Mastering Autodesk® Revit® MEP 2011

Autodesk®
Official Training Guide

Don Bokmiller

Marvin Titlow

Simon Whitbread

Wiley Publishing, Inc.

Senior Acquisitions Editor: Willem Knibbe
Development Editor: Susan Herman
Technical Editor: Simon Whitbread
Production Editor: Christine O'Connor
Copy Editor: Kim Wimpsett
Editorial Manager: Pete Gaughan
Production Manager: Tim Tate
Vice President and Executive Group Publisher: Richard Swadley
Vice President and Publisher: Neil Edde
Book Designers: Maureen Forys, Happenstance Type-O-Rama; Judy Fung
Proofreader: Nancy Bell
Indexer: Nancy Guenther
Project Coordinator, Cover: Lynsey Stanford
Cover Designer: Ryan Sneed
Cover Image: Pete Gardner/DigitalVision/Getty Images

ISBN: 978-0-470-62637-5

ISBN: 978-1-118-00690-0 (ebk.)

ISBN: 978-1-118-00692-4 (ebk.)

ISBN: 978-1-118-00691-7 (ebk.)

For general information on our other products and services or to obtain technical support, please contact our Customer Care Department within the U.S. at (877) 762-2974, outside the U.S. at (317) 572-3993 or fax (317) 572-4002.

Wiley also publishes its books in a variety of electronic formats. Some content that appears in print may not be available in electronic books.

Library of Congress Cataloging-in-Publication Data is available from the publisher.

10 9 8 7 6 5 4 3 2

Dear Reader,

Thank you for choosing *Mastering Autodesk Revit MEP 2011*. This book is part of a family of premium-quality Sybex books, all of which are written by outstanding authors who combine practical experience with a gift for teaching.

Sybex was founded in 1976. More than 30 years later, we're still committed to producing consistently exceptional books. With each of our titles, we're working hard to set a new standard for the industry. From the paper we print on, to the authors we work with, our goal is to bring you the best books available.

I hope you see all that reflected in these pages. I'd be very interested to hear your comments and get your feedback on how we're doing. Feel free to let me know what you think about this or any other Sybex book by sending me an email at `nedde@wiley.com`. If you think you've found a technical error in this book, please visit `http://sybex.custhelp.com`. Customer feedback is critical to our efforts at Sybex.

Best regards,

Neil Edde
Vice President and Publisher
Sybex, an Imprint of Wiley

To my wife, my family, and my friends, with much gratitude.
—Don Bokmiller

To my loving wife and son and my supportive friends: thank you all very much.
—Marvin Titlow

Acknowledgments

I would first like to thank my wife, Shelley. Without her loving and prayerful support, this simply would not have happened. I could fill the rest of the pages in this book with words attempting to express my gratitude, and it still would not be enough to convey how much you have supported me. Also to my mom, my brother Steve, my sister Joli, and the rest of my family, thank you so much for your kind encouragement throughout this adventure.

I have had the great opportunity to work with many wonderful people who have influenced my career and provided wisdom, guidance, and friendship. Thanks to Dave Sammons and Mike Taylor for sharing so much insight on what it takes to collaborate and work as a team and for being such good friends. Thank you, Kevin Austin, for giving me the opportunity to learn, research, and grow as a professional. Thanks to Jim Bish for knowing pretty much everything. Thank you to everyone else at Inlet Technology; my time there has proved to be invaluable. I also want to thank my friends and colleagues at Clark Nexsen, where I have been given the opportunity to grow and learn in such a wonderful working environment, which I could never take for granted. Thanks Johan, Tom, Scott, Willie, Mickey, and Bobby. You guys are the best! Thanks also to Bob Middlebrooks for getting me started with Revit and for being a great mentor.

I cannot bypass the opportunity to thank all the incredible people at Sybex. Thank you, Willem, for taking a chance on an inexperienced author. Thank you, Susan and Christine, for being such great editors who, despite my inexperience, never made me feel out of place. I'm sure there are many others who have worked hard to bring this book together. Thank you, all! Thanks also to Marvin, Simon, and Steven. It has been a pleasure to work with you.

—Don Bokmiller

To my wife, Tammie, thank you for your support through the long nights of patiently waiting while I was doing research and writing on this book. I would also like to thank my son, James Isaac, who really inspired me the most by reminding me that through hard work and dedication anything is possible.

To all my friends, thank you for all your support. Also, I dedicate this book to my good friend and colleague Dan Farmer who went on to meet the Lord during the writing of this book. He always had kind words and was supportive of the tasks that I took on for Michael Brady Inc. To Mike Brady and Louis Cortina, thank you for the time needed to write this book; also to Jeremy Nichols and Eric Baker who kept production going when I needed to find a secluded corner to write in and to Stanley Wolfson for the opportunities to share my plumbing knowledge of BIM with ASPE.

A special thank-you to Don, Simon, Susan, Willem, Pete, and Steven. All of you taught me so much about writing; I could never repay you for this experience.

—Marvin Titlow

About the Authors

Don Bokmiller is a CAD/BIM manager at Clark Nexsen Architecture and Engineering in Norfolk, Virginia. He has worked in the AE design industry since 1996 when he started out as a CAD technician in the electrical department. He eventually became one of a few CAD managers as the company grew, while also participating as an electrical designer on several projects. When Revit Systems came along, he participated in the Autodesk Beta program and has continued to do so for each release. His current position is to optimize the company's use of Revit MEP. Don has also worked as an application specialist for Inlet Technology, an Autodesk reseller, where he supported clients of various size and company structure on their use of Revit MEP. He has taught classes and given presentations to local engineering organizations. Don is an AUGI member and attends Autodesk University whenever possible.

Marvin Titlow has been working for mechanical, plumbing, and fire protection engineering firms since 1988. He started working for Michael Brady Inc. in 2004 as a plumbing/fire protection designer. In 2006 he began working with Revit MEP in beta and soon discovered that this was going to be a game changer. In 2008 he was promoted to technologies/BIM manager. Marvin has taught hundreds of plumbing designers how to get started with Revit MEP through teaching seminars and webinars for the American Society of Plumbing Engineers, as well as taught numerous mechanical engineering firms in Knoxville, Tennessee, how to use Revit MEP. Marvin has also had the pleasure of lecturing on the topics of BIM and Revit MEP at the Tennessee AIA convention in 2007 and at AUGI in 2008.

Simon Whitbread has been using Revit since release 5.1 and has over 25 years experience in Building Services and Architectural industries. Since the early 1990's he has been involved in developing and managing CAD and IT Systems for a number of companies while also teaching a variety of CAD classes to students at all levels of age and experience. He led the Revit implementation at JASMAX, one of New Zealand's leading architectural practices; trained and supported Revit MEP customers at an Autodesk reseller; and now provides Revit implementation, training, and support for Beca, a multinational, multi-service company.

Contents at a Glance

Contents

Introduction

Hello, and welcome to *Mastering Revit MEP 2011*. This book has been in the making for some time now, and we have been working diligently to bring you the first of its kind — a book that takes you through the core features and functionality of Revit MEP 2011 from both the design and documentation perspectives.

Revit MEP started out as Revit Systems in 2006 and in just a few years has been on a fast-track development pace in order to bring it up to speed with the Revit Architecture and Revit Structure platforms. The 2011 release of Revit MEP is the most improved and enhanced version of the software yet. When Revit Systems was first released, it was primarily to allow MEP engineers to join the move toward BIM that was being taken by architects and structural engineers. The features and functionality were, in the opinion of most, limited to provide a complete MEP project. The development team has been listening to the needs of users and has delivered tools and features in this release that have been desired by many from the beginning. We now have conduit and cable tray tools, oval ductwork, customizable demand factors and load classifications, and many other new features.

The primary focus of this book is, of course, on the MEP disciplines, but there is plenty of information that applies to Revit in general. The idea behind the format is to take you through the major points of the design process and requirements for completing a building design and project submittal. This book focuses on building engineering, but it may also be helpful for other types of engineering projects such as process piping design or any others that require a combination of data and model components.

The book is written in five parts, the first of which covers general functionality that is useful for all disciplines. You will find suggestions throughout the book for including features and components in your project templates. The first part does not cover every pick and click available in the software; it approaches the use of Revit from a best-practices standpoint, which we hope will inspire you to think about ways to make Revit MEP 2011 work best for you. Any topics not covered were not omitted to imply that they are unimportant but simply because you can find information about these features in the documentation provided by Autodesk and in the Revit MEP 2011 Help menu.

The next three parts of the book are MEP-specific and have been written to cover the key design areas of each individual discipline. Again, we focus on best practices by relating our professional experience with not only the software but also the design industry. In an effort to tie it all together, the fifth part of the book contains information on how to optimize your Revit experience by learning the tools and features available for creating the various components that make up an MEP model.

Who Should Buy This Book

This book is intended for readers who are at least somewhat familiar with Revit MEP. It is not intended to be a "how-to" book by simply explaining picks and clicks; it is more for readers who are looking to find ideas on how to make the software work for them. Engineers, designers, and CAD technicians will all find useful information related to their workflows. If you are looking to move further with your Revit MEP implementation, you should find this book to be a useful resource. Even if you find that you know the topics discussed in this book, we hope you will be inspired to think of new ways to improve your Revit MEP experience.

What's Inside

Here is a glance at what's in each chapter.

Part 1: General Project Setup

Chapter 1: Exploring the User Interface The ribbon interface is designed for optimal workflow. In this chapter, you will discover the features of the user interface that allow you to work efficiently. There are some new features in Revit MEP 2011 that improve the user interface dramatically.

Chapter 2: Creating an Effective Project Template The key to success with Revit projects is to have a good template file. Chapter 2 takes you through the major areas of a template file, offering ideas for settings that will make starting a project as simple and efficient as possible.

Chapter 3: Worksets and Worksharing This chapter guides you through the process of setting up a project file in a multiuser environment. The features of a worksharing-enabled file are explained in a manner that promotes ideas for project workflow efficiency.

Chapter 4: Best Practices for Sharing Projects with Consultants Revit has many features that make project collaboration easy to manage. In this chapter, you will learn about ways to use the power of Revit MEP to coordinate your design and documents with other members of the project team.

Chapter 5: Schedules The best way to extract the data contained in your Revit project model is to use the power of schedules. In this chapter, you will learn the tools available for scheduling model components and how to use schedules to manage data within your projects. The new panel schedule template feature is also covered in this chapter.

Chapter 6: Details Although creating a model with computable data is the primary reason for using Revit MEP, you do not want to model every minute detail of the design. The tools for creating detail drawings of your design are examined in this chapter. You will also learn how to use existing CAD details along with strategies for creating a library of Revit details.

Chapter 7: Sheets When it comes time to submit a project, you need to have a set of coordinated construction documents. In this chapter, you will learn the ways you can create and manage your project sheets. You will also learn about how you can print and export your project sheets for submittal or coordination with clients.

Part 2: Revit MEP for Mechanical

Chapter 8: Creating Logical Systems In Chapter 8, you will learn how to set up logical systems and how each system is affected by the type of systems you have created. From mechanical systems to fire protection systems, all have a certain role to play in BIM.

Chapter 9: HVAC Cooling and Heating Load Analysis Mechanical design must first start with understanding how your building will perform in different weather conditions and climates. In Chapter 9, you will learn that properly produced building loads can ensure that the mechanical design has been sized for maximum efficiency, saving energy and money while reducing the impact of the environment.

Chapter 10: Mechanical Systems and Ductwork Understanding how to successfully route ductwork can lead to error reduction and better coordination. In Chapter 10, you will learn how to locate mechanical equipment and how to use the proper routing methods for ductwork.

Chapter 11: Mechanical Piping Routing mechanical piping can be a daunting task. In this chapter, you will learn how to route and coordinate your piping and how, through these techniques, you can speed up production and take full advantage of what Revit MEP 2011 has to offer.

Part 3: Revit MEP for Electrical

Chapter 12: Lighting In this chapter, you will learn how to place lighting fixtures into your projects, including site lighting. The use of lighting switches is also discussed, along with the relationship between lighting fixtures and the spaces they occupy. This chapter also covers the basics for using Revit MEP for lighting analysis.

Chapter 13: Power and Communications In Chapter 13, the basics for placing power and communication devices into a model are covered. You will also learn how to place electrical equipment and connections for use in distribution systems. The new conduit and cable tray tools are also explored in this chapter.

Chapter 14: Circuiting and Panels Creating systems for your electrical components is just as important as it is for mechanical components. In this chapter, you will learn how to set up your projects to your standards for wiring, create circuits within your model, and create panel schedules to report the loads. The new tools for load classification and demand factors are also covered in this chapter.

Part 4: Revit MEP for Plumbing

Chapter 15: Plumbing (Domestic, Sanitary, and Other Piping) In this chapter, you will learn how to modify plumbing fixture families and create custom systems to speed up plumbing design. You will also learn how to use the new copy/monitor features in ways you may not have thought possible.

Chapter 16: Fire Protection Fire protection systems protect building and lives. You will learn how to lay out a fire pump system and assemble components to help in your design process. You will learn how to coordinate with other disciplines and how to effectively enter into the BIM arena through the use of Revit MEP 2011.

Part 5: Managing Content in Revit MEP

Chapter 17: Solid Modeling The foundation for custom content creation is having the ability to create the forms required to build component families. In this chapter, you will learn how to use the tools available in Revit MEP to create model geometry. You will also learn how to make geometry parametric, increasing its usability.

Chapter 18: Creating Symbols and Annotation Because so much of MEP design information is conveyed with schematic symbols, it is important to have the symbols and annotative objects commonly used for projects. Revit MEP has the tools needed to create schematic symbols for use in component families or directly in projects. In this chapter, you will learn how to use these tools and how to create constraints within families for display of the symbols in your projects.

Chapter 19: Parameters Parameters are the intelligence within a BIM project. This chapter explores how parameters can be used in both projects and families for applying computable data to your Revit models. The creation of shared parameters and their use is also covered.

Chapter 20: Creating Equipment Equipment families are an important component of a Revit model because of the space they occupy within a building. In this chapter, you will learn how to use solid modeling tools to create equipment. You will also learn how to add connectors for systems and how to create clearance spaces for coordination with other model elements.

Chapter 21: Creating Lighting Fixtures Lighting fixture families are special because they can hold photometric data that allows for lighting analysis directly in your Revit model. This chapter covers how to create lighting fixture families and add the data needed for analysis. You will also learn how lighting fixture families can be represented in project model views using detail components, line work, and annotation within the family file.

Chapter 22: Creating Devices This chapter examines the process for creating MEP system devices and how to use annotations to represent them on construction documents. In this chapter, you will also learn how parameters can be used to control and manage symbol visibility.

Sybex.com

Sybex strives to keep you supplied with the latest tools and information you need for your work. Please check the website at `www.sybex.com/go/masteringrevitmep2011`, where we'll post additional content and updates that supplement this book if the need arises. Enter **Mastering Revit MEP 2011** in the search box (or type the book's ISBN—9780470626375), and click Go to get to the book's update page.

The Mastering Series

The Mastering series from Sybex provides outstanding instruction for readers with intermediate and advanced skills, in the form of top-notch training and development for those already working in their field and clear, serious education for those aspiring to become pros. Every Mastering book includes the following:

- Real-World Scenarios, ranging from case studies to interviews, that show how the tool, technique, or knowledge presented is applied in actual practice
- Skill-based instruction, with chapters organized around real tasks rather than abstract concepts or subjects
- Self-review test questions, so you can be certain you're equipped to do the job right

Part 1

General Project Setup

Chapter 1

Exploring the User Interface

Revit MEP 2011 is the second release of Revit utilizing the ribbon-style interface. Having a good knowledge of where tools are located and how to access commands easily is the best way to be efficient in your use of Revit MEP 2011. Some improvements and changes have been made to the ribbon portion of the interface, so there is a slight learning curve when transitioning from Revit MEP 2010.

If you are transitioning to Revit MEP 2011 from a release prior to Revit MEP 2010, then the ribbon-style interface will be totally new to you. The ribbon works well with Revit because it allows tools to be organized in one area of the interface, which gives you more screen real estate for viewing the model. Although the user interface is customizable, you are limited in the amount of customization and number of features that you can change. At first this may seem a bit restrictive, but like any software, once you work with it enough, it will become more familiar, and your efficiency will increase.

Improvements have been made to the user interface for Revit MEP 2011. Some features have been added to improve workflow and efficiency, and typical workflow features that were previously accessed through buttons in the interface are now available as part of the interface itself.

Knowing your way around the Revit MEP 2011 user interface is the first step to success in reaping the benefits of utilizing a building information modeling (BIM) solution for your building projects. In this chapter, you will learn to

- Navigate the ribbon interface
- Utilize user interface features
- Use settings and menus

The Ribbon

Making the transition to a ribbon interface has been difficult for some people. If you are familiar with the Revit MEP user interface prior to the 2010 version, transitioning to the ribbon-style interface may indeed take some getting used to. Once you understand the way that the ribbon is set up and how you can customize it to better suit your workflow, though, you will see that it is an optimal interface for a BIM and design application.

Tabs

The ribbon portion of the user interface consists of several tabs, each organized by panels that relate to the topic of the tab. Each panel contains one or more buttons for the relevant features available in Revit MEP 2011. You can access a tab by simply clicking the name at the top of the ribbon. Although each tab is designed to provide a unique set of tools, some of the features of Revit are repeated on different tabs. The panels and tools for each tab are described here (not all panels are shown for each tab):

Home The Home tab is the main tab for MEP modeling tools. This tab is divided into panels that are specific to each of the main disciplines. This tab is where you will find the tools to build an MEP model. Each of the discipline panels has a small arrow in the lower-right corner that provides quick access to the MEP settings dialog box for that discipline.

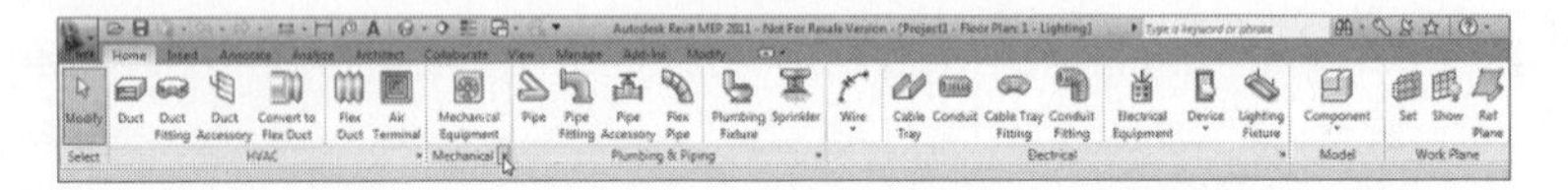

Insert The Insert tab contains the tools for bringing other files into your Revit projects. The tab is organized by panels for linking and importing files and also contains tools for loading Revit families. The small arrow at the lower right of the Import panel is for accessing the Import Line Weights dialog box, where you can associate imported CAD color numbers to a Revit line weight. The Insert tab also contains the Autodesk Seek panel, which provides a search window for content available on the Autodesk Seek website.

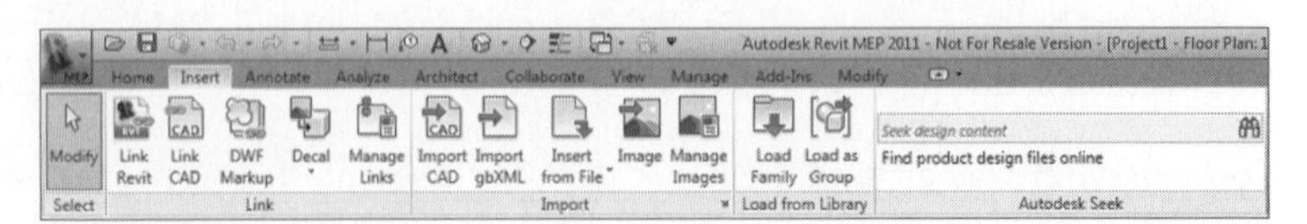

Annotate On the Annotate tab you will find the tools needed to add annotations to your model views along with drafting tools for creating details. The Dimension and Tags panels can be extended by clicking the arrow next to the panel name, revealing the tools for establishing settings. The Symbol button is used for placing annotation families onto views or sheets. The small arrow at the lower-right corner of the Text panel provides access to the Type Properties dialog box for creating or modifying text styles.

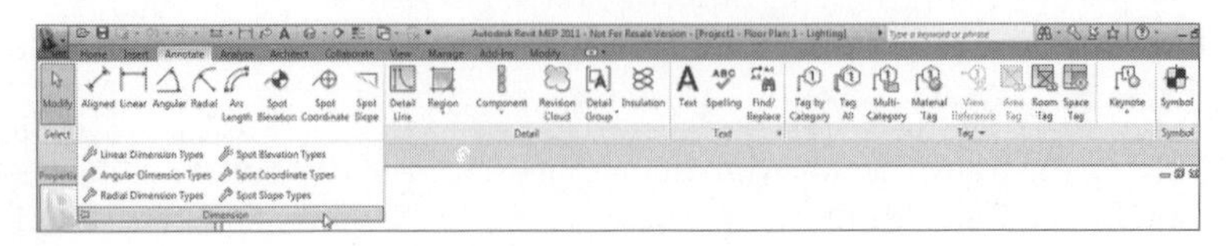

Analyze Tools for model analysis and systems checking are located on the Analyze tab. There are some other tools on this tab that allow you to add color to your ductwork and piping based on defined criteria. The Spaces & Zones panel on this tab contains the tools for placing Space objects and Space Separator lines.

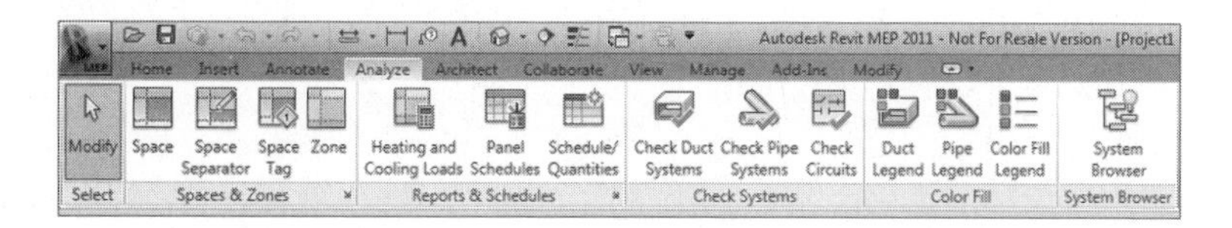

Architect Revit MEP 2011 is capable of modeling much more than just the components of an MEP engineering system. The Architect tab contains tools that allow you to model architectural components or to create conceptual mass objects. The Room & Area panel contains the Legend button, which allows you to apply color fill to your views based on defined criteria.

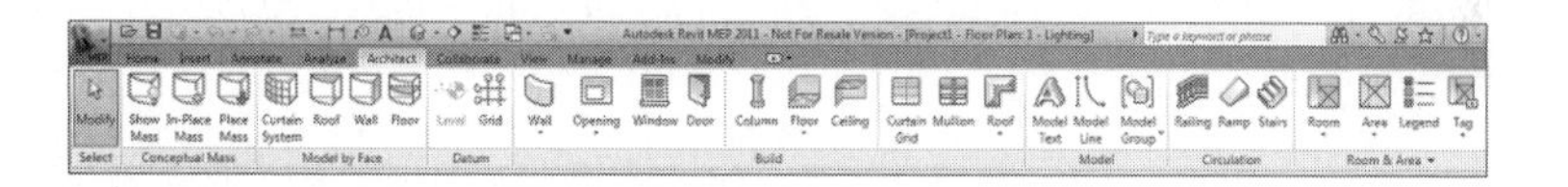

Collaborate To keep your model coordinated with other disciplines and work in a multi-user environment, you need tools that allow you to do so. These tools can be found on the Collaborate tab. The Worksets panel has a drop-down for switching worksets, and the Coordinate panel contains tools for copying and monitoring objects from linked files. The Coordinate panel has a new tool for locating face-hosted elements that have lost their association to their host. You can check for clashes between model objects by using the Interference Check tool on the Coordinate tab.

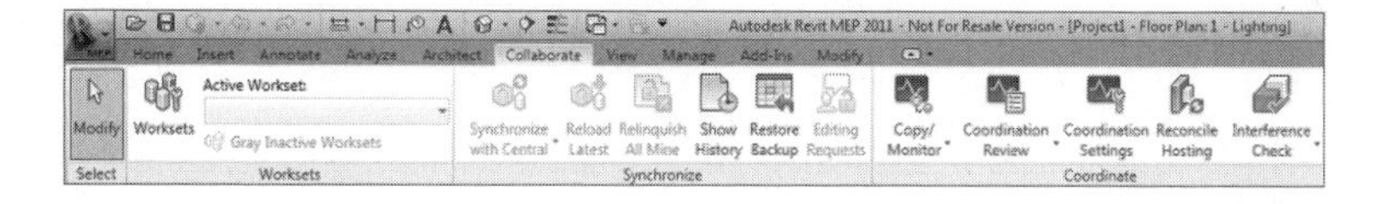

View You can use the tools available on the View tab to create views. There are also tools for managing the views you have open in the drawing area. On the Graphics panel there are tools for creating view templates and filters. The Sheet Composition panel has tools for creating sheets as well as for adding matchlines or revisions. The User Interface button allows you to toggle the visibility of key user interface features.

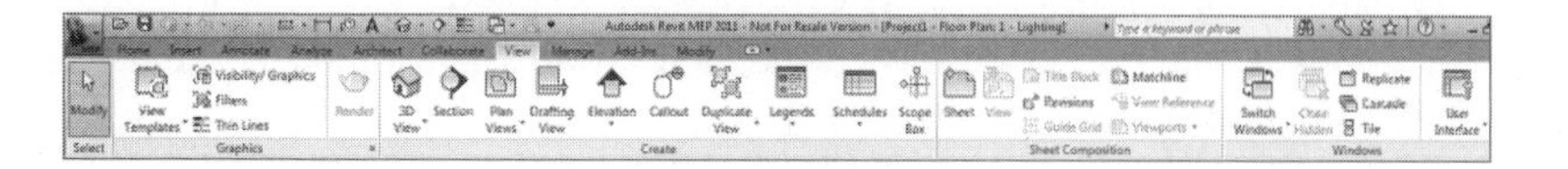

Manage On the Manage tab, you can find the tools needed to establish project settings. The Inquiry panel has tools that can be used to locate specific objects in your project model and display any warnings associated with your project. Along with the settings that can be accessed from the tools on the Settings panel, the Additional Settings button is a drop-down list of even more options. New to Revit MEP 2011 is the Panel Schedule Templates button, which allows you to define and edit the layout of panel schedules.

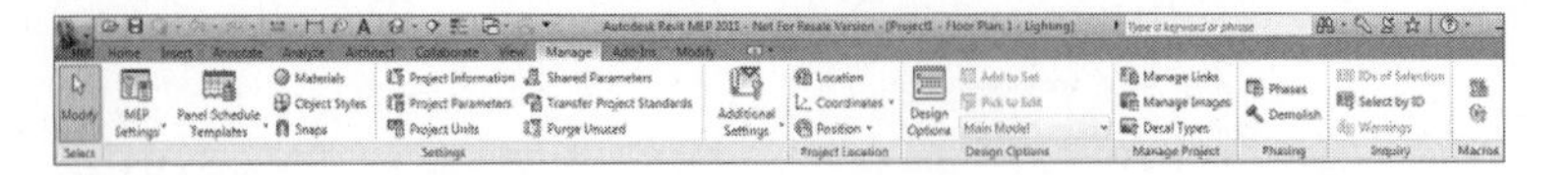

Modify The Modify tab is in a new location for Revit MEP 2011. It is now located at the end of the tabs, moving it closer to the center of the user interface for easier access. The Modify tab has the tools needed to make changes to components or line work in your project views. The tools on the Modify panel have been arranged with the more commonly used tools having larger buttons. Some of the tools that used to be drop-downs have been separated into individual buttons to reduce the number of mouse clicks required to access them. You can toggle the Properties palette on and off from this tab.

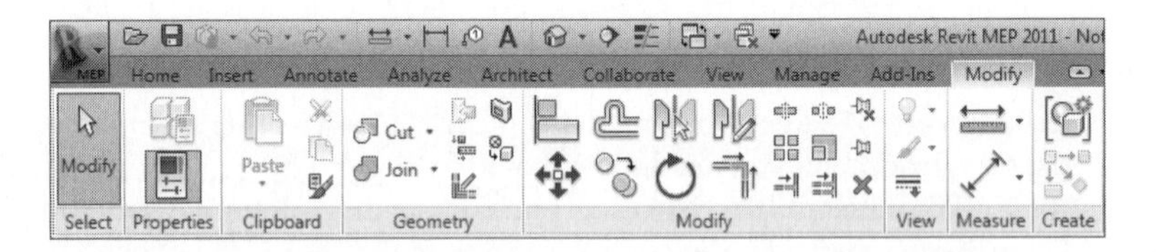

If you are running any external applications or macros, the Add-Ins tab will appear on your ribbon as the last tab. The buttons or other features provided by the external application will appear on the tab as configured.

The Modify button appears in the Select panel on every tab. This button allows you to exit from any active command, giving you an alternative to using the Esc key multiple times or selecting another tool.

Contextual Tabs

In addition to the tabs provided by default on the ribbon, there are tabs that appear when you select objects in your project. These contextual tabs appear with tools specific for modifying the selected object. Contextual tabs appear in the location of the Modify tab and are identified by their green color and a name that applies to the selected object. A contextual tab for a selected object is an extension of the Modify tab, which is why the base Modify tab is so compact compared to the other tabs. This allows for the selection-specific tools to appear on the right side of the Modify tab. Figure 1.1a shows the contextual tab for an air terminal selected in the model of a project. The standard tools on the Modify tab are available at the left but not shown in this figure for clarity.

When you select an object in the model that is part of a system, an additional contextual tab appears with tools for editing the system. These tabs are completely separate from the standard Modify tab and only contain tools for system editing. If you select an object on a system, the system tab will appear along with the contextual Modify tab, as shown in Figure 1.1b, but if you select an actual system, only the system tab appears. The panels and buttons on contextual tabs cannot be removed or rearranged on the ribbon. The buttons cannot be added to the Quick Access toolbar.

FIGURE 1.1
(a) Contextual tab for an air terminal; (b) contextual tab for duct system

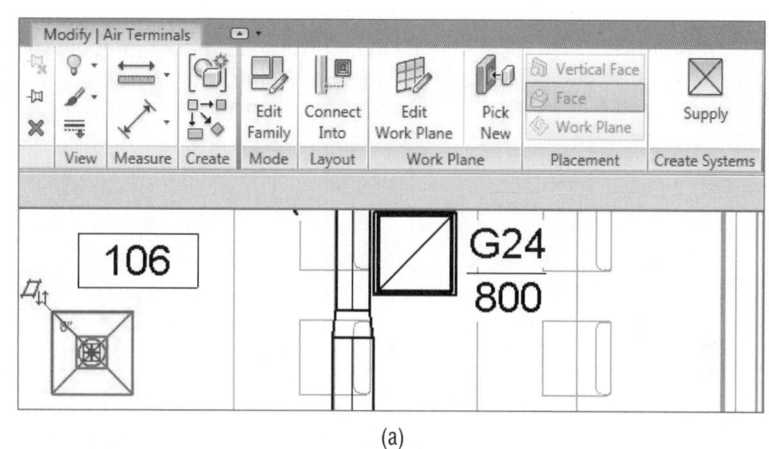

(a)

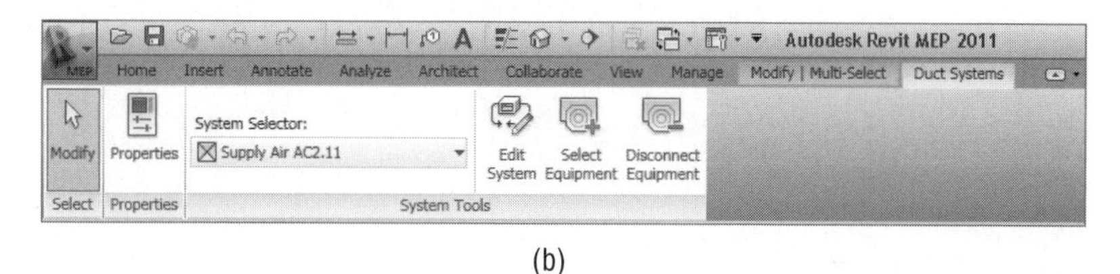

(b)

Family Editor Tabs

The tabs in the Family Editor environment differ from those in the project file environment. When you open a family file, the tabs on the ribbon contain some familiar tools, but many of them are specific to the creation and modification of family components. The tabs available in the Family Editor environment are as follows:

Home The Home tab in the Family Editor environment contains tools for creating solid geometry and lines, adding system connectors, and creating and managing references.

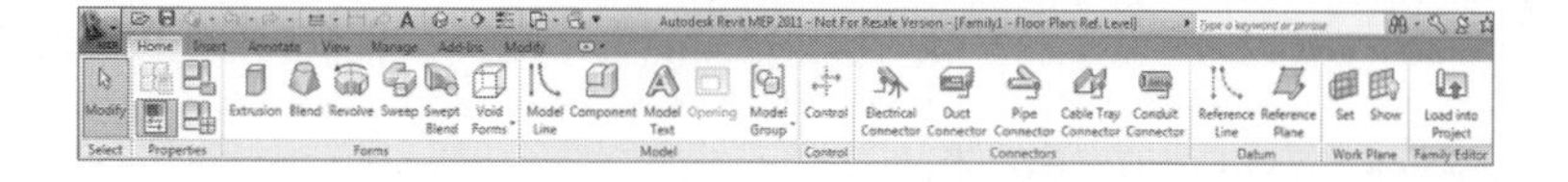

Insert The Insert tab in the Family Editor environment contains tools for bringing other files into your family file. The tools for linking are visible but disabled, because importing is the only available method for bringing a CAD file into your Revit family file.

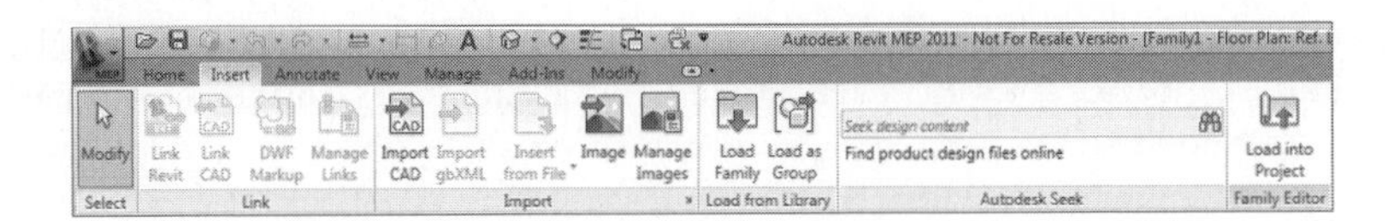

Annotate On the Annotate tab within the Family Editor environment, you will find the types of annotation tools that can be used in a family file. The Dimension panel can be expanded to establish dimension styles within the family file.

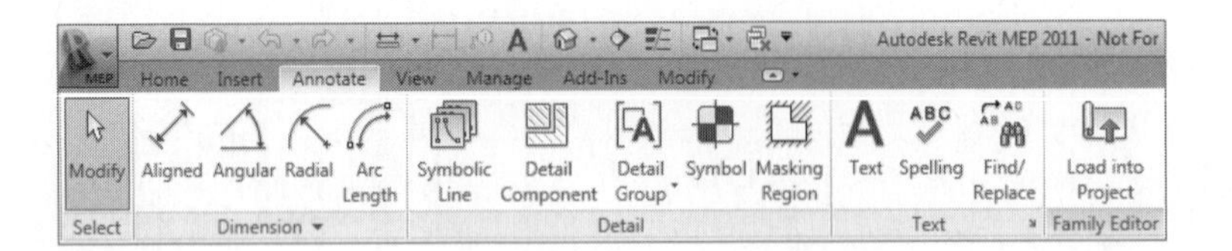

View The View tab in the Family Editor environment is limited to tools for managing the family views. Section views can be created and camera positions can be established for 3D views also.

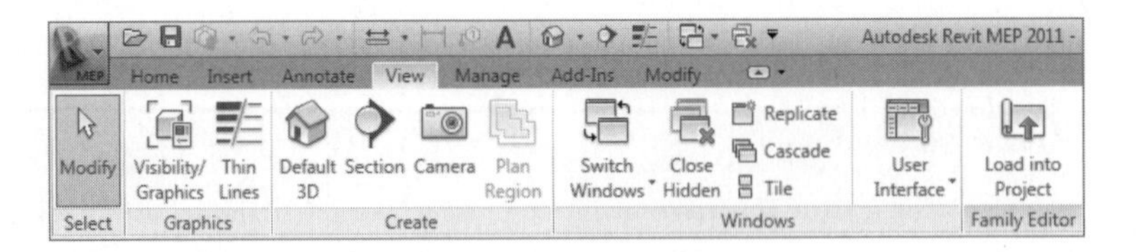

Manage In the Family Editor environment the Manage tab is populated with tools for establishing settings within the family file. The MEP Settings button allows you to establish load classifications and demand factors while the Additional Settings button drops down for access to general settings.

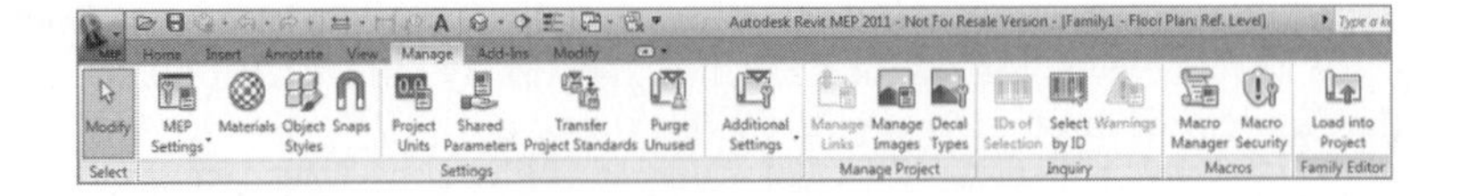

Modify The Modify tab in the Family Editor environment is the same as the one found in the project file environment. This tab is also compact, allowing for a contextual tab when objects within the family are selected.

The Load Into Project button is available on each tab in the Family Editor environment.

Customizing the Ribbon

You can customize the ribbon interface to better suit your workflow. For example, you can rearrange the order of the tabs by holding the Ctrl key and clicking a tab name to drag it to a new location.

You can move panels on a tab to different locations on the tab by clicking a panel name and dragging it to a new location. Figure 1.2 shows the Mechanical panel being dragged from its location on the Home tab. The panels to the right will slide over to fill in the space left by the moved panel.

FIGURE 1.2
Moving a tab panel

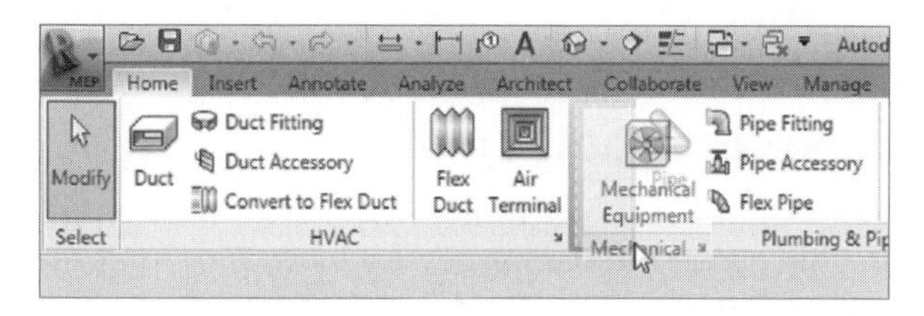

You cannot move a panel from one tab to another, however. If you attempt to drop a panel onto another tab, it will return to its original location on its normal tab.

You can remove panels from a tab and place them in another location on your screen. You can dock floating panels together by dragging one panel over the other, and you can move the docked panels as a group by clicking and dragging the gray grip that appears when you hover the mouse pointer over a floating panel. If you use dual monitors, you can even drag a panel to the second monitor. The panel's new position is maintained when you restart the software but will not appear until a file is opened. Keep in mind that moving tools to another screen may actually hinder your workflow efficiency.

BACK TO NORMAL

You can return a floating panel to its default location by clicking the small button in the upper-right corner of the panel.

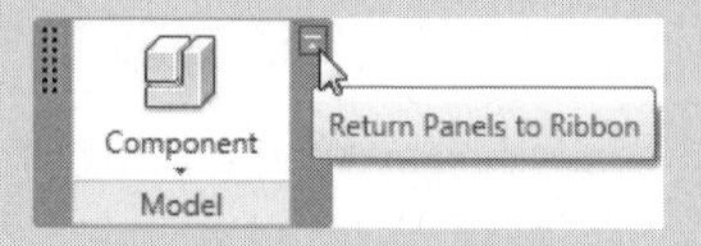

If you want to return the entire ribbon interface to its default settings, you can do so by locating and deleting the `UIState.dat` file. Deleting this file will also remove any customization done to the Quick Access toolbar.

You can control the visibility of the ribbon tabs by clicking the small button to the right of the tabs. This button will cycle through the different display options. You can also click the small arrow next to the button to display and select a specific option.

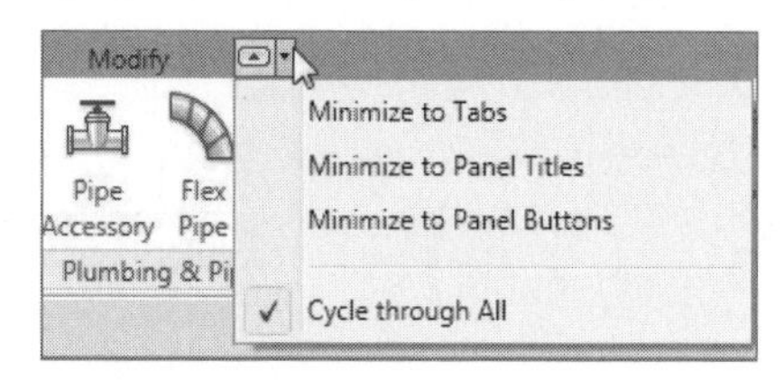

You can establish the switching behavior of the tabs on the ribbon to determine what tab is displayed when you exit a tool or command. When you click a tool, the contextual Modify tab for that tool appears. The interface will stay on the Modify tab when you exit the tool, or you can set it to return to the previous tab. These settings are located on the User Interface tab of the Options dialog box, which is discussed later in this chapter.

Quick Access Toolbar

As you are working and switching between tabs, you may find yourself taking extra steps to switch tabs in order to access the desired tools. The Quick Access toolbar (QAT) is a place where you can put frequently used tools for instant access.

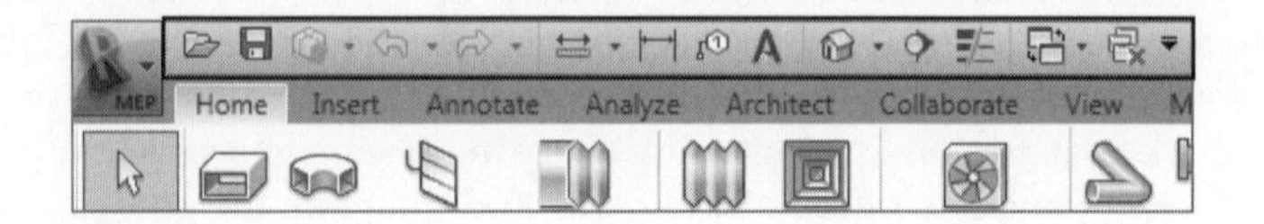

You can add tools from any of the standard tabs to the QAT by simply right-clicking the button or drop-down and selecting the Add To Quick Access Toolbar option. The tool will be placed at the end of the QAT. To manage the tools available on the QAT, you can click the small arrow button at the far right of the QAT, as shown in Figure 1.3. Each button on the QAT is listed, and removing the check mark next to it will turn off its visibility in the QAT.

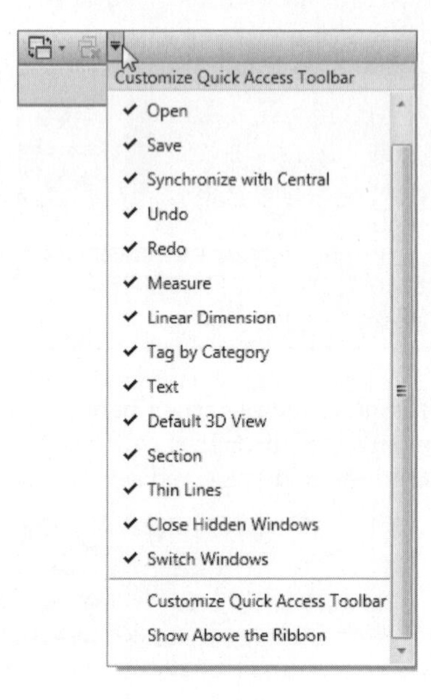

FIGURE 1.3
Quick Access toolbar customization menu

The option at the bottom of the list allows you to set the location of the QAT either above or below the ribbon. Setting it below the ribbon moves it closer to the drawing area for easier access. Moving it to this location does not take the place of the Options Bar, however. If you add several buttons to the QAT, you may want to move it below the ribbon so that it does not crowd out the filename on the title bar.

Clicking the Customize Quick Access Toolbar option in the drop-down opens the dialog box shown in Figure 1.4. In this dialog box, you can change the order of the buttons, create separator lines, or delete buttons.

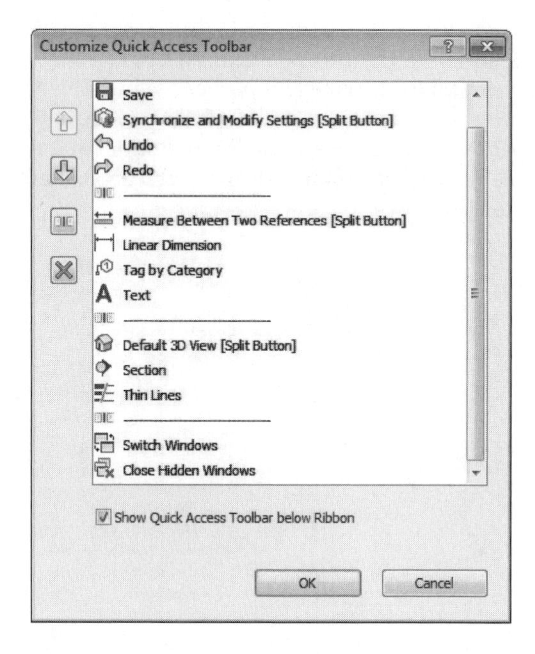

FIGURE 1.4 Customize Quick Access Toolbar dialog box

You can also right-click a button on the QAT for quick options such as removing the button, adding a separator line, or accessing the customization dialog box.

User Interface Features

The Revit MEP 2011 user interface is full of features designed to help you design and model efficiently. Some items are new, some have been modified, and some are the same as they have always been. The title bar at the top of the screen will still inform you of what file you are in and what view is currently active in the drawing area.

Options Bar

Despite the functionality of the ribbon with its contextual tabs, the Options Bar is still an important part of the user interface. This should be the first place you look when a tool or object in the project is selected. Though the number of options that appear may be limited, they are usually very important to the task you are engaging in. When placing Space objects, for example, pay close attention to the Upper Limit and Offset options to ensure proper space height.

You can dock the Options Bar at the top of the screen below the ribbon, which is the default location, or at the bottom of the screen just above the status bar. Right-click the Options Bar to change its docked position.

Properties Palette

New to Revit MEP 2011 is the Properties palette. This feature reduces the number of mouse clicks necessary to access the properties of a model object or project component. You can dock the Properties palette to the sides, top, or bottom of the screen, or it can float. If you dock the Properties palette to the same side of the screen as the Project Browser, the two features will split the docked space, as shown in Figure 1.5.

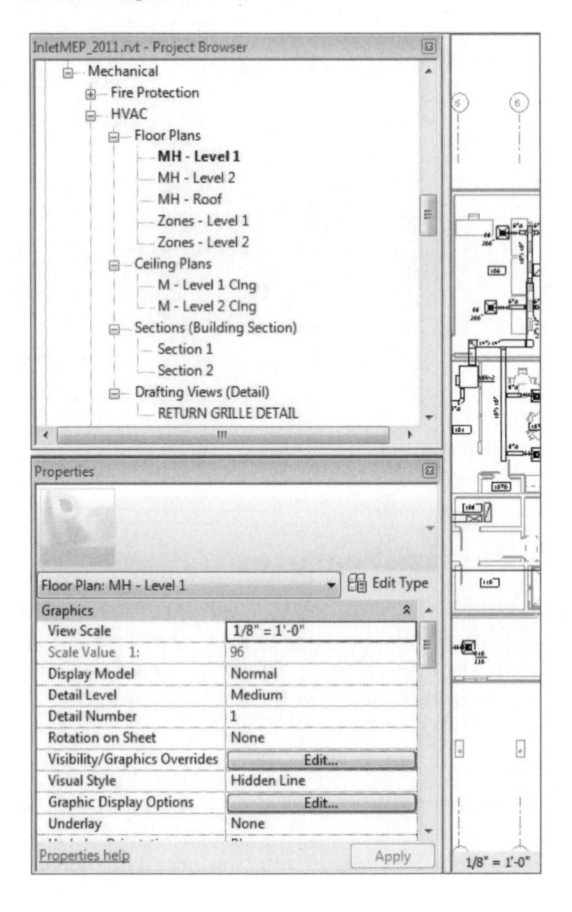

FIGURE 1.5
Project Browser and Properties palette docked together

If you do not have the Properties palette turned on, you can access it by clicking the Properties button located on the Modify tab or contextual Modify tab of a selected object. The Properties palette will remain on until you close it.

When no object is selected in the model or in a drafting view, the Properties palette will display the properties of the current view in the drawing area. You can select a view in the Project Browser to view its properties in the Properties palette.

The top section of the Properties palette acts as the Type Selector when an object is selected or a tool is chosen for placing an object. When an object is selected, you can switch to the properties of the current view by using the drop-down located just below the Type Selector, as shown in Figure 1.6.

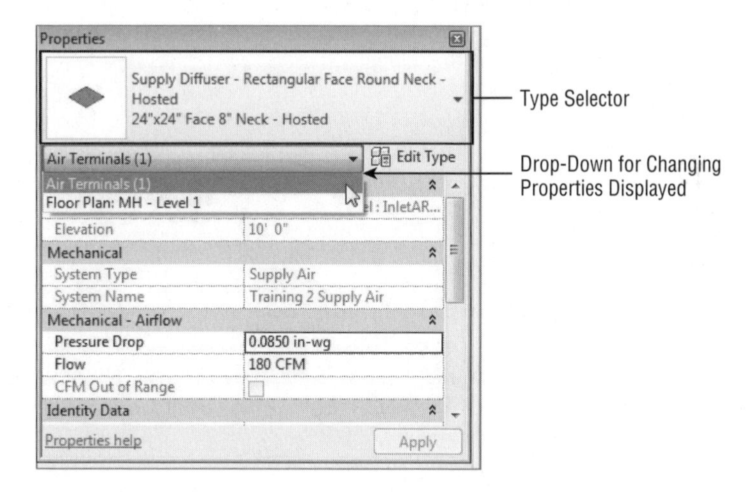

FIGURE 1.6
Properties palette

The properties shown in the Properties palette are instance properties. You can click the Edit Type button to display the Type Properties dialog box for the selected item or view. When viewing the instance properties of an object or view, the scroll bar at the right side of the palette will hold its position when you move your mouse pointer away from the palette. The scroll bar will remain in position even when other items are selected in the model.

When you make a change to a parameter in the Properties palette, you can click the Apply button at the bottom-right corner of the palette to set the change, or you can simply move your mouse pointer away from the palette, and the change will be applied.

View Control Bar

The View Control Bar is often overlooked but contains tools that are very important to the display of the contents in the drawing area. There are a few new features for the View Control Bar in the Revit MEP 2011 release. The Sun Settings button has been added so that the Sun Path can be turned on or off in the view. You can access the Sun Path settings by clicking the Edit button in the Graphic Display Options parameter of a view's properties.

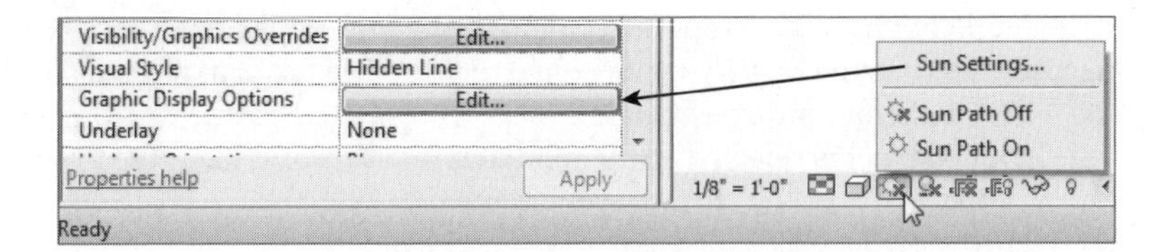

Another change to the View Control Bar is the addition of styles to the Visual Styles tool. The Consistent Colors setting has been added so that the colors of objects will appear in a shaded view with the same consistency regardless of its orientation to a light source. So when you are viewing the model in a 3D view, the color of all sides of an object looks the same. The Realistic setting allows the rendering materials of objects to be displayed in an editable view. This setting works only if you have Hardware Acceleration turned on.

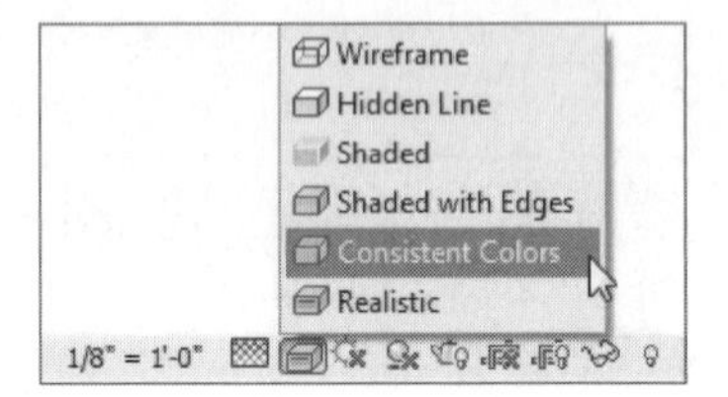

Take some time to explore the new visual styles by completing the following exercise. If you do not have Hardware Acceleration turned on, you can do so by clicking the Options button on the Application menu and going to the Graphics tab. You must do this prior to opening the file. If activating Hardware Acceleration causes problems because of your video driver, you can choose to not activate it and skip step 4:

1. Open the `Ch1_Project.rvt` file found at `www.wiley.com/go/masteringrevitmep2011`.
2. Click the Visual Styles button on the View Control Bar. Set Visual Style to **Shaded With Edges**. Zoom, pan, and orbit the view, and make note of the variations in color based on the model orientation.
3. Click the Visual Styles button, and change the style to **Consistent Colors**. Notice that the colors remain a consistent shade when you zoom, pan, and orbit the view.
4. Click the Visual Styles button, and change the style to **Realistic**. Notice that the render material defined for the objects is now displayed.
5. Click any object in the model. If you do not already have the Properties palette active, click the Properties button on the Properties panel of the contextual tab. Take some time to become familiar with the behavior of the Properties palette.

Status Bar

Improvements have been made to the status bar for Revit MEP 2011. The status bar will still report the information about a selected item or prompt with instructions for multilevel commands, but it now also has an active workset indicator and design option indicator.

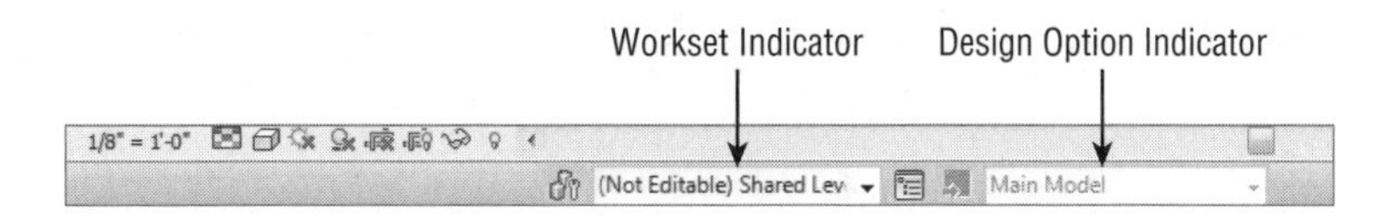

You can access the Worksets dialog box by clicking the Worksets button next to the active workset window, and you can switch between active worksets by clicking the window and selecting the desired workset. These are the same tools as found on the Collaborate tab, but having them on the status bar eliminates the need to switch tabs on the ribbon to access them. This also eliminates the need to add the tools to the Quick Access toolbar. The Design Options window displays the active design option, and you can access the Design Options dialog box by clicking the button.

The Editable Only and Press And Drag check boxes are located on the status bar. The Editable Only check box is for filtering a selection by only those objects that are editable in a worksharing environment, and the Press And Drag check box allows you to drag objects in the model without having to select them first. So, you can click an object and immediately drag it when the box is checked. When deselected, you must first click to select the object, release your mouse button, and then click the object again to drag it. In some cases, you may want to deselect this box to avoid moving objects inadvertently when working in a crowded area of the model.

Info Center

The Info Center is the portion of the title bar that gives you quick access to the Help menu or information about Revit MEP 2011. The search window allows you to search for information about a topic, and you can choose the locations to be searched by clicking the arrow next to the binoculars button.

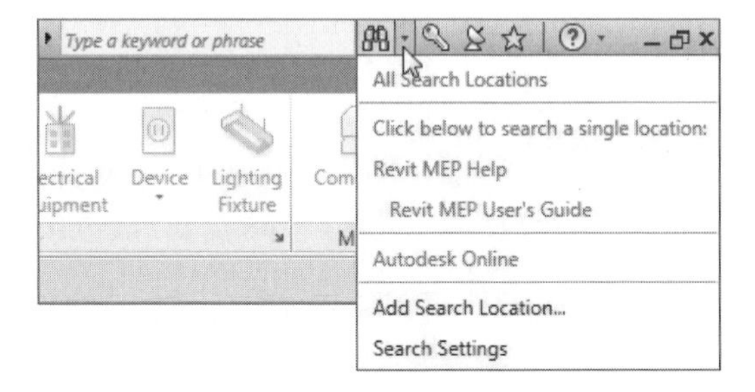

You can easily access the Subscription Center by clicking the button with a key on it. Any topics found in the search window or listed in the Subscription Center panel can be added to your Favorites list by clicking the star icon to the right of the topic. Clicking the star button on the Info Center will list all your favorite topics. The Help menu is displayed by clicking the question mark button. Revit MEP 2011 has a web-based Help menu. You can access additional information by clicking the arrow next to the Help button, as shown in Figure 1.7.

The Communication Center button is the one with the radar device on it. The Communication Center provides information about product updates and announcements.

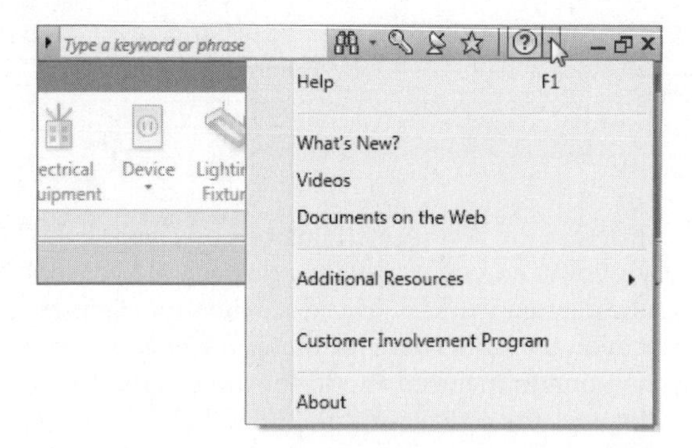

FIGURE 1.7
Additional Help options

User Interface Control

Many components of the Revit MEP 2011 user interface can be turned on or off for workflow efficiency or to maximize screen real estate. The User Interface button, located at the far right of the View tab, allows you to select which user interface components are visible, as shown in Figure 1.8.

This is not only a way to display or remove interface components for more screen real estate but is also a way to access the Recent Files screen, which cannot be accessed from the Switch Windows drop-down button.

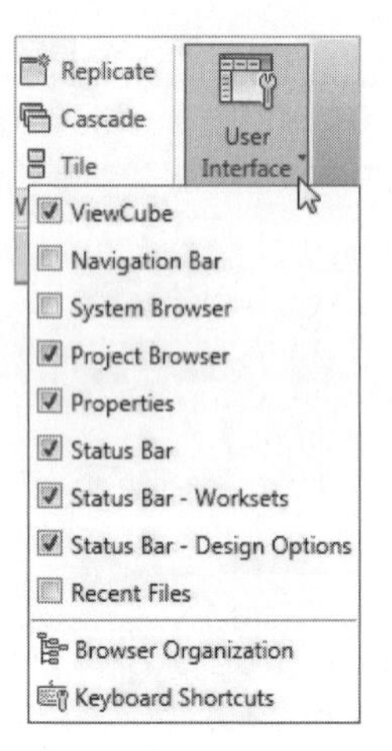

FIGURE 1.8
User Interface button options

Menus and Settings

Some changes and new features are available for user interface settings and for context menus when right-clicking. Along with options for file export, printing, and opening and saving files, the Application menu has an Options button where you can access the settings to establish the behavior of the interface as well as the location of directories and files used for working on projects. You access the Application menu by clicking the Revit logo button in the upper-left corner of the user interface; the Options button is located in the lower-right corner of the menu, as shown in Figure 1.9.

FIGURE 1.9
Application menu

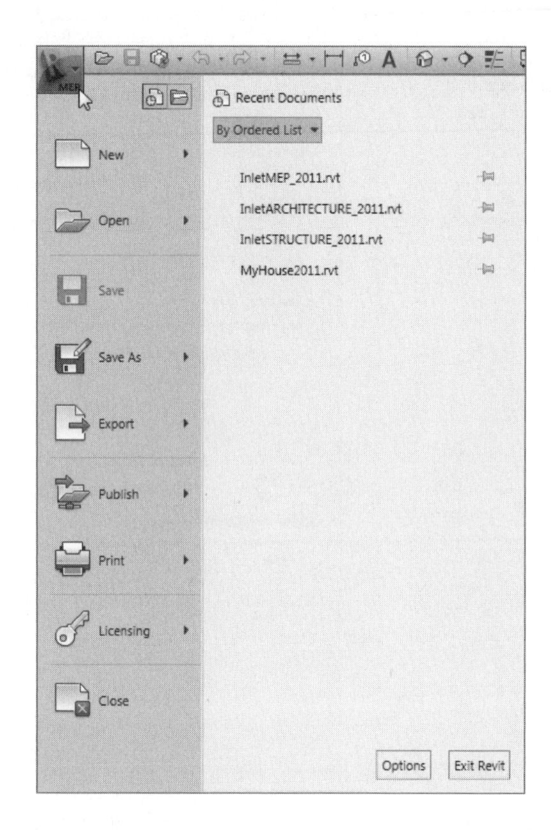

Clicking the Options button opens the Options dialog box, which has several tabs for different settings within Revit MEP 2011. The User Interface tab shown in Figure 1.10 is where you can set some general behavior for the interface. For example, you can choose between a light gray or dark gray theme for the interface from the Active Theme drop-down.

The Tab Display Behavior section in the center of the tab is where you define how the ribbon tabs will behave after an action is completed. There are settings for the behavior in both the project and Family Editor environments. The check box for displaying a contextual tab on selection allows you to have the ribbon tab switch immediately to the contextual tab when an object is selected. If that check box is not selected, the contextual tab will still appear on the ribbon, but it will not automatically become the active tab.

In addition, you can define the level of information provided by tooltips from the Tooltip Assistance drop-down. If you are still learning the function of different tools within Revit MEP 2011, you may want to set this option to High so that you will receive detailed descriptions of how tools work when you hover over them with your mouse pointer. If you find the tooltips interfere with your workflow, you can set this option to None.

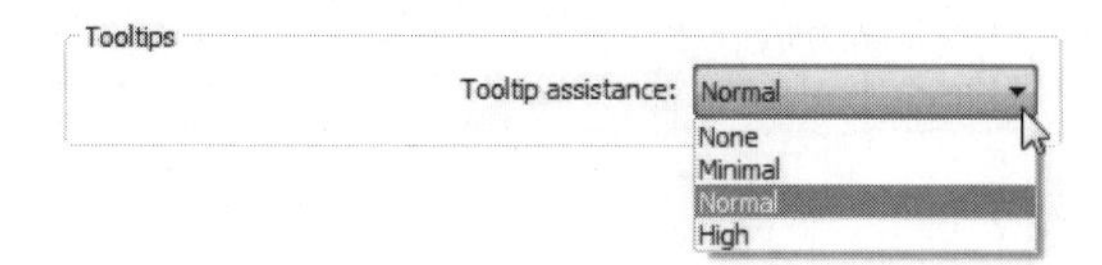

FIGURE 1.10
User Interface tab of the Options dialog box

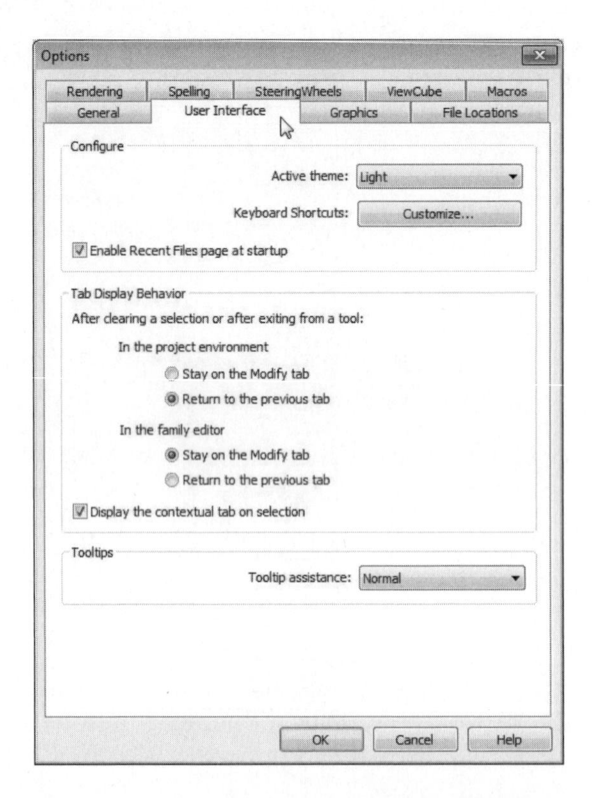

> **WHAT DO I TYPE?**
>
> With Tooltip Assistance set to at least Minimal, you can see the keyboard shortcut for a tool by hovering your mouse pointer over it. The keyboard shortcut is shown in parentheses next to the name of the tool.

Keyboard Shortcuts

On the User Interface tab of the Options dialog box is a button that enables you to customize your keyboard shortcuts. Clicking this button activates the Keyboard Shortcuts dialog box. In this dialog box you can filter the commands to make the list easier to manage and edit. You can even filter by specific tabs or menus, as shown in Figure 1.11.

> **SHORTCUT ACCESS**
>
> You can access the Keyboard Shortcuts dialog box by clicking the User Interface button on the View tab. There is even a keyboard shortcut for the Keyboard Shortcuts dialog box, KS.

You can sort the list in ascending or descending alphabetical order by clicking the desired column. Once you have located a command that you want to create a keyboard shortcut for, you can select the command to activate the Press New Keys window at the bottom of the dialog box. Input the desired keys that will activate the command. You can input up to five characters for a keyboard shortcut. Reserved keys cannot be used for keyboard shortcuts; you can find the reserved keys using the filter in the Keyboard Shortcuts dialog box. Click the Assign button to apply the shortcut to the selected command.

FIGURE 1.11
Keyboard shortcut list filter options

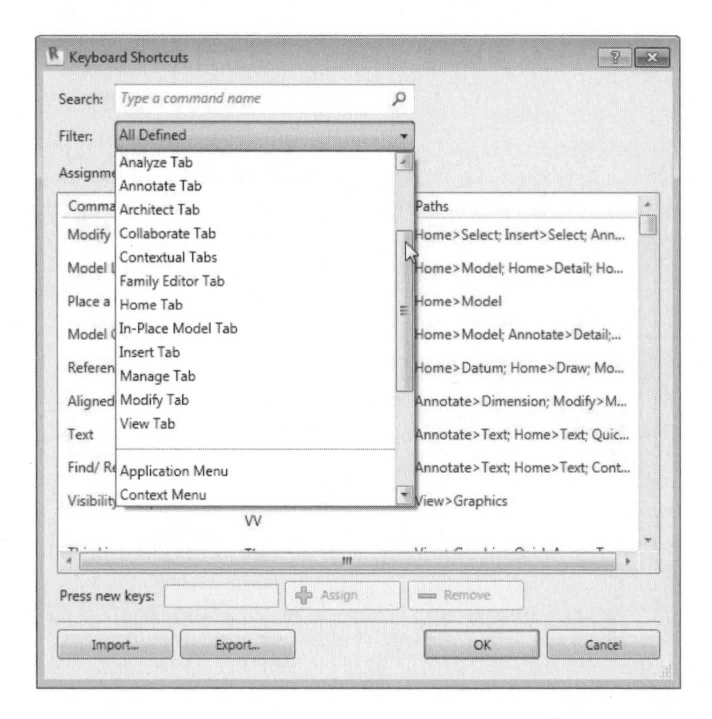

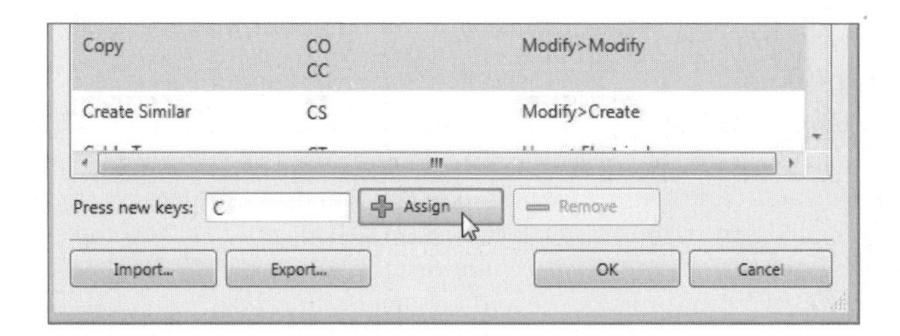

You can create multiple shortcuts for a single tool, and you can use the same shortcut keys for multiple tools. When you create a shortcut that is used for multiple tools, you must use the status bar to determine which tool to use when working in your project. When you type the shortcut, the first matching command will be displayed on the status bar. You can use the up or down arrows to cycle through available commands for the shortcut. Once the desired command is displayed on the status bar, you can activate it by hitting the spacebar. You can remove a keyboard shortcut from a command by selecting the specific shortcut and clicking the Remove button.

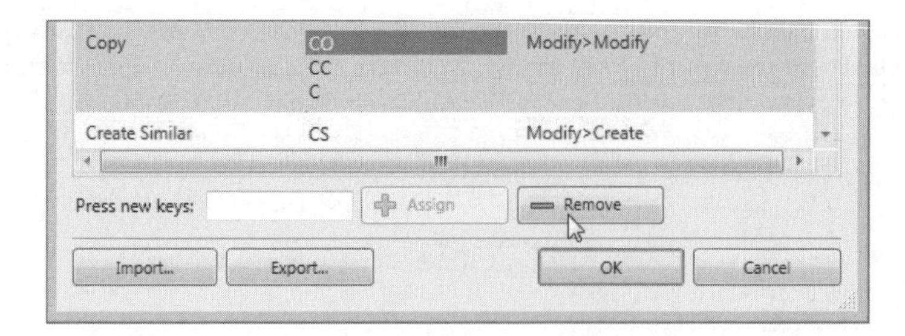

You can export your keyboard shortcut settings by clicking the Export button at the bottom of the Keyboard Shortcuts dialog box. This will save your settings as an `.xml` file that can be edited using a spreadsheet program. Using a spreadsheet is another way to manage and share your keyboard shortcuts. The `.xml` file can then be imported into Revit using the Import button, allowing you to set a standard for keyboard shortcuts in a multiuser work environment.

COMMON SHORTCUTS

Clark is responsible for teaching Chris and Allyson Revit MEP 2011. He has established keyboard shortcuts that fit his workflow best and allow for efficient use of the software. Because these shortcuts are what he is most familiar with, he wants to share them with his students so that during class they will all be using the same shortcuts.

Clark exports his shortcut settings to a file, which he imports into Revit MEP 2011 on the training computers. Prior to importing the custom settings, he exported the default settings so they can be used later if necessary. During class it is noted that the settings provided are preferred, but each student can further customize them if it results in improved efficiency.

Graphics

The Graphics tab of the Options dialog box allows you to set the selection, highlight, and alert colors used in the drawing area. The drawing area's background color can be inverted if you are more comfortable using a black background. Settings for temporary dimensions are also available to make them more readable, as shown in Figure 1.12. You can set the background for temporary dimensions to transparent or opaque (this setting is not shown in figure).

Settings for the appearance and behavior of the SteeringWheels and ViewCube can be found on their respective tabs in the Options dialog box.

Context Menus

Though the ribbon interface is designed for efficient workflow, context menus can be the easiest, most effective way to access settings or make changes to components of your Revit projects.

A *context menu* is a menu that appears when you right-click in open space, on an item in the Project Browser, or on an object in the drawing area. New to Revit MEP 2011 is the addition of the Repeat and Recent Commands options on the context menus. You can now repeat the last command by using the Enter key or by clicking the option on the context menu. The Recent Commands option displays a list of recently used commands for easy access during repetitive work. Figure 1.13 shows a context menu and the recent commands used during a working session. The last five commands used are displayed in the Recent Commands list.

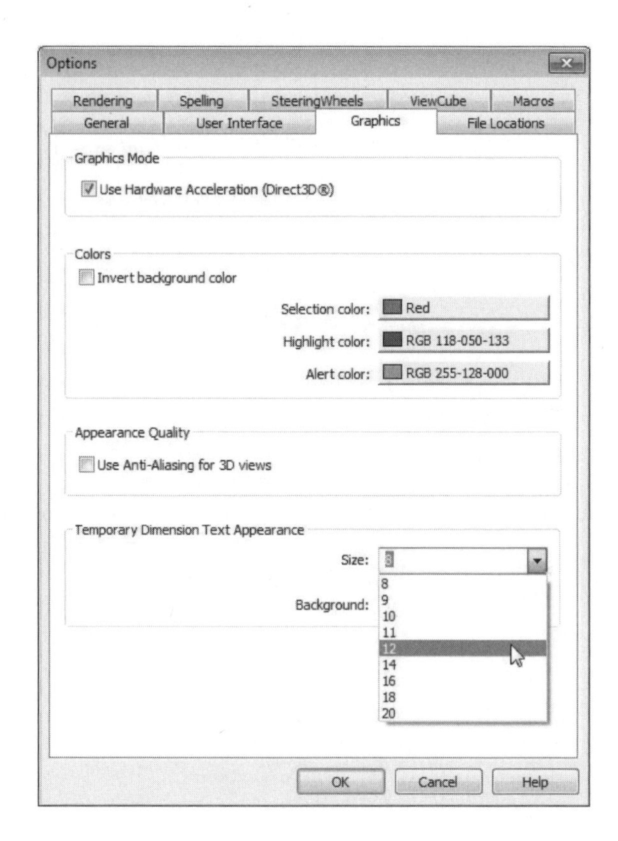

FIGURE 1.12
Temporary dimension appearance settings

The options displayed on a context menu depend on the object selected when the menu is accessed. Another new feature to Revit MEP 2011 is the ability to define the selection set when the Select All Instances option is chosen. Figure 1.14 shows that you now have the option to select only the objects in the active view or to select them throughout the entire project. This makes the Select All Instances feature much more useful without worrying that objects that should not be selected are not inadvertently included in the selection set.

There is no longer a View Properties or Element Properties option in the context menus because of the addition of the Properties palette. The Properties option at the bottom of a context menu allows you to toggle the Properties palette on or off.

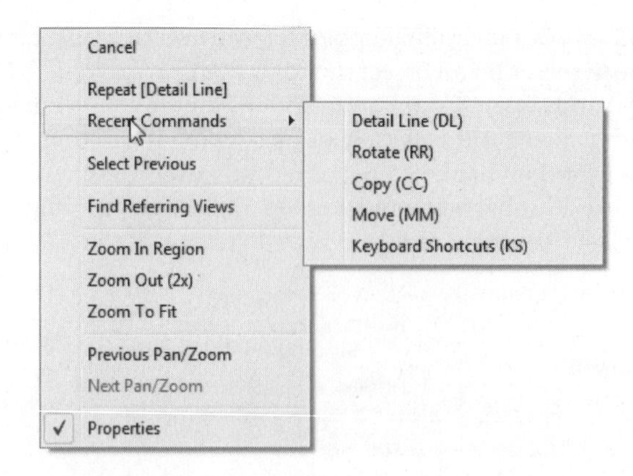

FIGURE 1.13
Context menu showing recent commands

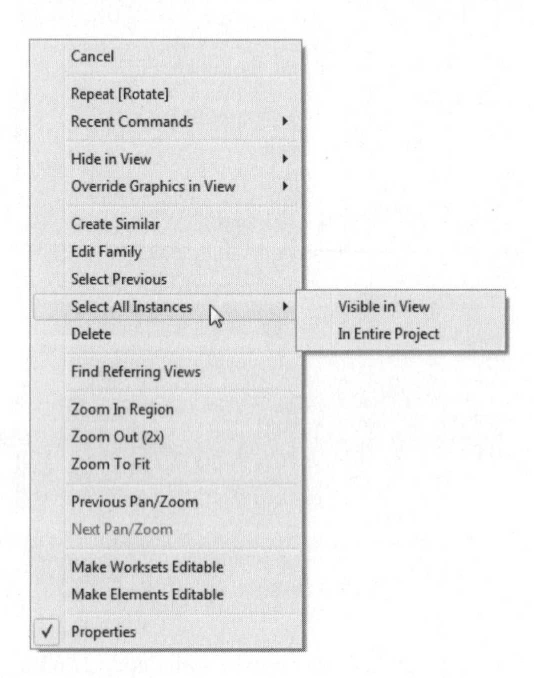

FIGURE 1.14
Selection set options

You can also right-click an element in the Project Browser for a context menu. Right-clicking a view will activate a menu with options for applying or creating a view template from the view. You can also save the view to a new file, as shown in Figure 1.15.

When you right-click a family in the Project Browser, you get a context menu with options to edit, rename, or reload the family. You can right-click a family type in the Project Browser to access its type properties or to select the instances in the project without having to locate one of the instances in the model, as shown in Figure 1.16.

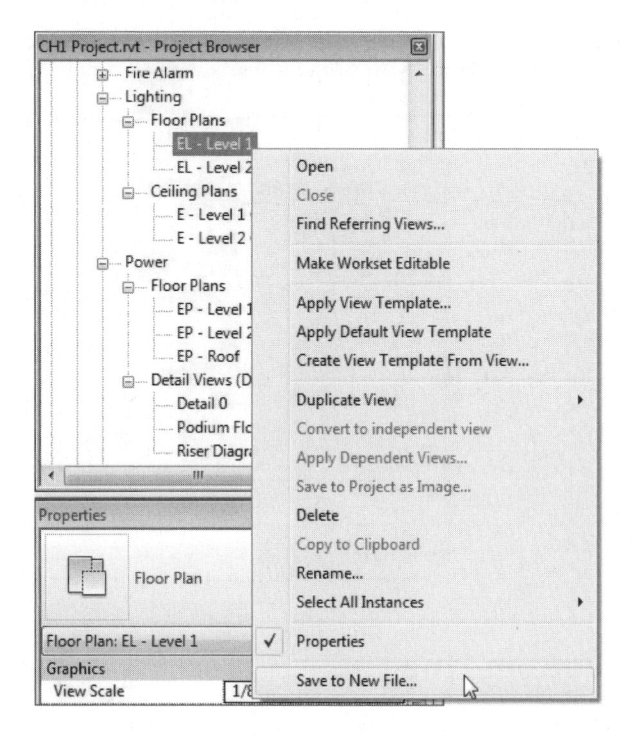

FIGURE 1.15 Context menu for a view selected in the Project Browser

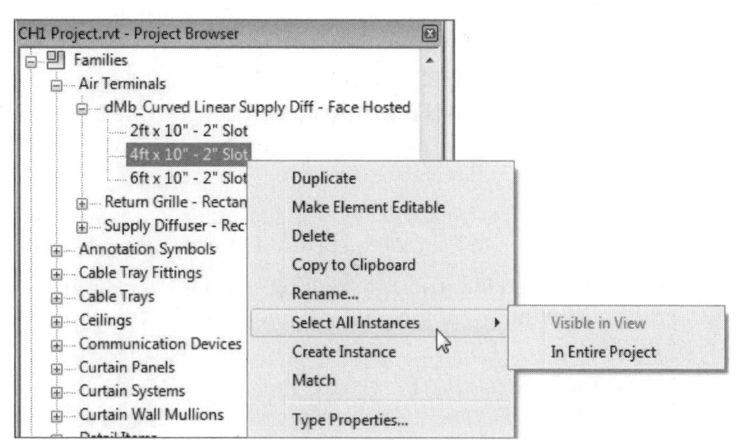

FIGURE 1.16 Context menu for a selected family type

You can also use a right-click context menu to manage linked files through the Project Browser. Figure 1.17 shows the options available on a context menu when a linked Revit file is selected in the Project Browser.

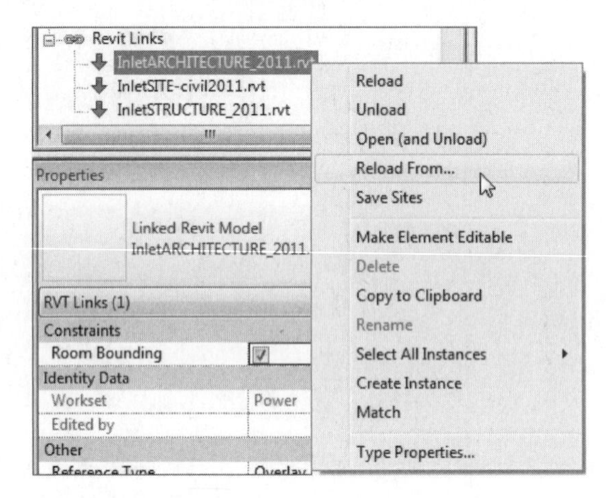

FIGURE 1.17
Context menu for a linked file in the Project Browser

The Bottom Line

Navigate the ribbon interface The ribbon is an efficient user interface style that works well in Revit. The ability to house numerous tools in a single area of the interface allows for maximum screen real estate for the drawing area.

Master It Along with the standard tabs available on the ribbon interface, contextual tabs also are available while you're working on a project. Explain what a contextual tab is and how it may differ throughout your workflow.

Utilize user interface features Many features are available in the Revit MEP 2011 user interface that allow for quick access to tools and settings. The use of keyboard shortcuts can also improve workflow efficiency.

Master It It is important to workflow efficiency to know how to access features of the user interface. What tool can be used to activate or remove user interface features?

Use settings and menus Establishing settings for your user interface is another way to create a working environment that is the most efficient and effective for your use of Revit MEP 2011.

Master It The use of keyboard shortcuts has been part of design and drafting software for a long time. The ability to customize the shortcuts to best suit your workflow is key to improved efficiency. How can the settings for keyboard shortcuts be accessed? How can the settings be shared among users?

Chapter 2

Creating an Effective Project Template

A lot of work goes into putting together a building design project. It is not only the coordination of design intent but also the coordination of the means to communicate the design intent. Anyone interested in saving time and money to achieve the goals required by a project will immediately begin to ask how they can simplify or automate the numerous tasks.

Project templates are the cornerstone to improving efficiency when working on a Revit project. Revit MEP is a design and documentation tool, and those who are paid to do design work should not have to spend time on anything other than achieving their design goals. A well-developed project template will enable you to focus more on design without having to spend time developing and defining settings or standards each time a task is required, because they'll already be created for you.

The first consideration for creating a Revit project template should be the requirements for the delivery of the design. There is no need for certain settings or features in a template if they are not used on a project-to-project basis. Some clients may have certain standards that require a unique template altogether. Because project templates are the culmination of company or client standards, they should be managed by one person or a small group of people. Project templates are fluid documents that require updates, so allowing global access to them will make them difficult to manage. However, input as to what should be included in a template can be made by anyone who works on projects and understands the need for features or functionality.

Revit MEP 2011 comes with template files that can be used for starting a project right away. You may choose to use these templates for a project or as a starting point for building your own template.

Whatever its use, project templates are the starting points that allow you to work seamlessly without breaking the momentum of collaboration and coordination efforts of your projects. The goal of creating a template is not to include every single item or standard that you use but to determine what is most often needed.

In this chapter, you will learn to

- Set up views and visibility
- Establish project settings
- Define preloaded content and its behavior
- Create sheet standards

Understanding Templates

You set standards so that project documents will look the same within a construction document set and so different sets of documents appear to have come from the same place. Some companies care about how their drawings look more than others, but there should be a uniformity regardless. This not only applies to the content that makes up a model but also applies to the organization of model views, the naming of views and schedules, and the overall drafting conventions used.

Once it has been determined how views should be displayed for each type of view and each discipline, you can establish those settings in your project template so that each project will begin from the same starting point. Because every project is unique is some way, it is possible to modify the default settings as needed, but that does not eliminate the need for baseline settings.

You want to be able to begin working on a project without having to spend time setting up how the project displays in your views. Having preset views and visibility settings will increase productivity on your projects by eliminating the need to do repetitive tasks just to get started.

Understanding the Project Browser Organization

How you keep your project organized within Revit's Project Browser will go a long way toward efficiency in the workflow of a project. Having a consistent Project Browser organization will make it easier when working on several projects. You always want to know where to find specific types of views and also what a view is by its name.

The first area of organization in the Project Browser consists of the views within the project. When transitioning from a traditional CAD program, it can be difficult at first to comprehend that each view is not a separate file. The views are created to determine how you are looking at the model. Views have many properties that determine their appearance and what discipline or system they belong to. You can organize your views based on any of these properties in order to group like views together. Figure 2.1 shows the different types of view organization available by default. You access this dialog box by right-clicking the Views heading in the Project Browser and selecting Properties.

Notice that the organization of views in the Project Browser is a system family. This means that you can create additional types by duplicating one of the default types and then changing how the views are grouped and listed.

When you click the Edit button for the Folders parameter in the type properties of the Browser – Views family, you can modify how the views are organized and listed in the Project Browser. You can choose up to three levels of grouping for your views and one sorting option. Each grouping option is determined by a view property. Once a property is chosen, you can select another level of grouping by using a different property and then another if necessary. When all the grouping options have been established, you can select the sorting option for the views. So, creating grouping options is similar to creating a folder structure for files, and the sorting option is how the views will be listed in the "folder."

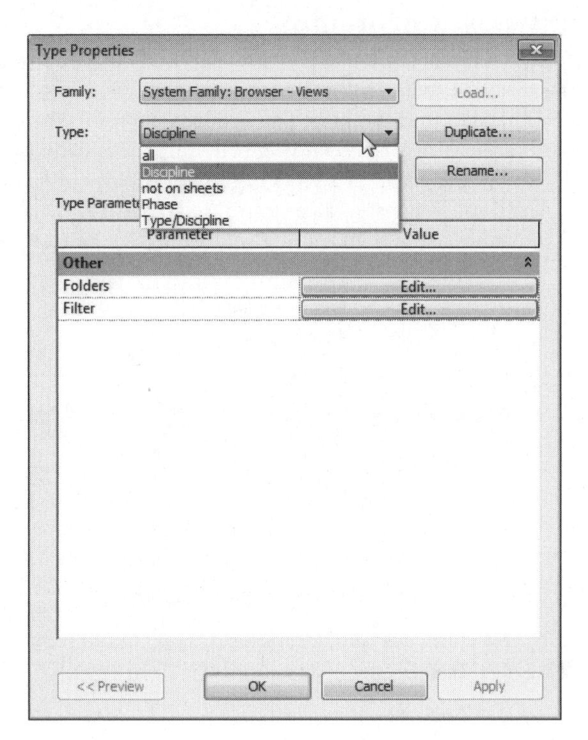

FIGURE 2.1
Default view organization types

Determining Which Views Are Grouped Together

Every view within a project belongs to a design discipline. The discipline property of a view is the most common property used for the first level of grouping. Revit MEP uses only two disciplines for MEP engineering, Mechanical and Electrical. The Sub-Discipline parameter can be used as the second level of grouping to further distinguish views used for fire protection, plumbing, or other types of engineering systems. This parameter exists in the default templates that can be installed with Revit MEP 2011. If you choose to create a template or project without starting with one of the default templates, you will have to create a parameter for this type of use.

Using a Sub-Discipline type parameter makes it possible to separate views into their specific engineering systems. You can have all your electrical views listed under the Electrical discipline, but that may make it difficult to quickly find views you want. Having a second level of grouping that puts all the lighting, power, communications, and other systems views in their own group will create a more organized environment in the Project Browser.

The Family And Type value of a view is another commonly used level of grouping. This is what defines your views as plan views, sections, elevations, 3D views, or ceiling plans. When you begin to annotate your views in order to place them on construction documents, it is important to place the annotation in the views that will be used on your sheets, so being able to easily locate the proper view for annotation is important. Each grouping option can be set to utilize all the characters of the parameter value or only a specific number of leading characters.

Sorting Views within Groups

Once you have established what types of views will be grouped together, you can determine how the views will be sorted in their respective groups. View Name is most often used because ultimately you have to find the view you are looking for in the Project Browser. Views can be sorted in either ascending or descending order alphabetically or numerically.

Figure 2.2 shows the setup of a view organization that utilizes the Discipline, Sub-Discipline, and Family And Type properties as a grouping structure with the views sorted by view name. Notice how the views are shown in the Project Browser because of this organization.

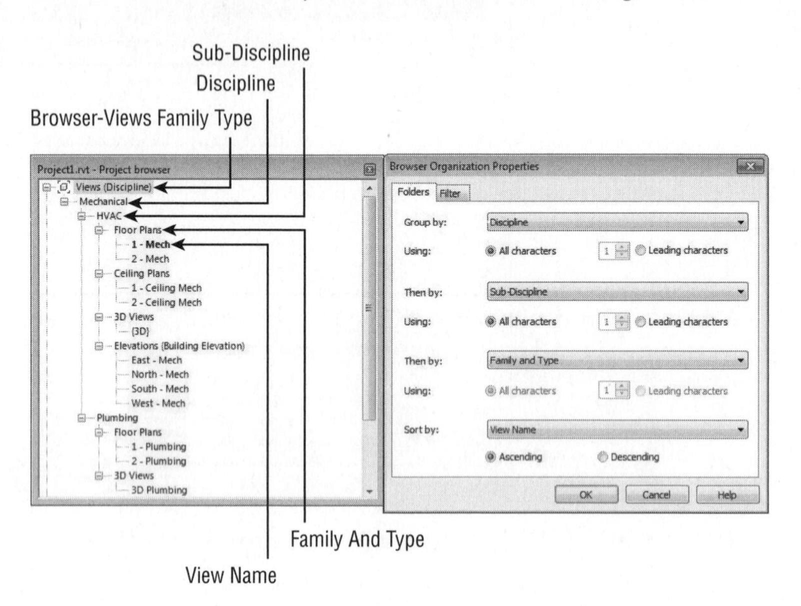

FIGURE 2.2 Sample view organization structure

Schedule views and legends are organized separately from model views. The only control you have as to how they are listed is by your naming convention because they will be listed alphabetically. The sheets used in your project are the only other element that you can sort and group in the Project Browser. Sheet organization options are discussed later in this chapter.

Determining the Number and Types of Views Needed

From a production standpoint, the purpose of a project template is to eliminate or reduce the need for repetitive tasks such as setting the scale of a view or turning off the visibility of certain model objects each time a new view is created. The ability to create a view and begin working without spending time setting it up can help reduce the time it takes to complete a project. It is also helpful in reducing drafting errors and maintaining a consistent look among construction documents.

We know that certain elements should be displayed in certain types of views and that some elements should display differently depending on the type of view. For example, you may want to show plumbing fixtures as halftone in a mechanical HVAC plan but display them normally in a plumbing plan.

The way that objects are displayed by default is set in the object styles of a project. We'll discuss object styles later in this chapter. The way that a view displays the model and specific objects within the model is controlled by the properties of the view.

When you select a view in the Project Browser, its properties are displayed in the Properties palette; if you are not using the Properties palette, you can right-click a view and select Properties. Another common method for accessing the properties of a view is to right-click in the drawing area and select Properties.

ACCESS TO VIEW PROPERTIES

One useful feature of the Project Browser is that it allows you to access the properties of your views (and any other element in the project) without having to open them in the drawing area. Parameters can be changed on many views without taking the time to open each view.

Some properties of a view are a result of the type of view that has been created. View types include floor plans, reflected ceiling plans, sections, elevations, and 3D views. These are all views used for displaying the model. There are also drafting and detail views that are used for displaying 2D details or diagrams. Some detail views are a combination of model display and detail components.

When building a project template, you should consider what types of views will be necessary. As with any component of a template, there is no need to create every type of view imaginable just because it might be used. Only choose to create views that you know will be used on nearly every project. Otherwise, you might end up creating more work for each project by having to clean up all the unused items. The types of views you create will also depend on your workflow for a Revit project. If all the design disciplines share a common Revit model, then you will want to have the views that each discipline requires. Obviously, if you create a separate project file for each discipline, then there is no need for all the discipline views in each template.

Number of Levels

Another important consideration is the number of levels to include in a template. A view should be created for each level, so it is important to decide on how many to start a project with. If you work primarily on two- or three-story buildings, it may be best to have only three or four levels established in your template, including one for the roof. More levels can be added to a project as needed. The number of levels is typically determined by the architect, so it is possible to not include any levels in your template and copy/monitor the architectural levels once the project file has been created. However, you must have levels in your template if you want to create views.

Plan Types

Views should be created in your template for each of the types of views to be used. These are generally determined by the Sub-Discipline property of the views. Whether your template is multidiscipline or single, a floor plan should be created at each level for each subdiscipline type.

When creating reflected ceiling plan views, it is a good idea to create only one ceiling plan view for each level. This promotes coordination among disciplines because everyone will be viewing the same ceiling plan and because all components of the model that occupy the ceiling are visible. Because this type of view does not belong to any one specific discipline, you may want to assign it to the Coordination discipline to distinguish it as unique.

Reflected ceiling plan views that are needed for construction documents can be created for the specific discipline that requires them once the project has been created.

Creating a Working View

A very useful type of view to create is what is typically called a *working* view. These are views that are the same as the views that will be used on construction documents, but the settings are different so that more or less of the model can be seen. They also allow for different graphic representations without having to constantly change back and forth within the view that goes on a sheet. For example, ductwork plans are typically shown in Hidden Line style, which can be difficult to navigate because of performance. A working view can be created that is set to Wireframe, which is easily navigable, and the modeling can be done there.

Choosing How to Display Each View

Once you have established the types and how many views to include in your template, you can set the properties that determine the display characteristics of the view. These settings will be the default, or *baseline*, settings because the need to change them occurs regularly while working in a project. In fact, the need is so common that a set of tools are available on the View Control Bar of the user interface for quick access to changing them. It may not seem necessary to set these properties since they can be so easily changed, but it is good to start with the best options for these settings.

You can choose the default settings by editing these parameters:

- View Scale
- Detail Level
- Visual Style

For the most part, model plan views are set to 1/8″ = 1′-0″ (1:100) scale and are displayed at Medium detail with a Hidden Line visual style. Consider using the Shaded w/Edges visual style at a Fine level of detail for 3D views. This gives the viewer a better sense of the model and will display any pipes or conduit in full 3D.

When creating views that will include piping, setting the detail level to Fine enables you to see the actual pipe, pipe fittings, and accessories instead of their single line representation. Many users prefer to model their piping systems in Fine detail because it is easier to see where connections are made and to discern differences in pipe sizes. For piping plans that are set to Fine detail, the Visual Style option should be set to Wireframe, because this will help improve performance when zooming or panning in the view.

Some other parameters you may want to edit include the following:

- Underlay
- Underlay Orientation
- Phase properties

You can use the Underlay and Underlay Orientation parameters to display other levels of the model as an underlay to the current view. Doing so will cause the underlay to display as halftone, while any detail or annotation graphics display normally. You can choose which level to underlay using the Underlay parameter, while the Underlay Orientation parameter determines how the model is being viewed. Any level of the model can be used in any other view. Although these are very useful settings for seeing how things line up in your model, it is not necessary to set an underlay in your default view settings.

Phase properties of a view are very important when working in phased projects. They add another level of visibility that can cause frustration if not set properly. Although these are instance parameters, they really should be used to determine the types of views you create. If you do a lot of renovation work, it is good to have default existing and demolition phase views in your template. Items placed into your model will take on the phase of view in which they are placed.

Plan views have type parameters only for setting what family to use for callout tags within the view and for setting what reference label is used when other callouts reference the view. Drafting views have the same type parameters as well as one for the section tag. Once you have determined what families will be used for your sections and callouts, you can assign them to the appropriate views.

Section and elevation model views have a unique parameter that allows you to not display the section or elevation marker at specified view scales. This eliminates the need for controlling the visibility of these markers with Visibility/Graphics Overrides settings. Figure 2.3 shows the parameter for hiding a section view marker at a specified scale.

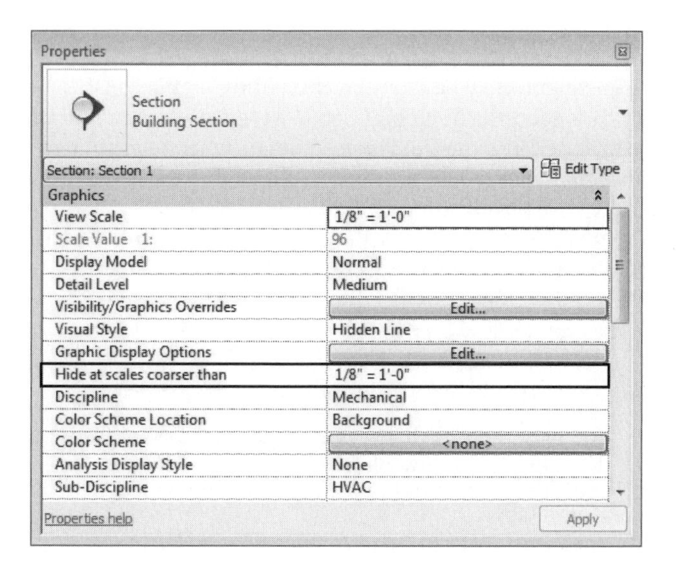

FIGURE 2.3
Parameter for hiding a section view

Visibility Settings for Template Views

There are many parameters for view properties that can be set by default. It is best to keep things simple by setting the most common parameters that will determine the general style of the view. Visibility settings are the most important to any view because you want to see what you expect to see in a particular type of plan. It can be frustrating and time-consuming to have to turn off unwanted model elements every time you open or create a new view.

There are two primary areas of visibility control within a view: the Visibility/Graphics Overrides settings, which allow for turning categories of elements on or off, and the View Range settings, which determine the field of view when looking at the model. Setting default values for these parameters is a key element of a good project template. Since working in a Revit project is a collaborative exercise, it is good to have the objects in your views display as expected or to not have objects showing up from other disciplines as they are placed in the model.

View Range

The three major components of View Range settings are the Top, Bottom, and Cut Plane of a view. For a floor plan, Top defines the elevation at which the model is being viewed from. Bottom is the extent to which the model is being viewed from the Top setting. In other words, it is how far you are looking. Cut Plane is an imaginary plane that cuts through the architectural and structural elements. The portions of these elements that are above the Cut Plane elevation are not visible.

Though you may have levels established in your template, there is no way of knowing what their actual dimensions will be until the building is modeled. However, there is a way to set default View Range settings that ensures the initial view of the model will correctly show the building elements. For example, to create a first-floor plan view, follow these steps:

1. For plan views, select View Properties ➤ View Range, and set Top to Level Above with an Offset setting of 0′-0″ (0mm).
2. Set Cut Plane to 4′-0″ (1220mm).
3. Set Bottom and View Depth to the Associated Level with an Offset setting of 0′-0″ (0mm).

Adjustments may be required depending on the construction of the building, but these settings are a good starting point because they will display all the visible model components from floor to floor. Since the Cut Plane setting is what determines the visible architectural and structural components, you don't need to worry that the actual floor object of the level above will interfere with visibility.

Figure 2.4 shows the View Range dialog box with the settings described in the previous steps for a first-floor plan view.

For a ceiling view, do the following:

1. Set Top to Level Above with an Offset setting of 0′-0″ (0mm).
2. Set Cut Plane to 4′-0″ (1220mm).

   You may choose a higher Cut Plane if you do not want to see items such as doors or windows in your ceiling plans. The Bottom setting is irrelevant in a ceiling plan because it is always behind your field of view.
3. Set View Depth to Level Above with an Offset of 0′-0″ (0mm).

FIGURE 2.4
Typical View Range settings for a plan view

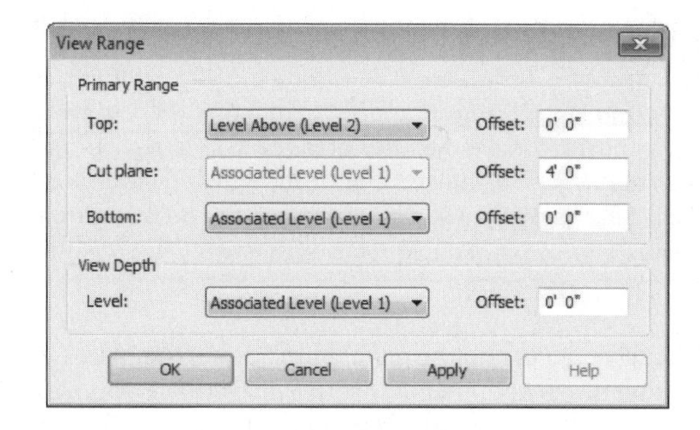

These settings ensure that for your ceiling plans you will see all visible model elements from the cut plane up to the top of the view range. So if there are ceilings with varied elevations on a level, they will all display, and any items you want to see above the ceiling for coordination will also be visible.

VISIBILITY/GRAPHICS OVERRIDES

In the Visibility/Graphics Overrides settings of a view, not only can you turn components on or off, but you can also change their color, line type, or transparency. Items that might ordinarily display with normal lines can be set to Halftone. You can apply settings to the subcategories of components as well.

One of the ideas behind establishing default visibility settings is that you do not want certain items showing up in specific views. For example, lighting fixtures are typically shown on a separate plan from receptacles and power devices, but since all of these components are being placed into one model, they will show up in every view (depending on the View Range settings) unless they are turned off.

It is a good practice to make a list of all model components that you would like to see in a particular type of plan and then turn off all others. If there are components that you are not quite sure of, it is best to leave them on, because seeing items encourages coordination, whereas not seeing them may lead to a design conflict in the model. Be sure to select the box in the lower-left corner of the Visibility/Graphics Overrides dialog box to display categories from all disciplines.

Annotation components are specific to the view that they are placed in, so it is not crucial to set up default visibility for them. After all, it is not likely that you will be placing air terminal tags in your lighting plan, for example, so there is no need to turn them off. However, there are some annotation categories that you may want to adjust such as setting the Space Tags category to Halftone. Many families contain nested annotations, so you should also check the subcategories of Generic Annotations to set any necessary visibility.

If your workflow consists of using linked files from other disciplines or consultants, those links will react to whatever visibility settings you apply to the view. This usually works well when a project begins, but as the model is more fully developed, you will find yourself constantly managing the visibility of the linked files in views. Consider linking files into your template that will act as placeholders for the actual files to be linked. Doing so will enable you to establish default visibility settings for the links, reducing time spent managing visibility once the project is in design.

PLACEHOLDER LINKS

Having linked files in your project template is a very useful feature for not only visibility but also for model positioning. If the Auto – Origin To Origin positioning option is used for a placeholder link, then the file used to replace it will also be positioned at its origin. Placeholder links can be very small files. In fact, they do not need to contain any information at all. When a project is set up, all that is needed is to use the Reload From option to replace the placeholder link with the actual model file.

Worksharing cannot be enabled in a Revit template (`.rte`) file, so if you want to establish default worksets and visibility settings for them, you will need to create a Revit project file (`.rvt`) that is to be used as a project template. The file can then be copied to establish the central file for a new project.

This scenario requires careful management of the file because of the nature of a worksharing environment. Some companies have written applications to make it easy for their users to set up a project this way without damaging or misplacing the project template file. You should set your template up this way only if you are absolutely certain it will be managed by personnel who have extensive Revit experience.

VIEW FILTERS

View filters are very useful for distinguishing similar components by line type or color. If you have a standard for the color or line type of certain elements, you can create the necessary view filters and apply them to the appropriate views. With view filters already established, you will see the components as expected as soon as you begin to model them.

If there is a variance in how filtered items are displayed from project to project, you may not want to apply the filters to your views, but having them established in your template will still save you some time. Be sure to name your filters in a manner that clearly indicates what the filter does so they will be easy to find and use.

Visibility Settings Shortcut: View Templates

Revit allows you to create a template within your template to expedite the process of establishing the visibility settings of your views. View templates enable you to define preset properties of view types that can be applied to any view in one simple step. These properties are not just for visibility but for any of the main view properties.

You can create a view template by clicking the View Templates button on the View tab of the ribbon and then selecting the View Template Settings option. On the left side of the dialog box is a list of any view templates that exist. You can sort the list by view type by using the drop-down in the upper-left corner. The right side of the dialog box lists the commonly used view parameters. Here you can set the default values for these parameters to establish the view template. There are also Edit buttons for setting the Visibility/Graphics Overrides settings and Graphic Display Option, which is used for displaying shadows in views. Any project parameters that you create that apply to views will also be available for editing.

The buttons in the lower left of the dialog box are for duplicating, renaming, or deleting a view template. There is no button to create a new template, so you must duplicate an existing one before you can begin. In fact, if no view templates exist in your file, you cannot access the View Templates dialog box from the ribbon.

A more common method for creating a view template is to first create a view and then to establish its desired properties, including the visibility settings. This is a more preferable method because you can see what the view will look like as you make adjustments to the properties. Creating a view template this way in your project template will require you to temporarily load a model to be used as the visual reference. Once you have the properties of a view established, you can right-click the view in the Project Browser and create a view template based on that view. You can also use the View Templates button on the View tab and select the Create Template From Current View option, as shown in Figure 2.5.

FIGURE 2.5
View Templates button

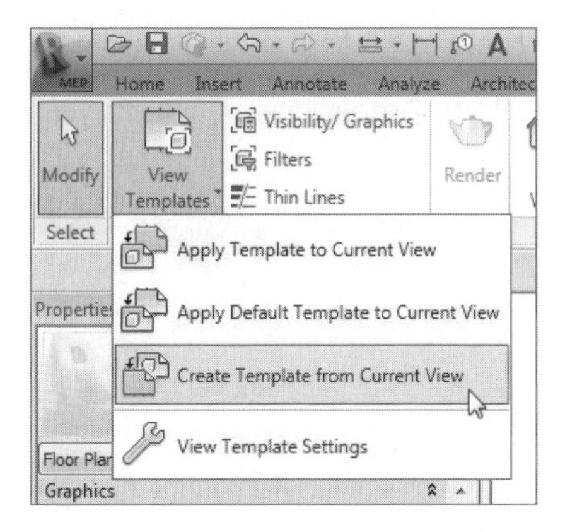

View templates are useful while creating your project template, but they are also very useful after the project has been created. One of the properties of views is Default View Template. You can assign a default template so that as changes are made to a view, you can revert to the default settings at any time. You can also apply any view template to any view, especially newly created ones, and quickly have the desired settings. When you work on a large project that has many levels or many dependent views, having view templates can save a significant amount of time.

In the View Templates dialog box, there is a check box next to each property that you can use to determine whether the settings of the template for that property will be applied to the view. This enables you to apply only the desired portions of a view template if necessary.

Schedule Views

Preset schedules in a project template are another feature that increases productivity and coordination. This topic is included in this section of the chapter because by displaying the data within the components that make up the model, schedules are actually views of the model.

Schedule views can be saved as their own project file that can be loaded into a project as needed, so it is not necessary to have every schedule that you might use in your project template. The types of schedules you should include are ones that you know will be in every project. If you are going to create a schedule in your template, you must use parameters that are available in the template file. The parameters will be available either because they exist in components that are loaded into your template or because they are set up as project parameters. This is most easily achieved by using shared parameters because it is important that the parameters in your content are the same as those in the schedule. Careful consideration should be taken as to what parameters need to be in the template prior to setting up a schedule. For more information on creating schedules, see Chapter 5; for more about creating and managing parameters, see Chapter 19.

Some types of schedules can be included in your template that are useful for managing the project. Consider adding a View List schedule to your template. This enables you to quickly view information about your views to determine whether you have all the views required for a project or whether a view has not yet been placed onto a sheet. This also is a good way to change the parameters of views without having to locate them in the Project Browser. Figure 2.6 shows a small sample of a View List schedule as it would look in a project file. There is much more information that can be added to a View List schedule, including any custom parameters that you create for views.

FIGURE 2.6
Sample View List schedule

VIEW LIST					
View Name	Title on Sheet	Sheet Number	Sheet Name	Scale Value 1:	Detail Level
EL - Level 1	FIRST FLOOR - LIGHTING	E3	FIRST FLOOR PLAN - LIGHTING	96	Coarse
EL - Level 2	SECOND FLOOR - LIGHTING	E4	SECOND FLOOR PLAN - LIGHTING	96	Medium
EP - Level 1	FIRST FLOOR - POWER	E5	FIRST FLOOR PLAN - POWER	96	Fine
EP - Level 2	SECOND FLOOR - POWER	E6	SECOND FLOOR PLAN - POWER	96	Medium
FA - Level 1	FIRST FLOOR - FIRE ALARM			96	Medium
FA - Level 2	SECOND FLOOR - FIRE ALARM			96	Medium
ISO DIAGRAM	DOMESTIC WATER ISO DIAGRA	P4	DIAGRAMS AND DETAILS	96	Medium
MH - Level 1	FIRST FLOOR PLAN - HVAC	M2	FIRST FLOOR PLAN - HVAC	96	Medium
P - Level 1	FIRST FLOOR PLAN - PLUMBIN	P2	FIRST FLOOR PLAN - PLUMBING	96	Medium
P - Level 2	SECOND FLOOR PLAN - PLUMB	P3	SECOND FLOOR PLAN - PLUMBING	96	Fine
Riser Diagram	POWER RISER DIAGRAM	E8	DIAGRAMS AND DETAILS	48	Fine
SANITARY ISO	SANITARY ISO DIAGRAM	P4	DIAGRAMS AND DETAILS	96	Medium
Zones - Level 1				96	Medium
Zones - Level 2				96	Medium

Other types of schedules you should consider for your template are any schedules that are used for analysis or schedule keys for applying values to parameters. These types of schedules are usually Space schedules since spaces hold much of the analytical information. Even if you do not use Revit for energy or engineering analysis, there is a type of Space schedule that can be very useful for managing names and numbers.

Revit's ability to generate spaces automatically is a great feature; however, it will generate spaces where there is no room object in the linked model. It can be tedious and time-consuming to search through a large floor plan looking for any spaces that should not be in the model. When these spaces are found in a plan view, deleting them removes them only from the model, not from the project. A simple Space schedule can be used to find all the spaces that should not exist and gives you the ability to delete all of them from the project with one click.

Figure 2.7 shows a sample Space schedule that reports the room name and room number of each space. The blank rows indicate spaces that have been placed where there is no room object. These rows can be highlighted in the schedule and removed from the project by clicking the Delete button on the ribbon.

If you are using the Panel Schedule feature within Revit MEP 2011, you can create templates for panel schedules and store them in your project template. Click the Panel Schedule Templates button on the Manage tab, and select Manage Templates. In the dialog box, you will see a list of any templates in your file. You can create a new one by clicking the Duplicate button at the bottom of the dialog box. See Chapter 5 for more information on creating and editing panel schedule templates.

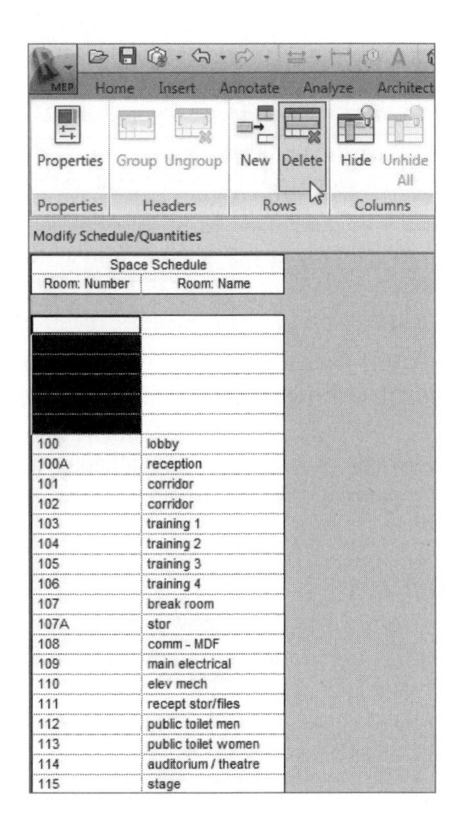

Space Schedule	
Room: Number	Room: Name
100	lobby
100A	reception
101	corridor
102	corridor
103	training 1
104	training 2
105	training 3
106	training 4
107	break room
107A	stor
108	comm - MDF
109	main electrical
110	elev mech
111	recept stor/files
112	public toilet men
113	public toilet women
114	auditorium / theatre
115	stage

FIGURE 2.7
Space schedule

Establishing Project Settings

Many settings can be preset in a project template to make it easier to begin a project. Some settings relate to how system families will behave, while others determine things such as how objects will print or what text will look like. There are also settings for values that Revit will use in calculations. Having these set properly in your project template ensures that when you begin a project, you will see the model and data correctly and to your standards.

Object Styles

The Object Styles settings within Revit determine how elements will display by default if no overrides are applied to them in a view. You can set the defaults for model and annotation components as well as for the layers of any linked or imported CAD files. Settings can be applied to subcategories as well.

Even though the Visibility/Graphics Overrides settings are often used in views, it is important to set the standards for how elements will display. The need for overrides comes from having several types of an element within one category. For example, you can set the style for the Ducts category, but if you want to display different types of duct in unique ways, you will have to use overrides.

You can access the Object Styles settings for your template from the Manage tab of the ribbon. The Object Styles dialog box lists all the model, annotation, and imported categories on separate tabs for easy access. These are the types of settings you can apply to a category:

Line Weight Projection line weight is the thickness of the lines of an object if it falls within the view range of a view and is not cut by the cut plane.

Cut line weight is the thickness of the lines of an object that is cut by the cut plane. Keep in mind that the cut plane does not apply to MEP objects and affects only architectural and structural elements, which is why the column is grayed out for MEP categories.

Line Color This setting is for establishing the default color of objects within a category.

Line Pattern This setting determines what type of line will be used for objects in a category. This setting does not utilize a line style but applies a line pattern directly to the objects.

Material You can apply a material to a model category. This is primarily for rendering purposes but can also be useful for material takeoffs; however, the material applies to the entire category, so you cannot establish unique materials for different types of pipe or duct using this setting.

Drafting Line Settings

In the same way that the Object Styles settings define how model, annotation, and imported objects are displayed and printed, it is necessary in a template file to define the various line styles that will be used for any drafting or detailing that may be done in your projects. A line style is defined by its weight, color, and pattern. You can create different combinations of these settings to define lines that are used for specific drafting purposes or that match your standards.

Line Weights

The first settings to consider when creating line styles are the available line weights in your template file. You can access the Line Weights settings by clicking the Additional Settings button on the Manage tab of the ribbon.

With Revit you can establish 16 different line weights. Typically, line weight 1 is the thinnest line, and line weight 16 is the thickest. The Line Weights dialog box has three tabs that give you access to the settings for lines depending on what type of view or what objects the line weights are applied to. The first tab is for model objects. Model line weights are dependent on the scale of the view they are in. You can define a thickness for each of the 16 line weights as it appears in a specific view scale. This gives you the freedom to show lines that are usually very thick as much thinner when the view scale is larger. Figure 2.8 shows the Model Line Weights tab of the Line Weights dialog box. Notice that line weight 14 is half as thick in a 1/16 scale view as it is in a 1/8 scale view. This keeps items from printing as blobs when using larger scale views, without having to manually adjust the line weights of objects using Visibility/Graphics Overrides.

You can add or delete view scales for line weight settings by using the buttons on the right side of the dialog box.

The Perspective Line Weights tab lists the thickness settings for the 16 line weights as they would appear in a perspective view. These settings do not apply to the default 3D view types but only to views that are generated from an explicit camera position. There is only one setting for each line weight because these line weights are consistent at any view scale.

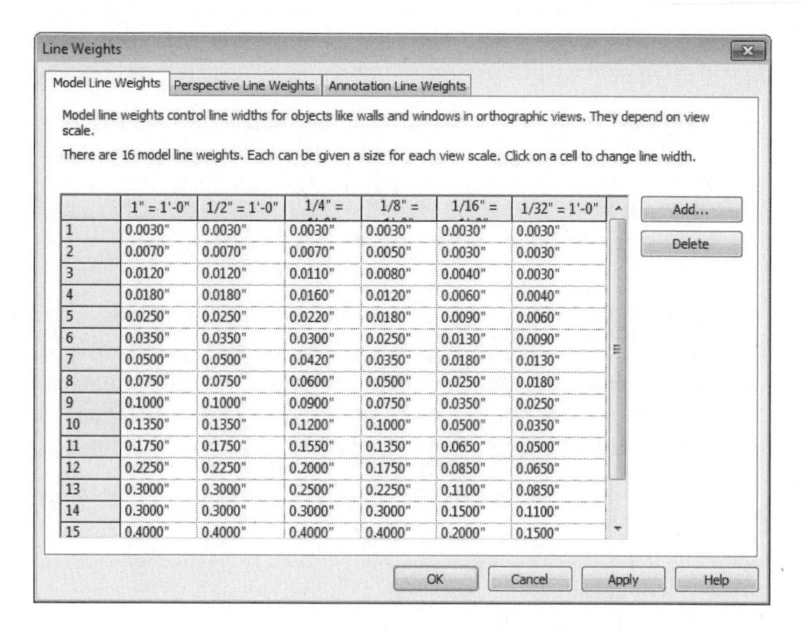

	1" = 1'-0"	1/2" = 1'-0"	1/4" =	1/8" =	1/16" =	1/32" = 1'-0"
1	0.0030"	0.0030"	0.0030"	0.0030"	0.0030"	0.0030"
2	0.0070"	0.0070"	0.0070"	0.0050"	0.0030"	0.0030"
3	0.0120"	0.0120"	0.0110"	0.0080"	0.0040"	0.0030"
4	0.0180"	0.0180"	0.0160"	0.0120"	0.0060"	0.0040"
5	0.0250"	0.0250"	0.0220"	0.0180"	0.0090"	0.0060"
6	0.0350"	0.0350"	0.0300"	0.0250"	0.0130"	0.0090"
7	0.0500"	0.0500"	0.0420"	0.0350"	0.0180"	0.0130"
8	0.0750"	0.0750"	0.0600"	0.0500"	0.0250"	0.0180"
9	0.1000"	0.1000"	0.0900"	0.0750"	0.0350"	0.0250"
10	0.1350"	0.1350"	0.1200"	0.1000"	0.0500"	0.0350"
11	0.1750"	0.1750"	0.1550"	0.1350"	0.0650"	0.0500"
12	0.2250"	0.2250"	0.2000"	0.1750"	0.0850"	0.0650"
13	0.3000"	0.3000"	0.2500"	0.2250"	0.1100"	0.0850"
14	0.3000"	0.3000"	0.3000"	0.3000"	0.1500"	0.1100"
15	0.4000"	0.4000"	0.4000"	0.4000"	0.2000"	0.1500"

FIGURE 2.8
Model Line Weights tab

The Annotation Line Weights tab is used for defining the thickness of annotation objects. These settings are also independent of view scale. Another way of thinking of this is that the settings you apply determine how thick the lines will print. This is the easiest place to start when establishing your line weights for your template. Most people use only about six or seven different line weights, so it is not necessary to come up with a thickness for all 16 line weight options.

Once you have established the settings for your line weights on the Annotation Line Weights tab, you can then apply the same settings to the Model Line Weights and Perspective Line Weights tabs. For the Model Line Weights tab, you will need to decide how to reduce the thickness of your heavier line weights at larger scales. The line weights that you define will be available for use in your template and subsequent project files for setting object styles or overriding the visibility of categories in views.

LINE COLORS

The next consideration for a line style is the color of the lines. The main thing to remember when setting the color for objects or for lines is that the color has no bearing on how thick the lines will print. You can have a red line with a line weight of 3 next to a blue line with a line weight of 3, and they will both print the same way.

Color can be useful for distinguishing systems or objects in a crowded engineering plan or detail. If you have a set of standards for the color of specific types of components, you can set the color for line styles that you create to represent those objects in the same way you would set the color for model objects in the Object Styles settings. Be aware that using colors in Revit may cause printing issues. If your print settings are set to print color, then all colored lines and objects will print as expected. However, if you are printing to a black-and-white printer and your settings are set to Grayscale, any colored objects or lines will print at the grayscale equivalent of that color, which could produce unexpected results. This has more to do with managing your print settings, but it is important to consider when deciding to apply colors to line and object styles.

LINE PATTERNS

The final consideration for creating a line style is the pattern of the lines. To access the line patterns available in a Revit file, click the Additional Settings button on the Manage tab, and select Line Patterns. Revit line patterns consist of dashes, dots, and spaces.

> **TEXT IN A LINE PATTERN**
>
> It is not possible to include text in a line pattern in Revit. One common workaround to this shortcoming is to use tags. You can place a tag multiple times along a line that represents something such as a pipe or wire. Though the tags must be placed individually, they will at least move with the object and be removed when the object is deleted.

The line patterns in Revit are independent from the view scale in which they are drawn, and there is no setting to apply a scale to an individual line. Because of this, you will need to create multiple line patterns for the same kind of line at different scales. This will allow for various lengths of the dashes and the spaces between dashes and dots as needed to display the line pattern properly. In Figure 2.9, you can see the Line Patterns dialog box and that additional Center line patterns have been created at 1/4, 3/8, and 5/8 scales to be used in different view scales.

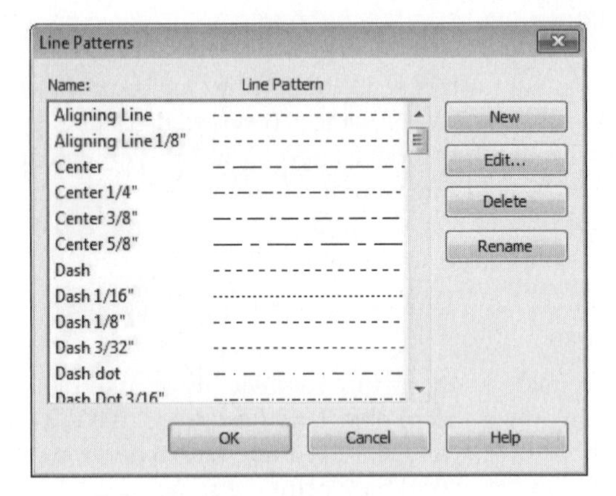

FIGURE 2.9
Line Patterns dialog box

You can modify the settings for a line pattern by clicking the Edit button on the right side of the dialog box. In the Line Pattern Properties dialog box, you will see the components that make up the line pattern. In the Type column, you can select a dash, dot, or space. Spaces can be used only after a dot or a dash. In the Value column, you assign the length of the dashes or spaces. Dots have a static length value that cannot be changed. You can enter up to 20 of these components, and when you have reached the point where the pattern will repeat, you are finished.

Variations of a line pattern for different view scales can be made by first looking at the settings for a line pattern. Figure 2.10 shows the settings for a Dash Dot Dot line pattern that is used in 1/8 scale views. To create the same pattern for use in 1/4 scale views, you would click the New button on the right of the Line Patterns dialog box, give the pattern a name such as Dash Dot Dot 1/4", and then put in the dashes and dots with values that are half of those in the Dash Dot Dot 1/8" pattern.

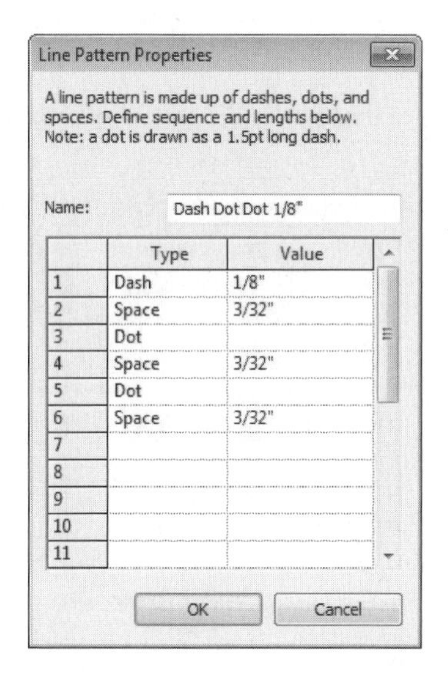

FIGURE 2.10
Line Pattern Properties dialog box

It is a good practice to test your line patterns as you create them. You can do so by creating a drafting view and drawing some parallel lines. As you create new patterns, you can assign them to the lines to see how the variations display and make adjustments to dash or space values as necessary. Figure 2.11 shows the multiple variations of a Center line pattern. The variations you create do not have to match the view scales. You can create variations that slightly modify the lengths of dashes and spaces so that the line pattern is more usable in certain situations. The 3/8" and 5/8" patterns in Figure 2.11 are examples of line patterns that may be used at any view scale depending on the length of the line.

LINE STYLES

With line weights and patterns defined in your template file, you can create line styles that can be used for model or detail lines in your projects. Line styles are separate from object styles because they apply only to lines created by using the Detail Lines or Model Lines tools or when creating the boundary of a region. The line styles you create will appear under the Lines category on the Model Categories tab of the Visibility/Graphics Overrides dialog box when changing the appearance of a view. You can access the Line Styles settings by clicking the Additional Settings button on the Manage tab and selecting Line Styles. Revit comes with some line styles that are coded into the program and cannot be removed or renamed, although you can change the settings for these lines. Figure 2.12 shows these lines and their default settings.

FIGURE 2.11
Line pattern variations

Center 1/2"
Center 1/4"
Center 1/8"
Center 1/16"
Center 3/8"
Center 5/8"

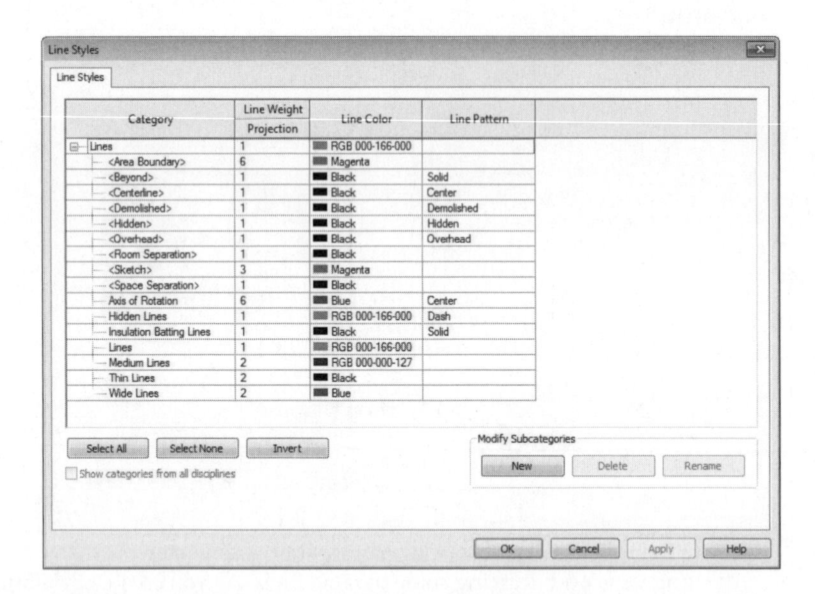

FIGURE 2.12
Line Styles dialog box

To create a line style, simply click the New button at the lower right of the dialog box, and enter a name. You can then assign a weight, color, and pattern to the line style. In some cases, it is useful to create line styles that match line patterns you have created to ensure consistency between model objects and drafting items. For example, if you create a line pattern that will be used for a domestic hot water pipe, you may want to create a line style called Domestic Hot Water that uses that pattern. That way, the domestic hot water pipe in your model can be given the pattern using a filter, and your diagrams can use the domestic hot water line type that matches. Otherwise, you would have to override each line in your diagram to the appropriate pattern.

Line styles are very useful, and having a good set of line styles can save time when drafting in a project. You can change a line from one style to another by using the Type Selector that appears in the Properties palette when you select a line. This is useful for creating details because you can draw all the line work with the most common line style and then easily change any lines that need to be different.

Export Settings

To share your project with consultants, it is often necessary to export your Revit views to CAD. Once you establish line weights and styles, it is a good time to consider your settings for exporting views. This task can be a bit tedious and time-consuming, but the settings you define can be saved and used any time. The settings you establish for exporting do not need to be applied to your template, but it is a good idea during your template creation to consider your export settings.

You can access the export settings from the Application menu, as shown in Figure 2.13. You can create settings for various types of CAD file formats and also for IFC export.

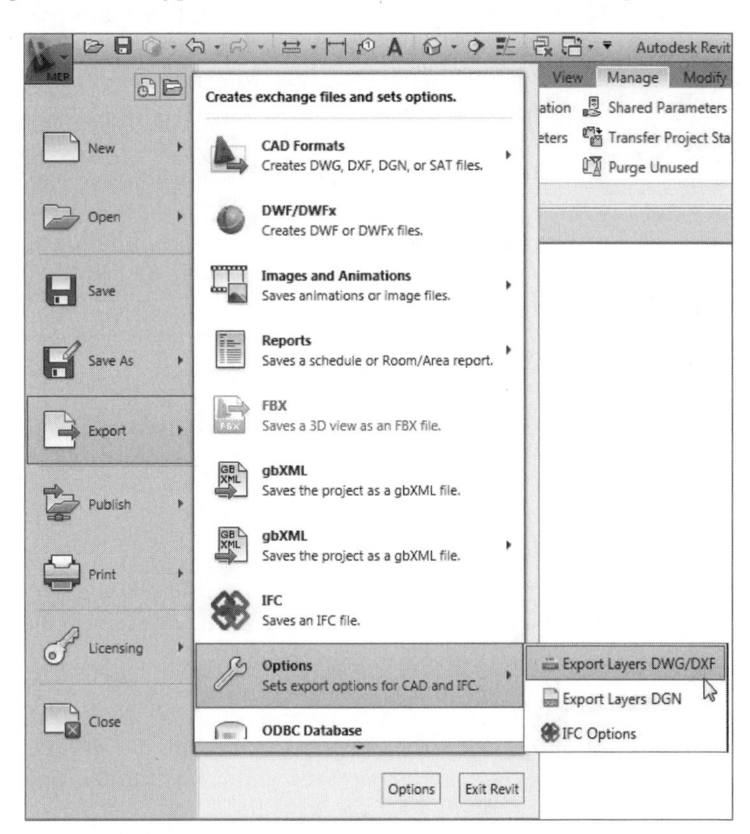

FIGURE 2.13
Export settings

IFC

Just as with CAD programs, there is a need for a standard file type that can be read by any BIM application. Industry Foundation Class (IFC) files can be shared by BIM applications, with the goal of being a seamless translation of data from one application to another. For more information on IFC, visit `www.buildingsmart.com/bim`.

When you click the option for exporting layers, a dialog box will appear that lists every category and subcategory within Revit. Alongside the list are columns that allow you to assign an associated AutoCAD layer name and color ID to each category and subcategory. The color ID is as it is defined in AutoCAD. 1 is for red, 2 is for yellow, and so on. It is important to note that when you export a view with the settings you establish, the lines created in the CAD file will be the color you assign. They will not be colored BYLAYER. There are also columns for assigning layers and colors to objects that are cut by the cut plane of a view.

You can choose an industry standard for export settings by clicking the Standard button on the right side of the dialog box and choosing the desired standard, as shown in Figure 2.14. This is a good starting point if you are creating settings that match your standards.

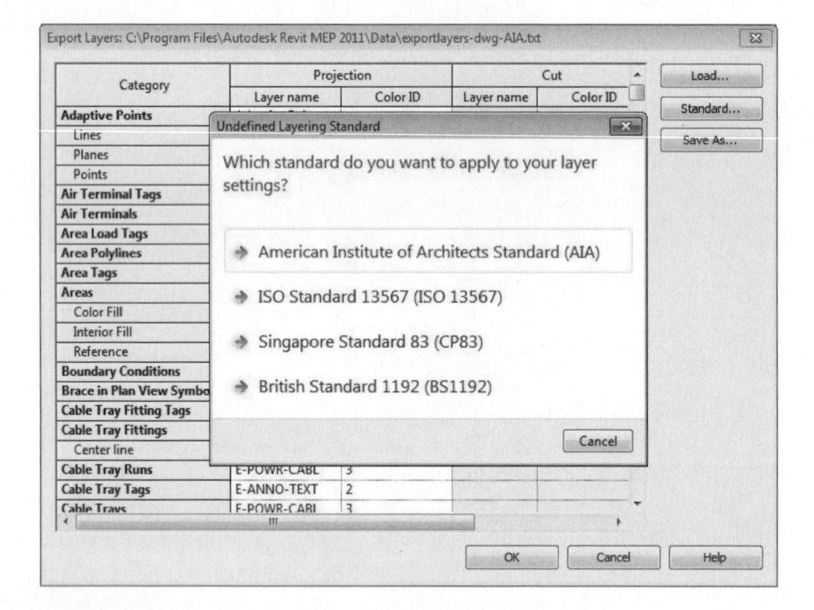

FIGURE 2.14
CAD export standards options

After you have assigned a layer and color to each category and subcategory that you require, you can save the settings for future use or for sharing with consultants. Click the Save As button to save your settings as a text file. You can import this file into other projects prior to exporting views, and you can share the export settings among consultants to ensure consistency between construction documents and shared `.dwg` files.

The process for creating export settings for `.dgn` files is the same except that you associate a level number with each category instead of a layer name.

Annotation Styles

The goal of most Revit projects is to create a set of coordinated construction documents. Annotation that is consistent from drawing to drawing and from discipline to discipline is a major part of that coordination. One of the benefits of using computers to do drafting is that it makes it easier to apply drafting standards. In the days of manual drawing, there was no guarantee that one person's lettering would look exactly like another. Some may argue that you still don't have that guarantee by using computers. That is why it is important to establish your annotation standards in your Revit project template.

TEXT

Text is a Revit system family, and the various text styles that you need can be created as types of that family. There is no dialog box for text settings, so to create the styles you want, you should start by accessing the properties of the default type and duplicating it for each type. This is done by clicking the small arrow located on the Text panel of the Annotate tab, as shown in Figure 2.15.

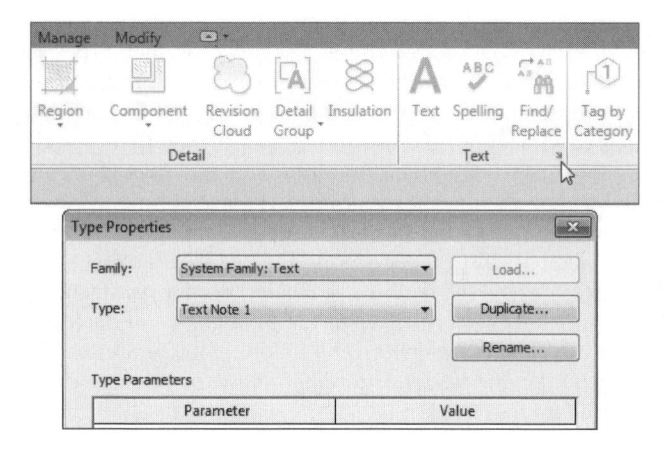

FIGURE 2.15 Accessing the type properties of the Text system family

To create a new text type, duplicate the default type, and provide a name for the text type you are creating. It is a common practice to name text types using the size and name of the font applied to the type. The type properties of a text type allow you to set the behavior and general appearance of the text type.

It is very important to understand how Revit handles text types before choosing how to set up your text family types. Each text family type that is created is unique to the file in which it was created. Figure 2.16 shows an example of a text type called STYLE 1 that was created in a project and a text type called STYLE 1 that was created in an annotation family. When the annotation family is used in the project, the text within the annotation family maintains its settings as defined in the family file. It does not take on the settings for STYLE 1 as they are defined in the project, even though both family types have the same name.

FIGURE 2.16 Text types with the same name

The easiest way to manage this behavior is to use the default font, which is Arial. Although this may be a deviation from your normal CAD standards, it gives a huge return in time savings. If you choose to use a different font, you will have to change the text and labels in every family or detail that you bring into your Revit project in order to maintain consistency in your construction documents. This includes all the preloaded content that comes with your Revit MEP installation.

By setting the properties of a text style, you can create several variations of text for use in your projects. Remember that these are type properties, so you will need to create a new text type for each variation in settings. The most common types are based on text height, background, and leader arrowhead style. You can also create types based on other properties, which are usually variations of your standard text types. For example, you can create a standard 3/32″ Arial text type for normal use and then create another type that is 3/32″ Arial – Underlined for use where underlined text is required. This will save you time in having to edit text to make it underlined by giving you the option to switch from one type to the other.

The Show Border and Leader/Border parameters allow you to create a text type that has a border automatically placed around the text that is offset a certain distance from the text. Even if you do not show the border, the Leader/Border parameter determines the distance between where the leader starts and the text. The thickness of the leader and the border are determined by the Line Weight parameter. The Color parameter determines not only the color of the text but also the border and leader. Figure 2.17 shows some sample text types, named based on the settings used for each type.

FIGURE 2.17
Sample text types

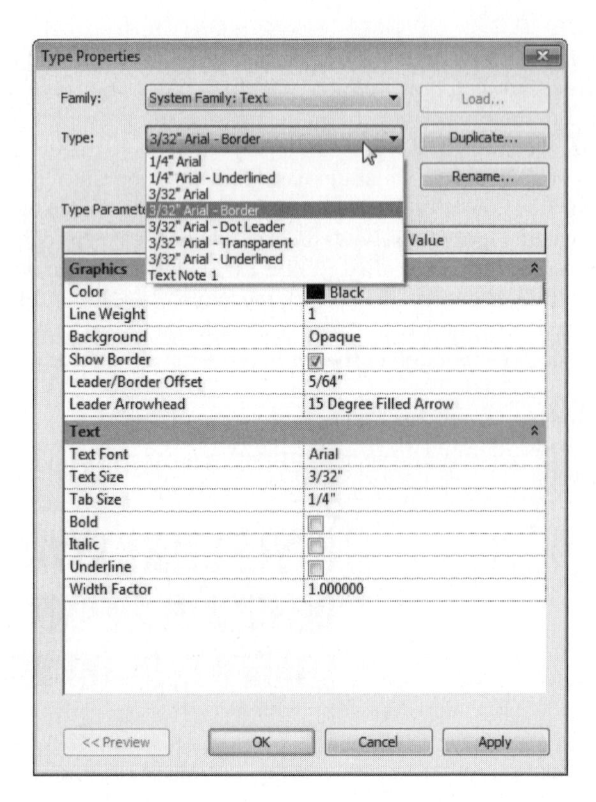

ARROWHEAD STYLES

One of the key considerations for setting the text standards in your template is the type of arrowhead your text types will use. You have the ability to modify the look of arrowhead types to suit your needs if necessary. You can find the settings for arrowheads by clicking the Additional Settings button on the Manage tab and selecting Arrowheads. Arrowheads are a Revit system family, and although you can create different types, the Arrow Style parameter determines the shape of the arrowhead. You cannot create your own arrow styles.

By adjusting the settings, you can control the size of an arrowhead type. The Tick Size parameter controls the overall length of the arrowhead, while the Arrow Width Angle determines the angle of the arrow from its point, which ultimately sets the width of the arrow. The dimension styles you establish will use these same arrowhead styles.

DIMENSION STYLES

Dimension styles are another key factor in creating consistently annotated construction documents. To establish the settings for your dimension styles, click the small arrow on the Dimension panel of the Annotate tab, as shown in Figure 2.18. Each type of dimension can have its own unique settings. If you do not use dimensions very often in your projects, you may want to consider leaving the default settings. When the need arises to show dimensions, you could establish the settings or transfer the dimension settings from a file in which they have already been established, such as a linked consultant's file or a previous project.

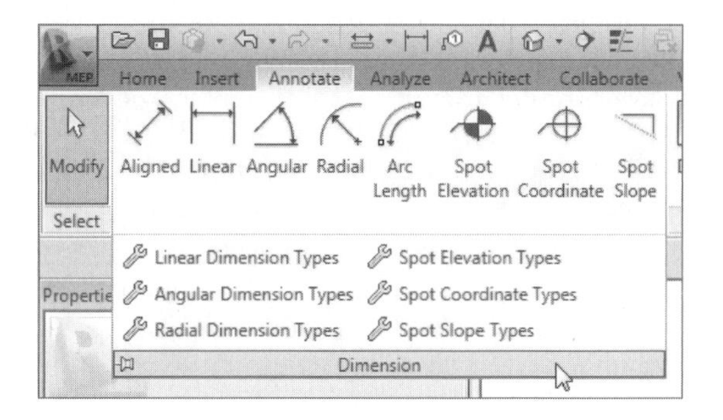

FIGURE 2.18
Accessing Dimension Types settings

> **TRANSFER PROJECT STANDARDS**
>
> The Transfer Project Standards feature of Revit can be very useful and save you time on your projects, but you should use it with caution. If you find yourself using this feature often on projects, you should consider establishing the settings you are transferring in your template to reduce the need for transferring.

Each dimension style is a system family, and you can create types within that family in the same way text types are created. Dimension styles have more parameters than text, however. Many of these parameters are for controlling how the graphics of the dimension will display. One of the key parameters in determining a dimension family type is the Dimension String Type parameter. This defines how the dimension will behave when a string of dimensions is placed.

You have options for creating a Continuous, Baseline, or Ordinate dimension string. If you use all or some of these types, you will need to create a separate dimension type for each one.

The Tick Mark parameter determines which arrowhead style will be used. There is a drop-down list when editing this parameter that lists all the arrowhead styles defined in the file. The line weight of the dimension can be controlled independently from the line weight of the tick mark used by setting the parameters for each. The line weight for tick marks in dimensions should match that of the line weight for leader arrowheads in your text types for consistency. Because of their relatively small size, using a heavy line weight may cause your arrowheads to look like blobs, so choose wisely. The Interior Tick Mark parameter is available only when you have set the dimension Tick Mark to an arrow type. This determines the style of arrowhead to be used when adjacent dimension lines are too close together to fit the default tick marks.

Other parameters control the lengths of witness line components and gaps and also the text used in the dimension style. Some of the settings for text within a dimension style are the same as those in a text style, such as font and text height. The Read Convention parameter allows you to set the direction in which the text will be read for vertically oriented dimensions. With the Units Format parameter, you can set the rounding accuracy of dimension types independent from the default project settings.

Project Units

Whether you are creating a template using metric or imperial units, you will need to establish what units of measurement are used and their precision. These settings will determine the default reporting of data not only in views but also in schedules and parameters. Click the Project Units button located on the Manage tab to access the settings for units.

You can set the default units for any graphical or engineering measurements. Figure 2.19 shows the Project Units dialog box. The drop-down list at the top contains the different discipline specific groups of units, with the Common discipline containing units that are used by all regardless of discipline.

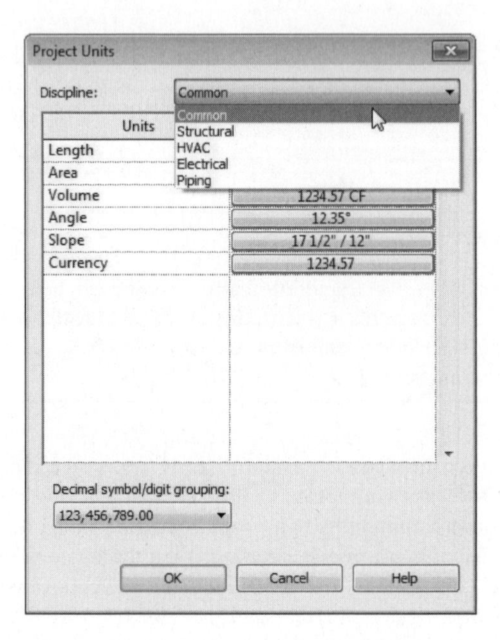

FIGURE 2.19 Project Units dialog box

Clicking the button in the Format column next to a unit will activate the Format dialog box for that unit. Here you have a drop-down list for the different unit options for that unit type. Once you have chosen a unit of measurement, you can determine the accuracy by selecting a Rounding option. The rounding increment will display to the right of the drop-down as an example of the option chosen. If you are using a decimal measurement, you can select Custom from the Rounding drop-down and designate the rounding increment manually. The Unit symbol drop-down offers the option for displaying the measurement unit next to the value if desired. There are check boxes in the bottom half of the dialog box for suppressing zeroes or spaces or for digit grouping depending on the type of unit you are formatting. When you click OK to finish formatting a unit type, you will see a sample in the button in the Format column next to the unit.

Setting the accuracy of certain unit types is important not only for how units display but also for the availability of model elements. For example, if you set the rounding of the Pipe Size unit to the nearest inch, then when you go to place a pipe in the model, only pipe sizes of 1″ increments will be available, even if you have pipe sizes defined at smaller increments in your project.

Project Phases

Although each project is different, you may want to establish phases in your project template if you are required to use them on many of your projects. The most common use of phasing is for renovation projects where the existing portion of the project is modeled. Phasing can be difficult to manage, so having the settings established in your template can be very beneficial.

Click the Phases button on the Manage tab to access the settings for phases in your template. The Phasing dialog box has three tabs for setting up the phases and their behavior. The Project Phases tab is where you establish what phases exist in your file. Revit starts each file with an Existing phase and a New Construction phase by default.

You can add phases using the buttons on the upper-right side of the dialog box, inserting them before or after the phase selected in the list. The list of phases starts from the earliest and ends with the latest. So if you were to insert a Demolition phase, you could select New Construction and use the Insert Before button, or you could select the Existing phase and use the Insert After button to place the Demolition phase between Existing and New Construction. The order of phases is very important because when views are set up, you will establish what phase they belong to. Any items placed into the model will be part of the phase that is set for the view in which they are placed.

During the course of a project, it may be decided that a phase is no longer necessary. You can use the Combine With buttons at the right of the dialog box to transfer the items from one phase to another.

The Phase Filters tab of the dialog box lists the different viewing options that can be applied to any view in order to display items from various phases, as shown in Figure 2.20.

The filter names describe what will be shown in the view that they are applied to. The New, Existing, Demolished, and Temporary columns define how the items that belong to the phases will be displayed. If you create a custom phase, the New column controls how the items are displayed in a view set to that phase. Any items placed in a phase prior to that are considered Existing. When you demolish an item, you can assign in which phase the demolition occurs; otherwise, the item is considered to be demolished in the phase that is applied to the view you are working in. You can create custom settings by creating a new filter using the New button at the bottom of the dialog box. It is a good practice to name the filter so that it is evident what will be shown when the filter is applied to a view.

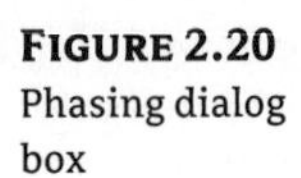

FIGURE 2.20
Phasing dialog box

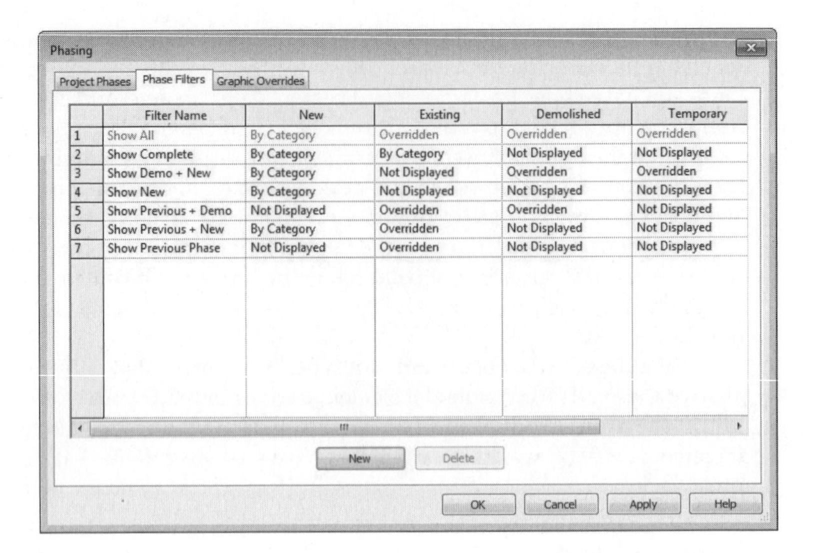

	Filter Name	New	Existing	Demolished	Temporary
1	Show All	By Category	Overridden	Overridden	Overridden
2	Show Complete	By Category	By Category	Not Displayed	Not Displayed
3	Show Demo + New	By Category	Not Displayed	Overridden	Overridden
4	Show New	By Category	Not Displayed	Not Displayed	Not Displayed
5	Show Previous + Demo	Not Displayed	Overridden	Overridden	Not Displayed
6	Show Previous + New	By Category	Overridden	Not Displayed	Not Displayed
7	Show Previous Phase	Not Displayed	Overridden	Not Displayed	Not Displayed

The options for the display of a phase are defined on the Graphics Overrides tab of the Phasing dialog box. Overrides to each Phase Status will affect only the objects when they are shown via a phase filter. They do not override the object styles defined in your project in views in which the objects are created. In other words, if you apply an override to the existing status, the overrides will not apply when you are working in an existing view, only when the existing phase is displayed in a view of another phase.

Figure 2.21 shows two pipes in a view that is set to the New Construction phase. The pipe on the bottom was modeled in a view that is set to the Existing phase. The Phasing dialog box shows the overrides for existing items. Notice in the properties of the view that the Phase Filter property applied to the view is Show Previous + New. These settings result in the existing pipe being displayed as dashed and halftone in the New Construction view.

If you establish phases in your template file, it is helpful to create views for each phase with the proper Phase and Phase Filter properties to maintain consistency throughout the project and to see expected results when modeling.

Defining Preloaded Content and Its Behavior

When you begin a project using a template file, you want to be able to start modeling right away without taking the time to load components and set up system families up front or having to stop periodically during the design and modeling process. Determining what content is loaded or defined in the template will give you more time to focus on the model and design decisions and will ensure consistency of standards between projects.

Annotation Families

Loading annotation families is especially important for consistent standards when working in a project that is shared by multiple disciplines. Even if you are creating a template for just one discipline, there are many annotations that are used on every project and should be included in your template(s).

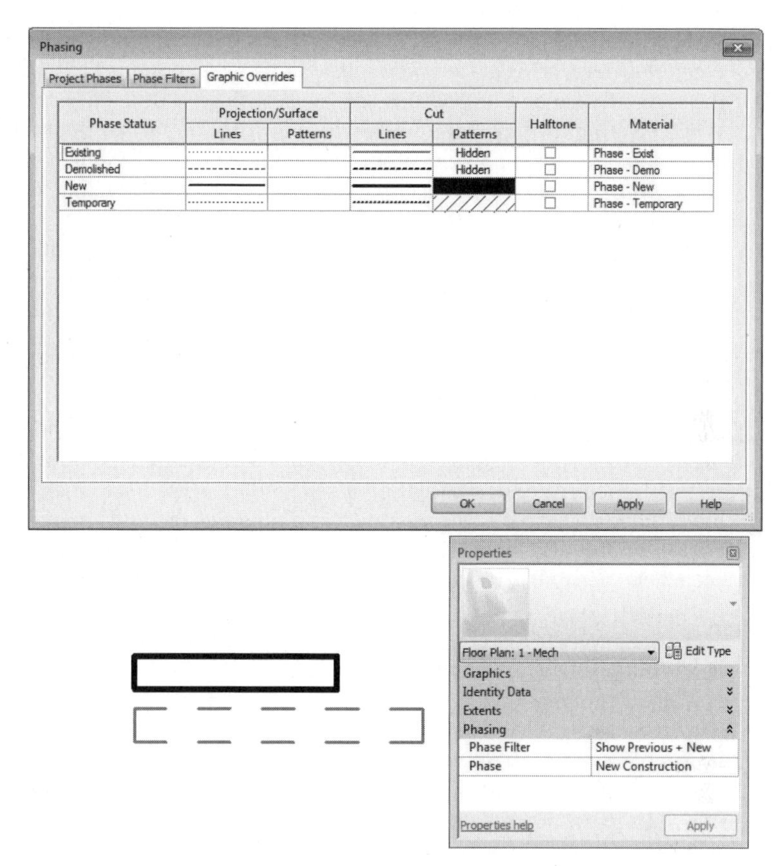

FIGURE 2.21 Existing and New pipes shown to demonstrate phase display

The symbols used for sections, callouts, and elevations should all be defined in your template file. To set up these standards, click the Additional Settings button on the Manage tab. For section tags, you can define what annotation family is used for the head of the section as well as the tail. There is also a setting for how the section tag will display when broken using the Gaps In Segments grip on a section line. These settings apply to the Section Tag system family. You can then create different types of sections by defining what section tag is used in the type properties of a section.

Elevation marks are created in a similar fashion. You first define the different types within the Elevation Tag system family by defining which annotation symbol is used and then apply the Elevation Tag types to various types within the Elevation system family. For callout tags, you can define what annotation is used and also the radius of the corners of the callout box that is drawn around a room or area of the model. The type properties of a callout are where you define what callout tag is used and what label to use when a callout references another view.

View titles are another type of annotation that should be defined in your template to match your drafting standards. To create custom viewport types, you will need to first establish the annotation family to be used as a view title. These annotations do not require a line for the title line because title lines are part of the Viewport system family and are generated automatically when the viewport is placed on a sheet. You also do not need to include a callout tag within the view title annotation because the tag to be used is defined in the type properties of a view. When creating the label that will be the view title, be sure to extend the limits of the label to accommodate a string of text; otherwise, your view titles will become multiple lines with only a few words. You can load several annotations into your template file in order to create multiple types of viewports.

To access the properties of a viewport, you need to place one on a sheet. Once placed, you can click the viewport and access its type properties. There you can duplicate the selected type and name it with a descriptive name. In the type properties of a viewport, you can define the annotation used for the title as well as the color, line weight, and pattern of the title line. There are also options for displaying the title or the title line, giving you the ability to create viewport types that do not display a title.

Another type of annotation to consider for your template are any tags that are commonly used. Pipe and duct size tags, wire tags, and equipment tags should all be loaded into their respective templates or into a template shared by MEP disciplines. In some cases, you may have more than one tag for a category such as a pipe size tag and a pipe invert elevation tag. For categories with multiple tags, it is very helpful when working in a project to have the default tag set. Click the small arrow on the Tag panel of the Annotate tab, and select Loaded Tags to access the Tags dialog box. In the Tags dialog box, you can define which tag will be used by default for each category. You can change this at any time during the project, but it is nice to start with the most commonly used option.

General annotations such as graphic scales or north arrows should also be included in your project template. You may choose to include a north arrow in your view title annotation, but keeping it separate gives you the freedom to rotate and place it anywhere on a sheet.

The same annotation families that are used in your model components can also be loaded into your template for use in creating legends. Legend components are limited in their placement options, so it may be easier to use annotation symbols; however, this method also results in having numerous annotation families loaded into your project that are only there for the legend view.

If you use a generic tag for plan notes, it should be loaded into your template. If you use the keynoting feature within Revit, you should have a keynote tag loaded as well as a keynote data file location defined. You access the Keynoting settings by clicking the small arrow on the Tags panel of the Annotate tab. In this dialog box, you can browse to the keynote data file to be used and set the path options. You can also define the numbering method to display either the specification section number and text from the data file or the By Sheet option, which numbers the notes sequentially as they are placed.

Component Families

The types of component families that you load into your project template really depend on what discipline you are creating the template for. If your template is used for a single MEP discipline, then there is no need for components that are used by other disciplines, unless you use those types of components regularly in your projects.

TOO MUCH OF A GOOD THING

It may be tempting to load your entire component library into your project template so that you would never have to pause what you are doing to load a family. Another reason may be so that you are certain the proper content is loaded. Although this may seem logical, with proper training on how to load families, it is not really necessary. Having an entire library of components in your template will increase the size of the file greatly and will result in every one of your projects starting at that size, not to mention that most of that content will be unused.

The most effective use of preloaded components is to have components loaded that are used on every, or nearly every, project. The following are examples of the types of components to consider for each discipline:

Mechanical Projects with HVAC systems will require air terminals and some type of distribution equipment. Consider loading an air terminal family for each type of air system into your template. Even if the types of air terminals used for the project end up being different from those in the template, at least you will have something to start with for preliminary design. The same is true for equipment. Load an equipment family that you most commonly use. Figure 2.22 shows an example of components loaded into a template for HVAC systems.

FIGURE 2.22
Sample HVAC components in a template

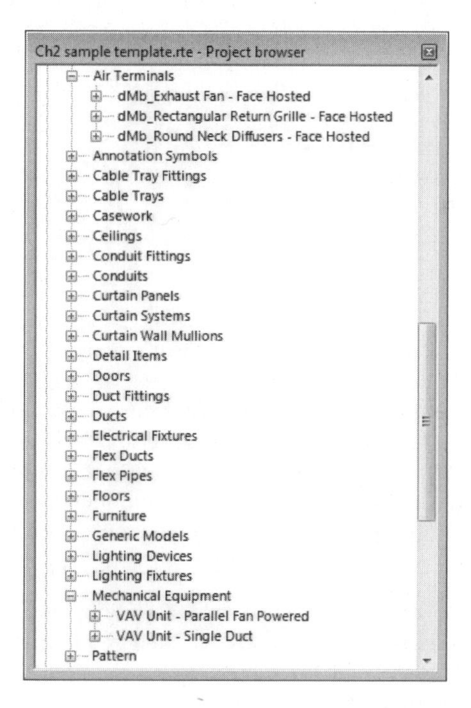

Electrical Projects with lighting, power, and communications systems require fixtures and receptacles along with distribution equipment. Having the common types of these components that you use on every project loaded into your template will make it easy to begin laying out a preliminary design while decisions are being made for specific object types. The components used in a preliminary layout can easily be changed to the specified components once decided upon. Figure 2.23 shows some examples of components loaded into a template for electrical systems design.

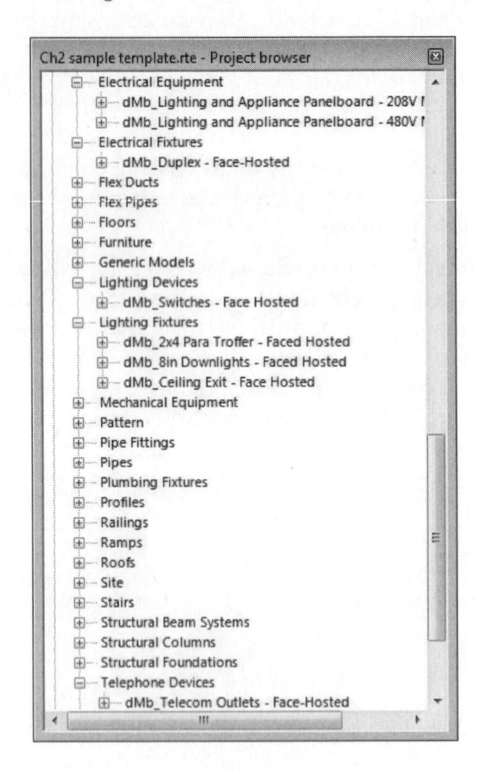

FIGURE 2.23
Sample electrical components in a template

Plumbing The types of plumbing components you load into your project depend on your workflow and how you coordinate with the architectural model. If you work in an environment where the architects typically show the plumbing fixtures, you do not need to have plumbing fixtures loaded into your template. One item that is useful is a plumbing fixture connector, which acts as a plumbing fixture and provides a connection point for piping. If there are valves or other types of components that you use regularly, they should be loaded into your template. Figure 2.24 shows some examples of plumbing components loaded into a template.

The basic idea is to keep it simple. Your template will be a fluid document that will change as your needs change or as you discover new requirements. If you find that you are having to load a particular component on many projects, then you should consider adding that component to your project template file.

Other components that you will need in your template are all of the duct, pipe, cable tray, and conduit fittings that will be used by their respective system families.

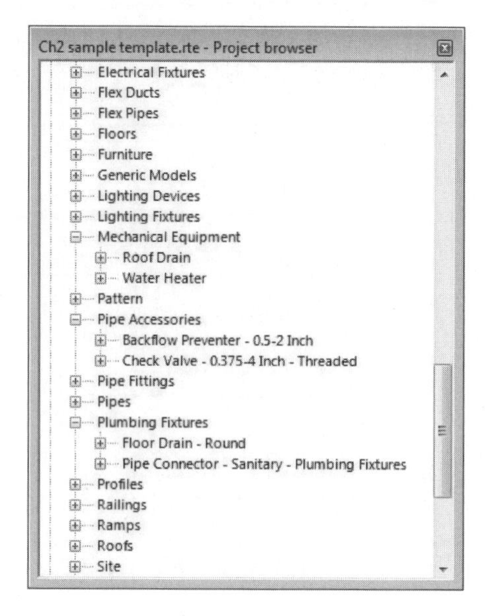

FIGURE 2.24 Sample plumbing components in a template

System Families

Along with having components preloaded into your project, it is important to define your system families. This will establish the default behavior for any types of system families that you define. If you start your template from scratch without using another template as a basis for your file, you will need to attempt to draw duct, pipe, cable tray, or conduit before the system family will appear in the Project Browser.

Once you have the desired system family in the Project Browser, you can right-click it to access its properties. The properties for MEP system families are primarily the same for each system. The idea is that you need to define what types of fittings are used. You can create variations of a system family to utilize different fittings. System family types should be named descriptively to indicate their use. Additional fittings can be added at any time to create new family types, but it is best to start with the basics for your project template.

An MEP system family will not be usable without fittings defined. To establish the fittings for a system family, right-click the family in the Project Browser. In the Type Properties dialog box, you will see options for assigning fitting component families to the system family. The fitting components you have loaded into your project will be available in the drop-down of the Fitting parameter for each specific type of fitting. Use the Duplicate button to create a new type of system family with its own unique fittings. Figure 2.25 shows an example of the settings for an Oval Duct system family. Notice that the family type has been named to indicate the elbows and preferred junction types defined in the family type.

Now that we have covered the importance of settings in your project template, practice setting the properties of both views and model objects by completing the following exercise:

1. Open the `Ch2_Template Settings.rte` Revit project template file at `www.wiley.com/go/masteringrevitmep2011`.

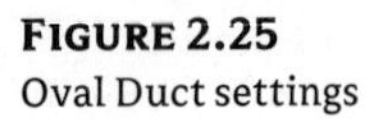

FIGURE 2.25
Oval Duct settings

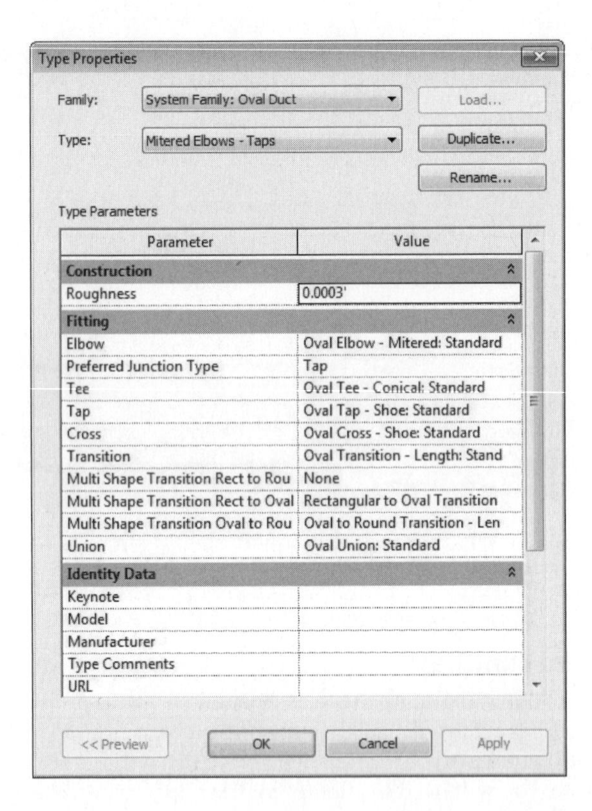

2. Access the properties of the 1 – Mech floor plan view. Set the Sub-Discipline parameter to HVAC. Set the Phase Filter parameter to Show Previous + New.

3. Click the Edit button in the View Range parameters. Set Top to Level Above, and set Offset to 0′ 0″. Verify that the Cut Plane setting is at 4′ 0″ and the Bottom setting is Associated Level (Level 1) with an Offset setting of 0′ 0″. Click OK.

4. Click the Apply button in the Properties palette to apply the changes.

5. Access the properties of the 1 – Mech Existing floor plan view. Apply all of the same settings as in steps 2 and 3 except set the Phase Filter parameter to Show All. Click the Apply button in the Properties palette.

6. Open the 1 – Mech view, and access Visibility/Graphics Overrides. On the Annotation Categories tab, set Grids to Halftone. Click OK.

7. Click the Object Styles button on the Manage tab. Set the Projection Line Weight for the Ducts category to 5. Set Line Color for Ducts to Blue. Click OK.

8. In the Project Browser, expand Families, and then expand Ducts. Right-click Standard under Rectangular Duct, and select Properties. Click the Duplicate button in the Type Properties dialog box. Name the new duct type **Mitered Elbows – Taps**, and click OK. In the Type Properties dialog box, set the Elbow parameter to Rectangular Elbow – Mitered: Standard. Verify that the Preferred Junction Type parameter is set to Tap.

9. Choose settings for each of the fitting types from the available items in the drop-down list for each parameter. The Multi Shape Transition Oval To Round parameter can be left as None. Click OK to exit the Type Properties of the duct family.
10. Use the Save As command to save the template in a location that you can access. Close the file.
11. Click the arrow next to New on the Application menu, and select Project. In the New Project dialog box, click Browse to the location where you saved the file in step 10. Verify that Project is selected, and click OK.
12. Open the 1 – Mech Existing view. Click the Duct button on the Home tab. If the Properties palette is not visible, click the Properties button on the ribbon. Click the drop-down at the top of the Properties palette, and set the duct type to Rectangular Duct Mitered Elbows – Taps. Draw a duct from left to right in the view, and then change direction to draw a duct toward the bottom of the screen, creating a 90-degree bend. Notice that the duct is blue and that a mitered elbow fitting was used.
13. Open the 1 – Mech view. Notice that the ductwork drawn in step 12 is displayed as halftone because it was modeled in an existing view and is therefore existing ductwork. Click the Duct button on the Home tab, and draw ductwork in the 1 – Mech view. Notice the ductwork is blue because it is new ductwork drawn in a New Construction phase view.
14. Click the Grid button on the Architect tab, and draw a grid across the view. Notice that the grid line and bubble are halftone.
15. Open the 1 – Mech Existing view. Notice that the ductwork drawn in the 1 – Mech view does not appear in this view. Notice that the grid drawn in the 1 – Mech view also appears in this view.

MEP Settings

You can use another group of settings to establish standards in your project template. The MEP settings of a Revit project file are used to determine the graphical representation of systems as well as the available sizes and materials of system families used. Click the MEP Settings button on the Manage tab to access the settings for a discipline. The settings you establish may determine the type of template you are creating. Some of the standards defined for the MEP settings may be unique to a project type or client's requirements, which would result in a unique template for those settings.

Mechanical Settings In the Mechanical Settings dialog box, you can establish the display of graphics that represent your design and also define values used in calculation. The left side of the dialog box lists all the settings. When you select a setting from the list, the options will appear in the right side of the dialog box. You can choose to include or exclude certain duct or pipe sizes for each type of pipe or duct. Figure 2.26 shows an example where the odd duct sizes have been excluded from use in projects and by Revit when sizing ductwork.

Be aware that the options you choose for Pipe Settings will apply to all plumbing and mechanical piping, so you may have to experiment with different settings to achieve the desired results for both disciplines. This is another reason for creating separate template files for each discipline.

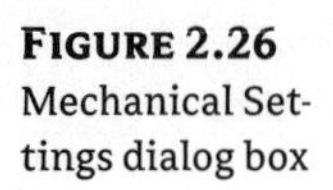

FIGURE 2.26
Mechanical Settings dialog box

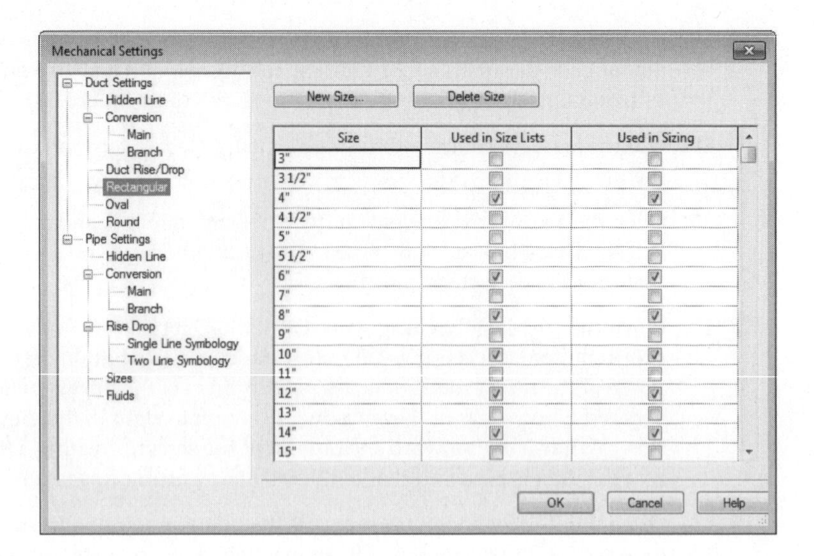

Electrical Settings In the Electrical Settings dialog box, the options you choose will define the graphical representation of items such as tick marks or circuit names. Here you can also define voltages and the behavior of distribution systems. There are settings for conduit and cable tray display and sizes as well.

Other MEP settings include Load Classifications, Demand Factors, and Building/Space Type Settings. These should be established to the extent that they are used in your projects.

Creating Sheet Standards

Drawing sheets are a key element of your project template. Because these are often the delivered product, it is important that they are put together in a consistent manner. The ability to easily manage the information included in your sheet views is one of the benefits of establishing sheet standards in a project template.

Titleblocks

The sheet border of a construction document is called many things by many different people and organizations. In this chapter and throughout this book, the graphics and information that make up the border of a construction document, or sheet, is referred to as a *titleblock*.

If you generally work as a consultant to the primary design discipline, then it may not be necessary for you to have a titleblock family for use in your template. In this type of environment, you would normally acquire the titleblock to be used from the primary design discipline and load it into your project file(s).

If you work in an environment in which you are the primary design discipline and have a titleblock that is unique to your standards, it should be included in your project template. You can have multiple titleblocks loaded into your template. Your clients may require their own standards for a titleblock. This is another reason why you might create multiple templates. If you include all the titleblocks that you work with into one template, it is possible that the wrong one could be applied to a sheet. This is not to say that you need a separate template file for each variation of titleblock you may have. For example, if you have a company standard titleblock that comes in two different sizes, it would be reasonable to include both sizes in your template. When a project begins and the size of sheets is determined, the unused titleblock should be removed to avoid confusion or it mistakenly being used.

For more information on the creation of titleblocks and how to include the desired information within them, see Chapter 7 and Chapter 21.

Sheet Organization

Whether you are creating a single-discipline template or one for use by multiple disciplines, it is important to establish the organization of your drawing sheets in order to make sheet management consistent from project to project.

Setting up sheet organization is similar to setting up the organization of views. You can set up the Browser – Sheets system family with different types to organize your sheets in any manner desired. To access this system family, right-click Sheets in the Project Browser, and select Properties. Use the Duplicate button in the type properties of the system family to create a new organization method. It is helpful to name the type based on how it organizes the sheets so that it is clear to all users what the type is used for. Once you have created a new family type, you can click the Edit button of the Folders parameter to set the organization settings. Figure 2.27 shows a sample of the settings for an organization type that is based on the discipline of sheets. In this example, the sheets are numbered using a discipline prefix so the sheets are organized by the first character of the sheet number in order to group each discipline's sheets together.

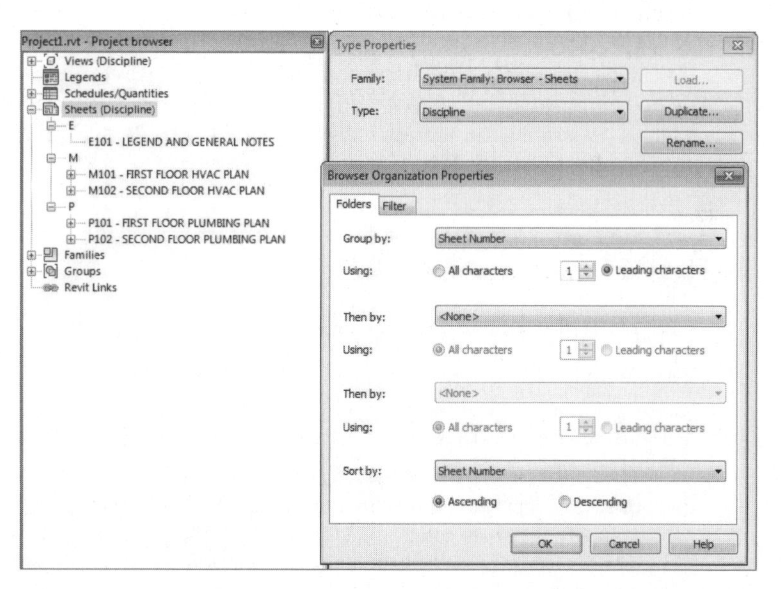

FIGURE 2.27 Sample sheet organization settings

Additional parameters can be applied to sheets that can be used as a basis for organization. These parameters can be included as project parameters in your template so that when a new sheet is created, it will have the parameters applied to it. If you create a drawing list that appears on your construction documents, you can use these parameters to organize the order of the list, which may not always match the order in the Project Browser. For more information on creating a drawing list or adding parameters, see Chapter 5 and Chapter 19.

Preset Sheets

You know some types of sheets will be included in every project that you do. These sheets can be created in your template file to ensure that they are properly set up for each project. Any views that are required for those sheets can be applied to the sheets also. So when a project has begun, any modeling done in the views will be shown on the sheets, streamlining the process of creating construction documents. Having multiple titleblocks in your template can complicate this slightly. Unless you want to set up each type of sheet with each available titleblock, it is best to use the most commonly used titleblock for your preset sheets. If the project requires a different titleblock, the default one can be switched using the Type Selector.

With preset sheets, you can establish consistent locations of sheet-specific items such as plan notes, key plans, and graphic scales. Preset detail sheets give you a place to put drafting and detail views as they are loaded or created. General notes or legends that are used on every project can be already in place when a project begins.

Having a set of sheets also is useful for determining how views will be organized and how many sheets will be required for a project. Once a project begins, you can print the sheets to create a mockup set, sometimes referred to as a *cartoon set*, to better determine how many sheets need to be added to or removed from the project, as well as how the building model will fit onto plan view sheets. With this available at the start of a project, decisions can be made early on to increase efficiency and improve workflow.

As with any element or standard you create in your project template(s), it is important to recognize how much is needed on a project-to-project basis. Consult with the people who work on projects on a daily basis to determine the types of things they would like to have in their projects from the start to improve their workflow.

We hope throughout this book you will find many ways or ideas to improve your processes. Any one of these can be considered for use in your project templates. Remember that your project templates are the foundation for your success with Revit MEP 2011 when it comes to an efficient and effective workflow.

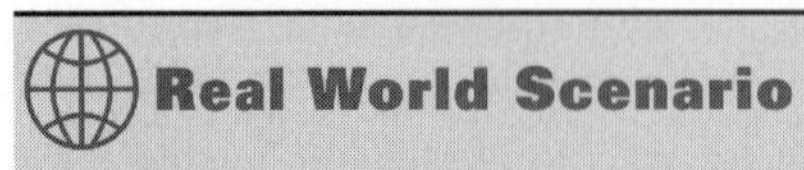

TIPS FOR MANAGING TEMPLATES

The management of a project template is an ongoing process. As standards are developed and enhanced, new workflows are discovered, and new clients are acquired, the need for template maintenance grows. Here are some items for consideration when managing your templates:

Take caution when transferring project standards Once a standard has been developed in an actual project, the easiest way to establish it in a template is to use the Transfer Project Standards feature of Revit. This is a powerful but potentially dangerous tool. Be sure that when you use it, you transfer only the standards required; otherwise, you may overwrite standards in your template with ones that are specific to a project.

Take caution with Purge Unused The Purge Unused command is another powerful but potentially dangerous tool, especially when used while working on a project template. After all, nearly everything in a template file is "unused." Exercise extreme caution when using this tool in a template file. Some items such as text styles can be removed only with this tool, so its use is sometimes required.

Keep it simple It can be tempting to include anything and everything in your project template(s). More often than not this will cause more work than it saves. Along with the potential for the wrong types of components being used in the model, there will be a lot of work cleaning up unwanted or unneeded items. Having to weed through a bunch of components just to find the one you are looking for will hamper your productivity.

Keep it safe It is usually best to limit the access to your project template(s). The more people who have access to them, the more potential for them being lost or damaged occurs. Keep an updated archive of your template(s) in a secure location so that when the time comes, you can replace it with minimal interruption of work.

The Bottom Line

Set up views and visibility The settings for views are crucial to being able to visualize the design and model being created and edited in a project. Establishing the default behavior for views and visibility of objects can increase not only the efficiency of working on a project but also the accuracy of design.

Master It The properties of a view determine how objects and the model will appear in the view. Along with Visibility/Graphics Overrides, what other view property determines whether items are visible in that view? For a floor plan view, describe the three major components of this property.

Establish project settings Many project settings can be established in a Revit template to determine the display of objects in views and on construction documents. There are also settings that define the makeup of the project itself.

Master It Phase settings for a project are very important to defining what portions of a building design occur in certain phases. Explain why having phases established in a template might cause a need for a separate template file for phased projects.

Define preloaded content and its behavior The more items you have available for immediate use when a project begins, the more your focus can be on design decisions and less on loading required items. In a multiuser environment, preloaded content ensures that improper variations do not occur, causing inconsistencies in the project documentation.

Master It Having system family types defined in your template is just as important as having the appropriate components loaded. Explain why certain component families are required in order to create and define MEP system family types.

Create sheet standards As with other template elements, standards for sheets are a useful component to have established.

Master It Having a predefined organization for drawing sheets in your template will ensure consistency from one project to the next. True or false: You must have all the required sheets for any project built into your template in order for them to be organized properly? Explain.

Chapter 3

Worksets and Worksharing

Revit MEP provides users with the ability to design and model complex engineering systems. More often than not, this is a team effort. It is important to fully understand how a Revit MEP file can be set up in a way that allows for multiple designers and engineers to work on these systems at the same time. Unlike traditional CAD solutions that require many files to represent the design intent, the entire mechanical, electrical, plumbing, and fire protection systems can be modeled in one Revit MEP file. Doing so will require careful setup of the project file in the early stages of the design.

Harnessing the power of Revit MEP to create a single model containing all the MEP systems will provide for quick and easy collaboration and coordination. The same is true for any Revit MEP file that requires more than one person working on the same model at the same time. Creating a central file in which all users' changes are accounted for will ensure that these changes can be coordinated as they occur. This empowers the team to make design decisions early on in the design process. It will also give you control of how entire systems are displayed in views and ultimately on construction documents.

To take advantage of the functionality created by enabling worksharing, it is necessary to know how the features work and how to manage them to best suit your company workflow and standards.

In this chapter, you will learn how to

- Create a central file by dividing the model into worksets
- Allow multiple users to work in the same file
- Work with and manage worksets
- Control the visibility of worksets

Understanding Central Files

It can be difficult to grasp the idea that several users can work in the same file at the same time. When worksharing has been enabled in a Revit MEP file, the file is "transformed" into a *central file*. This file is the repository for the model and all of its associated components. In truth, each user will not be opening and working in this specific file but rather in a copy of this file that communicates the actions and model changes of each user back to the central file. A copy of a central file is called a *local file*. These files can be created by copying and pasting the central file to a folder that is connected to the location of the central file via some networking means. When the copied file is opened for the first time, Revit will warn the user that saving the file will make it into a local file copy of the central file.

A typical practice for working with a central file is to store it in a job folder on a network with the local copies residing on the hard drives of the users' workstations. This prevents the users from having to work across the network while designing and modifying the model. This workflow reduces network traffic to occurring only when the local file is synchronized with the central file.

DANGERS OF OPENING A CENTRAL FILE

Once multiple users have begun work on the Revit project, it is important that the central file is not opened by any user. Having the central file open will prevent others from synchronizing their changes and can even keep them from accessing control of certain model elements. Avoid opening the central file unless absolutely necessary, in other words, for situations such as relocation and regular auditing.

This workflow can be achieved regardless of the number of users. Each user works in their respective local file, saving changes to the design and periodically synchronizing those changes with the central file. So, the big question is, "How do several users all work in one file at the same time?" One analogy to explain the workflow is to consider a jigsaw puzzle. You could dump out the pieces onto a table and gather your friends or family around and begin work on the puzzle. Each person could scoop up a handful of the pieces (local file) and work to put the puzzle (central file) together. The next sections explain the workflow in more detail.

Creating a Central File

Like most data files on a computer, Revit project files can be accessed by only one user at a time. To allow for multiple users, the project file must be made into a central file. This process starts when worksharing is enabled for a project. Any user can enable worksharing, and it needs to be done only once, by whoever is setting up the project. To enable worksharing, click the Worksets button on the Collaborate tab.

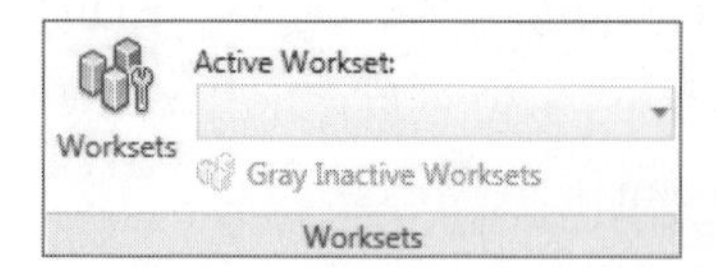

This can be done at any point in a project, but it is best to do so as early as possible because it will keep you from having to go through the model and assign elements to specific worksets after they have already been placed.

Worksets are the divisions of a model into four categories:

- User Created
- Families
- Project Standards
- Views

Each category contains a number of worksets depending on the makeup of the Revit project. Each view of the model, component, and project standard (things such as settings and types) has its own unique workset. Every family that is loaded into the project is also assigned its own unique workset. This partitioning is done by Revit automatically when worksharing is enabled to keep users from modifying the same model elements at the same time. The User Created category allows you to create your own set(s) of model components in order to control their visibility as a group or to control access to those components.

For example, if you are working on a project that will contain the elements from multiple disciplines, you may decide to create worksets that correspond to those disciplines. In other words, there would be a workset for all the electrical components, mechanical components, plumbing, and so on. However, the real world often dictates that multiple designers and drafters will be working on a project, so it may be necessary to divide the model into more specific categories based on the roles of the people assigned to the project. Perhaps there will be one person working on the HVAC systems while another works on the mechanical piping. Likewise, there may be a designer doing the lighting as another develops the power systems. In this type of scenario, it is wise to use worksets to divide the model into these categories to allow for control of these specific systems.

Figure 3.1 shows the dialog box that appears the first time you click the Worksets button in a project. This dialog box informs you that you are about to enable worksharing and contains some important information.

FIGURE 3.1
Worksharing dialog box

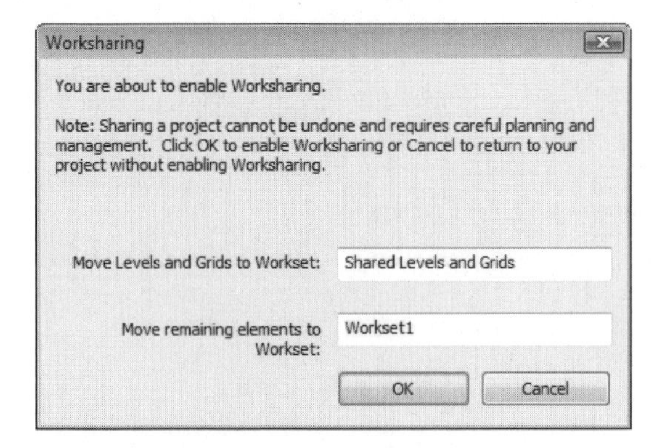

The first thing to notice is the note that warns you that if you continue with the process of enabling worksharing, it cannot be undone. In other words, once you have enabled worksharing for a project, it will always be enabled. So, the file will always behave as a central file. The importance of this warning is that if you really do not need multiple users working on the file at the same time, then enabling worksharing is not really necessary. At this point, there are several methods/points of view. One viewpoint is that all project files, regardless of how many people are working on the project, are workset enabled. This method means that all projects can be accessed in the same way, and users always work on a local file. The benefits of this are as follows:

- There is less network traffic.
- There is more than one location for backup files.
- Users, especially new to the software, have less to think about.
- There is less reliance on users renaming files.

The second piece of information that this dialog box provides is an explanation of the worksets that the components in the model (if any exist at the time you click the Worksets button) will be assigned to. So if you have already begun to build the model and later decide to enable worksharing, Revit will place all the model components into two default worksets. Revit will put any levels and column grids onto a workset called Shared Levels And Grids. This is done because these types of elements are typically used by all disciplines and are typically visible in all model views. Everything else that exists in the model will be placed into a workset called Workset1. You have the option to rename these default worksets to something more appropriate to meet your project needs. It is recommended that you rename Workset1 because this name gives no indication of what types of components belong to the workset. If you enable worksharing in a file that contains no model, the default worksets will still be created; they will just be "empty."

When you click OK on this dialog box, Revit will go through the process of creating worksets for each view, project standard, and family that is in the project file, and if a model exists, Revit will assign each component to the default workset. Once worksharing has been enabled, you have access to the Worksets dialog box, shown in Figure 3.2 by clicking the Worksets button on the Collaborate tab.

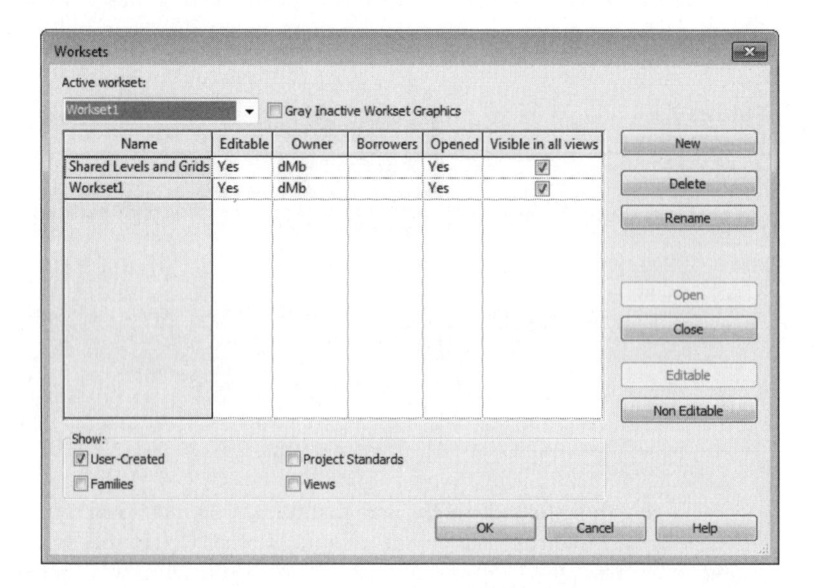

FIGURE 3.2
The Worksets dialog box

The first thing to notice is the drop-down in the upper-left corner of the dialog box. This indicates which workset is active. When a workset is active, any components that are placed into the model are part of that workset. This is similar to AutoCAD; when lines are drawn, they are on the layer that is set current. The Name column of the dialog box lists all the worksets in the project depending on which category is being shown. The radio buttons below the columns allow you to switch between categories of worksets you are viewing. This keeps you from having to view the entire list of worksets at one time.

The Editable column in the dialog box has two options: Yes or No. When a workset is set to Editable (Yes), then the user has what is known as *ownership* of that workset. Having ownership of a workset means you are the only person who can modify the model elements that belong to that workset. Ownership can be taken only by the user; it cannot be assigned by someone else. If a workset is set to Non-Editable (No), it does not mean that workset cannot be edited. It simply means that nobody "owns" that workset, so anybody working in the model can modify elements that belong to that workset. You will notice that by default the user who first enables worksharing and creates a central file is the owner of the default worksets. That user must save the file before they can relinquish their ownership of those worksets. This is a critical step when a central file is created because the goal is to allow multiple users to access the model.

The next column in the Worksets dialog box is the Borrowers column. This will display the username of anyone who has worked on elements in the model that belong to that particular workset. As you can see in Figure 3.3, the username is listed in the Borrowers column for the Lighting workset because that user made a modification to a model element (in this case the lighting fixture shown to the left) that is in that workset.

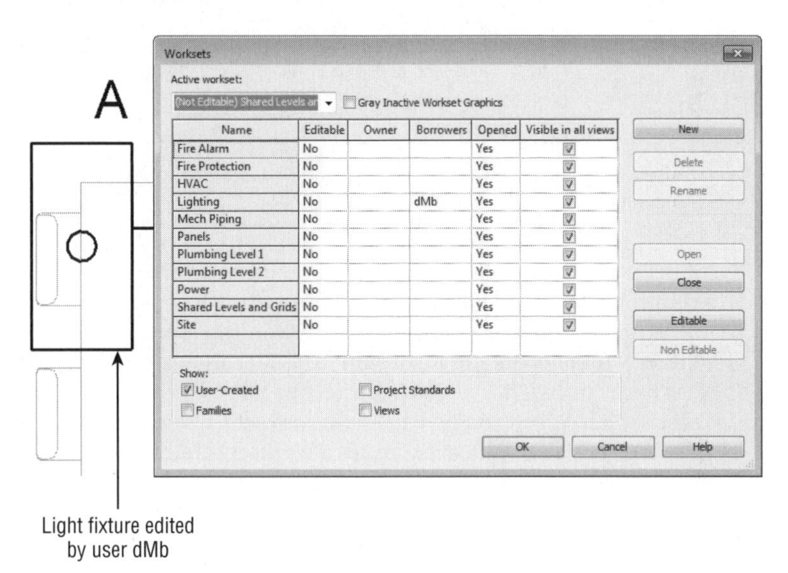

FIGURE 3.3 Username displayed in the Borrowers column of the Worksets dialog box

This is the core functionality of worksharing. Revit is keeping track at all times of who is manipulating the model and what parts of the model they are working on. The central file is keeping track of the work done in the local files. Two users cannot "borrow" the same model element at the same time. This is how multiple users can work on the same model at the same time. If you select a model element such as a light fixture and some other user has already selected and modified that light fixture, Revit will warn you with a dialog box that you cannot modify that model element until the other user relinquishes their control of it. Going back to the jigsaw puzzle analogy, if you were to pick up a puzzle piece to place it somewhere, no other person at the table could do anything with that piece until you let go of it. Revit will list all the usernames of anyone borrowing elements on a workset.

The next column of the dialog box indicates whether a workset is open. If a workset is not open, then the items that are in that workset are not displayed in the model and are not able to be modified. When you open a Revit project that has worksharing enabled, you have the option in the Open dialog box to set which worksets will be open with the file. This can save file opening time and improve file performance. For example, you could choose to not open the Lighting workset if you are working only on the plumbing systems and have no need to coordinate with the lighting portion of the model at that time. This may help with model regeneration time and improve your productivity.

The last column of the dialog box is new to Revit MEP 2011. The Visible In All Views column allows you to control the behavior of the visibility of the worksets. You will see that when you create a workset, you can determine whether it is visible in all views. In previous versions of the software, once you made that decision, it could not be reversed. This new column allows you to reverse that decision if necessary.

Creating a New Workset

To create a new workset, click the New button in the upper-right corner of the dialog box. This will display the New Workset dialog box that allows you to name your new workset.

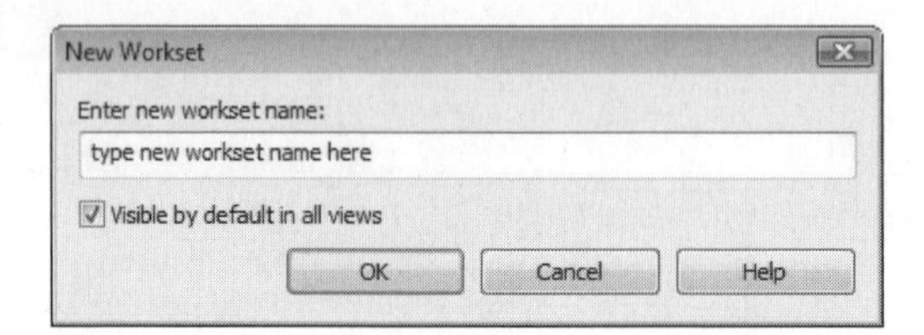

This dialog box has a check box allowing you to set the default visibility of the workset in your model views. This is an important setting that we will discuss further in the next chapter because it has an effect on what can be seen when your model is linked into another file. With the new functionality of Revit MEP 2011 that allows you to change this setting after the workset has been created, it is not as important as it used to be to make the decision at the time of workset creation. When you give the new workset a name and click OK, it is added to the list of User Created worksets. Whoever creates a workset is the owner of that workset by default. Once you have created the worksets necessary for your project and you click OK in the Worksets dialog box, you must save the file to complete the creation of a central file. Revit will warn you with a dialog box if you attempt to relinquish your control of the worksets prior to saving the file as a central file.

After enabling worksharing and establishing your worksets, click the Save button to create the central file. This will open the Save As dialog box. If you click the Options button in the Save As dialog box, you will see that because worksharing has been enabled, Revit will save your file as a central file, as shown in Figure 3.4.

When your file has been made into a central file, a backup folder is automatically created. This folder contains data files that will allow you to restore a backup version of your project if you should ever need to roll back the project to an earlier date or time. Having several users on a project who are saving frequently can limit how far back in time you can go to restore your project. For this reason, it is recommended that you archive your project file at intervals logical to the submittal of the project.

The first thing you should do after the file is saved is click the Relinquish All Mine button on the Collaborate tab. This will release ownership of all worksets and borrowed elements. The file should then be saved again so as to have the central file in a state where no user has ownership of any worksets. However, if you have a central file open, you will notice that the Save button is inactive. To save a central file when you have it open, you must click the Synchronize With Central button on the Collaborate tab. This will save the file and create a backup folder of the project. Once the central file has been saved, it should be closed before any more work on the project file continues. From this point on, all work should be done in the local files of the project.

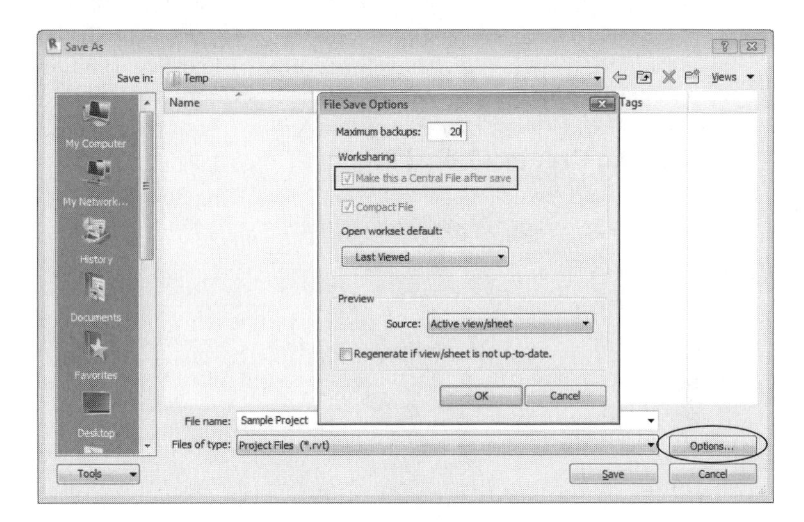

FIGURE 3.4 Options for saving a file after enabling worksharing

BEST PRACTICE FOR NAMING CENTRAL FILES

Regardless of your company file naming standards, it is a good idea to put the word *Central* somewhere in the name of your central file. Because the central file is so important, you want to clearly identify it in your project folder structure. Using this naming will also help confirm that the correct file is being shared with consultants.

To summarize, these are the steps to create a central file for a project.

1. Select New ➤ Project from the Application menu.
2. Click OK in the New Project dialog box.
3. Click the Worksets button on the Collaborate tab.
4. Verify that the Shared Levels And Grids and Workset1 worksets will be created, and then click OK on the Worksharing dialog box.
5. Create any additional worksets needed for your project in the Worksets dialog box.
6. Click OK. You will be asked whether you want to make one of your User Created worksets the current workset. You can choose to do so, but it has no effect on the file creation.
7. Save your file to the desired location. Your file will be saved as a central file. You are the owner of all of the worksets until you relinquish them.
8. After you have saved your file, click the Relinquish All Mine button on the Collaborate tab. The Save button is no longer active, so you must use the Synchronize With Central button to save your changes.

Working with Local Files

Now that the central file has been created, each person who will work on the project must create a local file copy of the project. This local file can reside on the users' local workstations or on the network, although local C: drives are generally the best location. The important thing is that the local file copies must be connected to the central file via the network in order to maintain the ability to synchronize them and for Revit to manage borrowing of elements.

Ways to Create a Local File

The Create New Local option in the Open dialog box shown in Figure 3.5 is selected by default when you attempt to open a central file.

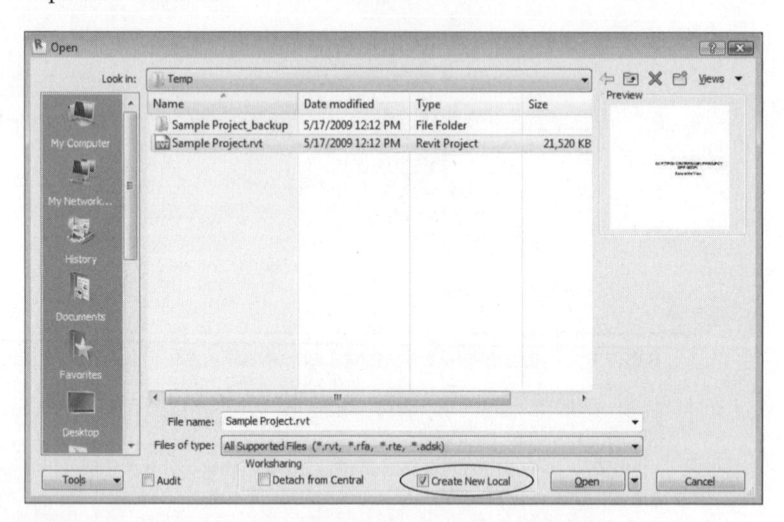

FIGURE 3.5
The Revit MEP 2011 Open dialog box

Selecting the Create New Local option can prevent someone from accidentally opening the central file by automatically creating a local file when they click Open. This is the easiest way to create a local file. Using this method will create a local file copy of the central file in the folder specified in your Revit settings. The filename will be the same as the central file with _username added to the end. Figure 3.6(a) shows the File Locations tab of the Options dialog box indicating the destination folder for project files, and Figure 3.6(b) shows the local file saved to the destination folder.

There are utilities available that automate the process of creating a local file, and many companies are developing their own routines that conform to their specific standards such as file naming and project folder structure. If you do not have such a utility, another way to create a local copy of a central file is to simply browse to the location of the central file and copy/paste it to the desired location using Windows Explorer (see Figure 3.6). Once you have placed the local copy, it is important to rename the file, removing the word *Central* from the filename to avoid confusion. It is suggested that you use the word *Local* along with your username in the filename to keep track of who owns the local files. If you use this method when you open the copied file, you will receive a warning dialog box stating that the central file has moved or been copied and that saving the file will make it into a local file (which is the intent). Click OK on this warning, and you will have a local file once you save.

FIGURE 3.6
(a) Options dialog box; (b) file saved to destination folder

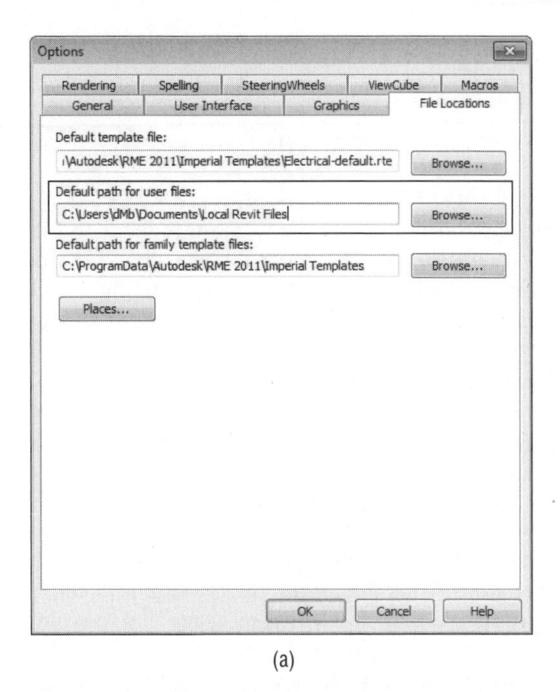

(a)

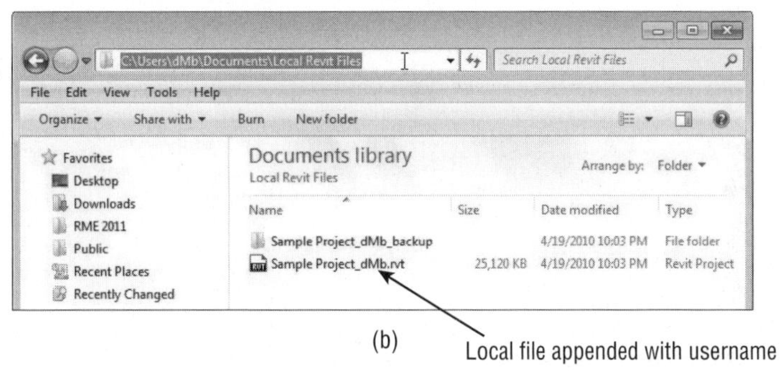

(b)

Synchronizing a Local File with the Central File

The central file is in constant communication with all of its local file copies in order to maintain coordination of the model elements. As you are working in the model, it is aware of what elements you have control of. You have two options for saving your work. The first option is to save your local file by simply clicking the Save button or selecting Save from the Application menu. When these changes are saved, they are saved to your local file only. This gives you the freedom to save at frequent intervals without accessing the central file. This also allows for a more efficient workflow because minor changes and reworking of the design can be finalized before pushing them out to the entire design team.

The second option for saving changes is to synchronize with the central file. Synchronization occurs when a user wants to save their work to the project and clicks the Synchronize With Central button on the Collaborate tab.

Clicking this button saves any changes made in the local file to the central file. It also updates the local file with any changes that have been synchronized to the central file by other users. So when you click this button, you are updating your local file to the most current status of the project. Another very important thing happens when you synchronize with the central file. Any items that you are borrowing in the model or any worksets that you have taken ownership of are relinquished when you synchronize. If you need to relinquish your control over items in the project but are not ready to save your changes yet, you can do so by saving your local copy and then clicking the Relinquish All Mine button on the Collaborate tab.

Synchronizing and relinquishing will become a regular part of your daily workflow once the project gets rolling. There are some occurrences in this workflow that are important to be aware of. If two or more users try to synchronize with the central file at the same time, Revit will synchronize them in the order that they happen. So whoever hit the button first, their changes will be saved first. While the central file is being updated, the others who have attempted to synchronize will see a dialog box with a message that the central file is currently busy. They can cancel their attempt to synchronize or wait until the central file is available. Clicking the Cancel button means that no synchronization will occur. If you wait, the synchronization will happen as soon as the central file becomes available.

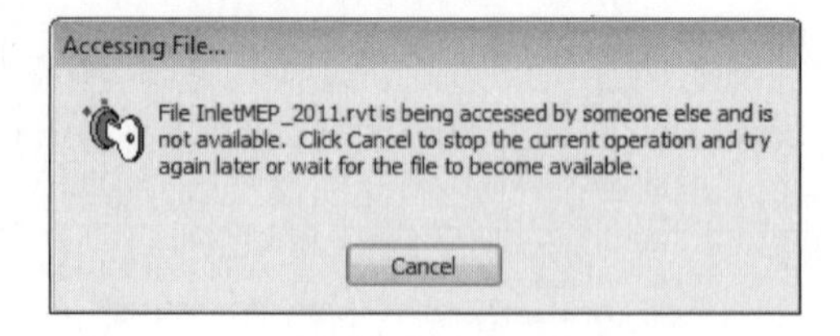

Local files can be deleted and new ones created at any time. It is recommended that you create a new local file every day as you begin work. This will help keep the number of warning and error messages that are saved to the central file down and will ensure that you start your Revit session with an updated copy of the central file. If you do not choose to make a new local file each day, it is important to update your local file with any changes that other users may have made while you were gone. You can do this by selecting Reload Latest from the Collaborate tab. When another user synchronizes with the central file, keep in mind that your file will not update with those changes automatically. You must either synchronize your file or reload the latest version of the central file.

When you start working on a project and want to create a new local file, you can browse to the location of the central file and open it with the Create New Local option selected. If you already have a local file for the project, a dialog box will appear with options for creating the new local file, as shown in Figure 3.7.

Depending on the changes made since you left, it is possible for your local file to become unreconcilable with the central file. It is for this reason that it is a good practice to create a new local file each day.

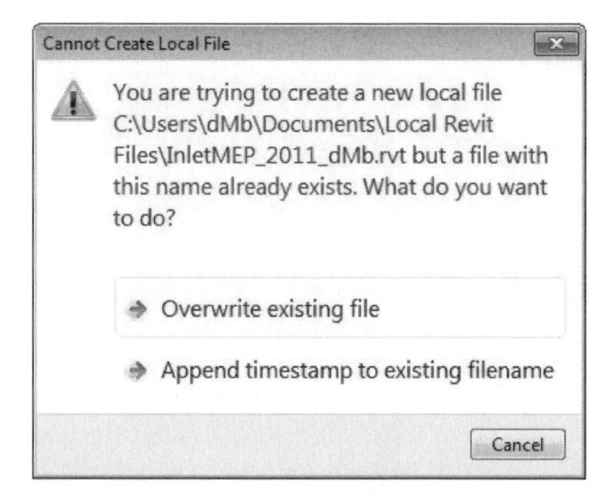

FIGURE 3.7
New local file creation options when a local file already exists

When the time comes to end your session of Revit MEP, it is best to synchronize your local file with the central file. If you forget or you choose not to by closing the program, you will see a dialog box with options for closing your session of Revit MEP. Figure 3.8 shows the dialog box with options for closing a local file that has not been saved or synchronized with the central file.

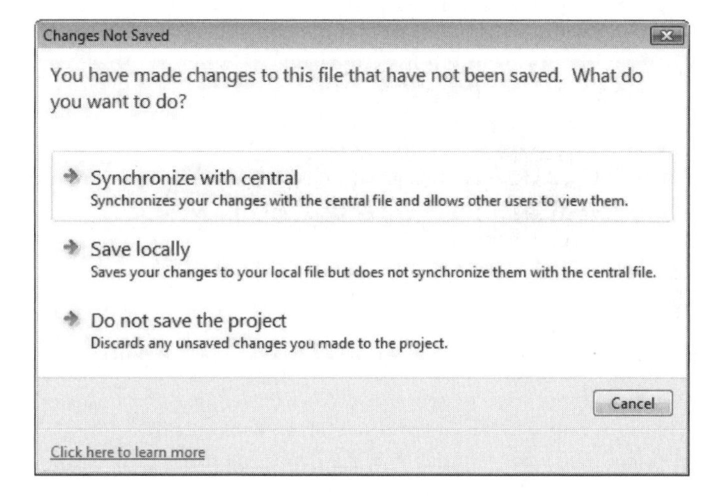

FIGURE 3.8
Ending Revit MEP session options

Clicking the Cancel button in this dialog box will return you to your session of Revit MEP. Clicking the Synchronize With Central option will open a dialog box with settings for your synchronization, as shown in Figure 3.9.

Notice in Figure 3.9 that you have the option to save your local file before and after it is synchronized with the central File. There is also a setting to control whether you will relinquish any worksets or borrowed elements.

If you click the Save Locally option in the dialog box shown in Figure 3.8, you will be given another dialog box with your options for relinquishing any borrowed elements or owned worksets. You can choose to relinquish any items even though your changes are saved only to your local file and not the central file, or you can choose to keep ownership of the items. Keeping ownership will prevent others from editing those items even after you close your Revit session.

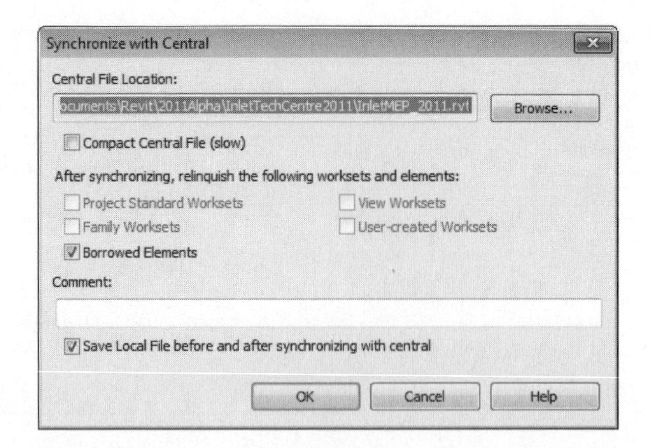

FIGURE 3.9
Synchronization settings

Selecting the third option in the dialog box in Figure 3.8 will open the same dialog box mentioned earlier, providing options for relinquishing control of worksets and borrowed elements. So even if you open the file to make changes that you don't intend to keep, be aware that you may affect others' use of the model by borrowing elements. It is important to relinquish worksets and borrowed elements when you are closing your Revit MEP session to allow others to work without running into the obstacle of not being able to edit an element because you have ownership.

Managing and Using the Power of Worksets

Any work done in a local file will transfer to the central file when synchronized. This is true for worksets also. When the central file is created, you may not know what worksets will be necessary, so they may be created in the local file and will appear in the central file after synchronization.

To make a model element part of a workset, that workset has to be set as the current workset. You do not need to be the owner of a workset in order to place a model element in that workset.

You do not have to access the Worksets dialog box in order to set a workset as current. The Workset panel of the Collaborate tab contains a drop-down with a list of all the worksets in the project. The workset visible in the drop-down window is the one that is set current. Remember that the words *(Not Editable)* next to the workset do not mean that changes cannot be made to the workset; they simply indicate that the workset is not owned by anyone.

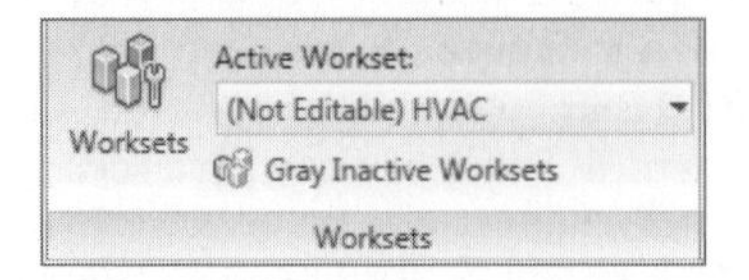

With the ribbon-based user interface, it can be cumbersome to switch to the Collaborate ribbon each time you want to set a workset as current or check to confirm the correct workset is current before placing an element in the model. For this reason, the current workset drop-down list is available on the status bar, so it is visible at all times. You can also add the Workset drop-down to the Quick Access toolbar if you choose.

Taking Ownership of Worksets

There are two primary reasons for taking ownership of a workset. Both result in that no other user can edit any model elements that are in that workset. If you are concerned that someone else might inadvertently move or edit an element that is part of your workset, you can take ownership of the workset. This scenario can be detrimental to the efficiency of the team if other users need to make changes. There are other alternatives to protecting elements in a model such as pinning them in place.

Another reason for workset ownership is to protect a group of elements such as a system or area of the model from changes while design decisions are being made. This most commonly occurs when you have to work on the project offline, such as when meeting with a client.

To take ownership of a workset, all you have to do is open the Worksets dialog box and change the value to Yes in the Editable column. It is best to synchronize your local file with the central file after taking ownership of a workset.

Real World Scenario

Using Workset Ownership Away from the Office

Rusty is an electrical designer working for an engineering firm that is consulting with an architect in another state. He is nearing his project deadline, and the owner is ready to decide on some of the different lighting options that have been proposed. While Rusty is away meeting with the architect and client, he wants to make sure that no unnecessary work is done to the lighting model since he will most likely bring back a new design that would make those changes obsolete. So, he takes ownership of the Lighting workset and then takes a copy of the project with him on his laptop.

During the next two days Rusty makes several modifications to the lighting design while the owner and architect provide input. When Rusty returns to his office, he connects his laptop to his network, opens his local file, and synchronizes his changes with the central file. Since he had ownership, no other work was able to be done to the lighting design, and his remotely made changes are now part of the central file.

Model Elements and Their Worksets

When a workset is set to be the active workset, any component placed into the model will be part of that workset. Annotative elements such as text, tags, or dimensions do not become part of a User Created workset but rather the workset of the model view that they are drawn in. Each project view, including sheet views, has its own workset. When you are working in a view adding annotation or detailing, you are borrowing that view workset. This prevents others from annotating the same view or changing the properties of that view until your changes are synchronized.

Oh No I'm Not!

Users often forget to select the View And Family Worksets check box when looking at the Worksets dialog box to see whether they are borrowing an element. Approached by fellow workers who cannot access items in the project, they insist that because the User Created list of worksets is clear, there must be a problem with the software!

The User Created worksets are the most managed worksets since the Views, Families, and Project Standards worksets are created automatically and there is rarely a need to delete or edit them. Deleting a family or view will also remove its workset from the project.

To delete a workset, you must first take ownership of it. You can do so only if there no other users borrowing elements on that workset. You have to close the Worksets dialog box after taking ownership before you can delete the workset. You can then reopen the Worksets dialog box, select the workset from the list, and click the Delete button on the right side of the dialog box. If there are any elements in the model that are in the workset you are deleting, you will get a dialog box that gives you options for what to do with those elements. Figure 3.10 shows the options of either deleting the elements from the model or moving them to another workset.

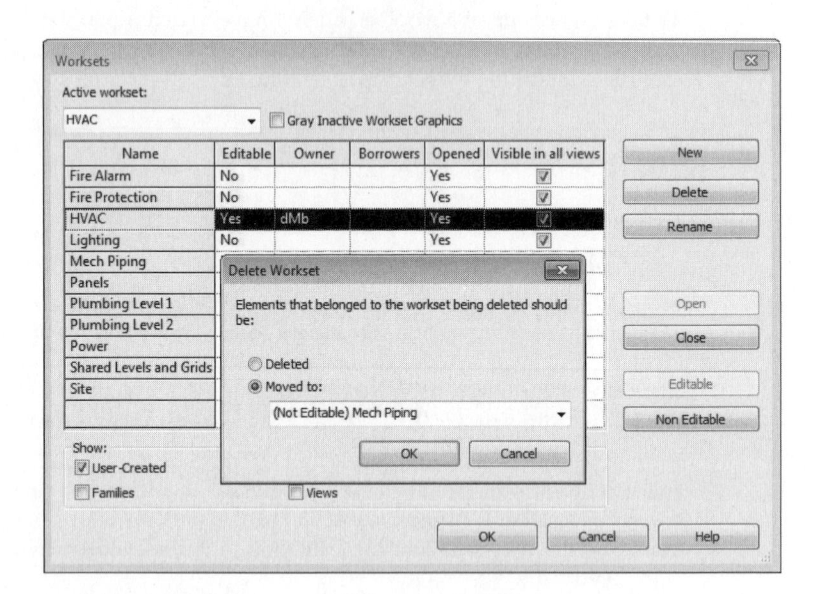

FIGURE 3.10
Options for deleting a workset

If another user owns the workset you are trying to move your elements to, you will need to have them relinquish their ownership of that workset before you can move your items to it.

When you place your cursor over an object in the model, you will see a cursor tooltip that lists some information about that object, as shown in Figure 3.11. When worksharing has been enabled in a project, the first item in the list of information on the tooltip is the name of the workset to which the object belongs.

This is the easiest way to determine which workset the object belongs to. This information can also be found in the properties of a model object or project component. The Workset parameter of an object is an instance parameter listed under the Identity Data group. If you need to change the workset of an object, you can edit the setting for the parameter in the Properties palette. The parameter and its value will be grayed out and inactive if you are not borrowing the selected item.

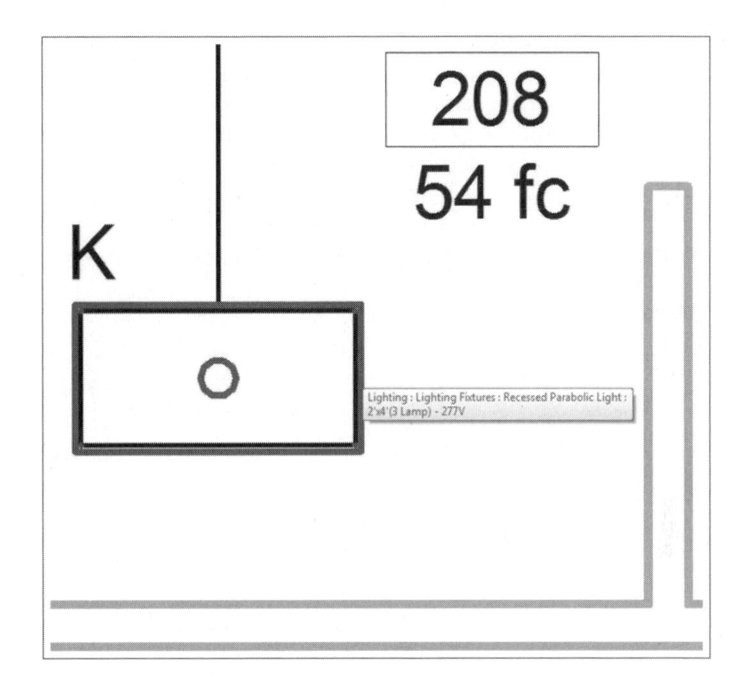

FIGURE 3.11
Model object information cursor tooltip

> **SELECTING MODEL ELEMENTS**
>
> Clicking an element and selecting it does not mean that you are borrowing that element. You must modify it in some way, either by changing one of its properties or by moving it before you will be borrowing the element.

Once you have borrowed an item in the model, the Workset parameter will become active, and you can change its value by selecting from the drop-down list of workset names, as shown in Figure 3.12.

You can change the workset of multiple items by selecting them all and then accessing the Workset parameter in the Properties palette, but you must be borrowing all of the selected elements. Since going through the model and selecting elements to modify them just so they will be borrowed is impractical, another method to activate their Workset parameters is to take ownership of the workset to which the items belong. This will enable you to access the Workset parameter by simply selecting an item and using the Properties palette. Keep in mind that nobody else will be able to edit items on that workset until you relinquish your ownership, even if they are not items that you are working on.

As mentioned earlier, each model component has its own workset. So if you select an item in the model that is a family and you select the Edit Family option from the Modify tab, once you load the family back into the project, you will see a dialog box that notifies you that the family is not editable and asks whether you would like to make it editable. This will give you ownership of the workset for that family. Because this does not occur until you load the family back in, it is possible that two or more users could edit a family, but only the first one to load it back into the project would gain access to the workset; therefore, only their changes would occur in the project.

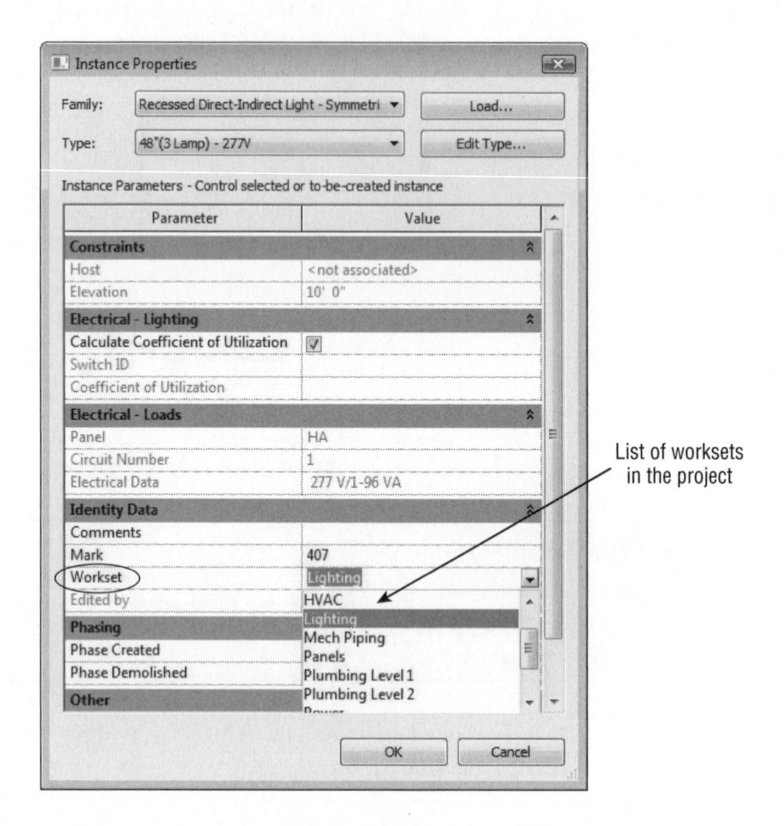

FIGURE 3.12 The Workset parameter in the properties of a model object

If you are the owner of a workset and want to release your ownership, you can do so by accessing the Worksets dialog box and changing the editable value from Yes to No. If it is a User Created workset that you are releasing, you will remain the borrower of any items that you have modified on that workset until you synchronize with the central file. If you attempt to use the Workset dialog box to release your ownership of a family workset, you will be notified that the local file must be synchronized with the central file in order to release that workset.

UNDO, UNDO...UNDO!!!

We tend to rely on the good old undo button to clean up our messes, but beware: the undo function will have no effect on workset ownership or borrowing of elements. Once you own a workset or borrow an element, it is yours until you release it.

Visibility and Worksets

Using the power of visualizing the model is one of the key advantages to using a building information modeling (BIM) solution for project design. We also use visibility settings to define how our views will appear on the construction documents. With the model divided into worksets that denote the engineering systems, we can harness this power to help us design more efficiently using "working" views. A working view is one that may show more or less model information than would be represented on a construction document.

The first feature of visibility control is simply turning items on or off. This applies not only to model categories but to worksets as well. When worksharing has been enabled in a project, an additional tab appears in the Visibility/Graphics Overrides dialog box, as shown in Figure 3.13.

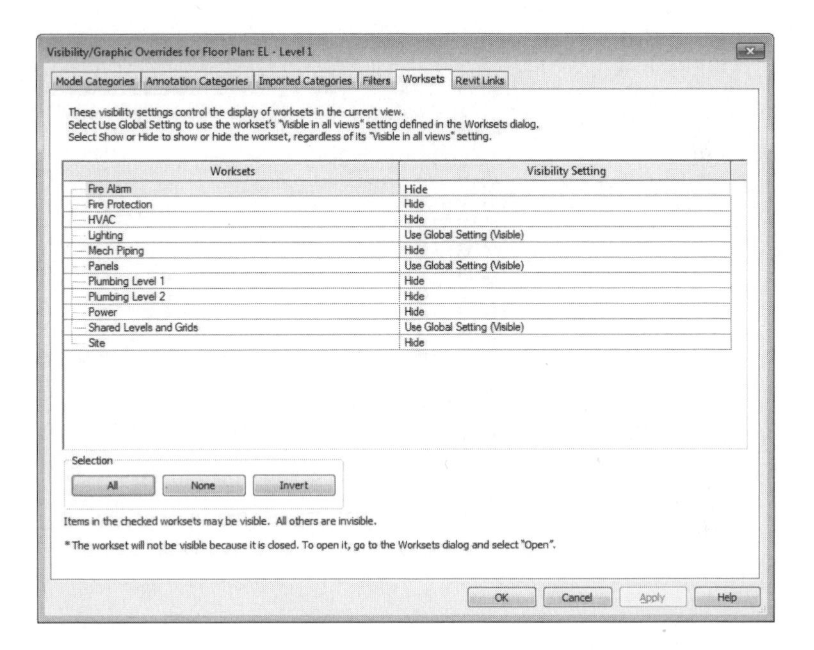

FIGURE 3.13
Visibility/Graphics Overrides dialog box

This tab of the dialog box provides a list of the User Created worksets in the project and allows for setting the visibility behavior of the worksets in the current view. There are three options for workset visibility in a view.

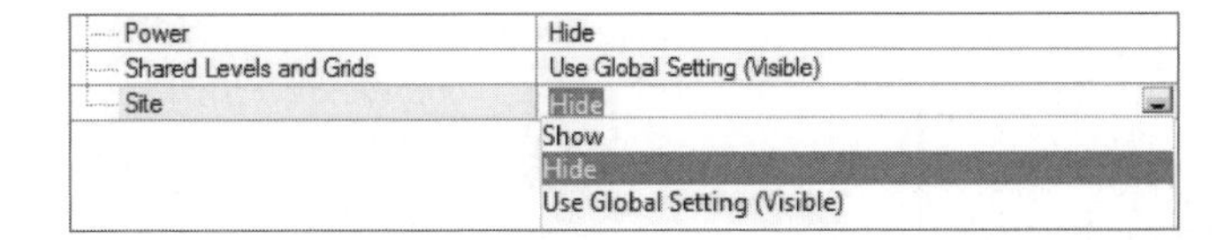

When you hide a workset, none of the items in that workset are visible in the view, even if the categories of the individual items are set to visible. The visibility settings for the workset take precedence over the settings for model categories.

The Use Global Setting option means that the visibility of the workset is determined by the Visible In All Views setting established when the workset was created or the status of that setting in the Worksets dialog box. The status is shown in parentheses.

When setting up your project worksets, it is important to consider this functionality. Thinking about what types of systems or groups of model components might need to be turned on or off will help you make decisions on what worksets will be created. Doing so early on will save time from having to do it later when the model contains many components that would have to be modified and will make controlling visibility easier from the start.

Another way to control a workset's visibility is to determine whether the workset will be opened when the project is opened. In the Open dialog box, you can click the arrow button next to the Open button to access options for which worksets to open with the file, as shown in Figure 3.14.

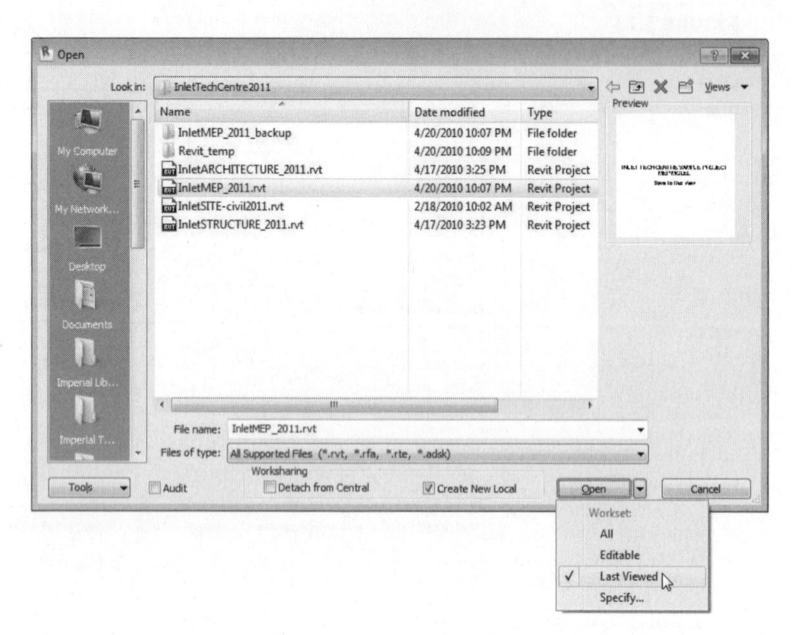

FIGURE 3.14
Workset options on the Open dialog box

Selecting the All option will open all the User Created worksets when the file opens. This does not mean that all the worksets will be visible in every view; it just means that the worksets are open, or "loaded," into the project. Selecting the Editable option will open (load) all worksets that are editable (owned) by you. If another user has ownership of a workset, it will not be opened with this option. The Last Viewed option is the default option.

FILE LINKS ON THE RECENT FILES SCREEN

Whatever option you choose from the Open dialog box will be applied to the thumbnail link of the file on the Recent Files screen. Using the link will open your file with the Last Viewed workset settings applied.

The final option for opening (loading) worksets in a file is the Specify option. This allows you to choose which worksets will be open when the file is opened. Selecting this option opens a dialog box similar to the Opening Worksets dialog box, as shown in Figure 3.15, where you can choose the worksets to open.

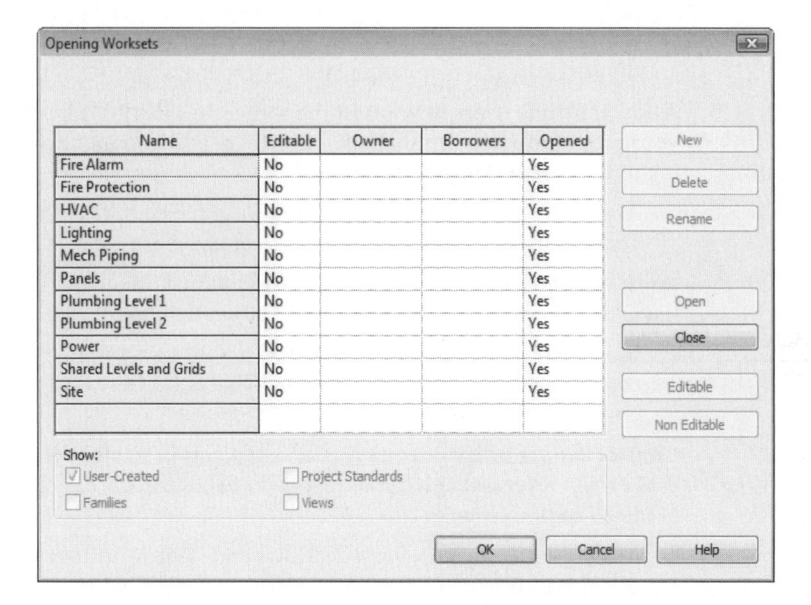

FIGURE 3.15
The Opening Worksets dialog box

This is a powerful option because it can greatly improve file opening time and overall performance of your model. If you need to work on a system in the model and do not need to see other systems, you can choose to not open their respective worksets, decreasing the amount of time for view regeneration and limiting the amount of behind-the-scenes calculations that are occurring when you work on your design. If you decide later that you want to see the items on a workset that you did not open when you opened the file, you can open the workset via the Worksets dialog box. When selecting the Visibility/Graphics Overrides dialog box for a view and there are worksets that are unopened, the unopened worksets will appear in the list with an asterisk next to them. So if you are not seeing a workset that is set to be visible, you can quickly see that it is not visible because it is not open.

One way to utilize this functionality for improved performance is to create a workset for each Revit file that you will link into your project. This gives you the option of not opening the workset of a linked file without having to unload the linked file.

WORKSETS AND VIEW TEMPLATES

There is no option for direct visibility control of worksets in view templates; however, you can control the visibility of the worksets within a linked file using view templates.

The Bottom Line

Create a central file by dividing the model into worksets Setting up your Revit project file correctly will help users visualize and coordinate their systems easily.

Master It You are working on a project with a mechanical engineer, a plumbing designer, and an electrical engineer. Describe the types of worksets that the model can be divided into to accommodate the different systems for each discipline.

Allow multiple users to work in the same file Revit MEP provides functionality to set up your project in a manner that allows users to edit and manage their systems without conflicting with other systems in the model.

Master It Describe how to create a local file copy of a central file and how to coordinate changes in the local file with other users who are accessing the central file.

Work with and manage worksets Working in a project with multiple users means it is likely that you will need to coordinate the availability of worksets.

Master It Describe how you would isolate a system in the model so that no other user could make changes to that system. What is the best way to release a system so that others can work on it?

Control the visibility of worksets Visualization is one of the most powerful features of a BIM project. Worksets give you the power to control the visibility of entire systems or groups of model components.

Master It You are facing a deadline and need to add some general notes to one of your plumbing sheets. Because of the intricate design of the HVAC system, your project file is very large and takes a long time to open. What can you do to open the file quickly to make your changes?

Chapter 4

Best Practices for Sharing Projects with Consultants

Since the beginning of engineers and architects working together on building projects, an integrated project delivery (IPD) has been the ultimate goal. The goal of consultants working together as seamlessly as possible is made more achievable by using a building information modeling (BIM) solution such as Revit MEP 2011. Many are quick to believe that BIM equals IPD, but the truth is that developing a building information model is only part of the process of IPD. You can build a 3D model and even put information into it, but it is how this information is shared and coordinated that defines an IPD.

Revit MEP enables you to interact with the project model in a way that you can realize the design as a whole, even if you are only working on pieces or parts. You do not have to work for a full-service design firm to participate in a project that results in a complete model. Architectural, structural, and MEP engineering systems can all come together regardless of company size, staff, or location. It is the sharing of computable data that makes for an integrated project delivery. Owners and contractors can also participate in the process. In fact, anyone who has any input into the design decisions should be included in and considered part of the project team.

Zooming in from the big picture a bit, you need to focus on your role as a project team member. You can be an effective player in the process by ensuring that the data you are sharing is usable, accurate, and timely. With a good understanding of how Revit manages your information, you can develop some good practices and standards for collaborating with your project team.

In this chapter, you will learn to

- Prepare your project file for sharing with consultants
- Work with linked Revit files
- Coordinate elements within shared models
- Work with non-Revit files
- Set up a system for quality control

Preparing Your Files for Sharing

In Chapter 3, you looked at ways to set up your project file that allows for your design team to work in a collaborative manner within their specific design disciplines. Although one goal is to make yourself more efficient and accurate, you must also recognize the importance of setting up and developing the project file in a way that you can easily share the design with your consultants. It is a good idea to consider whether the workflow or process you are setting up will have an adverse effect on the other project team members.

Consulting In-House

The process for sharing information among consultants can be applied to your Revit project even if all the design team members work for the same company. Many full-service firms are finding it effective to treat the differing disciplines as "consultants" in order to achieve an IPD. Using this workflow helps reduce inaccurate assumptions that may occur and provides an atmosphere that encourages increased communication.

A Revit project file does not have to be set up as a central file in order to share it with others. If there is no need to establish worksets in your project, then you can simply save your file as the design develops and share it when necessary. This is the simplest type of project file to share because there is no concern for receiving a local file by mistake or a file that is not updated with the most current design changes. As with all files that will be shared among the team members, it is a good practice to name the Revit project file in a manner that clearly indicates what systems it contains, such as `MEP Model.rvt` or `HVAC.rvt`. This will make linked files easy to manage and reload when updates occur.

If your Revit file has been set up with worksets, there are other considerations for sharing the file with your consultants. You should take care to realize that the decisions you make when setting up your worksets can have an effect on those team members who need to visualize and coordinate with your design elements.

The first decision comes when you create your worksets. In the New Workset dialog box, you have the option of naming your workset, as shown in Figure 4.1. The name you choose will have no effect on anyone you share your model file with but should be descriptive so that others will know the types of items included in the workset. Using a descriptive name is also helpful to your consultants, who now have the ability to see what worksets are in your file. The other part of this dialog box offers you the option to make the newly created workset visible in all your model views. If you choose not to make the workset visible in all views, then you will have to make it visible in each view of your model via Visibility/Graphics Overrides settings.

When you are creating your worksets, you may want to consider the Revit files that you will be linking into your project. It can be very useful to create a workset for each linked Revit file so that you can control not only their visibility but also whether they are loaded into the project. Since you have the ability to choose which worksets to open when you open your file, you can determine which linked files will be present and visible when your file is opened. This can increase your productivity by improving file opening and model regeneration times. It also helps by not having to load and unload the links, which can cause problems with hosted elements. When you do not open a workset that a linked file is part of, the file is still technically loaded into your file; it's just not visible or "present" in your session.

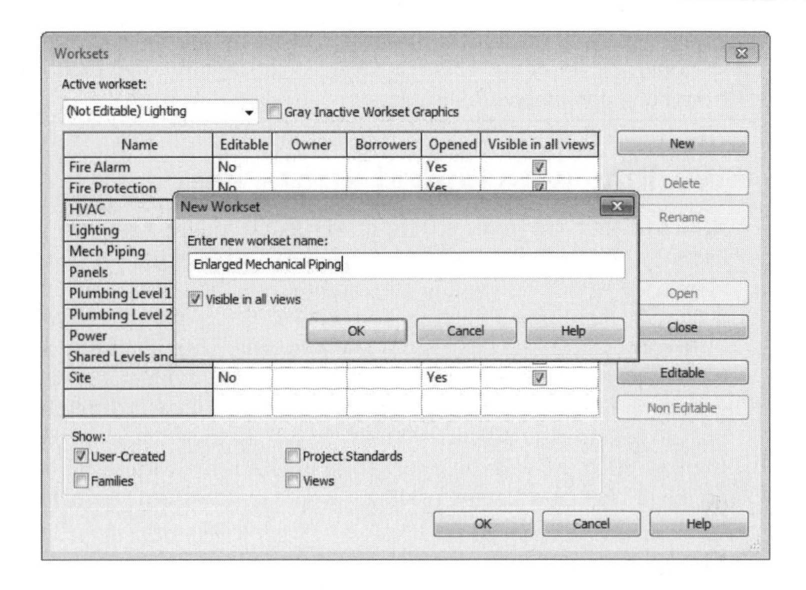

FIGURE 4.1
New Workset dialog box

The makeup of your Revit project template file has a lot to do with the preparation for sharing your file with others. Consider carefully the content that you put into your template. It is tempting to dump your whole library into the template in order to have everything right at your fingertips, but this makes for large files that are less easily shared. Once you have worked on a few projects, you will be able to determine what is really necessary to have in your template file.

There is no denying that Revit files can get to be very large. Purging any unused items from your file prior to sharing is helpful in keeping file sizes to a minimum, but use this feature with caution.

PURGE WISELY!

Do not use the Purge Unused feature in Revit when you are working in a template file. After all, nearly everything in a template file is "unused."

Because of the size of the files, sharing them among consultants via email is not a recommended practice. Most people do not have the capacity to receive large files via email. Using an FTP site is the most common practice for file sharing. Also, certain third-party applications are designed specifically for project collaboration. It is usually the primary design team that establishes the site and gives all teams access to upload and download files. Each consultant should have at least one designated person to manage the transfer of files for their team. This person should be responsible for making sure that the most current files have been shared. The frequency of updated files being posted to the site should be determined early on during a project kickoff meeting (you are having those, right?) and can change depending on the needs of the project. At a minimum, you should prepare to post your file once a week during active development. This will keep your consultants up-to-date with the progress of your design. If you have enabled worksharing on your project, it is important that you send your central file to your consultants, not a local file. This will ensure that they are receiving a file with updated information from all your team members.

Working with Linked Revit Files

Revit MEP not only gives you the power to model and document your design but also to collaborate with the designs of your consultants directly in your project file. This concept is not new to the world of digital documentation. It is the same workflow that occurs in the world of 2D CAD: you receive a file from your consultant, overlay it into your file, and watch as it updates during the changes to the design. The difference with Revit is in how much more you are able to coordinate by receiving a model that is not only a graphical representation but also a database with computable information. Some of the terminology is different when working with Revit. Instead of external references (*Xrefs*), we call the files that are loaded into our project *linked* files. In this section, we will cover some best practices for loading, managing, and viewing links that are Revit `.rvt` files. Later we will cover files of other CAD formats.

Linking Revit Files

When you are ready to bring another Revit file into yours, you should first do some maintenance on that file. Copy the file you have received to your network. If the file you are receiving is a central file, you should open the file using the Detach From Central option, as shown in Figure 4.2.

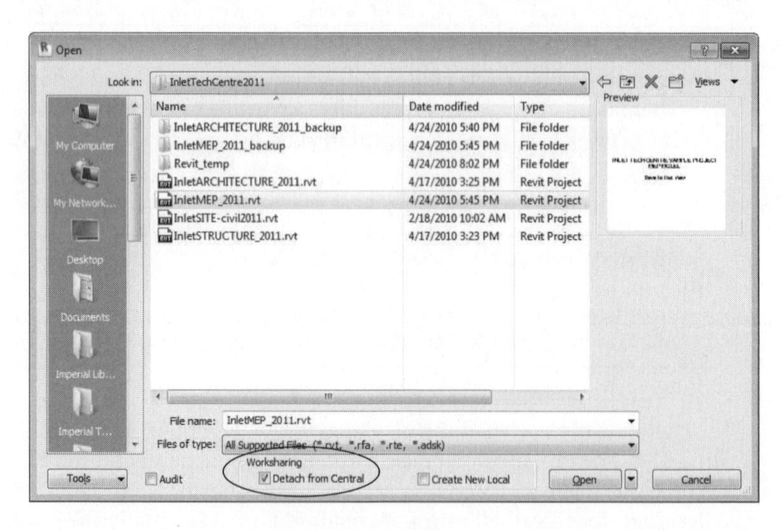

FIGURE 4.2
Using the Detach From Central option

When you open a file with the Detached From Central option selected, it becomes a new, unnamed project, and you are the owner of all the worksets in the file. A warning dialog box appears when you select this option that provides confirmation. Once the file is open, you can save it to the location in which you want to load it from by using the Save As command. Using the same location as your file will make it easier to manage all your linked files. If the filename does not suit your standards, you can give it any name you want, but the location you save it to and the filename that you use should remain the same throughout the life of the project. It is best to avoid overwriting a file with the same name using Save As, so when you download the file and put it on your network, you may want to rename it prior to opening it. You should also avoid downloading it directly to the directory in which the working linked file resides, especially if you are not changing the filename. You do not want to overwrite the linked file before you get a chance to clean it up.

You then need to relinquish your ownership of all the worksets and save the file again. This will ensure that the file you have received from your consultant is not attempting to access other files or coordinate with local files from another network. If the file you are linking is not a central file, you do not need to open it since there will be no worksets or local files associated with it.

There is no need to keep an archive of all the files being sent to you. Revit files can be very large, so keeping an archive will quickly use up storage space. When worksharing has been enabled in a file, Revit keeps a backup log of the file. The Restore Backup button on the Collaborate tab allows you to roll back changes to the file. So if you require a previous version of your consultant's model, you could ask them to save a rolled-back version of their file and send it to you. If you would like to keep archived versions of the project, it is best to use media such as a CD or DVD to save storage space on your network or hard drive.

Once the maintenance has been done on your consultant's file, you are ready to link it into your file. Let's first take a look at linking in the architectural model. Here are the steps:

1. The first thing to do when bringing in any type of file into your project is to make sure you have the appropriate workset active (if you have enabled worksharing). Considering that every discipline will need to use the architectural model for background and coordination, you should place it on a workset that is common to all users. The Shared Levels And Grids workset is a good option if you have not created worksets specifically for linked files. This workset can be visible in all views without causing too much extra work to control the visibility of elements on that workset.

2. With the appropriate workset active, click the Link Revit button on the Insert tab. Browse to the location of the file you saved after receiving it from your consultant. The position of the linked file is crucial to coordination, so before you click Open, you must determine the positioning of the file by using the Positioning drop-down menu at the bottom of the Import/Link RVT dialog box. Figure 4.3 shows the six options for placing the linked file into your project.

The three Manual options allow you to manually place the model into your file using an insertion point from the origin, the center, or a specified base point of the file you are linking. These options are rarely used because they require you to manually locate the model, which can be difficult without a common reference point in both files.

The automatic options are more commonly used, and each has a specific function:

Auto – Center To Center This option will link the file from its center point to the center point of your file. The center point is determined by the location of the graphics within the file. This option should be avoided unless you have a coordinated center point with your consultants and the makeup of the model will prevent that point from changing.

Auto – Origin To Origin This is the best option to use unless a shared coordinate system is required. The origin of the linked file will not change if the model changes size or shape, or even if it is moved or rotated, so you will be able to stay coordinated with the building when major position changes occur.

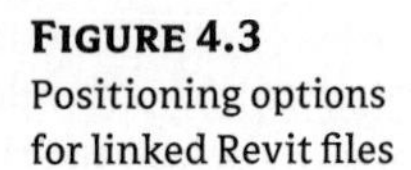

FIGURE 4.3
Positioning options for linked Revit files

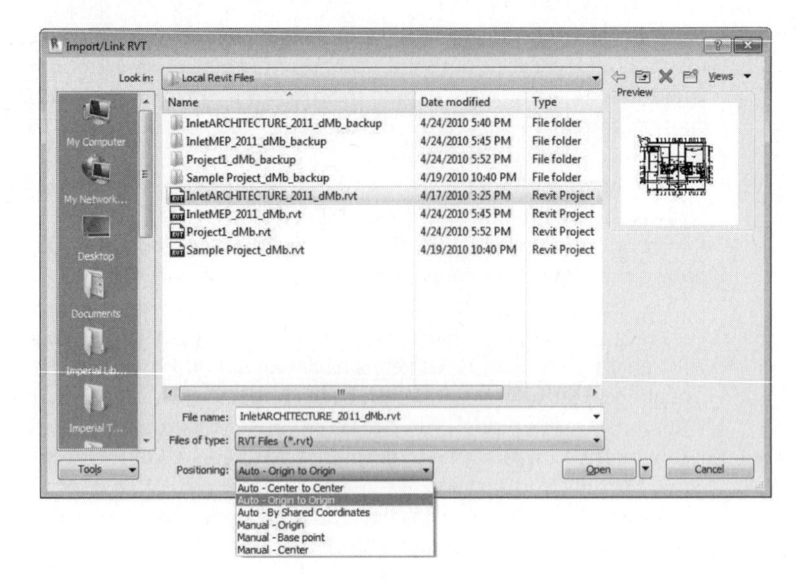

Auto – By Shared Coordinates This option should be used when a project requires that common location coordinates be used. This is most commonly used on projects that have multiple buildings because it enables you to keep the relationship between buildings coordinated. If you link in a file that is using shared coordinates, you can set the coordinates of your file to use the same coordinates.

Be careful when linking in a Revit file because the Auto – Center To Center option is the default. Forgetting to change the option could result in a serious headache further along in the project, because we know that architects *never* move or rotate the building, right? Make it a habit to use the Auto – Origin To Origin option, unless shared coordinates are required.

3. Once you have chosen a positioning option, you can choose options for loading worksets, similar to when you open a file that contains worksets. Just click the pull-down menu next to the Open button. You can bring in all the worksets or specify which ones you want to load. If you specify worksets to be loaded, they are the only ones that will be visible when the file is loaded.

4. After selecting workset options, click the Open button to link the file into your project. The file will appear in your file at the location determined by your positioning option. If you selected a manual option, you will need to click at a position in the view window to place the linked model.

5. Once the linked model is in the proper position, you want it to stay there, so before you do anything with the linked model, it is a good idea to pin it in place. This will prevent users from accidentally moving the whole building when clicking and dragging items during design. Click the model, and select the Pin button that appears on the Modify panel of the Modify RVT Links tab.

6. Repeat the process for all the files you need to link into your project. Once you have all your Revit links in place, you should synchronize your file with the central file or, if your file is not using worksharing, simply save your file.

Managing Revit Links

Throughout the design process you will want to keep up with the many changes occurring in the files you have linked into your project. Revit does not always automatically update your file to changes made in the linked files. The only time it updates automatically is when you open your project file after the linked files have been updated. If you are in a working session of Revit and one of your linked files is updated, you will have to manually reload the link to see the update.

Updating files from your consultants is the same process as when you first received the files except you do not have to place them into your file again. Download the new file from your consultant, and apply the maintenance procedures discussed in the previous section. Be sure that you overwrite the existing file with the new one and avoid changing the name.

After you have updated the files in your project folder, you can open your file. If you are working in a local file, you should synchronize it with the central file as soon as it is finished opening. This will ensure that the central file will contain the updated linked file information.

Sometimes new versions of the files you are linking become available while you are working in your project file. If another user has updated the linked files and synchronized them with the central file, you will not see the updated links even when you synchronize with the central file. You will need to reload the links also.

COMMUNICATION

Working with Revit should increase the amount of communication between design team members. If you update a linked file, let your team members know. One method to achieve this is to create a view in your project used by everyone where notices can be placed and then always save with this view current. This way, when someone opens the project, they immediately will read what needs to happen next to get totally up-to-date.

To reload a linked Revit file, click the Manage Links button on the Manage tab. This will open the Manage Links dialog box that allows you to manage the files you have loaded into your project (see Figure 4.4).

If your file is not synchronized with the central file, you will get a dialog box with options for syncing prior to managing links. The Revit tab of this dialog box lists all the Revit files linked into your file. Here you can determine whether the file is an attachment or an overlay. This is the same behavior found in external references in AutoCAD. An *attachment* will need to travel with the file, while an *overlay* is only for the file it is linked to. If a file you are linking in has other Revit files linked to it as overlays, you will see a dialog box indicating that the overlaid files will not link in with the file you are linking. This allows you to have the architectural file linked into your file and your file linked into the architectural file without duplicating either model. In other words, your file does not come back to you when you link in the architectural file. The default setting for linking files is Overlay, because this is the best choice.

The Save Locations button in the lower-left corner of the dialog box is used when your project is using shared coordinates. If you move a link within your project, you will receive a warning dialog box with options for saving the new location, or you can use this button.

Other buttons allow you to unload, reload, and remove links from your file. Once a link is removed, you cannot get it back by using the Undo feature; you will have to load the link back into your project. The Reload From button allows you to reload a link that may have moved to a new location on your network. You can even reload from a different file altogether.

FIGURE 4.4
Manage Links dialog box

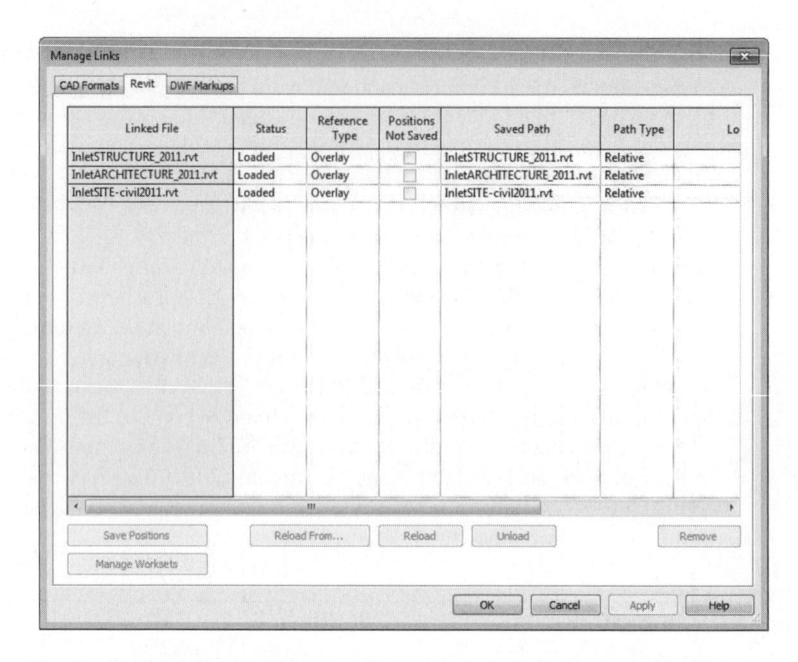

You can also use the Project Browser to manage the files linked into your project. All linked Revit files are listed at the bottom of the Project Browser. Right-clicking a link in the Project Browser provides a menu with quick access to link management tools, as shown in Figure 4.5.

FIGURE 4.5
Quick access to link management tools from the Project Browser

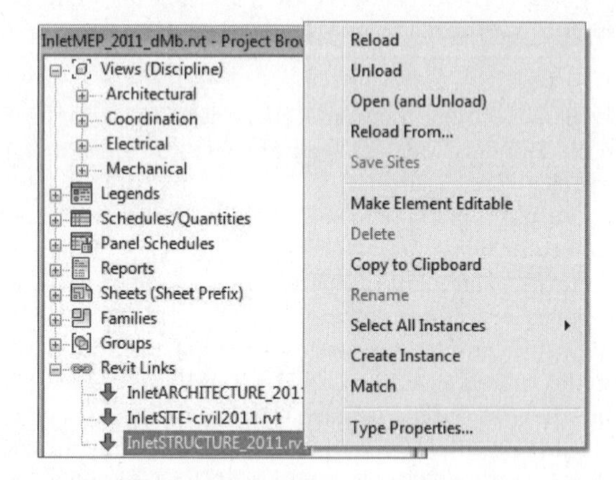

Notice that one option is to Select All Instances. You can have multiple instances of a linked file in your project by loading the file and copying it.

Visibility of Revit Links

When a Revit file is linked into your project, it is treated as one element, a linked model. Linking in the architectural model does not mean that you now have hundreds of walls, doors, and windows in your project. The model is seen as a single linked file. You do, however, have control over the visibility of the individual components that make up the linked model, as well as the entire link itself.

When you link in a Revit file, a new tab appears in the Visibility/Graphics Overrides dialog box. This tab lists the linked Revit files and allows you to turn them on or off. If you have multiple copies of a link, they are listed beneath the name of the link as seen when you click the + next to the link name. Each copy can be given a unique name for easy management. There is also a check box for setting the link to halftone (this is not necessary for views set to Mechanical and Electrical disciplines because Revit automatically applies halftone to the architectural elements in these discipline views). This is useful for turning links on or off in views as needed during design, and the settings can even be applied to view templates.

You also have the ability to control the visibility of components within the linked Revit file. By default, the components in the Revit link will react to the settings for visibility that you apply to your view. So if you were to turn off the Doors category in your view, the doors in the Revit link would also turn off. The same is true for any settings you may apply for line weight or color. Figure 4.6 shows that this behavior is indicated in the Visibility/Graphics Overrides dialog box as By Host View.

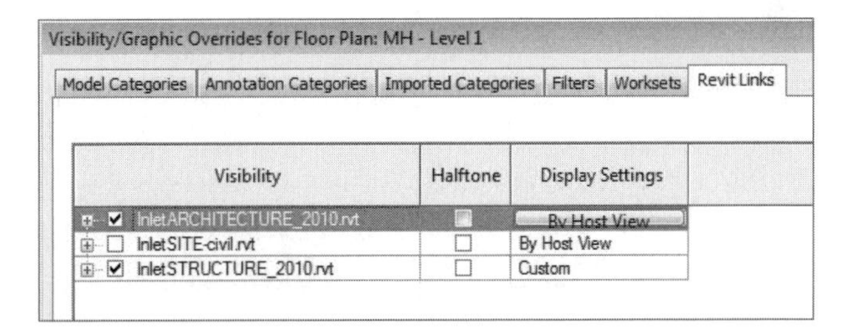

FIGURE 4.6
Linked Revit file Display Settings options

Clicking the By Host View button for a link will open the RVT Link Display Settings dialog box. This is similar to the Visibility/Graphics Overrides dialog box, but it is specific to the selected Revit link. The first tab of this dialog box allows you to set the behavior of the visibility of the link. You can set the link to behave By Host View, By Linked View, or Custom.

If you choose the By Linked View option, the option to select a view from the Revit link becomes active. Choosing a view from the link will set the link to display in your view with all the settings from the view that resides in the linked file. You do not have access to the views contained in a linked file, but you can set the link to display in your file the same way it appears in those views. This means that the link will display as set in that view despite any visibility settings in the host view. For example, in Figure 4.7, you see a view with the grids turned off in the host view, yet they are displayed because the linked file is set to display from a linked view.

It is not likely that your consultants have set all of their views in a manner that works exactly for how you would like to see the link, so you also have the Custom option for visibility. Selecting Custom activates all the setting options on the Basics tab of the RVT Link Display Settings dialog box. With this option not only can you choose a linked view, but you can also determine other behaviors for the display of the link. You do not have to choose a linked view if you only want to control the visibility of certain components. The Custom option is the same as accessing the Visibility/Graphics Overrides dialog box, but for the link only. Once you have selected Custom, you can use the other tabs in the dialog box to set the visibility of the components within the link. When you select a tab, the items are grayed out. You must set that specific category to Custom as well by using the pull-down at the top of the tab, as shown in Figure 4.8.

FIGURE 4.7
Example of linked file view settings, showing (a) the View Grids option turned off, and (b) By Linked View selected

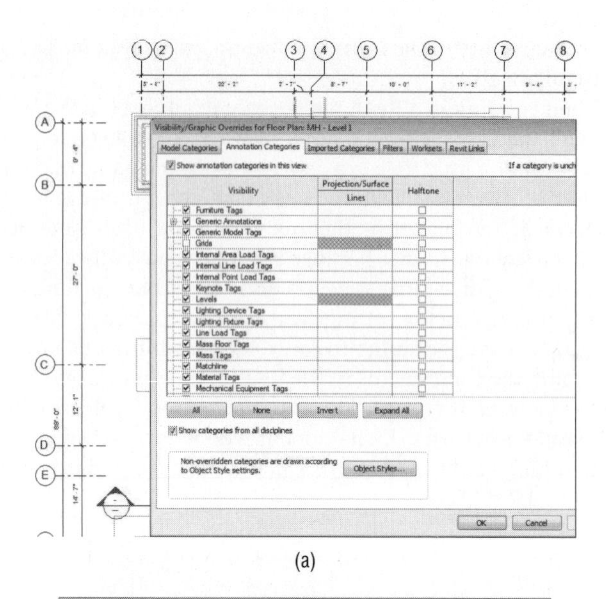

(a)

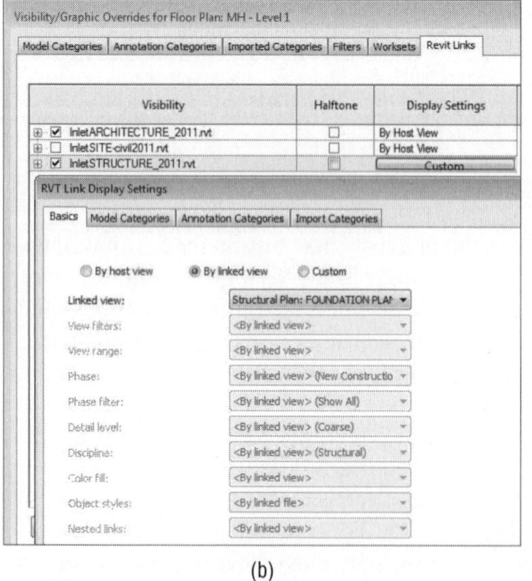

(b)

After setting a category to Custom, you can control the visibility of any Revit model category within the linked file. You can control elements in the Model categories as well as Annotation and any Import items such as linked CAD files. Figure 4.9a shows that the Grids category is displayed because the linked structural file is set to display by a linked view, despite the fact that the Grids category is turned off for the host view. A better solution might be to set the structural link to a Custom display and turn off all annotation categories except for Grids, as demonstrated in Figure 4.9b. Doing this will keep the structural link as a halftone automatically and will only display the items you want to see in your view.

Notice that none of the other annotations such as Dimensions is showing from the structural link.

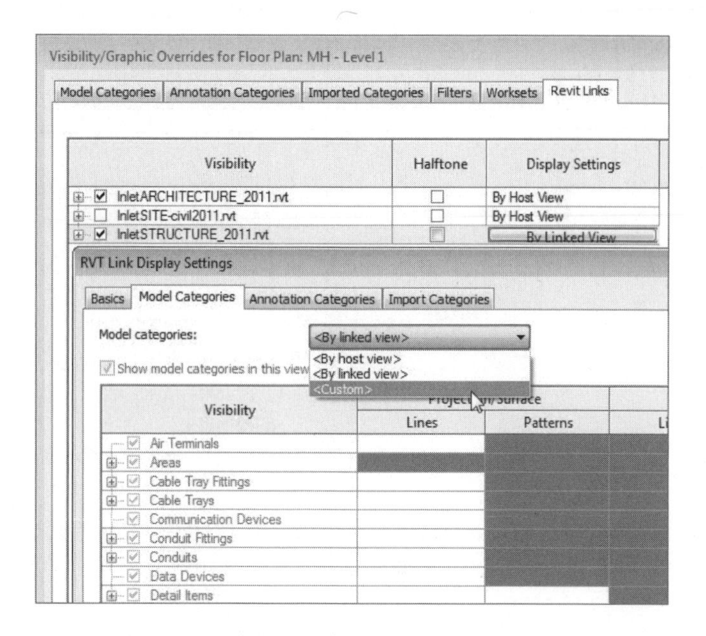

FIGURE 4.8
Setting a model category within a linked file to Custom

A new feature for Revit MEP 2011 is that you can control the visibility of worksets within a linked Revit file. If you choose not to make your worksets visible in all views within your project, then your consultants will have to turn on the visibility of the workset when your file is linked into their project file. A workset can be turned on or off by accessing the linked Revit file in the Visibility/Graphics Overrides dialog box. Clicking the button in the Display Settings column for the linked file opens the RVT Link Display Settings dialog box. This dialog box has a new tab for controlling the visibility of worksets, as shown in Figure 4.10. Setting the display to Custom allows you to turn on or off any worksets within the linked file.

LINKED FILES FOR PROJECT TEMPLATES

Mick has been tasked with creating his company project template. He works for an HVAC design firm and has to link the architectural and structural models whenever they do a project. His various consultants do things differently, but there are some common settings he would like to add to his template to improve efficiency. Since every project is different, he cannot have linked models in his template file. Or can he?

Mick decides to link some "dummy" files into his template to act as placeholders for when he will receive actual models from his consultants. He creates a file called `Arch.rvt` and one called `Struct.rvt` and links them into his template. The files are empty, so they do not add a lot of bulk to his template, but they do give him the ability to create visibility settings for linked files in his view templates.

When Mick uses his template to start a project, he can overwrite his placeholder links with the actual models received from his consultants using the Reload From option, and all of his view settings will remain. Mick is happy.

FIGURE 4.9
(a) Grids turned off in our view and (b) with the Structural link set to Custom and only displaying the Grids category

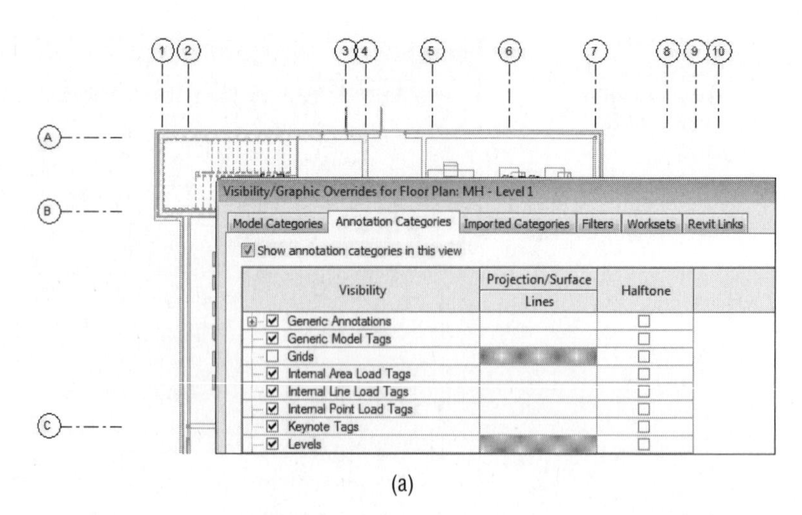

(a)

(b)

The output of Revit documents can vary depending on the type of printer you use. The most common issue is that the automatically halftone linked files do not print dark enough. Revit provides you with the ability to control how dark the linked files appear and print. You can find the Halftone/Underlay settings on the Settings menu located on the Manage tab, as shown in Figure 4.11.

In the Halftone/Underlay dialog box, you can turn off the automatic halftone of the linked file by deselecting the Apply Halftone check box. You can also override the line weight of the background Revit link. If you are using an underlay in your view, you can change the line pattern to any of the patterns you have loaded in your project. Figure 4.12 shows the Halftone/Underlay dialog box with sample settings applied.

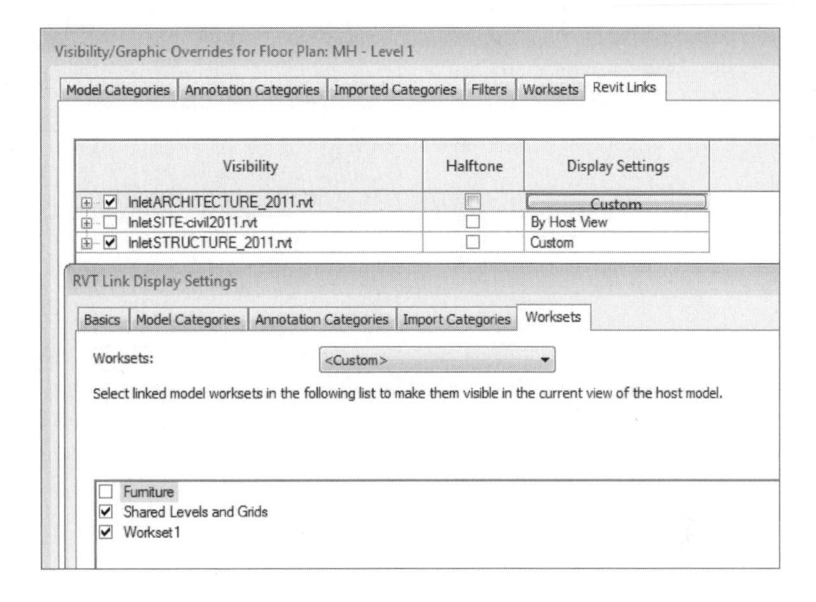

FIGURE 4.10
Visibility control of worksets within a linked file

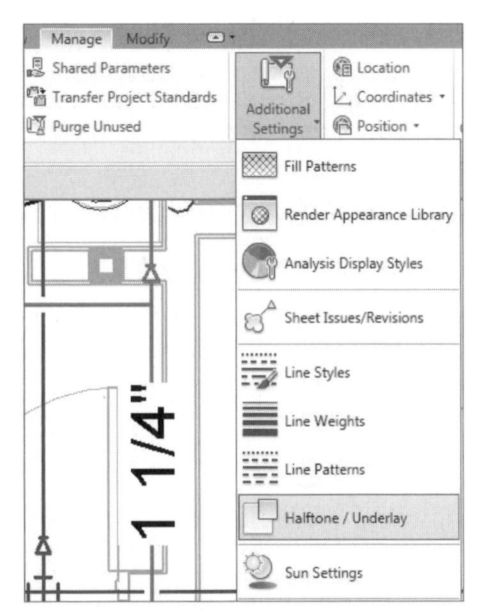

FIGURE 4.11
Halftone/Underlay settings

The slider on the bottom half of the dialog box lets you manually adjust the percentage of halftone applied to all halftone items in the view, both linked items and those in your model. Notice that only the elements within the link that are halftone are affected. The plumbing fixtures are part of the linked architectural model, but since they have been overridden to not be halftone, they are not affected by the adjustment. The settings that you use apply to the entire project, so you will only need to set them once. This is a helpful tool for getting your documents to print properly.

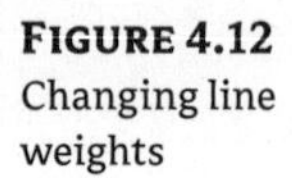

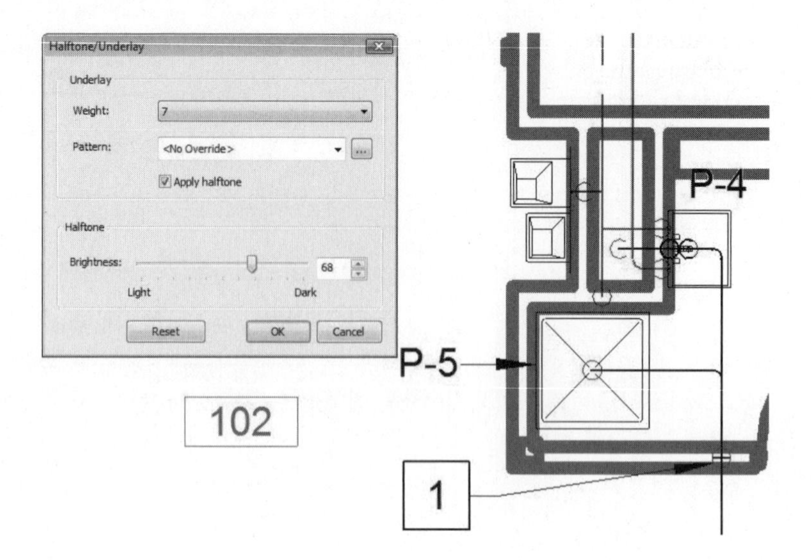

FIGURE 4.12
Changing line weights

> **PRINTING OPTIONS FOR HALFTONE**
>
> You can use the Replace Halftone With Thin Lines option in the Print Settings dialog box if you choose not to adjust the halftone settings in your model but still want darker backgrounds.

Coordinating Elements within Shared Models

One of the key benefits to using a BIM solution such as Revit MEP is that you can see changes in the design more readily because you see more of the other disciplines' work than you do with a traditional workflow. The ease of cutting section views and inspecting the model in 3D allows you to notice what is going on not only with your design but with the project as a whole. You may not have time to look over the model inch by inch, though, so there are some items that you can keep track of in the event that the design changes in a way that affects your systems.

Monitoring Elements

Revit gives you the capability to monitor certain elements within a linked file in order to be made aware of any changes. This functionality is limited to a few categories because, after all, you wouldn't want to know every time the architect moved a chair or changed the color of a door. The types of components that can be monitored are items that if changed can have a major effect on your design, especially late in the game. The architectural categories that can be monitored for coordination are Levels, Grids, Columns, Walls, and Floors. These items are often hosts for components in your model, so being aware of changes to them provides you with a higher level of coordination.

The Copy/Monitor function in Revit not only lets you monitor an element of a linked file but also allows you to create a copy of that element within your file. With the use of face-hosted elements, the need for copying items so that you could have something for them to be hosted by is not absolutely necessary. There are items, however, that are very important to coordinate with, such as levels and grids.

Your template file may contain levels in order to have preset views, but it is not likely that your levels will be the same on every project. After pinning a linked file in place, the next thing you should be in the habit of doing is monitoring the levels. This mainly applies to the architectural link since architects are the ones who will determine floor-to-floor heights. If your file has fewer levels than the linked file, you can copy the additional levels, and they will then be monitored automatically.

To coordinate with the levels in the linked file, follow these steps:

1. Open an elevation or section view of the model to see the level locations.
2. Set your levels to monitor the levels in the link by clicking the Copy/Monitor button on the Collaborate tab. Your two options for monitoring are selecting items in your file and selecting a link.
3. Click the Select Link option, and click the link in your view. When you select the link, the ribbon will change to display tools for Copy/Monitor.
4. There is no need to copy levels that already exist in your file, but you will want to modify them to match the name and elevation of the levels in the linked file. If you do not, then they will monitor the linked levels relative to the location they are at. Figure 4.13 shows an example of project and linked file levels.
5. Click the Monitor button on the Tools panel, and select the linked level you want to monitor.
6. Next, select the level in your file that you want to coordinate with the linked level.
7. Repeat this process for each level that you have in your file. If there is a level in the link that does not exist in your file, you can use the Copy button to create a monitored copy of the level.

   It is not necessary to monitor all levels, because your consultants may create levels for use as coordination points rather than for view creation. Only monitor levels that you need to create views for or associate your model components to, as in Figure 4.14.

   When you use the Copy button in the Copy/Monitor function, the copied element will automatically monitor the object it was copied from, so there is no need to repeat the monitor process after copying an item.

8. After you have finished coordinating your levels to the levels within the linked file, click the Finish button on the Copy/Monitor tab. You can now create views for any additional levels you may have added.

New to Revit MEP 2011 is the ability to Copy/Monitor MEP elements within a linked file. Clicking the Coordination Settings button on the Coordinate panel of the Collaborate tab opens the dialog box for establishing how MEP elements will be copied from a linked file into your project. Figure 4.15 shows the dialog box. Notice that only four categories of MEP elements can be copied and/or monitored. These settings are specific to the current project.

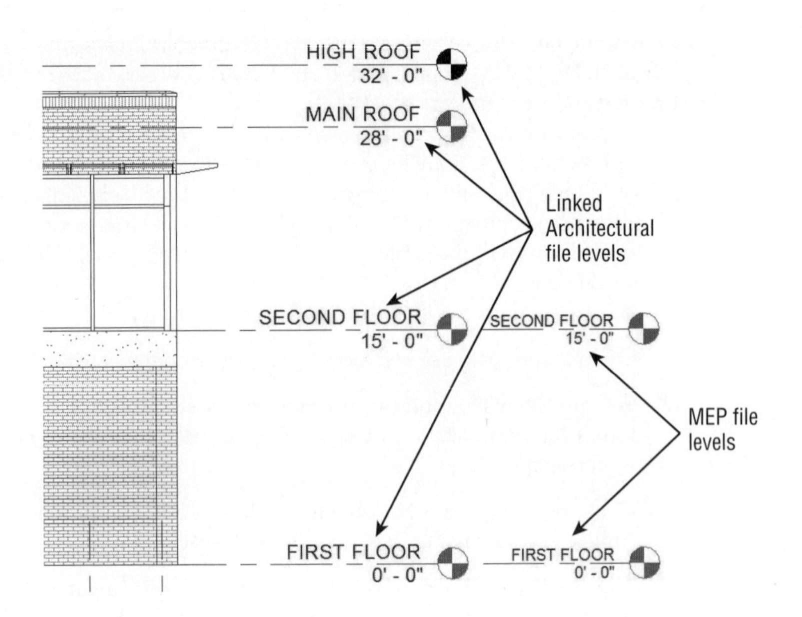

FIGURE 4.13
Project levels and linked file levels

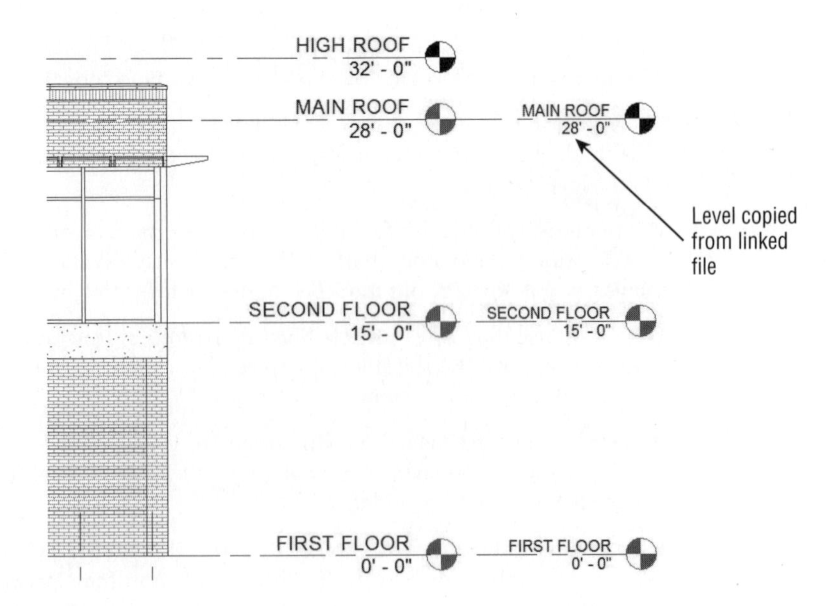

FIGURE 4.14
Level copied from linked file

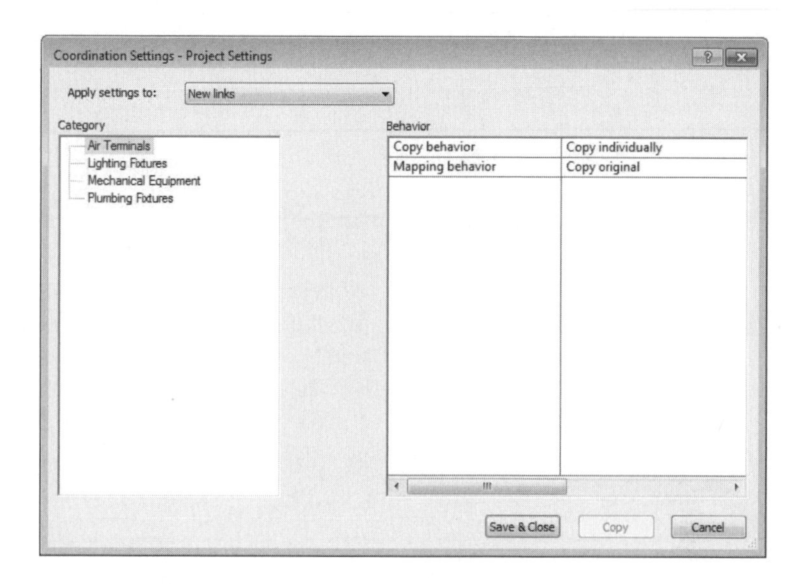

FIGURE 4.15 Coordination Settings dialog box

You can apply the settings to any of the linked files within your project by using the drop-down at the top of the dialog box.

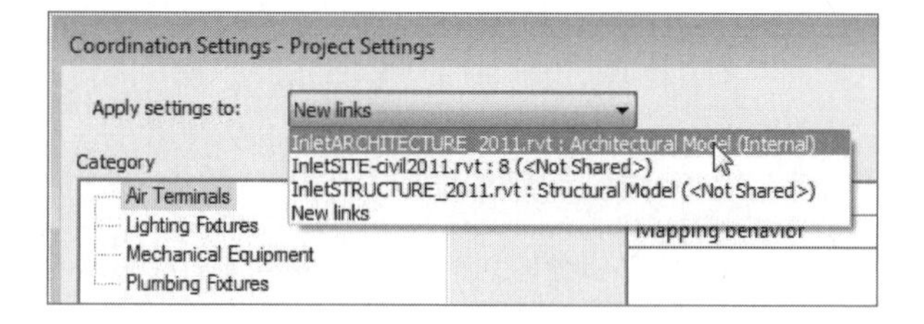

There are three choices for the Copy Behavior setting of each category. Items can be batch copied, which will allow for multiple elements of the category to be copied. You can also choose to ignore a category.

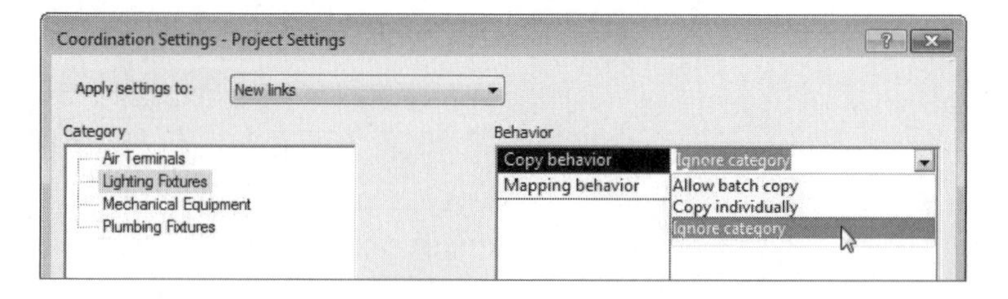

If you are applying settings for new links, then the Mapping Behavior setting is set to copy the original components, by default. If you have selected a specific linked file, you can set the mapping behavior for a category as long as the category is not set to be ignored.

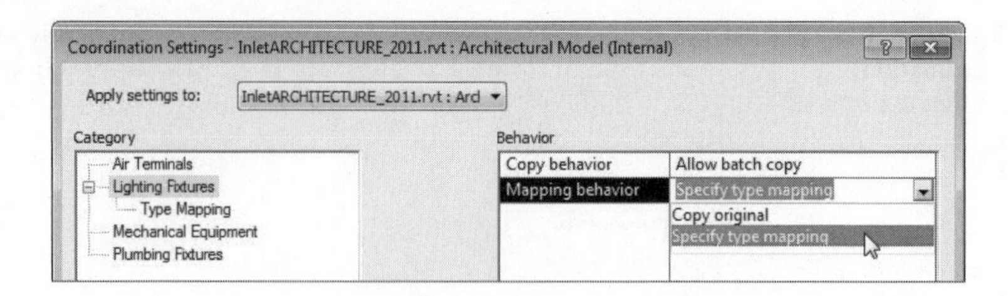

When you choose the Specify Type Mapping option, the Type Mapping settings appear beneath the category in the list at the left of the dialog box. When you click the Type Mapping option below the category, the dialog box changes to show the items in that category that are in the linked file and a column for defining what family within your project to use, as shown in Figure 4.16.

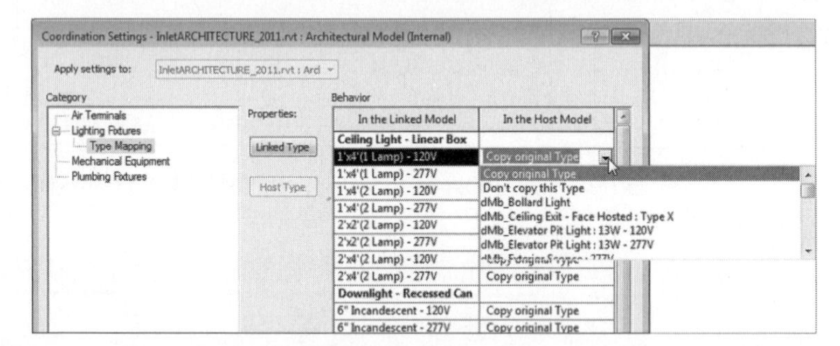

FIGURE 4.16
Type mapping options

The Linked Type and Host Type buttons allow you to view the Type Properties dialog box for the selected families. Changes cannot be made to the properties, but you can see them in order to determine the right families to use, as shown in Figure 4.17.

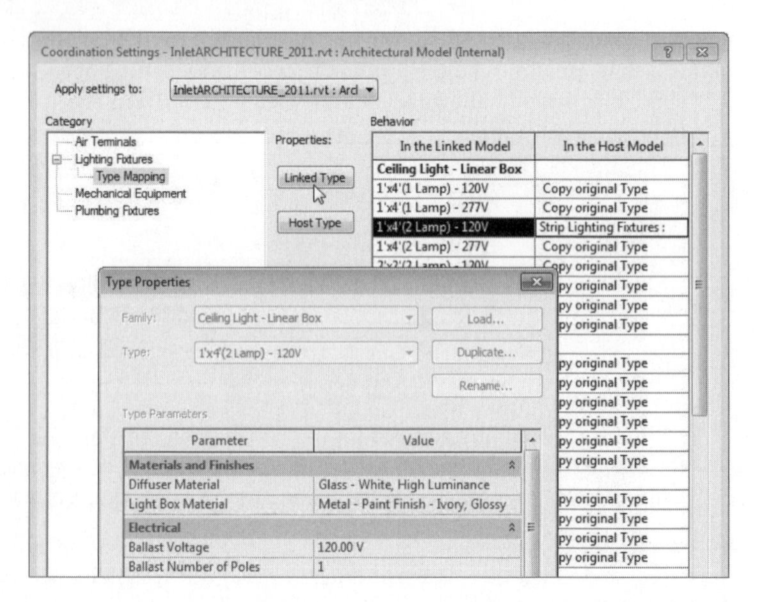

FIGURE 4.17
Type properties shown in the Coordination Settings dialog box

The main purpose for monitoring items is to be alerted when a change has been made to them. This is done by the Coordination Review feature within Revit. When you update your linked files, if there is a change to an item you are monitoring, a warning will appear during the loading of the link. This warning lets you know that a Coordination Review needs to be executed.

Responding to Change Alerts

To perform a Coordination Review, you can click the Coordination Review button on the Collaborate tab. The command is also on the Modify RVT Link tab that appears when you select a link in the view window. The dialog box that appears contains a list of the issues that require coordination, as shown in Figure 4.18. Expanding each issue will reveal the specific elements involved.

FIGURE 4.18 Coordination Review dialog box

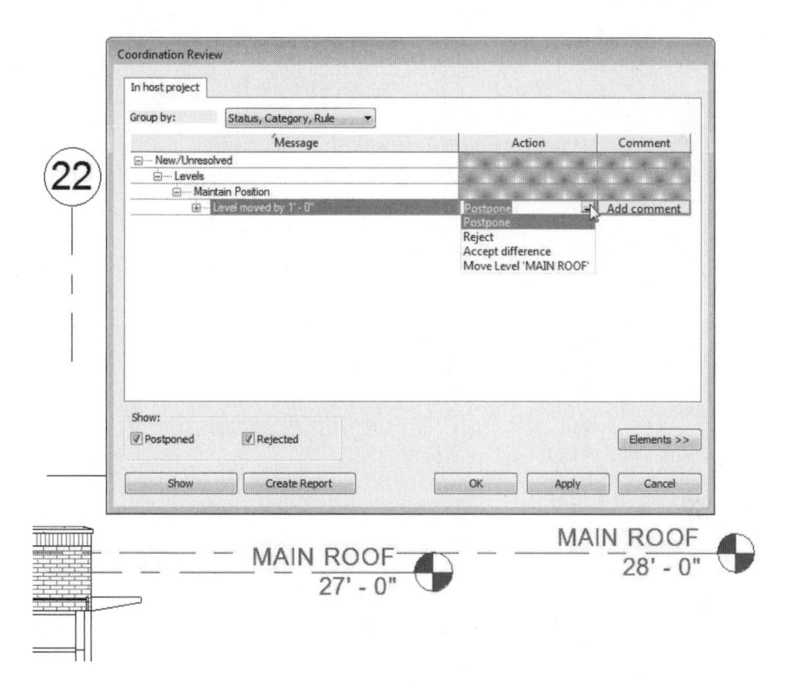

The Action column on the left provides four options for dealing with the coordination issue:

Postpone If you choose to postpone the action, Revit will warn you of the issue each time your file is opened or the linked file is reloaded until another action is taken.

Reject This option should be accompanied by a description in the Comment column to the right. If a change has been made that you must reject because of design constraints or some other reason, you can select this option. When you send your file to your consultants, they will receive a notice that a Coordination Review is required and will see that you rejected the change. Most likely a change that is rejected will generate communication between team members, but adding a comment is a good way to document the action choice.

Accept difference This choice can be used when the change does not affect your design and when the effort to coordinate outweighs the benefit. For example, you may be monitoring a wall in the architectural model to see whether it moves. On the day the project is due, you receive an alert that the wall has moved 1/32". You could decide that the change is insignificant enough to allow for the difference.

Modify your element to match the item it is monitoring In Figure 4.19, you can see you have the option to move the level so that it is coordinated with the linked Revit model. When this option is chosen, the change is made automatically by Revit. Select the option, and click the Apply button to see the change occur. If you are not in a view that shows the items requiring coordination, you can select one of the items in the list and click the Show button in the lower-left corner. Revit will search for a view that shows the selected item and open that view.

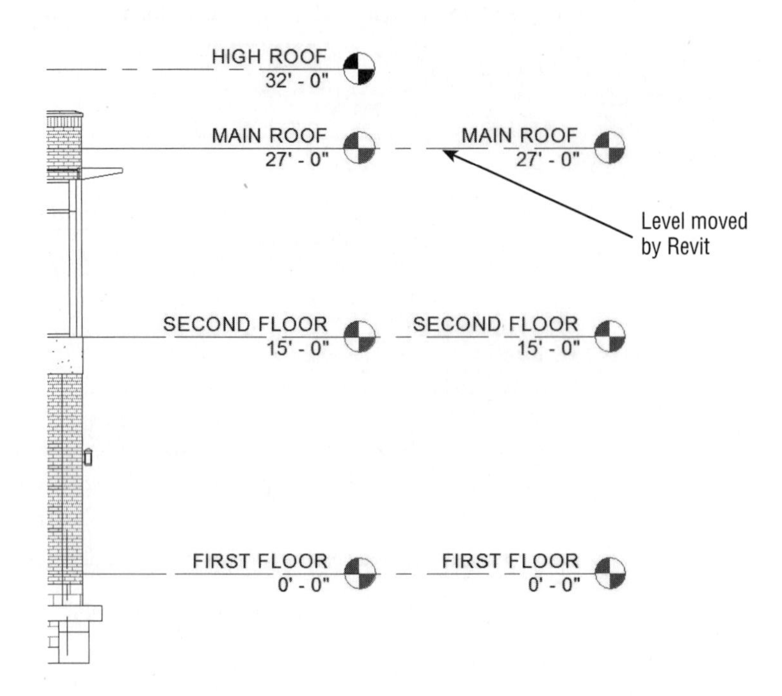

FIGURE 4.19 Level adjusted by Coordination Review

If the change made has too great of an effect on your model, you can simply undo the action and revisit the Coordination Review dialog box to select another action.

Reconciling Hosting

The use of face-hosted families allows you to attach your components to the 3D faces within a linked file. When the host of one of these types of families is deleted, the family will remain in the model, but it will no longer be associated to the linked file. This behavior can cause objects within your model to be floating in space when they should be attached to a surface. The most common reason for an element losing its host is that the host was deleted and redrawn, instead of just being moved. For example, if you have some air terminals hosted by the face of a ceiling and the ceiling needs to move up, the architect may delete the ceiling and create a new one at the new elevation. This would cause your air terminals to be floating below the new ceiling.

Revit MEP 2011 has a new feature that allows you to identify and reconcile any families that have lost their host. Clicking the Reconcile Hosting button on the Coordinate panel of the Collaborate tab turns on the Reconcile Hosting palette, as shown in Figure 4.20.

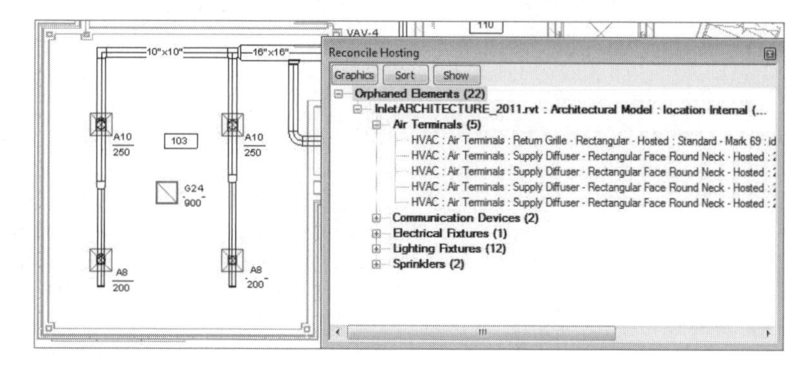

FIGURE 4.20 Reconcile Hosting palette

You can dock this palette to the interface or have it float. When the palette is active, any objects that are listed will be green in the model view. You can click the Graphics button on the palette to change the visual display of orphaned elements when the palette is active.

The Show button allows you to locate an item selected from the list if it is not readily visible in the current view. The Sort button lets you sort the list by either the linked file and then the categories or by the categories and then the linked file.

You can right-click an item listed in the palette and choose the option to delete it or choose the Pick Host option, which allows you to place the element on a new host. Be aware that when you use the Pick Host option from the palette, the only hosting option will be to a vertical face. So if you have an orphaned element that requires a horizontal host, the best option for reconciling would be to select the object in the model view and use the Pick New button on the contextual tab.

Tips for Maintaining Project Coordination

Proper setup of your project file is essential to maintaining coordination throughout the life of the project. You can find an example of how to do this on the companion website. Create a project file by following these steps:

1. Open the `Ch4Sample.rvt` file.
2. Click the Link Revit button on the Insert tab.
3. Select the `ArchModel.rvt` file, and choose the Auto – Origin To Origin positioning option. Click Open.
4. Select the linked model in the active view window, and click the Pin button on the Modify RVT Links tab.
5. Open the North elevation view.
6. Click the Copy/Monitor button on the Collaborate tab. Click Select Link.
7. Click the linked model in the view window.
8. Click the Monitor button on the Copy/Monitor tab.

9. Click the First Floor level at the right, and then select the First Floor level of the linked model.
10. Repeat step 9 for the Roof level.
11. Click the Finish button on the Copy/Monitor tab.

Working with Non-Revit Files

Revit files are not the only type of files that you can link into your project. The Revit platform is compatible with many types of CAD formats as well. As you move into using a BIM solution for your projects, try to use CAD files as little as possible. A large number of linked CAD files in your project can cause the file to perform poorly. Choose the files that will save you time overall on the project. CAD formats that can be linked into Revit include the following:

- DWG files
- DXF files
- DGN files
- ACIS SAT files
- SketchUp files

You can also link image files and DWF markups. There are two options for bringing non-Revit file types into your project. You can link them or import them. Linking is the preferred method because it will update as the file changes without having to repeat the process of inserting the link. Linked CAD files will show up in the Manage Links dialog box with options for pathing, reloading, and unloading. If you share your project that contains CAD links with a consultant and you do not give the consultant the CAD files, Revit will display the CAD link as it last appeared in your file.

LINKING VS. IMPORTING

Be careful when choosing to import a file rather than link it. When you import a file, the graphics are not the only thing brought into your project. All text styles, line styles, layers, dimensions, and so on, are brought in as well. This can really clog up your file and impede performance. Imported files will not update when changes are made. New versions of the imported file must be put into your project manually as well.

Linking in CAD Files

Because there is no Revit platform for civil engineering, it is likely that you will receive the files from your civil engineering consultant in some type of CAD format. The process for linking in these types of files is similar to linking a Revit file.

1. On the Insert tab, click the Link CAD button. The dialog box for linking a CAD file contains more options than the one for linking a Revit file, as shown in Figure 4.21.

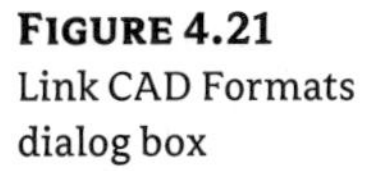

FIGURE 4.21
Link CAD Formats dialog box

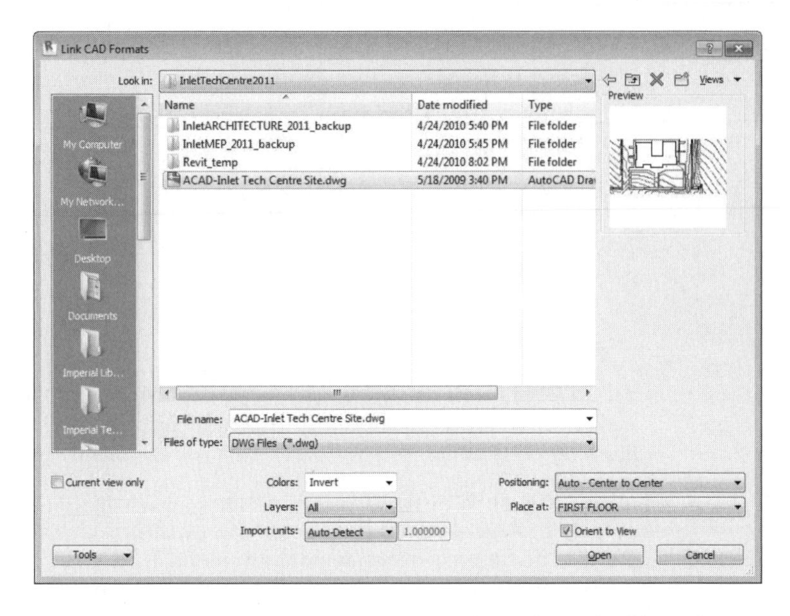

The check box in the lower-left corner provides the option for linking the file to the current view only. This will keep you from having to turn off the CAD link in any other view.

The Colors drop-down list allows you to invert the colors of the CAD file, preserve them as they exist, or turn all the line work to black and white.

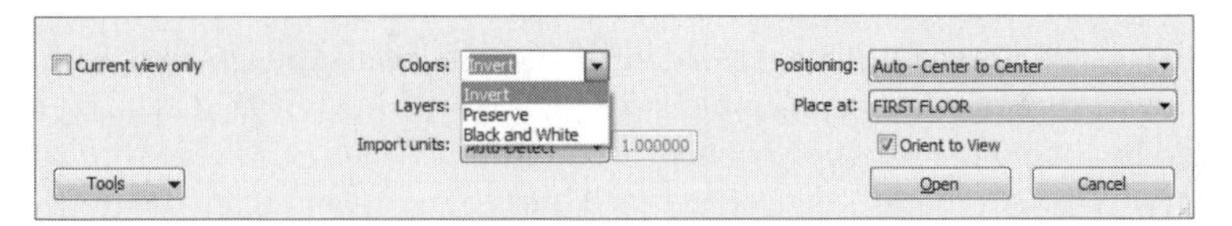

2. Determine which layers of the CAD file you want to load. This is similar to loading specific worksets of a linked Revit file. You can load all the layers within the file, you can load only those that are visible within the drawing (layers that are on and thawed), or you can specify which layers you would like to load.

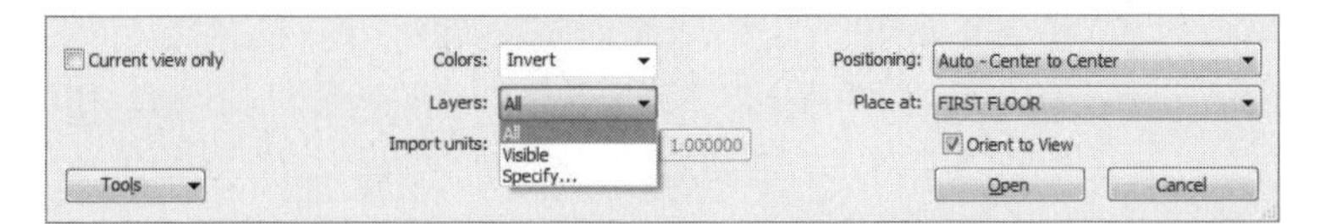

Choosing the Specify option will open a dialog box from which you can select the desired layers within the CAD file to load.

Layers within the linked CAD file can be removed after the file is in your project by clicking the link in the view window and clicking the Delete Layers button in the Import Instance panel.

3. Define the units for the file to be linked in. The Auto-Detect option means that Revit will use the unit setting that the CAD file is using.

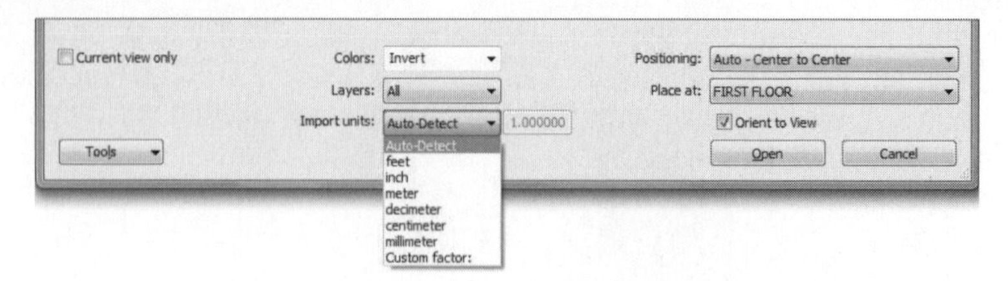

4. Position the CAD file. The positioning options for a linked CAD file are the same as for a linked Revit file. If you are using shared coordinates for your project, these coordinates need to be communicated to the civil engineering consultant in order for the site plan to align properly. If not, you can place the CAD file manually to line up the graphics.

Typically the architect will provide a building outline to the civil engineering consultant that will appear in the linked site plan. Whatever method is chosen, the file will automatically be pinned in order to prevent accidental movement. The Orient To View check box in the lower-right corner will position the file appropriately based on your view orientation, either Plan North or True North.

If you choose to use the Import option instead of linking, the options for placing the file are the same except for the automatic orientation to your view. For the sake of good file hygiene, it is best to link CAD files instead of importing them, especially if you are bringing in the file only for temporary use.

You can link any other type of CAD file needed in your project such as details or diagrams. These types of drawings can be linked to a drafting view that can be placed on a sheet, or they can be linked directly onto a sheet. To remove a linked CAD file, you can simply select it in the view window and hit the Delete key. This will remove it from the project completely.

You can view the properties of elements within a linked CAD file by selecting the link and clicking the Query button in the Import Instance panel that appears when the link is selected. When you click this, each element within the linked file can be selected, as shown in Figure 4.22.

Clicking an item will open the Import Instance Query dialog box with information about the selected item. The Delete and Hide In View buttons in this dialog box apply to the entire category of the selected item.

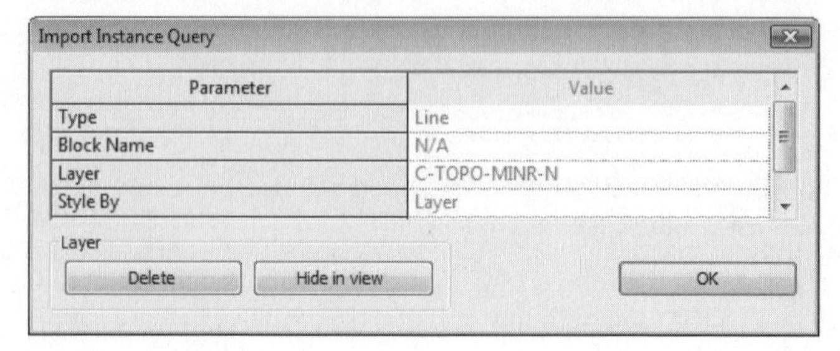

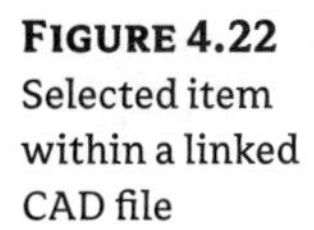

FIGURE 4.22
Selected item within a linked CAD file

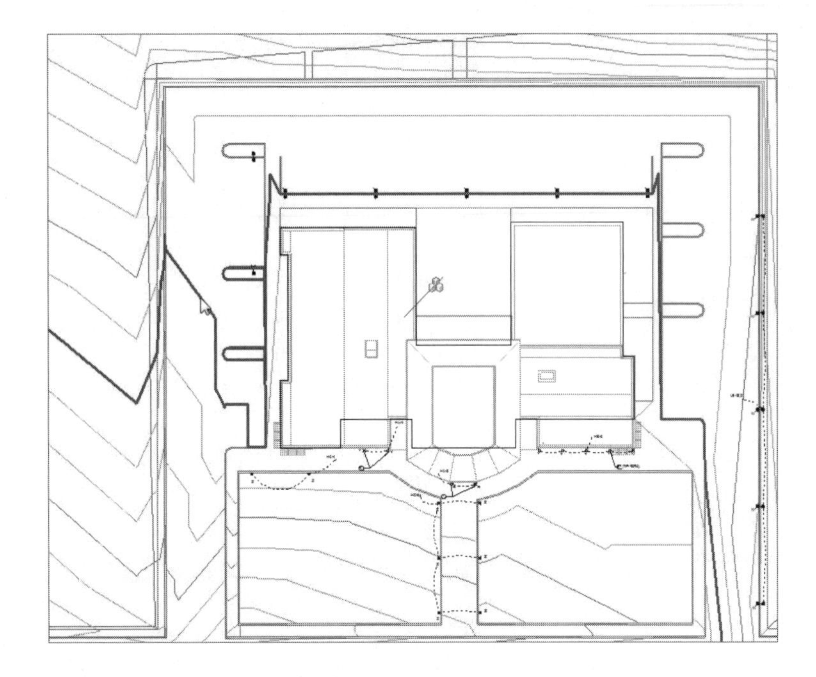

You can control the visibility of items in the linked CAD file from the Visibility/Graphics Override dialog box. The Imported Categories tab in this dialog box contains a list of all linked and imported CAD graphics. The layers within the linked files are listed and can be turned on or off, and their color, line weight, and line pattern can be overridden.

> **A NOTE ABOUT NOTES**
>
> When you link a CAD file that contains text, the text will take on the type of text your Revit project is using (Arial by default). This may affect the formatting of some notes and callouts within the linked CAD file.

Exporting Your Revit File to a CAD Format

Sharing your project file with consultants not using Revit is equally as important as receiving their file. You can export your file to several different CAD formats for use by your consultants. The most important thing to do in order to export your file is to establish the translation of model elements to CAD layers or levels. This is set up in the export options found on the Application menu under the Export tool (see Figure 4.23).

When selecting one of the file format export settings, the dialog box that appears contains a list of all the Revit model categories and their subcategories. You can assign a CAD layer or level and a color to each category and subcategory depending on whether the elements are displayed as cut or in projection. See Figure 4.24 for an example.

FIGURE 4.23
Export options

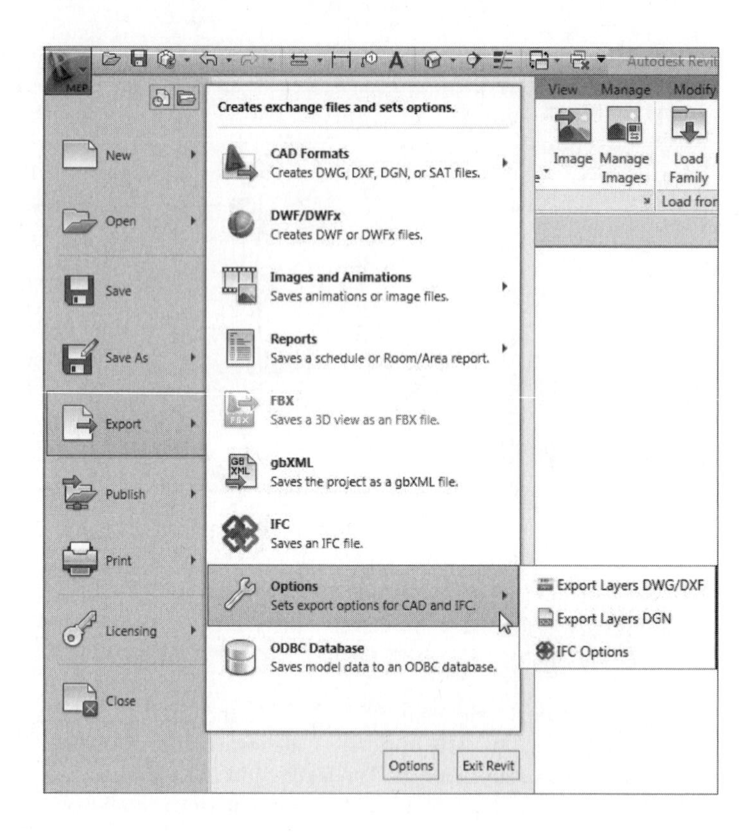

FIGURE 4.24
Layer export settings

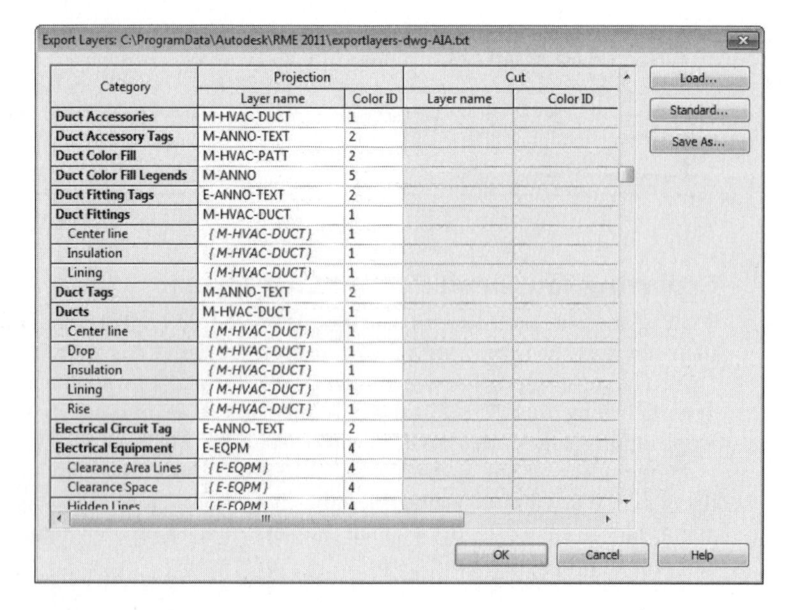

Once you have provided a layer and color for the elements that you need to translate to CAD, you can save the settings to a file that can be used again on future project by clicking the Save As button in the upper right of the dialog box.

With your export settings established, choose the appropriate CAD format from the Export tool on the Application menu. The Export CAD Formats dialog box provides you with options for selecting multiple sheets or views to export and the properties of the exported files.

Click the Export button once you have chosen the views or sheets to export, and a dialog box will appear to browse to the file export location. At this point, you can determine the software version you want to export as. If you are exporting a sheet view, you have the option to turn each view on the sheet into an external reference. If you are exporting a model view, selecting this option will cause any linked files in the view to be exported as separate CAD files to be used as external references.

Using Image Files in a Revit Project

Image files may be necessary for your Revit project to convey design information. They can also be used for presentation documents to add a more realistic look to the model and for logos on sheet borders. Sometimes renovation projects require information from older documents that are not in Revit or a CAD format. You can save the time required to re-create this information by inserting scanned images of the documents. As with linking CAD files, be careful not to use too many images in your project, for the sake of model performance.

You can insert an image file directly into any type of view except for 3D views. Click the Image button on the Insert tab, and browse to the location of the image. Images are imported into your project, not linked, and will travel with the file when shared. Figure 4.25 shows placing an image into a sheet border. An outline of the image appears at your cursor for placement.

FIGURE 4.25
Placing an image into a view

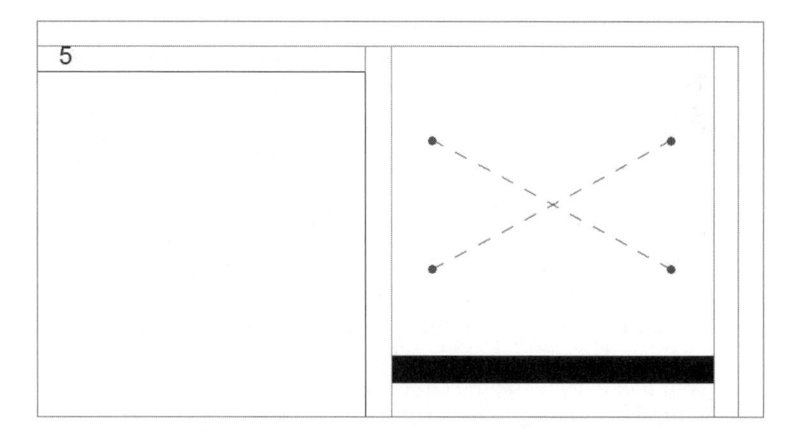

Once the image has been placed, the grips at the corners of the image allow you to resize it by clicking and dragging them. The aspect ratio of the image will be maintained when dragging a corner point. To access these grips after placement, just click the image. You can place text and line work on top of the image as necessary.

If you need to show an image in a 3D view for rendering, you can bring the image into your file by creating a decal. The Decal button is located on the Insert tab. Clicking this button activates the Decal Types dialog box, shown in Figure 4.26. You will first need to create a decal type within your project by clicking the Create New Decal button in the lower-left corner.

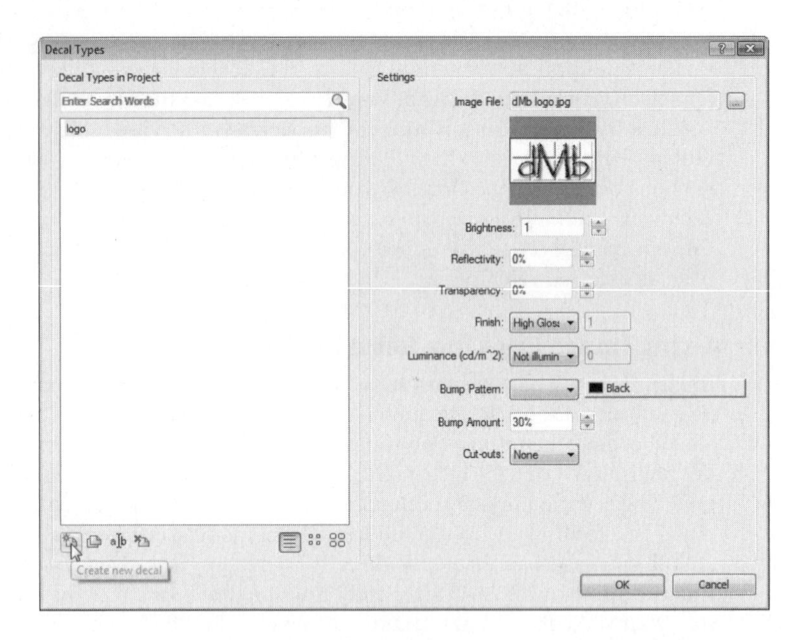

FIGURE 4.26
Decal Types dialog box

After naming your decal type, you can browse to the location of an image you want to use. The dialog box offers other settings for the rendering appearance of the image. Once the decal type is created, you can use the Place Decal button to insert the image into a 3D view. Settings for the decal size appear on the options bar during placement. You can place the decal on any one-directional planar 3D face, such as a wall, curved or straight. It will appear in the view as a box that defines the size of the image, as shown in Figure 4.27.

The decal will display its associated image only in rendered views. This allows you to show things such as signs or logos without having to model them. See Figure 4.28 for an example. Valid image file types for decals are BMP, JPG, JPEG, PNG, and TIF.

Image files brought into your project will not appear in the Manage Links dialog box. You can use the Manage Images button on the Insert tab to see what images are used in your project. Decals will not appear in the Manage Images list.

If you are working with a client or consultant who is not using Revit, you can still share your model by using the Drawing Web Format (DWF) export capabilities within Revit. You can export your model in either 2D or 3D format depending on your client's needs. DWF files are similar to `.pdf` files in that they cannot be edited, but they contain more than just graphical information. This will help you include your clients or consultants in the BIM process of your project by providing them the *I* in BIM — information.

To create a DWF file for sharing, click the Application menu, and select the Export tool. There you will see the DWF option (see Figure 4.29).

In the DWF Export Settings dialog box, you will see a preview of the current view and options for the output. If you want to create a 3D DWF, you need to be in a 3D view when you select the Export function. If your view discipline is set to Coordination, only the elements within your file will be exported; linked files will not be included in the DWF.

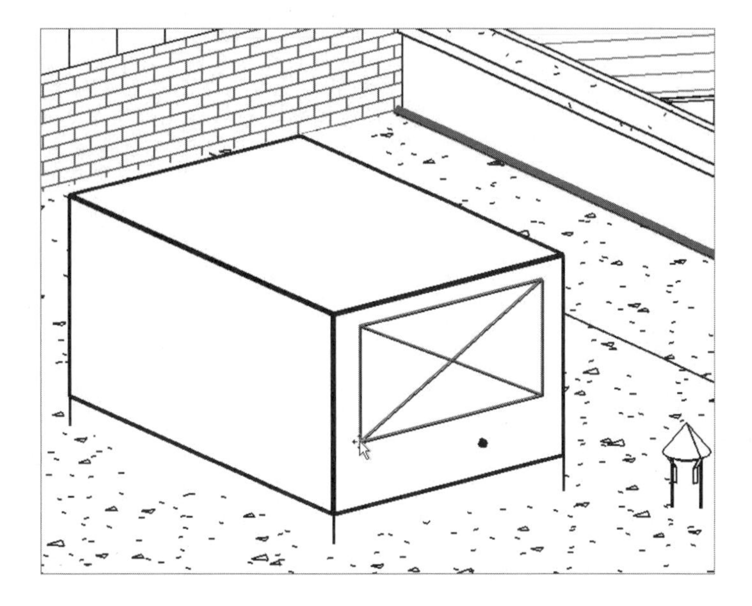

FIGURE 4.27
Placing a decal in a 3D view

FIGURE 4.28
Decal shown in a rendered view

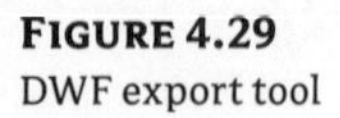

FIGURE 4.29
DWF export tool

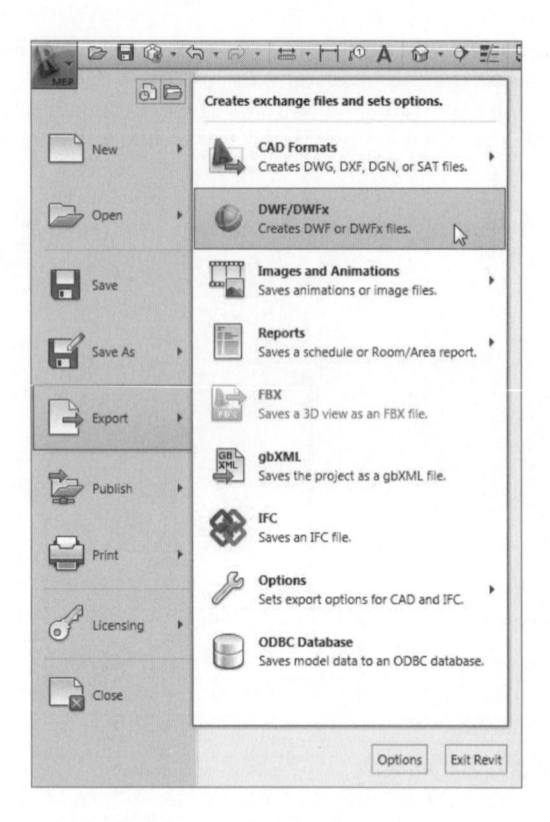

Options for Quality Control

The topics in this chapter discuss how you can work with all the files in your project to better coordinate your design with your consultants. One of the best things you can do is make sure that the file you are sharing is as complete and accurate as possible.

You can use some of the core functionality of Revit MEP to set up a system for quality control and project review. During visual inspection of the project, you may notice issues that require coordination. Unless you are the person responsible for modeling, there needs to be a workflow in place to document and manage the issues.

Consider creating an annotation family that acts as a comment tag for whoever may be reviewing the project. This tag can contain as little or as much information as you want. The sample shown in Figure 4.30 has parameters to assign a number, date, status, and owner of a comment along with space for a brief description. The project reviewer can place these tags throughout the project to point out areas that require modification or coordination. Comments can be made regarding issues within your file or issues about your consultant's files. When you share your file, your consultant can view the comments without sending additional information or files.

These tags can be scheduled, and the schedule can be used as a means of tracking the comments. The schedule also allows users to easily locate the comment tags using the Show feature within native Revit schedules, as shown in Figure 4.31.

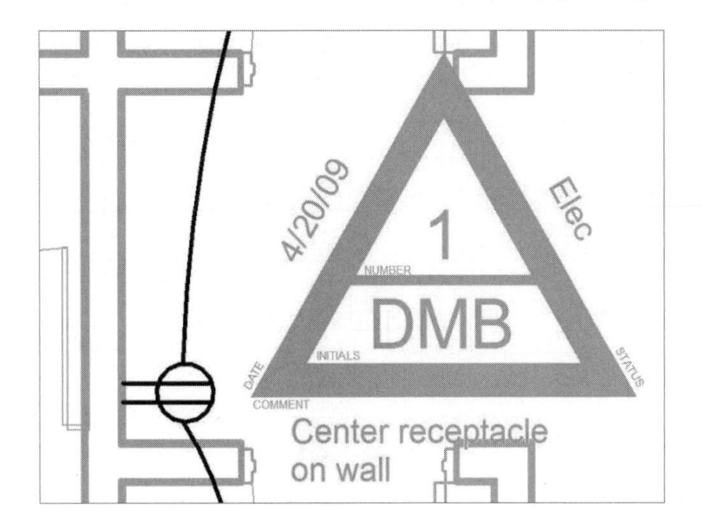

FIGURE 4.30
Sample review comment tag

Review Comments				
Status	Number	Reviewer	Comment	Date
ELEC	1	DMB	Center receptacle on wall	4/20/10
ELEC	2	DMB	Are there enough lights in this room?	4/20/10
FIRE PROT	9	SAB	Move riser to avoid conflict with Struct framing	4/22/10
FIRE PROT	10	SAB	Sprinkler is in same tile as light fixture, relocate	4/22/10
HVAC	3	MAT	Raise air terminals in this space to match light fixtures	4/20/10
HVAC	4	RXB	Coordinate shaft opening with Structural	4/21/10
HVAC	5	DMB	Increase duct size for new calculated airflow	4/21/10
PLUMBING	6	RXB	Arch removed a fixture, reroute sanitary piping	4/21/10
PLUMBING	7	RXB	Arch removed a fixture, reroute sanitary piping	4/21/10
PLUMBING	8	RXB	Arch removed a fixture, reroute sanitary piping	4/22/10

Repeat [Properties]
Recent Commands
Group Headers
Ungroup Headers
New Row
Delete Row(s)
Hide Column(s)
Unhide All Columns
Show...
✓ Properties

FIGURE 4.31
Revit schedule Show feature

As comments are addressed by users, the tags can be deleted from the views. An empty schedule would indicate that all comments have been taken care of. Another option would be to add a parameter, such as a check box, to the comment tag to indicate when a comment has been addressed. Using this workflow can be an efficient and productive method for dealing with coordination issues during the design process.

Another option to consider for quality control is using DWF markups in lieu of the traditional "red line on paper" method. The benefit of using DWFs is not only in the reduction of printing but also that your markups, and how they are handled, can be documented and archived. You can generate a DWF file of multiple sheets, which can be digitally marked up using the free Autodesk Design Review software that can be installed with Revit or is available for download from `www.autodesk.com`. The marked-up DWF sheets can then be brought into Revit and overlaid on their corresponding sheet views using the DWF Markup command on the Insert tab, as shown in Figure 4.32.

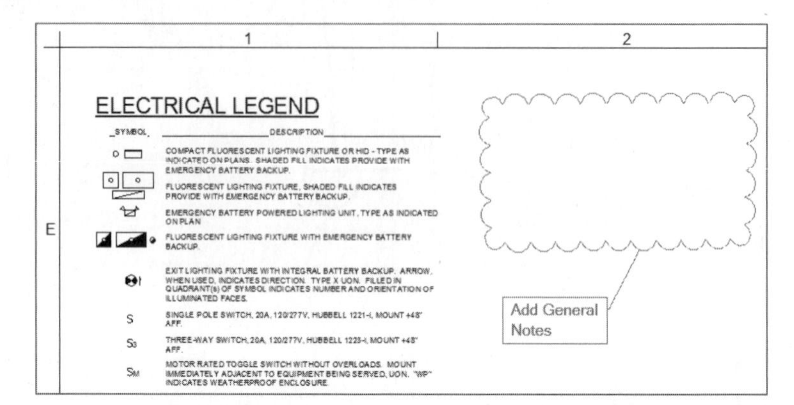

FIGURE 4.32
Revit sheet with DWF markup

As changes are made to your Revit file, you can edit the properties of the marks from the DWF, categorizing them as complete. The Revit file can then be exported to DWF again, and the new DWF file will show the marks as complete by highlighting them in yellow. The username and time of completion of the mark are stored in the DWF file also. This workflow enables you to keep track of when changes were made and who made them.

The Bottom Line

Prepare your project file for sharing with consultants Taking care to provide a clean, accurate model will aid in achieving an integrated project delivery.

Master it Describe the importance of making worksets visible in all views when your file will be shared with consultants.

Work with linked Revit files There are many advantages to using linked Revit files in your project. Revit provides many options for the visibility of consultants' files, allowing you to easily coordinate your design.

Master it How would you turn off a model category within a Revit link while allowing that category to remain visible in your model?

Coordinate elements within shared models Revit can alert you to changes to certain model elements within linked files. Managing these changes when they occur can reduce errors and omissions later in the project and help keep the design team coordinated.

Master it List the types of elements in a linked file that can be monitored for changes.

Working with non-Revit files Not all your consultants may be using Revit. This does not mean that you cannot use their files to develop and coordinate your design. You can also share your design by exporting your file to a format they can use.

Master it Describe the difference between linking and importing a CAD file and why the linking option is preferred.

Set up a system for quality control As a BIM solution, Revit provides functionality to keep your design coordinated with your consultants.

Master it What functionality exists in Revit that could allow a design reviewer to comment on coordination issues within a project?

Chapter 5

Schedules

Schedules of a parametric model provide the most immediate return of information from the data inherent in the model. With Revit MEP 2011 you can create schedules that are useful not only for construction documentation but also for data management, for object tracking, and even as tools for making design decisions.

By building a model, you can look at it from many angles or viewpoints. Schedule views are simply another view of the model. Instead of looking at the physical graphics that represent the design, a schedule allows you to view the data within the components in an organized and easy-to-manage format. Using schedules for data management is one way to increase efficiency, accuracy, and coordination within your Revit project. You can edit the properties of many objects quickly without having to locate the objects in the model.

Revit schedules can provide you with an accurate account of what objects are being used in your project model. The ability to track and manage objects can help with cost estimation and material takeoff. With this information readily available, you have the power to make decisions that affect cost and constructability.

Harnessing the power of creating schedules in a Revit MEP 2011 project can help you reap the benefits that come with easy access to any model or project information.

In this chapter, you will learn to

- Use the tools in Revit MEP 2011 for defining schedules and their behavior
- Schedule building components
- Create schedules for design and analysis
- Schedule views and sheets for project management

Defining Schedules

Mastering the scheduling tools in Revit MEP 2011 will enable you to easily extract any information from your projects. Schedules can be created and used on any of your Revit projects to establish consistency on construction documents and ease of data management for specified model objects.

Although there are different types of schedules, depending on their use and the items associated with them, the tools for creating a schedule in Revit MEP 2011 are similar for whatever type of schedule you are creating. Because schedules are essentially a view of the model, the tools are located on the View tab of the ribbon. Figure 5.1 shows the different types of schedules that can be created by clicking the Schedules button.

FIGURE 5.1
Schedule types

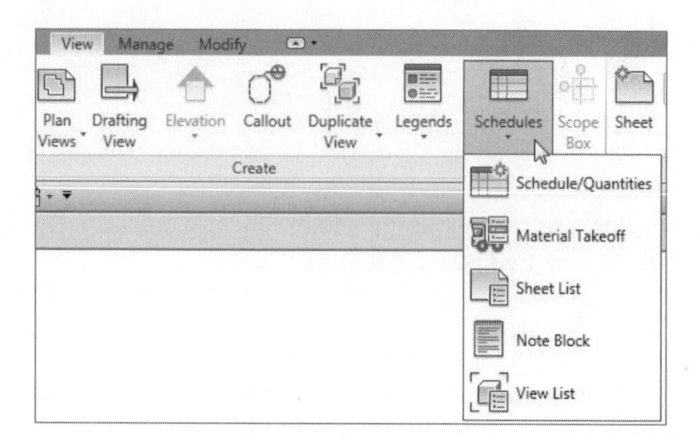

Clicking the View List or Sheet List option will take you directly to the View List or Sheet List Properties dialog box where you can begin creating your schedule. Clicking the Schedule/Quantities or Material Takeoff option will first take you to a dialog box that allows you to select the Revit object category that you want to schedule. The Note Block option takes you to a dialog box where you can select the annotation family that you want to schedule.

In the New Schedule dialog box, you can define whether you are creating a schedule of building components or a schedule key. You can also set the project phase of the schedule view. The project phase is an important property of schedules because only the objects in the model that belong to the same phase as a schedule will appear in that schedule. The name that you give your schedule is what will appear in the main header of the schedule when it is placed on a drawing. If your drafting standards dictate that the text should be in all-capital letters, you will need to retype the name even if it is correct. The same check box that appears in the Visibility/Graphics Overrides dialog box for showing the categories from all disciplines also is available in the New Schedule dialog box. This allows you to select items that are not MEP objects and may not even exist in your model. You can schedule items within any Revit files that are linked into your project. Figure 5.2 shows a sample of the New Schedule dialog box with settings to build a Space schedule.

One of the choices is to create a Multi-Category schedule. This type of schedule is for objects that are in different categories but have common parameters. Figure 5.3 shows an example of a Multi-Category schedule. Notice that the Finish Color parameter is used in the schedule even though not all objects have that parameter.

Once you have established the initial settings and chosen a category, clicking OK will open the Schedule Properties dialog box. This dialog box has five tabs across the top that have settings to define the behavior and appearance of your schedule. When scheduling space or room objects, there is an additional tab for building an embedded schedule within the schedule you are creating. It is a good workflow to move through the tabs sequentially, establishing the settings in each tab before moving to the next. The first two tabs let you determine what information will be scheduled, while the last three enable you to control how the data will be displayed and the graphical appearance of the schedule when placed on a drawing.

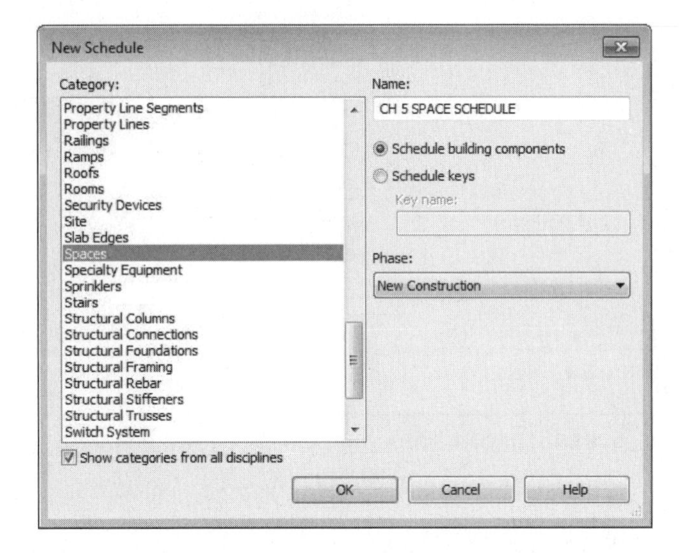

FIGURE 5.2
Sample schedule settings

FIGURE 5.3
Sample Multi-Category schedule

Multi-Category Schedule				
Family and Type	Type Mark	Description	Finish Color	Count
Check Valve - 0.375-4 Inch - Threaded: 3"		CHECK VALVE		2
dMb_8in Downlights - Faced Hosted: 1-32W Vertical - 120V	D	DOWNLIGHT, 8" DIA., 32W TT, VERTICAL LAMP	BRONZE	2
dMb_Duplex - Face-Hosted: Duplex Receptacle	R1	CONVENIENCE RECEPTACLE		5
dMb_Rectangular Duct Elbow - Mitered: Standard	SUPP	RECTANGULAR MITERED DUCT ELBOW		1
dMb_Round Neck Diffusers - Face Hosted: 12x12 - 6" Neck	A6	LAY-IN LOUVER FACE	OFF WHITE	3
dMb_Switches - Face Hosted: Single Pole - 120V	SW1	LIGHTING TOGGLE SWITCH, 120V	IVORY	3
Dry Type Transformer - 480-208Y120 - NEMA Type 2: 15 kV	TX1	POWER TRANSFORMER		1
Pipe Elbow: Standard		PIPE ELBOW		4
Pipe Types: Domestic Hot Water	DHW	DOMESTIC HOT WATER PIPE		7
Pump - Base Mounted: 100 GPM - 130 Foot Head	SP1	BASE MOUNTED PUMP		1
Pump - Base Mounted: 600 GPM - 120 Foot Head	SP2	BASE MOUNTED PUMP		1
Rectangular Duct Tee: Standard	SUPP	RECTANGULAR DUCT TEE		1
Rectangular Duct: Mitered Elbows / Tees	SUPP	RECTANGULAR DUCT		4

The Fields Tab

The Fields tab of the Schedule Properties dialog box contains many tools for choosing what information will appear in a schedule. Figure 5.4 shows the Fields tab of a newly created Space schedule.

The check box in the lower-left corner allows you to schedule items that are in any files that are linked into your project. The Available Fields list at the left side of the dialog box contains all the schedulable parameters for the category that was chosen to be scheduled. The Edit and Delete buttons below the list become active only when you select a project or shared parameter that you have created. Only Project parameters that are not Shared parameters can be deleted.

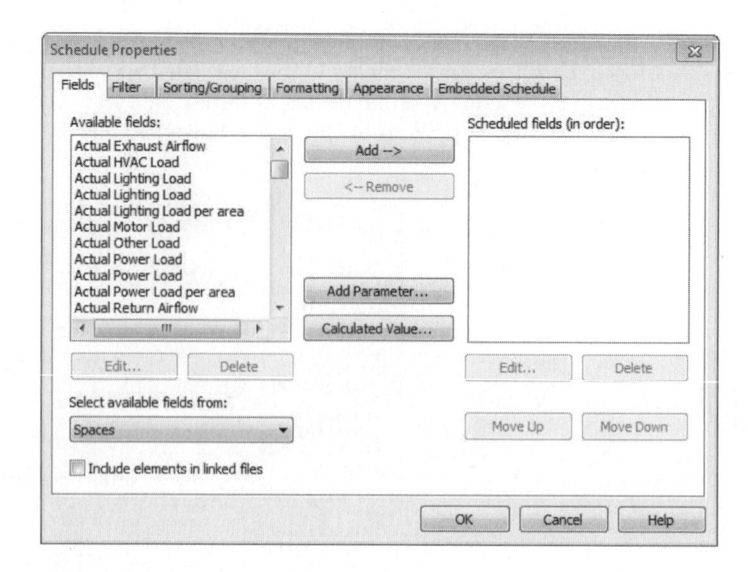

FIGURE 5.4
Fields tab of Schedule Properties dialog box

SCHEDULES AND PARAMETERS

The data within a Revit schedule comes from parameters assigned to the objects that are scheduled. It is very important that the parameters of objects are the same as those in the schedule. Because of the relationship between objects, parameters, and schedules, it is best to have a good understanding of how parameters work when you set out to build a schedule. For more information on the creation and management of parameters, see Chapter 19.

You can use the drop-down list in the lower left to change the list of available fields to parameters from another category. This is primarily for Space and Room schedules because the two objects are essentially the same thing. You cannot select fields from Mechanical Equipment if you are creating a Lighting Fixture schedule, for example.

To establish the columns in your schedule, you simply select the parameter from the Available Fields list and click the Add button at the top center of the dialog box. This will place the chosen parameter in the space at the right of the dialog box. Each parameter added to the schedule will be a new column, and the order they appear in the list determines the order of the columns in the schedule, from left to right. Once you have added a parameter to the schedule, you can use the Move Up and Move Down buttons to set the order. These buttons become active only when you select a parameter from the Scheduled Fields list at the right.

The Remove button at the top center allows you to remove a parameter from the schedule. This is not the same thing as deleting a parameter. Using either of the Delete buttons will remove the selected parameter from your project and any objects in your project.

If you want to schedule information about the chosen category and a parameter does not exist that contains that information, you can click the Add Parameter button at the center of the dialog box. Clicking this button will activate the Parameter Properties dialog box where you can choose the type of parameter and its properties. The box in the lower-left corner of the

Parameter Properties dialog box will be selected by default and is grayed out to ensure that the parameter you create will be added to all the objects in the category you are scheduling. Once you have created a parameter, it will automatically be added to the schedule.

Revit schedules have the ability to perform calculations on the data within the parameters that are included in a schedule. This information will then appear as a unique column in the schedule. To create a calculated value column in your schedule, click the Calculated Value button at the center of the Schedule Properties dialog box. Give the value a descriptive name so that when it is seen in the list or schedule, its purpose will be clear. You can create a formula by including parameters in a mathematical equation, or you can calculate a percentage.

The percentage calculation will return the percentage of a specific parameter as it relates to the project as a whole. Figure 5.5 shows the settings for percentage calculation of the volume of Space objects and the resulting column in the schedule.

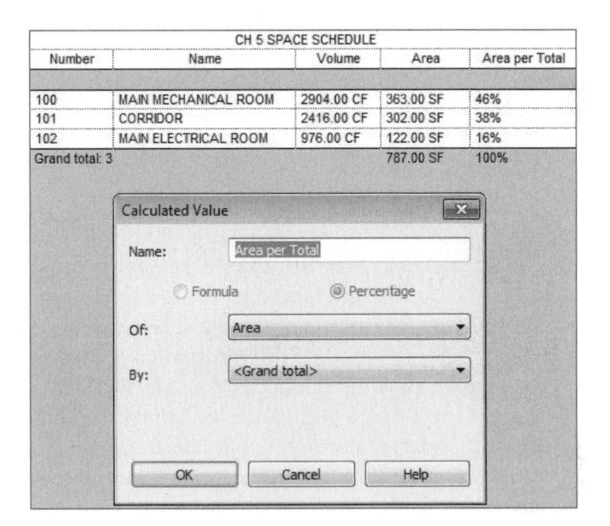

CH 5 SPACE SCHEDULE

Number	Name	Volume	Area	Area per Total
100	MAIN MECHANICAL ROOM	2904.00 CF	363.00 SF	46%
101	CORRIDOR	2416.00 CF	302.00 SF	38%
102	MAIN ELECTRICAL ROOM	976.00 CF	122.00 SF	16%
Grand total: 3			787.00 SF	100%

FIGURE 5.5
Percentage calculation in a schedule

When creating a calculated value based on a formula, it is important to use the proper Discipline and Type settings that coincide with the result of the equation. You will receive a warning stating that the units are inconsistent if the result does not match. Revit is case sensitive when it comes to typing in parameter names, so as you write your formula, it is crucial that you type the parameter name exactly. The small button next to the Formula field allows you to pick the parameter from a list and inserts it into your formula to ensure proper spelling and capitalization.

Once you have created a calculated value, it automatically appears in the Scheduled Fields list. If you use the Remove button to remove a calculated value from the schedule, it will be deleted from the project. You can use multiple operators and multiple parameters to create complex formulas if necessary. One example of a simple formula for Space objects is to divide the volume by the area to determine the ceiling height, as shown in Figure 5.6. This formula works only because the Space objects have an upper limit set to the level above and the ceiling objects are set to Room Bounding, therefore defining the volume of the Spaces.

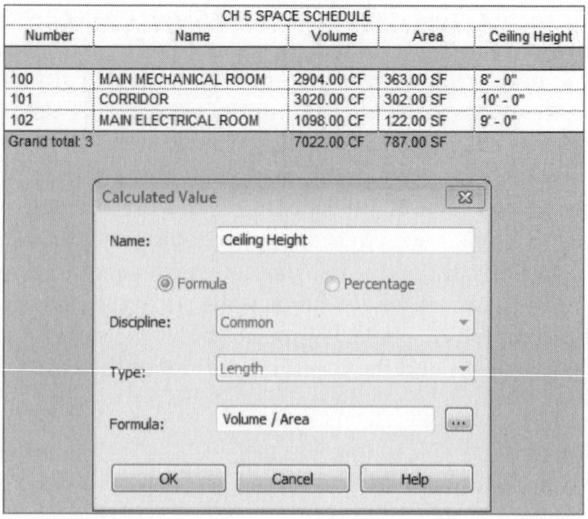

FIGURE 5.6
Space schedule with a calculated value for ceiling height

CH 5 SPACE SCHEDULE				
Number	Name	Volume	Area	Ceiling Height
100	MAIN MECHANICAL ROOM	2904.00 CF	363.00 SF	8' - 0"
101	CORRIDOR	3020.00 CF	302.00 SF	10' - 0"
102	MAIN ELECTRICAL ROOM	1098.00 CF	122.00 SF	9' - 0"
Grand total: 3		7022.00 CF	787.00 SF	

Calculated Value
Name: Ceiling Height
Formula
Percentage
Discipline: Common
Type: Length
Formula: Volume / Area
OK
Cancel
Help

The Filters Tab

The next tab in the Schedule Properties dialog box is the Filters tab. The settings on this tab allow you to filter out unwanted items in your schedule. You can set certain criteria for parameter values that objects in the schedule category must meet in order to be included in the schedule. This is very useful when scheduling a category that has many unique types of items, such as the Mechanical Equipment category. For example, if you wanted to create a pump schedule, you would have to schedule the Mechanical Equipment category. Without applying a filter to your schedule, all items categorized as Mechanical Equipment would show up in your schedule.

You can apply up to four conditions that an object must meet to make it into the schedule. To set a condition for a parameter, that parameter must be included in your schedule. If you want to use a parameter for filtering but do not want it to show up in your schedule, you have the option of hiding the column of the unwanted parameter. There are some parameters that you are not able to apply conditions to such as Cost or Family and Type.

Figure 5.7 shows a sample of a Mechanical Equipment schedule without any filter conditions applied to it.

FIGURE 5.7
Sample Mechanical Equipment schedule

Mechanical Equipment Schedule			
Type Mark	Description	Level	Family
CH1	CHILLER	Level 1	Scroll Chiller - Air Cooled - 60-120 Tons
EF-1	EXHAUST FAN	Level 1	Exhauster with Cabinet - Rectangular Conn
EF-1	EXHAUST FAN	Level 1	Exhauster with Cabinet - Rectangular Conn
EF-2	EXHAUST FAN	Level 1	Exhauster with Cabinet - Rectangular Conn
RF1	ROOFTOP FAN	Level 2	Centrifugal Fan - Rooftop - Upblast
SP1	BASE MOUNTED PUMP	Level 1	Pump - Base Mounted
SP2	BASE MOUNTED PUMP	Level 1	Pump - Base Mounted
VAV-1	FAN POWERED VAV	Level 1	VAV Unit - Parallel Fan Powered
VAV-2	FAN POWERED VAV	Level 2	VAV Unit - Parallel Fan Powered
VAV-2	FAN POWERED VAV	Level 2	VAV Unit - Parallel Fan Powered
VAV-2	FAN POWERED VAV	Level 2	VAV Unit - Parallel Fan Powered
VAV-3	FAN POWERED VAV	Level 1	VAV Unit - Parallel Fan Powered
VAV-4	FAN POWERED VAV	Level 1	VAV Unit - Parallel Fan Powered

To make this into a pump schedule, the filter settings shown in Figure 5.8 are applied. Notice the many options for filter rules that can be used to narrow down the qualifying objects. Using these conditions in combination with certain parameters gives you a very powerful tool for scheduling exactly the objects you want.

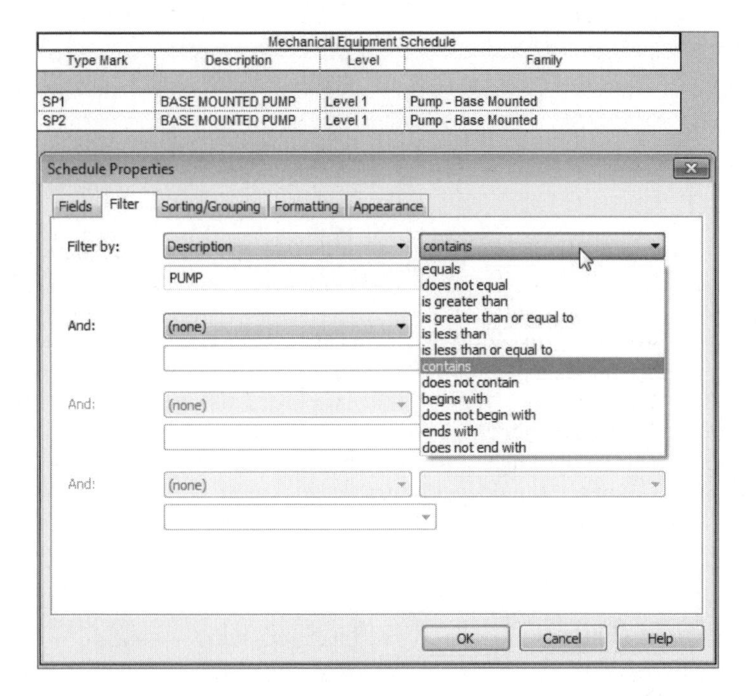

FIGURE 5.8
Filter settings for a pump schedule

As items are brought into your model, they will automatically appear in a schedule of their category as long as they meet the filter criteria. Remember that if you filter by a parameter that an object may not have or may not have a value for, you should input the data into the parameter when the object is brought into your project. This will ensure that your schedule is showing all that it should.

Another option for filtering a schedule is by setting the schedule phase, as mentioned earlier. The different options for the phase filter of a view can be applied to schedule views in the same way they are applied to model views. So if you were to set the phase of a schedule view to New Construction and the phase filter to one that shows only New Construction, then no items from a previous phase(s) would appear in the schedule.

The Sorting/Grouping Tab

The Sorting/Grouping tab of the Schedule Properties dialog box is where you can determine the order of the objects in a schedule and group like items together. You can assign up to four group conditions and provide a header or footer for each group. Sorting can be done based on parameters that are used in the schedule. When you sort by a particular parameter, all the objects in the schedule that have the same value for that parameter will be listed together.

Figure 5.9 shows a sample lighting fixture schedule that has been sorted by the Type Mark parameter. Notice that each instance of the different fixture types is listed in the schedule and that the Count column at the far right of the schedule confirms this.

FIGURE 5.9
Sample lighting fixture schedule

LIGHTING FIXTURE SCHEDULE								
			LAMPS					
TYPE	MANUFACTURER	CATALOG NO.	TYPE	QTY.	VOLTAGE	MOUNTING	REMARKS	Count
A	ACME LIGHTING	AL-12345	F32T8	2	120 V	RECESSED IN CEILING	NOTES 1, 3	1
A	ACME LIGHTING	AL-12345	F32T8	2	120 V	RECESSED IN CEILING	NOTES 1, 3	1
A	ACME LIGHTING	AL-12345	F32T8	2	120 V	RECESSED IN CEILING	NOTES 1, 3	1
B	REALLY COOL LIGHTS	RLC-432-EB2	F32T8/835/RS	4	120 V	RECESSED IN CEILING		1
B	REALLY COOL LIGHTS	RLC-432-EB2	F32T8/835/RS	4	120 V	RECESSED IN CEILING		1
B	REALLY COOL LIGHTS	RLC-432-EB2	F32T8/835/RS	4	120 V	RECESSED IN CEILING		1
B	REALLY COOL LIGHTS	RLC-432-EB2	F32T8/835/RS	4	120 V	RECESSED IN CEILING		1
B	REALLY COOL LIGHTS	RLC-432-EB2	F32T8/835/RS	4	120 V	RECESSED IN CEILING		1
C	DOWNLIGHT EMPORIUM	DE-120V-CAN-CF3	CF32TTBX	2	120 V	RECESSED IN CEILING	NOTES 1, 2,	1
C	DOWNLIGHT EMPORIUM	DE-120V-CAN-CF3	CF32TTBX	2	120 V	RECESSED IN CEILING	NOTES 1, 2,	1
C	DOWNLIGHT EMPORIUM	DE-120V-CAN-CF3	CF32TTBX	2	120 V	RECESSED IN CEILING	NOTES 1, 2,	1
D	JOES LIGHTING	JL77775-1	CF32TTBX	1	120 V	RECESSED IN CEILING	NOTES 1, 2,	1
D	JOES LIGHTING	JL77775-1	CF32TTBX	1	120 V	RECESSED IN CEILING	NOTES 1, 2,	1
X	EXIT LITES R US	ELRU-BR549	LED	-	120 V	SURFACE, CEILING		1
X	EXIT LITES R US	ELRU-BR549	LED	-	120 V	SURFACE, CEILING		1

A schedule will list each instance of the objects by default. To group items together that have all the same parameter values, you need to deselect the box in the lower-left corner of the Sorting/Grouping tab of the Schedule Properties dialog box. Figure 5.10 shows settings for the same lighting fixture schedule where the Itemize Every Instance box has been deselected. Notice that each type of lighting fixture is listed only once because the schedule is set to sort by fixture type. The Count column now indicates the total number of each type of fixture. The Grand Totals check box has also been selected to give a total number of lighting fixtures at the bottom of the schedule.

FIGURE 5.10
Lighting fixture schedule sorted and grouped by type

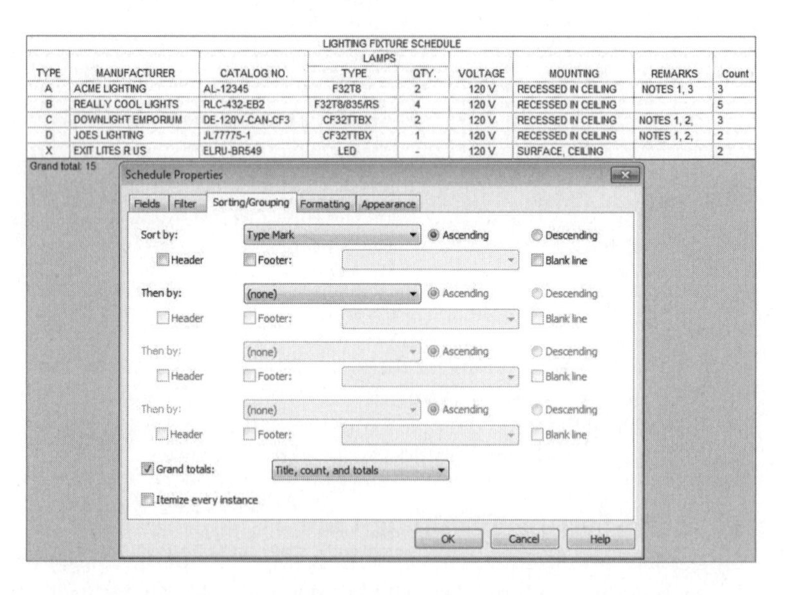

LIGHTING FIXTURE SCHEDULE								
			LAMPS					
TYPE	MANUFACTURER	CATALOG NO.	TYPE	QTY.	VOLTAGE	MOUNTING	REMARKS	Count
A	ACME LIGHTING	AL-12345	F32T8	2	120 V	RECESSED IN CEILING	NOTES 1, 3	3
B	REALLY COOL LIGHTS	RLC-432-EB2	F32T8/835/RS	4	120 V	RECESSED IN CEILING		5
C	DOWNLIGHT EMPORIUM	DE-120V-CAN-CF3	CF32TTBX	2	120 V	RECESSED IN CEILING	NOTES 1, 2,	3
D	JOES LIGHTING	JL77775-1	CF32TTBX	1	120 V	RECESSED IN CEILING	NOTES 1, 2,	2
X	EXIT LITES R US	ELRU-BR549	LED	-	120 V	SURFACE, CEILING		2

Grand total: 15

You can use the Sorting/Grouping settings with combinations of parameters to organize your schedules in the best way to easily read the data within them. Figure 5.11 shows the lighting fixture schedule again, with settings to sort first by the Manufacturer parameter and then by the Type Mark parameter with a footer to indicate the total number of fixtures per manufacturer. The Ascending or Descending buttons can be used to determine the order of items within a sorted group. Notice that fixture type A is listed before type D within the ACME LIGHTING manufacturer group because the Ascending option is used.

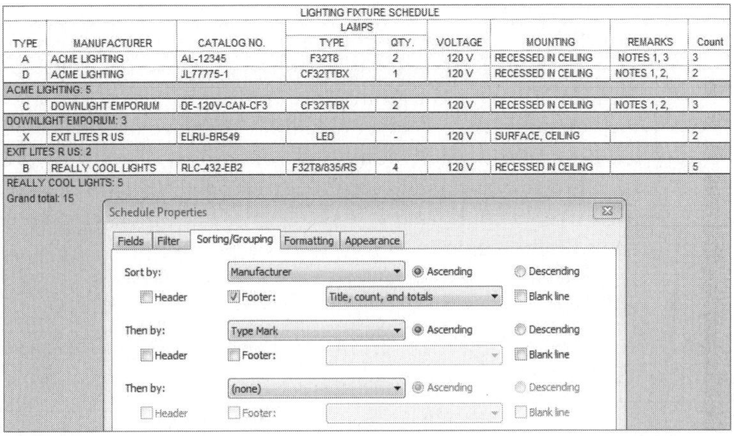

FIGURE 5.11 Lighting fixture schedule with two sorting options

LIGHTING FIXTURE SCHEDULE									
			LAMPS						
TYPE	MANUFACTURER	CATALOG NO.	TYPE	QTY.	VOLTAGE	MOUNTING	REMARKS	Count
A	ACME LIGHTING	AL-12345	F32T8	2	120 V	RECESSED IN CEILING	NOTES 1, 3	3
D	ACME LIGHTING	JL77775-1	CF32TTBX	1	120 V	RECESSED IN CEILING	NOTES 1, 2,	2
ACME LIGHTING: 5								
C	DOWNLIGHT EMPORIUM	DE-120V-CAN-CF3	CF32TTBX	2	120 V	RECESSED IN CEILING	NOTES 1, 2,	3
DOWNLIGHT EMPORIUM: 3								
X	EXIT LITES R US	ELRU-BR549	LED	-	120 V	SURFACE, CEILING		2
EXIT LITES R US: 2								
B	REALLY COOL LIGHTS	RLC-432-EB2	F32T8/835/RS	4	120 V	RECESSED IN CEILING		5
REALLY COOL LIGHTS: 5								
Grand total: 15								

The Formatting Tab

The Formatting tab of the Schedule Properties dialog box has tools for setting how the data will display within a schedule. Each parameter used in the schedule is listed at the left side. When you select a parameter from the list, you can apply the settings available on the right side. You can also apply the settings to multiple parameters.

The Heading setting establishes the name of the column in the schedule. The parameter name is used by default. Changing the column heading has no effect on the parameter itself, as shown in Figure 5.12, where the Description parameter is used but the column heading is called FIXTURE TYPE. Even if the parameter name is what you want for your schedule column heading, you will have to retype it if you want all capital letters.

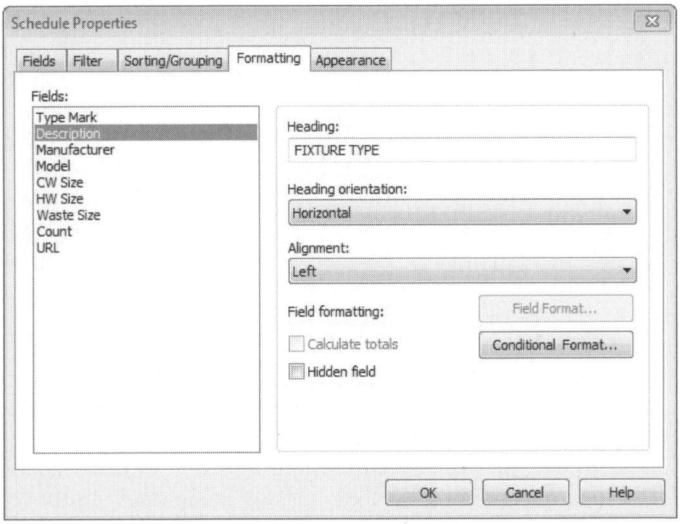

FIGURE 5.12 Column heading settings

On the Formatting tab there are also settings for the orientation of the column headings and the alignment of the data within a column. Headings can be either vertical or horizontal, and the data can be aligned to the left, right, or center of a column.

Two additional types of formatting are available depending on the type of parameter selected. The Field Format button becomes active when you select a parameter that is a measurement. In the Format dialog box that appears when you click the Field Format button, you can change the units and rounding accuracy of the data. You can also choose whether to show the unit symbol in the cell of the schedule. Whatever settings have been established for the project will be used by default. You can deselect the box at the top of the Format dialog box to overwrite the settings within the schedule. Figure 5.13 shows format settings for the Area parameter used in a Space schedule. The unit symbol has been set to None because the schedule column heading indicates the units.

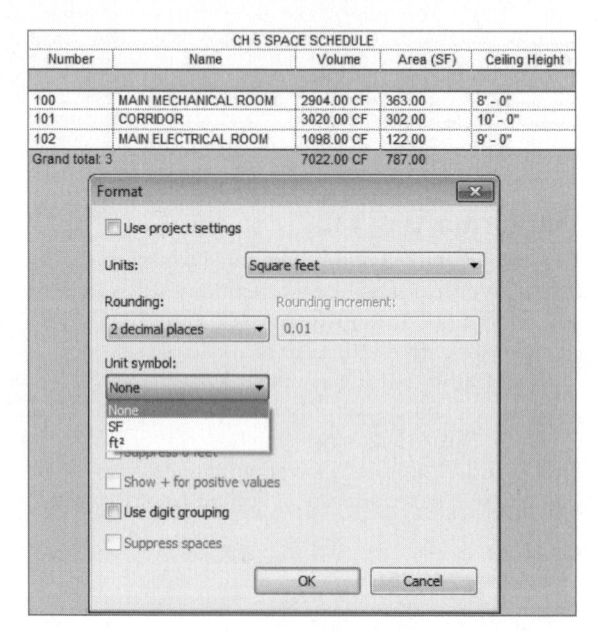

CH 5 SPACE SCHEDULE				
Number	Name	Volume	Area (SF)	Ceiling Height
100	MAIN MECHANICAL ROOM	2904.00 CF	363.00	8' - 0"
101	CORRIDOR	3020.00 CF	302.00	10' - 0"
102	MAIN ELECTRICAL ROOM	1098.00 CF	122.00	9' - 0"
Grand total: 3		7022.00 CF	787.00	

FIGURE 5.13
Field Format settings

You can change the background color of a cell based on conditions that you apply using the Conditional Format button. Clicking this button opens a dialog box where you can set a test condition for the value of a parameter and the background color of the schedule cell when the condition is met or not met. Figure 5.14 shows the Conditional Formatting dialog box for a calculated value in a schedule. Background Color is set to turn red when the result of the calculation does not meet the test condition. The different types of condition test options are shown in the Test drop-down list of the dialog box.

On the Sorting/Grouping tab is an option to display the totals for groups of objects. If you want to show the totals for an individual column, you can select the Calculate Totals box when you select a parameter on the Formatting tab. These totals will not appear if the Grand Totals option is not selected on the Sorting/Grouping tab.

Sometimes you need to include a parameter in a schedule for sorting purposes or for calculations, but you do not want to display the column in your schedule. If this is the case, you can use the Hidden Field check box on the Formatting tab to hide a selected column. This can be very useful because it allows you to hide information without having to remove the information from your schedule. It also allows you to create a schedule with more information than would normally be shown on a drawing but is useful for calculations or design decisions. When the time comes to put the schedule on a drawing, you can hide the unwanted columns.

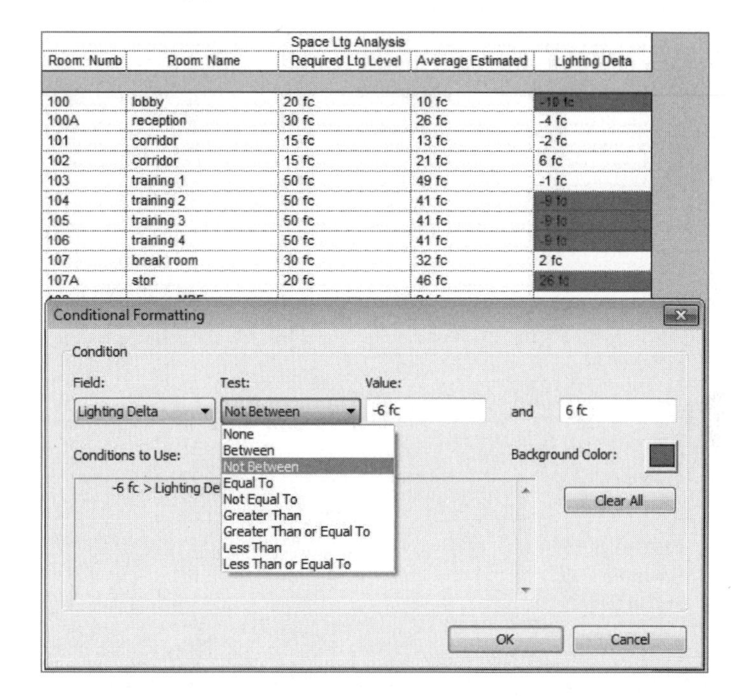

FIGURE 5.14
Sample Conditional Formatting dialog box

The Appearance Tab

The Appearance tab of the Schedule Properties dialog box contains the settings that define how the schedule looks when placed on a sheet.

Grid lines and an outline can be chosen from any of the line styles defined in your project. If you do not choose an option for the outline, the line style chosen for the grid lines will be used. The option to include grid lines within the headers, footers, or spacers is also available.

The check box to provide a blank row between the headings and schedule data is selected by default. If this is not how you normally display your schedules, you will have to deselect this box whenever you create a schedule. The Appearance tab also has settings for the text within a schedule, as shown in Figure 5.15. The font and text height that you choose for header text will be applied to all headings, subheadings, and the title of the schedule.

Editing a Schedule

Once you've created a schedule, you can edit it using the same Schedule Properties dialog box. When you double-click a schedule in the Project Browser, the schedule is displayed in the drawing area. You can right-click anywhere in the drawing area and select View Properties from the menu to access the Schedule Properties dialog box. The tabs of the Schedule Properties dialog box are listed in the Instance Properties list of a schedule view, as shown in Figure 5.16. Each tab has its own Edit button, but clicking any of these buttons will take you to the Schedule Properties dialog box. Whichever button you click from the list will take you to that tab within the dialog box.

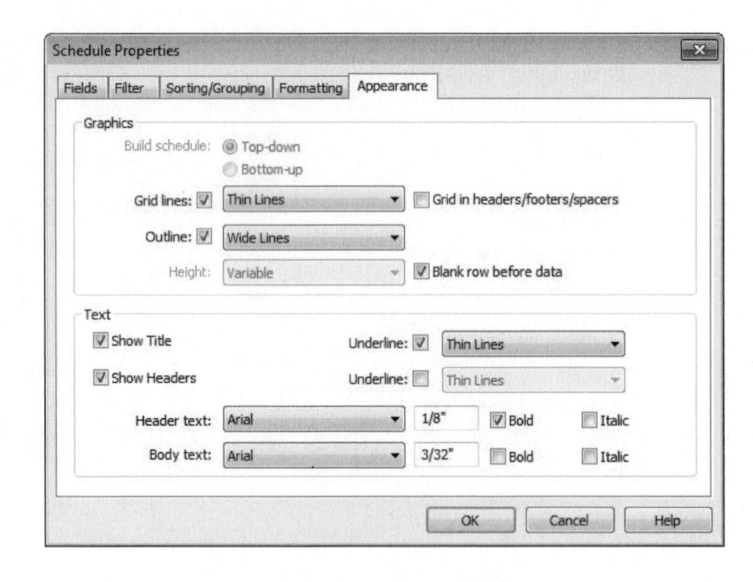

FIGURE 5.15
Appearance tab of the Schedule Properties dialog box

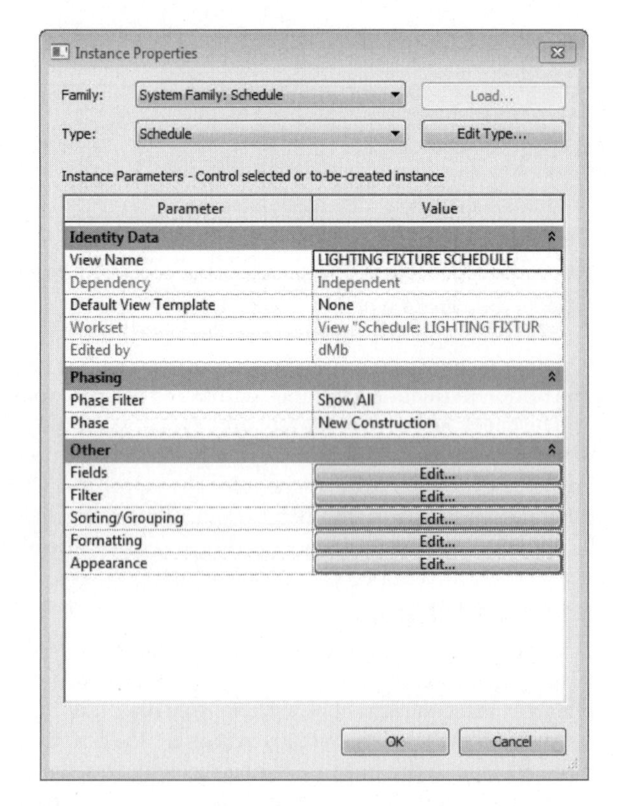

FIGURE 5.16
Schedule view properties

When you access the Schedule Properties dialog box, you can modify the schedule with any of the tools or settings that you would use to create a new schedule. One thing you cannot do is change what is being scheduled. For example, you cannot change a Lighting Fixture schedule to a Lighting Devices schedule.

Some of the formatting options that are defined can be modified in the schedule view, without having to access the Schedule Properties dialog box. In the schedule view, you can select anywhere within a column and use the Hide button on the ribbon to hide that column. There is also a Delete button on the ribbon to delete a row in the schedule.

DELETE WITH CAUTION

It is important to understand what will happen if you use the Delete button on the ribbon when you select a row in a schedule. If your schedule is scheduling objects from your model and you delete a row, all the objects from that row will be deleted from the project.

When you click in the cell of a column heading, you can change the text of the heading, eliminating the need to access the Formatting tab. You can change the title of the schedule with this method also. Columns can be grouped together under a common heading by selecting the columns and clicking the Group button on the ribbon. Columns must be adjacent to each other to be grouped. The key to selecting columns for grouping is to hold your mouse button down when you click the first column and then drag your cursor to highlight the desired columns to be included in the group. When you click the Group button, a new blank header will appear above the columns awaiting input for a heading. Columns groups can be removed by highlighting all the columns within a group and clicking the Ungroup button on the ribbon.

One of the best features of using schedules in Revit is that you can edit parameter values of objects right in the schedule. If you are changing the value of a type parameter, you will receive a warning letting you know that your change will be applied to all objects of that type when you finish editing the cell. Some cells will turn into a drop-down list when you click them. This lets you select from previously input values for that parameter as long as those values are being used in the project. If you click a cell and you cannot edit it, that means that either the parameter is a calculated value, it is a parameter of an object within a linked file, or the parameter does not exist in the object.

Because a schedule view is actually a view of the model, you can select a cell in a schedule and click the Highlight In Model button on the ribbon to see where the selected object exists in the model. This will take you to a view of the model where the object can be seen. You can continue to click the Show button for different views. If you have grouped objects in the schedule and you are not itemizing every instance of the objects in your schedule, then all the selected object types will be shown. This happens because when you select an object in a schedule, you are actually selecting that object, not just a list of the data within that object. The object (or objects) is not just highlighted but actually selected, so you can access its properties or edit it once you close the Select Element(s) In View dialog box.

Some schedules need multiple lines of input within one row. This can be accomplished by using your Ctrl+Enter keys when inputting information into a schedule cell or parameter value. Each time you press Ctrl+Enter, you will create a hard return in the value, therefore starting a new line of text. When you are in the schedule view, you will see only the first line of input in the cell. You will not see the other lines of input, so if you need to edit them, you will have to retype all the lines within that cell or parameter value. The multiple lines will display when the schedule is placed on a sheet.

Scheduling Component and System Family Data

Any object type that is placed into a Revit model can be scheduled. This seems like a very generic concept, but if you consider the possibilities that exist, it is easy to see why using a parametric model can be a great benefit to your design processes. Having readily accessible information about the components that make up your design will make coordination and decision making easier and more efficient.

It is important to know what kind of information about model objects can be scheduled because the types of schedules that you can create for model objects depends on the data within those objects. Some data is inherent in the objects based on how they used in the model, such as elevation or location. Most of the data used in component object schedules comes from parameters that are added to the objects either as project parameters or directly in the component family file as shared parameters. Family parameters cannot be scheduled.

Using the organizational and calculation tools within a Revit schedule can help you get the most from the data within the model components. In this section, we will cover some possibilities for scheduling building components.

Mechanical Equipment Schedules

The Mechanical Equipment model category covers a wide range of components. Chillers, water heaters, pumps, rooftop units, fans, and more all fall under the Mechanical Equipment category. As you have seen, using filters makes it possible to create specific schedules for different types of Mechanical Equipment components. The key to success with these schedules is developing a method that makes the filtering easy to use and manage.

When considering the information that is needed in your schedules, you can look at the parameters that already exist in every component and determine whether they can be used in lieu of creating a custom parameter. The Description parameter is one example of a parameter that is in every object. With the ability to change the heading in a schedule, this parameter can be used for any type of descriptive information that is scheduled about your components. You do not need to create a parameter for each type of information to be scheduled. For example, if your Mechanical Equipment schedule for pumps has a column for the mounting type, you could use the Description parameter to convey this information.

Every item that is placed into a model is given a Mark value. The Mark parameter is another useful parameter for scheduling mechanical equipment.

A WARNING ABOUT MARK VALUES

When you manually change the Mark value of an object, the next object that you place that is in the same category will have a Mark value sequential to the value you input. For example, if you place a boiler and give it a Mark value of B-1, the next Mechanical Equipment object you place will be given a Mark value of B-2 even if it is not another boiler. Using Mark values to identify equipment requires you to manage the Mark value of an object when it is placed into your model.

The best reason for using the Mark parameter for Mechanical Equipment schedules is that it makes it easy to filter the schedule for specific items. Figure 5.17 shows a Mechanical Equipment schedule for pumps and the filter settings based on the Mark parameter. This setup allows for having unique Mark values for multiple instances of the same type of component.

FIGURE 5.17
Pump schedule filtered by Mark parameter

Mechanical Equipment Schedule					
Mark	Description	Manufacturer	Model	Level	Cost
P-1	BASE MOUNTED PUMP			Level 1	
P-2	BASE MOUNTED PUMP			Level 1	
P-3	BASE MOUNTED PUMP			Level 1	
P-4	BASE MOUNTED PUMP			Level 1	
P-5	BASE MOUNTED PUMP			Level 1	
P-6	BASE MOUNTED PUMP			Level 1	

Schedule Properties
Fields | Filter | Sorting/Grouping | Formatting | Appearance
Filter by: Mark | begins with
P
And: (none)
And: (none)
And: (none)

Another way to easily filter your Mechanical Equipment schedules is to create a project parameter called Schedule Type that is applied to the Mechanical Equipment category. You can then assign values for this parameter in order to create schedule filter rules. This may be the preferred method because the Schedule Type parameter could be applied to all appropriate model categories and would give you a uniform method for filtering schedules.

If you have a schedule with a column for remarks or notes about an object, the Type Comments parameter can be a useful alternative to creating a custom parameter. The value of this parameter can refer to text notes associated with a schedule.

Lighting Fixture Schedules

The Type Mark parameter is a useful way to assign a unique identifier to each fixture type within your project and eliminates the need for a custom parameter. If you use a standard set of fixtures for each design, you might consider naming each family type within your fixture families and using the Type name to identify your fixtures in a schedule. The family type name shows up as Type in the Available Fields list, as shown in Figure 5.18.

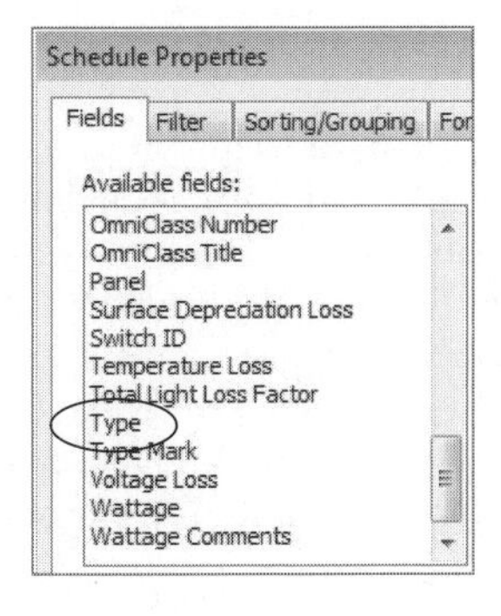

FIGURE 5.18
Family Type option in Available Fields list

The URL parameter can be used for any component and is particularly useful for lighting fixtures. You can input a link to the cut sheet of a lighting fixture into its URL parameter for quick access to the additional information that a cut sheet provides. Because a manufacturer's website can change without you knowing it, it is not recommended that you provide a link directly to a web page but rather to a downloaded file. Figure 5.19 shows an example of a lighting fixture schedule with a URL column. The cut sheet file opens when you click the small box at the right of the URL cell.

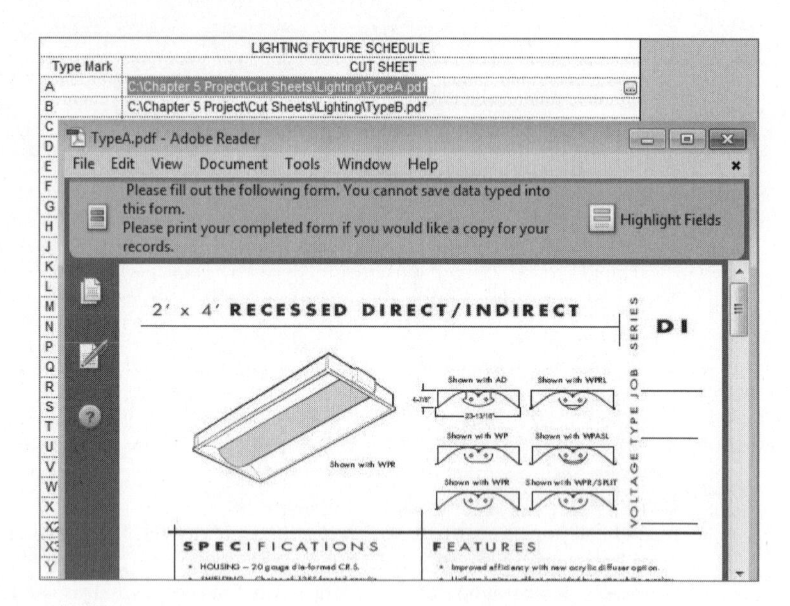

FIGURE 5.19
URL parameter used in a lighting fixture schedule

The paths you provide in URL parameters are not relative, so if you share or submit your Revit project file, the links will be inactive unless you also share the cut sheets and the URL parameters are repathed.

System Family Schedules

Schedules are useful not only for component objects but for system families as well. Duct, Pipe, Cable Tray, and Conduit schedules can be created for use on construction documents or just for keeping track of quantities and materials.

Duct Schedules

A lot of information can be taken from a ductwork model. How you organize a Duct schedule depends on the type of information you are looking for and what you intend to do with it. Determining the amount of sheet metal for ductwork is one way to utilize the power of scheduling in Revit MEP 2011. Unfortunately, the Material Takeoff option of the Schedules tool on the View tab is restricted to component families. However, with some calculated fields and the right parameters, you can create a schedule that works as a material takeoff.

Figure 5.20 shows a sample Duct schedule with calculated values to determine the total area of material for both round and rectangular duct. A duct material project parameter was created, and values were manually input for different duct types. The total area for each duct material is calculated at the bottom. This is just a small example of how you can get useful information from your model in schedule format. Keep in mind that because Revit is a parametric modeling tool, as ducts are removed, added, or changed, the schedule will reflect those changes automatically.

FIGURE 5.20
Sample duct schedule

Duct Schedule								
Duct Material	Width	Height	Width Area	Height Area	Rectangular Area	Diameter	Circumference	Round Area
Galvanized Steel					0.00 SF	12"	3' - 1 11/16"	9.42 SF
Galvanized Steel	12"	12"	3.42 SF	3.42 SF	13.67 SF			0.00 SF
Galvanized Steel	12"	12"	3.92 SF	3.92 SF	15.67 SF			0.00 SF
Galvanized Steel	8"	8"	4.00 SF	4.00 SF	16.00 SF			0.00 SF
Galvanized Steel	20"	20"	5.69 SF	5.69 SF	45.56 SF			0.00 SF
Galvanized Steel	10"	12"	10.83 SF	13.00 SF	47.67 SF			0.00 SF
Galvanized Steel: 7					138.56 SF			9.42 SF
Stainless Steel					0.00 SF	10"	2' - 7 13/32"	104.72 SF
Stainless Steel					0.00 SF	12"	3' - 1 11/16"	21.99 SF
Stainless Steel					0.00 SF	12"	3' - 1 11/16"	31.42 SF
Stainless Steel	12"	12"	2.13 SF	2.13 SF	17.00 SF			0.00 SF
Stainless Steel: 6					17.00 SF			158.13 SF
Grand total: 13					155.56 SF			167.55 SF

Another type of Duct schedule can help you keep track of what type of duct is used for different air systems. A schedule like this will enable you to see whether any errors have occurred in the model based on design criteria. If all return air ductwork is to be rectangular, for example, a schedule would quickly reveal any duct that does not meet the requirement.

PIPE SCHEDULES

Creating a custom material takeoff schedule for pipe is a little easier for pipe than duct because pipe types can be assigned a material without creating a custom parameter. Figure 5.21 shows a sample Pipe schedule that is organized to show the total length of each pipe size per material.

FIGURE 5.21
Sample pipe schedule for material takeoff

Pipe Schedule		
Material	Diameter	Length
Carbon Steel	3/4"	599' - 6 9/16"
Carbon Steel	1"	19' - 2 29/32"
Carbon Steel	1 1/2"	20' - 3 13/32"
Carbon Steel	2"	749' - 6"
Carbon Steel	3"	122' - 11 9/32"
Carbon Steel	4"	59' - 1 1/4"
Carbon Steel: 307		1570' - 7 13/32"
Copper	3/8"	0' - 7 7/8"
Copper	1/2"	824' - 11 5/8"
Copper	3/4"	16' - 4 7/8"
Copper	1"	50' - 4 13/16"
Copper	1 1/4"	164' - 5 5/32"
Copper	2"	152' - 3 1/32"
Copper: 221		1209' - 1 13/32"
Plastic	3/4"	4' - 0 11/32"
Plastic	1 1/2"	62' - 2 1/32"
Plastic	2"	67' - 7 1/8"
Plastic	2 1/2"	1' - 8 7/8"
Plastic	3"	133' - 11 29/32"
Plastic	4"	111' - 11 3/8"
Plastic: 147		381' - 5 11/16"
Grand total: 675		3161' - 2 1/2"

Although cost cannot be used in a calculated value, with this information readily available, you could easily estimate the cost of pipe for your project.

EXPORTING SCHEDULES

You can export Revit schedules by using the Export tool on the Application menu. The exported `.txt` file can be input into a spreadsheet for further computation or analysis. Unfortunately, spreadsheets cannot be imported into Revit schedules.

You can create pipe schedules to quickly input data into the properties of pipes without having to locate each pipe in the model. One example where this may be useful is with pipe insulation. You can create a schedule like the one shown in Figure 5.22 where pipes are grouped by their system and size.

FIGURE 5.22
Sample pipe schedule for parameter input

Pipe Insulation Schedule		
System Type	Diameter	Insulation Thickness
Domestic Cold Water	3/8"	1"
Domestic Cold Water	1/2"	1"
Domestic Cold Water	3/4"	1"
Domestic Cold Water	1"	1"
Domestic Cold Water	2"	1"
Domestic Hot Water	1/2"	2"
Domestic Hot Water	1 1/4"	2"
Fire Protection Wet	3/4"	0"
Fire Protection Wet	2"	0"
Fire Protection Wet	3"	0"
Fire Protection Wet	4"	0"
Sanitary	1 1/2"	0"
Sanitary	2"	0"
Sanitary	2 1/2"	0"
Sanitary	3"	0"
Sanitary	4"	0"

Even though Insulation Thickness is an instance parameter, you can apply a value to several objects at once by grouping the pipes and deselecting the Itemize Every Instance box on the Sorting/Grouping tab of the Schedule Properties dialog box. When you input a value for the Insulation Thickness parameter, it is applied to all pipes within that group.

SPACE SCHEDULES

Spaces are typically scheduled to analyze their data to determine the performance of systems, but creating a Space schedule can also help with model maintenance or quality control. A very simple Space schedule can be a very useful tool for removing unwanted or misplaced Spaces, reducing the risk of performing analysis of an incorrect model.

Spaces can be placed into a model manually or by using the Place Spaces Automatically tool on the Modify | Place Space contextual tab. The one drawback to the automated process is that Revit will place a Space object in any enclosed area bounded by objects defined as Room Bounding that is 0.25 square feet or larger. This often results in Space objects being placed in pipe and duct chases or column wraps and other small enclosures. In a large model, it could be time-consuming to search for all the unwanted Space objects, making the automation of space placement seem unreasonable. When these spaces are found, deleting them removes them from the model only. They still exist in the project and could be inadvertently used again or have unnecessary analysis performed on them.

Consider creating a simple Space schedule that will quickly identify any unwanted spaces and allow you to delete them all completely from the project with a few clicks. When you link in an architectural model, one thing you need to do is set the Room Bounding parameter so that it will define the boundaries of your spaces. Placing a space within the same boundaries as a room object will associate that space with the architectural room object. So unless your architect has placed room objects in areas such as chases or column wraps, you can easily see which of your spaces match up with the rooms.

Figure 5.23 shows the settings for a Space schedule that can be used to eliminate unwanted spaces. The schedule is sorted and grouped by room number so that all spaces without an associated room will be listed together at the top of the schedule.

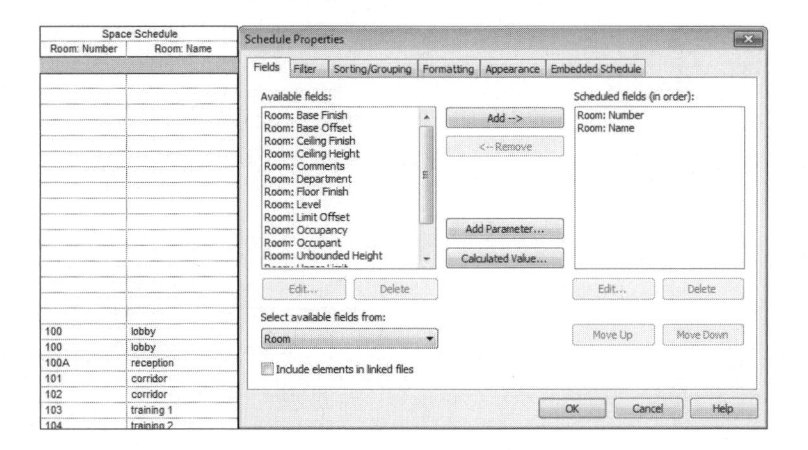

FIGURE 5.23
Space schedule

The unwanted spaces can be highlighted in the schedule and then removed from the project using the Delete button on the ribbon. You cannot use the Shift or Ctrl keys, but you can click and drag your cursor to select multiple rows within a Revit schedule.

The Schedule Properties dialog box for a Space schedule contains an additional tab for creating and managing an embedded schedule. This allows you to create an additional schedule that is set up "inside" your Space schedule. On the Embedded Schedule tab, you can select the box to include an embedded schedule within your Space schedule and choose the model category you want to schedule. When you click the Embedded Schedule Properties button in the lower-left corner, a new Schedule Properties dialog box appears. You can then set up the embedded schedule in the same way you would set up a regular one. The Appearance tab is not available in an embedded schedule because the settings are controlled by the host schedule.

Figure 5.24 shows the Fields tab of a lighting fixture schedule that is embedded into a Space schedule. The fixtures in the embedded schedule are sorted and grouped by the Type Mark parameter, and grand totals are shown.

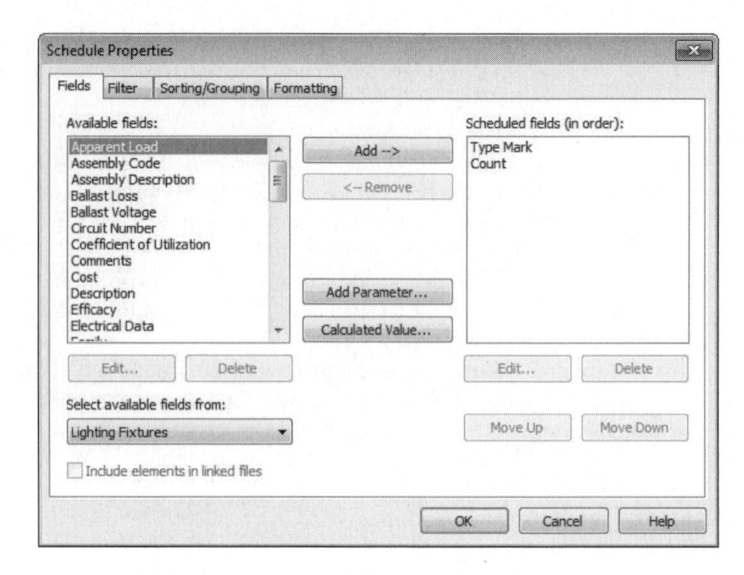

FIGURE 5.24
Embedded light fixture schedule settings

The information within the embedded schedule will be included in the host Space schedule, as shown in Figure 5.25. By creating this type of schedule, you can see how many elements are associated with each Space.

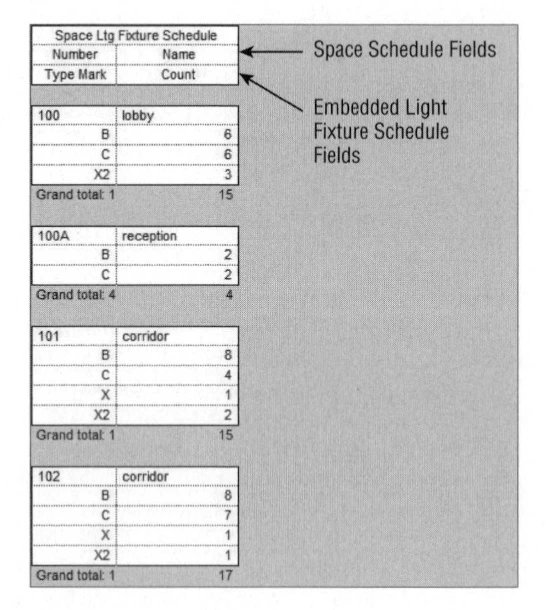

Space Ltg Fixture Schedule	
Number	Name
Type Mark	Count
100	lobby
B	6
C	6
X2	3
Grand total: 1	15
100A	reception
B	2
C	2
Grand total: 4	4
101	corridor
B	8
C	4
X	1
X2	2
Grand total: 1	15
102	corridor
B	8
C	7
X	1
X2	1
Grand total: 1	17

FIGURE 5.25
Space schedule with an embedded light fixture schedule

Creating a Schedule of Model Components

Understanding the basic features of a component schedule will help you begin to develop your own methods for scheduling your Revit projects. Create a simple component schedule by doing the following:

1. Open the `CH5_Schedules.rvt` file found at `www.wiley.com/go/masteringrevitmep2011`.
2. Click the Schedules button on the View tab, and select Schedule/Quantities.
3. Select the Air Terminals category. Ensure that the Schedule Building Components option is selected and Phase is set to New Construction, and click OK.
4. On the Fields tab, select Type Mark from the Available Fields list, and click the Add button. Repeat the process to add the System Type, Neck Size, CFM Range, Description, and Count parameters to the schedule.
5. On the Sorting/Grouping tab, select Type Mark in the Sort By drop-down. Deselect the Itemize Every Instance box in the lower-left corner.
6. On the Formatting tab, highlight all the parameters in the Fields list, and set Alignment to Center.
7. On the Appearance tab, select the Outline box, and change the line style to Wide Lines. Click OK to exit the Schedule Properties dialog box.
8. Click the Schedules button on the View tab, and select Schedule/Quantities.
9. Select the Spaces category. Ensure that the Schedule Building Components option is selected and Phase is set to New Construction, and click OK.
10. On the Fields tab, select Number from the Available Fields list, and click the Add button. Repeat the process to add the Number parameter to the schedule.
11. On the Filter tab, select Number in the Filter By drop-down and Begins With as the rule. Type 1 in the field below the Filter By drop-down.
12. On the Sorting/Grouping tab, select Number from the Sort By drop-down.
13. On the Formatting tab, select the Number and Name parameters, and set Alignment to Center.
14. On the Embedded Schedule tab, select the Embedded Schedule box, select the Air Terminals category, and click the Embedded Schedule Properties button.
15. Add the Type Mark, System Type, and Count parameters from the Available Fields list.
16. On the Sorting/Grouping tab, select Type Mark in the Sort By drop-down. Select the Grand Totals box in the lower-left corner.
17. On the Formatting tab, select the Count parameter, and select the Calculate Totals box. Click OK to exit the Embedded Schedule Properties dialog box. Click OK to exit the Schedule Properties dialog box.

Using Schedules for Design and Analysis

Scheduling building components to provide information on construction documents or to keep track of model components and materials are not the only uses for schedules in Revit. Using schedules to analyze the performance of MEP systems in relation to the building model can help you make design decisions. The ability to see and manipulate the information directly in Revit can improve the efficiency of your design processes.

When it comes to analysis, the focus is mainly on the Space objects in your model. Space objects hold a lot of energy analysis–related information either by direct input, from input from third-party analysis, or as a result of the characteristics of components associated with the spaces. Understanding the type of information you can retrieve from Spaces during the design process is the key to developing schedules that are most useful to your workflow.

Analysis can be as simple as checking to see whether the components you are using meet the engineering requirements or standards you are designing around. This often requires that you manually set the requirement information property of your Spaces. Selecting each Space and accessing its properties to input this information is an inefficient practice. A special type of Schedule can be created to improve the process of adding information to your Spaces.

A *schedule key* allows you to specify values for a parameter based on the value of a key parameter. One example is in lighting design where a certain lighting level is required for specific types of rooms. A schedule key can be created to associate a specific lighting level value with each Space key parameter value. The parameter associated with the key can be included in your Space schedule so that each space can be assigned a key and therefore a required lighting level. The required lighting level could be input manually into the parameter, but using a key ensures accuracy and consistency.

Click the Schedules button on the View tab, and select Schedule/Quantities to create a schedule key. When you select the category you want to schedule, the option to create a schedule key becomes available on the right side of the dialog box. You can then choose a name for the key parameter in the Key Name field. A style-based name will automatically be placed in the Key Name field, but you can change it to whatever you want. After you create the schedule, a parameter with this name will be added to all the objects in the category you are creating a schedule key for. Figure 5.26 shows an example of the settings for a Space schedule key. Notice the schedule has been named to identify it as a schedule key and not a component schedule.

FIGURE 5.26
Space schedule key setup

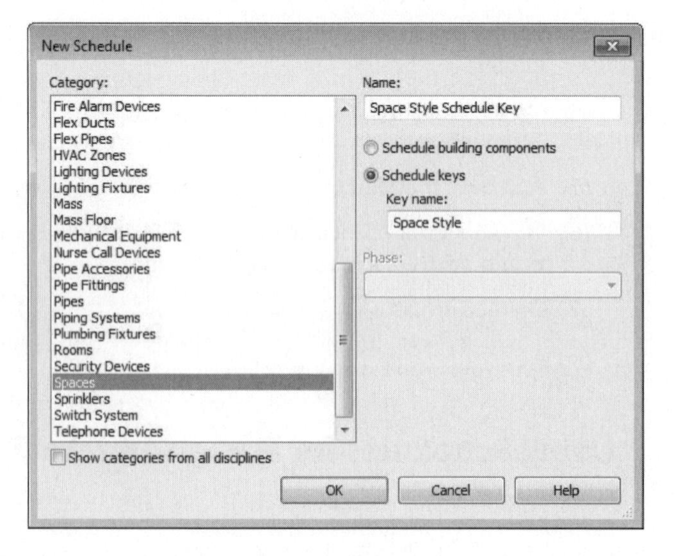

The Key Name parameter will automatically be included in the Scheduled Fields area on the Fields tab of the Schedule Properties dialog box. Here you can select parameters to assign values to based on the key value. In the case of a Space schedule key for lighting levels, the Required Ltg Level parameter is chosen, as shown in Figure 5.27.

FIGURE 5.27
Space schedule key Fields tab

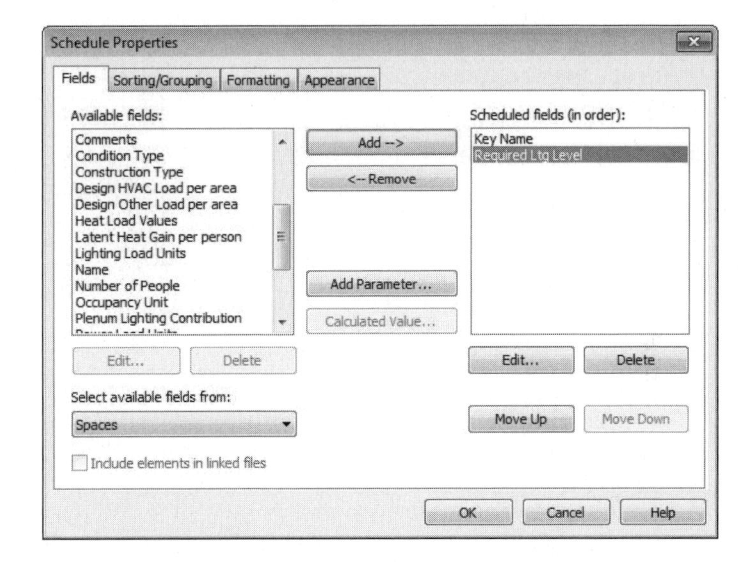

The schedule key is sorted by the Key Name parameter by default. Unless you will be placing the schedule on a drawing, there is no need for any other settings within the Schedule Properties dialog box. The schedule key will start out blank. Click the New button on the Rows panel of the ribbon to add a row to the schedule. You can then edit the name of the key and assign a value to the associated parameter. Repeat the process to create additional rows and build your schedule key. Figure 5.28 shows an example of a Space schedule key for lighting levels.

FIGURE 5.28
Sample schedule key

Space Style Schedule Key	
Key Name	Required Ltg Level
Auditorium	30 fc
Conference Room	30 fc
Control Room	50 fc
Corridor	15 fc
Janitor	15 fc
Kitchen	30 fc
Lobby	20 fc
Mech/Elec	20 fc
Office	50 fc
Stair	10 fc
Storage	15 fc
Toilet	10 fc
Training/Classroom	50 fc
Waiting/Lounge	30 fc

With this schedule key created, each Space now has a Space Style parameter that can be given a value from the schedule key to assign a value to the Required Ltg Level parameter. A schedule can be used to assign these values without having to access the properties of each Space. A drop-down list appears with all the key values when editing a key parameter value in a schedule, as shown in Figure 5.29.

FIGURE 5.29
Editing a schedule key parameter

Space Style Schedule		
Room: Numb	Room: Name	Space Style
100	lobby	(none)
100A	reception	(none)
101	corridor	Corridor
102	corridor	(none)
103	training 1	Auditorium
104	training 2	Conference Room
105	training 3	Control Room
106	training 4	Corridor
107	break room	Janitor
		Kitchen
107A	stor	(none)
108	comm - MDF	(none)
109	main electrical	(none)
110	elev mech	(none)

A key parameter can be included in a schedule that is set up for analysis to increase the functionality of the schedule. Additional parameters were added to the schedule shown in Figure 5.29 to determine whether the lighting levels in each Space meet the design requirements. A calculated value was added to show the difference between the required level and what is actually occurring in the Space. This type of schedule allows you to quickly see the performance of the design components used and you can adjust parameter values of the Spaces to investigate options. Notice in Figure 5.30 that Spaces that have been assigned a Space Style value have the corresponding Required Ltg Level parameter value from the key.

FIGURE 5.30
Space lighting level analysis schedule

Space Ltg Analysis Schedule							
Room: Number	Room: Name	Space Style	Required Ltg Level	Actual Ltg Level	Clng Reflectance	Calculation Plane	Lighting Delta
100	lobby	Lobby	20 fc	10 fc	0.8	2' - 6"	-10 fc
100A	reception	Waiting/Lounge	30 fc	26 fc	0.8	2' - 6"	-4 fc
101	corridor	Corridor	15 fc	11 fc	0.8	0' - 0"	-4 fc
102	corridor	Corridor	15 fc	18 fc	0.8	0' - 0"	3 fc
103	training 1	Training/Classroom	50 fc	50 fc	0.8	2' - 6"	-0 fc
104	training 2	Training/Classroom	50 fc	41 fc	0.8	2' - 6"	-9 fc
105	training 3	Training/Classroom	50 fc	41 fc	0.8	2' - 6"	-9 fc
106	training 4	Training/Classroom	50 fc	41 fc	0.8	2' - 6"	-9 fc
107	break room	(none)		33 fc	0.8	2' - 6"	
107A	stor	Storage	15 fc	46 fc	0.8	2' - 6"	31 fc
108	comm - MDF	Mech/Elec	20 fc	21 fc	0.8	2' - 6"	1 fc
109	main electrical	Mech/Elec	20 fc	19 fc	0.5	2' - 6"	-1 fc
110	elev mech	Mech/Elec	20 fc	31 fc	0.5	2' - 6"	11 fc

This same type of schedule can be created for airflow in spaces. The design airflow value can be checked against the actual airflow and the value calculated by a third-party analysis or from the Heating and Cooling Analysis tool directly in Revit MEP 2011.

The ability to see the performance-based data is essential to checking the quality of your design and for making appropriate changes to the design when necessary. Figure 5.31 shows an example of a Duct schedule that displays the properties related to the performance of the system rather than information about the physical properties of the ductwork.

FIGURE 5.31
Duct airflow schedule

Duct Airflow					
System Name	Friction	Flow	Pressure Drop	Velocity	Velocity Pressure
Aud Corridor Supply Air	0.06 in-wg/100ft	690 CFM	0.00 in-wg	690 FPM	0.03 in-wg
Aud Corridor Supply Air	0.06 in-wg/100ft	230 CFM	0.00 in-wg	518 FPM	0.02 in-wg
Aud Corridor Supply Air	0.07 in-wg/100ft	460 CFM	0.01 in-wg	662 FPM	0.03 in-wg
Aud Corridor Supply Air	0.38 in-wg/100ft	230 CFM	0.01 in-wg	1171 FPM	0.09 in-wg
Aud Corridor Supply Air	0.38 in-wg/100ft	230 CFM	0.00 in-wg	1171 FPM	0.09 in-wg
Aud Corridor Supply Air	0.09 in-wg/100ft	230 CFM	0.00 in-wg	659 FPM	0.03 in-wg
Aud Corridor Supply Air	0.09 in-wg/100ft	230 CFM	0.00 in-wg	659 FPM	0.03 in-wg
Aud Corridor Supply Air	0.09 in-wg/100ft	230 CFM	0.00 in-wg	659 FPM	0.03 in-wg
Aud Corridor Supply Air	0.09 in-wg/100ft	230 CFM	0.00 in-wg	659 FPM	0.03 in-wg
Aud Corridor Supply Air	0.09 in-wg/100ft	230 CFM	0.00 in-wg	659 FPM	0.03 in-wg
Aud Corridor Supply Air	0.09 in-wg/100ft	230 CFM	0.00 in-wg	659 FPM	0.03 in-wg
Aud Corridor Supply Air: 11		3220 CFM	0.03 in-wg		0.41 in-wg
Auditorium Supply Air	0.06 in-wg/100ft	660 CFM	0.00 in-wg	660 FPM	0.03 in-wg
Auditorium Supply Air	0.72 in-wg/100ft	3180 CFM	0.04 in-wg	2726 FPM	0.46 in-wg
Auditorium Supply Air	0.11 in-wg/100ft	2790 CFM	0.00 in-wg	1240 FPM	0.10 in-wg
Auditorium Supply Air	0.11 in-wg/100ft	2790 CFM	0.02 in-wg	1240 FPM	0.10 in-wg
Auditorium Supply Air	0.06 in-wg/100ft	330 CFM	0.00 in-wg	605 FPM	0.02 in-wg

Real World Scenario

USING THIRD-PARTY DATA

Shelley uses a third-party application to analyze her HVAC designs. Because the software she uses is compatible with Revit MEP 2011, she can import the data generated from the third-party analysis into her Revit projects.

When this data is imported, she wants to be able to see the values as they pertain to each space in combination with the types of air terminals used in her design. Shelley has created a Space schedule with an embedded Air Terminal schedule that allows her to view the results from the third-party analysis in each space alongside the airflow properties of her air terminals. She has added this schedule to her project template so that it is available for each project she does and now has a consistent workflow for analyzing the results.

Panel Schedules

Panel schedules are unique schedules in Revit MEP 2011. They are not created by using the scheduling tools but instead by a predefined format defined in a panel schedule template. The ability to create custom panel schedules is a long-awaited feature that is new to Revit MEP 2011. You now have tools to create a panel schedule template with settings for both the appearance of the schedule and the data that is reported.

Unlike in previous versions that have two separate tools for creating a panel schedule and viewing and managing circuits, Revit MEP 2011 combines all the functionality into the panel schedules. This means you will need to create a panel schedule to view and manage circuits even if you do not intend to use the schedule on your construction documents.

Once you have established panel schedule templates, they will be available for use on any project. If you want to create a custom panel schedule, the first step is to create its template. You can find the Panel Schedule Templates tool on the Manage tab.

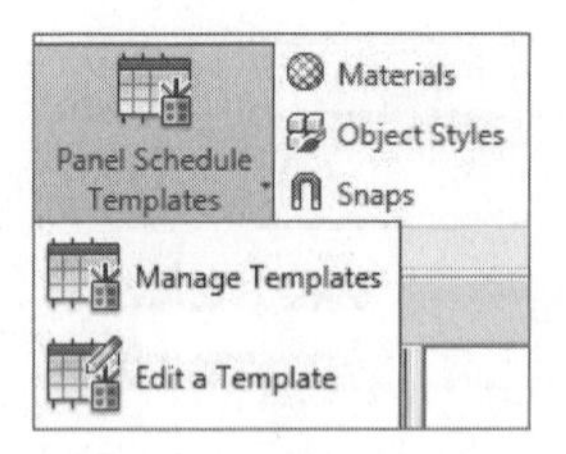

The Manage Templates option gives you access to your panel schedule templates for editing or duplication. Clicking this option activates the Manage Panel Schedule Templates dialog box. The first tab on the dialog box lists any templates in your file. The three types of panel templates that you can create are Branch Panel, Data Panel, and Switchboard. Once you have created a panel schedule template, the Make Default button enables you to establish a selected template as the default. Buttons at the bottom of the dialog box can be used to edit, duplicate, rename, or delete a template.

The second tab of the dialog box allows you to apply a panel schedule template to any panel in your project. When you make changes to your panel schedule templates, you can update the schedules for panels that are using the older version by selecting the panels from the list and clicking the Update Schedules button.

To build a custom template, start by clicking the default template in the list on the Manage Templates tab of the dialog box. Use the Duplicate button at the bottom of the dialog box to copy the default template. Select the newly created template in the list, and use the Edit button to begin customizing your panel schedule.

> **DEFAULT TEMPLATES**
>
> To see a good example of how a panel schedule template can be formatted, check out the default templates in the `Electrical-defaults.rte` project template file. These defaults can be transferred to any file using the Transfer Project Standards tool. If you want to start completely from scratch, the default panel schedule templates from other project templates are totally blank.

All the tools for editing and formatting a panel schedule template are located on the Modify | Panel Schedule Template contextual tab that appears on the ribbon when you click to edit a template. Every panel schedule template has a header area, a circuit table, a load summary area, and a footer. That does not mean that you have to use all three of these areas. You can access the settings for the overall appearance and behavior of a panel schedule template by clicking the Set Template Options button on the contextual tab.

There are three main areas for the overall settings. Figure 5.32 shows the General Settings options. The Total Width setting determines the size of the schedule when placed on a sheet. The height is determined by the row height settings within the schedule. There is no longer a need to adjust column widths after placing a schedule onto a sheet because the widths are defined directly in the schedule template view. You can deselect the boxes to remove the header, footer, or load summary, but the circuit table cannot be removed.

The Circuit Table options allow you to define the format for the columns that display circuit loads. Single phase panels can be created, and options for the third phase loads column are available. These columns will appear in the center of your schedule no matter what you do to format the other columns. The values in these columns cannot be edited, and you cannot replace them with another parameter.

The Loads Summary options let you set which load classifications will be listed when the loads summary is included in your schedule. A list of all load classifications that exist in your project is shown on the left, and the ones that appear in the schedule are listed on the right. If you choose the option for displaying only those loads that are connected to the panel, then all the classifications will move to the Scheduled Loads list. When you create a panel schedule, only the classifications that are connected will display; however, there will be blank rows in the loads summary for the unused classifications, as shown in Figure 5.33.

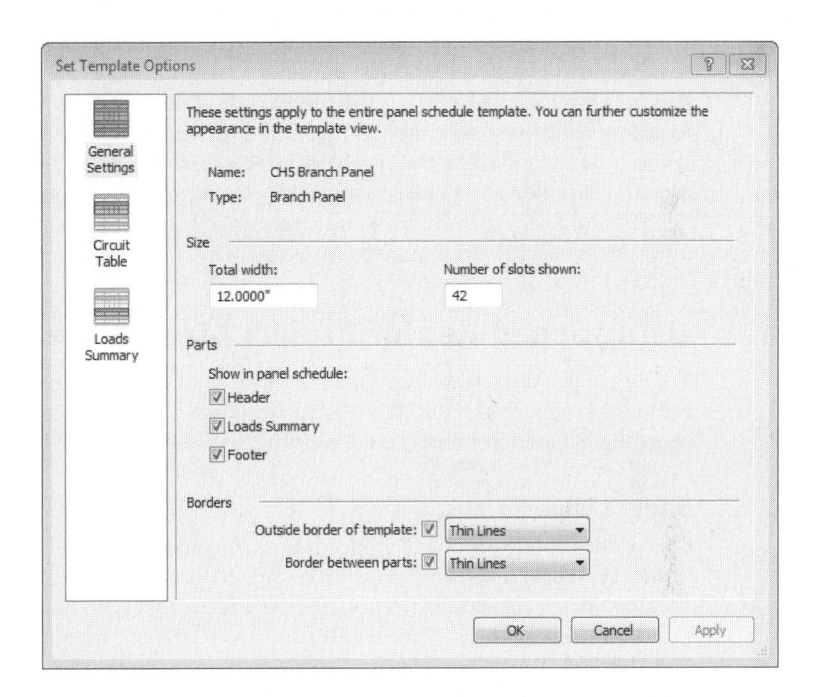

FIGURE 5.32
General Settings area for a panel schedule template

FIGURE 5.33
Loads summary

Load Classification	Connected Loa	Demand Factor	Estimated Dem	Panel Totals	
Receptacle	2520 VA	100.00%	2520 VA		
Appliance - Dwelling Unit	19 VA	100.00%	19 VA	Total Conn. Load:	2539 VA
				Total Est. Demand:	2539 VA
				Total Conn. Curren	7 A
				Total Est. Demand	7 A

The tools on the Modify | Panel Schedule Template contextual tab of the ribbon are the same kind of tools you would find in a spreadsheet program. You can merge cells, add, or delete columns and rows; set the borders and shading of cells; and set the alignment of column data. You can also adjust the font for a cell or group of cells. Many of these tools are also available on a menu that appears when you right-click a cell in the schedule view. You can adjust row and column heights and widths by clicking and dragging the border lines in the schedule view.

The Parameters panel of this tab has a drop-down for selecting a category and a drop-down for selecting parameters within a chosen category to be placed in a column. You can place unique parameters in individual cells in any area of the schedule except for the circuit table. When you click a cell in the circuit table and select a parameter, that parameter populates all the cells in that column and the corresponding column on the other side of the table. When you click a cell in the Circuit Table area, the only category available in the drop-down will be Electrical Circuits. In other areas of the schedule, you can choose parameters from the Electrical Equipment and Project Information categories.

You can set the units to be used in the circuit table with the Format Unit button on the Parameters panel. There is also a button for creating a calculated value in a cell. This tool works the same way as in a building component schedule. The Combine Parameters button allows you to put multiple parameters in a single cell. It is similar to the tool for building a label in an annotation or tag family. You choose the parameters for a cell, the options for a prefix or suffix, and how the parameters will be separated if necessary.

With the ability to customize and format panel schedules to match your standards, they become much more usable than in previous releases to track the loads within your electrical design.

Using Schedules for Project Management

You can apply the scheduling capabilities of Revit to other areas of your project to facilitate project management and organize construction documents. You can even use schedules as a way to coordinate plan notes and ensure coordination between notes and their callouts.

Sheet List

One of the unique types of schedules available on the Schedules button is a Sheet List schedule. A Sheet List schedule is built with the same tools as a model component schedule, but all the parameters are related to sheets. Creating a Sheet List schedule is useful for managing your project documentation. A Sheet List allows you to track what sheets have been created, revised, and checked. You can create custom parameters that apply to sheets for management by other means.

If you are required to submit a list of all the drawings in a submittal package, you can use a Sheet List schedule. Using a Sheet List schedule will ensure that all the sheets created in your project are listed. Some submittals do not require all sheets to be submitted, so a custom parameter that allows you to control what sheets are included in a submittal is a good way to manage your sheet list.

As with most schedules that appear on construction documents, it is a good practice to have two Sheet List schedules in your project: one that has all the parameters that you need for tracking and making changes and another that is actually included in your construction documents. This keeps you from having to hide and unhide columns when working in the schedule view. Figure 5.34 shows a Sheet List schedule with a parameter for whether a sheet is submitted and the corresponding Sheet List schedule that appears in the construction documents. The schedule on the bottom is filtered by the Submitted parameter so that only sheets that are selected appear in the list.

Even though you may have established how sheet views are organized in the Project Browser, it may not meet your requirements for listing the sheets included in a project submittal. You can use a custom parameter to sort your Sheet List schedule in the order you require. You can use a numbering system in combination with the items being listed alphabetically to arrange the sheets in your schedule, as shown in Figure 5.35.

Notice that different decade groups are used for each discipline so that sheets can be organized within a discipline. This allows the electrical site plan to be listed before the plan sheets, for example, without having to provide a unique number for each sheet.

FIGURE 5.34
Sheet list examples

Sheet List for Editing				
Sheet Number	Sheet Name	Sheet Issue	Current Revision	Submitted
E-100	ELECTRICAL LEGEND AND NOTES	02/13/10		☑
E-501	DIAGRAMS AND DETAILS			☐
E-601	PANELBOARD SCHEDULES			☐
EL101	FIRST FLOOR PLAN - LIGHTING	02/13/10		☑
EL102	SECOND FLOOR PLAN - LIGHTING			☐
EP101	FIRST FLOOR PLAN - POWER	02/13/10		☑
EP102	SECOND FLOOR PLAN - POWER	02/13/10		☑
ES101	ELECTRICAL SITE PLAN	02/13/10		☑
G-001	TITLE SHEET	02/13/10		☑
M-100	MECHANICAL LEGEND AND NOTES	02/13/10		☑
M-501	DIAGRAMS AND DETAILS			☐
MH101	FIRST FLOOR PLAN - HVAC	02/13/10		☑
MH102	SECOND FLOOR PLAN - HVAC	02/13/10		☑
P-100	LEGEND, NOTES AND SCHEDULES	02/13/10		☑
P-501	DIAGRAMS AND DETAILS			☐
PL101	FIRST FLOOR PLAN - PLUMBING	02/13/10		☑
PL102	SECOND FLOOR PLAN - PLUMBING	02/13/10		☑

SHEET LIST			
SHEET NO.	SHEET NAME	DATE	REVISED
E-100	ELECTRICAL LEGEND AND NOTES	02/13/10	
EL101	FIRST FLOOR PLAN - LIGHTING	02/13/10	
EP101	FIRST FLOOR PLAN - POWER	02/13/10	
EP102	SECOND FLOOR PLAN - POWER	02/13/10	
ES101	ELECTRICAL SITE PLAN	02/13/10	
G-001	TITLE SHEET	02/13/10	
M-100	MECHANICAL LEGEND AND NOTES	02/13/10	
MH101	FIRST FLOOR PLAN - HVAC	02/13/10	
MH102	SECOND FLOOR PLAN - HVAC	02/13/10	
P-100	LEGEND, NOTES AND SCHEDULES	02/13/10	
PL101	FIRST FLOOR PLAN - PLUMBING	02/13/10	
PL102	SECOND FLOOR PLAN - PLUMBING	02/13/10	

View List

A View List schedule is another unique type of schedule that can be created from the Schedules button on the View tab. Similar to a Sheet List schedule, a View List schedule can help you keep track of what views exist in your project.

WHY NOT USE THE PROJECT BROWSER?

It may seem a bit redundant to create a View List schedule when all views are organized in the Project Browser, but a View List schedule enables you to quickly view the properties of several views at one time. It also allows for quick comparison and editing of multiple views.

FIGURE 5.35
Sheet list sorted in order

Sheet List for Editing					
Sheet Number	Sheet Name	Sheet Issue	Current Revision	Submitted	Publish Order
E-100	ELECTRICAL LEGEND AND NOTES	02/13/10		☑	30
E-501	DIAGRAMS AND DETAILS			☐	33
E-601	PANELBOARD SCHEDULES			☐	34
EL101	FIRST FLOOR PLAN - LIGHTING	02/13/10		☑	32
EL102	SECOND FLOOR PLAN - LIGHTING			☐	32
EP101	FIRST FLOOR PLAN - POWER	02/13/10		☑	32
EP102	SECOND FLOOR PLAN - POWER	02/13/10		☑	32
ES101	ELECTRICAL SITE PLAN	02/13/10		☑	31
G-001	TITLE SHEET	02/13/10		☑	1
M-100	MECHANICAL LEGEND AND NOTES	02/13/10		☑	10
M-501	DIAGRAMS AND DETAILS			☐	12
MH101	FIRST FLOOR PLAN - HVAC	02/13/10		☑	11
MH102	SECOND FLOOR PLAN - HVAC	02/13/10		☑	11
P-100	LEGEND, NOTES AND SCHEDULES	02/13/10		☑	20
P-501	DIAGRAMS AND DETAILS			☐	22
PL101	FIRST FLOOR PLAN - PLUMBING	02/13/10		☑	21
PL102	SECOND FLOOR PLAN - PLUMBING	02/13/10		☑	21

SHEET LIST			
SHEET NO.	SHEET NAME	DATE	REVISED
G-001	TITLE SHEET	02/13/10	
M-100	MECHANICAL LEGEND AND NOTES	02/13/10	
MH101	FIRST FLOOR PLAN - HVAC	02/13/10	
MH102	SECOND FLOOR PLAN - HVAC	02/13/10	
P-100	LEGEND, NOTES AND SCHEDULES	02/13/10	
PL101	FIRST FLOOR PLAN - PLUMBING	02/13/10	
PL102	SECOND FLOOR PLAN - PLUMBING	02/13/10	
E-100	ELECTRICAL LEGEND AND NOTES	02/13/10	
ES101	ELECTRICAL SITE PLAN	02/13/10	
EL101	FIRST FLOOR PLAN - LIGHTING	02/13/10	
EP101	FIRST FLOOR PLAN - POWER	02/13/10	
EP102	SECOND FLOOR PLAN - POWER	02/13/10	

When printing a set of construction documents, you want to be sure that all your views that should be on a sheet have been placed onto the appropriate sheet. Not only will this save you the time it would take to visually inspect each sheet, but it can also save paper and wear and tear on your printer.

A View List is constructed with the same tools as a Sheet List, except only parameters that apply to Views can be used in the schedule. Unfortunately, not all view parameters are able to be scheduled, but there are many that are useful for project management. When working in a project that includes phases, a View List can be a useful tool for ensuring that views are set to the proper phase and have the proper phase filter applied.

To create a View List that only shows views that are placed on sheets, you can use a filter. If you use the Sheet Number parameter as a filter with the Does Not Contain rule and leave the value blank, the expected result would be a schedule that shows only those views that are placed on a sheet.

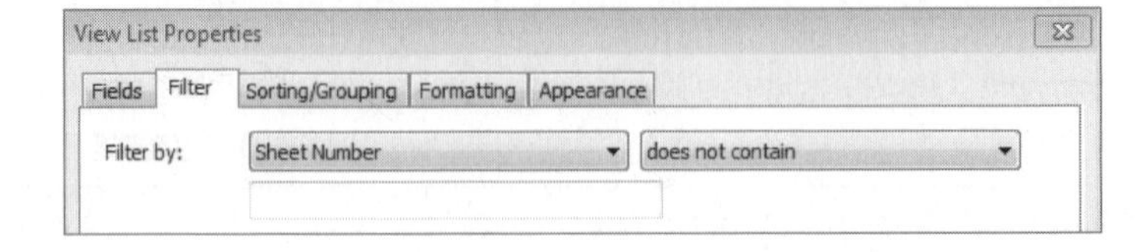

Normally you would apply a filter to remove all the scheduled items that do not meet your filter requirements. However, using these settings will actually remove all views that meet the requirements of the filter, causing your schedule to list all the views that have no value for the parameter.

Note Block

A Note Block is a schedule of an annotation family that is used in your project. Note Block schedules are very useful for managing plan notes on your construction documents as an alternative to keynoting. As note annotations are placed in a view, they can be given a description, and then a Note Block schedule can be created to list the descriptions of each instance of the annotation. Figure 5.36 shows the properties of a note annotation family used for plan notes.

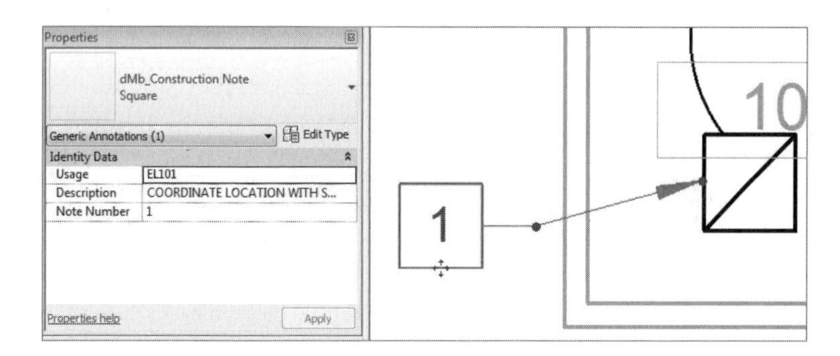

FIGURE 5.36 Note annotation properties

A Usage parameter has been created to determine what sheet the note belongs to. The Description parameter holds the contents of the note. Once a note annotation is placed in a view, it can be copied for each instance or to create the next note in the list.

A Note Block schedule is created by choosing the Note Block option from the Schedules button on the View tab. The annotation family that is used for plan notes is selected to be scheduled. The title of the Note Block should be a description of where the notes are located to make it easy to locate the Note Block schedule in the Project Browser. The Note Number, Description, and Usage parameters are added to the schedule. The schedule is then filtered by the Usage parameter so that only note annotations that appear on a specific sheet will appear in the schedule and the Usage parameter is hidden. The schedule is sorted by the Note Number parameter.

It is important to establish the Usage parameter value for the first note annotation placed into a view and then to copy that view whenever a new note is required. Copying the note annotation will ensure that the Usage parameter value is the same for each note in the plan view. The Note Block view and its associated plan view can be tiled so that as you add note annotations to the plan view, you can see them appear in the schedule. The Properties palette gives you instant access to the Description parameter of the note annotation so changes can be made easily.

One benefit of using a Note Block schedule over static text for plan notes is that if you need to change what a note says, you can deselect the Itemize Every Instance option in the schedule to see each unique note; then when you change the description of the note, it will change for all instances.

Another benefit is that if you need to delete a note, you can delete it from the Note Block schedule, and all instances will be removed from the associated plan view. This ensures that you do not have a note annotation in the plan view that references a note that does not exist in the plan notes. It also eliminates the need for having a note in the notes list with a value of Not used because the notes can easily be renumbered in the Note Block schedule when a note is removed. Renumbering the notes in the schedule will automatically update the annotations in the plan view.

When you place the Note Block schedule onto a sheet with its associated plan view, you will want to adjust the Appearance settings so that the schedule title and headings are not shown, giving the schedule the appearance of a list rather than a schedule, as shown in Figure 5.37.

Text was used on the sheet to provide a title for the list of notes. Notice that there is no square symbol for each note number in the list. Unfortunately, you cannot add graphics to a schedule at this time, but the benefits of coordination between the list of notes and the annotations on the plan far outweigh having a square around each number. A square symbol was added to the title as a legend to further clarify to the contractor what the note symbols are.

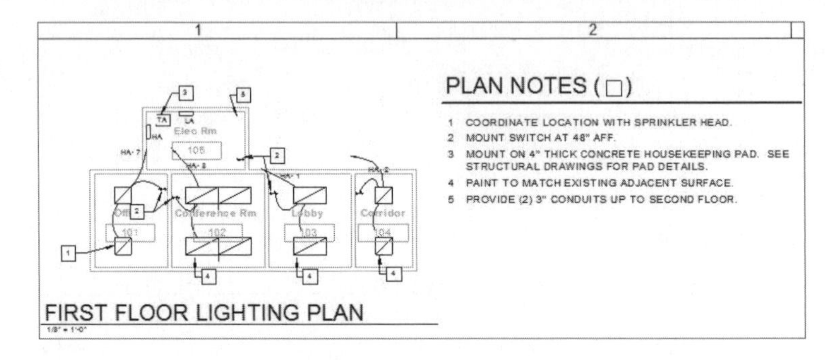

FIGURE 5.37
Note Block plan notes on a sheet

The Bottom Line

Use the tools in Revit MEP 2011 for defining schedules and their behavior The capabilities of schedules in Revit MEP 2011 can increase your project coordination and the efficiency of your workflow. The ability to track items within a model can help you to better understand the makeup of your design.

> **Master It** The information in schedules comes from information stored within the objects of a Revit model. Explain why editing the data of an object in a schedule will change the properties of the object.

Schedule building components Scheduling building components is the primary use of the scheduling tools in Revit. Schedules are used on construction documents to provide additional information about components so that drawings do not become too cluttered.

> **Master It** Understanding what information can be used in a schedule is important to setting up a specific component schedule within your Revit project. What types of parameters can be included in a schedule? What type cannot?

Create schedules for design and analysis Scheduling can go beyond counting objects and tracking their information. You can also create schedules that assist in making design decisions by providing organized analytical information.

Master It The information stored in Space objects often comes from their relationship with other objects. Some of the data for analysis needs to be manually input. Explain how using a schedule key can assist in adding data to a Space object.

Schedule views and sheets for project management Not only the components that make up a model, but also the views and sheets within your project can be scheduled. Specialized schedules for views and sheets are useful for project management.

Master It A Note Block schedule is a schedule of annotation family so a list of the information within the annotation can be generated. What are some of the benefits of using a Note Block instead of static text for plan notes?

Chapter 6

Details

With all the emphasis on using Revit as a design tool, it is easy to assume that it is weak when it comes to traditional drafting. Revit projects do not need to be modeled to the smallest detail just because they can be. Many of the details that are used to convey design intent with a traditional 2D drafting method can be used on a Revit project as well. In fact, it is best to keep your model as simple as possible and handle the more detailed information with, well, details.

The transition to Revit can be difficult because it may seem that you will have to abandon all the details you have accumulated over the years. With Revit, you are able to use the CAD details you have in their native format directly in your projects, or you can easily convert them to Revit format and begin to re-create your library. Actually, it provides you with an opportunity to update and organize your library of details.

Using the tools available in Revit MEP 2011 for details and diagrams will enable you to create a complete set of construction documents.

In this chapter, you will learn to

- Use Revit drafting and detailing tools efficiently
- Use CAD details in Revit projects
- Build a Revit detail library
- Create detail views of your Revit model

Drafting and Detailing Tools

Whether you are embellishing a model view or creating a detail or diagram on a drafting view, the features of the Annotate tab provide you with the necessary drafting tools to generate line work, patterns, and symbols. The Detail Line button enables you to draw line work that is specific to the view in which it is created.

When you click the Detail Line button, the Modify/Place Detail Lines contextual tab appears on the ribbon. This tab is very much like the Modify tab but also contains the Draw panel, which allows you to select from an assortment of line tools to create either lines or shapes. The Line Style panel lets you choose the type of line that will be used.

Line Styles

Line styles are an important part of your Revit projects because you can create styles that match the line patterns used for model components. This allows you to maintain a consistent look throughout your construction documents without having to override the graphic representation of each line. For example, you can create a line style to be used for lines that represent Domestic Cold Water Pipe with the same line pattern used in a view filter that applies to the actual Domestic Cold Water Pipe, as shown in Figure 6.1.

FIGURE 6.1
Line style matching model display

It may be helpful to think of line styles as family types for lines. You can draw a line with any line style and switch it to another style using the selector drop-down on the Line Style panel of the contextual tab that appears when a line is selected. This gives you the freedom to draw a detail or diagram with one line style, focusing on the content of what is being drawn. You can then go back and change selected lines to more appropriate line styles if necessary.

Regions

Filled regions are areas with a chosen pattern that can used to represent a material or designated area of a detail. When you create a filled region, you can choose a line style for the border of the region. When you are drawing a detail that requires a filled region, determine how its boundaries will interact with the rest of the detail, and choose appropriate border lines. Using the borders of a filled region requires less line work to be drawn because there is no need to have a line drawn over the border of a filled region.

Figure 6.2 shows a detail of a concrete floor penetration. A filled region is used to represent the concrete floor. The border lines of the region where the electrical box penetrates the floor are thicker than the rest. This is done so that additional lines do not have to be drawn to represent the box.

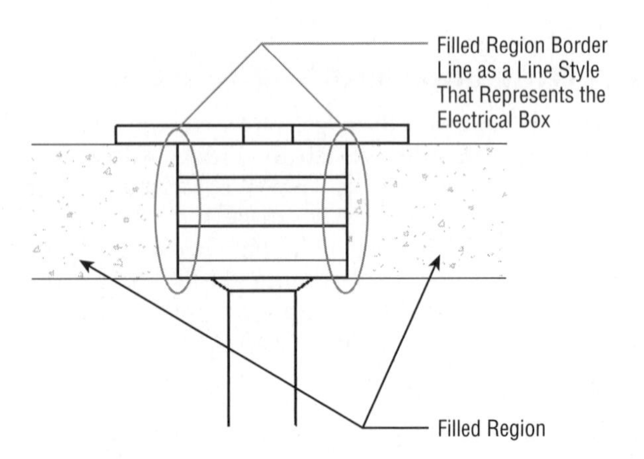

FIGURE 6.2
Filled region in a detail

Notice that at the outer edges of the filled region there are no border lines shown. You can use the <Invisible Lines> line style for borders of a filled region that you do not want to show. This is also useful for when you have two filled regions that are adjacent. One region can define the line of the detail, while the overlapping region border can be set to invisible. This is only really necessary if the overlapping lines cause the detail to look incorrect.

INVISIBLE LINES

The <Invisible Lines> line style is one of the line styles that comes with Revit MEP 2011. You can create your own invisible line style by setting the color of the line to white.

To draw a filled region, do the following:

1. Click the Region button on the Annotate tab, and select the Filled Region option. This will change the ribbon to a contextual tab for drawing the region. All visible elements in the drawing area will become halftone because drawing a region is done in sketch mode. The same tools for drawing detail lines are available when sketching a region.
2. Choose a line style for the borders of the region, and draw lines to define the shape of the region. Lines must form a closed loop to create a region.
3. When you are finished sketching the region shape, you must click the green check mark button on the Mode panel of the contextual tab to finish and exit sketch mode. If you want to abandon the creation of the region, you can click the red X button on the Mode panel.
4. When you exit the sketch mode for a filled region, you can change the pattern used for the region by using the Type Selector in the Properties palette.
5. Selecting a filled region activates grips that allow you to push and pull the edges to change the shape of the region. Edges can be aligned with other items in the view. An edge can be modified only to the point where it does not cause another edge to be too small or disappear.
6. To make changes to the borders of a filled region, you can click the Edit Boundary button on the Mode panel of the contextual tab that appears when the region is selected. This returns you to the sketch mode for making changes to the region's shape or to redefine the border line styles.

When you have a filled region that overlaps other detail lines, you can select the region and use the Bring To Front or Send To Back button on the Arrange panel of the contextual tab to change the draw order. If the region is defined as having an opaque background, it will mask out any items behind it as long as the view is set to the Hidden Line model graphics style. A Wireframe view will cause any line work behind a region to be displayed.

You can use two types of patterns for filled regions. Drafting patterns will change with the scale of the view, while model patterns are sized based on the dimensions in the model. To create a filled region pattern, click the Additional Settings button on the Manage tab and select Fill Patterns. Choose whether you want to create a drafting or model pattern at the bottom of the Fill

Patterns dialog box. Click the New button to create a pattern. You can choose between parallel lines or crosshatched lines, setting the angle and spacing of the lines for a simple pattern. If you choose the Custom option in the New Pattern dialog box, you can import a `.pat` file, as shown in Figure 6.3.

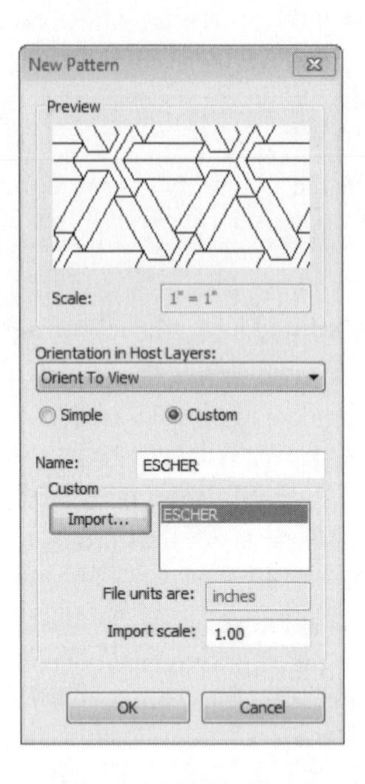

FIGURE 6.3
Custom pattern for filled region

Once you create a custom pattern, you must reimport it to change the scale. You can set the Orientation In Host Layers setting for a pattern when it is used as a cut pattern for objects such as walls, floors, and roofs. You can choose to keep the pattern aligned with the element, orient the pattern to the view, or keep it readable. These settings are described here:

Aligned With Element This option will cause the pattern to maintain its relationship to the host.

Orient To View This option causes all patterns to have the same alignment.

Keep Readable This option causes the pattern to behave in the same manner as text. The pattern will maintain its alignment until it is rotated past 90 degrees, where it will flip to keep its intended alignment.

These settings do not have any effect on filled regions that are drawn manually.

Once you have a pattern, you can assign it to a filled region style. Select a filled region, and click the Edit Type button in the Properties palette to edit a style or create a new one. Click the Duplicate button to create a new style, giving it a name that clearly defines what it is. It is usually best to name a filled region style with the same name as the pattern it uses. Click the Fill Pattern parameter to select a pattern used for the style. Other parameters allow you to set the transparency of the background, the line weight, and the color of the lines. These settings are for the pattern used by the style, not for the lines that define the border, as shown in Figure 6.4.

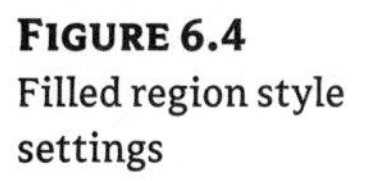

FIGURE 6.4
Filled region style settings

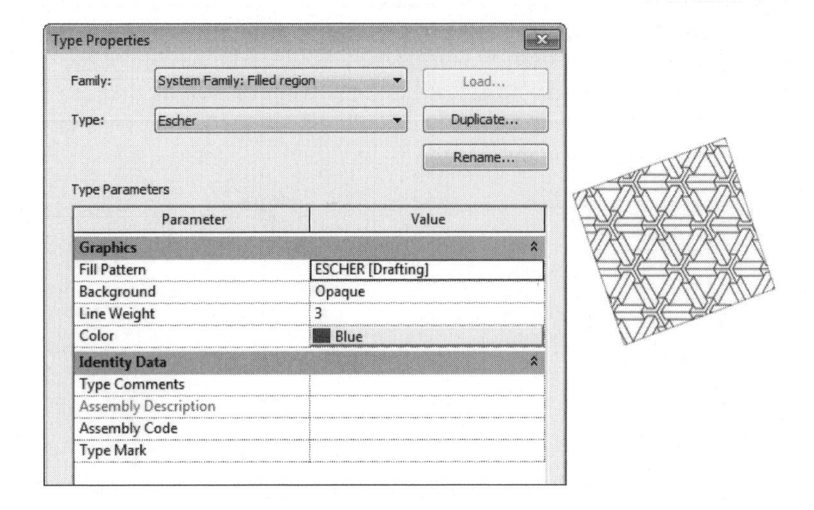

Another type of region you can use in details is a masking region. A masking region is a shape that will mask out any line work behind it. It is essentially a blank filled region. Masking regions will block out any detail lines or filled regions in a detail view that is set to the Hidden Line visual style. The tools for creating a masking region are exactly the same as for a filled region. These regions are very useful when creating a detail that displays model elements that may contain elements that you do not want to be seen in the detail. You can use the border lines of a masking region to define portions of a detail, in the same manner as with filled regions.

Detail Components

Many items within details are used repeatedly from one detail to the next. It can be tedious and time-consuming to locate a detail that has a specific item that you know you have already drawn so that you can use it in a detail you are currently drawing. *Detail components* are special annotation families that can be used to represent objects commonly used in details, saving you repetitive drawing time. When you install Revit MEP 2011 and its content, a Detail Components folder is included in the content library. The families in this folder are sorted by their construction specification category. Each folder contains families that you can use in your details to save the time it would take to draw them repeatedly. Many of the families have multiple versions, each of which is drawn from a different viewpoint so that they can be used in section detail views.

Detail component families are different from annotation families because their size is not dependent on the scale of the view they are drawn in. They are representations of building components drawn at their actual size. These items can be used when the building model has not been modeled to a level of detail to include the actual items. Certain architectural and structural elements do not need to be included in their respective models in order for these disciplines to achieve their project goals. When you use a section or callout of the model for a detail, you will need to include detail components for a true representation of the design.

It is a common practice to use generic styles for elements such as walls, floors, or roofs so as not to weigh down the model with unnecessary detail that can be handled with detail views. For example, though an exterior wall may have an outer layer of brick, the wall can be modeled generically and detail components can be used in a detail view to show the brick layer. Detail components are useful for MEP discipline details as well. Many structural member components are available in the library. Items such as angles, beams, and channels can be quickly and accurately represented when creating a detail. Figure 6.5 shows an example of detail components used to create a detail.

FIGURE 6.5
Sample detail using detail components

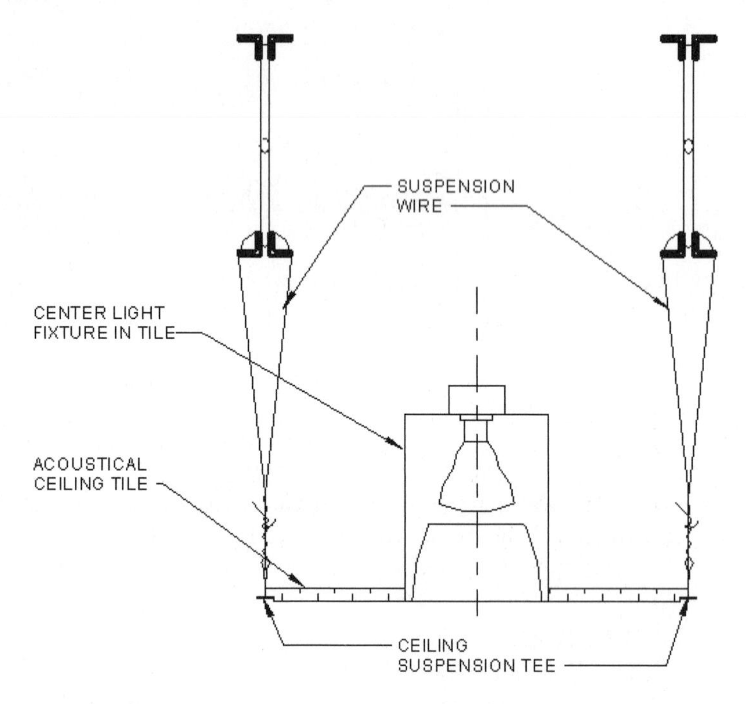

Even though the lighting fixture shown in the detail was included in the model, a detail component of a section view of a light fixture was used in the detail because the actual model component used was a simple cylinder. Every item shown in the detail is a detail component family except for the center line and annotations.

You can use detail component families to create repeating detail component styles. Repeating detail components allow you to quickly represent repetitive instances of components such as brick, glass block, or roof decking. They can also be used for MEP items such as pipe or conduit, as shown in Figure 6.6. In this example, the repeating detail was drawn horizontally to represent the rows of conduit.

To create a repeating detail component style, click the Component button on the Annotate tab, and select the Repeating Detail Component option. Click the Edit Type button in the Properties palette, and click the Duplicate button in the Type Properties dialog box. Name your style so that it can be easily identified for its use. Figure 6.7 shows the Type Properties dialog box for a repeating detail.

The Detail parameter allows you to select a detail component family that is loaded into your project for the repeating detail. There are four choices for the Layout parameter settings:

Fill Available Space This option will place as many instances of the detail component that will fit within the length of the path drawn for the repeating detail.

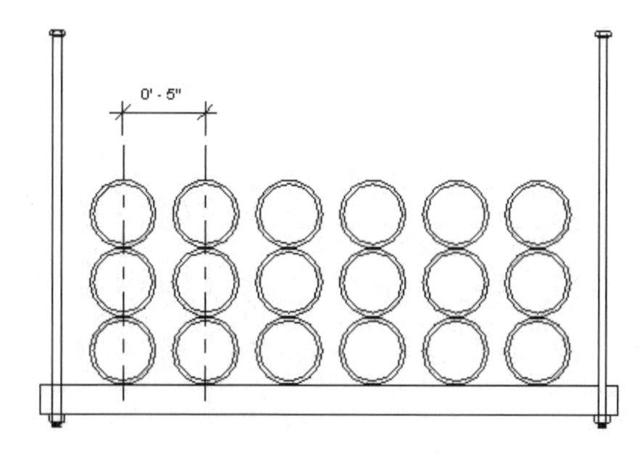

FIGURE 6.6
Repeating detail for conduit rack

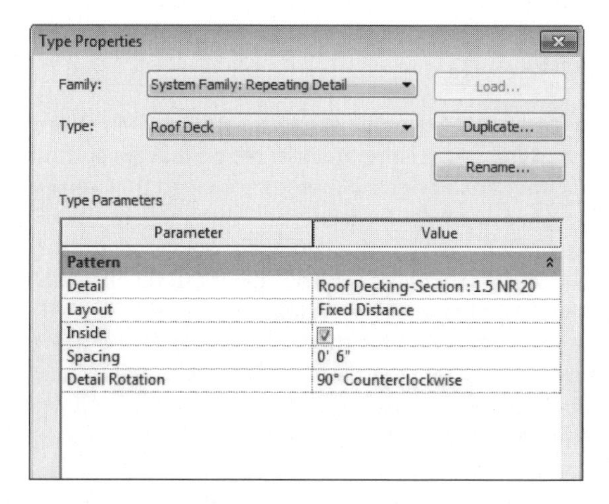

FIGURE 6.7
Repeating detail properties

Fixed Distance This option can be used with the Spacing parameter, which defines the distance between each instance of the detail component. The number of detail components with the spacing that completely fit within the length of the path will display.

Fixed Number This option will equally space the detail component along the length of the path drawn.

Maximum Spacing This option allows you to set the maximum distance between instances of the detail component. When the Maximum Spacing option is used, the space between instances may be less than specified depending on the length of the path. This ensures that a complete instance will occur at each end of the path. Additional instances are added as the path length causes the maximum spacing to be reached.

The Inside parameter determines whether the first and last instance of the detail fits within the length of the path drawn. The insertion point of the detail component determines the placement of the first and last instances. Figure 6.8 shows two repeating details with the same settings except that the top row's Inside parameter is set to No and the bottom row's is set to Yes. Notice that the bottom row has fewer instances of the detail component even though the path distance is the same.

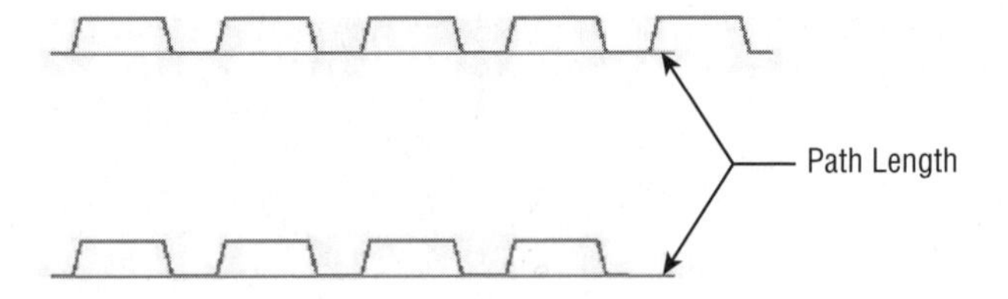

FIGURE 6.8
Repeating details

You can use the Detail Rotation parameter to rotate the detail component family so that it follows the path in the proper manner.

Because repeating detail components are system families, you cannot store them in your library, but you can create a project file with all the styles you use and then transfer them into your projects as needed. The detail component families used in the repeating details will be transferred along with the styles.

CAD Details

One of the primary benefits of a CAD application is the ability to save and reuse drawings. Because of this, most companies have spent years accumulating a vast array of CAD details. Making the transition to Revit does not mean that you can no longer use your library of CAD details. You can use CAD details and diagrams in your Revit projects, or you can re-create them in Revit format to build a new library of details for use on Revit projects. Converting your details to Revit will reduce the number of CAD files you will have to link into your projects, which will help improve performance.

The key to success when using CAD details in a Revit project is to link the CAD file. If you have any concern for standards and for file performance, do not import CAD files into your project. It can be tempting to import a CAD file, explode it, and then clean it up for use in your project. Although this might provide immediate results, it will have an overall negative effect on your project. When you explode an imported CAD file, Revit creates a text style and line style for each unique text style and layer within the CAD file.

The more of these unnecessary styles that you bring into your project, the poorer your file performance will be. It also opens the door for deviation from standards because nonstandard text styles and line patterns will be available for use in other areas of the project.

If your project contains line patterns that have come from imported CAD files, you can remove them to avoid improper use. The line patterns show up as IMPORT- patterns in the Line Patterns dialog box, as shown in Figure 6.9.

You cannot remove them using the Purge Unused tool, but you can remove them manually from the project in the Line Patterns dialog box.

When you link a CAD detail into your Revit project, it may not look exactly like it does in CAD. Variations in text styles from CAD to Revit can cause undesired display of text notes and leaders. CAD details can be easily converted to Revit format, which enables you to correctly display the detail.

Drafting Views

CAD details or diagrams can be linked directly onto a sheet in your Revit projects, but if you intend to convert a detail to Revit format, it is best to link it in a drafting view instead of onto a sheet. This allows you to save the view as a file for use on future projects.

FIGURE 6.9
Imported line patterns

Even if you are going to use CAD details in your project without converting them, linking them into drafting views will help keep your project organized. Drafting views can be organized in the Project Browser, which will help you keep track of what details exist in your project. You can specify the properties of drafting views to group them into their appropriate locations within the Project Browser for easy access and management.

It is best to set the scale of a drafting view that is used for linked CAD details to the same scale that the details are drawn in. This will ensure accurate display of the CAD details.

Converting Details

When you link a CAD detail into a drafting view by clicking the Link CAD button on the Insert tab, you have some options for its display. One of the choices is for the colors of the CAD lines. Many people use colors in CAD to determine the line weight of printed lines. Choosing the Preserve option in the Link CAD Formats dialog box will maintain the colors of the line work when inserted into Revit, as shown in Figure 6.10. This will help you to determine what Revit line styles to use when converting the detail.

There are a few ways to convert a CAD detail into Revit format. The most time-consuming and likely least desirable method is to import the CAD file, explode it, and convert the line and text styles to Revit lines and text. This should not be done directly in your Revit project but rather in a separate file in order to keep the imported styles from populating your Revit project. The file should then be purged of all unused styles before being saved to a library or used in projects.

The easiest method for converting CAD details to Revit format is to link the file into a drafting view and trace the line work. Although tracing seems like a tedious task, Revit has a drafting tool that expedites the process. With a file linked into a drafting view, click the Detail Line button on the Annotate tab. On the Draw panel, select the Pick Lines tool.

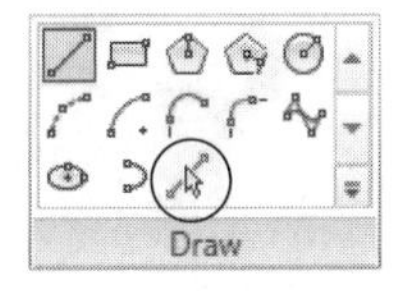

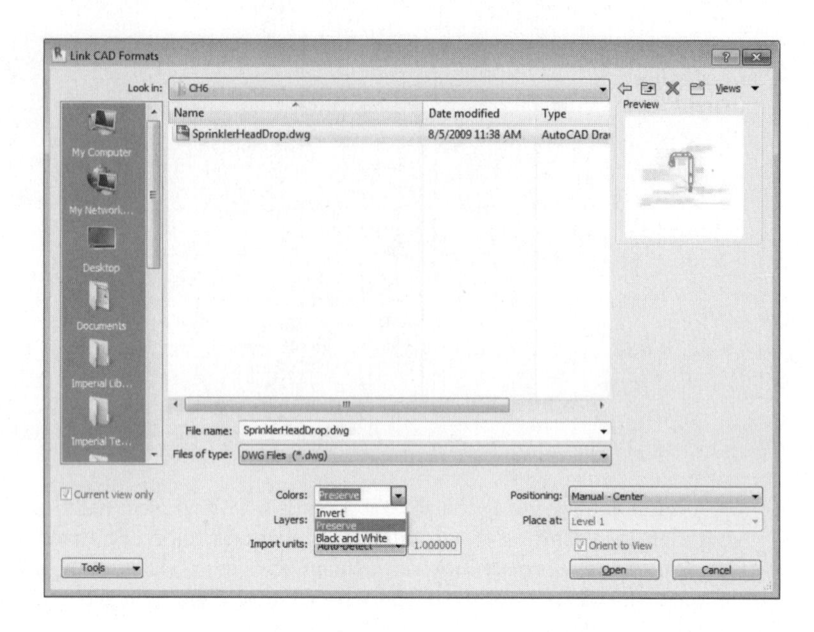

FIGURE 6.10
Color options for linked CAD file

The Pick Lines tool allows you to select lines within the CAD file and will place a Revit line in the same location that matches the selected line. With this tool you can pick the lines of a CAD detail and have a Revit version within a short period of time.

You cannot use the window selection feature to select the lines. Each line must be selected individually; however, when you combine this tool with the Tab selection functionality of Revit, you will save time and number of clicks. Placing your cursor over a desired line and hitting the Tab key on your keyboard will highlight any lines connected to the line that your cursor is on for selection, as shown in Figure 6.11. With the multiple lines highlighted, you can click to select them, which will place a Revit line at each location.

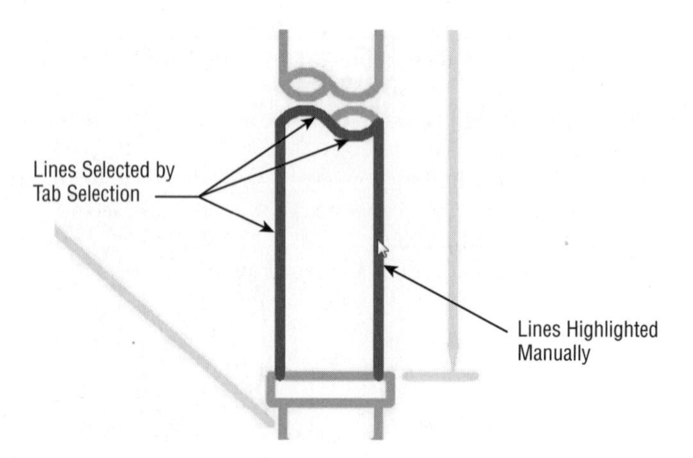

FIGURE 6.11
Tab selection of multiple lines

SHORT LINES

Revit cannot draw a line that is shorter than 1/32″ (0.8mm). When you are converting a CAD detail, you may come across lines that are too short for Revit to draw. Even though lines may highlight using Tab selection, they will not be drawn when clicked, and any lines connected to them will not be drawn. When you place your cursor over a line in a CAD detail, the cursor will indicate whether the line cannot be drawn in Revit. Adjustments may need to be made to your details to accommodate this limitation.

In some cases, you need to create a very small circle. You can use the Scale feature on the Modify tab to shrink a circle to a smaller size than you can create when drawing it initially.

When you are using the Pick Lines tool to convert a CAD detail, you may want to choose a Revit line style that is thicker than the lines being copied. This will help you keep track of which lines have already been converted, especially if your CAD detail is drawn with black lines. As you click to draw Revit lines, a padlock icon will display and allow you to lock the Revit line to the CAD detail line. This is unnecessary since the CAD detail will be removed. Once you have completed the line conversion, you can select lines and change their style to match the intent of the detail. Using this process is more efficient than constantly switching between line styles while converting the detail.

You cannot use the Pick Lines tool to copy the text within a CAD detail, so you will have to manually place notes and dimensions. Use the ability to override dimension text when necessary for details that are not drawn to scale. You will have to create any required filled regions manually as well.

Once a CAD detail has been converted to Revit format, you can use the Manage Links tool to remove the linked CAD file, leaving you with a native Revit detail that can be saved to a detail library for future use.

Reducing your reliance on linked CAD files is an important step toward reducing project load times, view regeneration, and other causes of poor file performance. Because details are such an important part of any project, having Revit details available for use is crucial to the success of a project. Now that you have learned some options for converting a CAD detail to Revit, practice by completing the following exercise:

1. Open the `Ch6_Details.rvt` file found at `www.wiley.com/go/masteringrevitmep2011`.
2. Click the Drafting View button on the View tab.
3. Name the view **`Sprinkler Head Drop Detail`**. Set the scale to Custom and the scale value to 1.

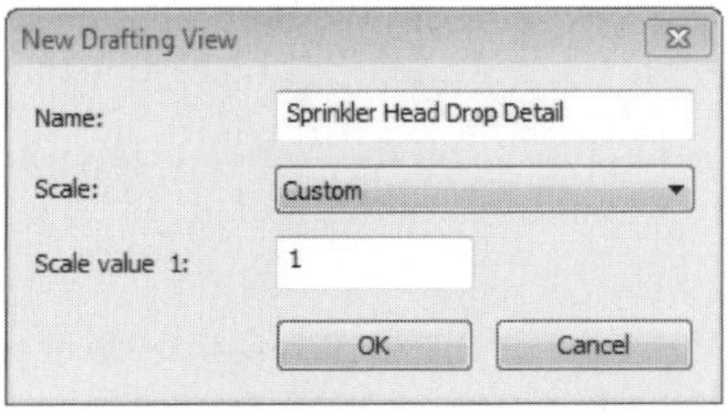

New Drafting View
Name: Sprinkler Head Drop Detail
Scale: Custom
Scale value 1: 1
OK Cancel

4. Download the `SprinklerHeadDrop.dwg` file found at `www.wiley.com/go/masteringrevitmep2011`. Save the file in a location that you can access during this exercise.

5. With the newly created drafting view open, click the Link CAD button on the Insert tab.

6. Browse to the location of the downloaded CAD file. Select the file, and set the Colors setting to Preserve. Choose Manual – Center for the Positioning option, and click Open.

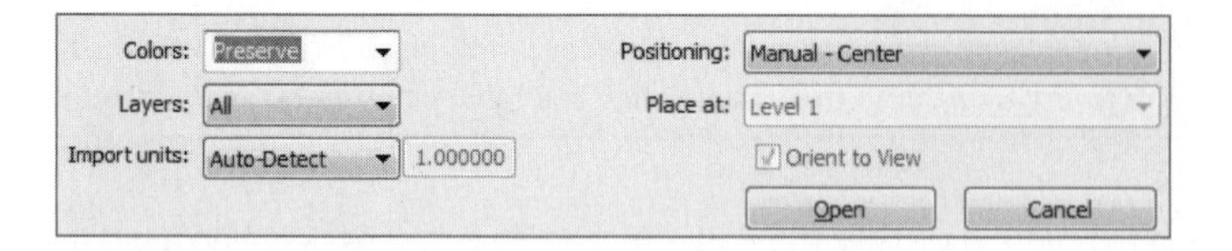

7. Click in the drawing area to place the CAD detail in the drafting view. Zoom to where you can comfortably see the entire detail.

8. Click the Detail Line button on the Annotate tab. Choose the Pick Lines tool from the Draw panel, and set the line style to Wide Lines on the Line Style panel of the ribbon.

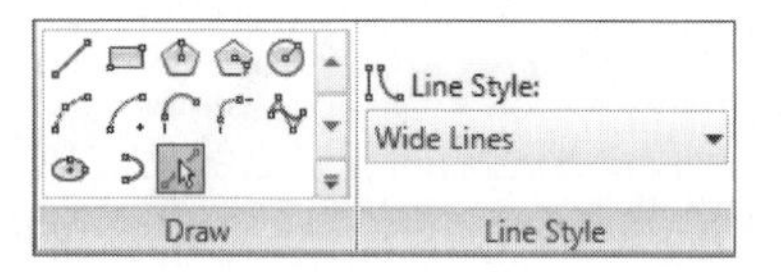

9. Place your cursor over the circle called out as the SPRINKLER BRANCH LINE in the detail. With the circle highlighted, click to place a Revit line.

10. Place your cursor over the vertical line at the top left side of the circle. Press the Tab key to highlight all three lines at the top of the circle, and then click to draw Revit lines.

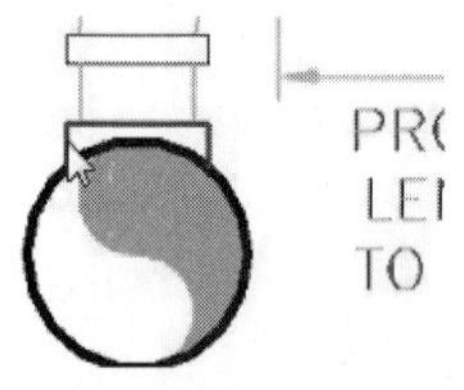

11. Continue placing lines by clicking the CAD detail lines, using the Tab key when applicable until you have completely duplicated the detail line work, including the detail title text line. Do not use the padlock icon to lock the Revit lines to the CAD detail.

12. Click the Region button on the Annotate tab, and select Filled Region. Select the Pick Lines tool from the Draw panel, and set the line style to Thin Lines in the Line Style panel.

13. Place your cursor over one of the arcs inside of the circle that is called out as SPRINKLER BRANCH LINE to highlight the line.

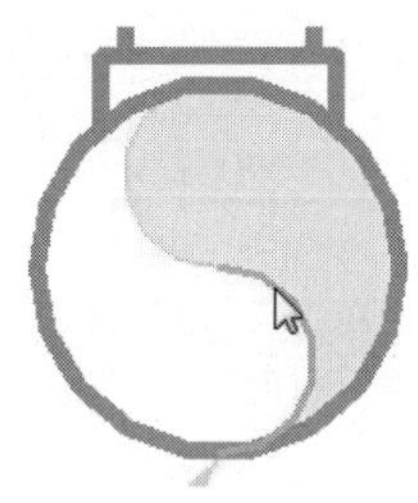

14. Press the Tab key to highlight the connected line work, and click to draw the region border with Revit lines. Click the green check mark on the Mode panel of the ribbon to exit the region's sketch mode.

15. Select the linked CAD file and move it off to the side, away from the newly created Revit line work. It is easiest to select the CAD file by clicking the text, since the line work has been covered by Revit lines.

16. The CAD file is drawn with standards where colors indicate line weight. In this example, Red = Thin Lines, Green = Medium Lines, and Blue = Thick Lines. Convert the Revit lines to their appropriate line style using the Line Style drop-down on the contextual tab that appears on the ribbon when you select a line. You can select multiple connected lines using the Tab key. You can also hold down the Ctrl key and use the window selection feature to select multiple lines that are not connected.

17. With the line work completed, place text and leaders to match those shown in the CAD detail. It may be helpful to move the CAD detail back into alignment with the Revit line work for placement of text notes and leaders.

18. Click the Aligned button on the Dimension panel of the Annotate tab to create a dimension in the same location as the horizontal LENGTH AS REQD dimension of the CAD detail. Click the dimension text to modify it. In the Dimension Text dialog box, choose the Replace With Text option, and enter **AS REQD**. Enter LENGTH in the Text Fields Above field.

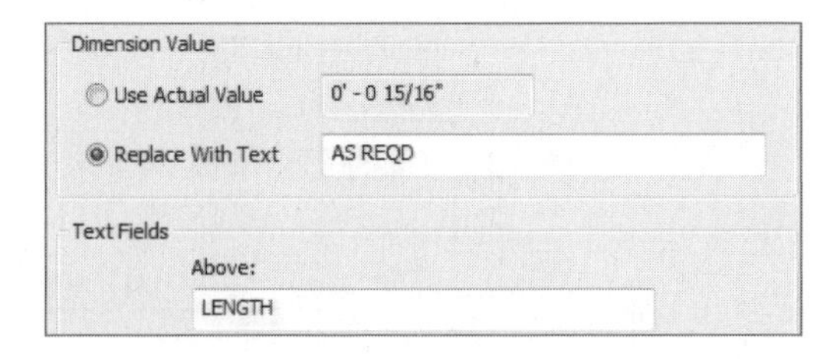

19. Repeat step 18 for the vertical dimension in the detail.

20. With the detail complete, use the Manage Links tool to remove the CAD file.

21. Locate the drafting view in the Project Browser. Right-click the view, and select the Save To New File option. Browse to a location to save the file.

Strategies for Creating a Detail Library

When you right-click a drafting view in the Project Browser, one of the menu options is Save To New File.

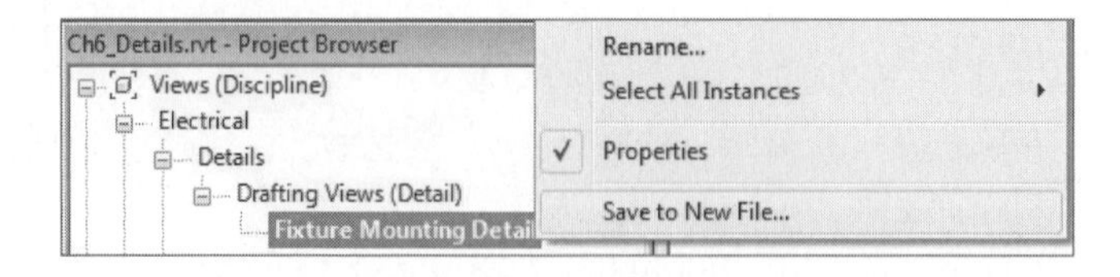

Saving a drafting view as a file allows you to put it in a location where you can build a library of details and diagrams that can be accessed for use on other projects. You can organize your detail library in any manner that suits your workflow or processes. One option is to create a drafting view that has several common details on it. The view is then saved as a file, and the file is opened. Within this new file, the drafting view is duplicated, and a new drafting view for each detail is created, leaving you with a single file containing multiple drafting views, each with its own unique detail. This will make it easy to locate specific details or allow you to place an entire group of details into your project. Figure 6.12 shows an example of a detail file containing multiple drafting views for individual details.

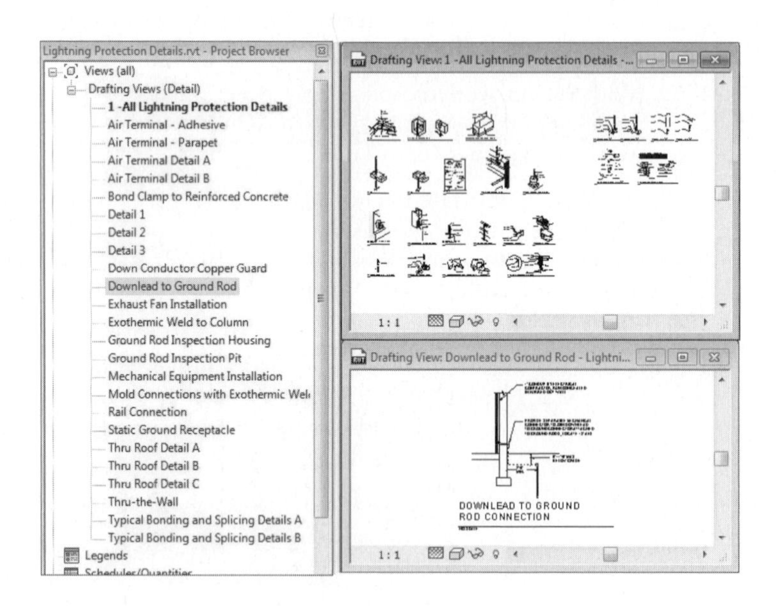

FIGURE 6.12
Sample detail file organization

When you create a drafting view for each detail, it allows you to easily bring the detail into your project without having to load a drafting view or an entire set of details. This will reduce the number of views in your project, making the Project Browser more navigable and keeping your project file size minimized. Figure 6.13 shows a sample Project Browser organization for detail views.

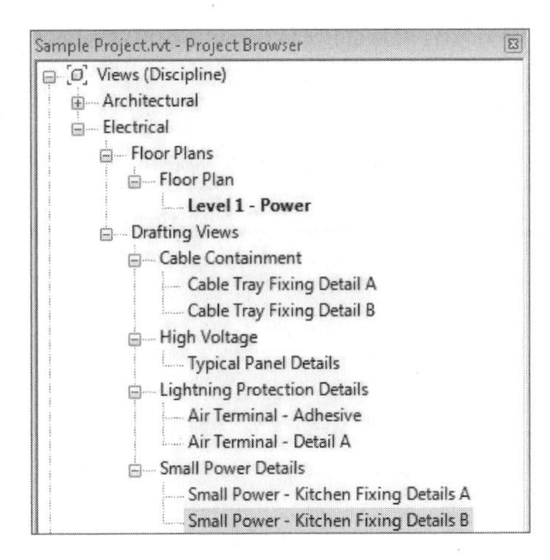

FIGURE 6.13
Sample detail view organization

You have two choices for inserting details on the Insert From File button located on the Insert tab. They are Insert Views From File and Insert 2D Elements From File. Each option is explained in the following sections.

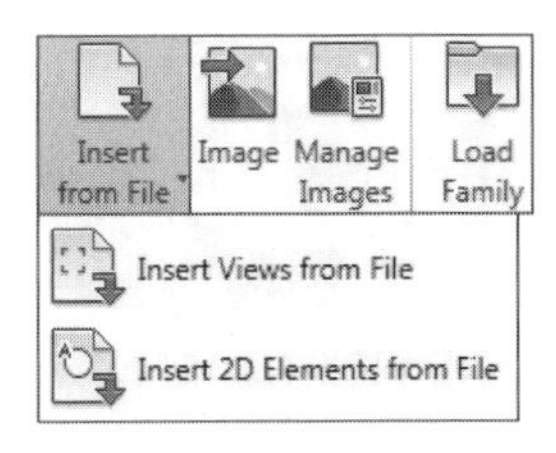

Insert 2D Elements

The Insert 2D Elements From File option allows you to bring in a detail without having to load the drafting view into your project. This is the preferred method if you have created a separate drafting view for each detail. When you choose this option, a dialog box appears for you to browse to the file containing the detail. When you open the file, the Insert 2D Elements dialog box appears, as shown in Figure 6.14.

In this dialog box you can select the view that contains the detail you want to place into your project. When you click OK to load the detail, you may get a dialog box alerting you of duplicate types within the detail file and your project file. The types defined in your project will be used. When you click OK, the detail is ready for placement in the current view of your project.

You should begin this process with a drafting view open in your drawing area. Because the detail is coming from a drafting view, you cannot place it directly onto a sheet using this process. To get the detail onto a sheet, it must first be placed onto a drafting view, and then the view can be placed onto a sheet.

FIGURE 6.14
Insert 2D Elements dialog box

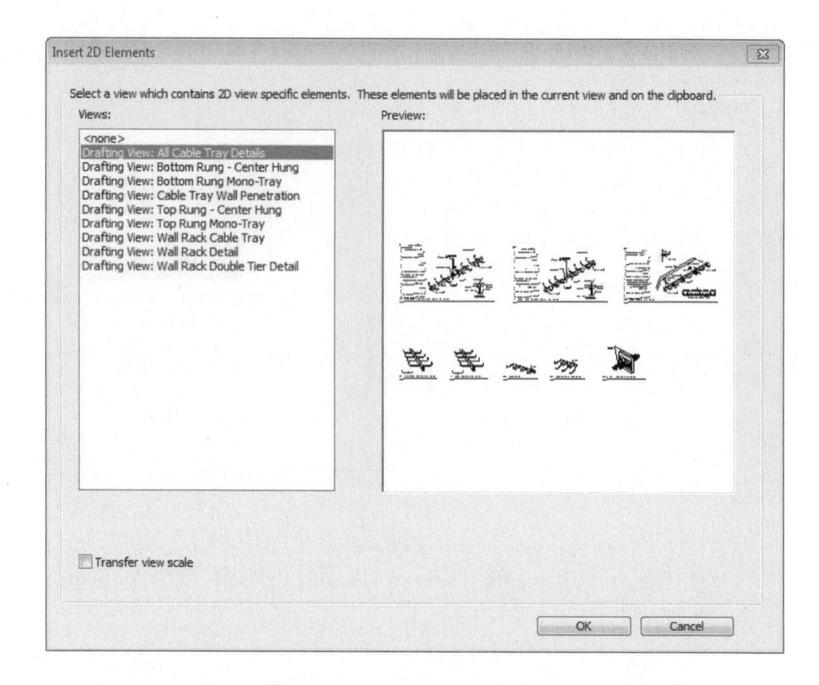

These elements can also be pasted into a callout of a plan or section so the detail relates directly to the model. In the properties of the view, the Display Model parameter can be set to Do Not Display.

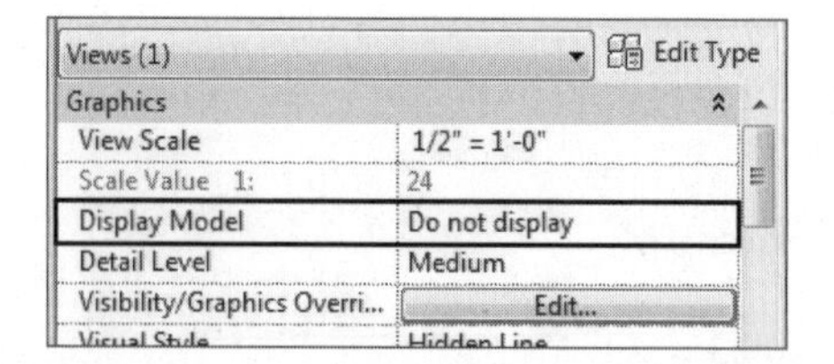

ORGANIZING DETAILS ON A DRAFTING VIEW

You can limit the number of drafting views required in your project by grouping details that will appear on a sheet into drafting views. This reduces the need for several drafting views on one sheet but may require some planning of sheet layouts prior to sheet creation.

When the detail is placed into your drafting view, it is just as if you drew the detail manually in the view. You can edit the lines and text if necessary to modify the detail for your specific project. If you change the detail significantly enough that you want to keep it as a new detail, you can copy and paste it into a new drafting view in the detail file where it came from.

Insert Views

The Insert Views From File option will load selected views from a chosen file. You should use this method if you want to have the drafting view in your project. Choosing this option brings up a dialog box where you can browse to the file containing the view you want to load. This does not have to be a file located in your detail library. For example, if you know of a drafting view in another project that has details on it that you want for your project, you can browse to that project and choose the drafting view from it.

The Insert Views dialog box that appears when you open a file shows any views available for loading into your project. You can select multiple views including drafting views, schedules, and reports to be added to your project. The Preview Selection option in the lower-left corner enables you to see the entire contents of the view prior to loading it. When you click OK in this dialog box, you may get another dialog box indicating that there are duplicate types between the file and your project and that your project types will be used. When you click OK, the selected views will be added to your project. The views will appear in the Project Browser based on their properties and your browser organization, so it is a good idea to give your detail library files the same view properties that you use in your projects. This will make the views easy to locate when they are inserted and will save you time in organizing your Project Browser.

Real World Scenario

IDEAS FOR DOWNTIME

Steve works for an MEP firm that is currently implementing Revit MEP 2011 into his office. The firm has determined that the best course of action is to train its personnel immediately prior to the kickoff of their first Revit project.

The training has taken place; however, there has been a change in the schedule of the project. Work on the project will not begin for a month, and Steve is worried that he will lose the knowledge he acquired during training.

One way he can maintain his familiarity with Revit MEP 2011 is to begin converting his vast library of CAD details into Revit format, because his company has invested in its implementation with the intention to use Revit MEP exclusively in the future.

By setting up a Revit detail library, Steve has been able to work with Revit MEP 2011 on a daily basis, keeping him familiar with the interface and some of the functionality so that when the project begins, he will not have forgotten all of his training. An additional benefit is that he will not need to rely heavily on CAD details for his projects.

Model Detail Views

Many project details are independent drawings that represent the construction of a component, but some details are taken from the design to further enhance the level of information given in plan, section, or elevation views. Details are often used to convey the installation of components rather than just to provide information about the component. It is important to a coordinated design to be able to convey the intent of placement of building objects and their relationship with other components.

You can use detail views in your Revit projects to provide additional information about the model that is not part of the model as a whole. For example, you may have a condition where a pipe is run in the space of a column wrap at several locations throughout the model. You may choose to show this one time in a detail view and then denote where it occurs in each floor plan, instead of modeling it at each occurrence. Unlike drafting views, detail views allow you to show model elements. Using detail views can help you provide the necessary level of information for the project without weighing down the model and spending time on repetitive modeling tasks.

Plan Callouts

The Callout tool on the View tab is one way to create a detail view from an area of your model. When you click the Callout button, you can choose from the Type Selector in the Properties palette what type of callout to create, either a Floor Plan callout that is used for enlarged plans or to isolate a specific area of a plan a Detail callout that creates a detail view.

Detail views are unique types of views in Revit. They have properties similar to plan, section, and elevation views, but they also have settings specific to the area of the model being shown. Detail views will take on the discipline of the view in which they are drawn. The Show In Parameter option allows you to display the callout in the parent view only, which is the view where the callout is initially drawn, or in any intersecting views, which means the callout will display in any views that show the area of the plan that the callout is taken. Unlike a regular floor plan view, you cannot adjust the View Range settings of a detail view.

A detail view will show the portion of the model within the boundaries of the callout box. This box determines the crop region within the detail view, as shown in Figure 6.15.

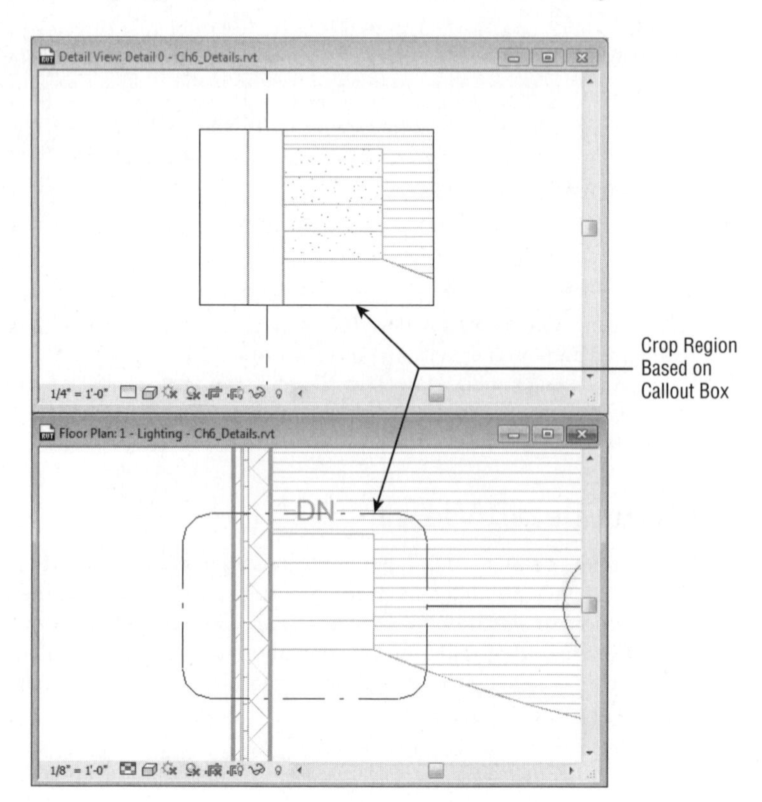

FIGURE 6.15
Callout box and crop region

A detail view taken from plan, elevation, or section will show any grid lines or levels that cross within the boundaries of the callout to help with location information. These datum annotation elements can be turned off if necessary, but you cannot edit them in a detail view. They will move to maintain their relationship to the crop region when the crop region is adjusted.

When you create a detail view of a floor plan area that contains objects represented by nested annotation symbols, the annotation symbols will not display in the detail view. This may require you to set the visibility of the actual model graphics in your families so that they can be seen at certain Detail Level settings. Figure 6.16 shows a floor box in plan view and the detail view of the same location. Notice that the annotation symbol does not show in the detail view, but the actual floor box is shown because the detail level of the detail view is set to Fine.

FIGURE 6.16
Detail view

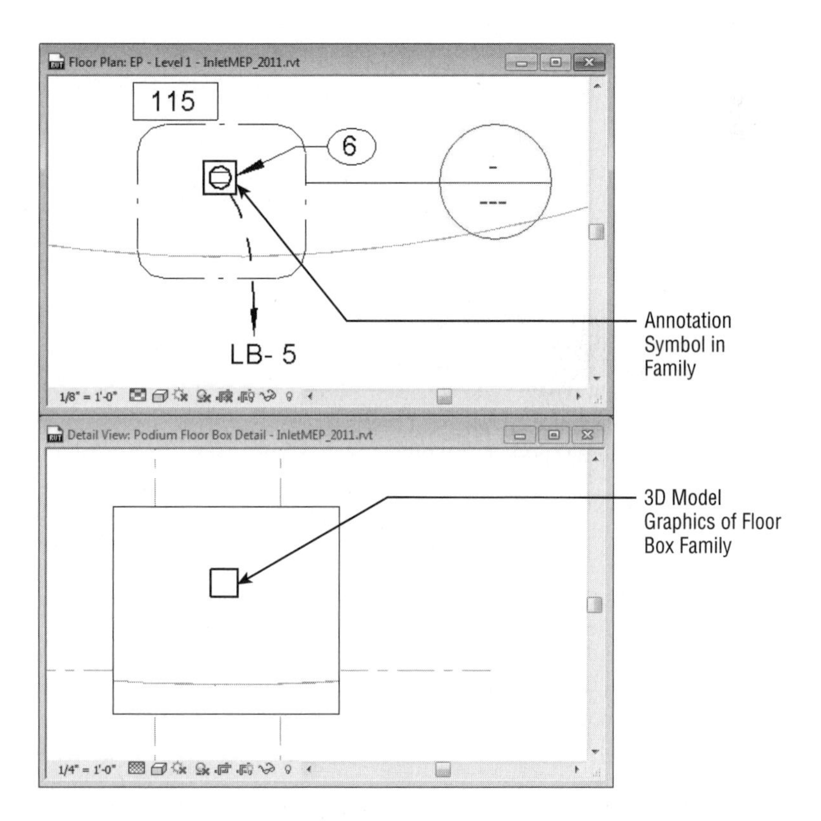

The main purpose of a detail view is to see a specific portion of the model and then add detail lines, regions, and other annotation objects to provide additional information about the design. Once a detail view has been created, you can use the same drafting tools as used in a drafting view to embellish the model information shown in the view. Filled regions can be added to show materials that may not be shown on the actual model objects. Masking regions can be used to block out items that you do not want to show in the detail. Figure 6.17 shows how the detail view of the floor box can be added to in order to provide more information about the object and its location.

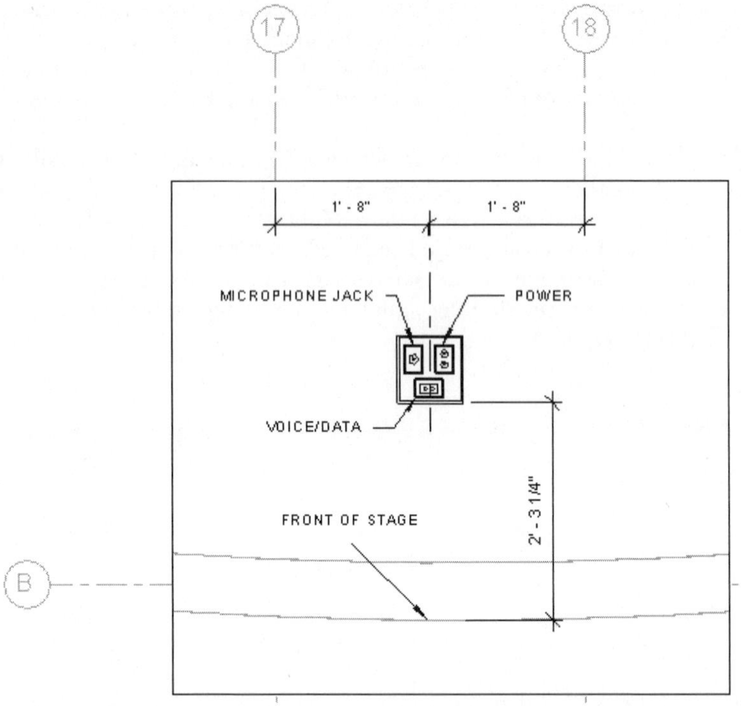

FIGURE 6.17
Detail view with added information

If you already have a detail drawn and you want to call out an object or an area of your model and reference that detail, you can use the option to reference another view. This option appears on the Options Bar when you click the Callout button on the View tab.

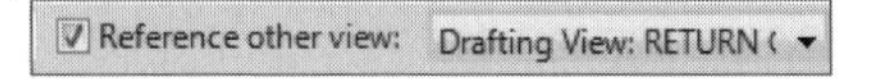

This will associate the callout with the view selected in the drop-down list next to the Reference Other View option instead of creating a new detail view. This option is very useful for details that occur in multiple locations throughout a project. Instead of creating a new detail view for each occurrence, you can create one and then reference that one view in every location required on your plans. When you delete a callout that references another view from a model view, the referenced view will remain in your project. With this functionality, you can combine your library of drafting details and callouts to convey design intent in a manner that is easily managed and coordinated.

Section Callouts

Another method for creating a detail view is to create a section view of an area of the model. One of the types of section views that can be created is a detail view, as shown in Figure 6.18. This type of section is created the same way as a building section, but the view created is a detail view instead of a model section view.

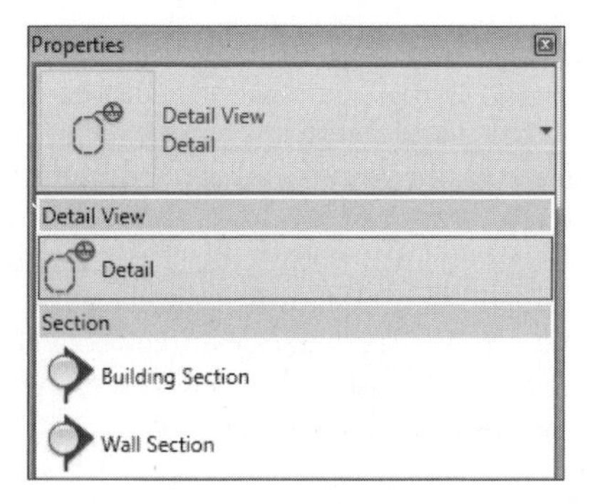

FIGURE 6.18
Section view types

Creating section detail views allows you to show project information without having to model elements that you ordinarily would not model to show in plan views. The same drafting tools used in drafting views can be used to provide additional information to the view in order to convey design conditions without having to add model components.

Section detail views can reference other views in the same way that plan callout detail views can. This allows you to create the detail drawing once and reference it as many times as needed in your project.

In some cases, you may be able to use a building section to convey more of the design intent without having to embellish the section with detail annotations. You can create different section tags to differentiate between model section and detail section marks in your views. Figure 6.19 shows how different section tags can be used for different types of sections. This will help you know what a section mark is used for when you come across it in a plan view. You can find the settings for section tags by clicking the Additional Settings button on the Manage tab.

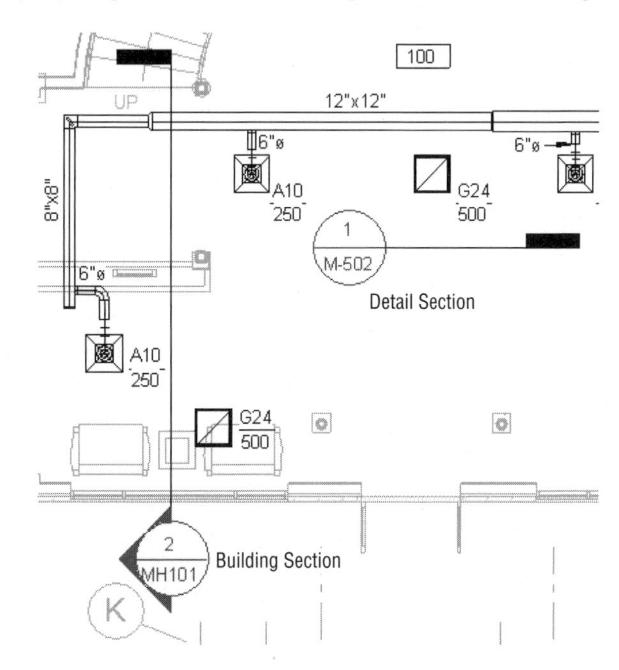

FIGURE 6.19
Section marks for different types of views

The Bottom Line

Use Revit drafting and detailing tools efficiently Revit MEP 2011 has many tools for creating the details and diagrams needed to enhance your model and provide the necessary level of information for your projects.

Master It Although the drafting tools in Revit MEP 2011 may be unfamiliar at first, learning to use them efficiently and effectively will help you spend more time focusing on design decisions instead of drafting efforts. Describe how filled regions can be used not only to display a pattern but also to provide line information in a detail.

Use CAD details in Revit projects Much of the detail information used in projects has already been drawn. When you transition to Revit, you can still use your CAD details.

Master It Using CAD details in a Revit project can be a quick way to complete your construction documents in a timely manner. However, using many CAD files for details can have a negative effect on file performance, so it is important to link CAD files whenever they are used. Explain why importing and exploding CAD files can adversely affect your project.

Build a Revit detail library Having a library of details makes it easy to save time on projects by not having to spend time drawing details that have already been created.

Master It Revit drafting views can be saved as individual files for use on projects as needed. True or false? A drafting view will be added to your project when you use the Insert 2D Graphics From View option of the Insert From File tool. Explain.

Create detail views of your Revit model Some details require the model to be shown in order to show installation or placement of objects.

Master It Callout views can be created from plan, section, and elevation views. Explain how detail views are different from drafting views.

Chapter 7

Sheets

It can be easy at first to view Revit MEP 2011 as nothing more than a modeling tool. There are many arguments about its ability to produce construction documents to the quality that we have grown used to with traditional CAD systems. Construction documents are extremely important because until we reach the day when clients and contractors are requiring virtual models as the official deliverables for project construction, drawing sheets will continue to be the standard for construction documentation.

Revit MEP 2011 has features that allow you to create construction documents from your model. You can include the schedules, details, and diagrams that you would normally provide for a coordinated and detailed set of drawings.

Many people view their construction documents as their product. They want their product to be unique to their company standards so that when others see them, they will know where they came from. You can develop standards for your construction documents that allow you to provide a unique, consistent look to your product.

Working with sheets in a Revit project is very similar to traditional CAD practices. The main difference is that the plan, section, and elevation information comes from views of your model, which allows for better coordination. Changes to the model are propagated throughout the project, so any sheets containing a view of the changed area are automatically updated. Actually, it is not that the sheet itself is updated; it is just that the view resides on the sheet. It may be best to not think of your sheets as separate from your project. They are simply the place where all the view information extracted from the model is coordinated.

Once you learn the tools available for generating a set of construction documents from your Revit model, you will see the benefits of using coordinated, computable data and 3D modeling to design your projects.

In this chapter, you will learn to

- Create a titleblock
- Establish sheets in your project
- Place views on sheets
- Print and export sheets

Creating a Titleblock

Whether you refer to it as a sheet border, drawing border, or some other name, a titleblock consists of the graphics that define the boundaries of a sheet. Some of your clients may have specific requirements for the size, shape, and information provided in the titleblock used for their projects. You may have already developed these for use in CAD projects, as well as a company titleblock for use when the client does not specify one. These CAD titleblocks can be used to develop titleblock families for use in your Revit projects. This will ensure a consistent look between your Revit sheets and CAD sheets, especially if the need should arise to combine CAD and Revit drawings in a document set.

Titleblocks are a unique family type within Revit. To create a new titleblock, you can choose one of the presized templates available in your template library. When you choose to create a new family, you can browse to the location of your titleblock family templates. Depending on which templates you chose to install with Revit MEP 2011, you will see templates available for standard sizes or one for creating your own unique size. Clicking New ➤ Title Block on the Application menu will take you to the location of your titleblock templates.

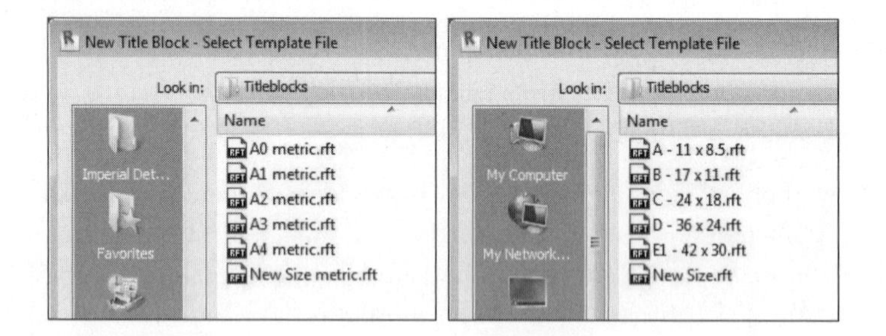

The templates for creating a titleblock contain a rectangle for defining the limits of the sheet size. If you are using the new size template, the rectangle can be modified to the required size by dragging the lines to the appropriate size. You can also select one of the lines and edit its dimension so you can input the precise size for the limits of the titleblock. If you are using a preset size template, the rectangle is already dimensioned to the appropriate size.

Using Existing CAD Graphics

Once you have chosen a template and the size has been determined, you can import a CAD titleblock for reference. When working in the Family Editor, you are not able to link a CAD file. It is important that you do not explode the imported CAD file; otherwise, all the styles in the CAD file will become Revit styles within your titleblock family. That may not sound like such a bad thing, but chances are there will be many unnecessary styles, and they will not conform to your naming standards. This excess information will be carried through to every project that the titleblock is used in. The idea here is to create a titleblock family that is in clean, native Revit format.

Click the Import CAD button on the Insert tab of the Family Editor to import your CAD titleblock. You may want to use the Preserve option for the colors in the CAD file so that you can draw an appropriate Revit line to match the line weight based on the CAD color. Align the imported CAD file with the rectangle in the family so that it fits properly, and pin it in place to avoid any accidental movement.

The Line button on the Home tab of the Family Editor allows you to draw the lines necessary to trace the imported CAD graphics. You can use the Pick Lines tool on the Draw panel of the Modify | Place Lines contextual tab to quickly generate Revit lines in the same location as the CAD graphics.

OBJECT STYLES

The templates for creating a titleblock have some predefined object styles that determine the lines used in the titleblock. You can create line styles, but they will not be available for use in the family. You must use object styles for the line work in a titleblock family. This distinguishes the lines from the line styles in your projects so that changes to line styles will not affect your titleblock. You can create any object styles you need for use in your titleblock family by clicking the Object Styles button on the Manage tab in the Family Editor.

Text and Labels

When you have finished creating Revit lines that match those of your CAD titleblock, you will also need the annotation objects that provide sheet and project information. Determine which annotation objects in the CAD file can be Revit text objects and which need to be labels. Any text that is constant from sheet to sheet and does not need to change during the course of a project can be text objects in your titleblock family. For example, if your titleblock provides information about who designed, drafted, and checked a drawing, you can place text in the titleblock as the titles for this information. Figure 7.1 shows imported CAD graphics in a titleblock family, with a text object placed in the same location as text in the CAD file.

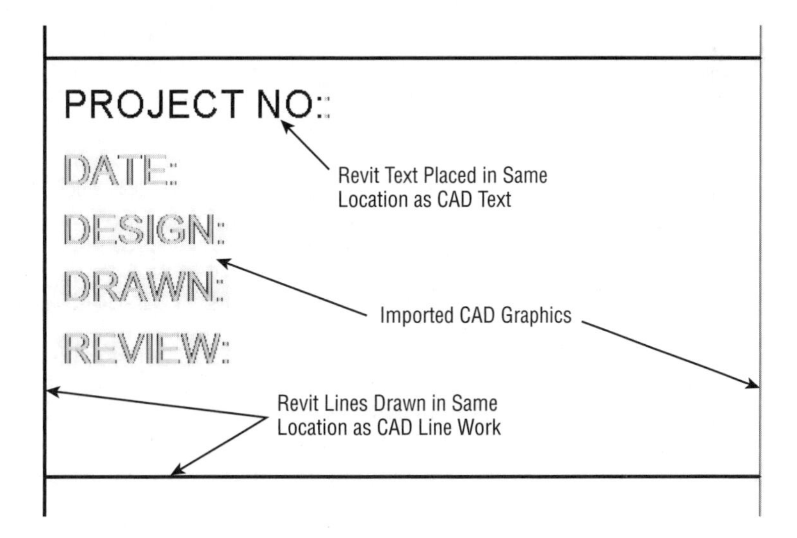

FIGURE 7.1
Revit text object used in a titleblock family

For more information about using the drafting tools available in Revit MEP 2011 and using CAD graphics, see Chapter 6.

You can create text styles within the titleblock family to match those used in the CAD version of the titleblock. Create different styles by duplicating the standard text style available when the family is created.

A titleblock typically contains information about the project in general, along with information specific to the items shown on the sheet. The key to coordinating the information that appears on your sheets is to coordinate where the information is taken from.

The most effective way to provide information on a titleblock is by using parameters. The use of parameters will allow you to make changes in a single location that will update all of your sheets in the project. Labels are used to show the information in a titleblock. Label styles should be created to match the text used in your CAD titleblock.

> **TIME FOR A CHANGE?**
>
> When you transition to doing projects in Revit, you may find that it is easier and more efficient to use some of the default settings. This is especially true with fonts. The default font used by Revit is Arial, and although this may not be in line with your standards, using it will save you time in re-creating annotations and managing them in families and projects.

Labels should be used in your titleblocks in the same way that attributes are used in CAD. Click the Label button on the Home tab to create a label. You can edit the default label style by clicking the Edit Type button in the Properties palette and then the Duplicate button in the Type Properties dialog box.

When you click in the view to place the label, the Edit Label dialog box will appear with a list of available parameters for the label. These parameters are all coded parameters that exist for either Sheet objects or project information. If one of the parameters matches the kind of information you are trying to include in your titleblock, select it from the list, and click the green arrow button to add it to the label, as shown in Figure 7.2. These parameters will be available in the properties of your project sheets or in the project information on the Manage tab. When their values are input or modified, the label in the titleblock family will update.

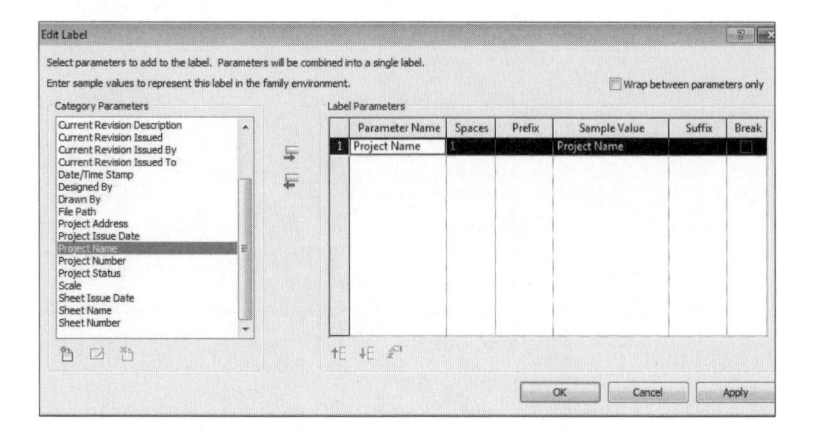

FIGURE 7.2
Adding a parameter to a label

When you exit the Edit Label dialog box, the label will be added to the view. You can edit the label in the view the same way as a text object. Clicking the label will activate grips for editing the location, rotation, and limits of the label.

If your titleblock requires a label that is not one of the coded parameters in the list in the Edit Label dialog box, you can click the Add Parameter button in the lower-left corner of the dialog box. In titleblock families, the only type of parameters that can be used are shared parameters. Click the Select button in the Parameter Properties dialog box to choose a parameter from your shared parameters file. If the parameter does not exist in your file, you can create one. Figure 7.3 shows an example of shared parameters used in titleblock families. See Chapter 19 for more information on creating and using shared parameters.

FIGURE 7.3
Shared parameters sample for titleblock

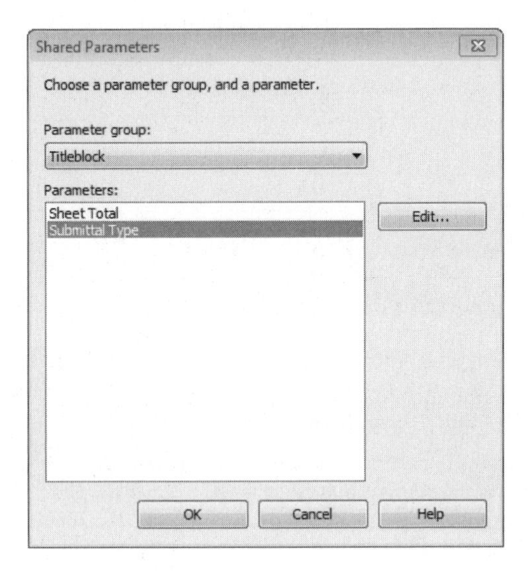

If your titleblock contains an area for a sheet revision schedule, you can create a schedule within the titleblock family that uses the parameters for revisions by clicking the Revision Schedule button located on the Create panel of the View tab in the Family Editor.

The Revision Properties dialog box looks the same as a regular schedule properties dialog box, with parameters available that are used for revisions. The schedule will automatically be populated with parameters, but you can use whichever parameters you choose. You cannot create parameters to be used in a Revision schedule. Figure 7.4 shows the default settings for a Revision schedule.

Once you have set the schedule properties, you can insert the schedule into the titleblock graphics by dragging and dropping it from the Project Browser. This is useful if your revision schedule is oriented horizontally. You cannot create a schedule that is oriented vertically in the titleblock because the schedule cannot be rotated or formatted in a vertical fashion. Unfortunately, you may need to change the way you provide revision information on your titleblocks to utilize the revision features in Revit.

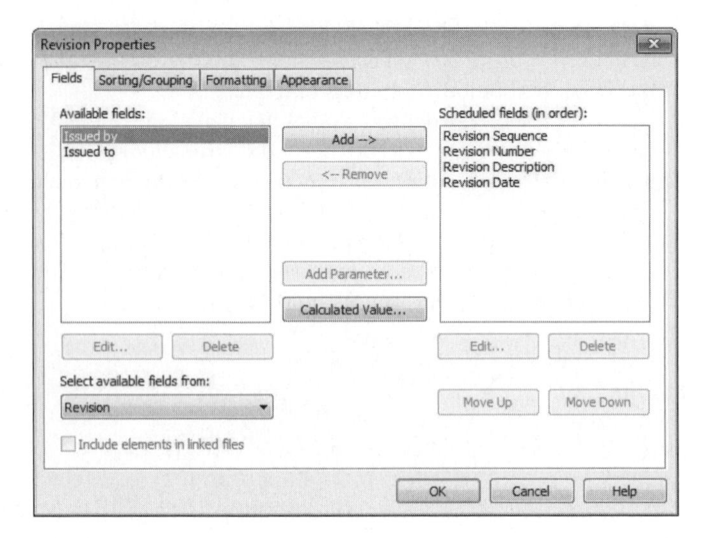

FIGURE 7.4
Revision schedule properties

Real World Scenario

VERTICAL REVISION SCHEDULE

Jo works for a company whose titleblock contains an area for revisions that is oriented vertically. To increase her efficiency in managing revisions, she has added labels to her titleblock family that are oriented in the proper location for showing revisions.

This works for her projects because when her company issues revised drawings, the only revision shown is the current one. So, she needs only one set of labels for the current revision. Otherwise, she would have multiple labels all showing the current revision information. After a revision has been issued, she can mark it as issued so that the next revision information will occupy the label, since the label parameters are for the current revision.

These are unique circumstances, and the workaround that Jo came up with is not really a schedule but a place on her titleblock to show the current revision. This method is valuable to her only as long as the current revision is all that is required. When her clients demand to see an actual schedule of revisions, her titleblock will need to be modified, or revisions will be managed manually with text.

Logos and Images

If your titleblock contains a company logo, you can include it in your Revit titleblock family as either an image or line work. Line work used to represent a logo can be made into an annotation family that can be loaded into the titleblock family. This will reduce the number of lines in the titleblock family, because many logos tend to contain numerous small lines. This also allows you to make changes to the logo without having to edit the titleblock family, except to reload the logo.

If you have an image file for your logo, you can load it into the titleblock family by clicking the Image button on the Insert tab in the Family Editor. You can resize the image to fit the titleblock graphics as necessary. Test the quality of the logo image by printing a sample sheet to ensure that the image used provides the expected results.

Working with Sheets in a Project

A Revit project is much more than a virtual 3D model with intelligent objects. All the documents that would be created for a building design in a traditional CAD environment can also be created in a Revit project. The nice thing about it is that these documents, or *sheets*, are contained in the same file with the model and its views. This allows for coordination directly with the building model and allows you to manage all the construction documents for a project in a single location.

Creating sheets in your Revit projects is a simple process of right-clicking the Sheets heading in the Project Browser and selecting the New Sheet option. The New Sheet dialog box lists any titleblock families that are loaded into your project. You can select which titleblock to use for the sheet and click the OK button. If your titleblock has not been loaded into your project, you can click the Load button to browse for the appropriate titleblock.

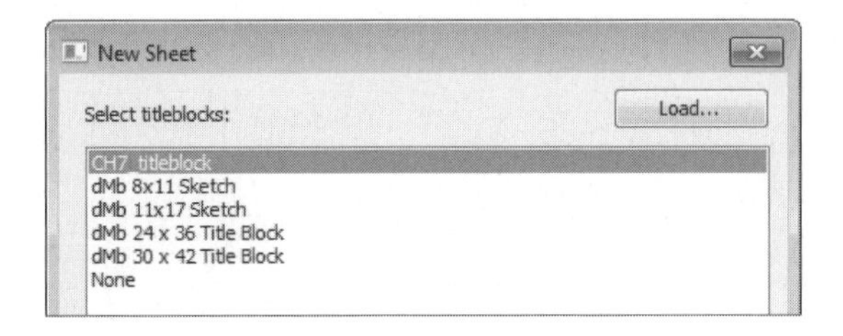

The bottom half of the dialog box allows you to select any placeholder sheets that you may have created. Creating placeholder sheets is discussed later in this chapter.

Available Titleblocks

It is a good idea to have your titleblock family loaded into your project template so that it is immediately available when sheets are created. If you have several titleblock families for various project types or clients, you can load them into your project template, and when sheets are created, the appropriate one can be selected. Be sure to delete the unused titleblock families from a project once the titleblock style has been established to avoid accidental misuse of an incorrect titleblock during sheet setup. Titleblock families are located under Annotation Symbols in the Project Browser.

Once you have chosen a titleblock and clicked OK, the sheet will be created and open in the drawing area. Depending on the parameters you have in your titleblock family and what data you may have already entered into the project, some of the information may already be filled out in the sheet. Revit will automatically apply a sheet number to the sheet starting with A101. When you change the sheet number, the next sheet you create will be the next number in the sequence. For example, if you give a sheet a number of M-1, the next sheet you create will automatically be numbered M-2.

You can change the values of the labels in the titleblock either by selecting the titleblock and editing them in the drawing area or by selecting the titleblock and changing the parameter values in the Properties palette. As with all parameters, if the label turns red when selected, that means it is a type parameter and must be edited in the type properties of the selected object.

Some of the labels may be properties of the project information and will not appear in the properties of the sheet. You can access the Project Information parameters by clicking the Project Information button located on the Manage tab.

As you give your sheets names and numbers, they will begin to appear in the Project Browser based on your sheet organization settings.

Project Browser Sheet Organization

You can organize the sheets that you create in your projects in the Project Browser for easy access and coordination. Right-click the Sheets heading in the Project Browser, and select Properties to establish the organizational structure for sheets in your project. The Type Properties dialog box will appear for the Browser – Sheets system family. There are a few preset types within the family that can be used, or you can create new types for custom organization.

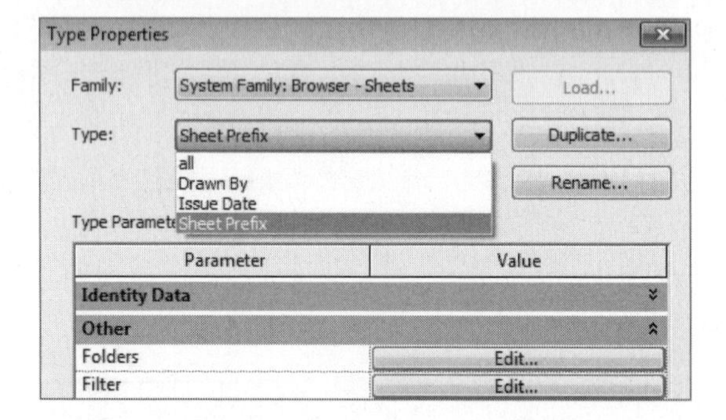

To create a new type of organization or to modify an existing one, click the Edit button in the Folders parameter. This opens the Browser Organization Properties dialog box, similar to the one that appears for organizing views. Your sheets can be grouped by parameters that apply to sheets, as shown in the example in Figure 7.5. You can apply up to three levels of grouping to organize your sheets.

FIGURE 7.5
Parameters available for sheet organization

Once you have chosen a parameter for grouping, you can specify the use of all the characters or a specific number for display in the Project Browser. The example shown in Figure 7.6 shows how the sheets appear in the Project Browser based on the settings for organizing by sheet number using one leading character. The sheet groups are listed alphabetically by default.

FIGURE 7.6: Sample sheet organization by sheet number

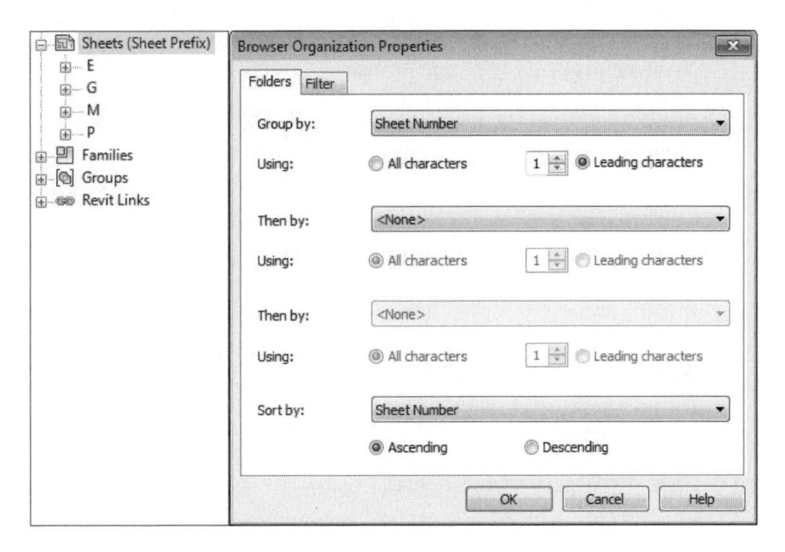

You can add other parameters to sheets so that they can be listed in the Project Browser as desired. Figure 7.7 shows that a parameter called Sheet Discipline has been created to organize the sheet order in the Project Browser.

FIGURE 7.7 Custom parameter used for sheet organization

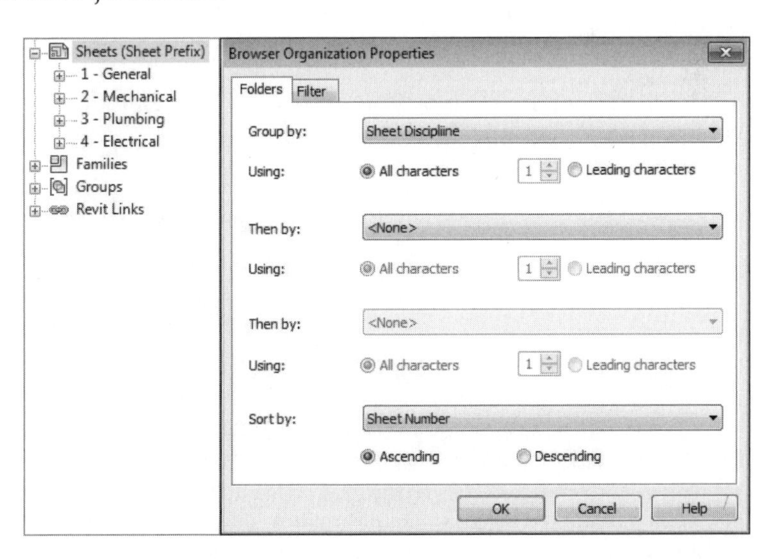

Additional levels of grouping are useful when you have a large number of sheets. Figure 7.8 shows an example of using a parameter called Sheet Sub-Discipline as a secondary level of organization. This makes it easier to find specific sheets once the sheet list becomes very large. Only one of the subdisciplines has been expanded in the Project Browser for image clarity.

FIGURE 7.8
Parameter used for two levels of sheet organization

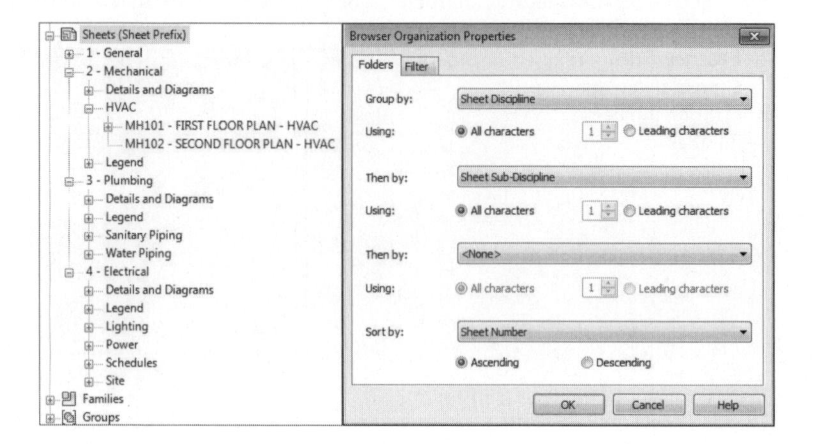

Placing Views on Sheets

In a Revit project, the model is developed to prove the constructability of the building, and views are created to display the model in a way that communicates the design. Annotations, dimensions, and detail lines are added to views to further enhance them so that the design intent is clear. These views eventually need to be put onto construction documents for official submittal of the project. You can add views to sheets by simply dragging and dropping them from the Project Browser onto a sheet that is open in the drawing area. You can even drag and drop a view onto a sheet name in the Project Browser. This will open the sheet for placement of the view.

A new feature to Revit MEP 2011 is the *guide grid*, which is a set of lines used to locate views on a sheet. Whether or not a sheet contains a guide grid is determined by the sheet properties. You can choose the grid style by using the Guide Grid parameter of the sheet properties.

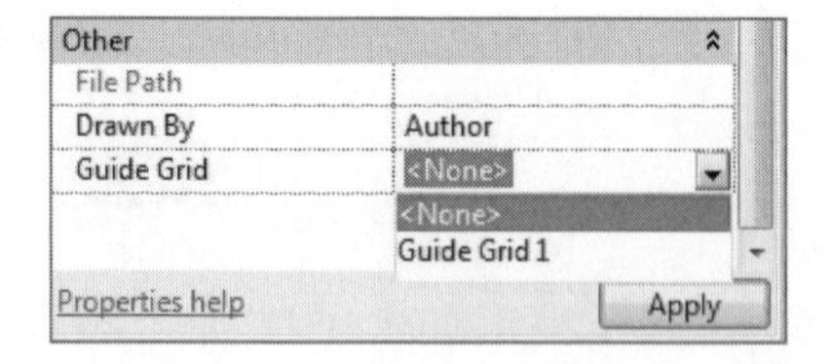

To establish the settings for a guide grid, you must have a sheet open in the drawing area. When you click the Guide Grid button located on the Sheet Composition panel of the View tab, you will be prompted to give the grid style a name. Consider using names based on the types of views placed onto the sheets or based on a unique feature of the building, such as its shape, that determines what portion of the building is shown when dependent views are used.

A standard grid of blue lines will appear over the sheet once you have given it a name. The size of the default grid is determined by the extents of your titleblock. You can select the grid by clicking one of the lines at the outer edge. This will activate grips that allow you to manually size the grid, as shown in Figure 7.9. You can move the entire grid by selecting the grid, placing your cursor over an outer edge line, and dragging it. You can also use the Move tool for more accurate movement of the grid.

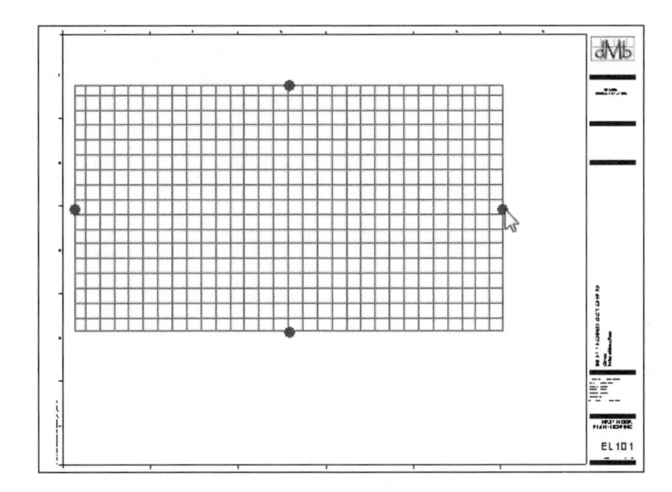

FIGURE 7.9
Grips for editing the size of a guide grid

When you select a guide grid, you can modify the spacing of the lines by accessing its properties. The Guide Spacing parameter enables you to set the distance between both the vertical and horizontal lines of the grid.

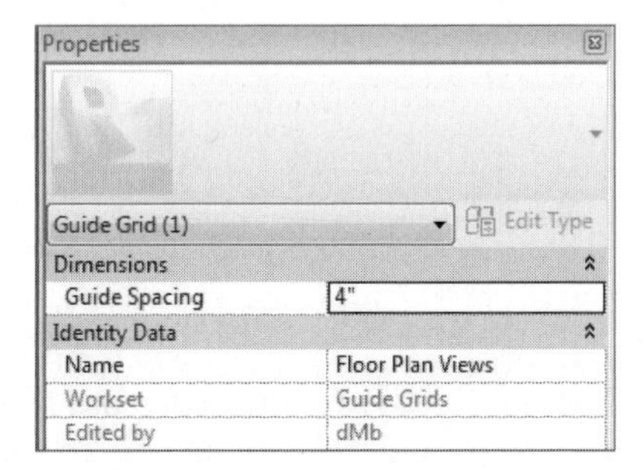

Unfortunately, you cannot snap or align to the grid lines when establishing its location on the sheet, and it will not snap to graphics of your titleblock. However, you can zoom in very close and drag the edges to the desired locations. There really should be no need to move the entire grid. Establish the spacing first, and if the grid must be moved, use the Move tool for accuracy. The key is that once you have established a grid style, it will appear in the same location on each sheet to which you apply it.

When you drag a view from the Project Browser onto a sheet, you click to place it on the sheet. It will not snap to the guide grid at this time. You can then select the view and use the Move tool to align it to the grid by selecting a point on a datum object such as a column grid line and moving it to one of the guide grid lines. The view will snap to the guide grid line when a point on a datum object is selected.

> **DATUM OBJECTS**
>
> Datum objects are annotation objects that can help determine the size and shape of the building and can be associated with model elements. Grid lines and levels are datum elements. If you are linking in architectural and structural models and using their grid lines or levels, you will not be able to snap them to your guide grid lines. You must have datum elements in your view to align to the guide grid.

Figure 7.10 shows how a view would be moved from a point on a column grid line in order to line up with a guide grid line.

FIGURE 7.10
Moving a view to a guide grid line

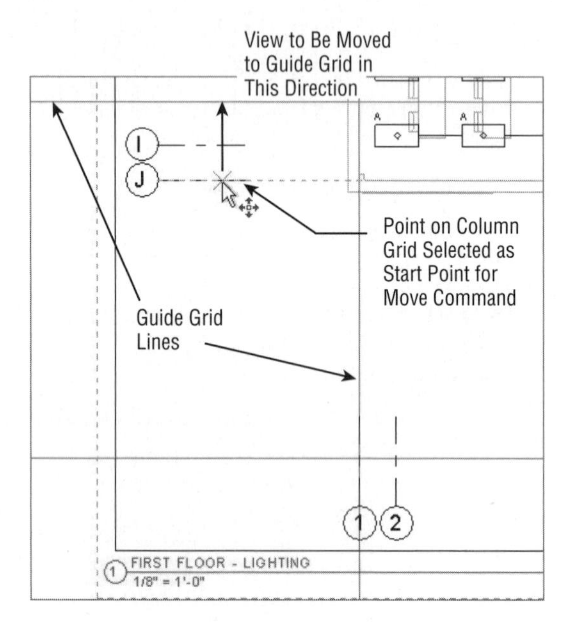

Guide grid lines will not print when they are visible in a sheet. You can control their visibility either by using the parameter in the sheet properties or by turning of the category in the Visibility/Graphics Overrides settings for the sheet. The Guide Grids category is located on the Annotations tab of the Visibility/Graphics Overrides dialog box.

When you place a view onto a sheet, you have the option to rotate it by selecting a rotation direction from the drop-down that appears on the Options Bar during placement. You can change the rotation of a view after the view has been placed by the same drop-down.

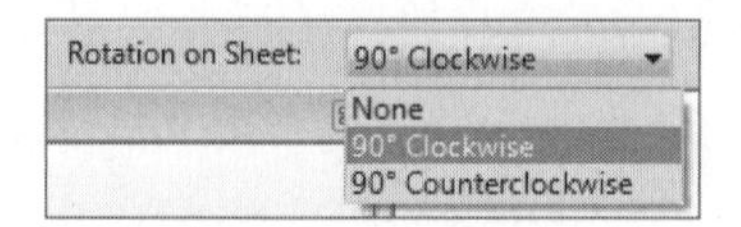

This does not allow you to continue rotating the view in the chosen direction. The settings on the drop-down are in relation to the normal orientation of the view. So, you can display the view normally, rotated 90° clockwise, or rotated 90° counterclockwise.

Viewports

Rotating a view will cause the view title to be rotated as well. The style of a view title is determined by the viewport type. Viewport types can be created so that the same kind of views can be represented with different title styles. For example, an enlarged plan is a floor plan view, but you may want to use a callout bubble for location of the view, where a normal floor plan does not require a callout bubble. Drafting views of details do not need a title if the detail title is included in the detail itself, so a viewport type without a title would be used.

When you select a viewport on a sheet, the properties that appear in the Properties palette are for the view. Clicking the Edit Type button will display the type properties for the viewport, as shown in Figure 7.11.

FIGURE 7.11
Type properties of a viewport

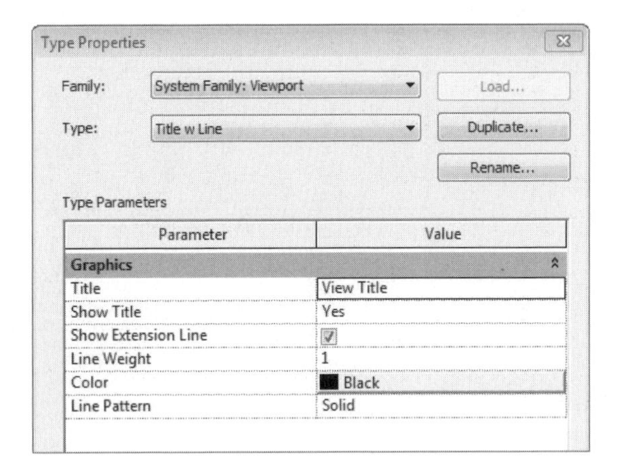

The Title parameter is where you define the annotation family that is used for the viewport title. Different view title families can be loaded into your projects for use with different viewport types. The Show Title parameter can be used to create a viewport type without a title for views that do not require one. Another choice is to show the title only when there are multiple viewports on a sheet.

TITLE FAMILIES

You can create view title families with any of the graphics you require for a view title. You do not need to include the title line in the family because it is generated by placement of the view. Labels used in a view title family can reference parameters for sheets, but you cannot create custom parameters. You should adjust the limits of the label text to allow for long titles, unless you intend for the title text to wrap, forming a multiline title.

To change the location of a viewport title, you can click the title line and drag the title to the desired location. To change the length of the title line, you must select the viewport. This will activate grips at the ends of the title line that can be dragged to change the line length.

By default, the title of a viewport will be the same as the name of the view in the Project Browser. Project views have a parameter called Title On Sheet that enables you to use a title that is different from the view name. This helps you keep your views organized as desired in the Project Browser while using the desired titles for views on your sheets, as shown in Figure 7.12.

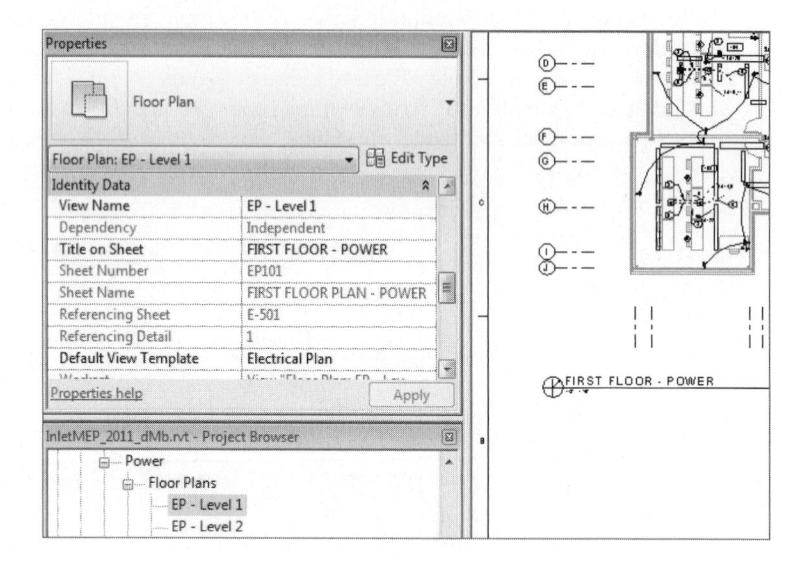

FIGURE 7.12 View properties for title when view is placed on a sheet

Annotations

Annotation objects can be added directly to sheets. You can place items such as north arrows, graphic scales, and other symbols directly on sheets in conjunction with title lines or titleblock graphics. To place these items onto sheets, you use the Symbol button on the Annotate tab. Annotation objects on a sheet will snap to titleblock geometry, but you cannot snap to title lines or schedule graphics other than the outline of a schedule. If you have an annotation, such as a north arrow symbol that needs to be part of your view titles, it can be nested into your view title annotation family.

You can also add text directly to a sheet using the Text tool on the Annotate tab. Plan notes or general notes can be typed directly on a sheet, or if you have a set of notes that is used on every project such as general notes, you may want to put them in a drafting view so that they can be saved as a separate file or included in your project template. Another option is to put the notes in a legend view, in cases where you need the notes on more than one sheet.

VIEWS ON MULTIPLE SHEETS

Plan, section, and elevation views can be placed on only one sheet. This is also true for drafting and detail views. Schedules and legend views can be placed on multiple sheets.

Schedules

Schedules are placed on sheets in the same way as views, by dragging and dropping them from the Project Browser. When you place a schedule onto a sheet, it may not look the same as it does when the schedule view is open in the drawing area. The columns of a schedule can be resized on a sheet using the triangular grips that appear at the top of each column separation line. The height of each row will adjust to accommodate the amount of text within a cell when the column width is modified. The small grip that appears at the right side of a schedule allows you to split the schedule if it is too long to fit on a sheet. Figure 7.13 shows that when you split a schedule, the headings are applied to each section.

FIGURE 7.13
Splitting a schedule

VAV SCHEDULE			
MARK	INLET SIZE	AIRFLOW	VOLTAGE
VAV-1	8"	1440 CFM	208 V
VAV-2	8"	1420 CFM	208 V
VAV-3	8"	1090 CFM	208 V
VAV-4	8"	1230 CFM	208 V
VAV-5	8"	1150 CFM	208 V
VAV-6	8"	690 CFM	208 V
VAV-7	8"	3180 CFM	208 V
VAV-8	8"	1800 CFM	208 V
VAV-9	8"	1500 CFM	208 V
VAV-10	8"	2000 CFM	208 V
VAV-11	8"	1020 CFM	208 V
VAV-12	8"	1650 CFM	208 V
VAV-13	8"	1090 CFM	208 V

VAV SCHEDULE			
MARK	INLET SIZE	AIRFLOW	VOLTAGE
VAV-1	8"	1440 CFM	208 V
VAV-2	8"	1420 CFM	208 V
VAV-3	8"	1090 CFM	208 V
VAV-4	8"	1230 CFM	208 V
VAV-5	8"	1150 CFM	208 V
VAV-6	8"	690 CFM	208 V
VAV-7	8"	3180 CFM	208 V

VAV SCHEDULE			
MARK	INLET SIZE	AIRFLOW	VOLTAGE
VAV-8	8"	1800 CFM	208 V
VAV-9	8"	1500 CFM	208 V
VAV-10	8"	2000 CFM	208 V
VAV-11	8"	1020 CFM	208 V
VAV-12	8"	1650 CFM	208 V
VAV-13	8"	1090 CFM	208 V

A schedule can be split multiple times if necessary. The dot grip at the bottom of the first section of the schedule can be dragged up or down to control at which row the split occurs. To put a split schedule back together, click and drag one of the blue arrow grips that appears in the center of a section onto another section of the schedule.

Schedules can be snapped to sheet guide grid lines by using the Move tool and snapping to the outline graphics of the schedule as the starting point. This can be very useful for schedules that appear on multiple sheets, such as a Room schedule.

Sheet Lists

A *sheet list* is a special kind of schedule that can be used to display project sheet information on a title sheet or anywhere within your project. You create sheet lists in the same way as other schedules, and the available parameters apply to sheets. You can also use a sheet list to keep track of the information that is required on your sheets.

When you begin a project, you can create a sheet list and input projected sheets to estimate the contents of your construction document set. This is done by creating placeholder sheets in the sheet list. A placeholder sheet can be created by adding a new row to the schedule. With the sheet list open in the drawing area, click the New button on the Rows panel to add a row. If you have a filter applied to your sheet list, the new rows may not appear if they do not meet the filter requirements. Creating a "working" sheet list that shows all sheets and information is a useful tool for sheet management. Another sheet list can be created to be placed on a title sheet for your construction documents.

The new row added to your sheet list will have a sheet number that is the next in sequence with the last sheet that was created. Adding a row to the schedule does not create an actual sheet object, so the placeholder sheet will not appear in the Project Browser. Any of the parameters used in your sheet list can be given values for the placeholder sheet.

When the time comes to create the actual sheet that the placeholder represents, you can choose the placeholder sheet from the New Sheet dialog box, as shown in Figure 7.14. You must also choose the titleblock option for the sheet to be created.

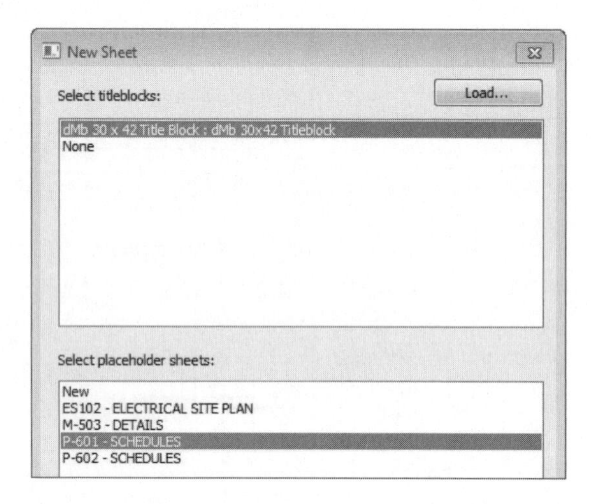

FIGURE 7.14 Choosing placeholder sheets in the New Sheet dialog box

When you have chosen the titleblock and placeholder sheet, clicking OK in the New Sheet dialog box will create the sheet. All the information that was entered into the sheet list for the placeholder sheet will populate the appropriate labels in the titleblock of the sheet. The sheet will now also appear in the Project Browser.

Using placeholder sheets is a good way to keep track of consultant sheets that are not part of your project file. For example, if you are working with a civil engineering consultant who is not creating a Revit file, you can create the civil drawings as placeholder sheets in your project to keep an accurate sheet count and sheet list.

Being able to create your project documents directly in the file that contains your model is one of the benefits of using Revit MEP 2011. Now that you have learned some of the skills required to create sheets, practice by completing the following exercise:

1. Open the `Ch_7 Project.rvt` file found at `www.wiley.com/go/masteringrevitmep2011`.
2. Right-click Sheets in the Project Browser. Select the New Sheet option.

3. Choose the D 22 × 34 Horizontal titleblock in the New Sheet dialog box, and click OK.
4. Select the titleblock, and in the Properties palette change Sheet Name to **FIRST FLOOR PLAN – HVAC**. Change Sheet Number to **M-101**. Click the Apply button in the Properties palette. Notice that the titleblock information is updated.
5. Right-click Sheets in the Project Browser, and select the New Sheet option. Choose the D 22 × 34 Horizontal titleblock, and click OK.
6. Repeat step 4, naming the sheet **SECOND FLOOR PLAN – HVAC**. You do not have to edit the Sheet Number option because it was created automatically in sequence with the previous sheet.
7. With the M-101 sheet open in the drawing area, click the Guide Grid button on the Sheet Composition panel of the View tab. Name the guide grid **Floor Plans, and click OK**.
8. Select the guide grid by clicking one of its outer edges. In the Properties palette, change the Guide Spacing parameter value to **4"**.
9. In the Project Browser, drag the 1 – Mech floor plan view to the drawing area. Click to place the view anywhere on the sheet.
10. Select the view, and click the Move button on the Modify | Viewports contextual tab. Click the horizontal "C" grid line to start the move, and click the first horizontal guide grid line near the bottom of the titleblock. Click the Move button again, and select a point on the vertical "1" grid line to start the move. Click the first vertical guide grid line at the left of the titleblock. Be sure to move horizontally so the "C" grid line stays aligned with its guide grid line, as shown in Figure 7.15.

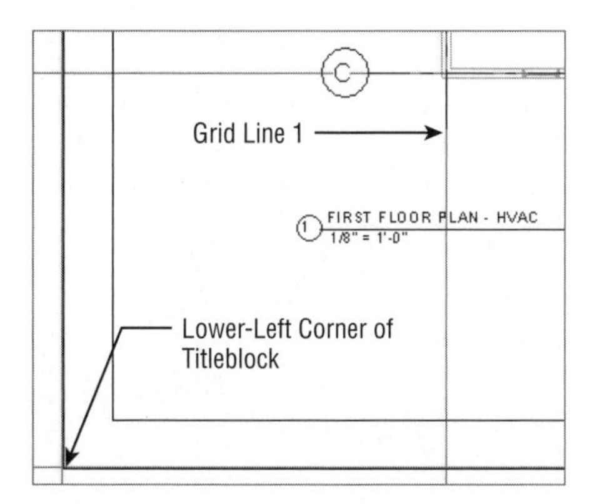

FIGURE 7.15 Alignment of grid lines to guide grid

11. Open the M-102 sheet, and drag the 2 – Mech floor plan view onto the sheet. In the Properties palette, change the Guide Grid parameter of the sheet to **Floor Plans**. Repeat step 10 to align the view to the guide grid.

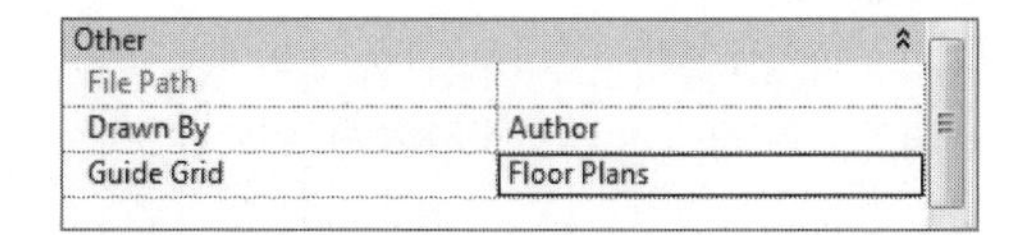

12. Open the Sheet List schedule by double-clicking it in the Project Browser. Click the New button on the Rows panel of the Modify Sheet List tab. Change Sheet Number to **M-501** and Sheet Name to **SECTIONS AND DETAILS** for the new row.

13. Right-click Sheets in the Project Browser, and select the New Sheet option. In the New Sheet dialog box, select the M-103 SECTIONS AND DETAILS placeholder sheet, and click OK.

14. Drag the Section 1 view from the Project Browser onto the sheet. Drag the DUCT DETAILS Drafting View onto the sheet from the Project Browser.

15. Select the detail view on the sheet. In the Properties palette, click the Edit Type button. In the Type Properties dialog box for the Viewport family, click the Duplicate button. Name the new family type **No Title, and click OK**. In the Type Properties dialog box, change the Title parameter to <none>, and deselect the box in the Show Extension Line parameter. Click OK in the Type Properties dialog box, and confirm that the detail viewport on the sheet has no title or line.

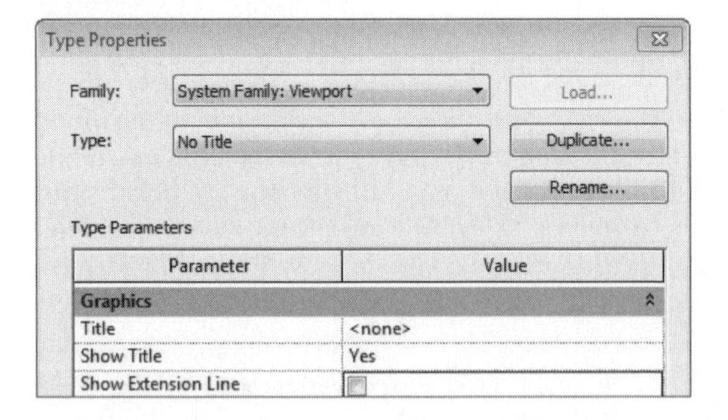

16. Right-click Sheets in the Project Browser, and select the Properties option. Click the Type drop-down, and change the type to Sheet Prefix. Click the Edit button in the Folders parameter. Set the first Group By drop-down to Sheet Number using one leading character.

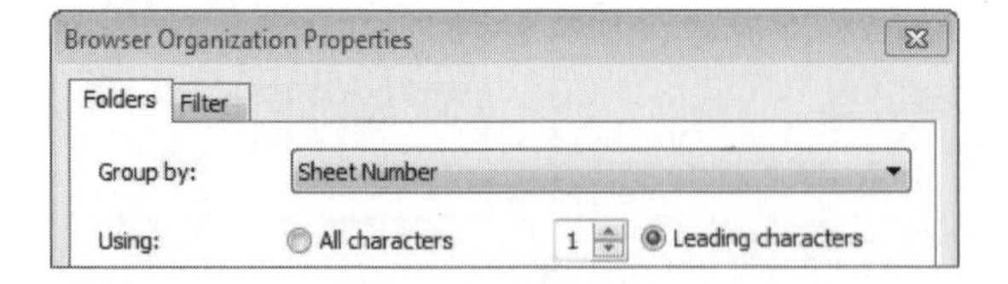

17. Click OK to close all dialog boxes. Notice how the sheets are organized in the Project Browser.

Sheet Revisions

Revisions are part of nearly every project. You can keep track of revisions to a Revit project by using a revision schedule in your titleblock families and managing the revisions with Revit.

When you make a change to your model and you want to issue it as a revision, you can draw a revision cloud directly in the view, or the cloud can be drawn on the sheet displaying the view. Either way, you should first establish the properties of the revision. To create a revision in your project, click the Revisions button on the Sheet Composition panel of the View tab.

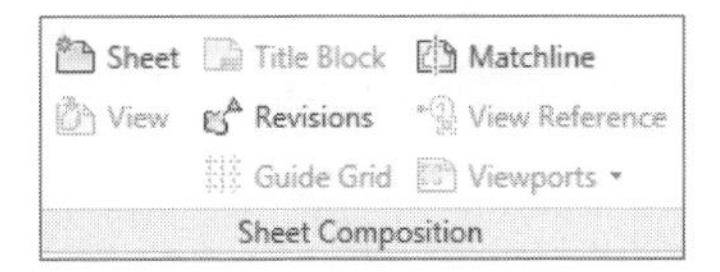

Do not confuse this with the Revision Cloud button on the Annotate tab, which is used to draw the actual cloud graphics. The Revisions button activates the Sheet Issues/Revisions dialog box where you can define the revision and establish the behavior of the revision schedule and graphics. Figure 7.16 shows the dialog box. Notice the available settings for the numbering, which can be numeric or alphabetic.

FIGURE 7.16
Sheet Issues/Revisions dialog box

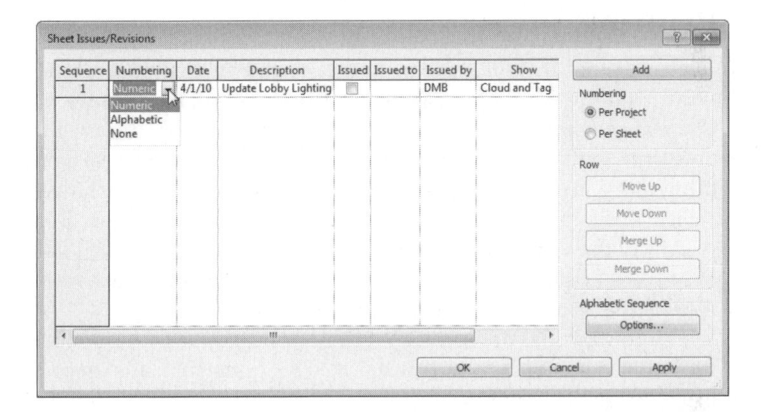

The settings on the right side of the dialog box allow you to number your revisions by sheet or by the entire project. Clicking the Add button in the upper-right corner will create a new revision sequence. The Move buttons in the Row section enable you to move a revision up or down the list to change the order in which they appear in the schedule. The Merge buttons allow you to combine the contents of one revision with another. Once you have created a revision sequence, it cannot be deleted. You must merge it with another revision. The Alphabetic Sequence setting can be defined to eliminate unwanted letters or establish a custom numbering system.

With the settings established for a revision, you can use the Revision Cloud button on the Detail panel of the Annotate tab to draw a cloud around the modified area. The properties of the cloud can be used to determine which revision sequence the cloud is a part of. Figure 7.17 shows a revision cloud that was drawn in a floor plan view and the resulting revision schedule on the sheet where the view resides. Notice that the cloud has been tagged using a revision tag that reports the revision number.

You can use the Issued parameter in the Sheet Issues/Revisions dialog box to establish that a revision has been issued. When a revision sequence has been marked as issued, you cannot add revision clouds to it. Sorting your revision schedule by sequence is a good way to organize your revisions. This allows you to use a custom numbering system without having to sort by it.

The visibility of revision clouds and their tags can be controlled via the Visibility/Graphic Overrides dialog box. Their categories are located on the Annotation Categories tab of the dialog box.

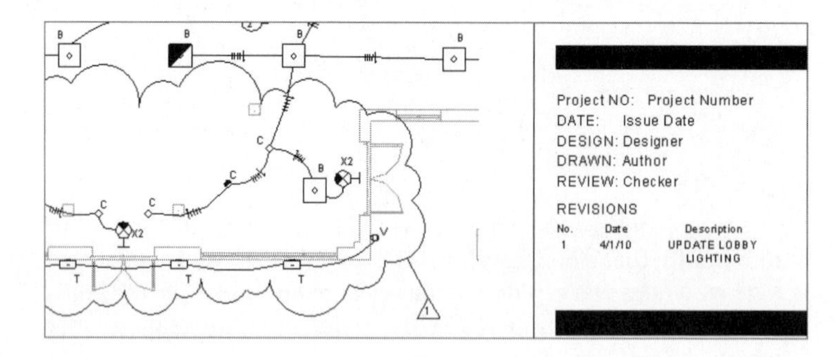

FIGURE 7.17
Revision cloud in a view

Printing Sheets

Though Revit MEP 2011 has many means and features that allow you to verify the coordination of your design and documents, printing is still a necessary part of a Revit project workflow. There are so many different types of printers and print drivers that it would be difficult, if not impossible, to describe print settings that would work for everyone on every project. The best thing that you can do to make your printing tasks easier and more efficient is to take a sample project and experiment with different print settings until the desired results are achieved.

Any of the views in your project can be printed except for schedules. A schedule would have to be placed on a sheet so that the sheet could be printed. Schedules can be exported to `.txt` files for use in and for printing by other software.

Printing options are located on the Application menu, and clicking Print will activate the Print dialog box, shown in Figure 7.18. This dialog box allows you to establish the printer to be used, define the name and location when printing to a file, and define which views or sheets are to be printed.

Clicking the Setup button located in the Settings section activates the Print Setup dialog box shown in Figure 7.19, which can also be accessed via the Application menu. This dialog box contains settings for how the print will appear on the paper or in the file.

The Name drop-down at the top is for saved print setups. Once you have established the settings that produce the desired print quality, you can save the setup for future use by clicking the Save button in the upper-right corner.

Vector or raster processing can be used for views set to Hidden Line. You may get varying degrees of quality depending on your printer drivers, so it is best to experiment with each option to determine the best one. For the Raster Processing option, you can choose the quality in the drop-down list in the Appearance section. The Colors drop-down in this section allows you to choose an option for color prints or black lines. The Grayscale option in this drop-down will convert any color lines to their grayscale equivalent when printing to a black-and-white printer.

In the Options section at the bottom of the dialog box, you can hide certain types of objects such as reference planes and viewport crop boundaries. There are also settings to hide section, callout, or elevation marks that reference views that are not placed on a sheet. This is very useful because it eliminates the need to hide those objects individually in the views prior to printing. The option to replace any halftone lines with thin ones is a useful setting if your printer driver causes halftone lines to be too faint when printed.

FIGURE 7.18
Print dialog box

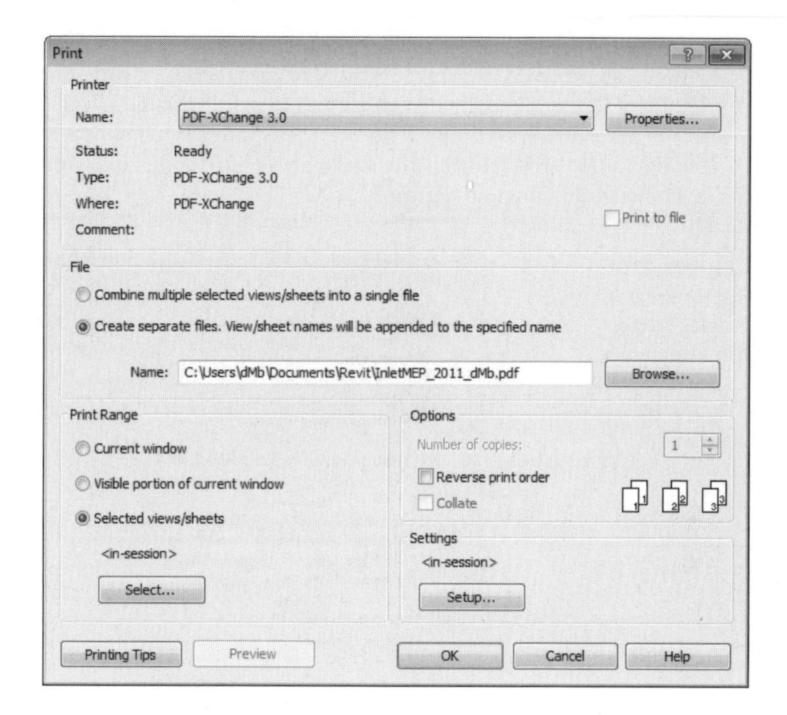

FIGURE 7.19
Print Setup dialog box

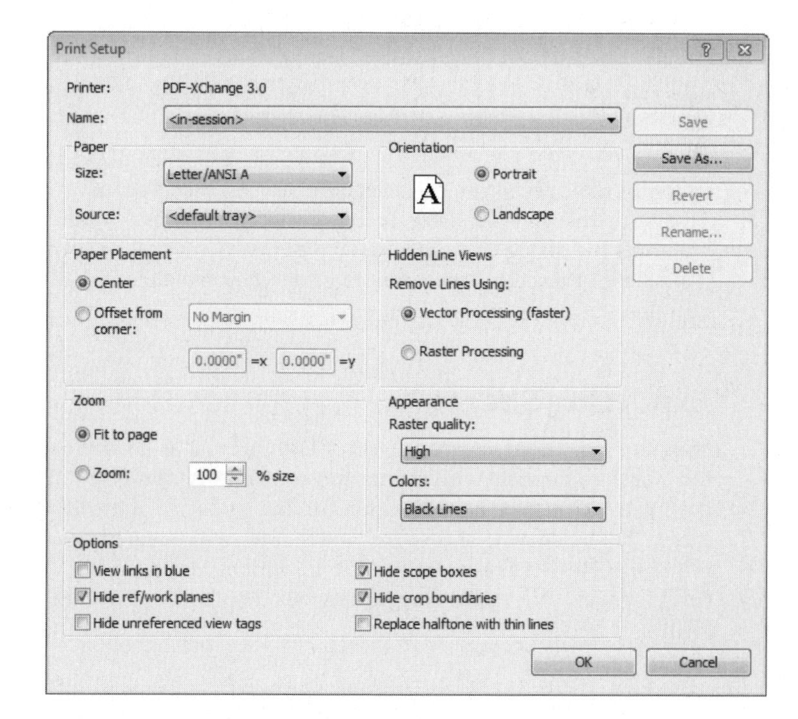

With the settings established, clicking OK will return you to the Print dialog box (if that is where you accessed the Print Settings from) where you can determine what is to be printed. In the Print Range section, you can choose to print what is currently visible in the drawing area or the currently open view, or you can select views and sheets to be printed. Clicking the Select button in this area activates the View/Sheet Set dialog box, which lists all the printable views and sheets, as shown in Figure 7.20.

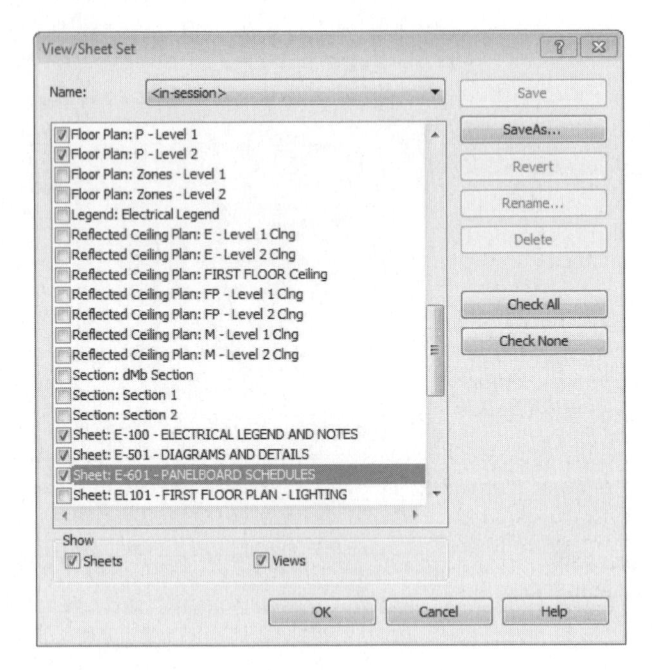

FIGURE 7.20
View/Sheet Set dialog box for printing

In this dialog box, you can select the desired sheets or views to be printed. You can filter the list of available views and sheets to show only views or only sheets using the check boxes at the bottom of the dialog box. Once you have chosen a set of views or sheets, you can save the selection set for future use. The Name drop-down at the top of the dialog box will list all the saved selection sets.

Exporting Sheets

You can export your sheets to CAD or DWF format as an alternative to printing directly from Revit or to collaborate with clients and consultants who may not have the ability to view Revit files. You can find the Export options on the Application menu. There are many options for export, but for sharing sheets, the two that we will focus on are exporting to CAD and DWF. Selecting CAD Formats from the Export options and choosing a file format activates the Export CAD Formats – Views/Settings dialog box. Figure 7.21 shows this dialog box for the DWG file format.

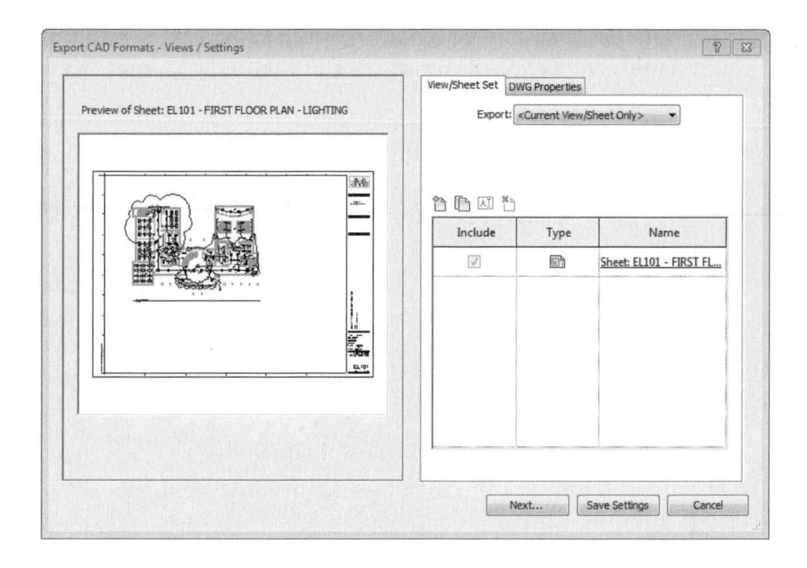

FIGURE 7.21
Export CAD Formats – Views/Settings dialog box

> **EXPORT SETTINGS**
>
> It is important to establish the settings for exporting to CAD formats so that your files will be accurate when translated to the chosen CAD file format. See Chapter 2 for more information on establishing settings for exporting to CAD.

The preview area of the dialog box shows the current view open in the drawing area. The right side of the dialog box is for establishing which sheets or views are to be exported. You can create a set of views and/or sheets by clicking the New Set button. Once you give the set a name, all the printable views and sheets will appear in the list. The drop-down at the top of the right side enables you to filter the list for specific types of views or sheets.

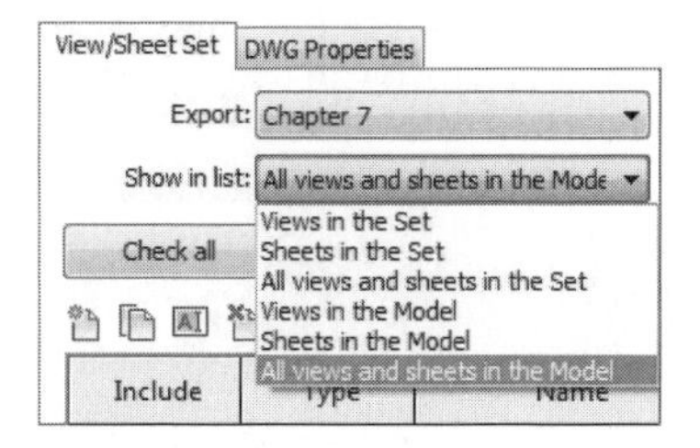

You can use the check boxes in the Include column of the list to determine which sheets or views will be exported. Once the list is established, you can click the Save Settings button so that the next time you access the dialog box it will return to the saved settings.

The DWG Properties tab contains settings for behavior of the CAD entities created by the export. You can establish how objects are colored, what line type scaling to use, and the units and coordinate system. The small box next to the Layers And Properties drop-down accesses the settings for exporting layers to determine what layers will be used for the Revit categories.

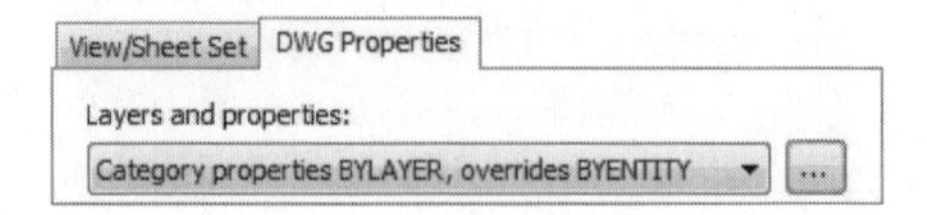

Once you have established the settings for export, clicking the Next button will allow you to browse to the location for the exported CAD file(s). The check box at the bottom of the window will create separate `.dwg` files for each view that is on a sheet when exporting Revit sheets. You can specify a prefix to the names of the files and the file type also.

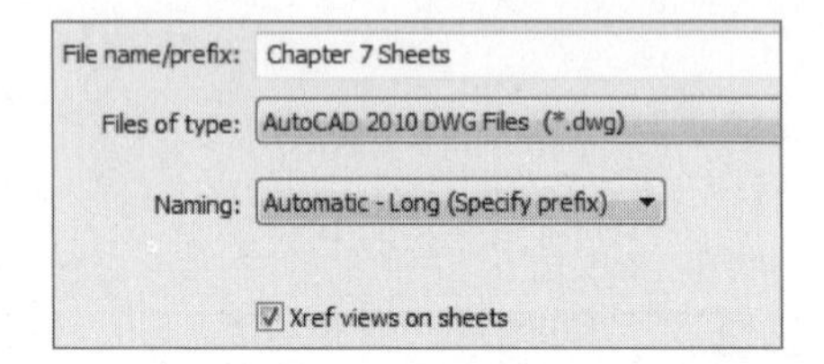

The option to export sheets or views to DWF works in the same way as exporting to CAD except additional options are available because of the capabilities of a DWF file to report element properties. In the DWF Export Settings – Views/Settings dialog box, the DWF Properties tab allows you to export the properties of your Revit model elements and to access print settings prior to export. The Project Information tab gives you direct access to the project information contained in your Revit project. You can edit the parameters in this dialog box to update any project information, such as issue date, prior to export.

When you click the Next button, you can browse to a location for the `.dwf` file. There is an option in the browse window to combine all the views and/or sheets to be exported into a single `.dwf` file.

If you are sharing your sheets or views with someone who is unable to open a `.dwf` file, you can export to DWFx format, which can be viewed using the free Microsoft XPS Viewer, available for download from Microsoft. The viewer is not able to view files containing 3D graphics, so it should be used only for 2D files.

The Bottom Line

Create a titleblock A titleblock can be the signature for your projects. Its design and layout can be an immediate indicator as to who has created the construction documents. A titleblock is also important for conveying general project and specific sheet information.

Master It To ensure that your Revit projects look the same as, or similar to, your other projects, it is necessary to have a titleblock family that looks the same as other file format titleblocks you use. Describe the process for creating a Revit titleblock family from an existing CAD format titleblock.

Establish sheets in your project The sheets that make up your construction documents can be organized in your Revit projects for easy access and for management of project information.

Master It A Sheet List schedule is a very useful tool for managing the information shown on your construction documents as well as for organizing the order of sheets for your project. Is it possible to create parameters for sheets that can be used in the sheet list? Explain.

Place views on sheets For a Revit project, the construction documents are created as a result of the model, whereas in traditional CAD environments the sheets are the main focus. You can put your construction documents together by placing the views you have created onto your sheets.

Master it Uniformity among sheets in a document set is important not only to the look of a project but also for ease in document navigation. Explain how guide grids can be used to place model views in the same location on individual sheets.

Print and export sheets Although we live in a digital age, the need to print sheets and views is still part of our daily workflow. With the ability to work with consultants from all over the world, the need to share digital information is crucial. Exporting sheets and views to a file format that is easily shareable increases our ability to collaborate with consultants.

Master It Printing sheets is often necessary for quality control checking of a project. How can you keep section and elevation marks of views that are not used on your sheets from printing?

Part 2

Revit MEP for Mechanical

Chapter 8

Creating Logical Systems

Creating and managing systems is the key to getting Revit MEP to work for you. Systems represent the transfer of information between families. They range from supply air to refrigerant to laboratory gases to anything else a building needs to operate.

In this chapter, you will learn to

- Create and manage air systems
- Create and manage piping systems
- Manage display properties of systems

Managing Systems

Systems are the logical connection between elements in the model. They are the link between the air terminal, the VAV (Variable Air Volume) box, and the air handler, and they represent an additional layer of information above the physical connections made with duct and pipe. Without systems, duct and pipe are only able to connect information between two points. Systems are needed to generate the bigger picture and allow you to manage the elements on a building-wide level. You can create systems to represent supply, return, and exhaust air as well as plumbing, fire protection, and hydronic piping. You can also create systems to represent other uses of duct and pipe outside the predefined types included with Revit MEP 2011.

Systems aid in the documentation of a model. Because elements across the entire model are linked together, tags and other properties can be managed quickly and accurately. The best example of this is using a pipe tag that includes not only the size but also the system name. Tagging any piece of pipe connected to the system, in any view, will immediately generate a complete and accurate annotation.

Why Are Systems Important?

Admittedly, it is possible to complete a project through construction documents in Revit MEP 2011 without using systems. Some workarounds are required to accurately create tags, and managing space air quantities will need to be done externally. Entering all the information needed to accurately represent the mechanical systems in a building may seem like a daunting task at first, but the benefits of having all the information in one place and directly in front of the user can lead to more accurate designs. Instead of flipping between a building load program and a duct sizing chart (or wheel) and trying to keep track of which terminal box is serving which space, systems can handle all of that for the user. By feeding Revit MEP 2011 the load information, calculated internally or externally, and creating an air system for a space, the CFM (Cubic Feet per Minute) required at each air terminal can be quickly determined with a schedule or a custom space tag. The airflow will then be assigned to the terminal box, and the space that it is serving can appear in a schedule. One program can handle all of these tasks, which the user would have to do anyway.

Using systems also carries a performance boost for Revit MEP. Even if systems are not being specifically set up, Revit MEP is using systems behind the scenes to keep track of all the information in the model. All elements get placed on default systems based on the type of connector: supply air, return air, hydronic supply, and so on. When the default system exceeds 50 elements, a warning message appears (Figure 8.1).

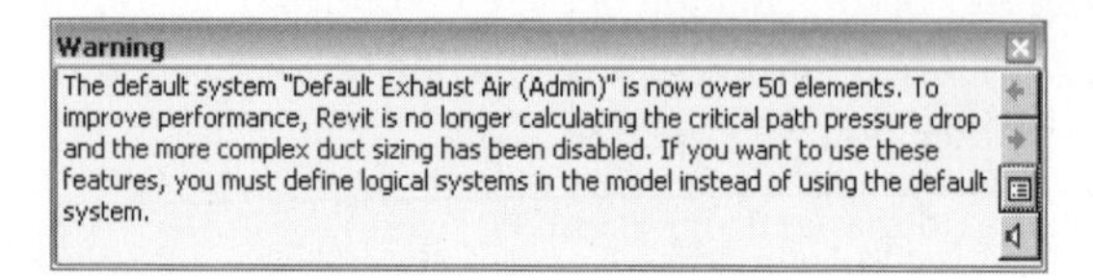

FIGURE 8.1
Fifty elements warning message

Using systems can eliminate this issue and a few others. For example, if an air terminal is not placed on a system, it cannot be rehosted to a new face. This can be a major problem at the 11th hour when an air terminal that has lost its host needs to be coordinated with the latest reflected ceiling plan.

Mechanical Settings

Before you can jump in and start creating systems, you need to set up several things to ensure systems work as they should. There is nothing wrong with the default settings, but every firm is different, and so are their standards and procedures. Most companies have developed standards over the years, and these are good guidelines to adhere to when using a new application such Revit 2011.

The Mechanical Settings dialog box, accessed from the MEP Settings panel's Manage tab contains some of the most critical settings for using systems in Revit MEP 2011 (Figure 8.2). This dialog box was briefly covered in Chapter 2, but a more in-depth look is needed so you can understand how these settings affect systems in Revit MEP 2011. All of these settings should be established in your company's project template, and changes to them should be discussed with the Revit team as well as the CAD manager because visibility and graphics can be dramatically affected by a minor change in this dialog box.

Several settings really affect systems graphically. For example, under Duct Settings ➪ Hidden Line, here you see Inside Gap, Outside Gap, and Single Line. By default, each one of these is set to 1/16″. By changing the numeric size of each one of these parameters, you can get a different look, which will help match your existing standards.

System Browser

The System Browser, shown in Figure 8.3, summarizes all the systems currently in the model. If that was all that it did, it would be a useful design tool. You could keep track of all the air and water in the building and see your system totals at a glance. But the System Browser in Revit

MEP 2011 takes this idea a step further; it is a live link to the components in the system as well as their parameters. You have full control to modify the airflows, equipment types, and diffuser selection, all from a single window.

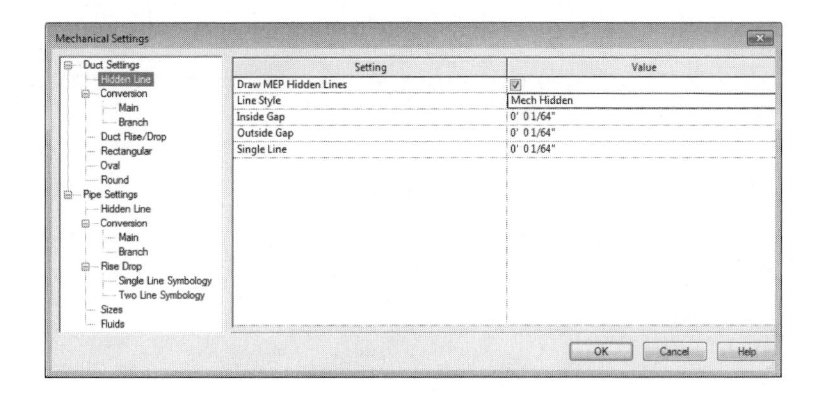

FIGURE 8.2
Mechanical Settings dialog box

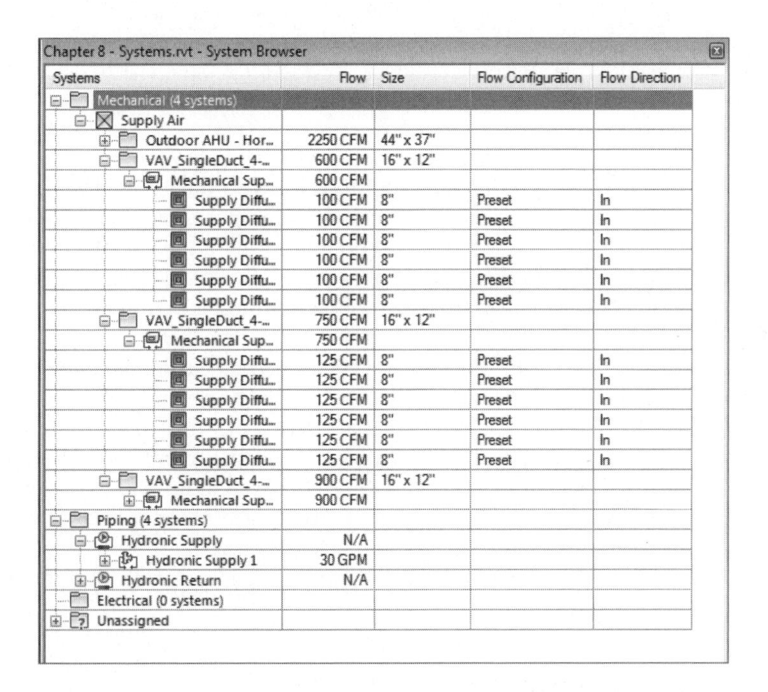

FIGURE 8.3
System Browser

You can access the System Browser from the Analyze tab or the keyboard shortcut F9. The first thing you will want to set up is the columns it displays. It can get very large, so a second monitor is very helpful. You can access the Column Settings dialog box by right-clicking in the System Browser, which gives you an expandable list of the information that can be referenced in the model (Figure 8.4).

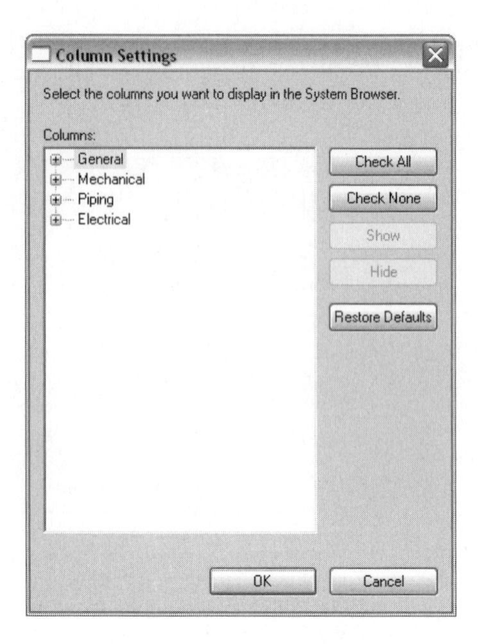

FIGURE 8.4
Column Settings dialog box

Obviously, not every parameter will be filled out for every part of the system, and some of the parameters will not be useful on a day-to-day basis. The columns you choose to display will rely on your personal preferences and how you model your systems. For example, Space Number and Space Name will populate only if the element and space touch. If spaces are bound by the ceiling and terminal boxes exist above the ceiling, they will not be associated with a space. You will need to use the Show command to find lost terminal boxes. To do this, right-click any element in the System Browser, and select Show.

Ideally, every connection on every piece of equipment would be associated with a system, and the Unassigned system category would be empty. This may not be realistic on a large job or where manufacturer content is being used. You may not model every condensate drain, but if the manufacturer has provided a connection point for it, there will be a listing in the System Browser. If your firm decides to use the System Browser to carefully monitor the systems and elements in the model, you may want to eliminate connectors that you will not be using to keep things clean.

To do this, you will have to create a duplicate family and remove the connectors you do not want to use.

Air Systems

Duct connectors in Revit MEP allow the user to connect ductwork to a family that may represent an air terminal, fan, VAV box, air handler, or chilled beam. Duct connections can also be used as a source for boiler combustion air and flues. There are many different applications of duct connections beyond a simple supply air diffuser. In this section, you will learn how to set up many different kinds of systems using duct connections.

Parameters

It takes a good understanding of the parameters besides height and width (or radius) before you can set up complex air systems. There are 14 parameters associated with a duct connection when it is not set to Fitting. Not all of the 14 are active all the time (Figure 8.5).

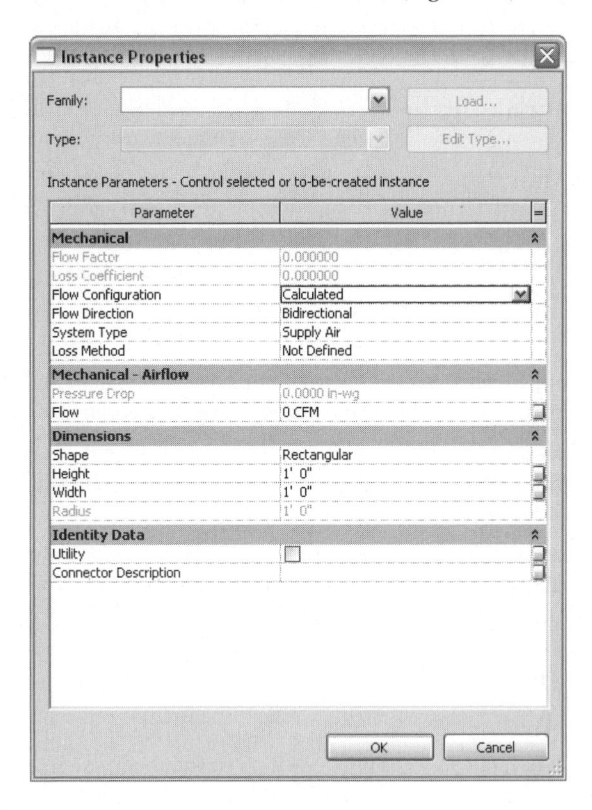

FIGURE 8.5
Parameters for duct connection

If the connector is set to Fitting, it has only six parameters. Revit MEP has several system types available for duct connections that facilitate system creation and view filters within a project (Figure 8.6).

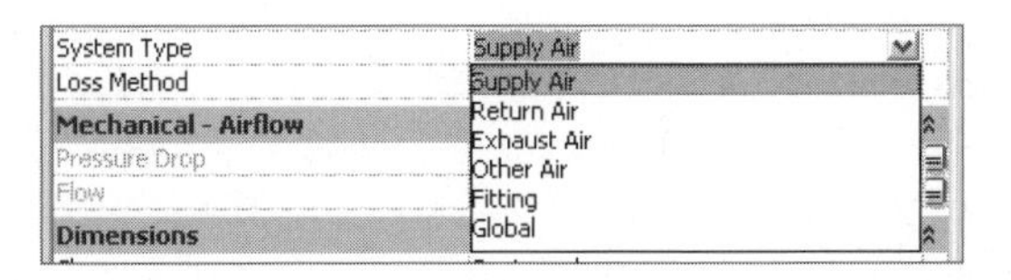

FIGURE 8.6
System types for duct connections

Starting at the top, here is the explanation of each duct connection parameter:

Flow Factor This parameter determines the percentage of the system flow that will be seen by the connector. It is available only when Flow Configuration is set to System. It is useful when using multiple devices, each of which is sized for part load.

Loss Coefficient Available only when Loss Method is set to Coefficient, this parameter is used in conjunction with the Flow parameter to determine the pressure drop.

Flow Configuration This parameter determines how the connector flow will be calculated.

Calculated This setting calculates airflow downstream of the connection and sets flow parameter to the sum of those flows. It is particularly useful for VAV boxes.

Preset No calculation is needed, and airflow is set to the Flow parameter.

System This setting is similar to Calculated, but the flow factor comes into play. It is best used for splitting the total system airflow between air handlers.

Flow Direction The flow direction can be In, Out, or Bidirectional. This direction is referring to the direction air is moving relative to the connector. For example, a supply air diffuser should be set to In because the air is flowing into the connection. An exhaust grille would be set to Out because air is coming out of the connection (to the system).

System Type Here the most appropriate system type is chosen for the application. Supply Air, Return Air, and Exhaust Air are all pretty self-explanatory, but they also have other uses.

Supply Air Air that is to be supplied to a space can also be used to model outside air, also known as air to be delivered to a space or air handler.

Return Air Air that is being returned from the space back into the system is called return air. It can also be used in place of relief air, but exhaust makes a little better candidate.

Exhaust Air Air that is destined to leave the space as well as the system is called exhaust air. This is the best option for modeling relief air.

Other Air Other air seems like a logical candidate for relief air or outside air; however, systems cannot be made with other air.

Fitting The Fitting system type is merely a pass-through connection; there is no effect on the airstream.

Global Global connections also cannot be made into systems on their own; however, they inherit the characteristics of whatever system connects to them. Fans are a good example of equipment that may use global connections.

Loss Method Not Defined, Coefficient, or Specific Loss are the options here, and Coefficient and Specific Loss each activate another parameter. Specific Loss should be used where the loss is known from a catalog or cut sheet. The pressure loss is taken literally as the entered value for Pressure Drop.

Pressure Drop This can be entered as a static value or linked to a family parameter. Units are handled in the Project Units dialog box (Manage ➪ Project Units).

Flow Values for the flow associated with the connection are highly dependent on flow configuration.

Shape The Shape settings of Rectangular or Round determine which dimension parameters are active.

Height Height is simply a dimension of the connector. It can be linked to a family parameter.

Width Width is simply a dimension of the connector. It can be linked to a family parameter.

Radius Radius is similar to Height and Width; however, be careful to use the radius and not the diameter when linking to a family parameter.

Utility This indicates whether the connector is exported on a site utility to an Autodesk Exchange file (ADSK).

Connector Description The option to assign a name to the connections shows up primarily when using the Connect To feature. It also appears when connections are in the same vertical plane.

Creating Mechanical Systems

Now that you have reviewed the parameters that mechanical systems consist of, you will now learn how to apply them in a simple exercise:

1. Open the `Chapter8_Dataset.rvt` file found at `www.wiley.com/go/masteringrevitmep2011`.
2. Next select Insert ➪ Load Family ➪ Imperial Library\Mechanical Components\Air-Side Components\Air Terminals, select `Supply Diffuser - Perforated - Round Neck - Ceiling Mounted .rfa`, and click Open (Figure 8.7).

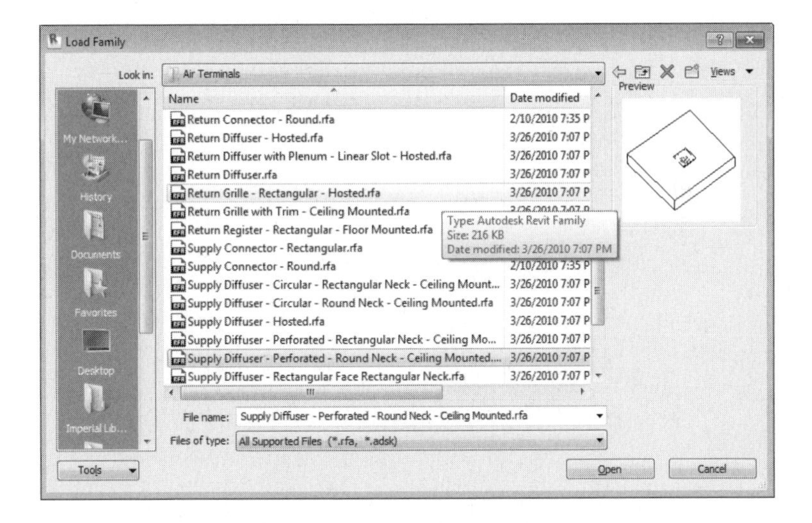

FIGURE 8.7
Selecting a supply diffuser

3. Now download `VAV_SingleDuct_4-16inch_break object styles.rfa`, located at `www.wiley.com/go/masteringrevitmep2011`, and insert it into your model (Figure 8.8).
4. After downloading all of your components, you will want to place the diffuser onto the ceiling plan of level one. Because this family is a face-based family, it will automatically attach to the ceiling surface. Make sure to select Place On Face located under Modify ➪ Place Air Terminal. If you don't, it will try to attach to the wall (Figure 8.9).

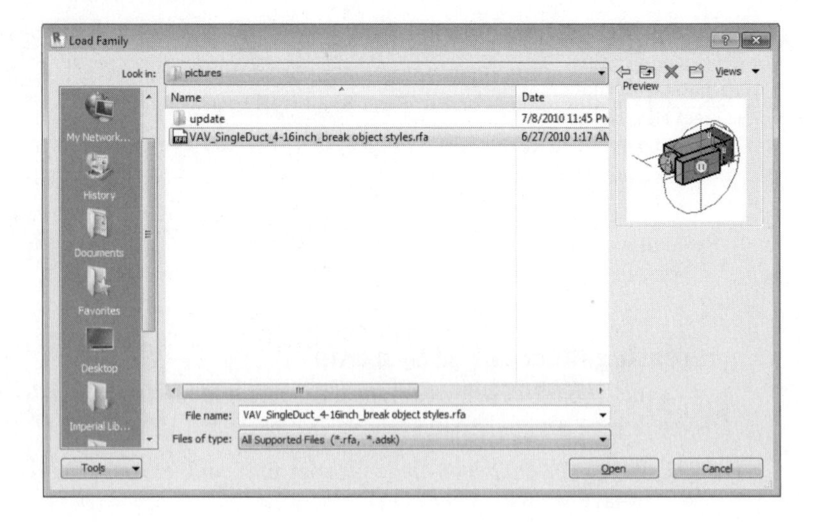

FIGURE 8.8 Inserting VAV single-duct family

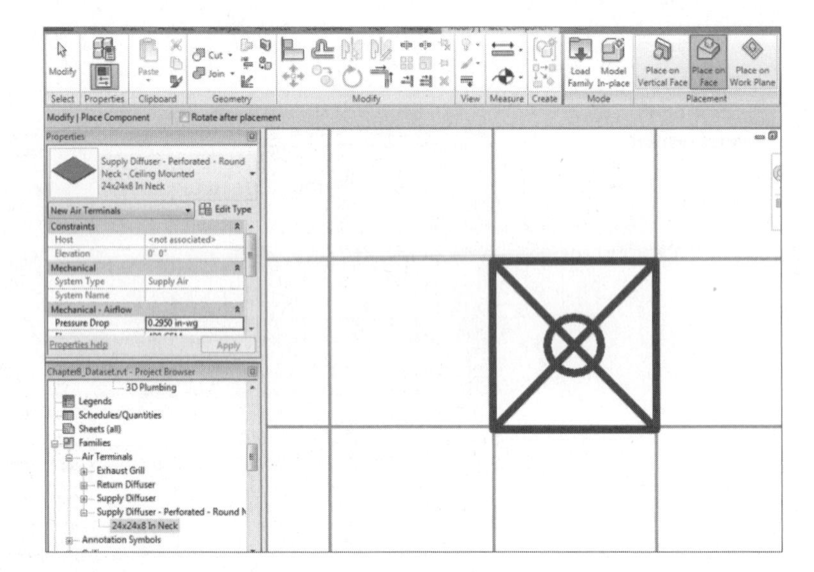

FIGURE 8.9 Placing ceiling diffuser into ceiling grid

5. Once the ceiling diffusers are located, place the VAV box above the ceiling at an elevation of 10′-0″. You can adjust the elevation by changing the elevation offset in the properties (Figure 8.10).

6. Now that the VAV box and diffusers are in place, you will want to make them part of a system. Select a diffuser, go to Home, Modify Mechanical Equipment, and select Create Supply Duct System (Figure 8.11).

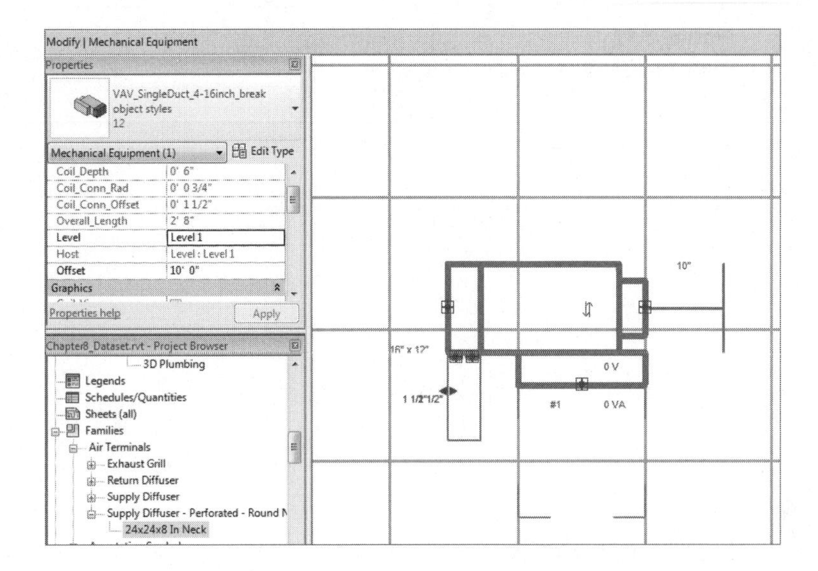

FIGURE 8.10 Changing elevation offset for the correct elevation of equipment

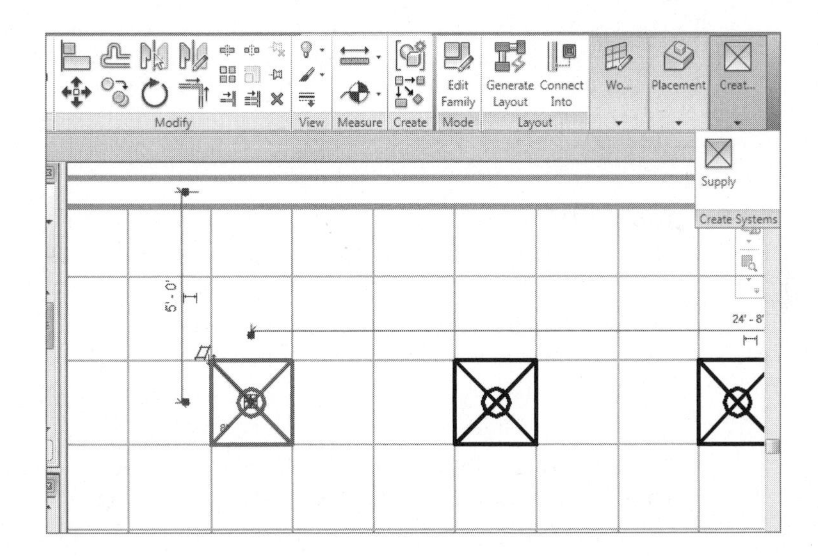

FIGURE 8.11 Creating a supply duct system

7. Once you have the system, you will need go to the Duct systems tab, then select Edit System, and edit the supply duct system name. Select Add To System, and select all the mechanical equipment that contains supply duct connections (Figure 8.12).

Now that you have created your system, you can route your ductwork, and it will take on the characteristics of the system parameters you set up. Understanding how to make mechanical systems will offer you the benefit of being able to create any duct system you may need for your design. For further instruction on how to route ductwork, refer to Chapter 10.

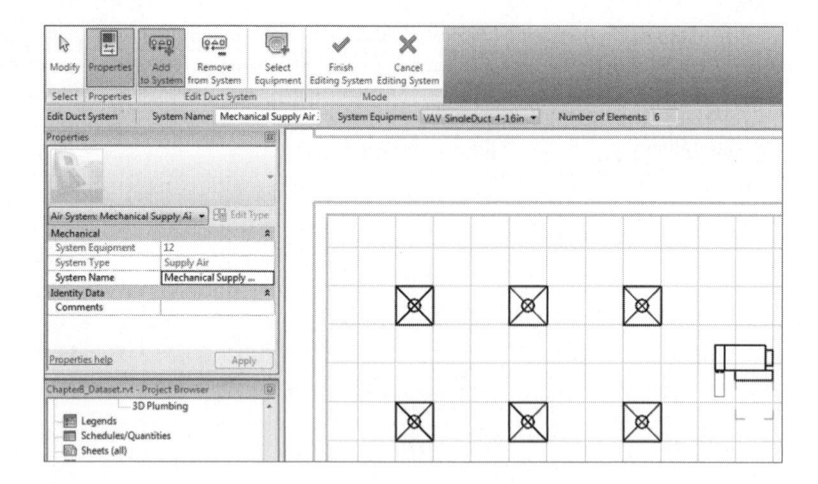

FIGURE 8.12
Editing supply duct systems

Piping Systems

Mechanical piping benefits greatly from systems in Revit MEP 2011. Graphics, annotations, flow, and pressure loss can all be handled with a small amount of setup on the front end. Pipes on different systems will not connect to one another even if they are at the same elevation. The pipes will show up as a conflict when collision detection runs, but that is a better result than a tee that connects sanitary piping with heating water. Pipe systems also allow filters to apply to all components in the system, including the fittings.

Parameters

Parameters for piping are similar to the parameters for air systems such as Flow Factor, Calculated, Preset, Flow Direction, and so on. The available pipe system types are as follows:

Hydronic Supply This pipe system can also be used for chilled water supply, cold water supply, steam, hot water supply, and process piping supply.

Hydronic Return This pipe system can also be used for chilled water return, cold water return, steam condensate, hot water return, and process piping return.

Sanitary This pipe system can also be used for grease waste, oil waste, storm drainage, acid waste, contamination drainage, and condensate drainage.

Domestic Hot Water This pipe system can also be used for different hot water systems such as 140-degree, 110-degree, and tempered water.

Domestic Cold Water This pipe system can also be used for filtered water, deionized water, and chilled water for remote drinking fountains.

Other This pipe system can be one of the most utilized pipe systems if you have a large piping project. This can be used for medical gas piping, vent piping, liquid propane piping, natural gas piping, and air piping.

Fire Protection This can be used for the building sprinkler piping, or it can used for the utility fire protection coming into your building to connect the base of your fire protection riser.

Wet Fire Protection This pipe system type normally is used for the layout of the piping from the riser to the sprinkler head layout.

Dry Fire Protection This pipe system is used for layout from the fire riser to the sprinkler head or standpipe to keep the system from freezing.

Pre-Action This pipe system can also be used for a deluge system.

Fire Protection Other This pipe system can be used for glycol antifreeze system and also can be used for chemical suppression system.

Creating Pipe Systems

You will need some pipe, some fittings, and something to which it will connect. Equipment is the source of the system, and the pipe and fittings connect everything. It seems pretty simple, but there are several things to consider when setting up the components of a pipe system.

Pipe types allow you to assign materials, connection types, identity data, and the fittings that will be used to connect it all together (Figure 8.13). Material, Connection Type, and Class are all set up in the Mechanical Settings dialog box. These should all reflect real-world values and the company standards and specifications. Even if you are not using Revit to size or lay out pipe automatically, not setting up the appropriate pipe types can cause headaches down the road. Pipe types should not be left at Standard and PVC. That hardly covers the necessary piping that a building requires, and more importantly plumbing and mechanical will be fighting over what fittings should be standard and what materials should be used. Mechanical piping and plumbing piping should have their own pipe types. Even if exactly the same materials and fittings are being used, there may be changes later, and splitting out pipe types late in a project will undoubtedly eat up a lot of time. This is one area of Revit MEP where you can have as many types as you want, so take advantage of it.

Pipe sizes and materials are set up in the Mechanical Settings dialog box (see Figure 8.14) and, similar to pipe types, should be separated for plumbing and mechanical piping. There are many instances where you would probably be fine using the same material and sizes, but when it is as simple as selecting the Add Material icon and selecting a source from which to copy, there is no reason not to create as many as are needed. All of these settings should also be determined using the company standards and specifications. It may seem tedious to set up, but if the inside diameter of a 6″ hot water pipe is not true and you are using a lot of it, your pipe volume calculations can be skewed.

Used In Size Lists vs. Used In Sizing

Used In Size Lists means that the various drop-down lists for available sizes will contain this size pipe; it can be easily accessed while routing or modifying pipe or in the path layout tool. For Used In Sizing, think of it as granting Revit permission to use a particular pipe size. If neither is checked, the size will not be available to the user.

FIGURE 8.13
Pipe type properties

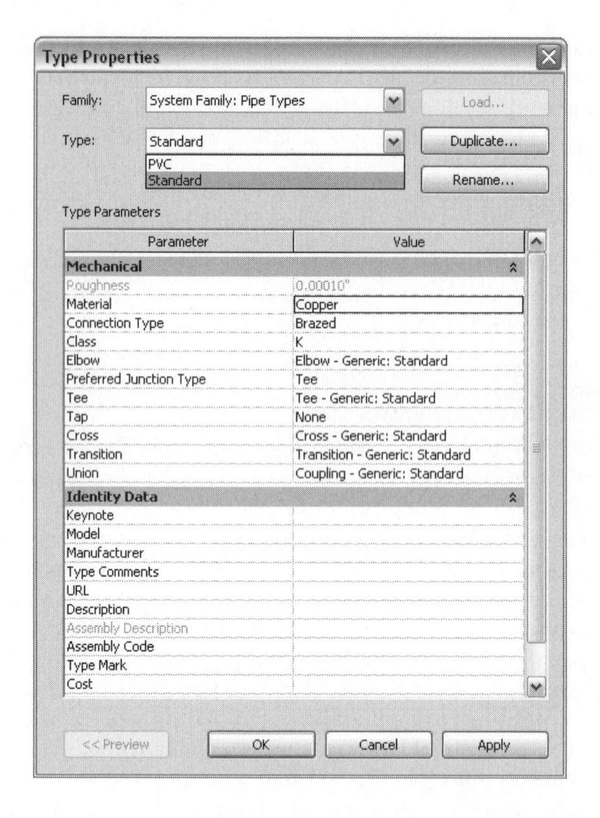

FIGURE 8.14
Pipe sizes

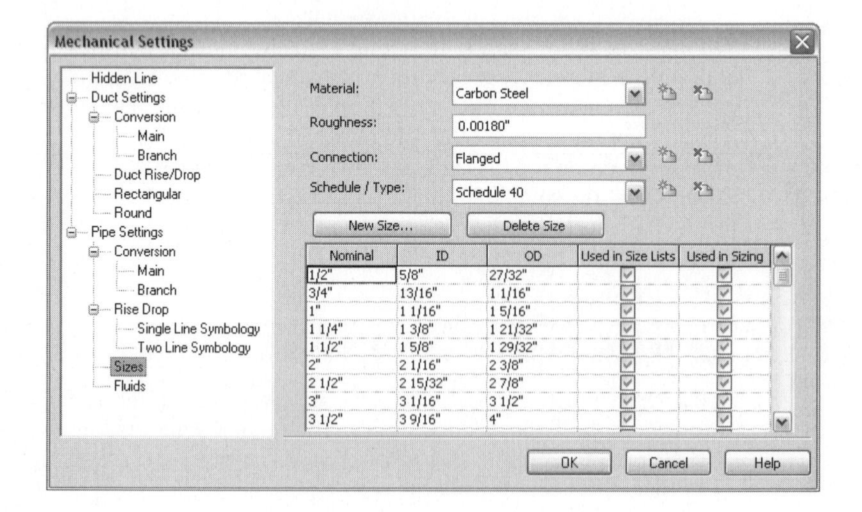

Fittings should be set up after sizes and materials for a couple reasons. First, the connection type needs to match. A solvent welded PVC fitting and a flanged steel fitting are vastly different. Second, the fittings need to be defined at all the available sizes for that type of pipe; if it goes down to 3/8″ or up to 36″, the fitting needs to accommodate. Fittings are going to be very specific to the type of system, which is another reason to have pipe types for plumbing and mechanical separate.

Hydronic systems have value even if the equipment is not piped together. Terminal box reheat coils are a good example because details generally cover their final connections. By simply adding all the terminal boxes to a hydronic supply and hydronic return system, you will be able to see the total flows for the entire model. You can use this flow summation to ensure that systems are adding up to what you expect and to compare flows between systems.

Quickly Adding Equipment to a Piping System

If the equipment you want to add to a hydronic system is all the same family and type, you can simply right-click one of the elements, select All Instances, and click the Hydronic Return/Supply/Other System button.

Now if your equipment has multiple types, this trick doesn't work as well. Start with the most common type, and use the Select All Instances trick. Once the initial system is created, edit it to add the remaining components. Unfortunately, the Select All Instances feature is not available within the Edit Systems menu. You can, however, drag a selection box to grab multiple pieces of equipment.

New to 2011 Revit MEP, the Select All Instances feature now has two options — Visible In View and In Entire Project. These new options offer more flexibility than the previous versions.

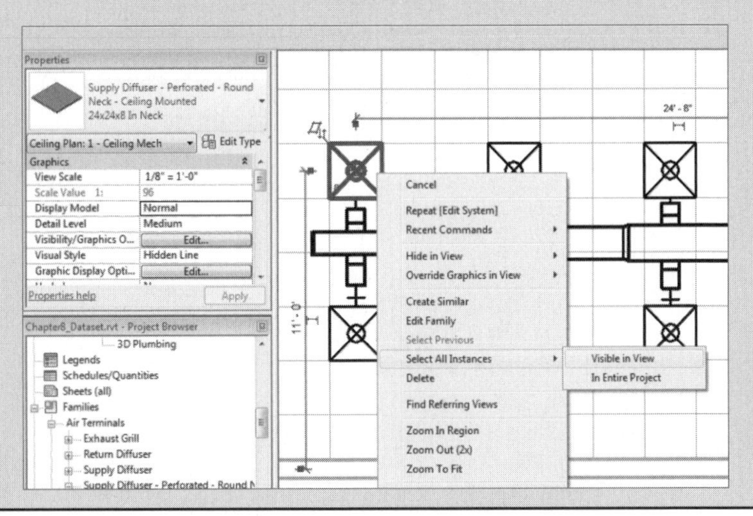

Fire Protection

Fire protection systems in Revit MEP are a sort of hybrid between air systems and hydronic systems. Sprinklers, from a system standpoint, are very similar to air terminals. They are designed to evenly distribute a fluid over a given area with pressure as the driving force of distribution. In the case of fire protection, water is the fluid, and the fire pump or city connection provides the pressure.

The key to a good manageable fire protection system is the families. Decide what type of system you will be using, and make sure appropriate families are developed before you or other users start laying out components. Revit MEP does have the ability to load a family in the place of another, but that tends to cause issues with system connections, pipe connections, and hosting. Sprinklers, standpipes, hose cabinets, and fire pumps may have to be created for the systems to work properly.

Display Properties of Systems

Now that you have air and piping systems, you will want to control how they are displayed. Display filters can be used to assign a color, line type, and line weight to a particular system.

System Filters

To set up system filters, you need to select the View tab on the ribbon and then select Visibility/Graphics ➪ Filters (Figure 8.15). Once selected, you will notice that in the Filters dialog box there are several filter names that are in Revit MEP by default.

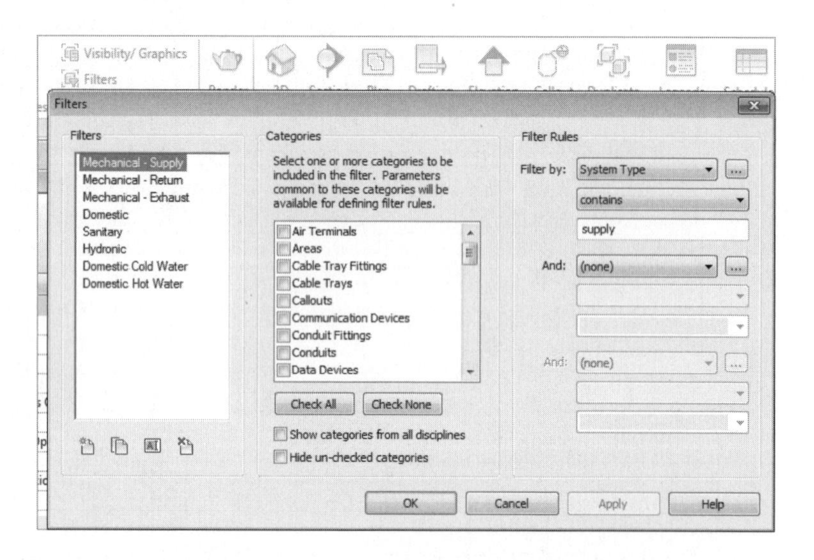

FIGURE 8.15
System filters

You will also notice that there are categories that have different elements selected. These are the items you want to have affected by the filter. Next you will notice under Filter Rules that by default System Type is selected. Because of the large number of filters that are normally created for a project, you will want to change the Filter By setting from System Type to System Name. This will help to properly identify and separate your systems. To create new system filters, the easiest method is to select an existing system filter and then click the Duplicate Icon. Once the system filter is duplicated, make sure the proper category elements are selected, and then rename the filter rule to the name you want to filter by.

To apply these filters to your views, you can go to the View tab on the ribbon and then select Visibility/Graphics. Next select the Filters tab (Figure 8.16).

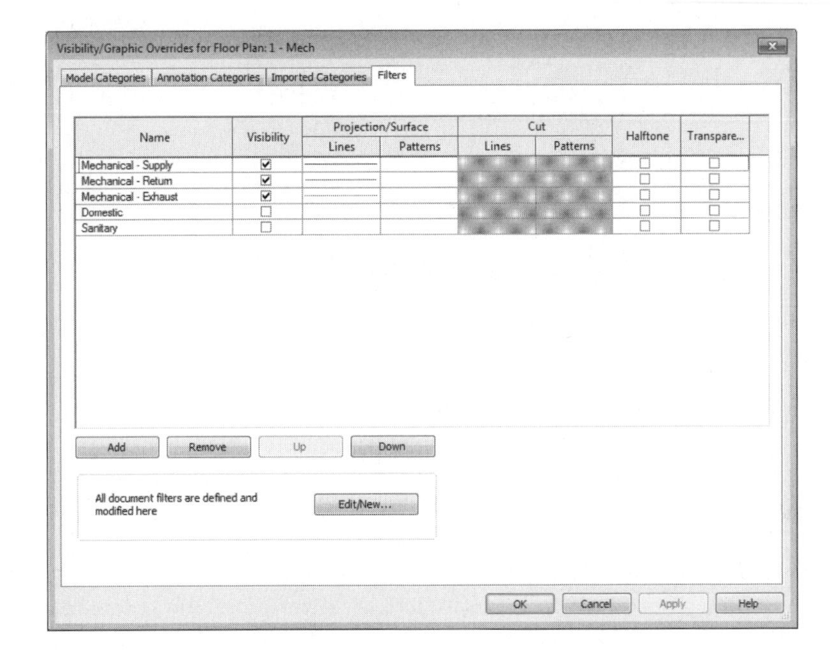

FIGURE 8.16
View filters

Once you have selected the Filters tab, select Add. This will bring up the filters that you had created, so select the filters that apply to your project. Once the filters are loaded, you can turn on and off the filters and adjust line weights, colors, and patterns.

Mastering filter options will give you the ability to create your models with the standards that your office has developed over the years producing CAD drawings.

Real World Scenario

FILTERS SAVE THE DAY

Jamie Abbott had been working on a medical facility in Revit 2011. When he completed the project, he submitted it to the state of Tennessee review board to make sure that the plans met the state requirements for standard of care for health facilities. He has just received the review comments, and the only comment on the review is that the reviewer would like to have the different duct systems shown with different patterns to help clearly tell the systems apart. To accomplish this, Jamie used system filters to distinguish the different duct systems. The following steps show how he accomplished his goal:

1. He pressed VG to bring up Visibility Graphics; then he selected Filters ➪ New/Edit. Next he duplicated Mechanical – Supply twice and renamed the new filters Mechanical Supply Air 1 and Mechanical Supply Air 2, as shown here.

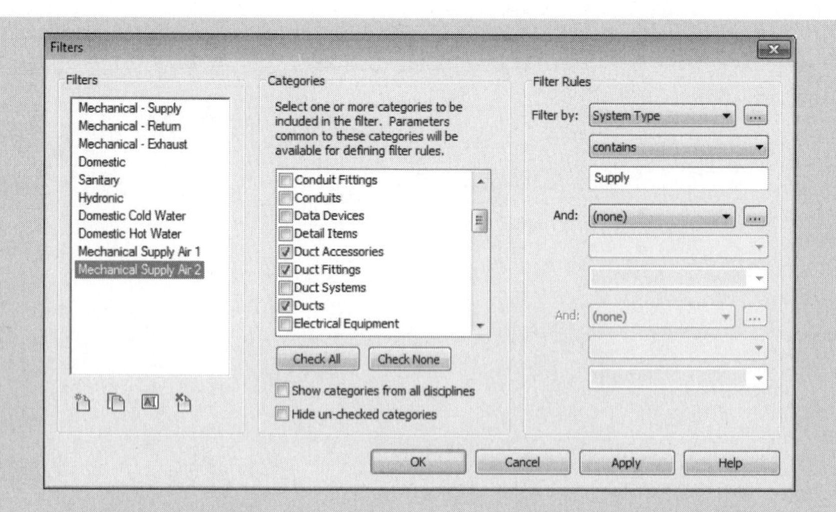

2. Next, he changed the filter rules from System Type to System Name and then changed the Contains statement to Mechanical Supply Air 1. He repeated the process for Mechanical Supply Air 2.

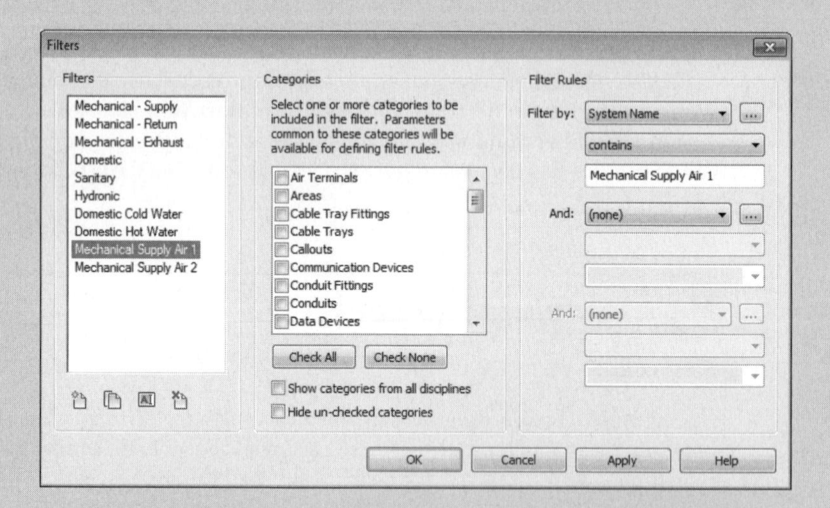

3. Once he had his filter rules created, he added them to his filters and then added patterns to help show the difference between the two supply systems.

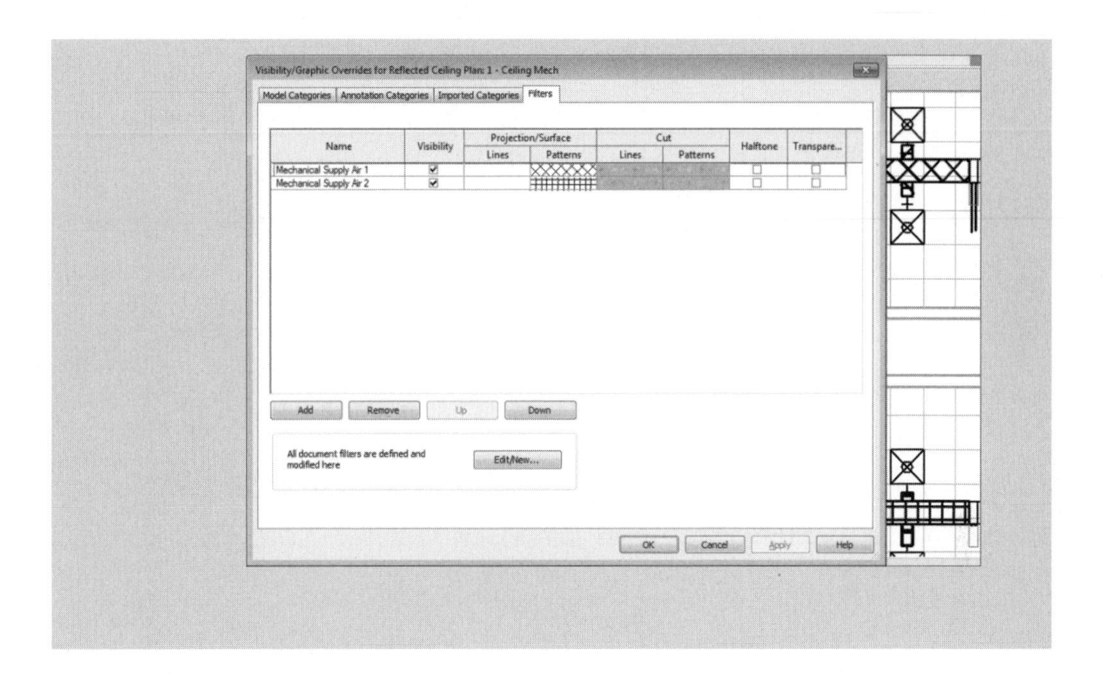

The Bottom Line

Create and manage air systems Knowing how to effectively manage air systems can help productivity through organizing systems so that items can be easily interrogated to verify that the systems are properly connected.

Master It True or False: Outside air cannot be modeled because there is no default system type to select from.

Create and manage piping systems Being able to understand how to change and manage piping systems, the user can create and maintain different systems effectively.

Master It A designer has been asked by the engineer to create a grease waste system to accommodate a new kitchen that has been added to the project. What would be the quickest way to accomplish this feat?

Manage display properties of systems Filters and Visibility settings can help the user show the intent of the layout.

Master It A plumbing designer has just created a Grease Waste system, and now the engineer has decided that the Grease Waste line should appear as a dashed line. How would the designer accomplish this?

Chapter 9

HVAC Cooling and Heating Load Analysis

When selecting and designing HVAC systems that will serve a particular building, modeling the building accurately is key. The majority of your time during the mechanical design phase of a project is spent on properly modeling the building in a load-simulating program, such as Trane TRACE 700 or Carrier's Hourly Analysis Program (HAP), among others.

Although these programs are essential to you, the mechanical designer, setting up the building accurately within these programs often can be a tedious task. Each space is set up individually, and typically, the physical construction and use of each space will be different. Alterations to the building design or space usage by the architect during this phase will cause you to return to any previously modeled spaces and coordinate the necessary changes. This is very time-consuming and often a point of contention between disciplines when changes occur later in the design phase.

Revit MEP enables you to model the building spaces accurately, as well as quickly and efficiently track any building design and construction changes on the fly, all within the project file. Revit MEP also gives you the option to either model the HVAC loads within the program itself or export the space load data via a gbXML file to an external simulation software program.

In this chapter, you will learn to

- Prepare your Revit MEP model for analysis
- Perform heating and cooling analysis within Revit MEP 2011
- Export gbXML data to load-simulating software

Modeling Spaces for Building Load Analysis

The key to any successful building load analysis lies with accurately modeling the spaces within the building. Components that need to be modeled for each space include but are not limited to building construction, such as walls, roof, floor slab, external shading and windows; internal loads, such as the number of people, the activity within the space, the heat gain from lighting, and the equipment operated within the space; and external factors, such as solar heat gain to the space, weather and typical outdoor temperatures, and infiltration. These are but a handful of factors that need to be addressed for each space that is being created, and each of these factors has several significant inputs that can affect the loads within the space.

Space Modeling

Before any HVAC loads can be run, spaces need to be created in the project file. Why? It seems redundant to create spaces, seeing as how the architectural model already has rooms created and defined, right? Not so.

In Revit MEP, spaces are not created *from* rooms. The same elements that define a room in the linked model define spaces in your MEP model. After you load your MEP project and link in the correct architectural model, you want to make sure that the elements that make up a room — walls, doors, ceilings, and so on — will define your MEP space accurately. Figure 9.1 shows a sample building model.

FIGURE 9.1
Sample building

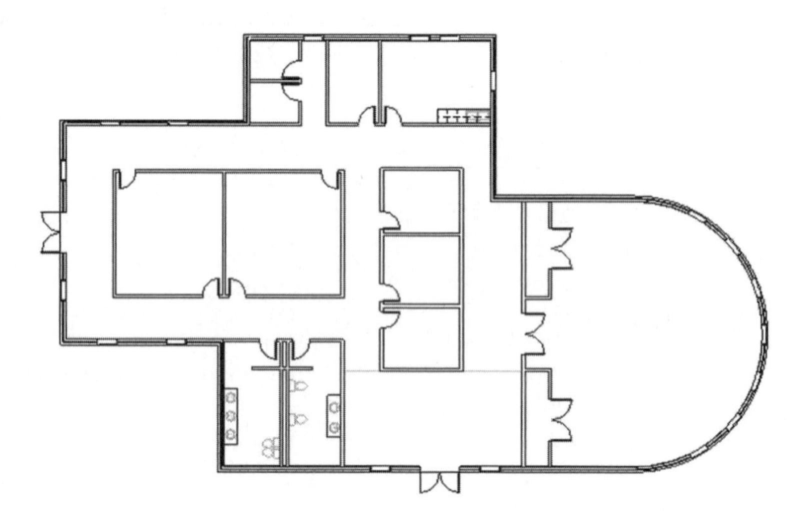

By selecting the architectural link in your project, the Edit Type button in the Properties palette is enabled. Clicking this will allow you to enable the selected link to be Room Bounding. This option will force the boundaries of the designed rooms to define the boundaries of the MEP spaces to be modeled. Figure 9.2 shows (a) the Edit Type button and (b) the type parameter Room Bounding.

If the link is not set to be room bounding, Revit MEP will offer a warning that the placed space is not in an enclosed region, and subsequently, if HVAC analysis is attempted, Revit will not be able to calculate load data for that space.

Placing Spaces

Revit allows you to locate the spaces within the architectural model in two ways: placing spaces manually or letting Revit locate and place spaces automatically. If your design building is complex, allowing the program to automatically place the spaces for you will save time.

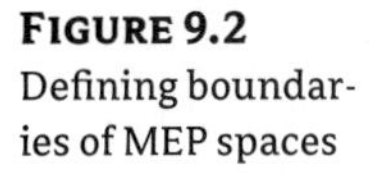

FIGURE 9.2
Defining boundaries of MEP spaces

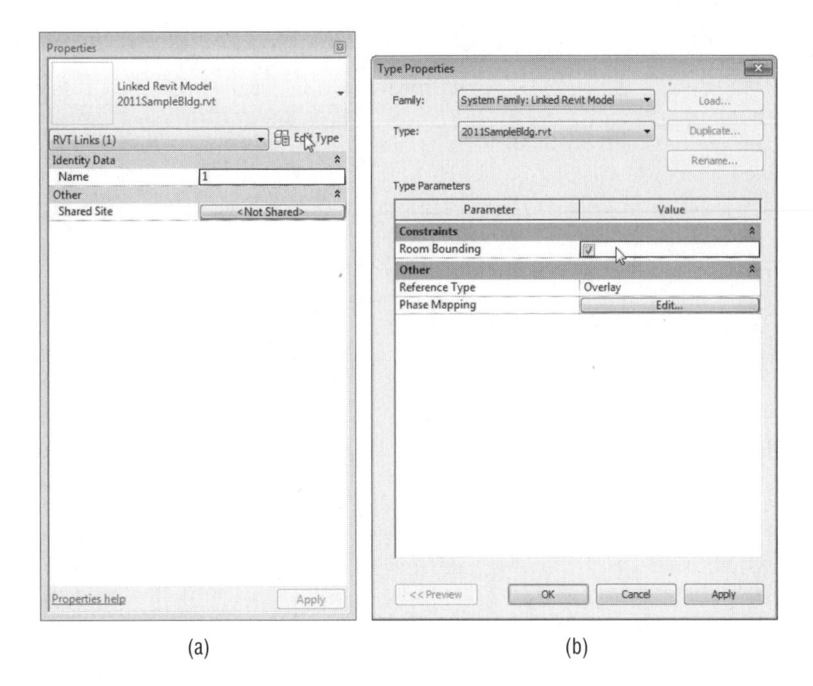

Be sure that all the spaces created are spaces that you intend to model. Check that areas such as utility chases, furrowed columns, or air spaces in wall constructions were not included when spaces were placed. To remove any unwanted spaces from the project file, open the System Browser, which is located on the View tab under User Interface, and set the view to Zones by right-clicking the System titlebar and choosing Zones from the View drop-down menu.

Select the desired spaces to remove by highlighting the space name, right-clicking, and selecting Delete, as shown in Figure 9.3. Click OK in the pop-up window to permanently remove the spaces and any associated space tags.

If you were to just select a space within an open view and delete it without using the System Browser, the space tag and visible space marker will be deleted, but the space information will still be present in your project. This will affect any building analysis performed, as well as any HVAC systems that will be set up in the building. Systems are covered in depth in Chapter 8 and Chapter 10.

Space Properties Schedule

When simulating the heating and cooling loads of a building, correctly modeling and accounting for space usage can be time-consuming. Often, design loads have to be revisited because of inaccurate or incomplete accounting of space usage, components, internal loading, or design changes of the space use. Creating a working schedule of the building space properties in Revit MEP will help you account for and coordinate these factors.

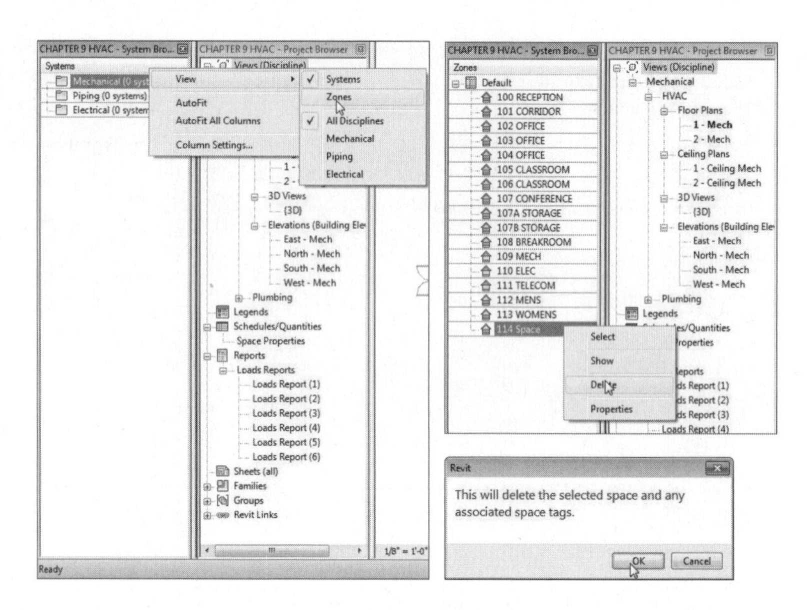

FIGURE 9.3
Deleting unwanted spaces from the project file

Similar to other schedules in Revit, you are given a choice of several fields to display in the Space Properties schedule to help track and modify data pertaining to each space. Setting up a working schedule such as this one for space properties will be an essential tool that will enable you to track how changes in certain properties will affect the spaces. Note that the schedule shown in Figure 9.4 is not intended to be provided to your client on a project sheet but rather to be used as a personal design tool within Revit to help you, the designer, organize the space data to better fit your personal workflow. Create this schedule as you see fit, using the many fields, sorting, and other formatting options available to you within the program to readily organize the data you need.

Useful fields to display, during the initial space creation, include the following:

Name Designate the space name in this field.

Number Assign each space a number using this field.

Space Type Describe how the space will be used in this field.

Occupiable This is a yes/no check box that shows whether this space will be occupied.

Number Of People Specify the space occupancy in this field.

Area List the space area in square feet or meters.

Construction Type Describe the space physical construction in this field. We will go over the various construction options later in this chapter.

Condition Type Describe type of space conditioning in this field (such as Heated and Cooled, Heated, Cooled, or Unconditioned).

FIGURE 9.4
Sample Space Properties schedule

Space Properties							
Name	Number	Space Type	Occupiable	Number of People	Area	Construction Type	Condition Type
RECEPTION	100	<Building>	✓	1.205679	371 SF	<Building>	Heated and cooled
CORRIDOR	101	<Building>	✓	3.986143	1226 SF	<Building>	Heated and cooled
OFFICE	102	<Building>	✓	0.33665	104 SF	<Building>	Heated and cooled
OFFICE	103	<Building>	✓	0.380517	117 SF	<Building>	Heated and cooled
OFFICE	104	<Building>	✓	0.342819	105 SF	<Building>	Heated and cooled
CLASSROOM	105	<Building>	✓	0.939224	289 SF	<Building>	Heated and cooled
CLASSROOM	106	<Building>	✓	1.017531	313 SF	<Building>	Heated and cooled
CONFERENCE	107	<Building>	✓	3.585021	1103 SF	<Building>	Heated and cooled
STORAGE	107A	<Building>	✓	0.162136	50 SF	<Building>	Heated and cooled
STORAGE	107B	<Building>	✓	0.162136	50 SF	<Building>	Heated and cooled
BREAKROOM	108	<Building>	✓	0.642562	198 SF	<Building>	Heated and cooled
MECH	109	<Building>	✓	0.301316	93 SF	<Building>	Heated and cooled
ELEC	110	<Building>	✓	0.137718	42 SF	<Building>	Heated and cooled
TELECOM	111	<Building>	✓	0.142964	44 SF	<Building>	Heated and cooled
MENS	112	<Building>	✓	0.453117	139 SF	<Building>	Heated and cooled
WOMENS	113	<Building>	✓	0.494751	152 SF	<Building>	Heated and cooled

SPACE NAMING

When creating spaces, Revit MEP assigns each one a generic space name and number. To reflect the architectural room names and numbers, add the fields Room: Name and Room: Number from the Rooms category in the Available Fields drop-down that's on the Fields tab in the Schedule Properties dialog box. Edit the space names and numbers to match the linked file. You can then hide the Room: Name and Room: Number columns to streamline your schedule and unhide them periodically to view any changes that may have occurred during the design. The Space Naming Utility is another essential tool that, if installed in Revit, will name and number the spaces automatically per the architectural link. The Space Naming Utility is available for download from Autodesk.

Modifying Space Properties

You will notice that the program defaults to a generic space naming and numbering convention when placing spaces. It also sets the Space Type and Construction Type values as <Building>, indicating that the spaces will be modeled generically, relying on default global building characteristics within the project.

How a space will be used is the overall factor driving the internal loads within a space. An enclosed office space is modeled differently than the conference room next door, and an office break room is modeled differently than a typical restaurant dining room. Lighting, population density, activity levels, equipment, and ventilation loads are all major factors that will vary with the space type.

If you open the Manage tab and then select MEP Settings ➪ Building/Space Type Settings, you will be able to view and modify both the space type options within Revit MEP as well as the global building types (see Figure 9.5).

FIGURE 9.5
Building/Space Type Settings

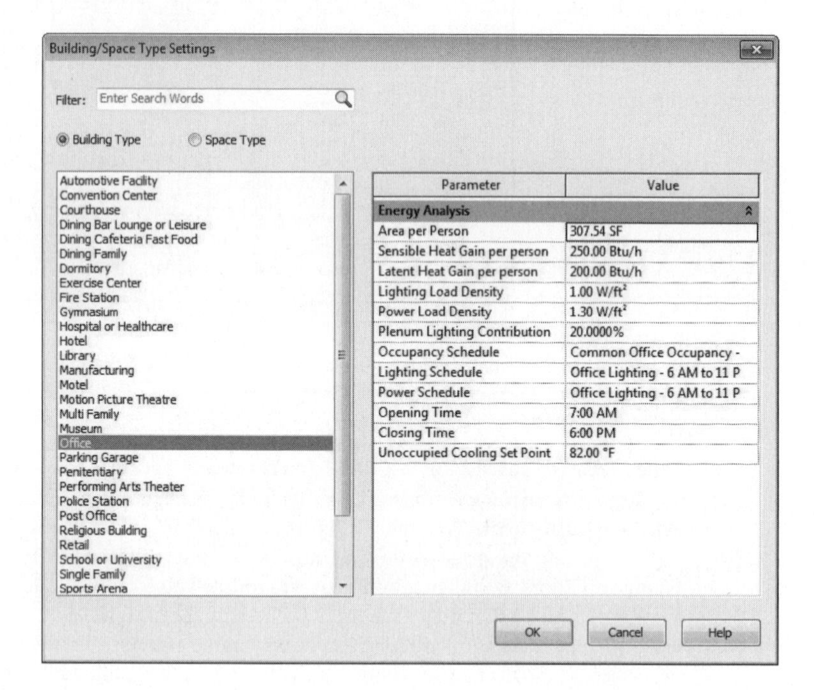

Under the Building Type category, you are able to choose a global usage for the building. This can be a good option if you want to do a quick takeoff of a typical building type for the overall model. You are given similar parameters to model as in the Space Type Settings window, as well as options to set building operating times — Opening Time and Closing Time — and an unoccupied space cooling temperature set point.

Revit MEP has several predefined options to model how each space is utilized, and each option has preset parameter and schedule values for the internal loads. Parameters such as Area per Person, Lighting Load Density, Power Load Density, and Heat Gain per Person should all be familiar to you, the mechanical designer. You may use the preset values within the program, or you are allowed to input more accurate design load values based on typical code-driven or industry-standard values.

Note that you can access the Space Type Settings window from the Properties palette when a space is selected in your view, as well as from within your space schedule by locating the cell in the Space Type column (if used in your space schedule) for the space you would like to modify and then selecting the ellipsis next to the current type name. If you click the People or Electrical Loads buttons on the space Properties palette, the default values within these windows are the values listed in the space type settings (see Figures 9.6 and 9.7).

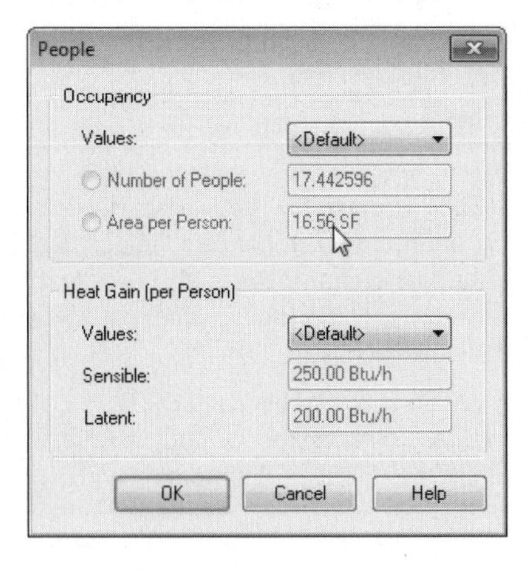

FIGURE 9.6
Default Occupancy and Heat Gain values

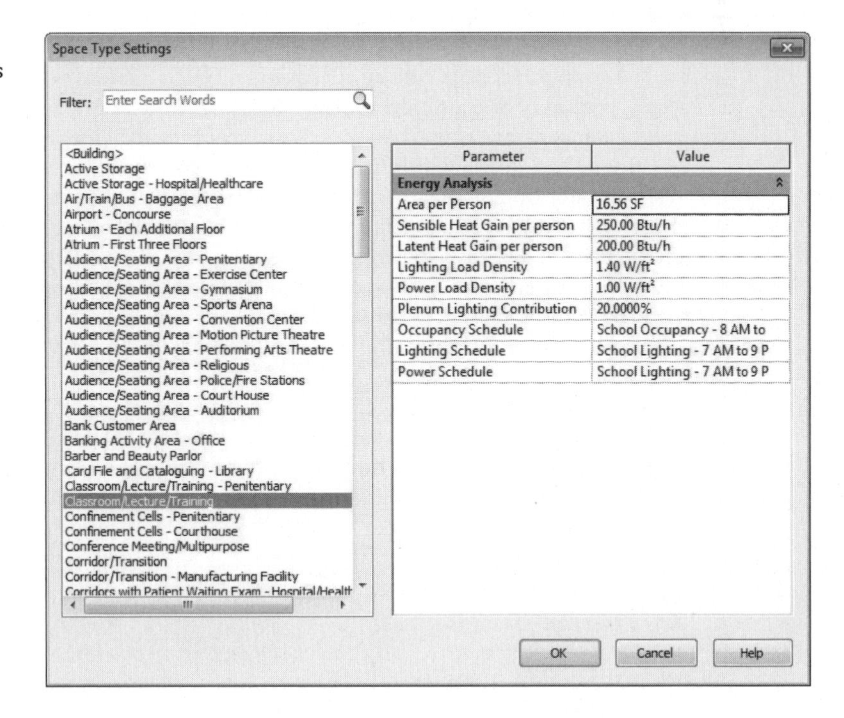

FIGURE 9.7
Space Type Settings window

Revit MEP also has default building and space schedule types that it uses to generate load data. You can modify the default schedules or create new schedules to fit your varying occupancy, power, and lighting needs. You can find Default Schedules, People Heat Gain, and Building and Space Type Data in the Revit MEP 2011 Help file, under Reference ➪ Energy Analysis Building & Space Type Imperial Data.

GOING ABOVE AND BEYOND

Doug has an issue. He has received the preliminary building model from the architect but has not been able to set up each individual space to get an accurate building load takeoff prior to the next project design meeting. Knowing that Revit MEP automatically defaults to a global building type, he adjusts the building type to Office, adjusts the required parameter values, and tells the program to lay out the spaces automatically.

He is able to quickly generate initial heating, cooling, and ventilation loads, and with that data, the mechanical design team will be able to analyze different system possibilities to condition the building and present options to the project team.

Zones

The next step in generating HVAC loads for your design building is to group similar spaces into zones. The main purpose of zoning spaces in an HVAC system is to provide controllability of the space air quality or condition. A single terminal unit, heat pump, or air handling unit can control the temperature and air quality of the spaces within its defined zone.

Creating HVAC zones within your building allows the designer to control the airflow to given spaces, shutting off airflow to areas that are not occupied or increasing airflow to spaces when the space load increases, via a signal from a temperature sensor or other sensing devices within the zone to a central control panel. Zoning also allows certain spaces to be controlled via a different system than the rest of the building, such as having a dining area and kitchen served via a dedicated outdoor air system that is cycled on with increasing occupancy to the space, while the rest of the building is controlled by a packaged VAV rooftop system on a typical office's 8 a.m. to 5 p.m. schedule.

Zoning spaces in Revit MEP is easy. On the Analyze tab, click Zone. This activates the Zone tool, and you are automatically prompted to add the first space to the zone. Clicking a space adds it to the zone, and then you have the option of adding other spaces to that zone. Click the Finish Editing Zone button on the ribbon to end the editing session.

To edit a zone that has already been created, simply select the required zone in the open view, and the Modify HVAC Zones tab appears. Click the Edit Zone button, and the zone creation ribbon appears. Here you are able to modify the zone by adding or removing spaces.

A second way to create a zone is to first select all the spaces you want to group together, click the Analyze tab, and select the Zone tool. Creating a zone this way automatically groups the selected spaces, without having to select each space individually with the tool active.

Selecting a zone displayed in an open view, the zone data appears in the Properties palette. From there, you are able to view the calculated heating and cooling loads and zone airflow (once Heating and Cooling Analysis has been run), the physical data of the zone (area, volume, and perimeter), and its characteristic data, which include the following:

Service Type This drop-down allows you to select the type of system that will be serving the spaces within the zone. Revit MEP offers four groups of systems: Constant Volume, Variable Air Volume (VAV), Hydronic, and Other. Within these four options, there exist several system variations. Revit MEP automatically defaults to the service type that is selected in the Project Energy Settings window. See Chapter 2 for information on how to establish project settings.

Coil Bypass This is where you are able to input any manufacturer's coil bypass factor for the unit serving the zone. This value indicates the volume of air that passes through the coil, unaffected by the coil temperature.

Cooling Information This button allows you to set the space cooling set point, coil leaving air temperature, and space humidity control.

Heating Information This button allows you to set the space heating set point, coil leaving air temperature, and space humidity control during heating.

Outdoor Air Information This button allows you to input the ventilation loads for the space: Outdoor Air per Person, Outdoor Air per Area, and Air Changes per Hour. You may specify individual Outdoor Air options or enter a value in all three options. Revit MEP only calculates heating and cooling loads with the largest calculated outdoor airflow, *not* a combination of the three values. To obtain the required combined breathing zone ventilation rate as defined in ASHRAE 62.1, add the ventilation CFM needed per occupant to the CFM required per area, and divide the result per the space area. Then enter this value in either of the first two options to force Revit into calculating the code-required ventilation rate.

OCCUPIABLE

Note that when you are creating spaces to use for heating and cooling analysis, make sure that the Occupiable box is selected in the space properties. This ensures that ventilation loads are computed as part of the space load.

Building Construction

So, now all the spaces within your design building are placed, the internal data has been identified, and the spaces have been grouped into zones ready for the heating and cooling data to be analyzed. The next step is to investigate the physical construction of your building's exterior and interior walls, roofs, ceilings, glazing, doors, and floor slabs. Each of these elements controls how heat enters or leaves your design spaces to or from the surrounding outdoor environment. Each element has a specific coefficient of heat transfer, or *U-value*, that is dependent on the element's material composition and assembly, as well as its thickness.

But the architect who created the building elements has already defined the wall or roof construction. That data should already be loaded in with the link, right? Well, no. Although the architectural model will have accurate wall and roof constructions modeled in the linked building, the U-values will not transfer from the link. In the MEP model, the building construction parameters have to be defined by you. This will set U-values for the different elements that bound your design spaces, allowing for heat transfer into and out of the spaces to be calculated and accounted for.

To set the building construction U-values, open the Manage tab, and click Project Information.

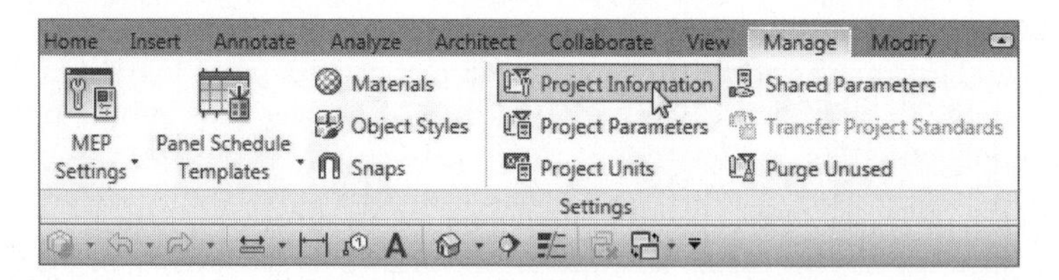

From this window, click the Edit button for Energy Settings, and then click the ellipsis next to <Building> in Building Construction. This brings you to the Building Construction window (see Figure 9.8).

Once here, you will be able to define the default construction characteristics. Clicking the drop-down arrow at each category exposes a wide array of common material constructions and assemblies that you can choose from to represent the design building walls, slab, roof, and glazing, each with a unique U-value associated to it.

Unfortunately, as of this version of Revit MEP, the constructions offered for each category in Building Construction are the only constructions that are available to use. The option to edit or customize wall or roof assemblies or even to alter the thermal properties of the materials that make up the assemblies is not present in this version, nor can you enter specific U-values, such as ASHRAE Standard 90.1 building envelope values. As the designer, you must choose the construction option that is closest to the actual U-value that has been calculated through material thermal takeoffs of the architectural design.

Just as all spaces are not going to be conditioned alike, some spaces have a different physical construction than the rest of the design building. For example, a utility space such as a mechanical or electrical room typically can be seen with exposed block wall construction, bare floor slab, open roof trusses or beams, and little or no insulation on the exterior walls. In Revit MEP, just as the individual space types can be defined differently from the overall building type, each individual space has the option to define its own construction U-values.

To alter the space construction, select a space in an open view; or, if you have created a Space Properties schedule, select the ellipsis next to <Building> in the Construction Type column of the space that you want to edit. The Construction Type window will open, and here you will be able to create individual space construction types as needed (see Figure 9.9).

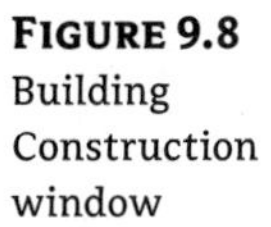

FIGURE 9.8 Building Construction window

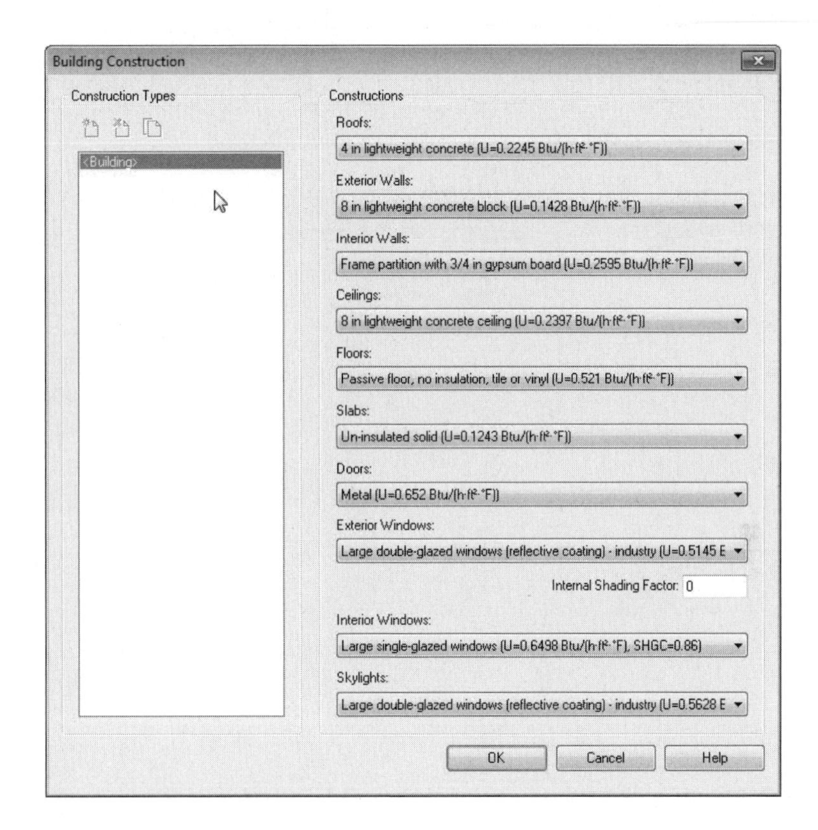

The same construction categories and options exist here that are available in the Building Construction window. Create as many different constructions as needed, and if there exists, in your design building, several spaces that utilize a different construction from the main building, you will be able to easily select the appropriate construction from your created list. Note that you are not able to alter the default <Building> construction type in this window. This is accessed from the Building Construction window, mentioned earlier.

Performing Heating and Cooling Load Analysis

Now you have all your design spaces created and respective parameters, conditioning systems, and space constructions defined. The next step then is to pull all this data together into a heating and cooling load analysis report to tell you how this particular building will perform throughout the year. You will then use that data to further refine your conditioning systems as well as size the equipment you will assign to your HVAC zones.

We've already discussed how the building and space construction can be modified to suit your building. But how do the different construction options affect the heat transfer into or out of the space? The engine that performs the heating and cooling load analysis in Revit MEP 2011 uses a Radiant Time Series (RTS) method to determine the building and space peak heating or cooling loads. This method takes into account the time delay effect of heat transfer through building envelopes, from the outside, and into spaces. A brief explanation of this method follows, but the RTS method of calculation is defined in detail in Chapter 30 of the 2005 ASHRAE *Handbook of Fundamentals*, as well as in the Load Calculation Applications Manual, also published by ASHRAE (visit `www.ashrae.org` for details).

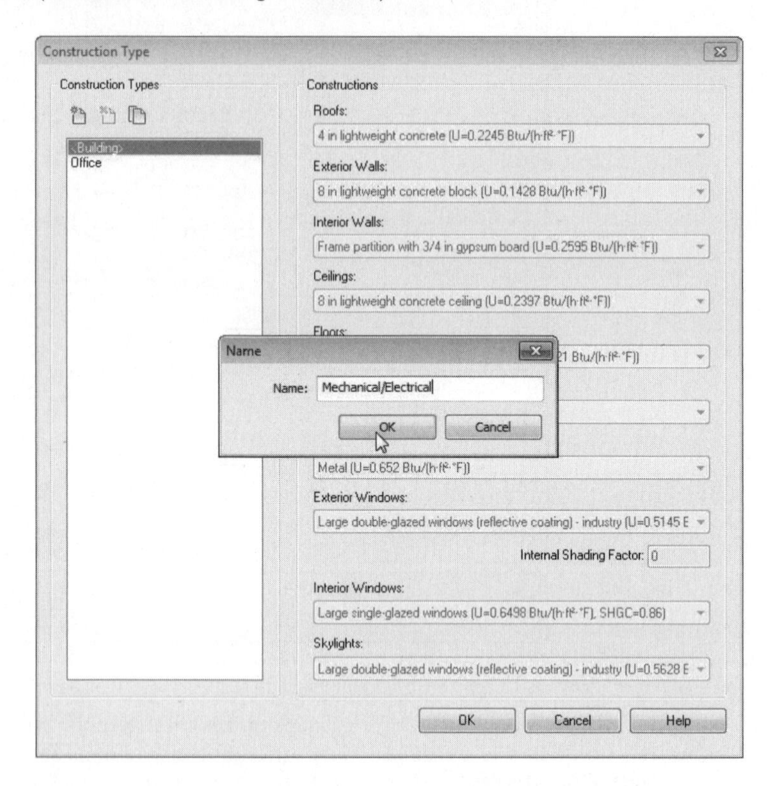

FIGURE 9.9
Construction Type window

The RTS calculation method determines cooling loads based on an assumption of steady periodic conditions, such as occupancy, design day weather, and cyclical 24-hour heat gain conditions. Two time-delay effects are addressed during cooling load calculations:

- Delay of conductive heat-gain through opaque massive exterior surfaces, such as exterior walls, the building roof, and floor slab on or below grade
- Delay of radiative heat gain conversion to cooling loads

Figure 9.10 shows a flow chart summarizing the RTS calculation method.

Exterior building elements conduct heat because of a temperature differential between indoor and outdoor air; solar energy is absorbed by exterior surfaces as well. Since each surface has a mass and an associated thermal capacity of the materials that make up its construction, a time delay occurs from when the heat input of the outdoor and solar loads becomes heat gain to the interior space. The majority of energy transferred to a space as heat is by a combination of convection and radiation — the cooling load immediately picks up the convective part of the energy transfer, and any radiant heat is absorbed into the surrounding space constructions and interior room finishes.

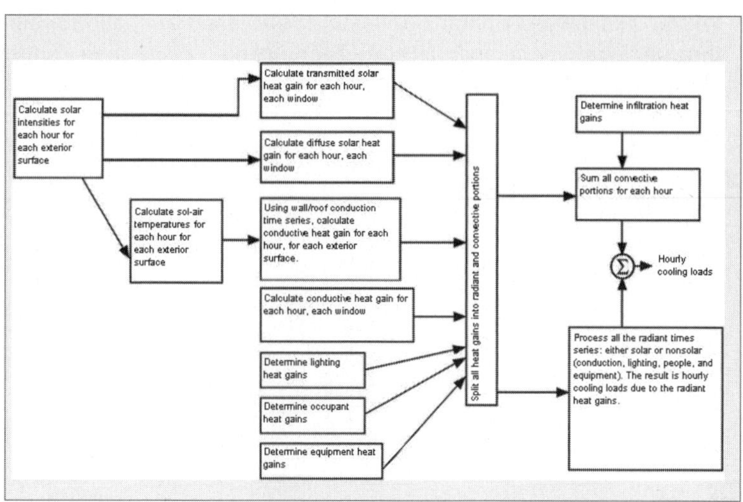

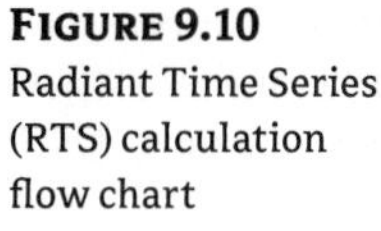
FIGURE 9.10
Radiant Time Series (RTS) calculation flow chart

The radiant heat is then transferred via convection to the space from those surfaces at a delayed time. Interior loads contribute to both the sensible heat gain of the space as well as a latent heat gain that is given off by people activity within the space. The latent heat gain contributes to the instantaneous cooling load of the space, while the sensible heat gain from internals is absorbed and retransmitted by radiation to the space.

The engine sums up the calculated cooling loads to determine a total cooling load per each design hour, and it selects the highest load, or *peak*, for the design of the air-conditioning system. Note that Revit MEP 2011 uses, for the standard calculation, the hours of 6 a.m. to 6 p.m. for the design day, *not* the full 24 hours, and only the months of April through November (October through May for southern hemisphere locations), *not* the full calendar year. The design day is derived from weather data for the location that you set during project establishment. We will revisit weather data later in this chapter.

Heating loads are calculated much the same way. The major differences are the obvious lower outdoor air temperatures in the heating design day, solar heat gains and internal heat gains are ignored, and the thermal storage effect of the building construction is not included. Negative heat gains, or *heat losses*, are considered to be instantaneous; therefore, heat transfer is dealt with as conductive. Latent heat gains are treated as replacing any space humidity that has been lost to the outdoor environment.

The worst-case load, as determined by the Revit MEP engine, is based on the design interior and exterior conditions and loads due to infiltration or ventilation. Solar effects are ignored — assuming night or cloudy winter day operation — and Revit does not recognize any internal heat gain from people, lights, or miscellaneous equipment to offset the heating load needed.

Load Analysis

You now have a complete building modeled and ready to analyze. The next step is to verify and specify, if needed, the building energy parameters, as shown in Figure 9.11. On the Analyze tab, click the Heating And Cooling Loads button. The Heating and Cooling Loads window will appear, and on the General tab, you will be able to view the building energy analysis project information that directly affects the heating and cooling analysis. You can also access these parameters through the Manage tab's Project Settings ➪ Project Information setting.

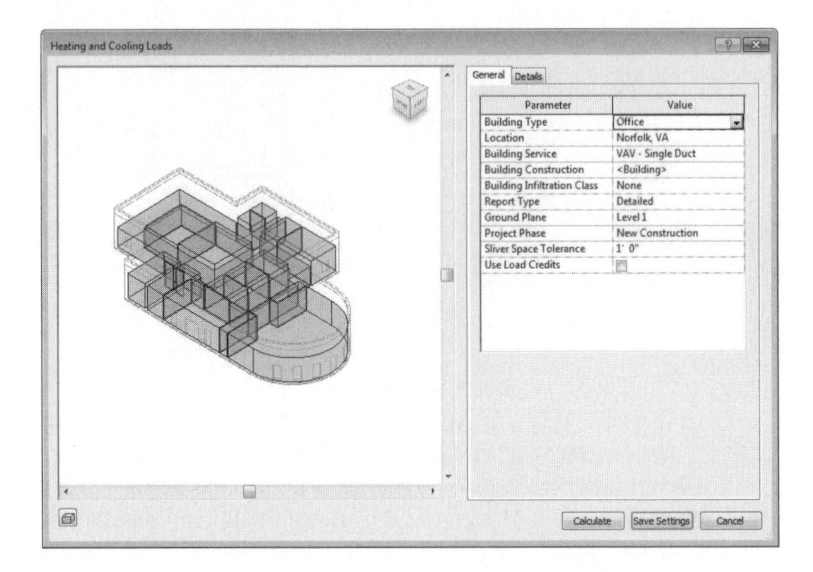

FIGURE 9.11
Heating and Cooling Loads window

In this window, you are able to set parameters such as the global building use type, building location, the global conditioning system, overall building construction and ground plane, the ground level reference, and building infiltration. You are also able to define the level of load analysis report detail you want to see.

Weather Data

We discussed earlier in the chapter building type, service type, and building construction, but another major component of energy analysis is the physical location of the building. The location determines environmental conditions such as outdoor air temperature and humidity and also ASHRAE climate zone data. To define the building location, click the ellipsis next to the city and state in the Location cell. You will be taken to the Location dialog box, as shown in Figure 9.12.

You have two ways to input the project design city: select the location from the default city list, or use the Internet mapping service, as shown in Figure 9.12. If you are connected to the Internet, this option allows you to locate your project using an interactive map through the Google Maps mapping service. The default city list contains major cities from which to select your project location. Select the major city from the list that is nearest to the project location address, or, if it is known, you can enter the latitude and longitude coordinates for a more exact location. Select Use Daylight Savings Time if it is appropriate for your project location.

The next step is to modify the cooling and heating design temperatures, if needed. You may use the default values that are associated with the closest weather station to your design city; or, by deselecting that option, you are able to modify the Dry Bulb, Wet Bulb, and Mean Daily Range temperatures as needed to fit your design location. You may specify the Heating Design Temperature value and set the Clearness Number value, which ranges from 0 to 2, with 1.0

being an average clearness. Clearness is defined in section 33.4 of the 2007 ASHRAE Handbook – HVAC Applications in the following way:

Greater than 1.2	Clear and Dry
1.0	Average
Less than 0.8	Hazy and Humid

FIGURE 9.12
Location Weather and Site

Outdoor Air Infiltration

The Building Infiltration class needs to be set next. You are given four choices to model the rate of outdoor air that enters the building, typically through leaks in the building envelope created at envelope openings such as windows, doors, and where perpendicular building surfaces join. In Revit, infiltration is defined with the following categories:

Loose	0.076 CFM/ft^2 of outside air.
Medium	0.038 CFM/ft^2 of outside air.
Tight	0.019 CFM/ft^2 of outside air.
None	Infiltration air is excluded from the load calculation.

Sliver Spaces

The next parameter to define is Sliver Space Tolerance. Sliver spaces in Revit are considered to be narrow areas that are bounded by parallel interior room-bounding components — parallel interior walls. These spaces include, but are not limited to, pipe chases, HVAC shafts, furrowed columns, and wall cavities. A sliver space is included in the heating and cooling load analysis only if identical parallel room-bounding elements enclose the space, the width of the sliver space is equal to or less than the Sliver Space Tolerance parameter, *and* if a space component has been placed in the tangent spaces on either side of the sliver space. If any one of these three requirements is not satisfied, Revit does not recognize any effects of the sliver space. If there are different geometries to the same sliver space, only the areas of the space that meets the previous criteria is counted in the load analysis. The sliver space volumes are added to the volume of the larger tangent analytical spaces.

Define the detail level of the heating and cooling analysis report. Three report detail levels are available: simple, which contains summary data for systems, zones, and spaces; standard, which expands the simple report to include psychrometric data as well as building level summaries and load summary data for each space; and detailed, which further expands the data displayed to include individual component contributions to zone and space loads.

Before you finish, define the ground plane, project phase, and whether to include heating or cooling load credits, which are negative load values that come from heating entering or leaving a space through a partition into another zone, for example.

Details

Before you analyze the building performance, you are given the opportunity to go through your created zones and spaces to make sure your desired settings have not been compromised and to make sure there are no warnings that would produce undesired effects on your loads. Switching to the Details tab, you can view the space and zone data that directly affects the heating and cooling analysis, as in Figure 9.13.

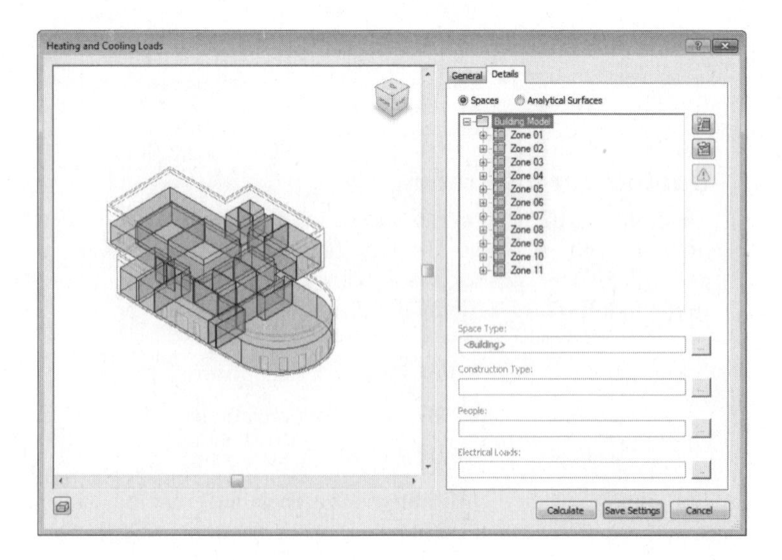

FIGURE 9.13
Building model details

Selecting a zone, you are able to expand the tree to view the associated spaces and verify or modify the zone service type, heating, cooling, and outdoor air information as needed. You can see whether a space is occupied, unoccupied, or a Plenum space by the symbol preceding the space name in the tree.

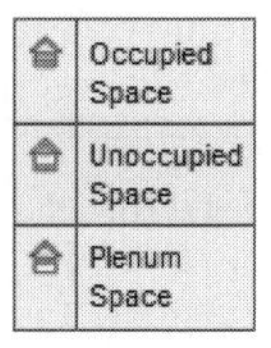

If a space is selected, you can set the space and construction type, as well as internal load information. If, when expanding a zone, a warning symbol appears alongside a space, you are able to investigate by clicking the Show Related Warnings symbol and then correct or ignore the cause of the warning (see Figure 9.14).

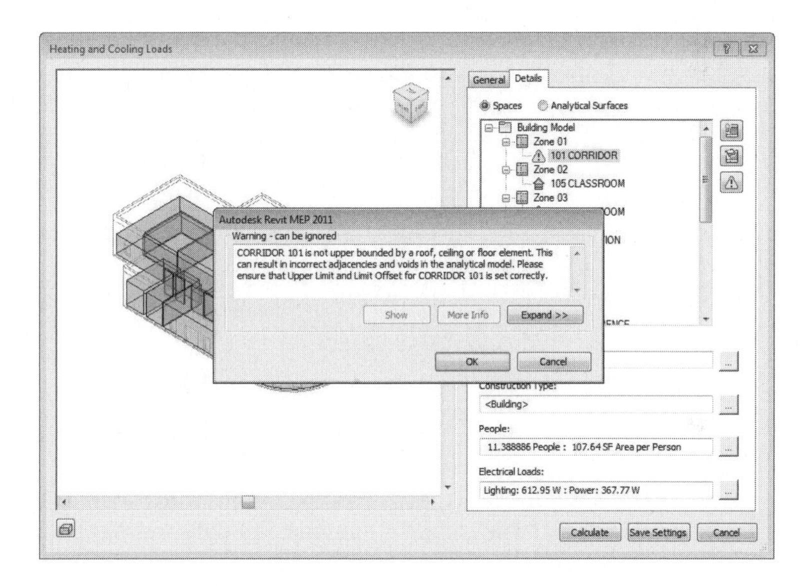

FIGURE 9.14
Warning

The Space Warning shown in Figure 9.14 suggests that a space exists without an upper bounding element. Click the Save Settings button, and you will be directed back to your model. Locate the space in question, and verify that the upper limit of the space is correct through the space Properties palette. You can also inspect the space visually by creating a section view through the space and verifying that the upper limit extends past a bounding element in the linked model.

Opening the Heating and Cooling Loads window and selecting Analytical Surfaces allows you to view and isolate the physical elements that bound the spaces to be analyzed. You are able to isolate every individual bounding element that has been defined for the space — roofs, exterior and interior walls, ceilings, floors, and any air gaps or sliver spaces — and view them for any modeling errors, before the simulation is performed.

Heating and Cooling Loads Report

Go ahead, click the Calculate button.

After the simulation is completed, you are directed to the Heating and Cooling Loads report, as shown in Figure 9.15. Depending on the level of report detail you selected prior to running the simulation, the tabulated results will be shown, broken into an overall project summary, a building summary, building level summaries, individual zone data, and individual space data.

FIGURE 9.15 Sample Heating and Cooling Loads Report

Project Summary

Location and Weather	
Project	Project Name
Address	
Calculation Time	Monday, May 31, 2010 3:25 PM
Report Type	Standard
Latitude	36.92°
Longitude	-76.24°
Summer Dry Bulb	97 °F
Summer Wet Bulb	81 °F
Winter Dry Bulb	23 °F
Mean Daily Range	14 °F

Building Summary

Inputs	
Building Type	Office
Area (SF)	4,395
Volume (CF)	43,657.18
Calculated Results	
Peak Cooling Total Load (Btu/h)	**173,779.6**
Peak Cooling Month and Hour	July 2:00 PM
Peak Cooling Sensible Load (Btu/h)	151,514.8
Peak Cooling Latent Load (Btu/h)	22,264.8
Maximum Cooling Capacity (Btu/h)	176,642.8
Peak Cooling Airflow (CFM)	6,901
Peak Heating Load (Btu/h)	**74,755.5**
Peak Heating Airflow (CFM)	2,644
Checksums	
Cooling Load Density (Btu/(h·ft²))	39.54
Cooling Flow Density (CFM/SF)	1.57
Cooling Flow / Load (CFM/ton)	476.56
Cooling Area / Load (SF/ton)	303.46
Heating Load Density (Btu/(h·ft²))	17.01
Heating Flow Density (CFM/SF)	0.60

The Project Summary area lists the project information (name, address, location), as well as calculated design date and time, Summer Dry Bulb and Wet Bulb temperatures, the Winter Dry Bulb temperatures, and the Mean Daily Range — values that should match the inputs you have entered with the weather data. The Building Summary area includes the global building type and its total analytical area and volume, as well as the overall calculated performance of the building — Peak Cooling Loads, Peak Heating Load, Airflows, and building checksums.

The Level Summary area includes the analytical area and volume of each level of the design building, if applicable, as well as each level's individual performance values, similar to the Building Summary. The Zone Summary lists each analyzed zone, listing its inputs, psychrometrics, and the calculated performance results. It also breaks down the various cooling and heating components and displays a list of the spaces that make up the zone along with a brief summary of the space performance. The Space Summary displays the space analytical areas and space volume, load inputs, and space type, as well as the calculated results for the space. It, too, contains a breakdown of the individual space components and how they contribute to the cooling and heating loads.

Revit MEP allows you to run a heating and cooling load analysis, make changes, and run subsequent analysis all while retaining the reports run for each analysis. Load reports are individually timestamped and can be accessed in the Project Browser under Reports. This enables you to easily flip to a previous report and quantify the changes in your design, without resorting to printing out each report as it is generated. Each report can grow to several hundred pages in length, depending on the size of your job and the detail level selected.

Now is a good time to set up and run a sample HVAC load analysis in Revit. Here's how:

1. Create a new Revit MEP 2011 file.
2. Link in the Architectural model file `ch9ArchModel.rvt` on the book's download page, `http://www.wiley.com/go/masteringrevitmep2011`. Use Origin To Origin placement.
3. Set the property of the linked file to Room Bounding by selecting the link, editing its Type Parameters, and checking the Room Bounding box. Click OK.
4. Place your spaces. Click the Analyze tab, and select Spaces to begin placing your spaces in the view. Either select the rooms you want to model, or click Place Spaces Automatically. Confirm the number of spaces created.
5. Create a working Space Properties schedule to manage space data. Click the Schedule/Quantities button.
6. Select Spaces as the category, and use New Construction as the phase. Click OK.
7. From the available fields, select the appropriate categories to match the Space Properties schedule shown in Figure 9.4.
8. Adjust the Name and Number fields to match the architectural link, either by manually entering the values or by using a space naming utility. Sort the rows in descending number order.
9. Notice that all the space types are set to <Building>, and the condition type is Heated and Cooled. If you examine the building type, you should see that it is set to Office. Select Manage ➪ Project Information ➪ Energy Settings to confirm. Click OK twice.
10. In the Space Properties schedule, select the space type cell associated with Office 102. Open the Space Type Settings dialog box by clicking the ellipsis in that cell.
11. Set the space type to Office – Enclosed. Notice the changes in the space parameter values. Verify that the Occupancy schedule is set to Common Office Occupancy – 8 AM to 5 PM and that the Lighting and Power schedules are set to Office Lighting – 6 AM to 11 PM. Click OK.
12. Notice the Number Of People value for Office 102 has changed. This reflects the different Area Per Person parameter value for the Space Type – Office vs. the Building Type – Office.
13. Set the remaining space type values appropriately.
14. Notice that the number of people in each space is taken out to six decimal places. Set the field format for this column to use fixed units with a rounding to 0 decimal places. Switch back to the 1-Mech floor plan view.

15. Click the Heating and Cooling Loads button on the Analyze tab. Switch the view to Details, and expand the Default tree.

16. Each space has a warning symbol attached to it. Select a space. Clicking the Warning button, you will see that the space is not upper bounded by a roof, ceiling, or floor element. You need to set the upper limits of each space to ensure accurate building analysis. Click OK and then Cancel.

17. In the 1-Mech floorplan, select all the spaces. In the Properties palette, set the upper limits of each space to Level 2, with a 0′-0″ offset.

18. Open the Heating and Cooling Loads window, and verify that all the warnings have been cleared from the spaces.

19. Click Calculate.

20. View the building performance report that has been generated.

21. Switching back to the 1-Mech floorplan, click the Zone button on the Analysis tab.

22. Create zones in the 1-Mech floorplan. Group the spaces as shown in Figure 9.16, and name the zones per the sample schedule.

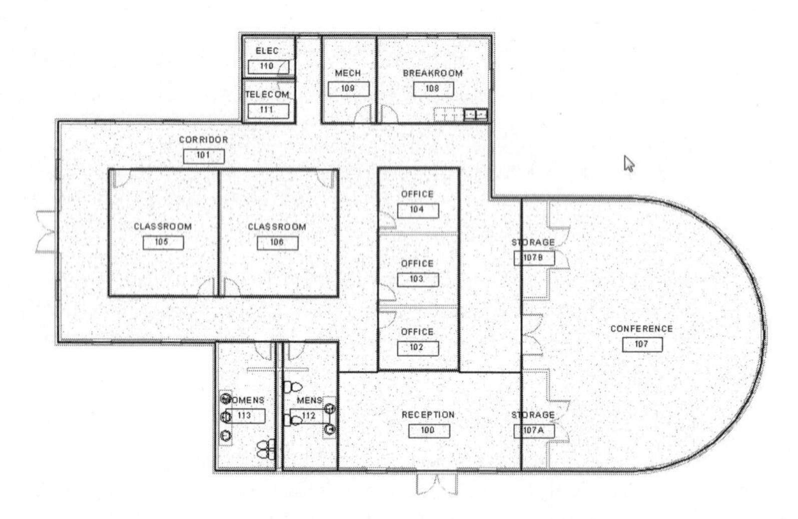

FIGURE 9.16
Grouping spaces in the floorplan

23. Select Zone 05 in the floorplan. Verify that the service type is VAV-Terminal Reheat and that the Cooling and Heating Information settings are both Default.

24. Click Edit for the Outdoor Air Information, and, for this zone, enter 5 CFM per person, and 0.06 CFM/SF in the appropriate boxes. Set the space occupancies equal to 1 person.

25. Run another building analysis.

26. Viewing the Zone Summary report for Zone 05, you should see that the Peak Ventilation Airflow is 20 CFM. The Ventilation Density value will equal 0.06 CFM/SF, the value you entered. The Ventilation/Person value, however, will equal 7 CFM/person, which

is not what was entered. Remember, Revit MEP takes the *largest* ventilation load, *not* the required combined ventilation load, when calculating the building performance.

27. To get the ASHRAE required ventilation, do the following:

   a. Calculate the required ventilation from area.

   b. Calculate the required ventilation from occupants.

   c. Add the two values, and divide by the total area.

   d. Enter this value in the Outdoor Air Information window for Zone 05.

28. Run the analysis again, and observe that Zone 05 Ventilation Airflow has changed to the airflow calculated in the previous step.

29. Using the space properties schedule, zone properties, and the heating and cooling loads window, adjust various parameters, and rerun the heating and cooling analysis.

30. Use the different reports generated to compare the effects of your changes to the building.

31. Close the file.

Exporting gbXML Data to Load-Simulating Software

Now we've explored how to set up a building for heating and cooling load analysis and how to perform the analysis in Revit MEP 2011. That's it, right? You're done? You have an accurate energy analysis of every component in the building heating and cooling systems?

Unfortunately, the engine built into this version of Revit MEP does not have the programming to perform a complete energy analysis of your building. Increasing LEED certification requirements need an equally increasing level of detail in the building analysis, comparison to baseline building design, and documentation to obtain the desired design certification. The solution to this dilemma is to export your Revit MEP model via gbXML format to a third-party simulation program, such as Trane TRACE 700 or Carrier's Hourly Analysis Program (HAP).

gbXML, or the Green Building XML schema, was developed in the late 1990s to enable interoperability between building design models and engineering analysis tools. Its use by major CAD vendors and several major engineering simulation vendors helps to streamline time-consuming building takeoffs — helping to remove some of the cost associated with the design of energy efficient buildings. Go to `http://gbxml.org/` for more details on the purpose and origin of gbXML.

To export your Revit MEP model, click ➪ Export ➪ gbXML on the Application menu. You will be directed to the Export gbXML window, shown in Figure 9.17. Even though the architectural model is linked into your project, the geometry will be exported to your analysis program.

This window looks nearly identical to the Heating And Cooling Loads window (shown earlier in Figure 9.11). If a heating and cooling load analysis has already been performed, the majority of the parameters in the General tab should already be defined per your design requirements. If not, these parameters have been discussed earlier in this chapter. Adjust the parameter values as needed.

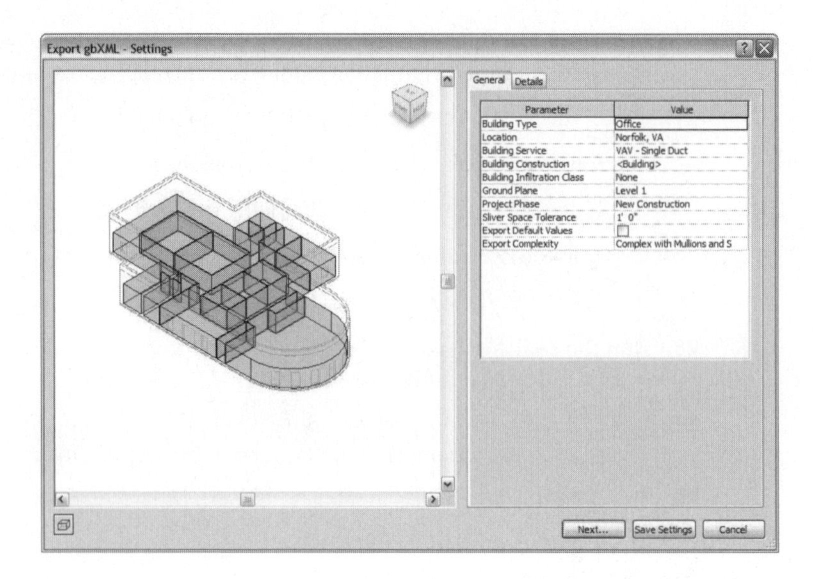

FIGURE 9.17 Export gbXML window

Two parameters that do not exist in the analysis window are Export Default Values and Export Complexity. If the Export Default Values box is selected, the default values for people, electrical loads, occupancy, lighting, schedules, and constructions will be exported, regardless of any user-overridden values. If this box is cleared, then only user entered values will be exported.

Export Complexity is simply that: the level of complexity of the information contained within the exported gbXML file. Five levels of complexity exist and are described here:

Simple Curtain walls and curtain systems are exported as a single opening. Simple complexity is used for the heating and cooling analysis and is most suited for exporting.

Simple With Shading Surfaces This is the same as Simple, but with shading surface information included.

Complex Curtain walls and panels are exported as multiple openings, each panel as a separate opening.

Complex With Shading Surfaces This is the same as Complex, but with shading surface information included; however, the shading surfaces (roof overhand, free-standing wall, and so on) are not associated with any room/space.

Complex with Mullions and Shading Surfaces This is the same as Complex With Shading Surfaces, but with mullions and curtain walls exported as simple analytical shading surfaces based on centerline, thickness, and offset.

The Details tab is identical to the analysis window, and as before, you are able to set or verify space energy parameters such as the building and space construction types, internal loads, and zone information like the service type and temperature set points. You are able to view and verify that all the analytical surfaces are correct and address any construction or space modeling warnings that are present before exporting.

Clicking Next, you are prompted to save the gbXML file. Browse to your project directory, and name the file accordingly. Your gbXML file is now ready to use.

Now you have complete model information set up to be read by a simulation program without issue and without any further manipulation. Right? Unfortunately, that is not the case.

A major issue with the exported data deals with the building construction. Revit breaks larger elements, such as the floor slab in large or complex-shaped space, into smaller polygonal elements for computation. This means that a single floor element is broken into several, individually tagged and defined, floor elements in the same space. Revit MEP also has a tendency, depending on how the architectural link is modeled, to assign exterior wall values to interior walls or roof values to ceilings that bound a space.

If your design space has an unusual or atypical geometry, such as a curved wall, the number of individual elements within a single type can grow to the hundreds. This is a prevalent issue when trying to import your gbXML file into an outside simulation program. There simply is not enough capacity of the third-party programs to handle large numbers of building surfaces — sometimes stopping at the first eight surfaces and causing you, the designer, to have to examine each space individually to delete and remodel any surfaces in error.

Also, importing errors may occur when parameters such as the construction data U-values fall outside a common range predefined by your simulation software. Often it will prompt a warning message, but other times, the gbXML data will fail to import completely. Most of the time, the remedy again is to check each individual building element construction or assembly to make sure that the appropriate U-values are modeled. Otherwise, you are forced to manually create the assemblies from scratch. In addition to checking the construction, make sure the internal load data has been properly read by your simulation program — manually adjust or enter the data as required.

Do not be discouraged. Yes, there are some drawbacks and trip-ups that may occur with exporting and importing gbXML data from Revit. As a designer, any effort you make to reduce the time-consuming process of modeling each space of the building in your simulation program will be worth the effort. Although there are some shortcomings, with Revit MEP 2011 you have the ability to examine the model geometry and data within the space prior to analysis, which can save time on your projects and increase coordination.

The Bottom Line

Prepare your Revit MEP Model for analysis The key element to a successful building performance analysis is the proper accounting of all variables that will influence the results.

Master it Describe the relationship between rooms and spaces — are they the same element? Describe an essential tool that can be created to maintain and track space input data and building construction for a heating and cooling load analysis.

Perform heating and cooling analysis with Revit MEP 2011 Before a piece of equipment can be sized or duct systems designed, the building heating and cooling performance must be known in order to accurately condition your spaces.

Master it How does project location affect building heating and cooling loads? Describe methods to determine project location in Revit MEP 2011.

What is a sliver space, and how does it affect the building performance?

Export gbXML data to load-simulating software Often, to complete the building analysis, the Revit MEP model has to be analyzed in greater detail by a third-party simulation program.

Master it What is gbXML? Why is it necessary to export your Revit MEP project?

Chapter 10

Mechanical Systems and Ductwork

Ductwork, like pipework, is a system family through which Revit MEP can calculate air flow rates and pressure drops on any correctly defined system. It also provides the graphics for the traditional "drawn" documentation and the variety of ways this can be represented on a drawing sheet at different stages of a project, regardless of whether this is "single line" at concept design or fully coordinated "double line" for construction issue.

There are three main types of duct: rectangular, round, and — new to Revit MEP 2011 — oval. These three types are the basis for your designs. They connect to air terminals and a variety of mechanical equipment. They also host duct fittings and accessories. By using the fittings and accessories available with the standard installation, the user can create supply, return, and exhaust systems with very little additional thought.

During your implementation, however, you should also consider the benefits of creating additional duct types that suit the way you work and your company standards.

In this chapter, you will learn to

- Distinguish between hosted and nonhosted components
- Use the different duct routing options
- Adjust duct fittings

Air Distribution Components

Air distribution components come in many different shapes and sizes. Depending on the design, they can be mounted in a variety of ways, including the following:

- Diffusers in a ceiling
- Duct-mounted sidewall diffusers
- Wall mounting
- Suspended

In each of these instances, the designer must decide whether to have hosted fittings (and if so, which type of hosting) or whether to not host at all. There are different — and similar — ways of placing these objects, and some may be conflicting.

An example of this is an installation where there are diffusers hosted in the ceiling as well as areas where the architect's design is for suspended fittings (see Figure 10.1). In this project, assume there is an external architect, and the architectural model is being linked. This means straightaway that you cannot use ceiling-hosted air terminals. Although you can see the linked ceiling, Revit only recognizes it as a face, and because of this, the air terminals will need to be created as face-hosted families.

FIGURE 10.1
Diffuser Hosting Methods

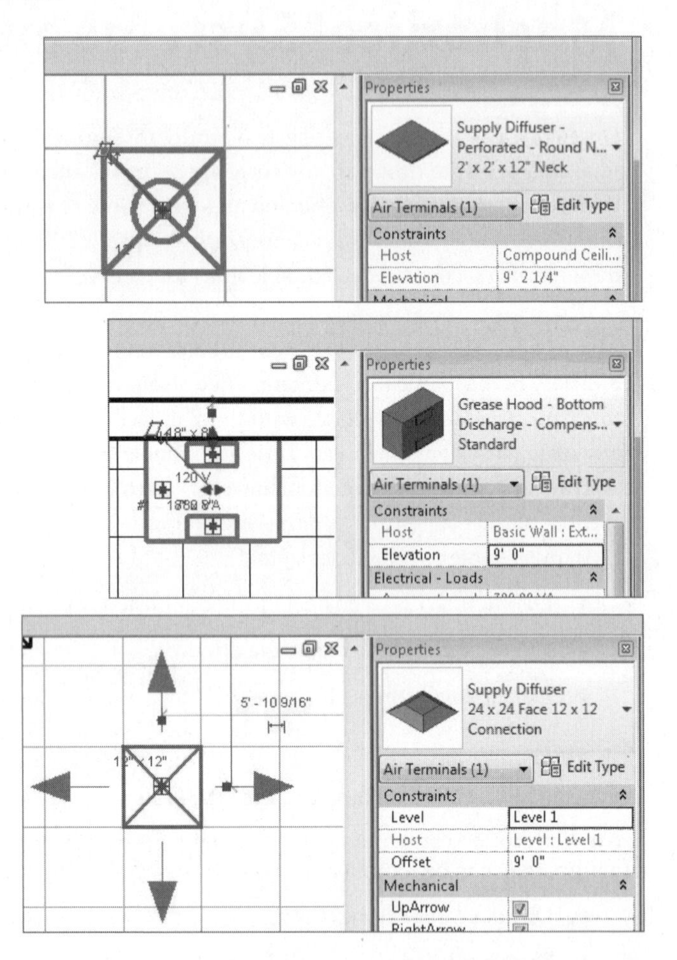

The most important point here is that once you commit to one type of family (hosted, non-hosted, wall, and so on), you can exchange for example, Air Terminals, with similarly created Air Terminals, not just those that are of the same category (see Figure 10.2).

The next problem is that the air terminals that are suspended are in fact the same type as those mounted within the ceiling tiles. The engineer wants to be able to schedule and filter them as one. How do you manage this?

New to Revit MEP 2011 is the ability to copy/monitor air terminals (as well as lighting fixtures, mechanical equipment, and plumbing fixtures). This means the services engineers can monitor the locations of air terminals that the architect has placed because the architect is responsible for placing these objects. As with the other items that can be copy/monitored, the services designer can choose to copy the original family type from the linked file or "map" it to one of their own choosing. In Figure 10.3, you can see, however, that once the air terminals have been copied/monitored, it does not matter what type of host association there is because this ceiling-hosted family has no host but is in the correct location specified by the architect.

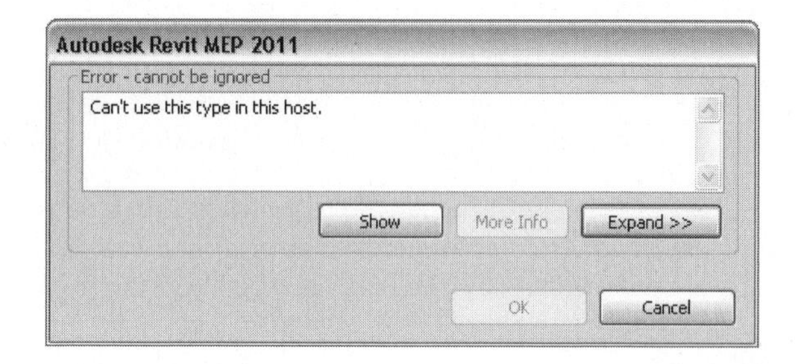

FIGURE 10.2
Hosting error message

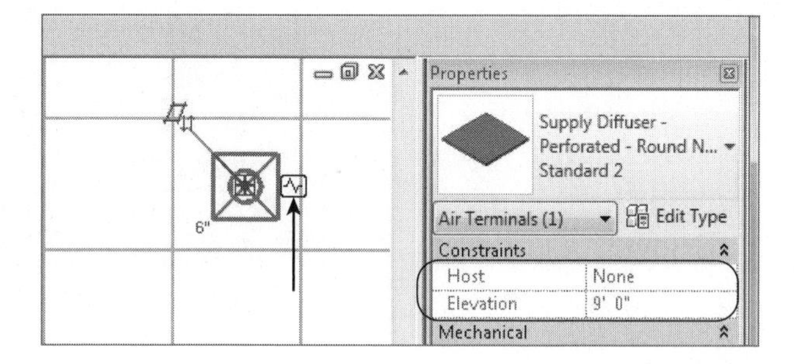

FIGURE 10.3
Copy/Monitor NonHosting

Mechanical Equipment Components

Mechanical equipment is the componentry that makes up the majority of large to medium-size plant objects for the mechanical designer. From air conditioners (ACs) and air-handling units (AHUs) to air curtains and heat pumps, these all provide the geometry and parameters associated with HVAC design. As with all components, choosing the hosting type is important. A level-hosted object cannot be exchanged for a face-hosted or ceiling-hosted one.

Air Conditioning/Handling Units

The heart of the mechanical air system, air conditioning/handling units can start life as generic "boxes" with intake and exhaust. Although basic in construction and with no manufacturer data attached, generic ACs/AHUs can have the same number of parameters as a more detailed family. Similar in dimension and performance, the concept box can be swapped out during the detail design period for a more detailed manufacturer or "construction-issue" family, or even set of families if the AC/AHU unit has been constructed from its manufacturer's component parts (see Figure 10.4).

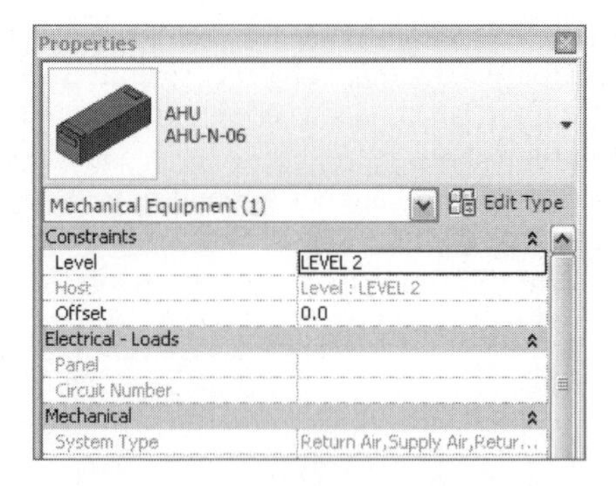

FIGURE 10.4
Basic AHU and type parameters

Placing the majority of AC/AHUs generally happens on the level that they are inserted on, with no offset. This can depend upon whether the unit is mounted on rails and whether those rails are part of the AC/AHU family (see Figure 10.5).

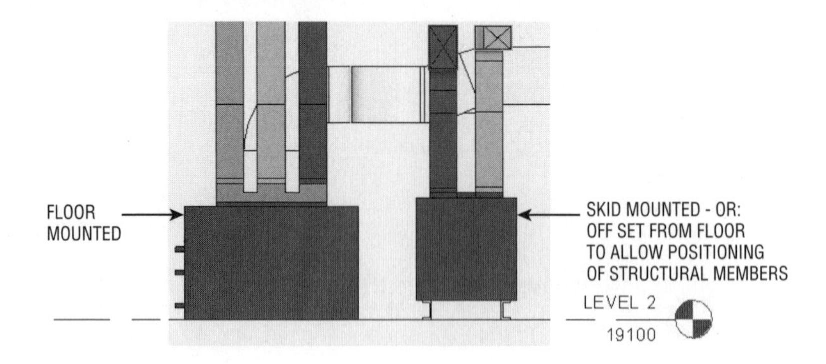

FIGURE 10.5
Floor-mounted and skid-mounted AC/AHU

MOUNTING OPTIONS FOR ACs/AHUs

There are two mounting options for ACs/AHUs.

FLOOR MOUNTED

In Figure 10.5, the mechanical plant on the left has been constructed to be mounted directly to the floor, which would probably be best created with a standard or default template (in other words, nonhosted).

SKID MOUNTED

With the example on the right side of Figure 10.5, the family can be created in a couple of ways:

- The family can be created in the same manner as the previous example, which allows for an offset to be applied. This allows for the structural engineer to locate elements in his file, and subsequent placement of the mechanical plant is then done in the MEP file.
- Alternatively, the Mechanical Equipment family has the structural elements built in, allowing the MEP designer to place the unit directly to the slab, premounted.

VAV Boxes

Generally created as nonhosted families, VAV boxes are usually mounted somewhere within the ceiling void, suspended from the underside of the slab above. In terms of placement, these are given an offset. One of the nice enhancements in Revit 2011 is that with the persistent properties box, the offset can be predefined, prior to the VAV (in this case) being placed. This makes for a much better workflow than previous releases where most objects were placed on the reference level and then subsequently moved to the correct invert level (see Figure 10.6).

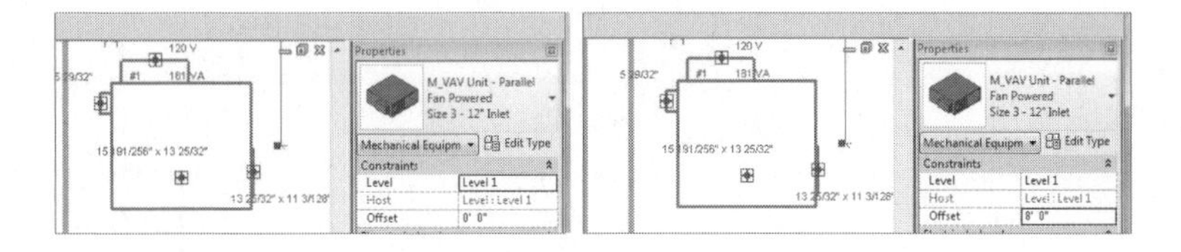

FIGURE 10.6
Changing the invert level

Connections to both AC units and VAV boxes can include heated and chilled water services and electrical for connecting to water and electrical systems (see Figure 10.7).

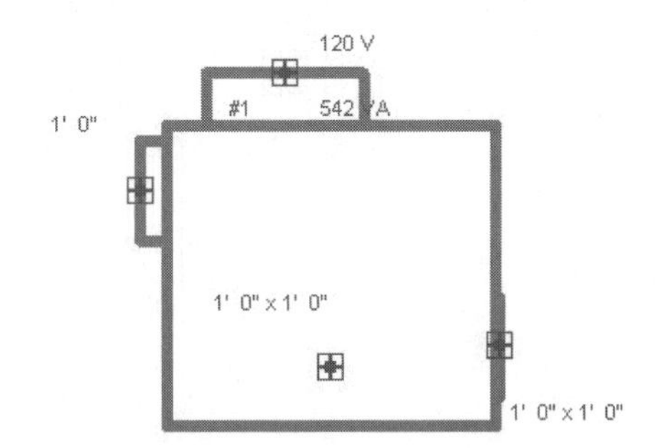

FIGURE 10.7
Typical mechanical and electrical connections to mechanical equipment

Ductwork

Ducts can be displayed in a variety of ways, including rectangular, round, or oval, as shown in Figure 10.8.

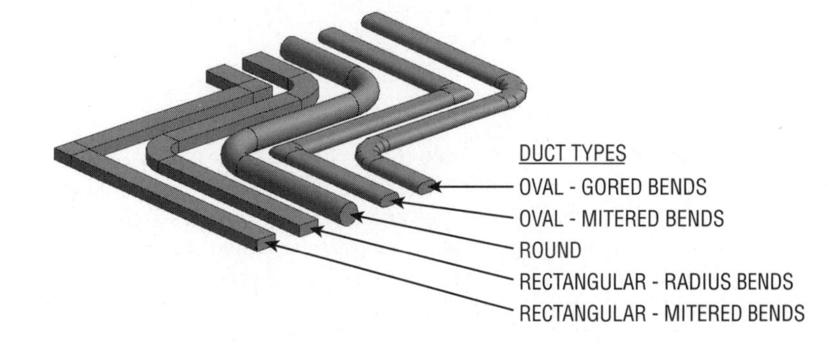

FIGURE 10.8
Ductwork

Ductwork is a system family and is the glue that holds systems together, but it also depends upon standard families to create a duct type, as described in the next section. Although ducts hold a huge amount of information, for the user, there are several important considerations to note.

First, ductwork and its associated fittings are systemless — they do not form part of a system until connected to points on mechanical equipment or air terminals. Figure 10.9 shows two ways in which duct can become associated with a system once connected to equipment.

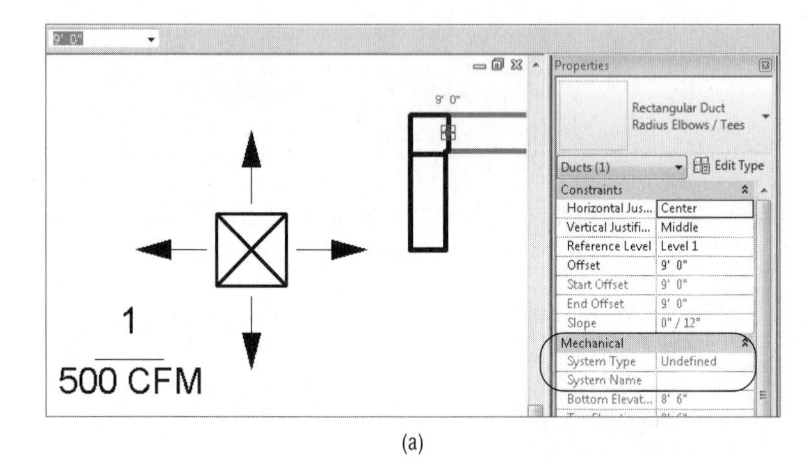

(a)

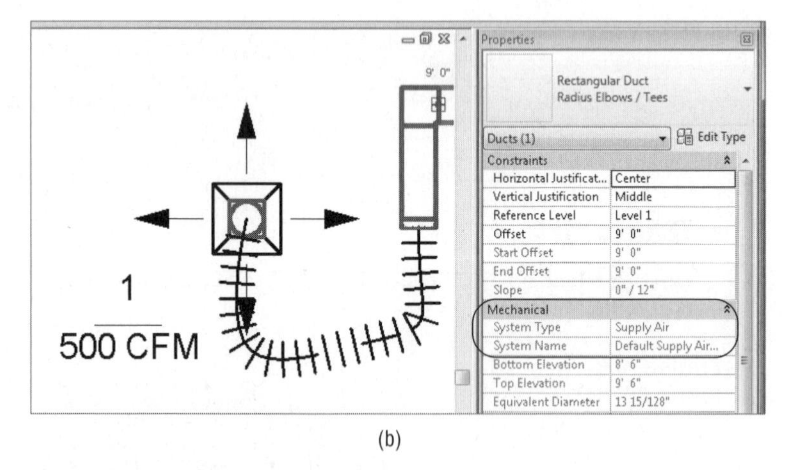

(b)

FIGURE 10.9 Ductwork unassigned to any system (a), ductwork assigned to default supply system (b)

Second, a named system will be more efficient in the design environment than the default setting (see Figure 10.10). This is because elements associated with systems aid the correct flow of data throughout the model, ensuring that analysis tools such as duct sizing work to the optimum. Ensure that systems are adequately named. This aids the designer and drafter in locating systems and objects related to them. An example of this is when you have a floor that is divided into zones; a system could be called Level 1 Zone A (see Figure 10.11).

FIGURE 10.10
System naming

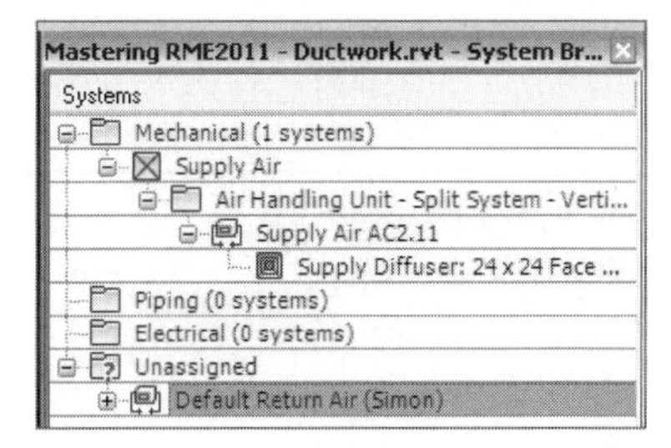

FIGURE 10.11
Ductwork assigned to named supply system

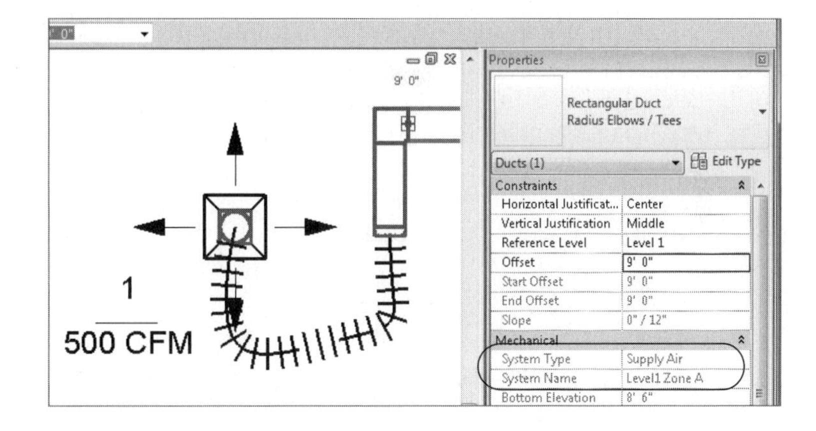

Third, when running an interference check, know that duct insulation will *not* form part of the calculation (see Figure 10.12) when only one interference has been detected.

FIGURE 10.12
Interference check

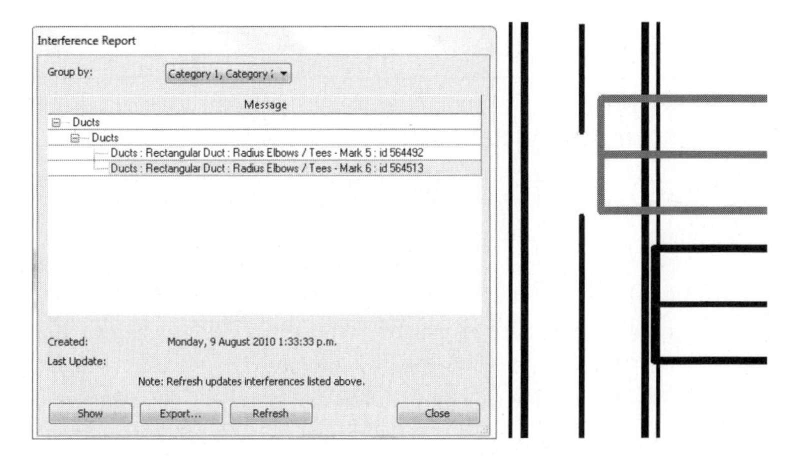

Duct Types and Routing

Creating new duct types is a way of managing how your duct runs work and connect into each other, such as whether bends are mitered or radiused. Although you can retrospectively change any of the fittings inserted, it is much easier to create a run in one go, using either the automatic or manual tools.

Creating New Duct Types

Although it may be tempting to create types such as Extract and Supply, try to keep these names more specific, such as Stainless-2D Radius/Taps or Galvanized-Mitered/Tees. This way, the user only has to think about what material the duct is. Although using a name that specifies Supply or Extract may be initially attractive, doing so can lead to ambiguity and misunderstanding among the modeler, designer, engineer, and potentially the client. Another downside to this is the need to create multiple types of bends, tees, crosses, and so on, for all the different material type *and* system types you are likely to use.

 **Real World Scenario**

SCHEMATIC LAYOUTS

Mike has been given the task of preparing a schematic ductwork layout for mechanical engineer Steve. Because the design architect has yet to supply a BIM model, he has some initial 2D AutoCAD plans and sections. Steve completes a rough design on paper along with calculations for duct sizes. When Mike starts the job, he links the AutoCAD section into a new section view. From here he can create building levels based on the section. With this done, floor plans are created, and the AutoCAD plans are linked into the relevant levels. Mike can now create a single-line, schematic layout for the project. At this stage, floor offsets, downstand beams, and coordination between services almost can be ignored, because the duct is being shown only as a single line. The main benefit of this is that the bulk of modeling for the duct system can be achieved at an early stage of the project and retained/modified as the project progresses.

Automatic Duct Routing

When using the automatic routing tools, as a rule of thumb, you should work on small sections — all the feeds to a VAV box is one good example. This means the computer has fewer objects to calculate, and the routing suggestions have less room for error. Before even starting this process, check the options under Mechanical Settings (see Figure 10.13) for default duct types, offsets, and length of flex duct, because these are used during the routing process.

Now you're ready to use automatic duct routing. Here's how:

1. Place your VAV box, ensuring it is located at the correct height. You can change it later if needed, but this may lead to you changing your duct route.
2. Place your air terminals. At this point, it is a good idea to consider the following, not from a design point of view but from a Revit one:
    - What type — if any — are the ceilings?
    - If there are ceilings, should you use face, ceiling, or unhosted families?

- Are you going to create your own placeholder ceiling to host your families?
- Should you use the new ability to copy/monitor the air terminals already placed by the architect?
- Choose the type of air terminal — is it top, side, or even sidewall entry?

3. Once you've made all these decisions, it's time to start laying out the equipment. Figure 10.14 shows a space where the upper limit of the space has been defined as the level above (1), the specified supply airflow has been entered manually (2), calculated supply airflow (3) is not computed because the analysis tools have not been used for this building, and the flow rates for the air terminals have been adjusted to suit the calculation (4).

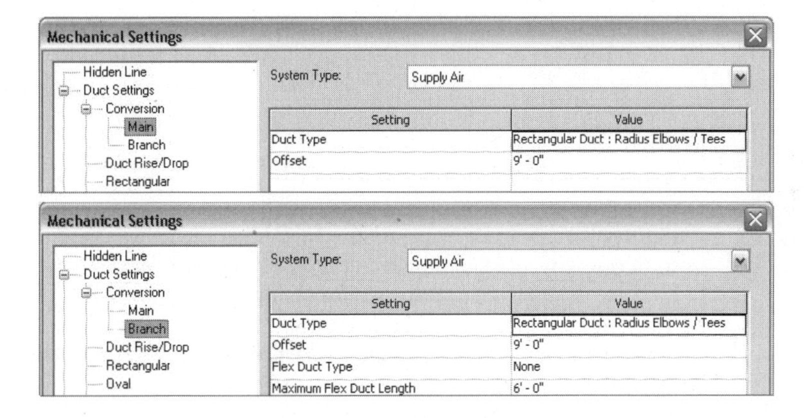

FIGURE 10.13
Mechanical settings

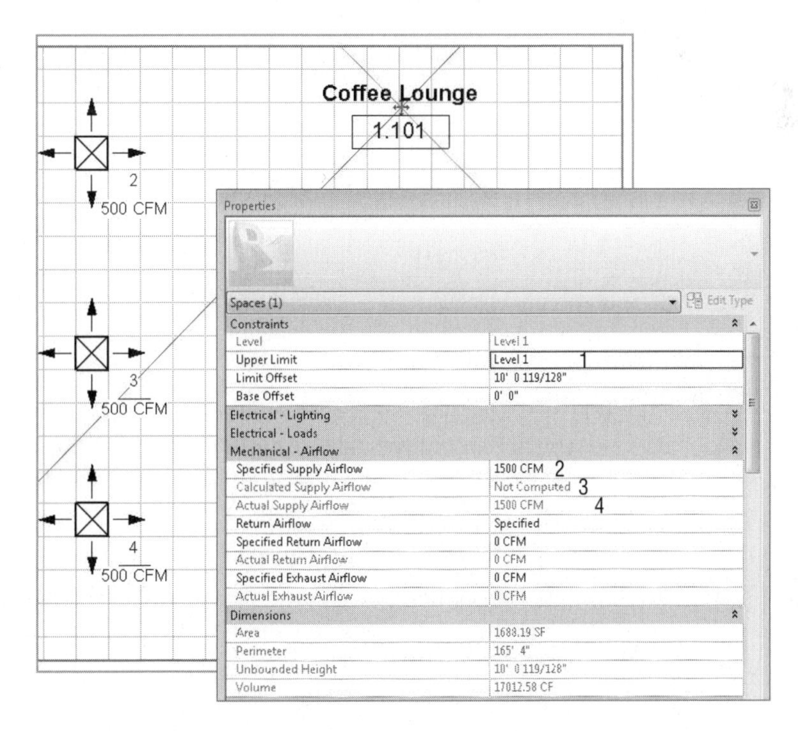

FIGURE 10.14
Space properties

4. Select one of the air terminals, and on the ribbon, click Supply on the Create Systems panel (Figure 10.15).

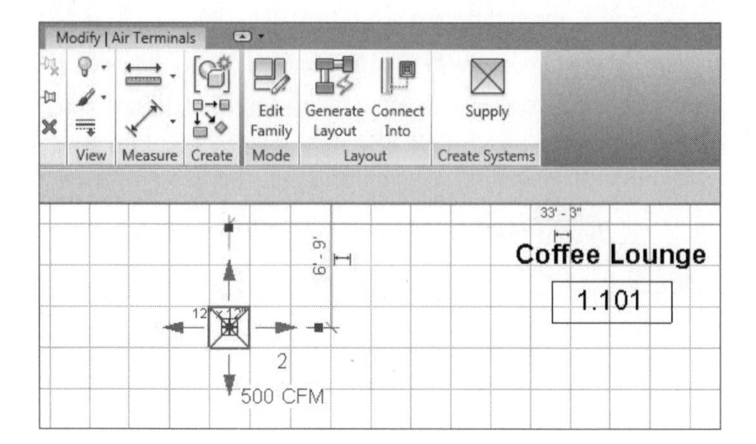

FIGURE 10.15
Creating systems

Figure 10.16 shows the option for choosing a suitable name for the system and then adding to the system. Objects that are not part of the current system appear in halftone, but because they are selected, their appearance changes to full weight.

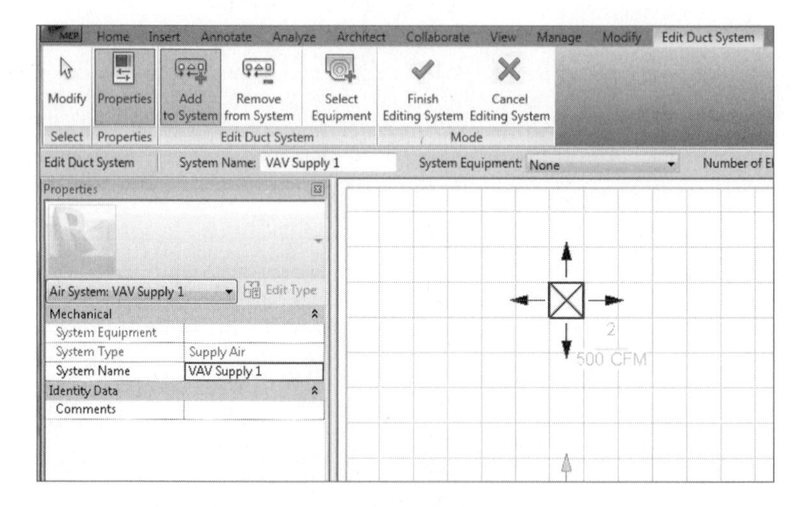

FIGURE 10.16
Add To System command

5. Once you have completed adding air terminals, click Select Equipment, and either select the equipment from the Options bar drop-down or select the actual VAV box indicated on the plan, as in Figure 10.17.
6. Complete this task by clicking Finish Editing System. This is where the fun begins!
7. Hover over one of the items in the system (but do not select it), and press the Tab key. All the items in the system, air terminals, and the VAV box should now be connected with a dotted line (see Figure 10.18).

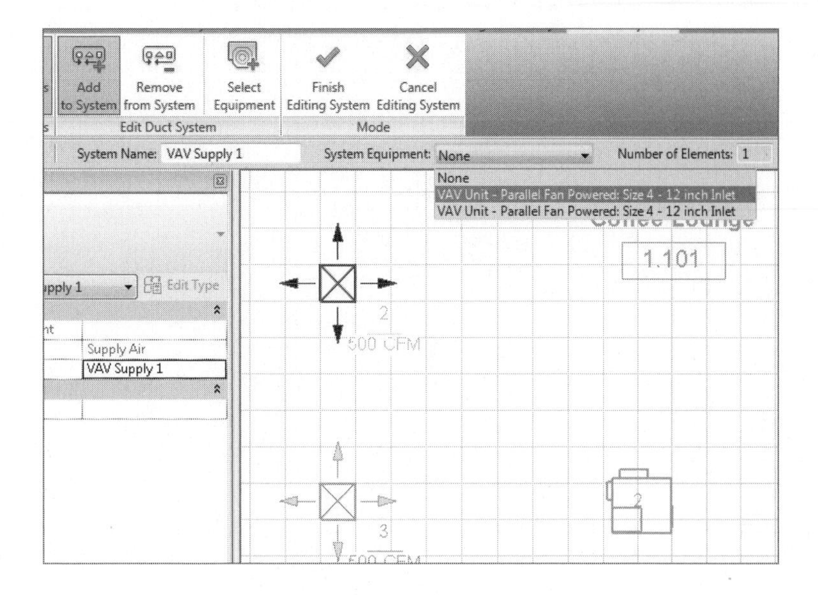

FIGURE 10.17 Selecting equipment

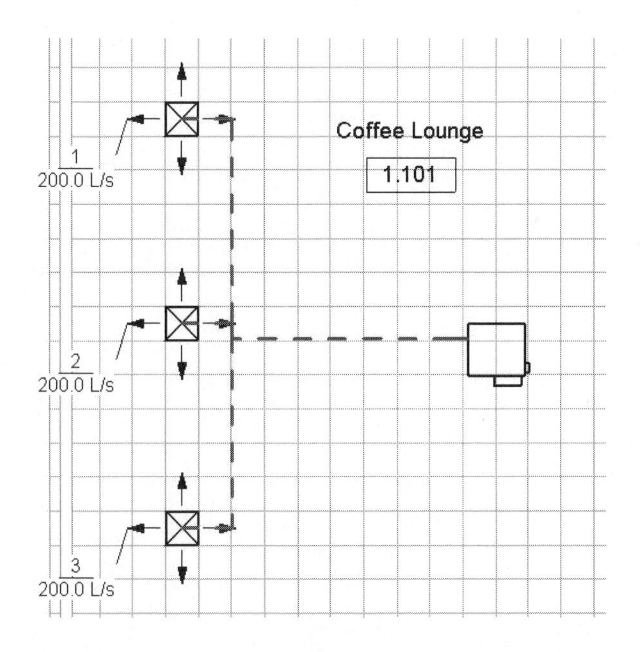

FIGURE 10.18 Tab-selecting the system

If you now click one of those objects in the system, the dotted lines turn red in color. This indicates that you have selected the system, not the objects themselves. The properties of the system will appear and can be edited in the Properties palette.

8. As shown in Figure 10.19, you can now see the Generate A Layout option.

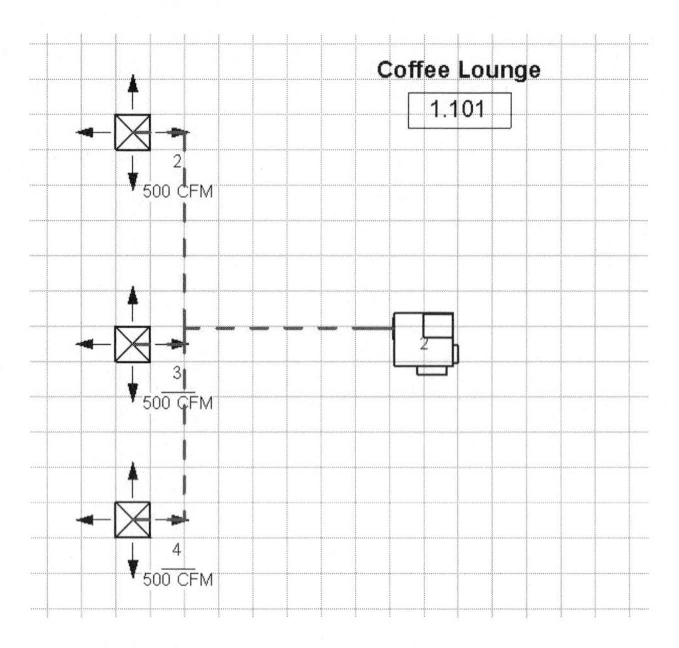

FIGURE 10.19
Generating a layout

9. There are now several options for automatically generating your duct layout, including the Network, Perimeter, and Intersection options. Each of these options can give you several solutions that can vary depending on predefined settings for the duct layout, which can also be accessed from the Options Bar. In Figure 10.20, you can click the Edit Layout button to make further changes to the layout.

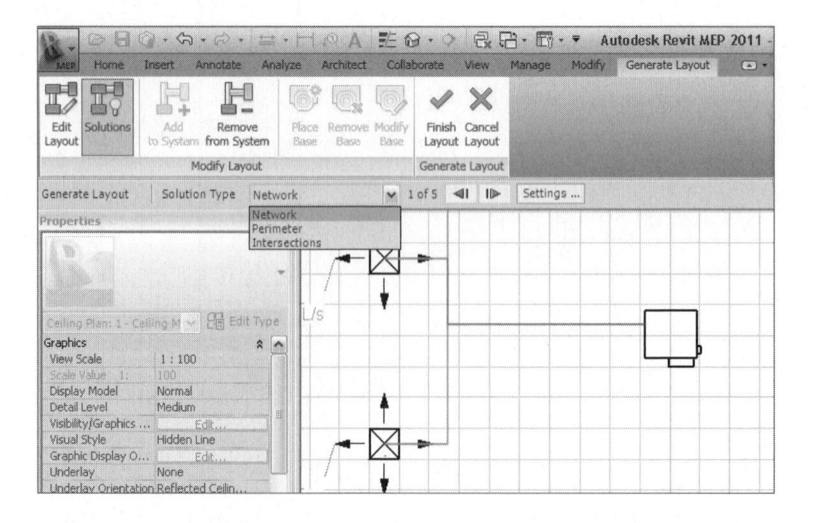

FIGURE 10.20
Ductwork routing solutions

10. Main duct runs are shown in blue, while branch runs are green. Once you have settled on your preferred layout, click Finish Layout, and the duct layout is created (see Figure 10.21).

It's worth pointing out here that the sizes used for this layout are based on the connection sizes and the settings for duct sizes located in the Mechanical Settings dialog box. There is another important consideration at this point. The designer/drafter must ensure that the default settings for the main and branch ducts are adequately high enough above the air terminals and associated equipment to generate the layout; otherwise, the layout tool will crash.

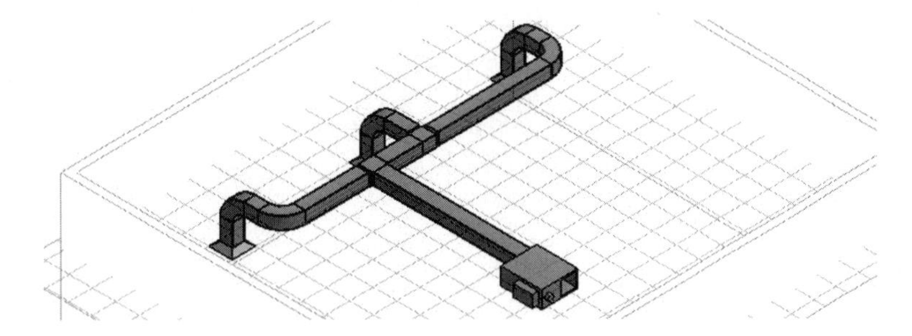

FIGURE 10.21
Completed duct layout

Manual Duct Routing

For the experienced design drafter, using the automatic tools may seem to be too limiting; however, through use, most users eventually settle for a variety, sometimes for no better reason than a "change is as good as a rest." Sometimes, however, the manual tool is much more efficient, especially when connecting different areas into a system or laying out runs back to a rooftop AHU.

To begin manual duct routing, do the following:

1. Click the Duct tool on the Home tab on the ribbon, and then choose the type and its various options from the Properties box, as shown in Figure 10.22.

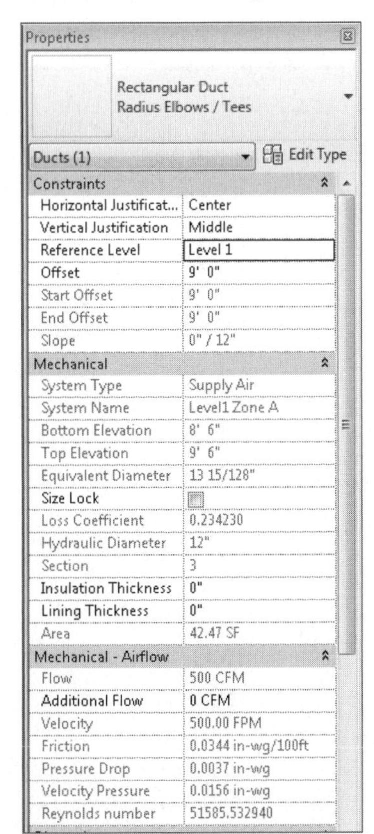

FIGURE 10.22
Duct types

Note the default settings for constraints, including Justification, Ref Level, and Level offset. The System type is undefined, and as the duct is unconnected, the airflow is 0.

You can adjust additional properties from the Options Bar and the Modify/Place Duct tab.

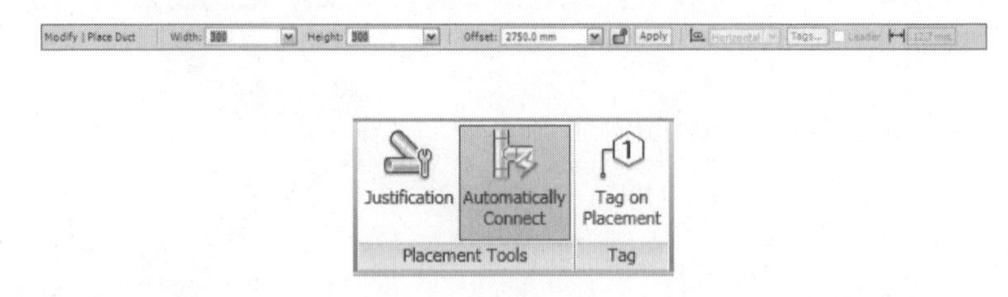

2. The duct can then be created/drawn to whatever path you choose. For vertical offsets, type the new height into the offset box on the Options Bar, but for angled setup/down, it is best to work in an elevation or sectional view.

For this example, the duct has been split into sections.

3. Hover over one end of the duct, right-click the connector, and select Draw Duct. Draw your duct, selecting the preferred route.

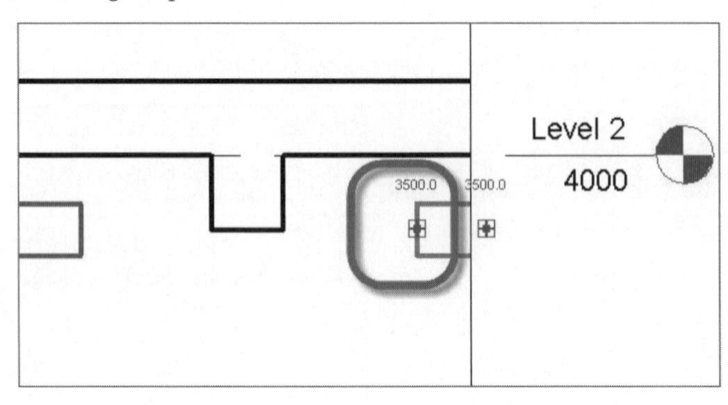

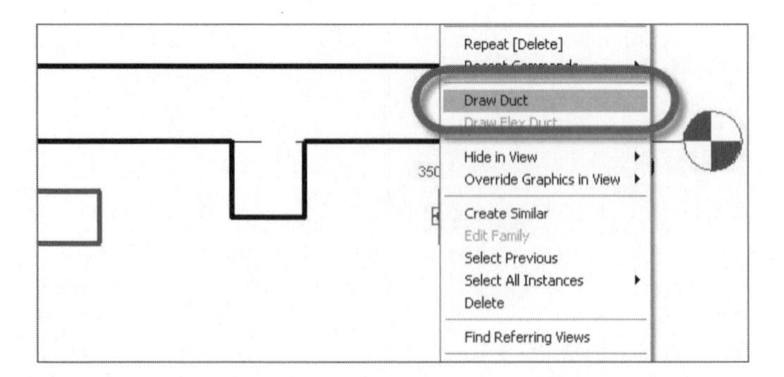

4. Using tools like the Trim command, you can complete the run as in Figure 10.23 and Figure 10.24.

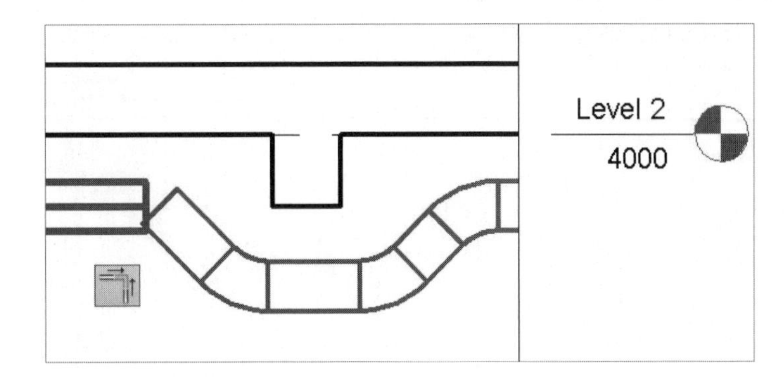

FIGURE 10.23
Trimming the duct

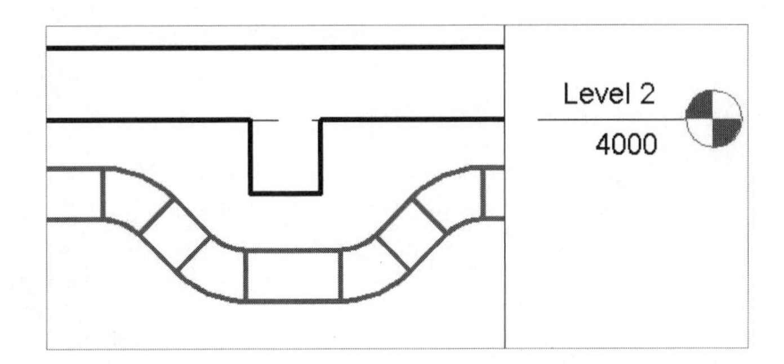

FIGURE 10.24
Completed duct

When creating ducts that set down/up or rise/drop, it is much easier to model these in a sectional or elevational view (as in Figures 10.23 and 10.24). To do this, first you need to create a suitable view. (Quite often individuals have a "personal section" that they move around the model. It can be opened when required, and then the duct — or whatever service being worked on — is modified, and the view is closed.)

ADJUSTING FITTINGS AND EXTENDING THE DESIGN

To extend the design (Figure 10.25), select one of the fittings (typically bends and tees). You will notice a small plus sign (+) adjacent. Clicking this will turn a bend into a tee and a tee into a cross, eliminating the need to delete fittings and inserting an appropriate one.

Duct Sizing

The most important factors to consider when using the duct sizing tools are that the ducts form part of a system and that this system should have a nominated name, not a default, which is created at the same time as the system being created. The system must also have a valid air flow, so either you specify the airflow of the air terminals or that flow is specified from the space and volume calculations.

Use the Tab key to select your system (Figure 10.26).

With the system selected, you are now able to use the Duct/Pipe Sizing Tool (Figure 10.27).

Figure 10.28 shows the options available when using the Duct Sizing Dialog tool.

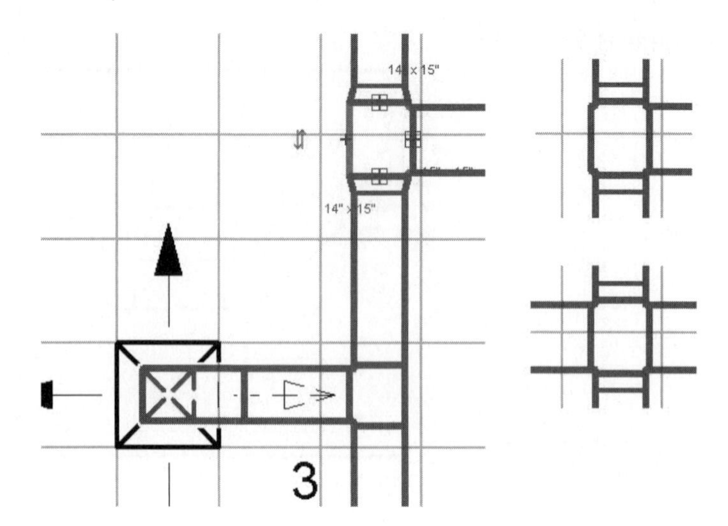

FIGURE 10.25
Extending the design

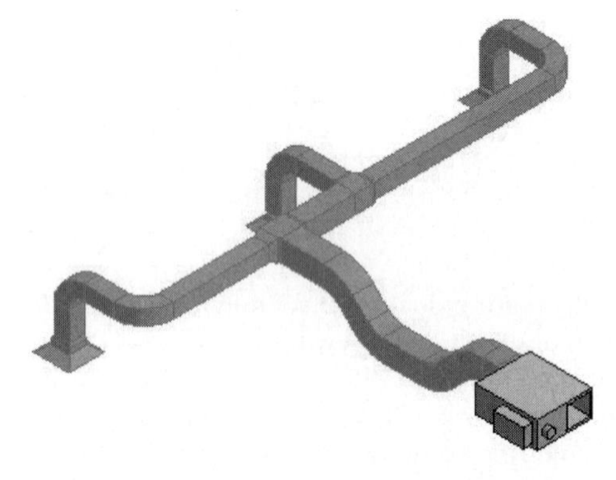

FIGURE 10.26
Tab-selecting the system

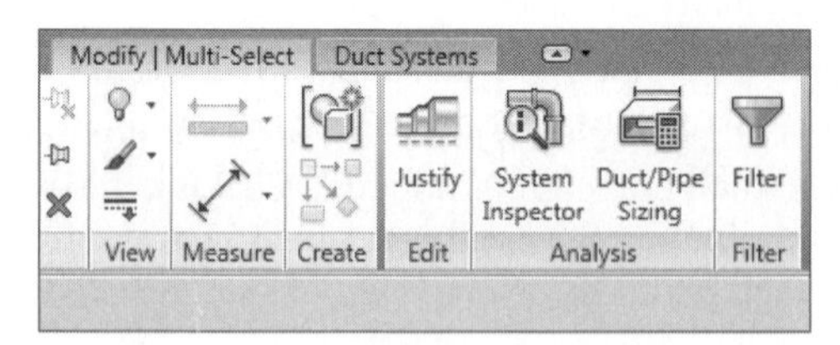

FIGURE 10.27
Duct/Pipe Sizing tool

Duct Sizing Methods

Although the sizing method you choose can be applied to an entire system, this is not necessarily the most efficient way to do things. If there are 5,257 objects in your system, it's probably not a great idea to ask your computer to process that amount of information — it may take a while! The logical choice is to split the task of duct sizing into manageable sections, such as a floor plan, a zone, or a group of air terminals fed from the same VAV box.

The various methods for sizing are as follows:

- Friction
- Velocity

- Equal Friction
- Static Regain

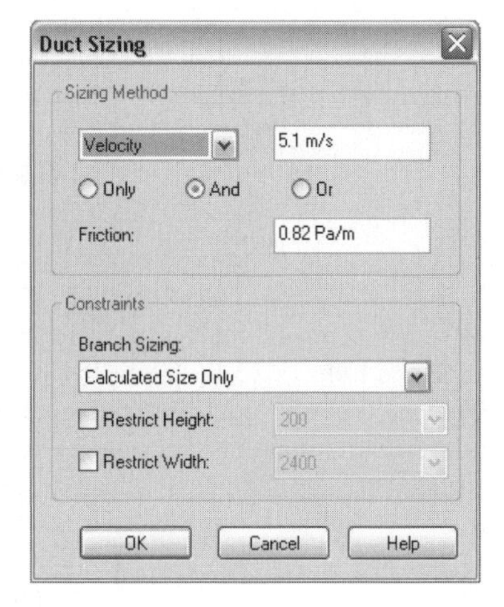

FIGURE 10.28
Duct Sizing dialog box

Friction and Velocity can be used independently of each other, when using the Only method. Alternatively, they can be used in conjunction with each other with the And and Or functions.

These allow the designer to force the sizing ducts to meet the parameters specified for both Velocity and Friction. With the Or method, the least restrictive of either of the parameters is used.

The Equal Friction and Static Regain methods use the ASHRAE Duct Fitting Database.

Air properties are set in the Mechanical Settings dialog box, as are the sizes of ducts used in the actual sizing calculations (Figure 10.29 and Figure 10.30).

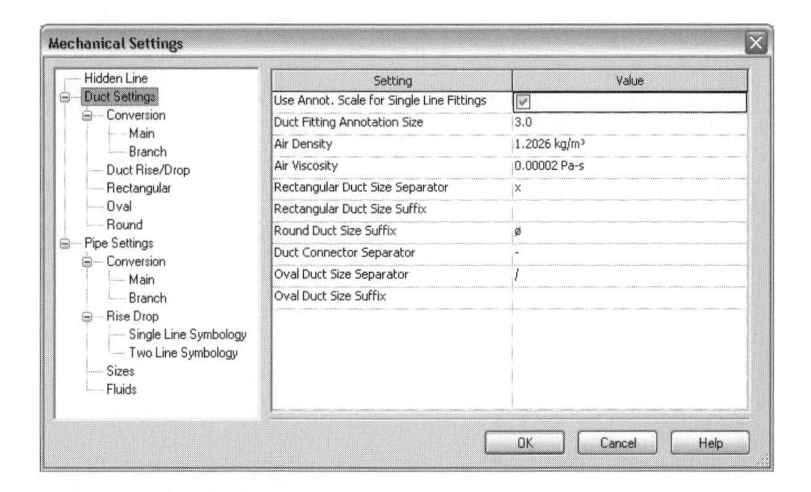

FIGURE 10.29
Mechanical Settings dialog box, air

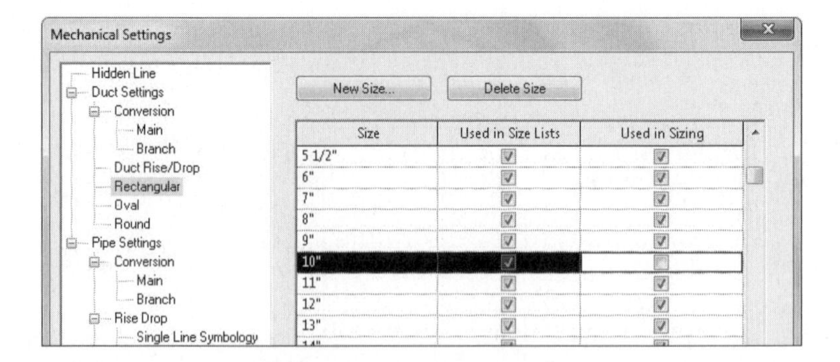

FIGURE 10.30 Mechanical Settings dialog box, duct sizes

During the sizing process, you can also apply constraints to the branch parts of the run. These are defined as follows: Calculated Size Only, Match Connector Size, and Larger Of Connector And Calculated (Figure 10.31). Depending on the stage of the design or your actual role (such as consultant), you could choose to select Calculated Size Only, because the design is still in early stages and equipment is Generic.

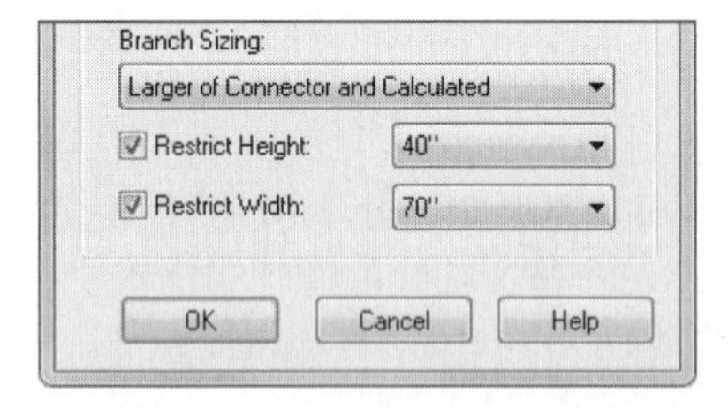

FIGURE 10.31 Duct sizing constraints

However, later in the project or as a contractor, when the equipment has been specified, you may want to select Match Connector Size and Larger Of Connector And Calculated to reduce the number of duct sizes used on the project, reducing your manufacturing costs.

This also allows you to specify a limit on the size of the ducts, which can further reduce costs or give you the ability to specify, say, a continuous duct height where you know access is a potential issue.

Factors That Do Not Affect Duct Sizing

When designing your system, you should be aware that Revit does not take into consideration external or internal insulation in sizing calculations or in clash detection. Duct insulation *does not* form part of any interference checking, so although you can see the clash, it does not get reported. In addition, duct lining can be shown on the plan but does not affect the airway size. These factors may be important, especially in areas where accessibility is an issue; the designer should be extra vigilant in these situations.

Lining and Insulation can be added only as instance parameters, so if the system depends upon a lined or insulated duct, the user will have to select all ducts for the system and then specify the insulation or lining required. This would also have to be done to elbows, tees, and so on. One tip here is to apply the required insulation in a schedule (Figure 10.32).

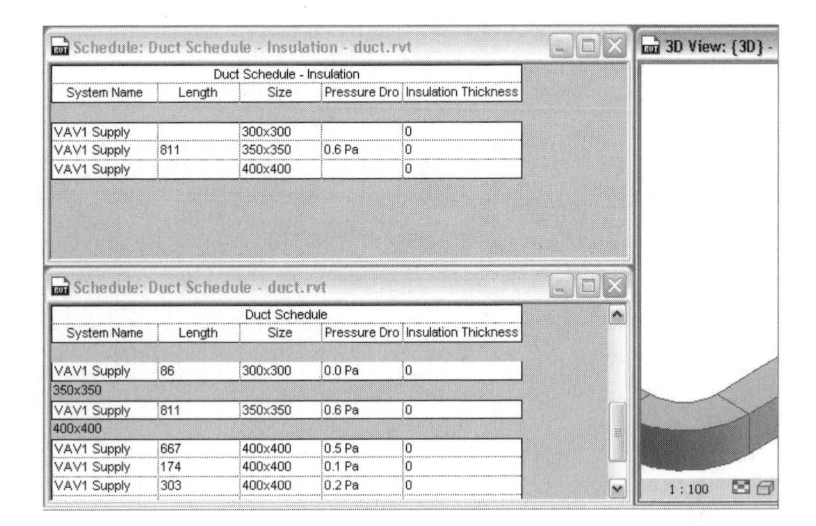

Duct Schedule - Insulation

System Name	Length	Size	Pressure Dro	Insulation Thickness
VAV1 Supply		300x300		0
VAV1 Supply	811	350x350	0.6 Pa	0
VAV1 Supply		400x400		0

Duct Schedule

System Name	Length	Size	Pressure Dro	Insulation Thickness
VAV1 Supply	86	300x300	0.0 Pa	0
350x350				
VAV1 Supply	811	350x350	0.6 Pa	0
400x400				
VAV1 Supply	667	400x400	0.5 Pa	0
VAV1 Supply	174	400x400	0.1 Pa	0
VAV1 Supply	303	400x400	0.2 Pa	0

FIGURE 10.32
Scheduling duct

In this example, there are two schedules, one showing the system you are modifying, with all the ducts (Figure 10.33). The other is a duplicate but doesn't show All Instances (Figure 10.34). When you type in the required insulation thickness, rather than it being applied to only one section of duct, it applies to all ducts of the same type in that system.

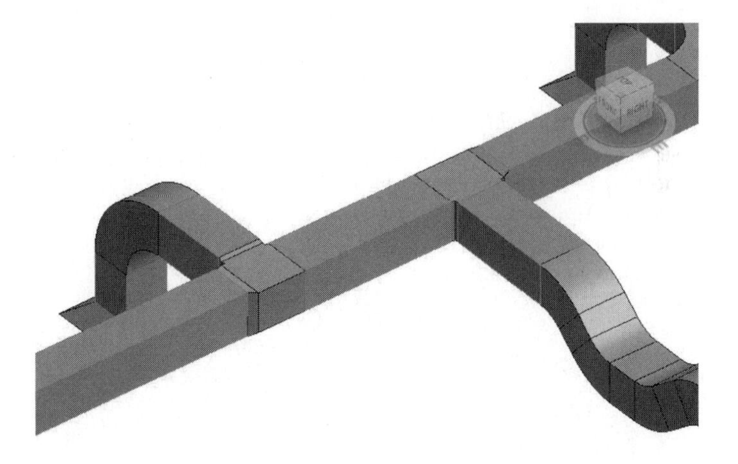

FIGURE 10.33
Scheduling duct

Duct sizing will work only where air terminals, ducts, duct fittings, and mechanical equipment are seamlessly connected, with no gaps, and where the equipment has a defined airflow. This airflow could be entered as part of the initial analysis or subsequent manipulation of the objects. If you are using Revit to do your documentation process without using the analysis and design tools, then all duct sizing will need to be done manually.

FIGURE 10.34
Applying insulation

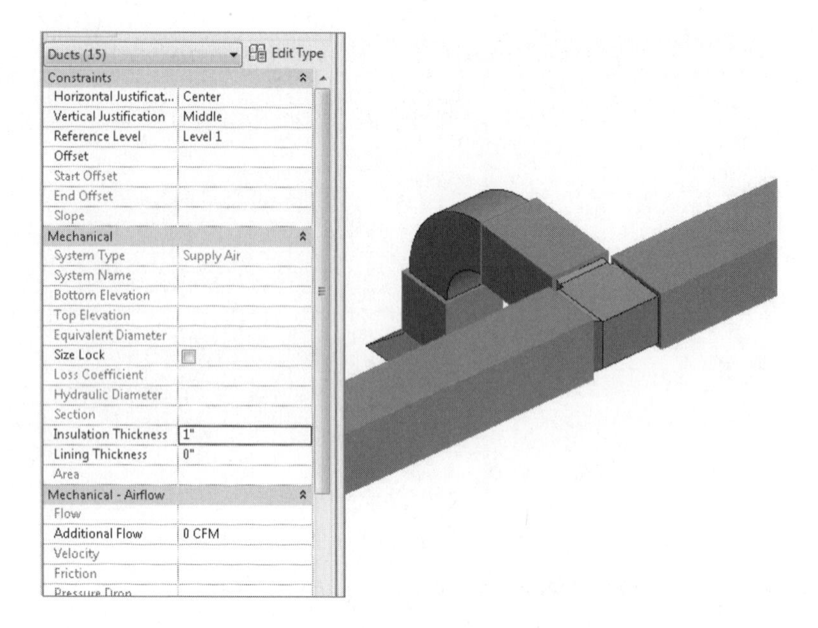

Real World Scenario

DUCTING AND COORDINATION

Multinational Consulting and Engineering company Beca Carter Hollings and Ferner, based in Auckland, New Zealand, planned to roll out Revit MEP to their engineers and drafting staff. Although there is a strict implementation process, offices around the country and farther afield take hold of the reins and give them a good shake. While the engineers are getting familiar with the design tools, Beca decides to develop Revit to provide documentation for projects and train its drafting staff to achieve this end.

With Revit Architecture rapidly becoming the documentation and virtualization tool of choice, the next logical step is for building services. With this process relatively new to Beca, the decision was made to complete documentation in new projects using Revit MEP in areas where a high level of coordination is required. In circulation and office areas where coordination, although important, is not critical for the consulting engineer, traditional 2D drafting methods are still employed.

On a new hospital project, Beca uses this technique to construct major plant areas. Very quickly the company discovers the value of this high level of coordination as the complex plant areas become extremely crowded. The image shown here indicates only the ductwork in this area (for a larger, color image, refer to the website file `cs1001.jpg`). Interference checking is used within Revit during the documentation process to ensure services are missing not only each other but also architecture and structure elements.

With design still being done in other software, modeling the plant areas with Revit MEP enabled Beca to concentrate on utilizing new software to its optimum without committing additional untrained staff to an untested environment.

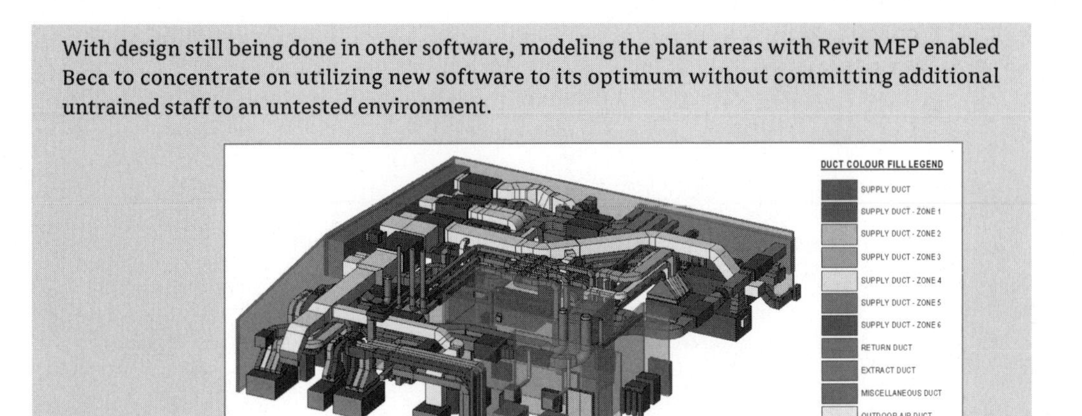

The Bottom Line

Distinguish between hosted and nonhosted components Deciding whether hosted or nonhosted components are used is crucial for the success of your project. It will play a large factor in performance and coordination with other companies.

Master It Should you choose hosted or nonhosted components for your project?

Use the different duct routing options When using Revit MEP 2011 for your duct layouts, the user must understand the functions of automatic duct routing and manual duct routing. Once these functions are mastered, then the user can lay out any type of ductwork system.

Master It When asked to submit a design proposal for a multifloor office building, the HVAC designer needs to show a typical open plan office and the supply and extract ductwork. How should the designer start this process?

Adjust duct fittings Duct fittings are needed in systems to make the systems function properly and to produce documentation for construction. Being able to add or modify fittings can increase productivity.

Master It You have just finished your modeled layout and given it to your employer for review. Your boss has just came back and has asked you to remove a couple of elbows and replace them with tees for future expansion. What would be your method to accomplish this quickly?

Chapter 11

Mechanical Piping

Mechanical piping is the lifeblood of a heating/cooling system. If not piped correctly, it can lead to problems in the field, and locating them may take months. There are simple two pipe systems and more complex multipipe systems. When using Revit MEP 2011 to lay out your systems, you will be able to very easily see your routing options and even calculate the total volume of fluid in your system.

In this chapter, you will learn how to

- Adjust and use the mechanical pipe settings
- Select and use the best pipe routing options for your project
- Adjust pipe fittings
- Adjust the visibility of pipes

Mechanical Pipe Settings

When setting up the mechanical piping to route, you will want to apply the proper pipe material. The purpose of this is to provide a more accurate layout by showing the correct pipe sizes and fittings for that system. You will have to adjust several areas to set this up properly. These areas are system pipes, fittings, pipe material, pipe sizing tables, and the fluids table, each of which is described here in more detail. Once you set these areas, you can then concentrate on the autorouting or manual routing of pipe.

System pipes These are the pipes that are hard-coded into Revit. You have a limited amount of freedom to adjust parameters for these system families.

Fittings These can be applied to the parameters of the system pipes, which will automatically populate the pipe system and the model. The pipe fittings must be loaded into the model for the system to work.

Pipe material This is set in Mechanical Settings ➪ Sizes. This allows you to duplicate a pipe material, rename it, and apply the piping specification of material and roughness of pipe wall.

Pipe sizing table This is set in Mechanical Settings ➪ Sizes. This allows you to duplicate and adjust inside pipe diameter, outside pipe diameter, and pipe sizes to the manufacturer's specifications.

Fluids table This is set in Mechanical Settings ➪ Sizes. This allows you to duplicate, rename, and adjust fluid type as well as the viscosity, temperature, and density.

Creating System Pipes

To modify and create new system pipes, in the Project Browser select Families ➪ Pipes ➪ Pipe Types, and then right-click the Standard pipe. Select Duplicate. This will create the additional pipe types you will need for your project. Next, right-click the duplicated pipe types, and rename them to the pipe type you require. We prefer to rename them to the pipe material type for takeoff purposes. (Refer to Figure 11.1.)

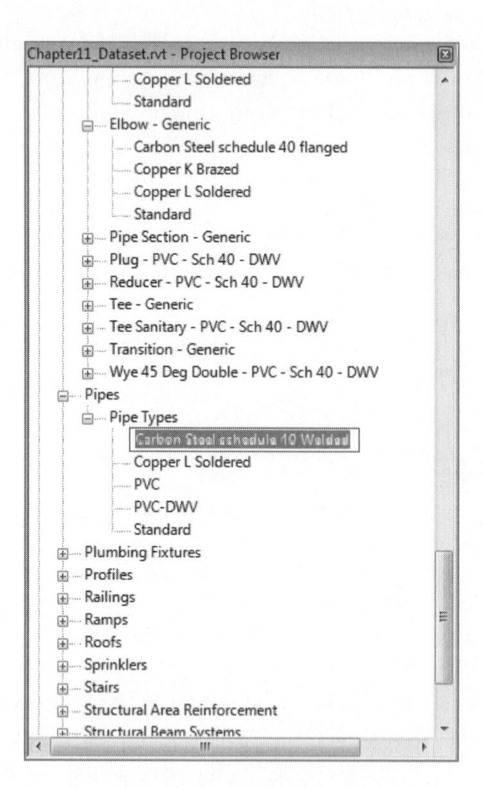

FIGURE 11.1
Renaming pipe types

NAMING PIPE TYPES

Try to avoid naming your pipe types as system type names (such as *hot water supply pipe*). This makes it harder if you use your model for an integrated project delivery where you are partnered with a general contractor and mechanical subcontractors. The contractors will want to take the information and use as much of it for shop drawings as possible. This is where the *I* in BIM becomes more than a catchphrase. The more accurate information about the material, the more accurate the material takeoffs will become for pricing and budgeting purposes.

Now that you have your pipe types created, you will want to change some of the parameter options. First right-click the pipe you want to edit and select Properties. This will open the Pipe Type parameters. Under Mechanical, you will find Material, Connect Type, and Class parameters. Change these settings to the appropriate types. (Refer to Figure 11.2.)

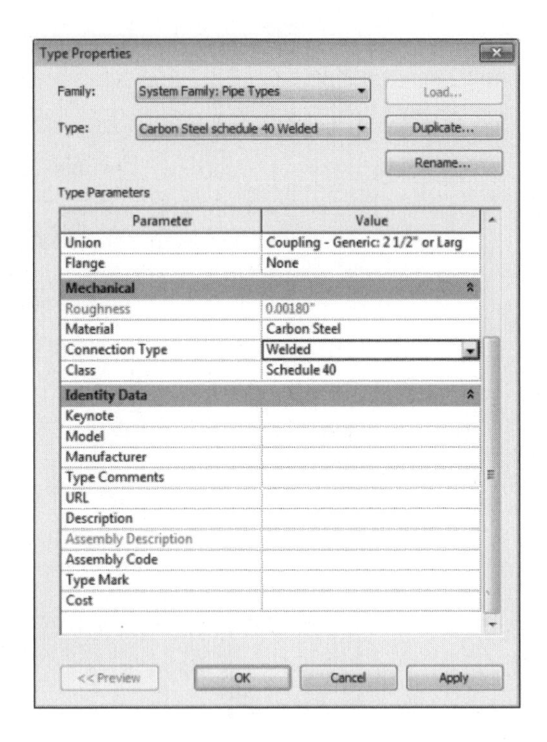

FIGURE 11.2 Changing the Mechanical parameters to the desired settings

Under the Identity Data parameters group, the following parameters are available: Keynote, Model, Manufacturer, Type Comments, URL, Description, Assembly Description, Assembly Code, Type Mark, and Cost. If you have a certain manufacturer, model, or other special note that you may want to denote on the plans, you can use these settings to further describe your pipe type. (Refer to Figure 11.3.)

Creating Fittings for System Pipes

Under the Fitting Parameters group, there are Elbow, Preferred Junction Type, Tee, Tap, Cross, Transition, Union, and Flange settings. Before you can adjust the rest of your system pipe type parameters, you need create the fittings that go with your system pipe types. To accomplish this, you must go to the Project Browser and select Families ➪ Pipe Fittings. Once you are under Pipe Fittings, you'll want to select the fittings that will be required for your pipe type parameters. You will right-click each of the family fittings and duplicate standard generic Elbow, Cross, Coupling, Tee, and Transition. After duplicating each one of these, rename them to the associated material of each pipe that you previously duplicated to create your pipe types. (Refer to Figure 11.4.)

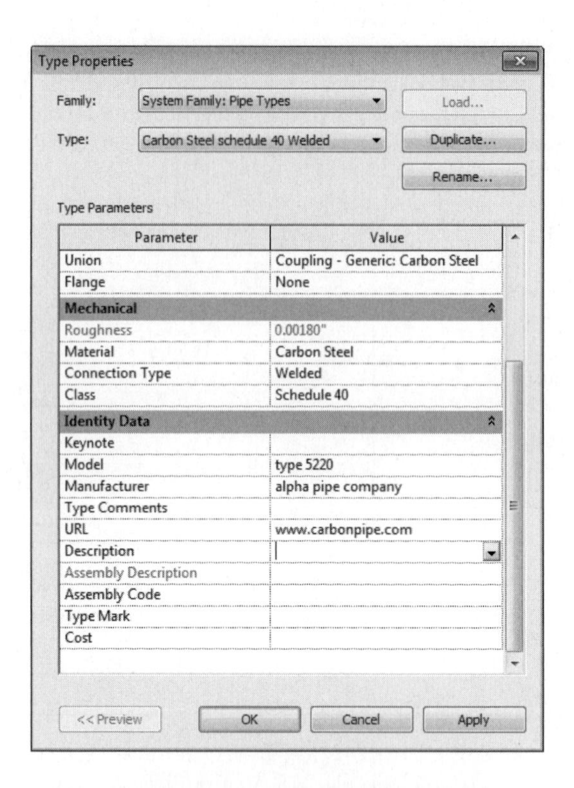
Type Properties
Family: System Family: Pipe Types
Type: Carbon Steel schedule 40 Welded
Load... Duplicate... Rename...
Type Parameters

Parameter	Value
Union	Coupling - Generic: Carbon Steel
Flange	None
Mechanical	
Roughness	0.00180"
Material	Carbon Steel
Connection Type	Welded
Class	Schedule 40
Identity Data	
Keynote	
Model	type 5220
Manufacturer	alpha pipe company
Type Comments	
URL	www.carbonpipe.com
Description	
Assembly Description	
Assembly Code	
Type Mark	
Cost	

<< Preview OK Cancel Apply

FIGURE 11.3
Identity Data parameters group

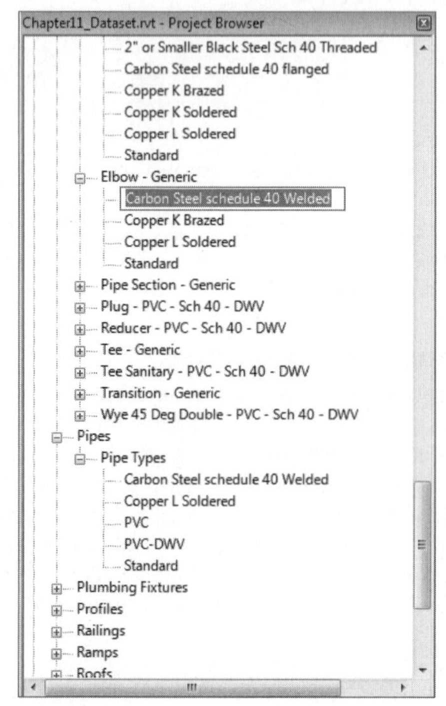

FIGURE 11.4
Duplicating and renaming fittings to match pipe types

IMPORTANCE OF SYSTEM PIPE FITTINGS

When possible, use fitting families that are true dimensions or that closely resemble the dimensions from the pipe fittings you have in your specifications, because the physical dimensions are so different because of flow characteristics of the pipe. The perfect situation is to use fittings that have been properly modeled from the manufacturer. As more manufacturers adopt BIM, the more accessible this information will become. Until that happens, you will want to review what options are available for pipe fittings that come with Revit. The way to achieve this is to take some time and familiarize yourself with the fittings directory located in the imperial library. For example, new to Revit MEP 2011, Autodesk has supplied flange fittings to help provide the end user with the ability to produce a more accurate model.

Once you have all your pipe fittings, go to the Mechanical parameters to find parameters for Material, Connect Type, and Class. Change these settings to the appropriate types. Also, you may need to modify the Loss method to K Coefficient From Table; also verify that the table is set to the appropriate setting. (Refer to Figure 11.5.)

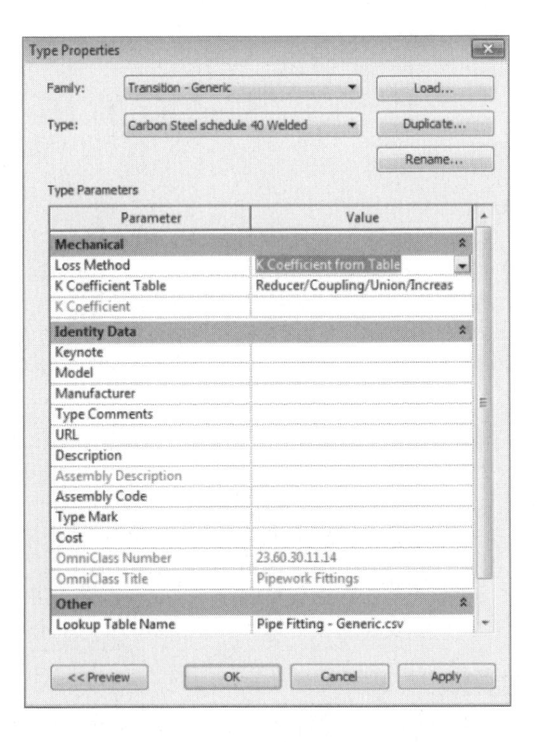

FIGURE 11.5
Setting the K coefficient to the appropriate setting

Once the pipe fittings have been created, go back to the Pipe Type parameters, and under Pipe Fitting select the proper pipe fitting for your pipe type. Doing this will allow you to filter and schedule piping for takeoff purposes. (Refer to Figure 11.6.)

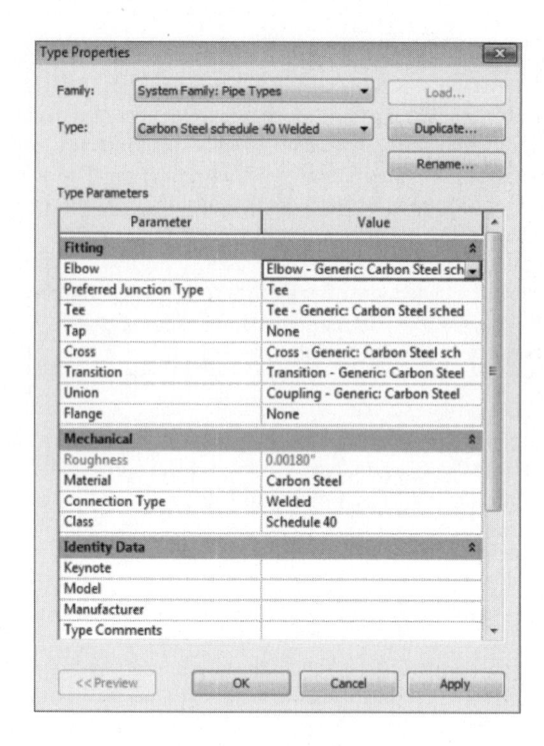

FIGURE 11.6 Selecting the proper pipe fittings for pipe types

Pipe Material

To get to the pipe material settings, go to Home ➪ Plumbing And Piping and then select the small arrow in the lower-right corner of the ribbon. This will open the Mechanical Settings dialog box (alternatively, go to Manage ➪ MEP Settings ➪ Mechanical Settings). Next, select Pipe Settings ➪ Sizes. If you want to create a new pipe material, you can duplicate an existing pipe material and rename it to the new pipe material as required. (Refer to Figure 11.7.) You can change connection type and schedule of pipe.

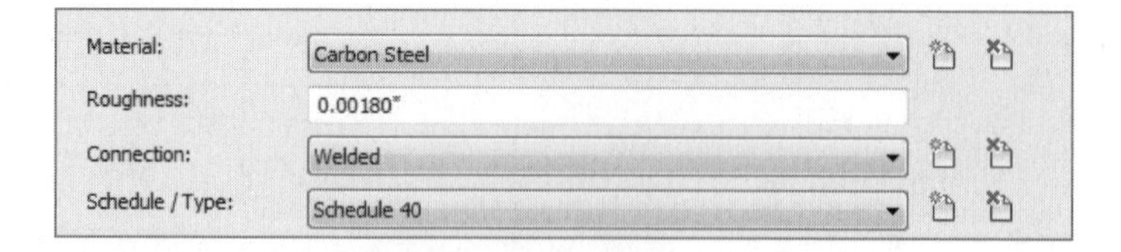

FIGURE 11.7 Duplicating pipe material and renaming to new material

Pipe Sizing Table

If you want to adjust the sizing table, go to Home ➪ Plumbing And Piping, and select the small arrow in the lower-right corner of the panel. This will open the Mechanical Settings dialog box. Next, select Pipe Settings ➪ Sizes. You can duplicate the schedule of pipe and apply the pipe wall thickness as required. You can also select and deselect the piping sizes to match your design standards. (Refer to Figure 11.8.)

FIGURE 11.8
Selecting default pipe sizes using the pipe sizing table

Nominal	ID	OD	Used in Size Lists	Used in Sizing
1/2"	5/8"	27/32"	☑	☑
3/4"	13/16"	1 1/16"	☑	☑
1"	1 1/16"	1 5/16"	☑	☑
1 1/4"	1 3/8"	1 21/32"	☑	☑
1 1/2"	1 5/8"	1 29/32"	☑	☑
2"	2 1/16"	2 3/8"	☑	☑
2 1/2"	2 15/32"	2 7/8"	☑	☑
3"	3 1/16"	3 1/2"	☑	☑
3 1/2"	3 9/16"	4"	☑	☑
4"	4 1/32"	4 1/2"	☑	☑
5"	5 1/16"	5 9/16"	☐	☐
6"	6 1/16"	6 5/8"	☑	☑
8"	7 31/32"	8 5/8"	☑	☑
10"	10 1/32"	10 3/4"	☑	☑
12"	11 15/16"	12 3/4"	☑	☑
14"	13 1/8"	14"	☑	☑
16"	15"	16"	☑	☑
18"	16 7/8"	18"	☑	☑
20"	18 13/16"	20"	☑	☑
24"	22 5/8"	24"	☑	☑

Fluids Table

You can add to the Fluids table information concerning temperature, viscosity, and density. To do this go, to Home ➪ Plumbing And Piping, and select the small arrow in the lower-right corner of the panel. This will open the Mechanical Settings dialog box. Next, select Pipe Settings ➪ Sizes. Then duplicate one of the fluid categories that is closest to the one you need, and modify it as required. (Refer to Figure 11.9.)

Pipe Routing Options

The size of mechanical piping can grow quite large because of the amount of fluids constantly transferred, so it is very important that routing is closely coordinated. Now using Revit MEP 2011, you can show your concerns with color-coded visual coordination and use interference checking to monitor conflicts with cable trays or sprinkler piping. There are a couple of routing options when you set out to design your piping model: the autoroute option and the manual routing option. The smaller a system, the more beneficial the autoroute feature. If you have a large system you are designing, then manual routing will benefit you more because of the nature of designs changing more often. Each of these is described in the following sections.

Automatic Pipe Routing

Ideally you have everything set up in your pipe types to begin routing piping. To start automatic pipe routing, select one of pieces of mechanical equipment that you have added to the mechanical piping system. Next press the Tab key until the autorouting feature highlights its suggested path. Then you can select Generate Layout ➪ Modify Mechanical Equipment. (Refer to Figure 11.10.)

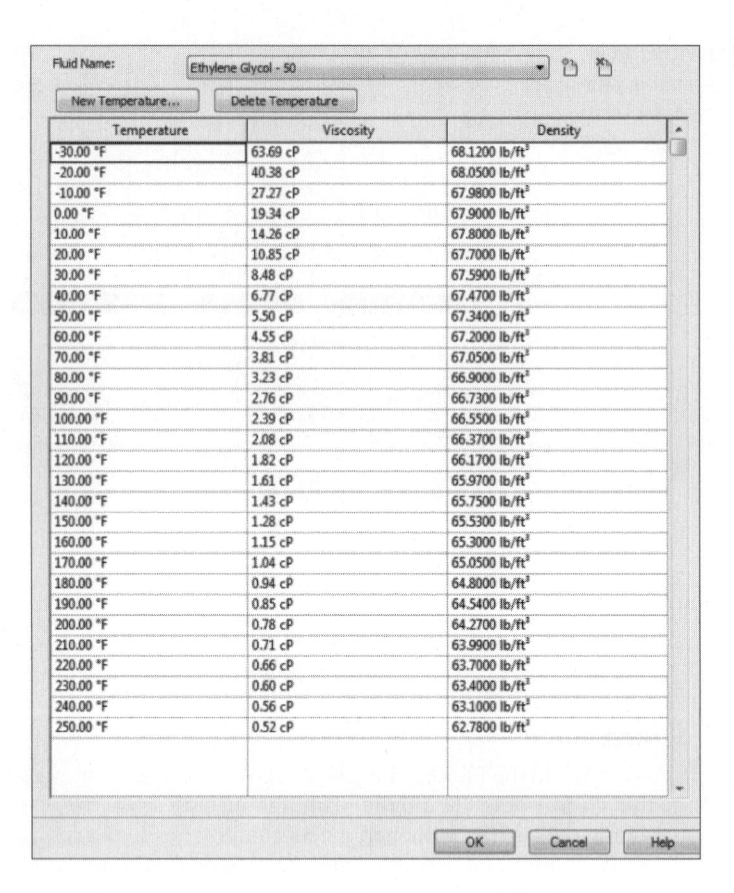

Temperature	Viscosity	Density
-30.00 °F	63.69 cP	68.1200 lb/ft³
-20.00 °F	40.38 cP	68.0500 lb/ft³
-10.00 °F	27.27 cP	67.9800 lb/ft³
0.00 °F	19.34 cP	67.9000 lb/ft³
10.00 °F	14.26 cP	67.8000 lb/ft³
20.00 °F	10.85 cP	67.7000 lb/ft³
30.00 °F	8.48 cP	67.5900 lb/ft³
40.00 °F	6.77 cP	67.4700 lb/ft³
50.00 °F	5.50 cP	67.3400 lb/ft³
60.00 °F	4.55 cP	67.2000 lb/ft³
70.00 °F	3.81 cP	67.0500 lb/ft³
80.00 °F	3.23 cP	66.9000 lb/ft³
90.00 °F	2.76 cP	66.7300 lb/ft³
100.00 °F	2.39 cP	66.5500 lb/ft³
110.00 °F	2.08 cP	66.3700 lb/ft³
120.00 °F	1.82 cP	66.1700 lb/ft³
130.00 °F	1.61 cP	65.9700 lb/ft³
140.00 °F	1.43 cP	65.7500 lb/ft³
150.00 °F	1.28 cP	65.5300 lb/ft³
160.00 °F	1.15 cP	65.3000 lb/ft³
170.00 °F	1.04 cP	65.0500 lb/ft³
180.00 °F	0.94 cP	64.8000 lb/ft³
190.00 °F	0.85 cP	64.5400 lb/ft³
200.00 °F	0.78 cP	64.2700 lb/ft³
210.00 °F	0.71 cP	63.9900 lb/ft³
220.00 °F	0.66 cP	63.7000 lb/ft³
230.00 °F	0.60 cP	63.4000 lb/ft³
240.00 °F	0.56 cP	63.1000 lb/ft³
250.00 °F	0.52 cP	62.7800 lb/ft³

FIGURE 11.9
Fluids table settings

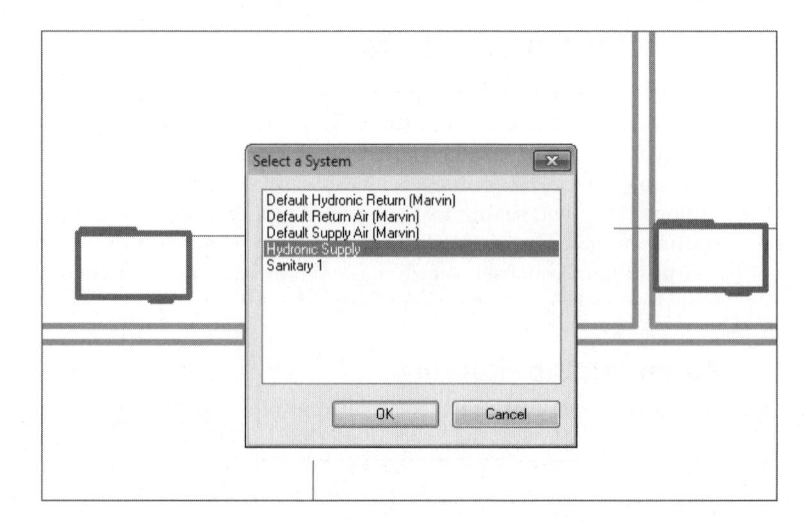

FIGURE 11.10
Generating layouts from systems

You have four options to select for generating the layout: Network, Perimeter, Intersections, and Custom, and each one has several routing solutions to choose from that consist of a main (blue) and branches (green):

Network This solution creates a bounding box around the components selected for the piping system and then bases several solutions on a main segment along the center line for the bounding box, with branches coming from the main segment.

Perimeter This solution creates a bounding box around the components selected for the system and proposes several potential routing solutions. You can specify the Inset value that determines the offset between the bounding box and the components. Inset is available only when the Perimeter option is selected.

Intersections This solution bases the potential routing on a pair of imaginary lines extending from each connector for the components in the system. Perpendicular lines extend from the connectors. Where the lines from the components intersect are potential junctions in the proposed solutions along the shortest paths.

Custom The solution becomes available once you begin to modify any of the other solutions.

When using the autorouting feature, the number-one mistake that everyone makes is to try to select every unit at one time. This will nearly always lead to failure. The process is to use the Add To System/Remove From System tool that is part of the General Layout ribbons. (Refer to Figure 11.11.)

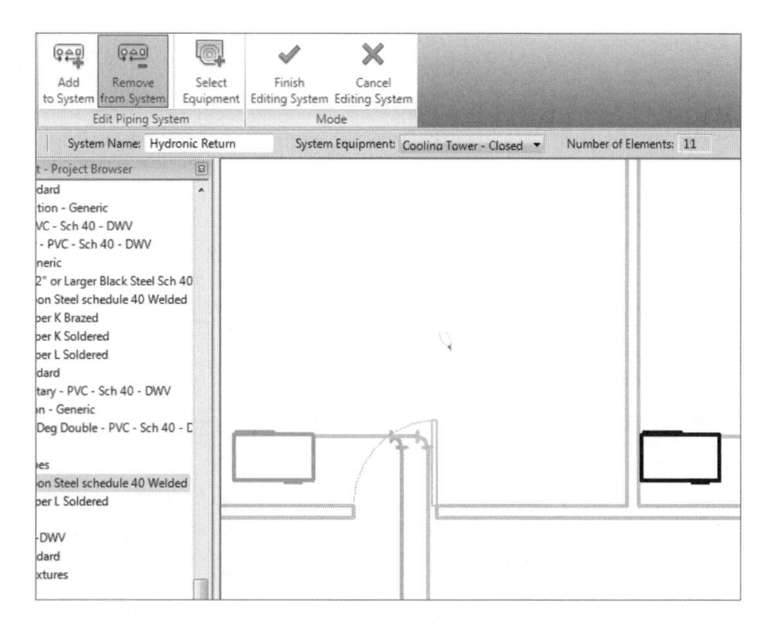

FIGURE 11.11 Adding to and removing from systems

By making the autorouting smaller, you can actually modify the runs better. Make sure this offset is at the same height as your piping, because this helps ensures that you are routing your piping at the correct offset. Once you have everything routed, connect the smaller piping systems together to make the overall system.

Manual Pipe Routing

Manual routing of piping is the next method. When routing, manually start the piping run at the elevation that you know will be most likely out of the way of other disciplines. Use the following steps to set up and place mechanical equipment, create a hydronic supply and return, and manually route pipe to all pieces of equipment.

1. Open the `Chapter11_Dataset.rvt` file found on `www.wiley.com/go/masteringrevitmep2011`.
2. Because of the normal size that models can grow, it is not uncommon for the architectural model to be linked into the mechanical model. Download the model `Chapter 11 base.rvt` file found at `www.wiley.com/go/masteringrevitmep2011`.
3. Go to Insert ➪ Revit Links, select the directory where you downloaded the Revit base to, select Auto – Origin To Origin, and then select Open. By selecting Origin To Origin, you are assured that the model will be properly located. (Refer to Figure 11.12.)

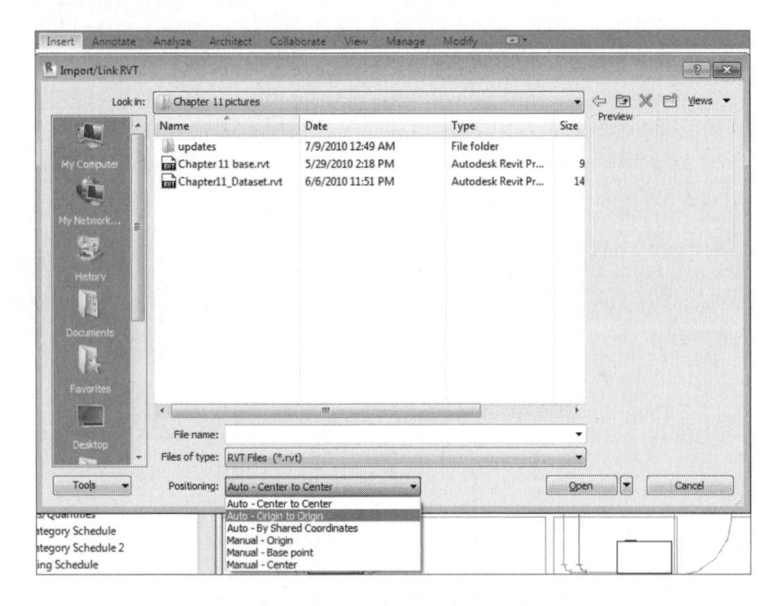

FIGURE 11.12 Linking Revit model, Auto – Origin to Origin

> **IMPORTANCE OF TEMPLATES**
>
> To improve productivity in piping layouts, make sure to create and maintain templates that contain your company standards. Properly created templates assure that everyone in your firm is on the same page when creating mechanical piping systems. For more information on how to create templates from existing projects, refer to Chapter 2.

4. Next go to Insert ➪ Load Family ➪ Imperial Library ➪ Mechanical Components ➪ Air-Side Components ➪ Heat Pumps, select `WSHP - horizontal.rfa`, and select Open. (Refer to Figure 11.13.)

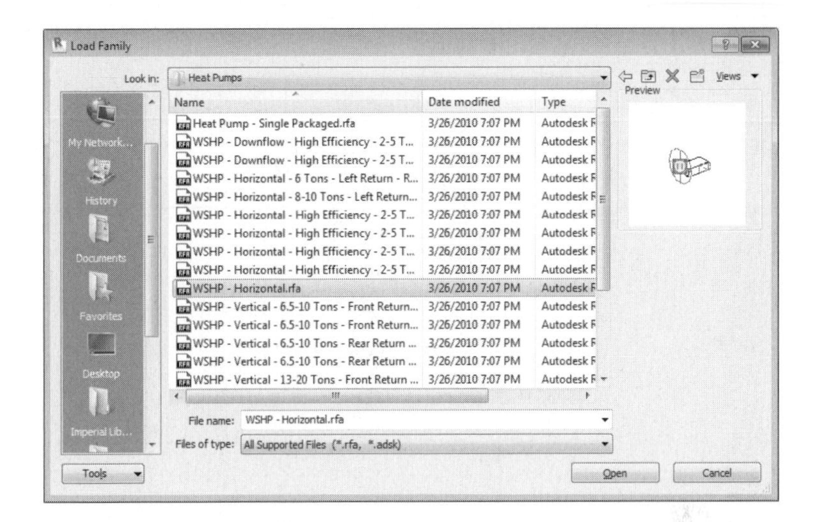

FIGURE 11.13 Water source heat pumps

5. Go to Home ➪ Mechanical Equipment, and start placing the water source heat pumps into the each room. After you have placed all the heat pumps into the model, right-click the water source heat pump you first spotted, and select All Instances ➪ In Entire View. Then right-click again and select Properties. Once the Properties dialog box is open, select Offset, and change the offset to 9′-0″ (274.32 cm). This will allow you to set the proper elevation of equipment. (Refer to Figure 11.14.)

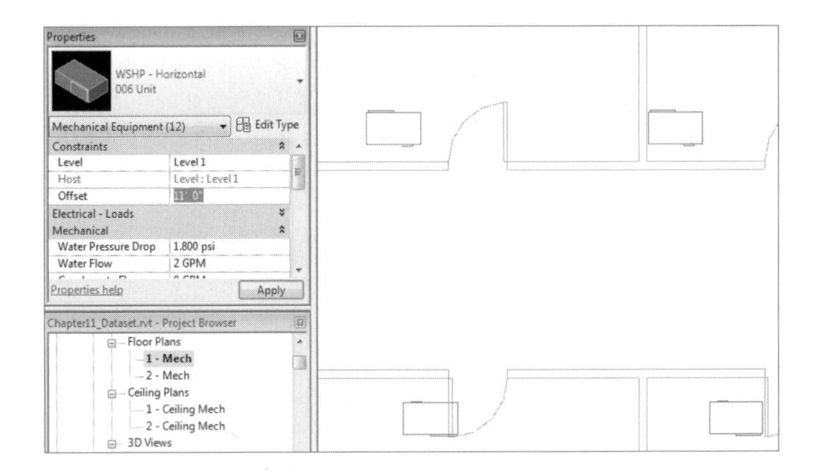

FIGURE 11.14 Verifying offset for proper elevation

6. Your system will require a boiler and closed circuit cooling tower. Go to Insert ➪ Load Family ➪ Imperial Library ➪ Mechanical Components ➪ Water-Side Components ➪ Boilers, select `Boiler.rfa`, and then select open. Repeat the process to load the `Cooling Tower - Closed Circuit - Counterflow - 37-211 MBH.rfa` file from the Cooling Tower directory located in Water-Side Components. Once these are located in the project, place them into the appropriate locations, as shown in Figure 11.15.

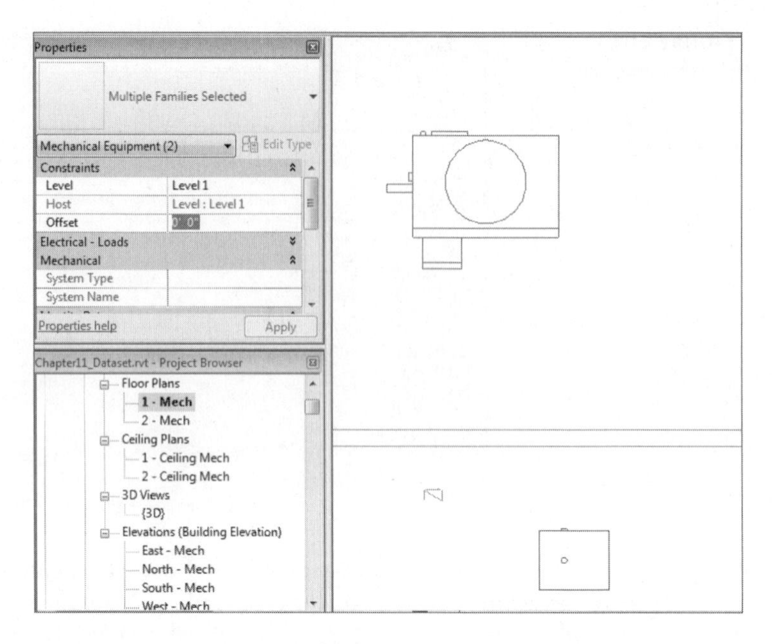

FIGURE 11.15
Placing mechanical equipment

7. Next you will need a centrifugal pump. Go to Insert ➪ Load Family ➪ Imperial Library ➪ Mechanical Components ➪ Water-Side Components ➪ Pumps, select `Centrifugal Pump - Horizontal.rfa`, and select Open. Place the pump where you want it to be located in the mechanical room. (Refer to Figure 11.16.)

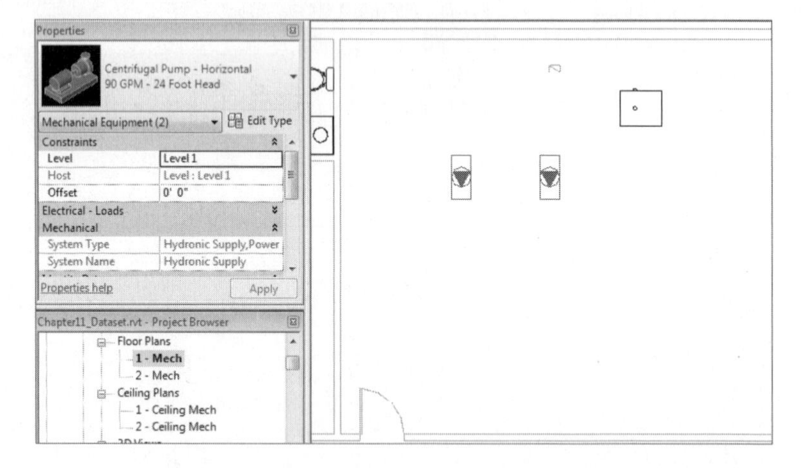

FIGURE 11.16
Placing the pumps in the mechanical room

8. You are now in a position to start routing piping. Route your mains first so you can make sure that most of the piping will fit before connecting all the branches to the mains. Select the proper piping material before starting your layout. This will help if you need to do a quantity takeoff for budgeting purposes. To route your pipe, select Home ➪ Pipe. Change the offset to 8′6″ (259.08cm) and pipe size to 2′ (60.6 mm), and route piping mains in the corridors and into the mechanical room. (Refer to Figure 11.17.)

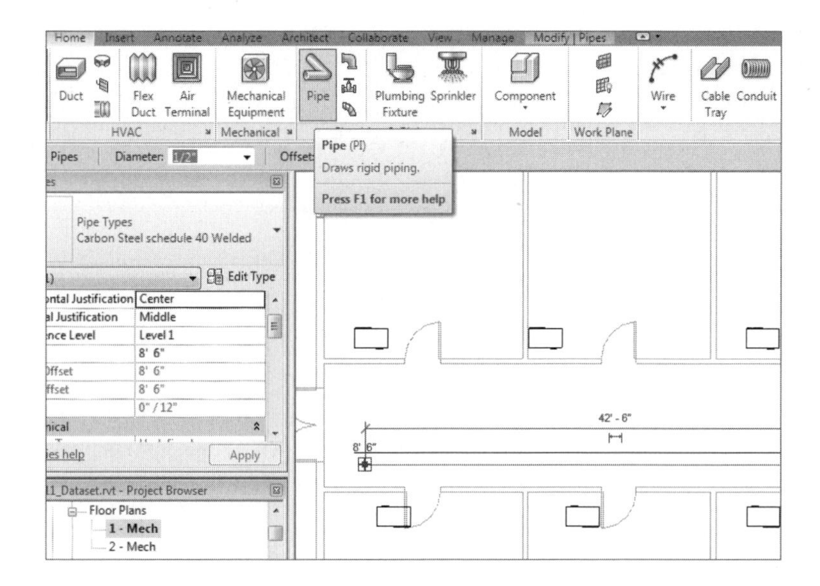

FIGURE 11.17
Routing pipe mains

9. Before you start connecting piping, you will need to make sure that your mechanical equipment is set up for both hydronic supply and return systems. To do this, select one of the pieces of mechanical equipment, then go to Home ➪ Modify Mechanical Equipment, and select Create Hydronic Return System. (Refer to Figure 11.18.)

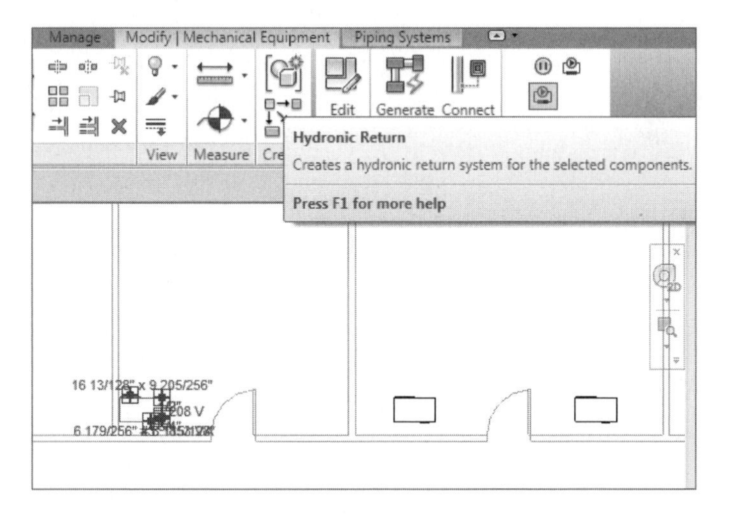

FIGURE 11.18
Creating mechanical piping systems

10. Once you have selected the hydronic return, go to the Piping Systems tab, select Edit System, and edit the Hydronic Return system name. Then select Add To System, and select all mechanical equipment that contains Hydronic Return connections. Repeat with Hydronic Supply. (Refer to Figure 11.19.)

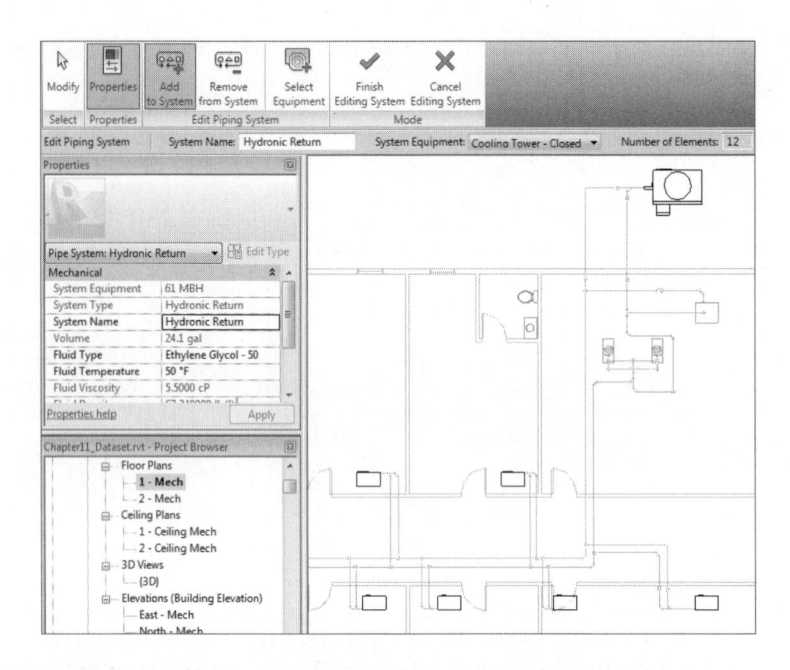

FIGURE 11.19
Adding equipment to systems

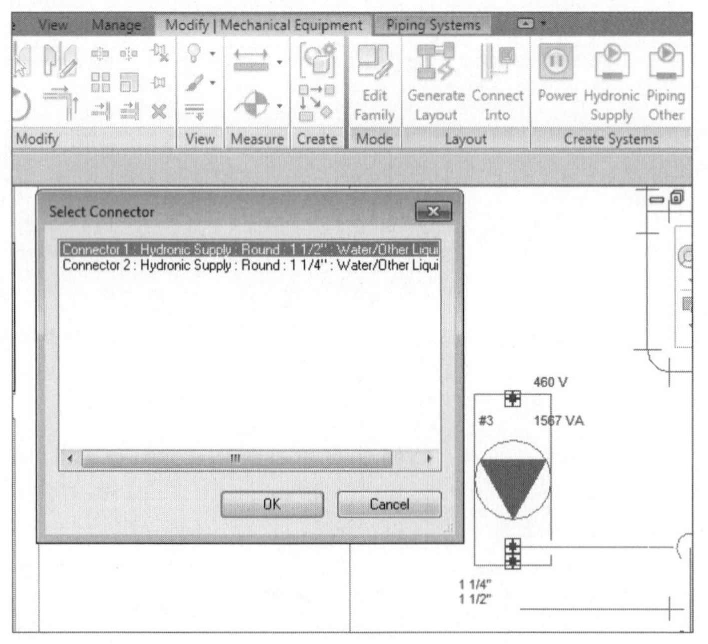

FIGURE 11.20
Using Connect Into

11. Cutting sections when routing piping can really improve the success of your coordination of your Revit model. To route piping, go to Home ➪ Pipe ➪ Modify Place Pipe, and be sure to select Connect Into (Refer to Figure 11.20.)

12. Repeat this process for all your piping systems Taking a few minutes to review any clashes between the structure, mechanical, and architectural models can really cut down on modeling time. The layout is now completed once you have coordinated between other services/structure and verified your layout.

Real World Scenario

COORDINATION IS KEY

In 2007, Michael Brady Incorporated used Revit Systems for MEP and Revit Architecture to design First Baptist Church in Farragut, Tennessee, and to produce contract documents. Because of the design timeline, the structural team used AutoCAD 2007 to lay out its structure.

Charles Merriman, the mechanical designer on the project, linked in the structural CAD plans and decided to mass objects above the ceiling space to represent the depth and location of the bar joists and beams. Although this was a little time-consuming, it helped show some of the conflicts he was facing with his design.

By using this technique, he was able to avoid some very large problems associated with the mechanical pipe routing and duct layout. When the project was in the construction phase, the general contractor began to ask for a potential change order. Charlie opened his Revit file and printed a 3D view of the area in question. Because of Charlie's earlier commitment to coordinating his design, the area in which the contractor thought was an issue was easily resolved.

Pipe Fittings

Without fittings, piping would not be worth a whole lot. Fittings help shut off flow, help regulate temperature, and help save lives. In Revit, most fitting families have the following functions:

End Cap These can be placed only at the end of pipe.

Tee, Tap, Wye, or Cross These can be placed anywhere along pipe runs.

Transitions, couplings, or unions These can be placed only at the end of pipe. They are used to join a smaller, larger, or same-sized pipe.

Flange These can be placed at the end of pipe or face to face with another flange.

Using Pipe Fitting Controls

Understanding pipe fitting controls can really make life easier if you are routing a lot of piping. Several fittings have this ability. When you are laying out your piping, turn 90 degrees to create a elbow. If you click the elbow, you will notice a plus (+) sign. If you click that sign, it will change from an elbow to a tee allowing you to add more piping and continue your pipe routing. If you click the minus (–) sign, it will downgrade the fitting. When you see the symbol on a fitting, it allows you to rotate the fitting, and the symbol allows you to flip the fitting.

Placing Fittings

When you need to add valves to your piping, select the Home tab, select Pipe Accessories, use the Type Selector, and select the type of valve you want to use. Most valves are "break into" types, so you can place them into a pipe run, and it will break into the piping, maintaining connections at either end. (Refer to Figure 11.21.)

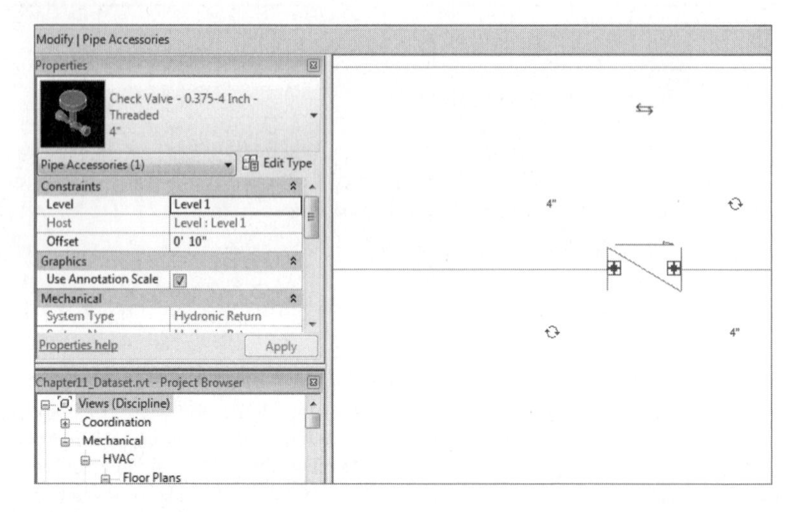

FIGURE 11.21 Fitting breaks into piping system

When creating or editing families, use the family parameter Part Type, which will define the ability to Breaks Into. (Refer to Figure 11.22.)

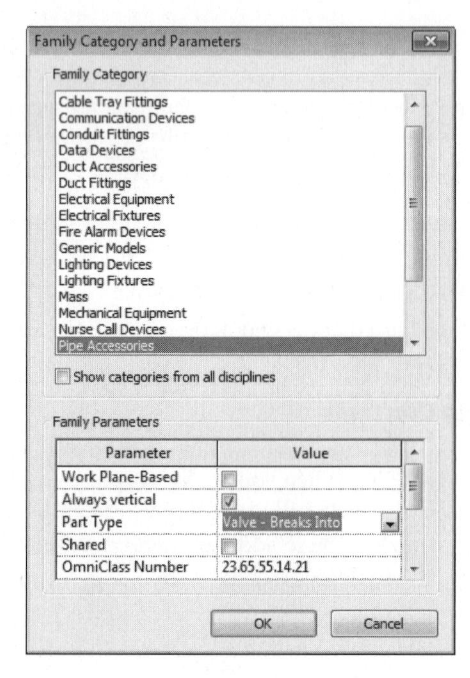

FIGURE 11.22 Fitting Breaks Into definition

Visibility of Pipes

You have created systems and have joined them together with pipes, fittings, and accessories. Now you want to have them represented according to your company standards in different views. This means changing colors, line weights, line types, and the actual ability to view them at all. If the systems were created correctly, you can apply filters to your views to add this additional control. To do this, go to Filters located under the View tab and duplicate, rename, and change system type to system name. (Refer to Figure 11.23.)

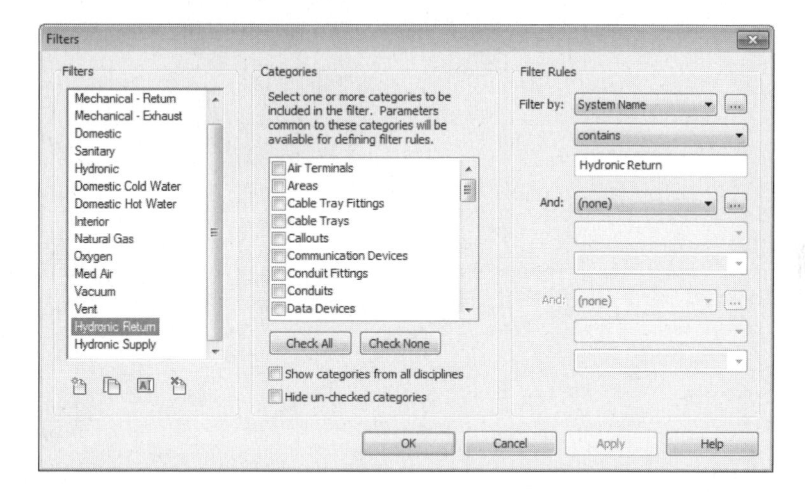

FIGURE 11.23
Modify filter settings

Once you have your filters created, go into each view where you want the different disciplines to show up. Press VG to bring open the Visibility/Graphics dialog box. Next go to the filters and the filters you want to control, and change the line types and colors to match your company standards. (Refer to Figure 11.24.)

FIGURE 11.24
Adjusting color and line types through Visibility/Graphics

lodel Categories | Annotation Categories | Imported Categories | Filters | Revit Links

Name	Visibility	Projection/Surface		Cut		Halftone	Transpare
		Lines	Patterns	Lines	Patterns		
Mechanical - Supply	☐					☐	☐
Mechanical - Return	☐					☐	☐
Mechanical - Exhaust	☐					☐	☐
Domestic	☐					☐	☐
Sanitary	☐					☐	☐
Vent	☐					☐	☐
Hydronic Return	☑					☐	☐
Hydronic Supply	☑					☐	☐

Add | Remove | Up | Down

All document filters are defined and modified here | Edit/New...

With the views defined, you can create a view template from the view, enabling you to ensure that all views of a specific discipline display with the same configuration. Once you have everything set the way you want the project to look, you can complete the documentation.

Your 3D views can also be color coded to make it easy to review piping. (Refer to Figure 11.25.)

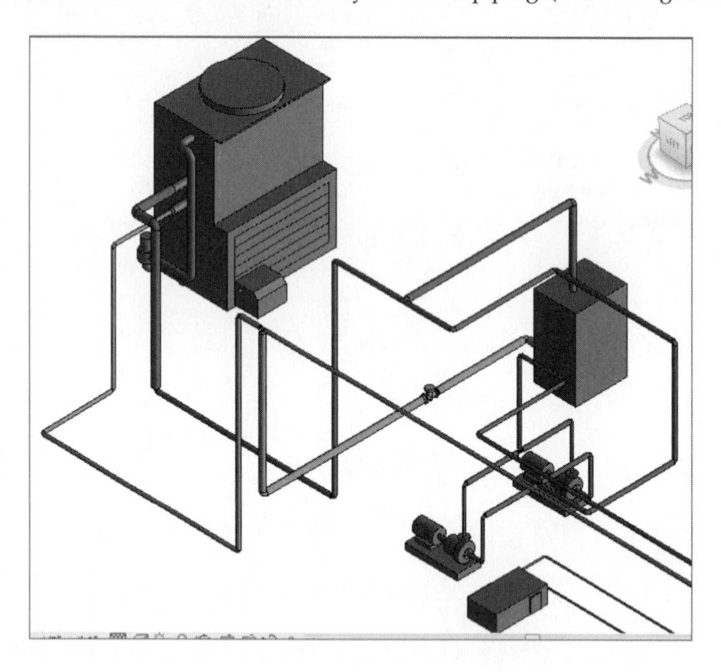

FIGURE 11.25 Color-coded 3D views for review

The Bottom Line

Adjust and use the mechanical pipe settings Making sure that the mechanical piping settings are properly set up is crucial to the beginning of any project.

Master It A designer has just been asked to model a mechanical piping layout, and the engineer wants to make sure the designer will be able to account for the piping material used in the layout. What steps must the designer take to the complete this request?

Select and use the best pipe routing options for your project When using Revit MEP 2011 for your mechanical layouts, one must understand the functions of automatic pipe routing and manual pipe routing. Once these functions are mastered, then the user can lay out any type of piping system.

Master It The engineer has just come back from a meeting with the owner and architect, and it has been decided that the owner wants to have a hot water system and a chilled water system rather than a two-pipe hydronic system. How would you modify your hydronic layout to accommodate the change?

Adjust pipe fittings Pipe fittings are needed in systems to make the systems function properly and to produce documentation for construction. Being able to add or modify fittings can increase productivity.

Master It You have printed off a check set for review and have noticed that there are no shutoff valves. Now you need to load the shutoff family. What directory should you look in for pipe fittings?

Adjust the visibility of pipes Being able to adjust the visibility gives the mechanical designer or user the ability to set up multiple views and control the graphics for documentation.

Master It The engineer has just come back from a meeting with the owner and architect, and it has been decided that the owner wants to have a hot water system and a chilled water system. You have just modified your hydronic layout to accommodate the change. Now the owner wants the pipes color coded, so it's easier to visualize the changes. Describe how this would be done.

Part 3

Revit MEP for Electrical

Chapter 12

Lighting

It may be difficult at first to see a good reason for taking the effort to include lighting systems in a 3D model. After all, lighting can be represented by symbols, can't it? Although that is true, a BIM project is much more than just creating a 3D model. The data from an intelligent lighting model can be used for analysis and can aid in design decisions.

Including light fixtures and their associated devices in a Revit model will allow you to coordinate your complete electrical design by providing electrical load information. They can also be used to develop presentation imagery by generating realistic light in renderings.

Creating a lighting model with Revit MEP enables you to develop your design while generating the necessary construction documents to convey the design intent.

In this chapter, you will learn to

- Prepare your project for lighting design
- Use Revit MEP for lighting analysis
- Compare and evaluate hosting options for lighting fixtures and devices
- Develop site lighting plans

Efficient Lighting Design

Let's face it, ceiling plans are one of the biggest coordination pain points for a design team. Nearly every MEP discipline has some type of element that resides in the ceiling. Using intelligent lighting families will help you, the electrical designer, stake your claim to that precious real estate. Using 3D geometry to represent light fixtures means you can detect interference with other model elements. This does not mean that your lighting fixture families will have to be modeled to show every trim ring, reflector, tombstone, or lens. The basic geometry is usually enough to satisfy the requirements for model coordination.

The intelligence put into your families is what will benefit you the most from an electrical standpoint. Photometric data, manufacturer, model, voltage, and number of lamps are just a few examples of the types of properties that can reside in your fixture families. An in-depth look at creating lighting fixture families occurs later in this book.

Spaces and Lighting

For the spaces in your model to report the correct lighting level, they must be modeled accurately. If the height of the space is short of the ceiling, the lighting fixtures will not be in the space and therefore will provide no light to that space. A ceiling can be defined as a room-bounding element, which means that it defines the upper boundary of the space. If you model your space so that the upper limit is higher than the ceiling height, you can be sure that you are getting accurate volume information for the space. When you are placing spaces into the model, set the upper limit to the level above the current level you are working on to ensure proper volumes. If you have a space that spans multiple levels, make sure you set the upper limit appropriately, as shown in Figure 12.1.

FIGURE 12.1
Space volume and ceiling relationship

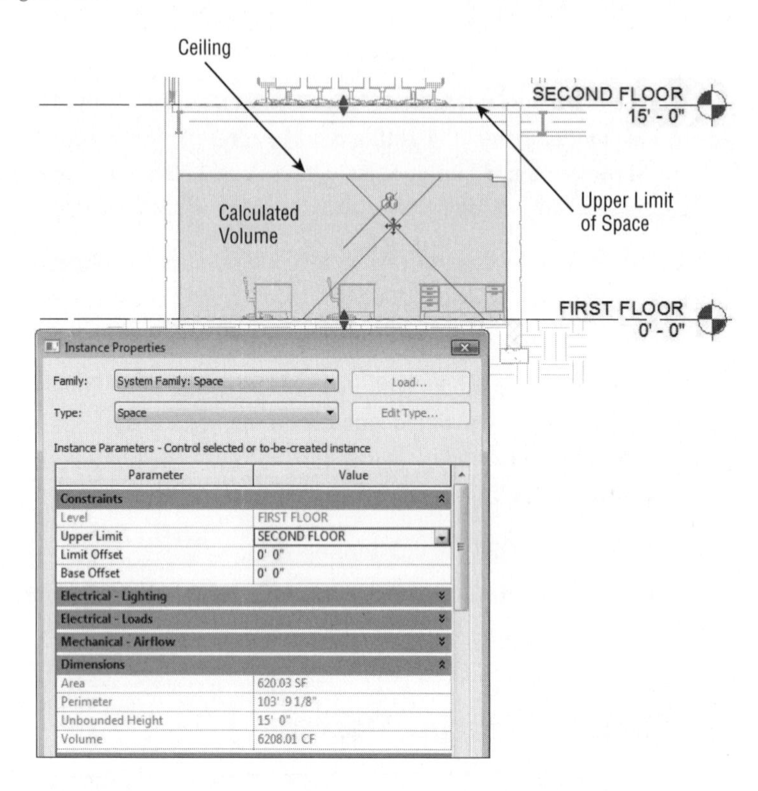

Real World Scenario

GOING ABOVE AND BEYOND

Rusty is getting some strange numbers from his lighting calculations. He checks his spaces and finds that the volumes of the exterior rooms are not being calculated to the ceiling, but all the way up to the next level. A quick phone call to his architect reveals that the mechanical designer has requested that all exterior room ceilings be set so they are not room bounding. This is necessary for accurate heating and cooling load calculations.

Rusty is glad that his file is separate from the mechanical file because all he has to do is set the upper limit of his exterior spaces to match the ceiling heights.

Space volume is important to the proper calculation of average estimated illumination within a room. The ability to calculate the volume of a space can be turned on or off to help with file performance. If you intend to use Revit MEP to analyze your lighting design, you need to ensure that this setting is turned on. Do this by clicking the Room & Area panel drop-down on the Architect tab.

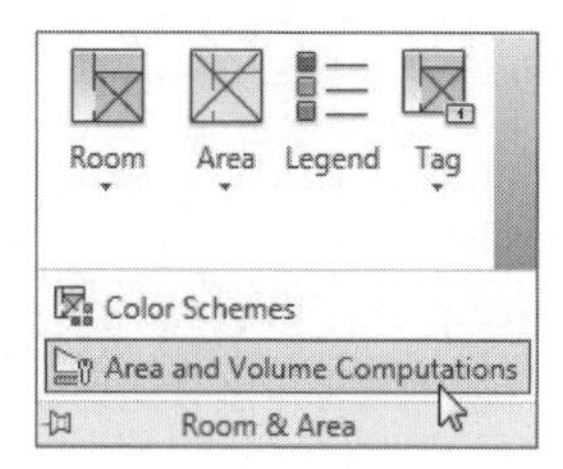

Select the Area And Volume Computations tool to access the settings for space volume computations. Choose the setting shown in Figure 12.2 when using Revit MEP for lighting analysis.

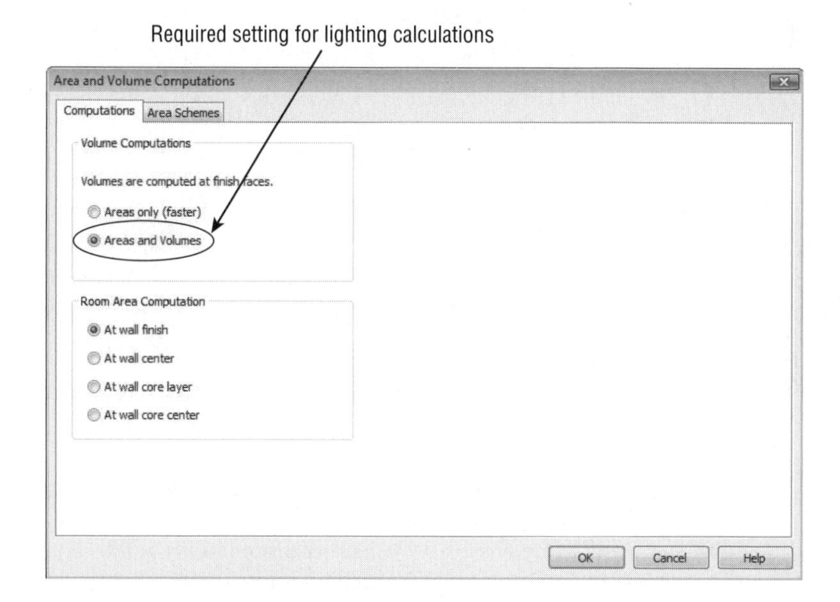

FIGURE 12.2
Area and volume settings

You can practice placing and manipulating spaces by doing the following:

1. Create a new project file.
2. Link in the architectural model file `ch12ArchModel.rvt` on the book's download page, `www.wiley.com/go/masteringrevitmep2011`. Use Origin To Origin placement.
3. Set the property of the linked file to Room Bounding by editing its type parameters and selecting the Room Bounding box.
4. Click the Space tool on the Analyze tab.

5. On the Options Bar, set Offset to 8′ 0″, and place a space in the large curved room at the right side of the building.
6. Create a section view, looking north, through the large room.
7. Open the section view, and select the Space object (you may have to hover your mouse pointer on the room edges). Notice that the upper limit is below the ceiling object in the room.
8. Check the instance properties of the space, and verify that the volume is computed under the Dimensions group. If not, go back and set your model to compute volume by adjusting the Area And Volume Computations settings on the Room And Area panel of the Architect tab.
9. Once again, with the Properties palette showing the properties of the space, change the Limit Offset parameter value to 10′ 0″, and examine the Volume parameter value of the space. You may need to click the Apply button on the Properties palette for the change to take place.
10. Change the Limit Offset parameter value to 12′ 0″, and notice that the Volume parameter value stays the same. This confirms that the ceiling is acting as a room boundary.
11. In a floor plan view, start to place another space in the room in the upper-right corner of the building. Set the upper limit of the space to the level above, and set the offset to 0′ 0″. Then click in the room to create the space.
12. Create a section view of this room, and select the Space object. Notice that the space extends beyond the ceiling. This confirms that the ceiling is not set to be room bounding.
13. In the section view, select the Space object, and use the grip arrow to stretch the top of the space beyond the roof. Notice that the space fills the entire volume up to the roof.
14. Close the file.

The Reflected Ceiling Plan

The first step in creating a well-coordinated lighting plan is to ensure that your reflected ceiling plan is properly set up to display the model in a way that allows you to see all items necessary to coordinate as you design. Set the view range of your view so that all ceilings are clearly visible, and turn on any worksets from other disciplines that may contain items in the ceiling. You can set the categories from other disciplines to halftone to see your layout more clearly. If you are linking in files from MEP consultants, use the visibility control options for linked files to achieve the desired result.

VIEW ORIENTATION

When you create a new ceiling plan view of a linked architectural model, you will have to manually set the Underlay Orientation parameter value of your view to Reflected Ceiling Plan in order to see the model in the proper orientation.

If you intend to display the ceiling grids on your lighting construction documents, you will have to make visibility adjustments to ensure that the building model displays correctly.

Remember, with a reflected ceiling plan, you are looking up at the model, so certain elements such as plumbing fixtures or windows may not display as desired until you adjust the view range and Visibility/Graphics Overrides. Stairs will also display differently in reflected ceiling views than they will in normal plan views.

Another method for displaying ceiling grids on construction documents of a Revit project is to create a ceiling view that shows only the ceiling objects. This view can be placed on a sheet in the same location as the lighting floor plan view. Revit will automatically snap the view to the same location as the floor plan view, so you can be sure of the alignment. This allows you to display the model correctly as a plan view and still be able to see the ceiling grids or surfaces.

There is often debate between architects and electrical designers as to whose model should contain the ceilings. Ceilings are not always required in early project submittals, so the architect may not get around to modeling them when the electrical designer needs them. This may prompt the electrical designer to create ceilings in their model in order to begin the lighting design, which can cause coordination issues once the architect begins designing the ceilings in the architectural model. Having duplicate information in multiple models can be a recipe for error. The electrical designer would have to keep the ceilings in the electrical model coordinated with the architect's ceilings and would have to ensure that all model views were displaying the proper ceilings. This extra effort defeats the purpose of using a BIM solution such as Revit MEP and hampers the effort to achieve an integrated project delivery, because it adds another level of manual coordination that creates more opportunities for error.

You may use the option of creating reference planes in your model to host your lighting fixtures temporarily. When the ceilings are placed in the architectural model, you could then rehost your fixtures to them. This is a better option than temporarily using nonhosted families because you cannot change a family from nonhosted to hosted, and you would have to replace each fixture once the ceilings are created.

Lighting fixtures that are modeled in the architectural model are another thing to consider. Many architects like to create lighting layouts for their design to get a feel for how the rooms will look with lighting fixtures in the ceiling. If the architect uses 3D families to represent the light fixtures, this can cause problems for the electrical designer when it comes to using the linked model for hosting. Lighting fixtures in the architectural model will most likely cut a hole in the ceiling where they are placed. When the electrical designer attempts to put a light fixture in the same location as the fixture in the linked file, they may not be able to because there won't actually be a ceiling there. There is also the chance that the fixture in the MEP file is hosted by a face of the fixture in the link. So if the architect deletes the fixture in their file, the fixture in the MEP file will not have a host and will not respond to changes.

Early in the project, the architect and electrical designer should agree on who will model the ceilings and which model they will reside in. They should also coordinate which types of families will be used if the architect intends to place lighting fixtures in the architectural model. If they need to be modeled in one file initially, they can be copied and pasted into another file if necessary. Another option is to use the Copy/Monitor tool to copy the lighting fixtures from the architectural model. The ultimate goal is to have one ceiling design that all disciplines can use for layout and coordination.

Lighting Worksets

When working in a model file with other MEP disciplines, it is best to create a lighting workset to distinguish model elements that would belong to that design system. It may even be necessary to create multiple worksets for lighting systems. Doing so will allow you to separate your lighting design into separate systems such as interior and exterior or by floor levels.

This will not only aid you in controlling the visibility of groups of model elements but will also enable multiple users to work on lighting at the same time without interfering with each other's designs.

Lighting Analysis

Because you are placing light fixtures for the purpose of a layout that is coordinated with other disciplines, you can also get design information that will help you make decisions on the types of lights to use. You can use the power of Revit MEP's scheduling capabilities to create a schedule of the spaces in your model that shows the lighting fixtures used and the lighting criteria that you are interested in. You can review this schedule as you place lights into the model to see whether you are making the right choices for lighting fixtures.

Figure 12.3 shows a simple version of this type of schedule. The last column is a calculated value that shows the difference between the required lighting level and the actual level. A difference greater than 5 footcandles causes the cell to turn red. Since there are no lights in the model yet, none of the spaces has the required lighting level, so every cell in the column is red. The goal as a designer is to achieve a schedule with no red cells in the final column.

FIGURE 12.3
Sample lighting analysis schedule

Space Lighting Schedule - Analysis

ROOM NO.	ROOM NAME	REQ LTG LEVEL	AVG EST ILLUMINATION	LOAD PER ARE	DIFF BETWEEN AVG AND REQ
100	lobby	20 fc	0 fc	0.00 W/ft^2	-20 fc
100A	reception	30 fc	0 fc	0.00 W/ft^2	-30 fc
101	corridor	20 fc	0 fc	0.00 W/ft^2	-20 fc
102	corridor	20 fc	0 fc	0.00 W/ft^2	-20 fc
103	training 1	50 fc	0 fc	0.00 W/ft^2	-50 fc
104	training 2	50 fc	0 fc	0.00 W/ft^2	-50 fc
105	training 3	50 fc	0 fc	0.00 W/ft^2	-50 fc
106	training 4	50 fc	0 fc	0.00 W/ft^2	-50 fc
107	break room	30 fc	0 fc	0.00 W/ft^2	-30 fc
107A	stor	20 fc	0 fc	0.00 W/ft^2	-20 fc
108	comm - MDF	20 fc	0 fc	0.00 W/ft^2	-20 fc
109	main electrical	20 fc	0 fc	0.00 W/ft^2	-20 fc
110	elev mech	20 fc	0 fc	0.00 W/ft^2	-20 fc
111	recept stor/files	20 fc	0 fc	0.00 W/ft^2	-20 fc
112	public toilet men	10 fc	0 fc	0.00 W/ft^2	-10 fc
113	public toilet women	10 fc	0 fc	0.00 W/ft^2	-10 fc
114	auditorium / theatre	30 fc	0 fc	0.00 W/ft^2	-30 fc

Prior to using this schedule, you should assign a target lighting level for all the spaces that you will analyze. Create a project parameter to be used for your targeted lighting level. This should be an instance parameter so that it can vary from space to space. Set the discipline of the parameter to Electrical and the type to Illuminance. Group the parameter in the Electrical-Lighting group so that it can be easily located. Name the parameter something such as Required Lighting Level so the intended use of the parameter is clear. You can create this project parameter in your project template file for use on every project if desired. Remember that you can use project parameters in schedules, but you cannot create an annotation tag for them. For more information on creating parameters, see Chapter 19.

Once you have established a parameter for the target lighting level of a space, you can create another type of schedule to associate standard lighting levels with certain types of spaces. This will not be a schedule of building components but rather a key that will assign a target lighting level to a space based on the type of space. This is known as a *schedule key*.

To create a schedule key, you use the same tool that you would use to create a regular schedule and do the following:

1. In the New Schedule dialog box, select Spaces as the category to be scheduled.
2. At the right side of the dialog box, name your schedule to indicate its use, and select the Schedule Keys option rather than the default Schedule Building Components option.
3. The key name that you choose will become an instance parameter of all the spaces in the model. The parameter will be located in the Identity Data group of parameters. Choose a name that clearly identifies the purpose of the parameter, as shown in Figure 12.4.

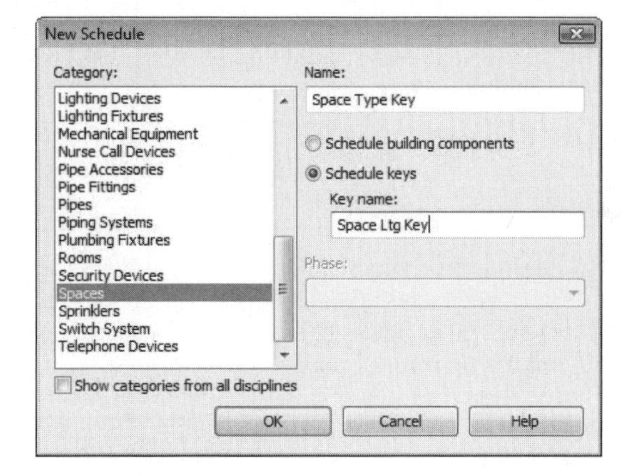

FIGURE 12.4
Creating a new schedule key for spaces

4. Click OK in the New Schedule dialog box to access the Schedule Properties dialog box. The key name is added as a schedule field by default, as shown in Figure 12.5.

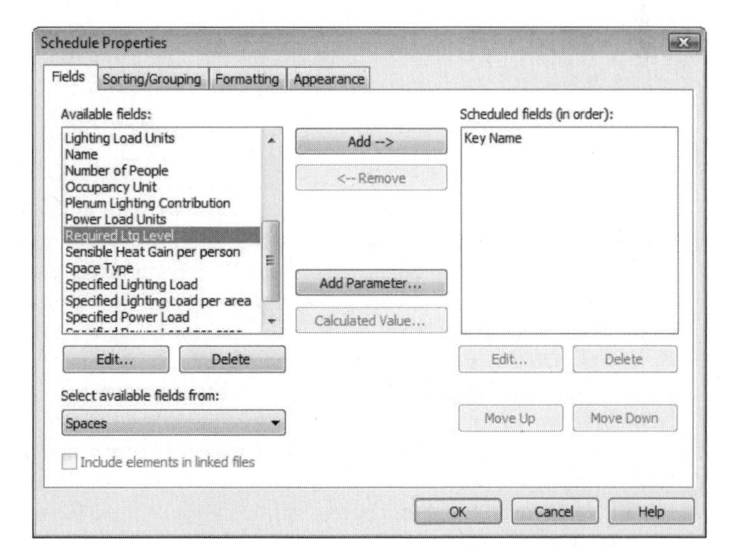

FIGURE 12.5
Schedule Properties dialog box

5. On the left side of the dialog box, select the parameter that you created as a target lighting level for your spaces, and click the Add button to include it in the schedule. These are the only two schedule fields required for this schedule, and there is no need to format them or adjust the appearance of the schedule, because it will only be used for analysis.
6. Click OK to create the schedule.
7. The schedule will not contain any data rows. At this point, you need to build your key for lighting requirements. Click the New button in the Rows section of the Modify Schedule/Quantities tab to create a row in your schedule.
8. Edit the name of the key in the schedule to that of a common type of building space.
9. Add the appropriate lighting level for that type of space in the second column of the schedule.

Space Style Schedule	
Key Name	Required Ltg Level
Corridor	15 fc

10. Repeat the process of adding rows, creating space types, and assigning lighting levels until your schedule contains all the space types you require for analysis of your project, as shown in Figure 12.6. You can create a comprehensive list in your project template file for use on future projects.

FIGURE 12.6
Schedule keys with required lighting levels

Space Style Schedule	
Key Name	Required Ltg Level
Auditorium	30 fc
Conference	30 fc
Corridor	15 fc
Janitor	15 fc
Kitchen	30 fc
Lobby	20 fc
Mech/Elec	20 fc
Office	50 fc
Storage	15 fc
Toilet	10 fc
Training/Classroom	50 fc
Waiting/Lounge	30 fc

The purpose for creating the schedule key is to maintain consistency throughout the model and to easily assign target lighting levels to spaces. You can now include the parameter created by the schedule key in your lighting analysis schedule and assign space types to all your spaces without having to select them in the model and edit their properties. Use the drop-down list in the parameter value to select an appropriate type for the space. When you select a type, the lighting level associated with that type will be input into the parameter for the target lighting level of that space, as shown in Figure 12.7. The value for the change calculated by the Lighting Delta conditional format automatically appears in the schedule when a key is assigned to a space.

You do not need to assign a type to a space in order to input a value for its targeted lighting level. Simply input a value into the schedule cell for the target lighting level. Notice in Figure 12.8 that room 100A has not been assigned a space type, yet a value has been given for the target lighting level.

FIGURE 12.7
Space lighting keys applied in a schedule

Space Ltg Analysis							
Room: Numb	Room: Name	Space Ltg Key	Required Ltg Level	Average Estimated Illumina	Ceiling Refle	Lighting Calculation Work	Lighting Delta
100	lobby	Lobby	20 fc	10 fc	0.75	2' - 6"	-10 fc
100A	reception	Lobby	20 fc	26 fc	0.75	2' - 6"	6 fc
101	corridor	Corridor	15 fc	13 fc	0.75	2' - 6"	-2 fc
102	corridor	Corridor	15 fc	21 fc	0.75	2' - 6"	6 fc
103	training 1	(none)		0 fc	0.75	2' - 6"	
104	training 2	Auditorium		41 fc	0.75	2' - 6"	
105	training 3	Conference		41 fc	0.75	2' - 6"	
106	training 4	Corridor		41 fc	0.75	2' - 6"	
107	break room	Janitor Kitchen		32 fc	0.75	2' - 6"	
107A	stor	Lobby		46 fc	0.75	2' - 6"	
108	comm - MDF	(none)		21 fc	0.75	2' - 6"	
109	main electrical	(none)		20 fc	0.75	2' - 6"	
110	elev mech	(none)		32 fc	0.75	2' - 6"	

FIGURE 12.8
Calculated values in a lighting analysis schedule

Space Ltg Analysis							
Room: Numb	Room: Name	Space Ltg Key	Required Ltg Level	Average Estimated Illumina	Ceiling Refle	Lighting Calculation Work	Lighting Delta
100	lobby	Lobby	20 fc	10 fc	0.75	2' - 6"	-10 fc
100A	reception	(none)	30 fc	26 fc	0.75	2' - 6"	-4 fc
101	corridor	Corridor	15 fc	13 fc	0.75	2' - 6"	-2 fc
102	corridor	Corridor	15 fc	21 fc	0.75	2' - 6"	6 fc
103	training 1	Training/Clas	50 fc	0 fc	0.75	2' - 6"	-50 fc
104	training 2	Training/Clas	50 fc	41 fc	0.75	2' - 6"	-9 fc
105	training 3	Training/Clas	50 fc	41 fc	0.75	2' - 6"	-9 fc
106	training 4	Training/Clas	50 fc	41 fc	0.75	2' - 6"	-9 fc
107	break room	Waiting/Loun	30 fc	32 fc	0.75	2' - 6"	2 fc
107A	stor	Storage	15 fc	46 fc	0.75	2' - 6"	31 fc
108	comm - MDF	Mech/Elec	20 fc	21 fc	0.75	2' - 6"	1 fc
109	main electrical	Mech/Elec	20 fc	20 fc	0.75	2' - 6"	0 fc
110	elev mech	Mech/Elec	20 fc	32 fc	0.75	2' - 6"	12 fc

With a target lighting level assigned to each space, you will now be able to determine how well the fixtures you have chosen are lighting the spaces. You can use the Space Lighting Analysis schedule to quickly see the types of adjustments required to meet the target levels. You can even include space parameters such as finishes and make adjustments to them for more accurate results.

CEILING HEIGHTS

Although ceilings can be bounding elements that determine the actual height of your spaces, you cannot include ceiling height in a Space schedule unless you create a parameter for ceiling height and manually input the data. One option for extracting the data from the model is to create a calculated value that divides the volume of a space by its area. This will give you an accurate dimension for the height of the ceiling if it is set to Room Bounding. In spaces with multiple ceilings at different elevations, the calculated value will result in an average height.

Hosting Options for Lighting Fixtures and Devices

Hosting fixtures and devices is important for coordination with other model elements and also reduces time spent modifying layouts to match design changes. There are a few different options for hosting, and you should choose the one that works best for the type of fixtures you are using and the file setup of your project. Face-hosted families are most commonly used because they work in many scenarios, but there may be times when you need to use an alternative hosting method.

Placing Light Fixtures in a Ceiling

Face-hosted lighting fixture families are most commonly used because they can be attached to ceilings in your model or ceilings within a linked file in your model. You can use face-hosted fixtures to represent recessed, surface, and pendant-mounted lights.

Cutting Holes in the Ceiling

Lighting fixture families can be made to cut the ceiling when they are placed. If you are attaching them to a ceiling that is in a linked file, the fixture will not cut the ceiling. This has no effect on lighting calculations and affects the appearance of the model only. If your architect is using your lighting model for a reflected ceiling plan, ceiling grid lines will be visible through lighting fixtures that cross them.

The default hosting for a face-hosted family is to a vertical face. To place lighting fixtures onto a ceiling, you need to select the Place On Face option. Recessed lighting fixture families should have an insertion point at one corner of the fixture. This will allow you to align the fixture to the ceiling grid on placement. If the family you are using does not have an insertion point at a corner, place the fixture on the ceiling, and use the Move or Align tool to line it up with the grid.

Using the Align tool is great for lining up your fixtures with ceiling grid lines, but it is important that you do not lock the alignment. Face-hosted families will not move with the grid lines when placed in a ceiling. Locking the alignment will cause constraint errors when the link is updated after the grid has moved. Either way, the lighting fixtures will stay attached to the ceiling if its elevation changes.

Once you have placed lighting fixtures onto a ceiling, you can copy them where needed. It is important to copy only the fixtures within the ceiling that they are hosted by. If you attempt to copy a face-hosted fixture from one ceiling to another that has a different elevation, you will receive a warning that the new instance of the fixture lies outside of its host. This will cause the fixture to be in the model without a host, which can result in an inaccurate model. By not being hosted by the ceiling, the fixture will not react to any changes in the ceiling elevation.

The Create Similar command is an easy way to get an exact duplicate of a fixture family from one ceiling to another. Use this command instead of Copy to duplicate a fixture family in another location. When you use this method, you will be required to set the hosting option to Place On Face before picking the location of the new fixture. Use the Pick New tool on the Workplane panel of the Modify | Lighting Fixtures contextual tab to move a lighting fixture from one ceiling to another.

Face-hosted lighting fixture families can be used in areas where a ceiling does not exist. Another choice for placement is to use the Place On Work Plane option. This will associate your fixture to a defined plane in the model. Because of the mounting behavior of face-hosted families, it is very important to draw your reference planes in the right direction. Drawing a reference plane from right to left will orient the plane properly for overhead lighting fixtures. Drawing from left to right will cause your lighting fixture families to be upside down in the model, as shown in Figure 12.9.

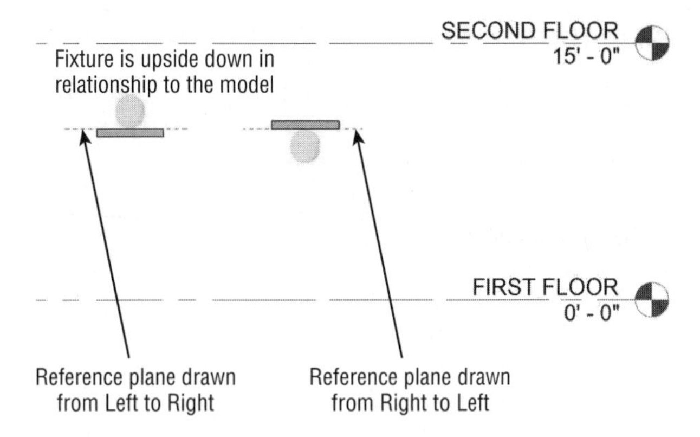

FIGURE 12.9 Lighting fixtures hosted by reference planes

Lighting Fixtures in Sloped Ceilings

Lighting fixtures can be hosted by sloped ceilings. Although this is an improvement to previous versions, there are some consequences. When the fixture is sloped, its symbolic line representation is no longer visible in a plan view. Figure 12.10 shows the same lighting fixture hosted by a sloped ceiling and a flat reference plane. The section view indicates that the fixture is hosted properly, but the plan view displays the fixture differently.

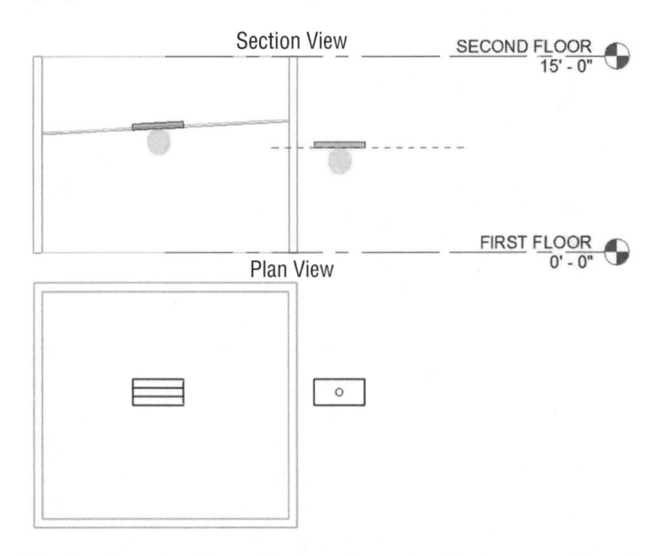

FIGURE 12.10 Lighting fixture in a sloped ceiling and in a flat ceiling

The fixture that is attached to the sloped ceiling displays the actual fixture graphics at an angle, while the fixture on the flat reference plane shows the symbolic lines used to represent the fixture. Because symbolic lines can be displayed only in a view that is parallel to the plane in which they are created, there is no way to display these lines when the fixture is sloped. You would need to either add a note to your documents that identifies the sloped fixtures or add the sloped fixture representation to your legend.

Another option is to use model lines in your lighting fixture family instead of symbolic lines. Model lines will display in plan view if the fixture is in a sloped ceiling. In this method, consider defining subcategories in the lighting family for Luminaire Body and Sloping Symbol. This will give you greater control over the look of your family. You wouldn't, after all, want the "symbolic" model lines showing up in any sections.

Ceiling Changes

Changes to ceilings will require some management of the lighting fixtures in your model. Face-hosted lighting fixtures will maintain their association with the ceiling when there is a change in elevation, but you should also be concerned with any lateral movement, especially with grid ceilings.

Lateral movement can have different effects on your lighting fixtures depending on how the grid was placed into the model. Grids that are placed by automatically locating the boundaries of a room will not affect your light fixtures when they are moved laterally. This is true as long as the movement does not cause your fixtures (which won't move laterally with the ceiling) to be located outside the boundaries of the ceiling. If the architect deletes a ceiling and then replaces it prior to giving you the updated file, your fixtures will not lose their association to the host, which is the linked file, but they will remain at the elevation of the original ceiling and will no longer respond to changes in the elevation of the new ceiling. You will have to use the Pick New tool to associate the fixtures with the new ceiling.

Ceilings that are placed into the model by sketching the shape of the ceiling will have a different effect on your fixtures when moved laterally. If the entire ceiling is moved, your light fixtures will remain in their location relative to the ceiling. If a boundary of the ceiling is edited by dragging it to a new location while in sketch mode, your fixtures will remain where they are located; however, any attempts to place new fixtures into the ceiling will cause them to appear outside the host, and you will receive a warning indicating that the fixture has no host. You can use the Pick New tool to then place the fixture in the ceiling.

Because ceilings tend to move around quite a bit in the early stages of a design, you may want to consider hosting your lighting fixtures to a reference plane until the major changes have settled down. At that point, you could move your fixtures to the ceilings using the Pick New tool.

Overhead Fixtures in Spaces with No Ceiling

Not every building area that you need to provide lighting for will have a ceiling. A space with no ceiling does not mean that you cannot use a face-hosted fixture family. Pendant-mounted fixtures can be face hosted to the floor or structure above, as shown in Figure 12.11.

Another option for using hosted fixtures in a space with no ceiling is to use a reference plane that defines the elevation of the fixtures. This is an effective method for lighting large spaces where one or two reference planes can handle all the lighting fixtures. It is not recommended that you use this method for multiple spaces because having many reference planes hosting items in your model will negatively affect your file performance.

You can also use lighting fixture families that do not require a host. You will have to manually set and manage their elevation. These types of fixtures should have a parameter that lets you define the mounting height, or you can use the Offset parameter.

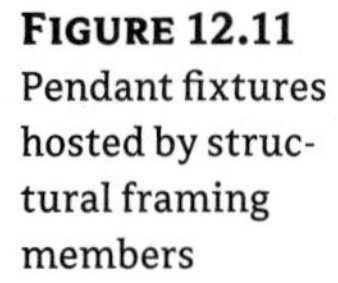

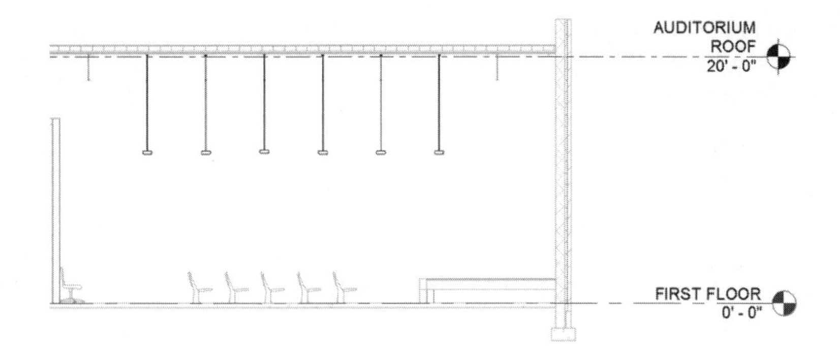

FIGURE 12.11 Pendant fixtures hosted by structural framing members

Wall-Mounted Lights

Lighting fixtures can be mounted to walls as well as ceilings. In fact, you can place a face-hosted lighting fixture family on any vertical surface that you need to. One thing to take note of is that any model element that is categorized as a lighting fixture does not have the ability to maintain its annotation orientation. What this means is you cannot use an annotation symbol nested in the family to represent the lighting fixture in a plan view when the fixture is mounted to a vertical face.

One option to overcome this shortcoming is to categorize your wall-mounted lighting fixture families as lighting devices instead of lighting fixtures. This will give you the ability to use a symbol but could also result in additional effort to control visibility and to schedule them along with all your other lighting fixtures.

Your best option, if you absolutely must use face-hosted families and represent wall-mounted lights with symbols, is to create line work in the family that represents the fixture and set the visibility of the line work so that it displays only in front and back views. This is necessary because with the face-hosted family placed on a vertical face, the direction that the fixture is seen from in plan view is from the back. The line work must be done with model lines; therefore, they will not react to changes in view scale. This technique is discussed in further detail in Chapter 21, but as shown in Figure 12.12, it is quite achievable.

The use of nonhosted families for wall lighting is perfectly acceptable. This will require that you manually maintain the association of the fixtures with the walls because you cannot lock the family to the linked wall. With a nonhosted family, you can use an annotation symbol to represent the lighting fixture in a plan view. This works well for exit lights because the actual fixture is typically not shown; rather, a symbol is shown.

Switches

Using face-hosted lighting switch families will keep your switches coordinated with the locations of their host walls. This does not mean that the movement of doors will not affect the hosting of your switches. Since you cannot constrain your switches to a distance from a door, if the door moves so that the switch is in the door opening, you will see the warning shown in Figure 12.13 that the switch has lost its association with the host.

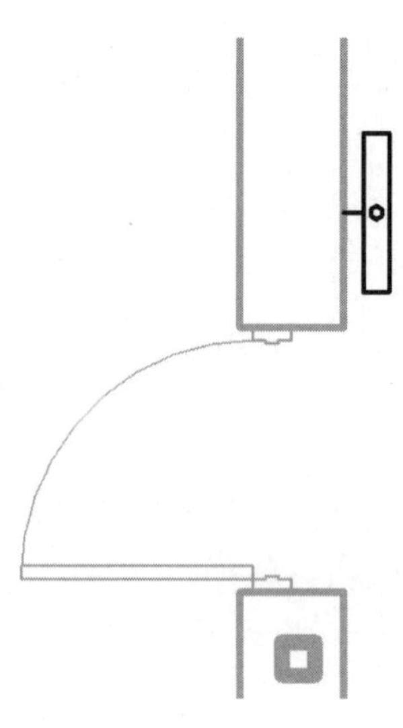

FIGURE 12.12
Face-hosted fixture mounted on a vertical wall with model lines to represent the fixture

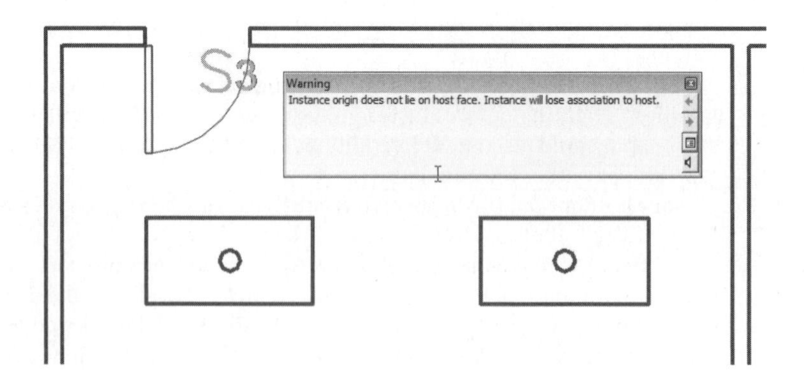

FIGURE 12.13
Warning that switch has lost association with host

You can associate your switches with the lighting fixtures that they operate. To create a switch system, do the following:

1. Select a lighting fixture, and click the Switch button on the Create Systems panel of the Modify | Lighting Fixtures contextual tab. This will change the ribbon to the Switch Systems tab.

On this tab, you have options to select the switch to be used for the system, view the properties of the system, and edit the system. Editing the system will allow you to add or remove elements and view the properties of the switch chosen for the system. Click the Select Switch button, and select the desired switch in the drawing area. Click the Switch Systems tab on the ribbon.

2. After clicking the Edit Switch System button, select the lighting fixtures to be included in the system, and click the Finish Editing System button; your switch system will be created.

3. To view the system, place your mouse pointer over any item that is part of the system, and press the Tab key. This will highlight the system elements and indicate their connectivity with dashed lines.

4. Click to select the system.

A switch system can contain only one switch, so for lighting fixtures controlled by multiple switches such as three-way switches, you will only be able to select one switch for the system. In the example in Figure 12.14, the three-way switch at the lower end of the room would control the lighting fixtures also but cannot be added as part of the switch system highlighted and indicated with dashed lines.

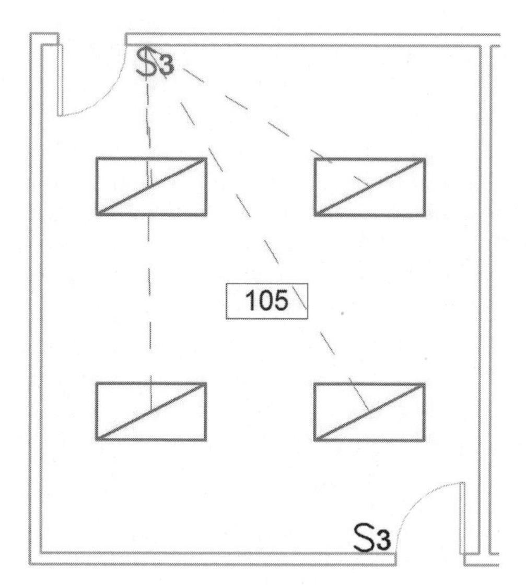

FIGURE 12.14
Switch system

Switches can be assigned an ID using the Switch ID parameter, which will help identify their relationship with lighting fixtures. When you select a lighting fixture and access the Switch Systems tab, you will see the ID of the switch associated with that light fixture in parentheses in the System Selector drop-down on the System Tools panel of the tab.

Creating switch systems is independent of any circuiting of the lighting fixtures and switches. You will still need to include the switches in the power circuit for the lighting fixtures if you want to remove them from the Unassigned category in the System Browser. Having as many elements as possible assigned to systems will help improve the overall performance of your model.

Site Lighting

Although you cannot do lighting analysis on site lighting within Revit MEP, a site lighting design can be useful to coordinate loads within panels and create a realistic view of the model from the exterior. Locations of poles, bollards, and other site lighting fixtures can be coordinated with other utilities within the project site. You can also create renderings to get an idea of the coverage of your lighting fixtures on the site.

The Site Plan

If you are working with a civil engineering consultant, it is likely that they are developing the site plan with some sort of CAD software. When they use a BIM solution, the 3D information such as elevation points and contours can be shared with Revit. This is necessary only if you are interested in creating topography within Revit to match the information in the site file. Otherwise, what you require from your civil engineering consultant is just line work that represents the layout of the site. Knowing the layout of parking lots, sidewalks, and major site elements should be enough for you to generate a site lighting design. In one sense, you are working with the site plan in the same manner that you would if you were using a typical 2D CAD system for your design. The difference is with Revit MEP you will be able to use the data within your design to help make decisions and to coordinate with other disciplines and project systems. Ask your consultant for a flat CAD file that you can use throughout the project, in other words, a file that they will update as changes are made to the project site.

The civil engineering and architectural consultants on your project may be sharing files also. At a minimum, the architect would share the building model so that the civil consultant could properly locate the building on the site. Your architect may choose to use the 3D data from the site file to generate a site plan within Revit. If so, you can use this information to create your site lighting layout. Although the architectural site model would give you topographical information, it also is only as up-to-date and accurate as the architect keeps it.

To get started, do the following:

1. Create a view associated with the ground level of your project. Because this is a site plan, the view does not have to contain only lighting system elements, so categorize your view in a manner that makes the most sense for your Project Browser organization. It may be best to create a subdiscipline under Electrical called Site to keep all your site-related views properly organized.

2. Set the View Range settings to properly display the building model as it would appear in a site plan. You can set the Top setting of your view and the Cut Plane setting to an elevation higher than the building so that it displays as seen from high above.

 Using a Plan Region should not be necessary unless you require the cut plane to be lower in a specific area of the site. If the building is represented in the linked site file, you may choose to not show the linked building model in your view. However, this workflow would mean that you are relying on the civil engineering consultant for an accurate representation of the building outline instead of getting that information from the architectural model.

3. Link the CAD file from your consultant into this view. Consider the option for linking the file into this view only if it is the only place that the site CAD file needs to appear. You can also create a workset for the linked site file to easily control whether it is loaded when your file is opened.

 Whatever you decide, link the file; do not import it. An imported site file will wreak havoc on your model, and you will not be able to automatically update the file when it changes.

4. Depending on the origin of the site file and your Revit file, you may have to manually place the site drawing into your view. The site file should contain an outline of the building, so you can align it with your linked Revit architectural model. If not, it may be necessary to

open the site CAD file and create an alignment point indicator that you can match up with inside your Revit file.

OPENING THE SITE FILE PRIOR TO LINKING

This practice is not uncommon. Many people do this to "clean up" the file prior to bringing it into their project. Unused layers and line work can be deleted to make the file easier to manage once it is linked into your project. Keep in mind that using this practice means you will have to do it every time your consultant gives you an updated version of the file. If you manually position the site plan into your view, you will have to use the Pin tool to pin it in place. CAD files that are linked in using the automatic positioning options are pinned in place by default.

Site Lighting Layout

With your site plan in place and a view that represents the building in relationship to the site, you can now begin to place site lighting fixtures in your model. If you have enabled worksharing in your project, it is best to create a workset for the site lighting plan, or at least for site elements in general. Lighting fixtures that are mounted to the exterior of the building should be included in this workset if they are to be displayed in the site plan view.

There are a limited number of site lighting fixtures that come with Revit MEP, so it is likely that you will have to get your site lighting families from manufacturers' websites or create them yourself. The chapters on content creation in Part 5 of this book will equip you with the skills necessary to create any lighting fixture that your project requires.

The topographic surface from a linked Revit site model will not provide you with a face to host your fixtures on. You can use the option to place your fixtures on a work plane and associate them with the ground level defined by the building. This will work fine for 2D plan view representation of the site, but if you need to show the site plan in section, elevation, or 3D, you will have to adjust your lighting fixtures to match the topographic elevation of their location. You will not have the ability to adjust the Elevation parameter of your lighting fixture families that are hosted by a work plan or level, so it may be best to use nonhosted families for site lighting. Notice in Figure 12.15 that the elevation of the lighting fixtures is not set to match the topography. Face-hosted fixture families were used in this example since the 3D view is used only for reference.

Site Lighting Analysis

Revit MEP can calculate the average estimated illumination of a Space object using the data from the photometric web file associated with a lighting fixture, but only because the Space object has a volume. Exterior lighting levels cannot be calculated. Most exterior lighting calculation applications are able to import CAD data, so you could export your Revit model to CAD and use the file in your analysis software.

You can, however, use Revit MEP for visual analysis of your site lighting layout. Lighting fixtures that contain photometric web files can display the pattern of light emitted from the fixture in renderings. Creating exterior renderings of your project will give you an idea of the coverage of your lighting fixtures on the site. This can help you determine whether you are using the right type of fixture or whether you need to adjust the number of fixtures used or the spacing of fixtures.

To see the exterior lighting in a rendered 3D view, you will need two things. You need to have a surface upon which the light will shine, and you need fixtures that contain a light source. If you are using a 2D CAD file for your site plan, you can place a "dummy" surface at ground level to act as your site surface. Your 3D view should be set with a sun position some time at night so the sun will not interfere with your lighting. You can display the sun position by using the Sun Settings options on the View Control Bar.

FIGURE 12.15
Site lighting fixtures in (a) 2D and (b) 3D views

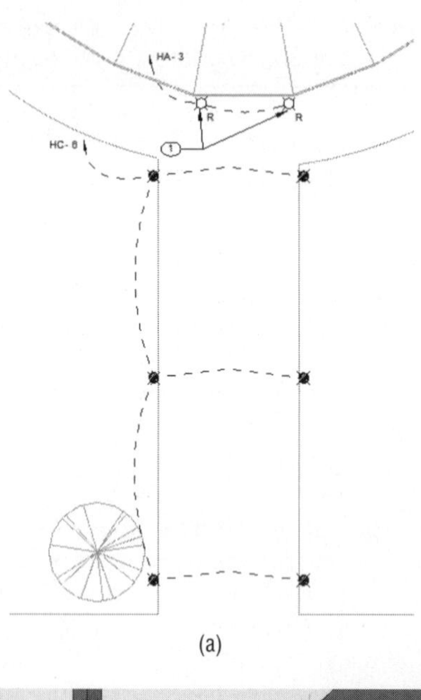

(a)

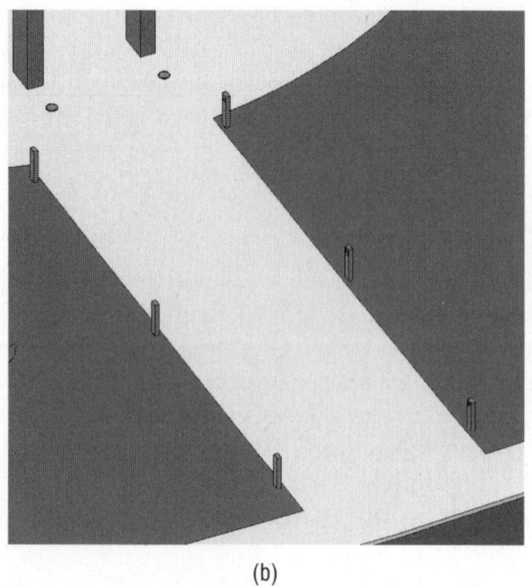

(b)

When you turn on the sun path, you may see a dialog box with options for displaying the sun path depending on the sun settings defined in the Graphic Display Options settings of the view, as shown in Figure 12.16.

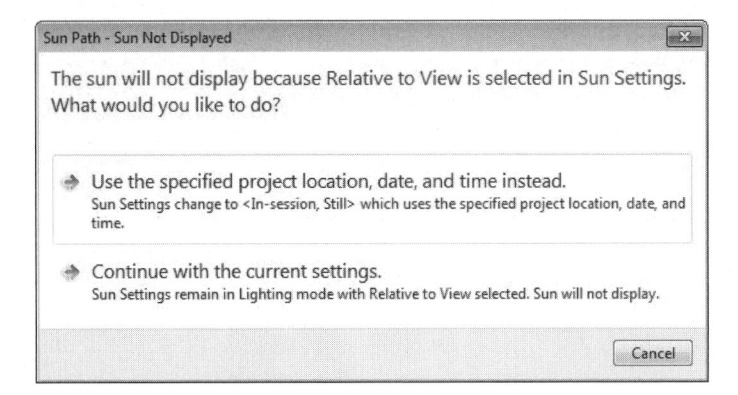

FIGURE 12.16
Options for sun path display

If you choose the option using project location and date, the sun path will appear in the view, as shown in Figure 12.17. You can adjust the position of the sun by dragging it along the path, or you can click the time shown and edit it manually. The date can also be edited by clicking the text.

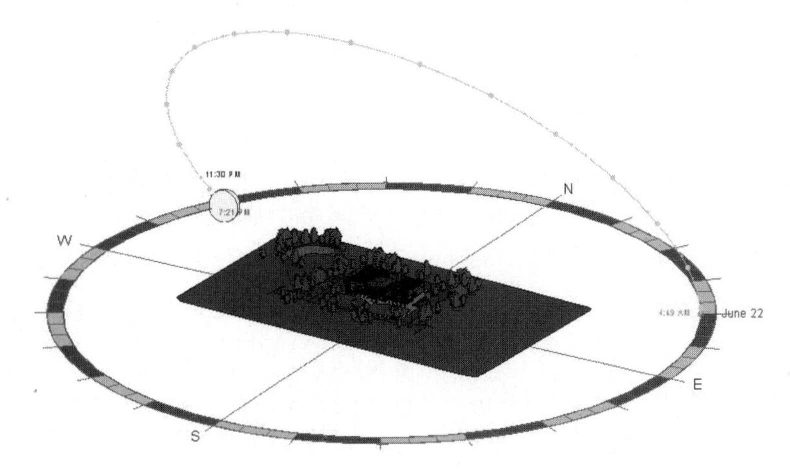

FIGURE 12.17
Sun path shown in a 3D view

To render your site lighting, do the following:

1. Click the Show Rendering Dialog button on the View Control Bar of your 3D view, and set the lighting scheme, as shown in Figure 12.18. It is best to do your renderings in draft mode when you are testing your design because of the amount of time it takes to render a view. Choose the Exterior: Artificial Only lighting scheme.

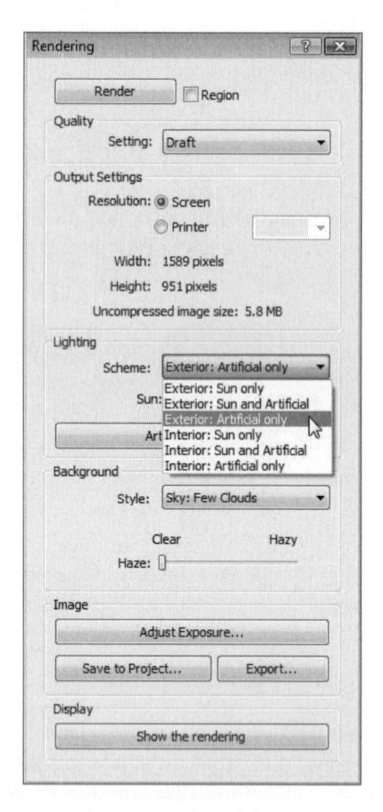

FIGURE 12.18
Render settings for draft view

2. You can group lights together and choose which groups to render. This will decrease your rendering times because Revit will take into account only the lights within the selected group. To create a light group, simply select a fixture, and use the Light Group tool on the Options Bar, as shown in Figure 12.19.
3. Click the Artificial Lights button in the Rendering dialog box to determine which groups of lights will be rendered, as shown in Figure 12.20. You can turn on or turn off entire groups of lights or individual fixtures to increase rendering times.
4. Click the Render button to generate a rendering of the view.

Once the rendering is finished, you will be able to see the lighting from your fixtures and how the light appears on the site. You can click the Adjust Exposure button to lighten or darken the image for more detail. The rendered view is a useful tool for the visual analysis of your lighting model. If you want to use the rendered view for presentation purposes, you can render the view at a higher level of detail. Figure 12.21 is an example of a draft-level rendering showing bollard lighting on a sidewalk.

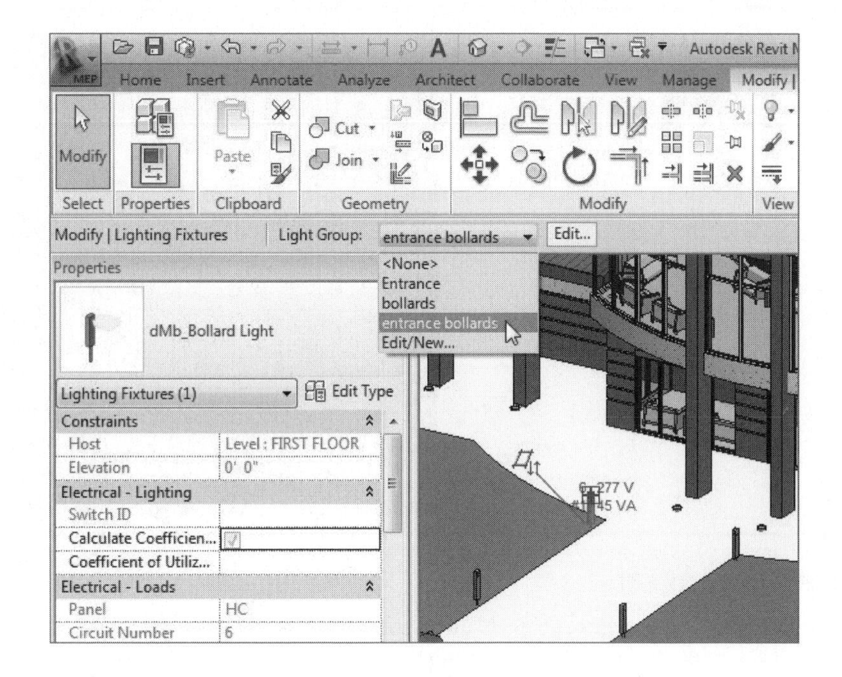

FIGURE 12.19 Selecting a light group

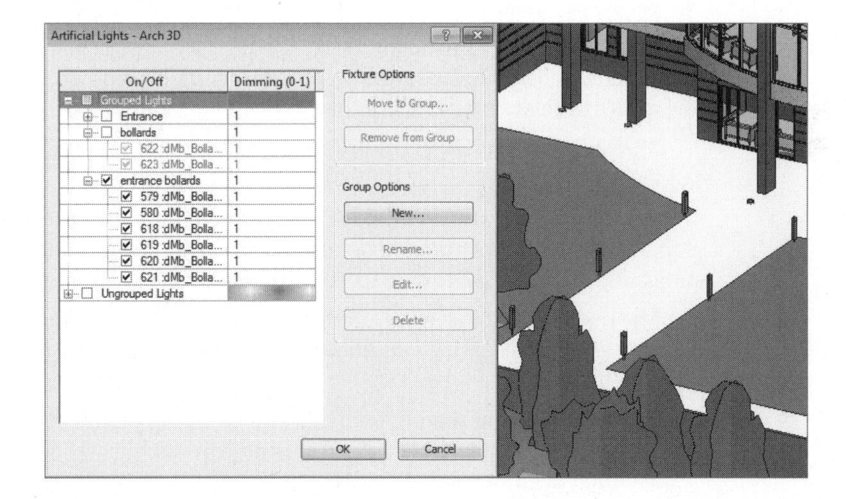

FIGURE 12.20 Choosing which groups of lights to render

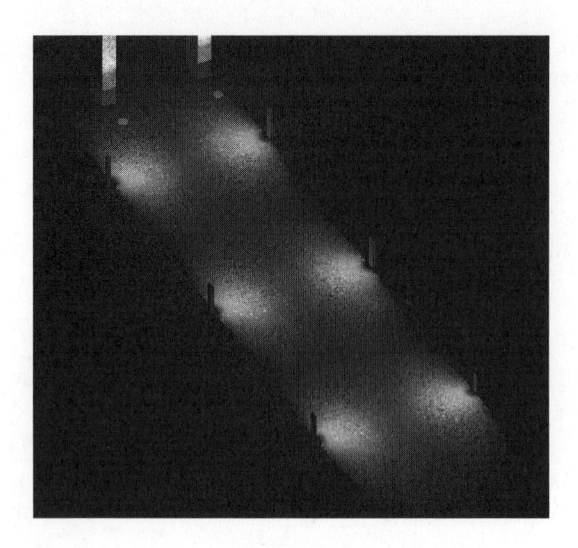

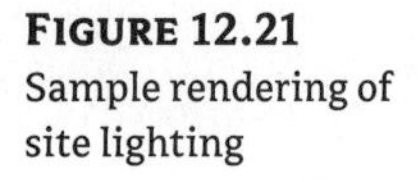

FIGURE 12.21
Sample rendering of site lighting

The Bottom Line

Prepare your project for lighting design The greatest benefit you can receive from a lighting model is coordination with other systems. Properly setting up the project file is key to achieving this coordination.

Master it Describe the relationship between ceilings and engineering spaces. How can you be sure that your engineering spaces are reporting the correct geometry?

Use Revit MEP for lighting analysis Though the design of electrical systems is usually represented schematically on construction documents, you can use the intelligence within the model to create a design tool that analyzes lighting levels.

Master it What model elements contain the data required to determine proper lighting layout?

Compare and evaluate hosting options for lighting fixtures and devices As a BIM solution, Revit MEP offers multiple options for placing your lighting model elements into your project. These options are in place to accommodate several workflow scenarios.

Master it What is the default hosting option for face-hosted families? Describe the limitations of representing wall-mounted lights with symbols and how they can be shown in a plan view.

Develop site lighting plans Creating a site lighting plan will allow you to coordinate with civil engineering consultants as well as with your architect. These plans are also useful for presentation documents and visual inspection of lighting coverage on the site.

Master it What is the benefit of using nonhosted lighting fixture families for site lighting?

Chapter 13

Power and Communications

Modeling power systems with a building information modeling (BIM) solution such as Revit MEP 2011 is just as important to project coordination as modeling systems that contain large amounts of physical data, such as HVAC systems. As with the lighting model, the key element to the power systems model is the data within the model elements. This information determines how systems can be put together. It can be extracted from the model for use in analysis and to aid in design decisions.

There is something to be said for the physical model of a power system as well. Receptacles and junction boxes are relatively small compared to other system components, but with a large number of them in a project, the potential for interference is increased.

Building communications systems have become more complex with the advances in modern technology. Voice and data networks and devices along with security and fire alarm systems are major design elements of new construction as well as with the renovation of existing buildings. Revit MEP 2011 has the tools necessary for you to communicate your design in a 3D model that contains the important data needed to ensure an efficient, effective design. Electrical equipment can be large and usually requires a clearance space around the equipment for service. Large conduit runs and cable tray are an important coordination consideration as well.

In this chapter, you will learn to

- Place power and systems devices into your model
- Place equipment and connections
- Create distribution systems
- Model conduit and cable tray

Methods for Modeling Power and Systems Devices

Power and systems plans can easily be created in your Revit model to show the locations of outlets and other types of electrical devices. You can use symbols, model elements, or a combination of the two to represent the design layout. The choice you make depends on what level of coordination you want to achieve and how much information you intend to extract from the design.

Revit MEP 2011 comes with a few basic device components to use for creating electrical layouts. Because you do not need to go into great detail to model these small elements, it is easy to create your own device families that match your company standards for electrical symbols. We cover creating electrical devices in Chapter 22, and we cover creating symbols in Chapter 18.

The Device button on the Home tab is used for placing a device into your model. This is a split tool with two parts. Clicking the top half of the button will invoke the command to insert an electrical device. Clicking the bottom half of the button reveals a drop-down of specific categories of devices, as shown in Figure 13.1. When you select a specific type of device from the drop-down menu, the top half of the Device button changes to match the category you have chosen. Clicking the top half of the button will invoke the command to insert that type of device. Selecting a device from a specific category will cause the Type Selector to only populate with devices in that category.

FIGURE 13.1
Drop-down menu from the Device button

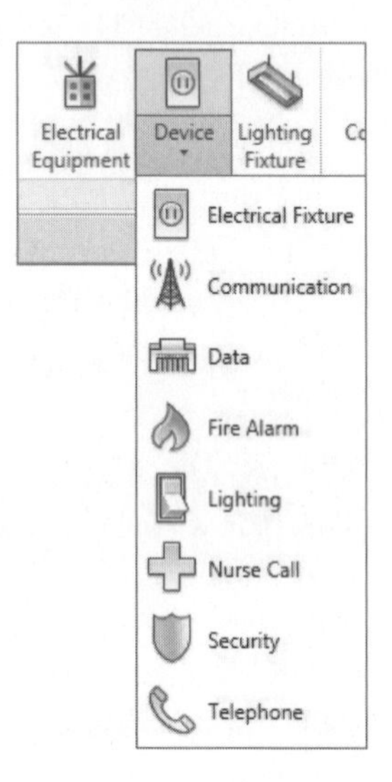

Select the Electrical Fixture category to insert power receptacles into your model. It is a good practice to go directly to the Type Selector after choosing a device category to insert. The Type Selector drop-down list contains all the families in the selected category that are loaded into your project. The list is organized by the family names highlighted in gray, with the family type names listed beneath each one. At the bottom of the list you will see a section with the most recently used families in that category. With tooltip assistance turned on, placing your mouse pointer over a family in the list will reveal a thumbnail view of the family. Figure 13.2 shows an example of the Type Selector drop-down for electrical fixtures loaded in the project.

If there is no family loaded into your project that matches the category you selected from the Device drop-down button, you will be prompted to load one. The default library contains an Electrical Components folder that has been reorganized for Revit MEP 2011. New content has also been provided with this release. Receptacle families are located in the Terminals folder found in the Electric Power folder. The Information And Communication folder contains families to be used for communications systems, including the Fire Alarm folder that has families for fire alarm design. All of these types of devices can be easily created because the level of model detail is not as important as the connector data. One family can be duplicated several times and given a different connector type in order to build a library of power and communications families.

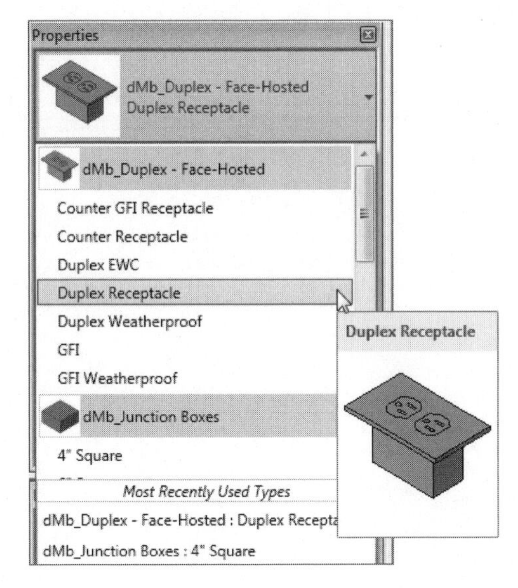

FIGURE 13.2
Type Selector drop-down

Using Annotation Symbols

Power and communications plans are typically shown schematically using symbols to represent the various devices and outlets. Because of this, annotation symbols are another key element of the families used to represent the design. Different symbols can be used within the same family to represent the different types of that family. This will help reduce the number of families required to maintain a complete library for design and reduce the number of families loaded into your project. Using multiple annotations within a family makes for easy modification of your design. Figure 13.3 shows a single communications family with three types utilizing different symbols.

Using Face-Hosted Families

It is best to use face-hosted families to model your receptacle and communications layouts. Unless otherwise noted, the examples discussed in this chapter will refer to face-hosted family types. This gives you the freedom to place the devices on any face within the building model. These types of components will move with their hosts, so you will spend less time moving your devices around to keep up with changes to the building model. If a device host is deleted from the model, the device will remain in place. This gives you the opportunity to relocate or remove the device manually and adjust any associated circuitry.

When you select a face-hosted device type to place in the model, the default placement is to a vertical surface. When you place your mouse pointer near a vertical surface such as a wall, the symbol used to represent the device will appear. You can press the spacebar to flip the orientation of the device to either side of the face that you are placing it on. Figure 13.4 shows a receptacle hosted to each side of the face of a wall. As you can see, this not only affects the symbol orientation but affects the model component as well.

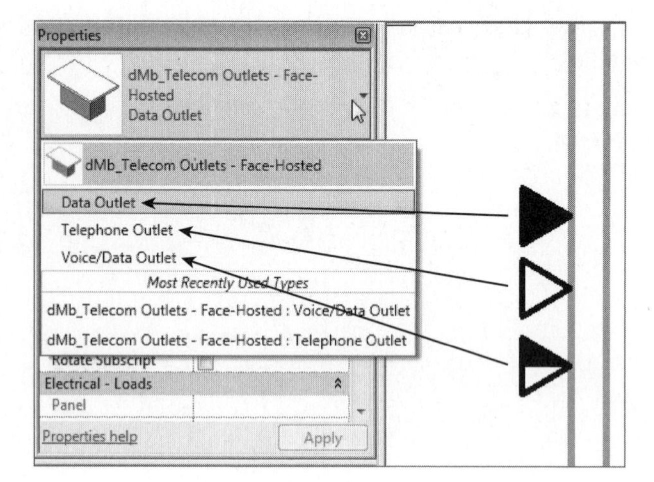

FIGURE 13.3
Multiple symbols in a device family

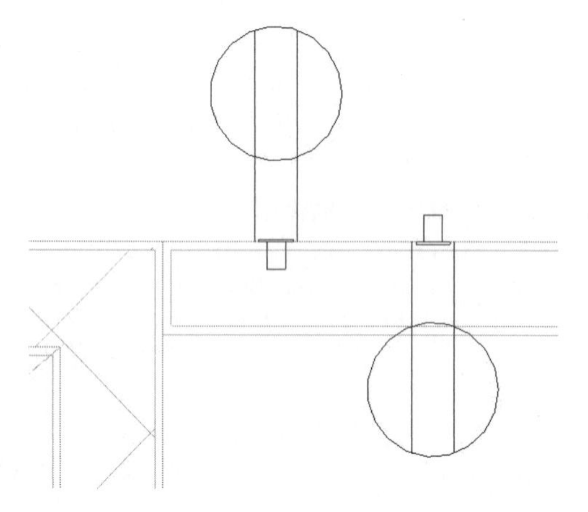

FIGURE 13.4
Orientation of receptacles on a vertical surface

If the device family you are using is not face hosted, pressing the spacebar will rotate the device in 90-degree increments parallel to the plane of the current view. After a face-hosted device has been placed, selecting it and pressing the spacebar will rotate the device 90 degrees parallel to the host face. This can cause your annotation symbol to "disappear" because annotations are visible only in views parallel to the view they are placed in. Figure 13.5 shows two systems outlets hosted by a wall. Both outlets were placed the same way, but the outlet on the right was rotated using the spacebar after placement. Notice that the annotation symbol of the outlet on the right is not visible in plan view.

Take care when placing devices on walls that contain multiple layers. Because of the structure of these types of walls, there are several vertical faces on which to host the device, as shown in Figure 13.6 with a power receptacle. You can use the Tab key to cycle through the different hosting options for the face you are trying to place the device on.

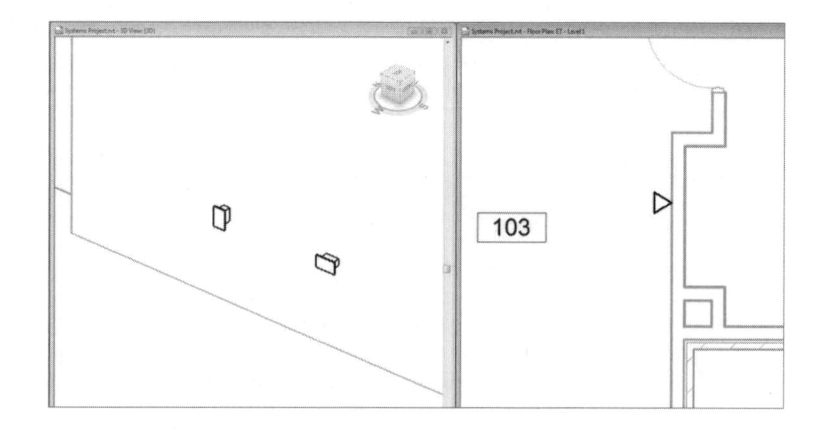
103

FIGURE 13.5
Outlet placement and rotation

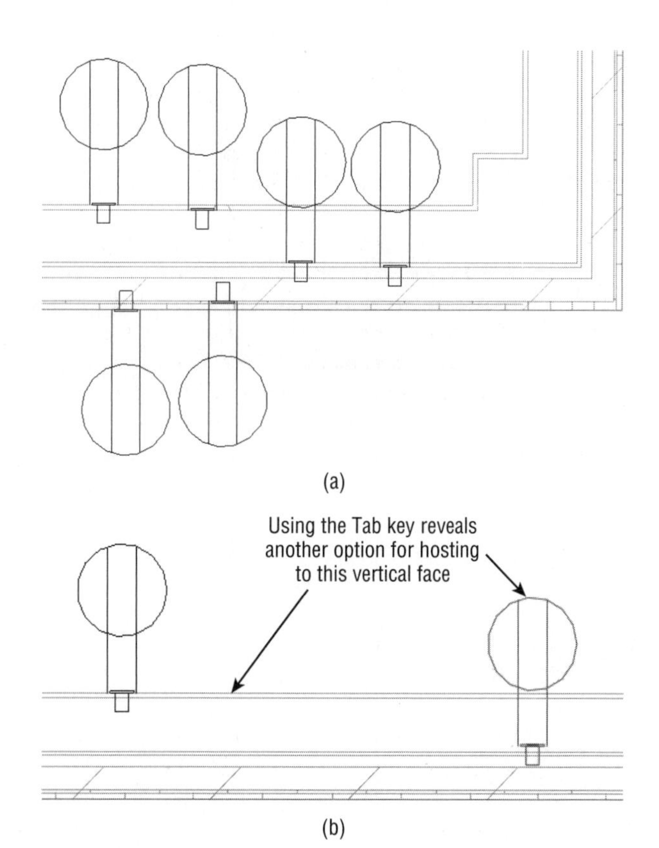
(a)
Using the Tab key reveals another option for hosting to this vertical face
(b)

FIGURE 13.6
(a) Multiple hosting options in a compound wall;
(b) using the Tab key for hosting options

Since any 3D surface can be used as a host, you can place devices on curved walls, and the component and symbol will follow the curve as you move your mouse pointer along the surface. There is no need to rotate the symbol or component after placement, saving time in laying out your design.

 **Real World Scenario**

ON THE FLOOR

Johan is working with an architect on a facility that will be used as a data management center. The main portion of the building is open office space on a raised floor. Because of the open nature of the floor plan, many floor boxes are needed for power and communications distribution. He is able to accommodate the design requirements by using face-hosted components that are hosted by the floor system. His plan views show the locations of the boxes with a standard symbol, and he is able to coordinate the locations with the office furniture layout.

Devices or outlets can be placed in floors or ceilings as well. To do so, you need to select the Place On Face option from the contextual tab after you have chosen your device. This enables you to place devices such as speakers onto the ceilings within your model, as shown in Figure 13.7.

FIGURE 13.7
Devices placed onto a ceiling face

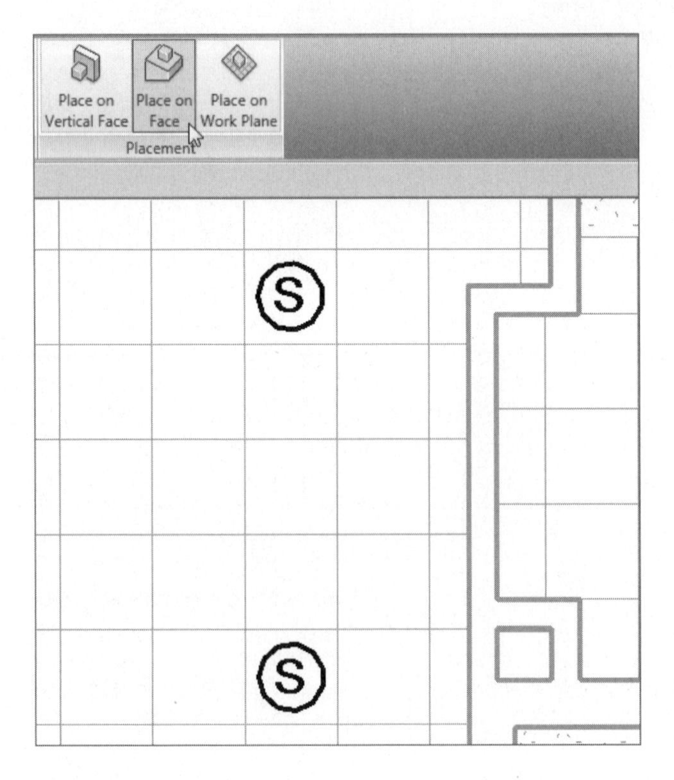

When you place devices or outlets onto floor objects, be sure that you are selecting the appropriate floor as a host. Often in projects the architect will model a floor object to act as a "placeholder" to show a floor in section or elevation views while the actual floor is being modeled in the structural project file. Use the Visibility/Graphics Overrides settings to turn on the appropriate floor object for hosting. This will ensure that if the floor object is removed from the architectural model, your devices or outlets will remain hosted. Use the tooltip or status bar to confirm the correct floor is displayed in your view, as shown in Figure 13.8.

FIGURE 13.8
Floor-hosted device in appropriate host

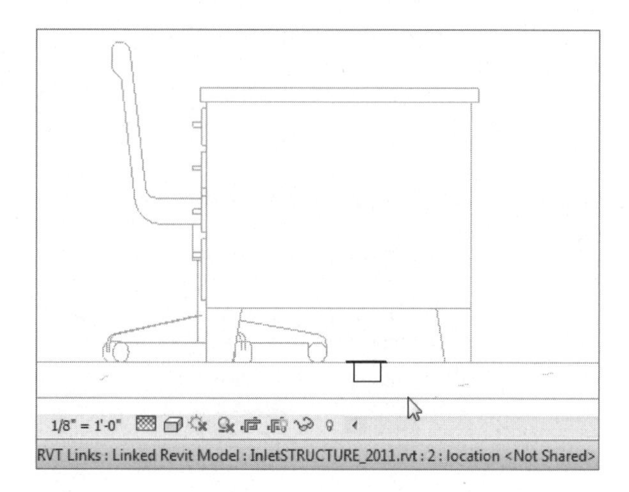

Avoiding Interference of Symbols

Because of the schematic nature of electrical construction documents, it can sometimes be difficult to place components without causing an interference between the symbols used to represent those components. Figure 13.9 shows two power receptacles placed in a model at the correct location for the intended design, but the symbols used for construction documentation are interfering with each other.

This scenario raises the question of whether it is more important to build the model accurately or to show the schematic symbols correctly for construction documentation. Using the functionality of families and nested annotations, Revit MEP gives you the ability to do both. Parameters within your component families can be used to offset the annotation symbols nested within them from the actual model components. In Figure 13.10 you can see how an offset parameter is used to shift the receptacle annotation of a family allowing for both accurate placement of the receptacles and correct symbolic representation.

This functionality can also be used for devices that are vertically aligned in the model. In plan view, these devices would appear to occupy the same space. A standard practice is to show one receptacle symbol at the location on the wall and another symbol offset from the wall at the same location. By using an offset parameter, the same result can be achieved. Figure 13.11a shows an instance parameter used to offset the symbol of a receptacle that is above another, and Figure 13.11b shows a section view and plan view of the receptacles. Using this functionality maintains both the model integrity and the schematic plan.

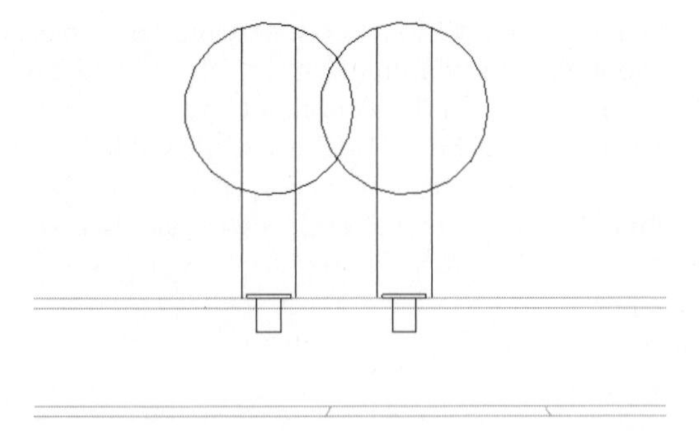

FIGURE 13.9
Receptacles with interfering symbols

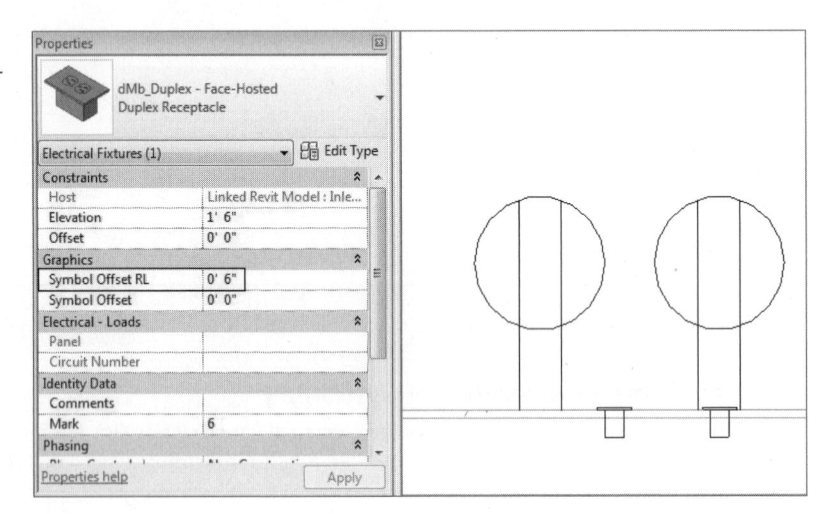

FIGURE 13.10
Receptacle annotation offset

Other devices such as junction boxes can be placed in the model in the same manner as receptacles or communication outlets. Creating different types within a junction box family is a good way to keep track of specific loads. Junction boxes for systems furniture can be hosted by the floor to show location if the furniture is not modeled in 3D. Boxes that are located above the ceiling can be hosted by the ceiling object and given an offset to maintain their distance from the ceiling if the ceiling height should change. When you place a junction box in a reflected ceiling plan and give it an offset, the box will offset from the front face of the ceiling. This means the box will move toward the floor, so if you need the box to be above the ceiling, you will have to give it a negative offset, as shown in Figure 13.12.

FIGURE 13.11
(a) Instance parameter settings for a symbol offset; (b) section and plan views of the receptacles

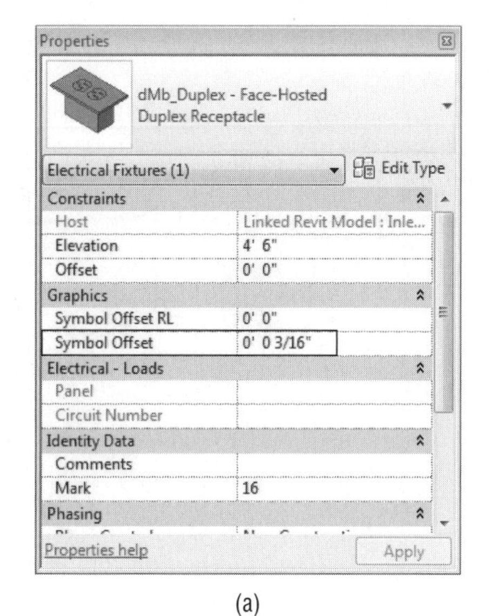

(a)

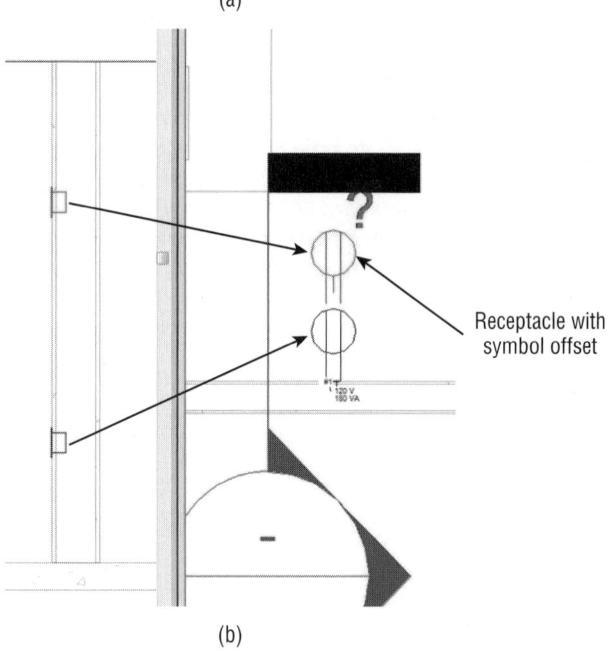

(b)

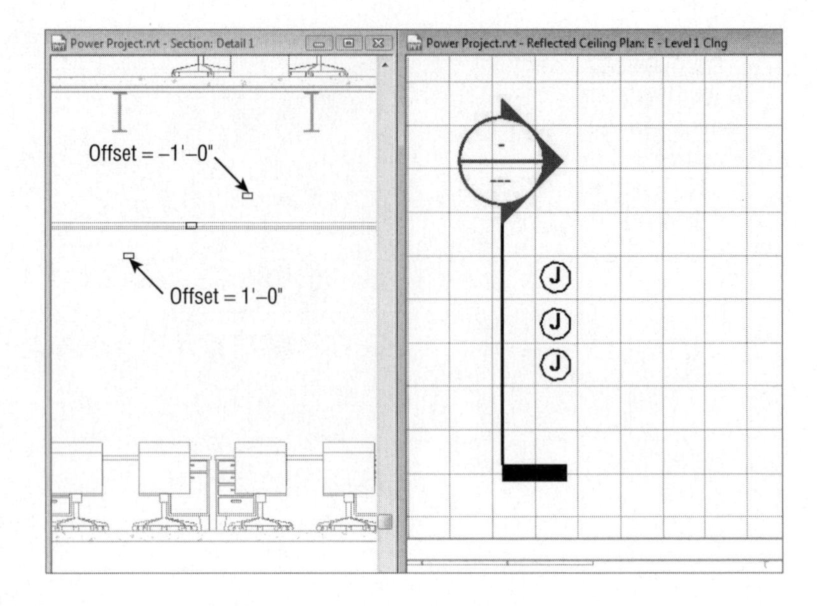

FIGURE 13.12 Ceiling junction boxes with offsets

Creating Circuits

When you place devices or outlets into your project, you can assign them to a system by creating a circuit for them. The type of circuit you can create is determined by the connector in the family. When you select a device in the model that contains a connector, the Create Systems panel appears on the contextual tab for that type of device. Clicking the button will create a circuit for that device. At this point, it is not necessary to select a panel for the circuit because you may not have placed the equipment yet, but having your devices on circuits will improve file performance. Figure 13.13 shows the available circuit types for a device that contains both a data and a telephone connector. A circuit can be created for each type of connector in the family.

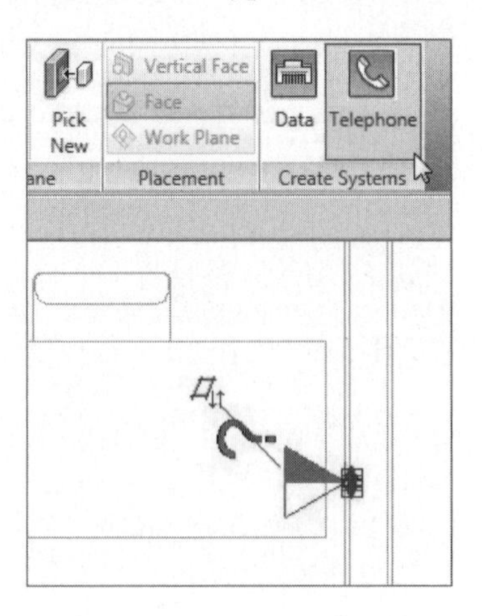

FIGURE 13.13 Circuit options for a multiconnector device

Some of your devices may require multiple connectors for different disciplines, such as a floor box with both power and data outlets. If your project contains several of these devices, it may be best to create a workset for multi-outlet devices so that their visibility can be easily controlled in both power and communications views.

Creating a Fire Alarm System Model

Fire alarm systems are often shown with the electrical construction documents, although on occasion they are handled separately. The same tools that allow you to model other electrical systems can be used for fire alarms. As with any unique system, it is recommended that you create a workset for the fire alarm system to allow for easy visibility control and multiple user access to the model. If the fire alarm system for your project is done by a consultant using Revit MEP, you can link in their model for coordination.

The number of fire alarm device families that are included with the installation of Revit MEP 2011 has increased from previous versions, but you may need to create devices that match your company standards. Fire alarm construction documents are usually schematic in nature, so the annotation symbols are the important part of your fire alarm system families. The `Manual Pull Station.rfa` family provided with Revit MEP 2011 can be copied and modified with the appropriate annotations to build your library of fire alarm devices. For more information on creating symbols and devices, see Chapter 18 and Chapter 22. Because fire alarm devices are typically mounted to walls and ceilings, face-hosted families are the best option.

Placing fire alarm devices into your model is the same workflow as power or communications devices:

1. Click the Device button drop-down on the Home tab, and select the Fire Alarm category. If no fire alarm devices are loaded into your project, you will be prompted to load one.

> **NO DEVICES LOADED**
>
> You can increase productivity by having the devices that you use most often on projects loaded into your project template, avoiding the need to interrupt your workflow by loading them as needed. On the other hand, it is possible to "overload" your template, so choose only those devices that you know will be used regularly.

2. Choose the desired device from the Type Selector, and choose a placement option for the host of your device. You can place pull stations, horns, strobes, and smoke detectors on any 3D surface in the model.

3. Use a reflected ceiling plan view to place devices on the ceiling. Be sure to turn on the visibility of other devices such as lighting fixtures, air terminals, and sprinklers to coordinate the location of your fire alarm devices in the ceiling. These other devices will likely be on worksets that contain other elements that are not necessary in your view, so you will have to control the visibility of the individual element categories.

Fire Alarm Riser Diagram

Because wiring is typically not shown on fire alarm drawings for the connectivity of the system, a fire alarm riser diagram is an important piece of the project. Although you cannot generate a riser diagram based on the devices placed into your model, you can use your model as a diagram, depending on the size and shape of the building. Figure 13.14 shows a floor plan with fire alarm devices placed throughout the building.

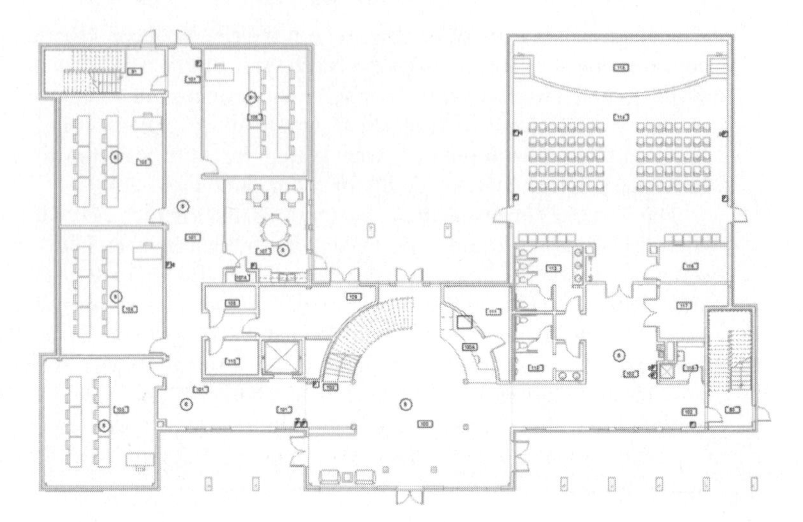

FIGURE 13.14 Sample fire alarm plan

With the devices in place, the view can be duplicated and used as a riser diagram. Right-clicking the view in the Project Browser and selecting the Duplicate With Detailing option will give you an exact copy of the view, including any annotation and detail lines. Using the Visibility/Graphics Overrides settings to turn off the linked models leaves just the fire alarm devices and room numbers shown. Room numbers and other items that do not need to be shown can be hidden in the view. Detail lines can then be added to the view to show the connectivity of the system, as shown in Figure 13.15. Using this method creates a diagram that is showing the actual locations of the devices, so as changes are made to the design, you will only need to edit the detail lines in this view to update your diagram.

This method has its drawbacks since the process would need to be repeated for every floor of a multistory building, resulting in several diagrams, so it is not a riser diagram in the true meaning of the term. It may also be difficult to use on a single-story building that is very large, where the devices are spread out, causing the diagram to be too large to justify the use of document space; however, if your project is the right size and shape, this method is a way to have your diagram tied directly to the model.

Fire Alarm Diagram Using Drafting Tools and Symbols

A more traditional approach would be to generate a diagram using drafting tools and symbols. This can be done while still utilizing some of the model information such as levels and an elevation of the fire alarm control panel or equipment. You can create a detail section of your control panel and then add symbols and detail lines to generate the riser diagram. In this section view, you can show the building levels and use constraints to maintain the relationship of your symbols to the levels so that if changes are made, your diagram will update appropriately. The same symbol families that are nested into your fire alarm device families can be loaded into your project file for use in your diagram.

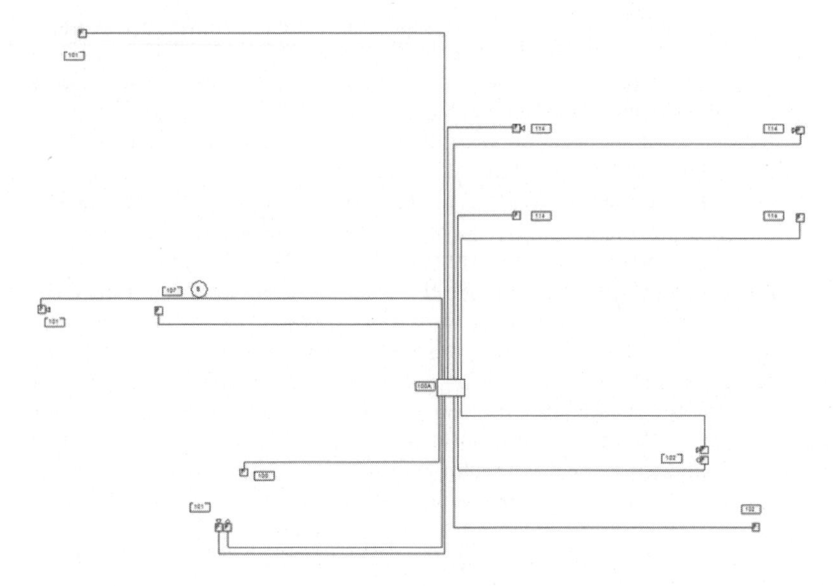

FIGURE 13.15 Fire alarm view with detail lines to be used as a riser diagram

By utilizing the drafting tools available in Revit MEP 2011 along with annotations and families, you can generate a fire alarm riser diagram without having to link in a CAD file and switch between applications to make changes. Figure 13.16 shows a very simple example of how Revit MEP 2011 can be used for fire alarm diagrams.

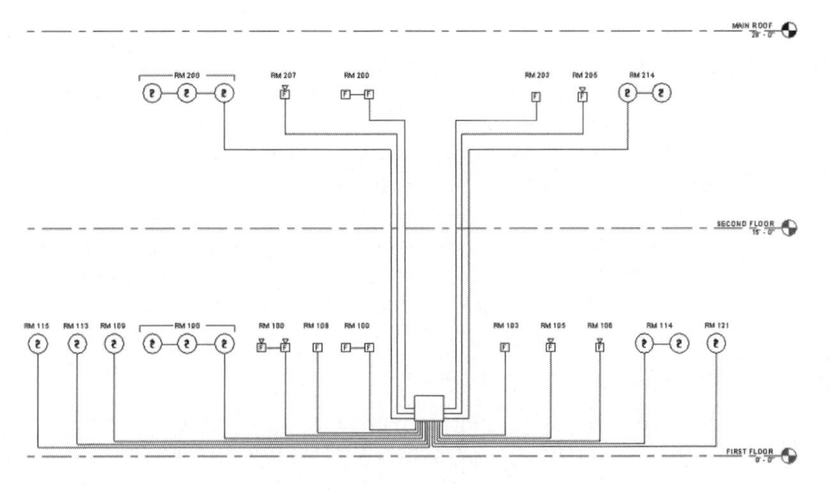

FIGURE 13.16 Sample fire alarm riser diagram using detail lines, text, and symbols

Equipment and Connections

Certain elements within a power design do not require a physical component in the building model. They are points of connection to the electrical system and are represented by a symbol to indicate to the electrical contractor where the connection is required. With Revit MEP 2011, you can represent these connection points with schematic symbols while building intelligence into the electrical model. Since the electrical information of a component is defined by its electrical connector, you can use symbols that contain connectors in order to account for the connections in your electrical model.

Equipment connections are the most common of these types of components. If you are linking an HVAC model into your project, you can use equipment connections that contain electrical information that matches the data in the component within the HVAC model. In this case, you would need to use an equipment connection family that contains an electrical connector. If you are working in a model that also contains the HVAC equipment, you can use an equipment connection symbol without an electrical connector (as long as the mechanical equipment has an electrical connector) and use the data from the HVAC equipment for your design.

With Revit MEP 2011, it is now possible to use the Copy/Monitor toolto copy and monitor mechanical equipment families from a linked file. This is a good way to coordinate the location of equipment; however, you are relying on the mechanical model to be using the appropriate family with the correct electrical data. If the equipment in the mechanical model does not contain an electrical connector, you will need to monitor it with one that does. This means you will have to load a similar mechanical equipment family that has an electrical connector into your project to be used for monitoring. Although this new functionality is a step in the right direction toward coordination between mechanical and electrical, it will require careful management of components used in the models and increased communication between consultants.

Figure 13.17 shows an equipment connection family used for an elevator. Since the elevator equipment has not been modeled, an equipment connection family with an electrical connector is used. Because the equipment connection family is used for many types of equipment, the load is an instance parameter. This allows for adjusting the load for each connection without having to create a new type within the family.

The mechanical equipment shown in Figure 13.18 is in the same model as the electrical components and has an electrical connector. In this case, a symbol without any electrical data is sufficient to represent the equipment connection.

In either case, the equipment connection can be attached to its associated model component by using face-hosted families. Constraints such as locked alignment can also be used but only if both constrained elements are in the same model. Aligning and locking a component to a linked component will cause a constraint error if the linked component moves and the linked file is reloaded.

Another example of an electrical design component that does not have any physical geometry is a motor connection. These can be represented in the electrical model in the same way as an equipment connection. The electrical data for the motor can come from either the equipment family or the symbol family used to represent the equipment.

Disconnect Switches

Disconnect switches may not require model components, depending on their use and who is to provide them for construction. Certain types of equipment come with a means of disconnect built into them, while others require that a disconnect switch be provided in addition to the equipment.

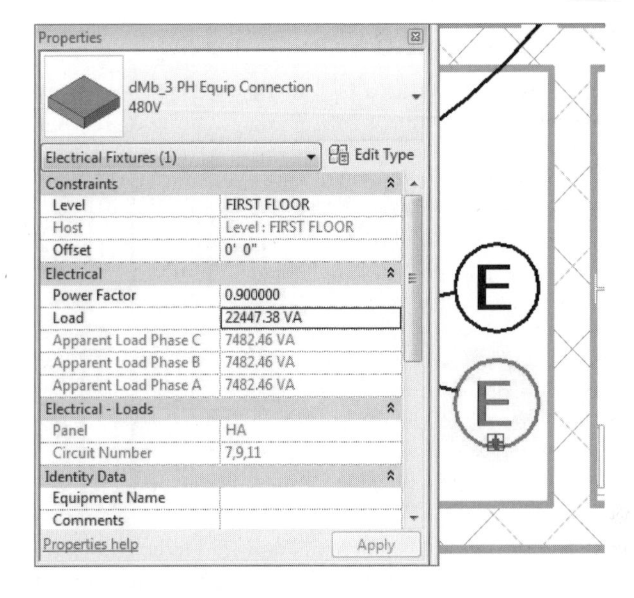

FIGURE 13.17 Equipment connection with a connector

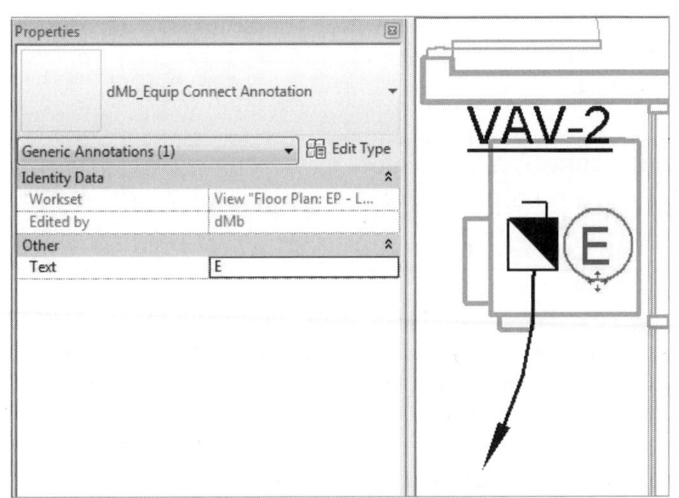

FIGURE 13.18 Equipment connection symbol

A disconnect switch that the electrical contractor is required to provide should be included as a model component in order to coordinate location and space requirements. Face-hosted families should be used because these types of disconnect switches may be mounted on walls or directly to the equipment that they serve. The disconnect switch family that comes with Revit MEP 2011 has several types based on voltage and rating. This family is found in the `Electrical Components\ Electric Power\ Terminals` folder of the library that installs with the software. The family is categorized as electrical equipment, so it is difficult to use a symbol to represent it, because face-hosted electrical equipment does not contain the parameter to maintain annotation orientation. So, each disconnect will display as its actual size based on the type chosen, as shown in Figure 13.19.

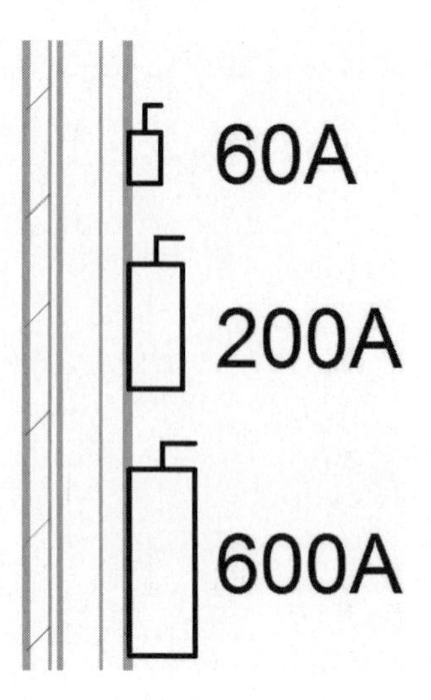

FIGURE 13.19
Disconnect switches in the building model

When a point of disconnect is required to be shown on the electrical plan but no equipment needs to be provided, a symbol can be used similar to the equipment connection symbol. There is no need for an electrical connector in your disconnect symbol family because a disconnect switch carries no load. However, wiring home runs for equipment are typically shown from the disconnect point, so you may want to include a connector point in your disconnect family to attach your wiring to.

CONNECTORS

Families that have connectors can be assigned to engineering systems. Having several components in your model that have not been assigned to systems can cause poor file performance.

Figure 13.20 shows a point of disconnect for a variable air volume (VAV) box. The electrical connector is in the VAV family, so the point of disconnect is represented as a symbol. As with the equipment connection, the disconnect symbol can be constrained to the equipment to provide coordination with changes to the model, as long as the equipment family is in the same file.

Distribution Equipment and Transformers

Distribution equipment not only plays an important role in the function of a building but also is an important element of a building model. These types of components require space for accessibility. Using accurately sized model components for distribution equipment allows you to coordinate early on with the architectural model for space requirements.

Transformers can be modeled at various sizes depending on their rating and voltages. This works well if you want to show the actual size of the transformers on your power plans. If you choose to represent your transformers with a symbol, one can be added to your transformer family. It is not necessary to use face-hosted families for transformers since they are typically located on the floor. An offset can be applied to wall-mounted transformers for placement at the proper height.

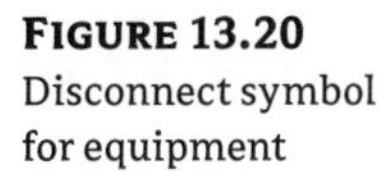

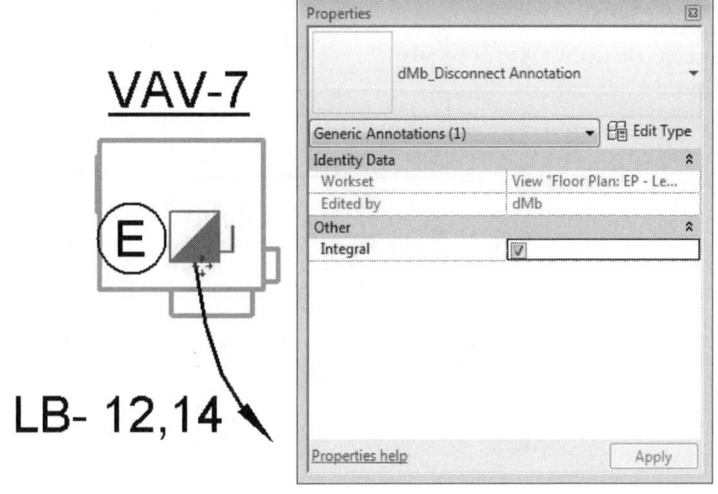

FIGURE 13.20 Disconnect symbol for equipment

To place a transformer, do the following:

1. Click the Electrical Equipment button on the Home tab to place a transformer in your model. The Options Bar allows you to rotate the transformer after placement, or you can use the spacebar to rotate the transformer prior to placing it in the model.
2. When you have placed the transformer in the desired location, it is best to assign distribution systems for the primary and secondary side. It is a good practice to do this at the time of placement so the transformer will be available as an option for service when you establish your distribution model. Selecting a transformer after it has been placed will activate the Distribution System drop-down list on the Options Bar. This drop-down contains a list of all the distribution systems defined in the electrical settings of your project. Only distribution systems that match the connector voltage of your transformer will appear in the list, as shown in Figure 13.21.

> **HOUSEKEEPING**
>
> The pads for your electrical equipment can either be added to the equipment families or be created as separate components. If they are to be created as separate components, you should coordinate with your structural consultant about who will create them.

3. To set the distribution type for the secondary side of the transformer, you need to access the element properties of the component. The Secondary Distribution System parameter is an instance parameter that contains a list of the distribution system types defined in the electrical settings of your project. There is no connector in the transformer family for the secondary side, but it is important to establish the distribution system for connectivity of components in your distribution model. Figure 13.22 shows the option for selecting a secondary distribution system for a transformer.

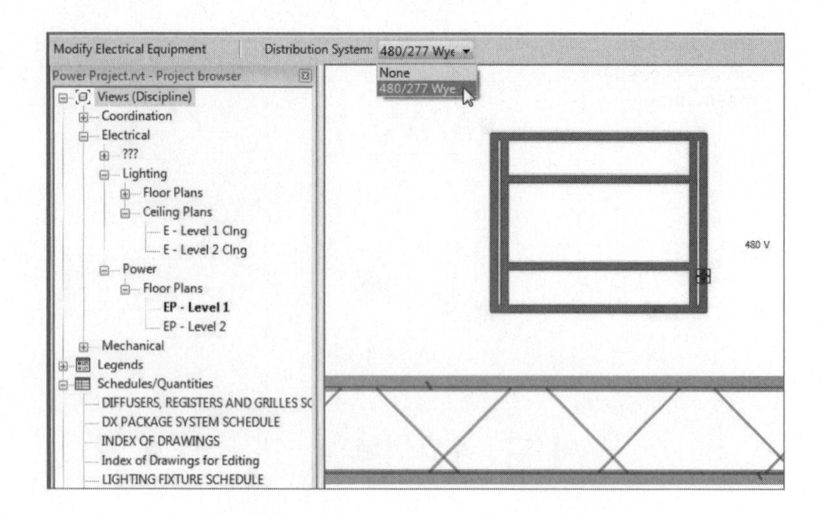

FIGURE 13.21 Distribution System drop-down for a transformer

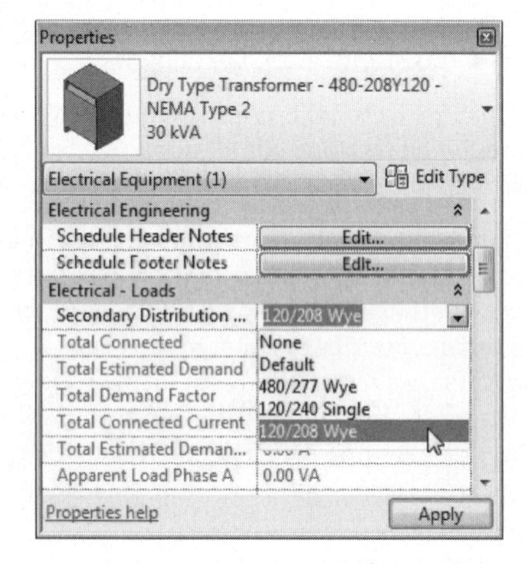

FIGURE 13.22 Secondary Distribution System parameter for a transformer

Switchboards

The various components that make up a switchboard are available in the Distribution folder within your Revit MEP 2011 library. Utility, metering, transformer, and circuit breaker sections are all available. Family types can be created within the families to meet the requirements of your design. If you do not know the exact equipment that will be used for your project, these families can act as placeholders until the equipment is chosen. Symbols can be used for various levels of model detail. The layout of switchboard sections can be used in elevation details if required.

To place a switchboard component, do the following:

1. Click the Electrical Equipment button on the Home tab. Select the component from the Type Selector drop-down, or load the family if necessary.

2. Use the spacebar to rotate the equipment prior to placement.

Figure 13.23 shows use of the spacebar to rotate objects adjacent to an angled wall or other objects. Begin by placing the object as normal (a); then hover your mouse pointer over the host object until it highlights (b), and press the spacebar (c). The object will rotate in line with the angled wall.

FIGURE 13.23
Using the spacebar to rotate objects

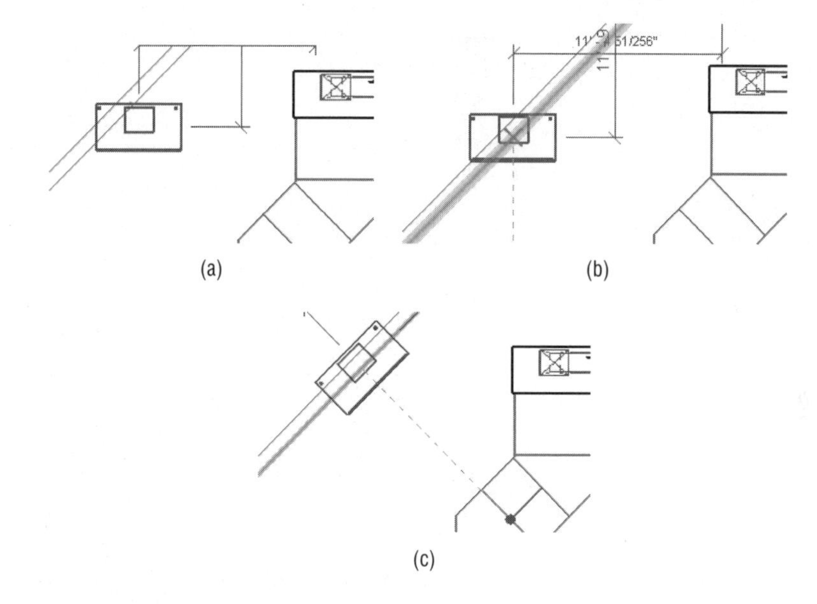

Components of the switchboard can be constrained together to move them as a whole unit if necessary.

3. Make sure each component contains a connector that will define it as part of a distribution system.

Figure 13.24 shows an example of a switchboard layout in (a) plan view and (b) section view.

The distribution system of a switchboard component can be defined after placement by selecting the element and using the Distribution System drop-down on the Options Bar or by editing the instance properties in the Properties palette.

Panels

Electrical distribution panels are called many things by different users, such as *panels, panelboards,* or *breaker boxes*. In this book, we will refer to them as *panels* and *panelboards*. The process for placing panels into your model is similar to placing other types of electrical equipment. You will want to use face-hosted panel families for coordination with architectural model changes. The panel families that come with Revit MEP 2011 are useful because they exist and are also easily customized to meet your company standards for panel representation on construction documents.

Many companies use specific symbols to represent panels based on their voltage. The panel families in Revit MEP 2011 simply represent the size of the panel by displaying the box. You can add a detail component to them that contains a filled region that represents your standard. You must use a detail component because a nested annotation symbol will not work for face-hosted electrical equipment. The detail component must be placed in the Front or Back elevation view of the family in order to display when the panel is mounted to a vertical surface.

FIGURE 13.24
Switchboard layout

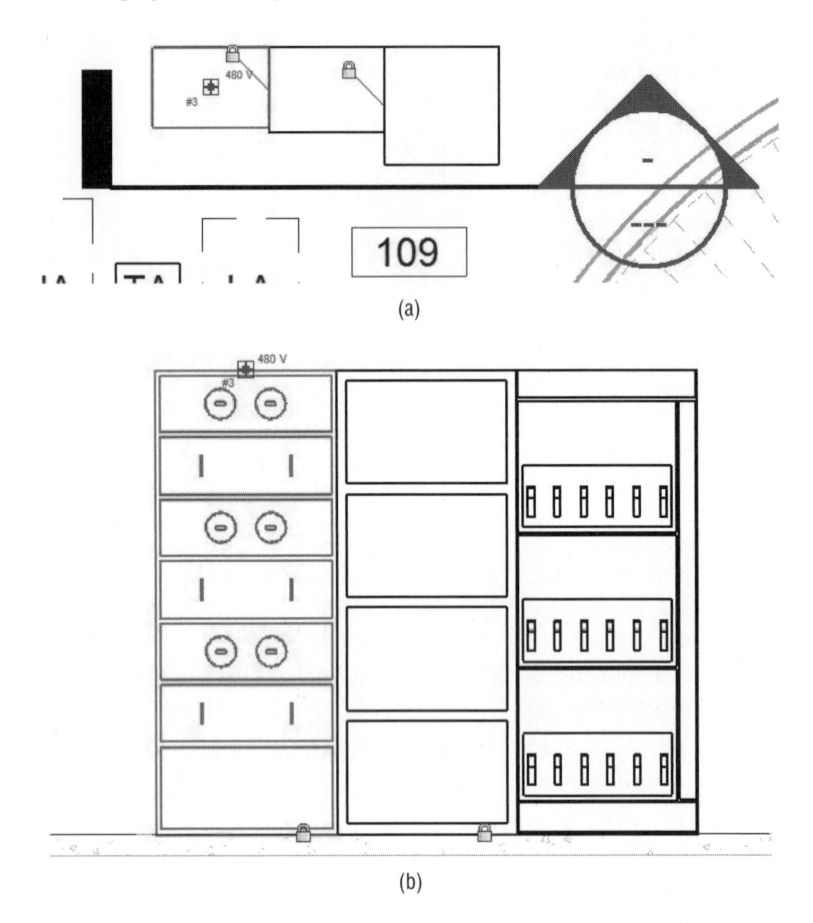

> **NESTED ANNOTATIONS**
>
> Annotation symbols cannot be placed in section or elevation views of families.

Figure 13.25 shows a panel family with a nested detail component family that is a filled region representing a 208V panel. This detail component is constrained to the parameters in the panel family and will change size with the panel.

Clearance space is an important issue when placing panels into your model. You could draw detail lines, or even model lines directly in your project view to represent the clearance spaces, but that would be difficult to manage when changes occur. Elements can be added to your panel families to represent clearances not only for 2D plan views but also for the 3D model.

This will allow you to check for interferences with objects that infiltrate that clearance space. In Figure 13.26 you can see a panel family that has line work to represent the clearance area in front of the panel for plan views and solid extrusions that represent the three-dimensional clearance areas.

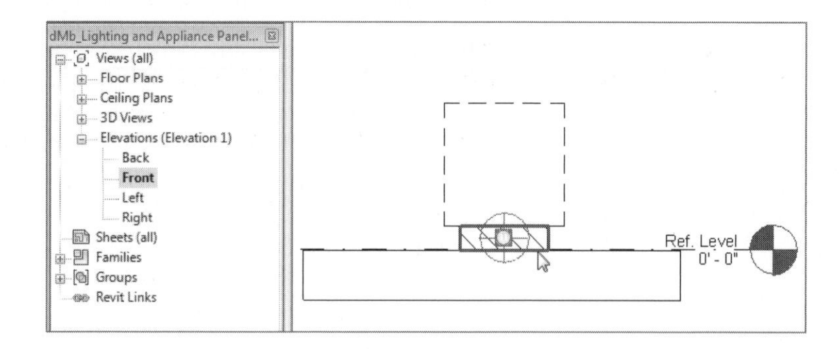

FIGURE 13.25
Detail component in a panel family

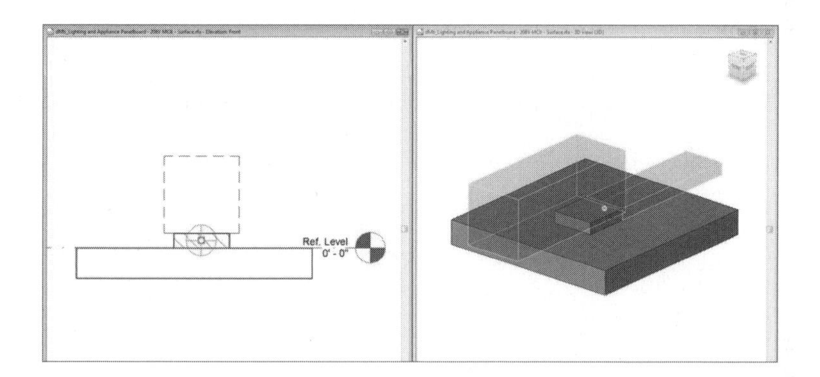

FIGURE 13.26
Panel family with clearance elements

Using subcategories for the clearance elements gives you the ability to control their visibility in your model views. The 3D clearance space does not have to be visible in your model in order for Revit to detect that an object is interfering with it.

Customizing your Revit families to meet your company standards makes them easier to use and can help utilize the model information for design decisions. To customize a panel with clearance space lines, do the following:

1. Open the `Ch13Panel.rfa` file found on `www.wiley.com/go/masteringrevitmep2011`.
2. Click the Object Styles button on the Settings panel of the Manage tab. Create a new subcategory under Electrical Equipment called **Clearance Lines** for the 2D lines that will show the clearance area in plan view. Choose a line weight of 4, blue line color, and hidden line pattern for this subcategory. Click OK to close the Object Styles dialog box.
3. Open the Front elevation view.
4. Click the Symbolic Lines button on the Annotate tab. Select Clearance Lines from the drop-down on the Subcategory panel of the Modify | Place Symbolic Lines contextual tab.

5. Draw a line from the upper-left corner of the panel object 4″ to the left. Draw a perpendicular line 3′-0″ up from the panel. Draw a line 2′-4″ to the right. Draw a line perpendicular down to the front of the panel. Draw a line to the upper-right corner of the panel to complete the clearance area. (You can use different dimensions according to your standards. The point is to draw the space in front of the panel.)

6. Click the Aligned button on the Dimensions panel of the Annotate tab to draw a dimension from the left reference plane to the Clearance Line parallel to it. Click the lock grip to constrain the Clearance Line to the reference plane. Repeat the process for the reference plane and parallel Clearance Line on the right side of the panel.

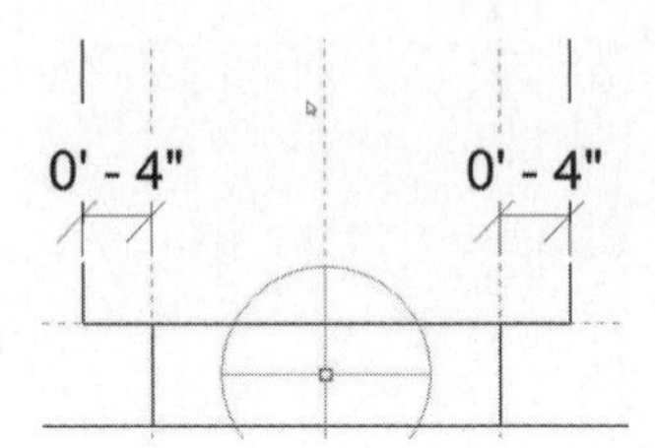

7. Dimension from the reference plane that defines the depth of the panel to the Clearance Line parallel to it in front of the panel. Lock the dimension.

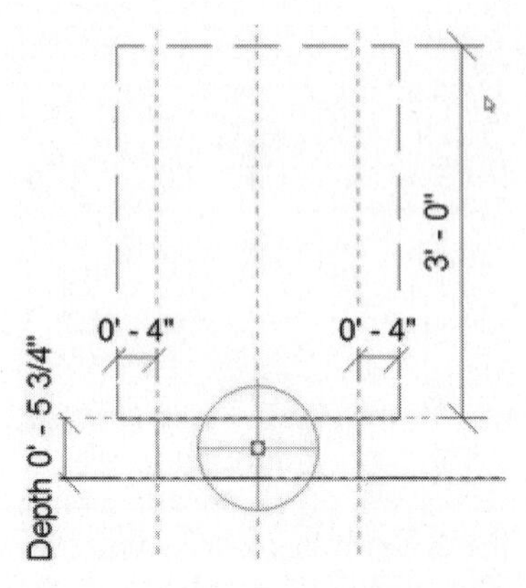

8. Use the Save As command to save the panel family to a location of your choice.

9. Create a new project file, and click the Wall button on the Architect tab to create a wall in plan view.

10. Load the new panel family into your project. Use the Electrical Equipment button on the Home tab to place an instance of the family onto the wall.

11. Use the View tab, and click the Visibility/Graphics button. Once in the dialog box, expand the Electrical Equipment category. Notice that the Clearance Lines subcategory now exists in the project. Turn on the display of electrical equipment, and click OK to close. You should now see your family and clearance line work.

Assigning a distribution system to your panels is crucial to the intelligence in your electrical model. This will enable you to create circuits for devices and lighting fixtures as well as model the distribution system. The Distribution System drop-down is available on the Options Bar when you place a panel into your model.

CREATING SIMILAR PANELS

If you use the Create Similar command to place a new panel in your model that is just like another one, you still need to select a distribution system for the new panel.

Panels can be named using your naming standard by editing the Panel Name parameter. When using the panel families that come with Revit or customized versions of them, you can control the size of a panel by establishing the number of poles. The Max #1 Pole Breakers parameter allows you to establish the number of poles available in the panel. An examination of the type parameters of one of these families reveals that the height of the panel is determined by the number of poles. The other parameters such as Mains, Enclosure, Modifications, and Short Circuit Rating that appear in the Electrical – Loads group of instance parameters are for information that will appear in the panel schedule. These parameters do not factor into the electrical characteristics of the panel or distribution system. They exist for reporting purposes only and have to be manually edited.

The Circuit Naming, Circuit Prefix Separator, and Circuit Prefix parameters allow you to control the naming of circuits that are fed from the panel. Because these are instance parameters, you can name circuits differently for each panel if necessary. You do not need to use these parameters in order to tag your circuits.

Other Equipment

Component families can be used to represent any type of electrical equipment required in your Revit project. Items such as generators, automatic transfer switches, starters, and variable frequency drives are available in the library. Figure 13.27 shows a diesel generator on the site plan of a project. The generator family displays differently depending on the Detail Level setting of the view. In Coarse detail, the generator displays as a simple box, while in Fine detail the shape of the equipment is visible.

For telephone device systems, you can use a family to represent the punch-down blocks that act as the termination point for the telephone wiring within the building. This would allow you to keep track of your telephone devices in different areas of the building. It is best to use a face-hosted family for this type of equipment because it is typically wall mounted. If your equipment family behavior is set to Other Panel, the Create Panel Schedules button appears on the Modify | Electrical Equipment contextual tab. You can create a panel schedule template that can be used for communications equipment to show connected devices. See Chapter 5 for information on creating panel schedules.

Server racks or audio equipment racks can be either wall mounted or free-standing equipment, so nonhosted families can be used to accommodate either condition. A server rack family can be used as the equipment for your devices with data connections, and an audio equipment rack family can be used for devices with communications connectors. Figure 13.28 shows server racks placed in the center of a room to allow for the required clearances for the equipment.

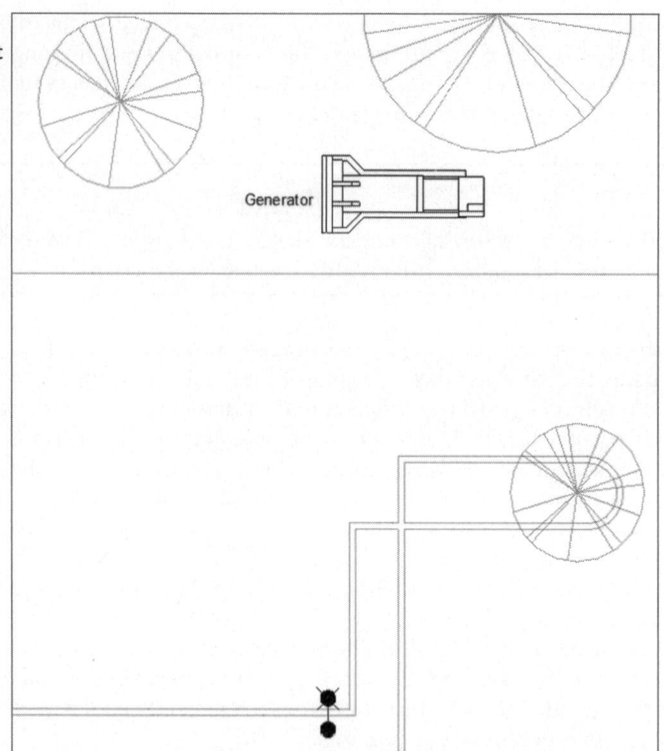

FIGURE 13.27
Generator in a Revit model

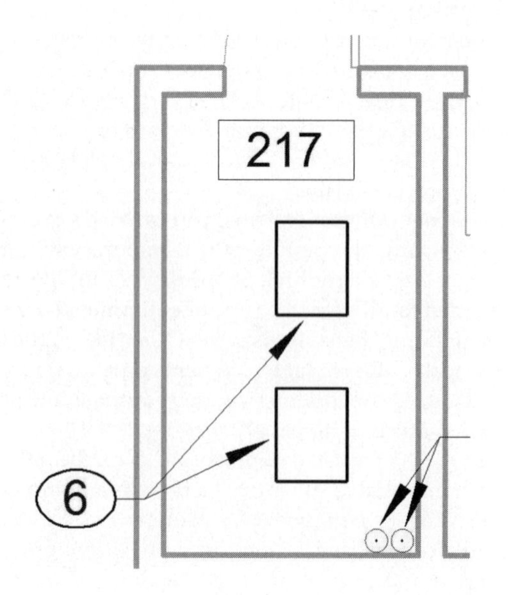

FIGURE 13.28
Server racks

ELECTRICAL COMPONENT FAMILIES

Detailed electrical equipment can be modeled using Autodesk Inventor and exported as an ADSK file. This file format can be used directly in your Revit project or in component families that you create. Electrical connection points transfer from the Inventor file to Revit during the export process.

Power Distribution Systems

Having electrical equipment components in your Revit project enables you to coordinate space requirements and interferences with other building systems, but they also allow you to build intelligence into the model by establishing distribution systems. Establishing the distributive relationship between electrical components gives you the ability to keep track of loads from the branch circuit panels all the way to the main electrical equipment.

This is where the importance of assigning a distribution system to your electrical power equipment as it is placed comes into play. With the equipment in place, you can easily create the relationship from component to component. You can start at any component and connect to related equipment upstream. Since you will be assigning the equipment from which a component is fed from, it is best to start at the branch circuit panels and work your way upstream through the system. No wiring or conduit needs to be modeled in order to establish the electrical connection between two distribution components.

To create a distribution system, follow this general process:

1. Start by selecting one of your low-voltage branch circuit panels. If you have not already done so, assign a distribution system to the panel from the drop-down on the Options Bar. The Modify | Electrical Equipment contextual tab will appear on the ribbon. Because the connector in the panel family is a power connection, a Power button appears in the Create Systems panel of the tab. Clicking this button will create a power circuit for the selected element. The active tab on the ribbon will change to the Electrical Circuits contextual tab. Click the Select Panel button to choose the equipment that feeds the selected component.

2. There are two ways you can select a panel, either by clicking the panel in the drawing area or by selecting it by name from the Panel drop-down on the Options Bar. The drop-down makes it easy when the panel you need is not shown in the current view. The only components that will show up in the drop-down or be available for selection in the drawing area are those that have connectors that match the distribution system of the selected component. This includes any panel or equipment in your project, not just in the current view.

In Figure 13.29, a power circuit is created for panel LA. Notice that transformer TA is the only equipment available to feed the selected panel, because the secondary distribution system of the transformer matches the distribution system of panel LA.

When the panel is selected and the Electrical Circuits tab is active, Revit MEP 2011 displays a red, dashed box around all elements that are part of the circuit. There is also a red arrow shown pointing from the item that the circuit was created for to the panel or equipment chosen to feed it. By placing your mouse pointer over an electrical component and pressing the Tab key, you can see whether a circuit has been created for it and the components of that circuit, as shown in Figure 13.30.

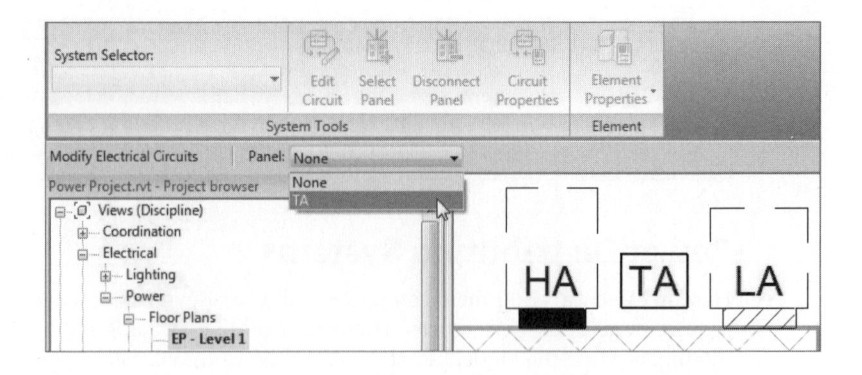

FIGURE 13.29 Selecting a panel for a distribution system

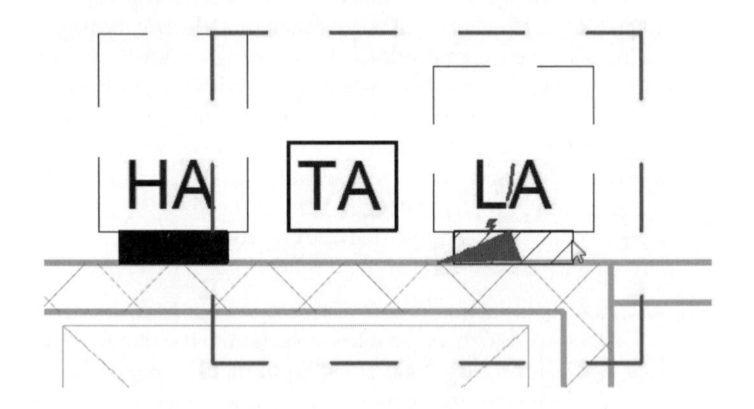

FIGURE 13.30 Using the Tab key to identify a circuit

3. Moving upstream from this point, you now need a circuit and panel for the transformer. In Figure 13.31, you can see that the steps for creating a circuit have been repeated for transformer TA. The list of available panels for the circuit is longer because there are more distribution components in the model that match the distribution system of the primary side of transformer TA. In this case, panel HA is the equipment chosen.

Also, notice in Figure 13.31 that naming your panels and equipment is important prior to creating distribution systems. The metering section of the switchboard was not given a name and shows up in the list as its type name along with its electrical characteristics.

You can repeat this process for equipment as far upstream as you want to keep track of load information. Obviously, you aren't going to model the utility company equipment or the local power plant, so when you get to the last component, you will not have any panel or equipment to select for its circuit. It is still recommended that you create a circuit for this equipment because you will be able to access the properties of the circuit and establish feeder information.

Power Diagrams

The most common way to communicate a distribution system is by a riser or one-line diagram. This schematic representation displays the connection relationships between electrical distribution equipment in a clear and easily readable manner (unless the project is large and complicated, in which case readability becomes debatable).

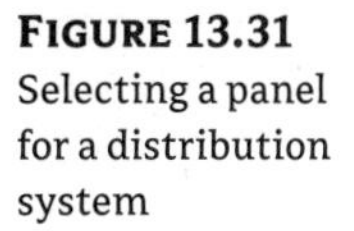

FIGURE 13.31
Selecting a panel for a distribution system

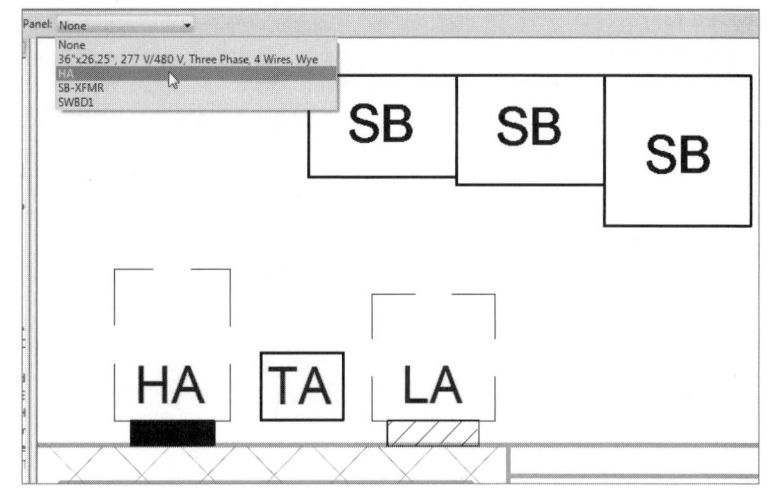

One of the first questions that comes up about using Revit MEP for electrical design is whether it can automatically create a power riser diagram from the components within the model. Unfortunately, at this point, Revit MEP does not have that capability. This does not mean that you cannot create a diagram directly in your Revit project. The drafting tools within Revit MEP 2011 provide you with all the tools you need to create your diagrams, or if you are more comfortable with using a CAD application for drafting tasks, you can link a CAD-generated diagram into your Revit project.

Tips for Creating Power Diagrams

It is best to have a library of commonly used symbols to facilitate quick and easy creation of your power diagram. Symbols for items such as panels, transformers, grounding, and generators can be easily created and stored for use on future projects. These symbols can contain parameters that make it easy for you to schedule and manage changes to the component data.

Consider creating a detail section view of your switchboard equipment or main distribution panel as the starting point for your diagram. In this view, you can turn off any linked files and worksets that do not need to be shown, leaving you with an elevation view of your equipment and the project levels, which are typically shown on riser diagrams.

Symbols can then be placed into this detail view to represent the electrical distribution equipment. Detail lines can be drawn in the view to represent the connection between elements. You can use the power of Revit to create line types specific to your needs for electrical diagrams to increase your efficiency when drafting. Although the symbols and line work are not tied to the model elements that they represent, you are able to coordinate data required in both places without having to switch between applications. Figure 13.32 shows an example of a power diagram created using an elevation view of the switchboard, symbols, text, and detail lines.

Temporary dimensions and alignment lines make drafting and editing diagrams in Revit MEP easy, and with a little practice, you can become proficient with the available tools. We are all looking forward to the day when power diagrams will be generated automatically from the model, but until then, you have the tools necessary to create them.

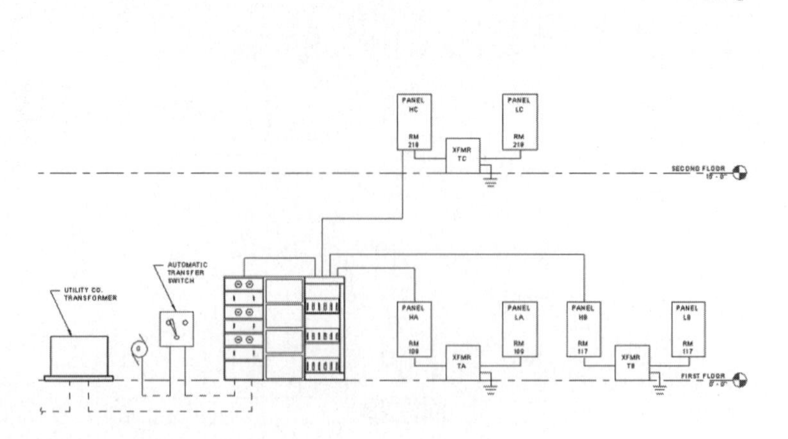

FIGURE 13.32
Power diagram created in Revit

The New Conduit and Cable Tray Tools

The tools for modeling conduit and cable tray are new to Revit MEP 2011. You can now create conduit and cable tray runs within your project model for coordination with other system components. Conduit and cable tray are system families that you can define different types for. Parameters can be used for tagging and scheduling these families to keep track of quantities and materials if desired. The techniques for modeling conduit and cable tray are similar to those for placing ductwork or pipe. It is not likely that you will model all the conduit required for your project, so it should be clear early on in the project what the size limit is for conduit that will be modeled.

There are two styles of system families for conduit and cable tray. You can create runs that utilize fittings or runs that do not. If you have a situation where you want to model a run of conduit that will be bent to change direction, you would use the style without fittings. This allows you to determine the length of the run and does not add components to the model that would not exist in the construction.

Because there are two different styles for conduit and cable tray, there are also two schedule types that can be created. The Cable Tray Runs category in the New Schedule dialog box is for cable tray style without fittings. This schedule style contains parameters for reporting the total length of a run, including any change in direction. The Cable Tray or Cable Tray Fittings schedule styles are used for the cable tray style that utilizes fittings. This type of schedule can be used to report data about the individual pieces that make up a run of cable tray. The same schedule types are available for conduit, as shown in Figure 13.33.

When you model a run of cable tray or conduit using the style without fittings, it does not mean that no fittings exist in the run. You still have to define what fittings will be used to transition or change direction, but the length of the fittings will be included in the total length of the run. Figure 13.34 shows the fittings assigned to a cable tray style without fittings.

A fitting family must be assigned to each parameter in order to model a run of conduit or cable tray; even if you are not modeling a condition that requires a tee, you still must have a fitting family assigned for tees. It may seem counterintuitive to assign fittings to a system family without fittings, but it is necessary in order for the runs to be modeled. When you model a run without fittings, the fittings used will have a value of Bend for their Bend or Fitting parameter. Runs modeled using fittings the parameter value will be Fitting. This can be useful information when scheduling conduit or cable tray.

FIGURE 13.33
Conduit and cable tray schedule categories

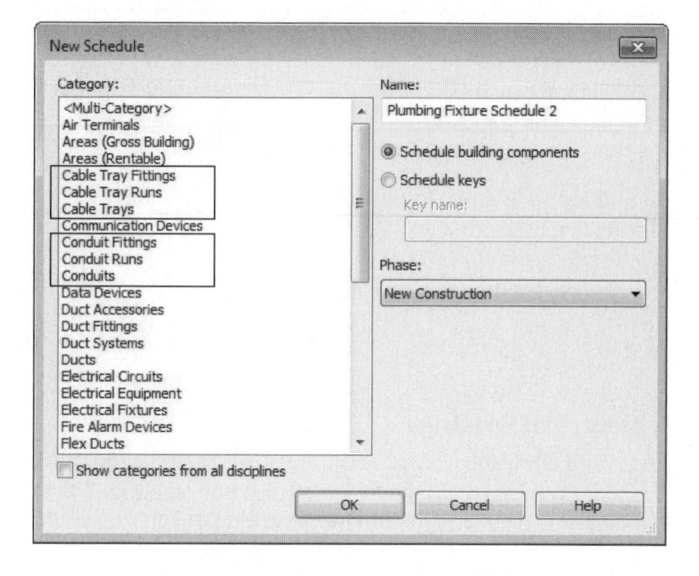

FIGURE 13.34
Fittings assigned to a cable tray family

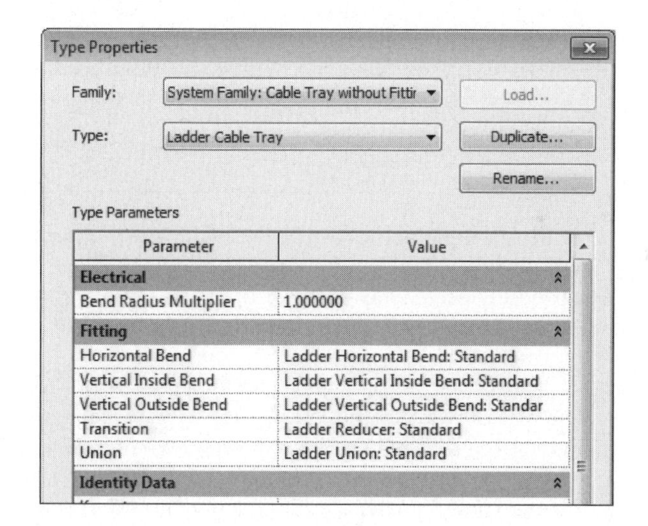

The bend radius for cable tray elbows is set to the width of the cable tray by default. The Bend Radius Multiplier parameter can be used to define a bend radius for different types within the system family. The bend radius of an elbow can be modified directly in the model by selecting the fitting and editing the radius temporary dimension that appears, as shown in Figure 13.35. The bend radius for conduit elbows is determined by a lookup table defined in the fitting family.

Conduit will display as single line in views set to Coarse or Medium detail level. Cable tray displays as a single line in views set to Coarse detail level. In views set to Medium detail level, cable tray will display as two-line geometry. Ladder-type cable tray will display as a ladder in views set to Fine detail level.

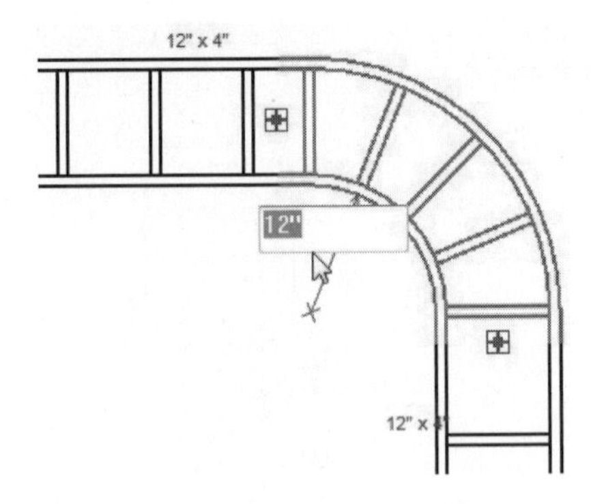

FIGURE 13.35
Editing a cable tray bend radius

Electrical Settings

Conduit and cable tray settings and sizes can be defined in the Electrical Settings dialog box, which is accessed via the MEP Settings button on the Manage tab. The general settings are used to define the visibility behavior of conduit or cable tray when shown as single-line graphics. A suffix and separator can be defined for tagging.

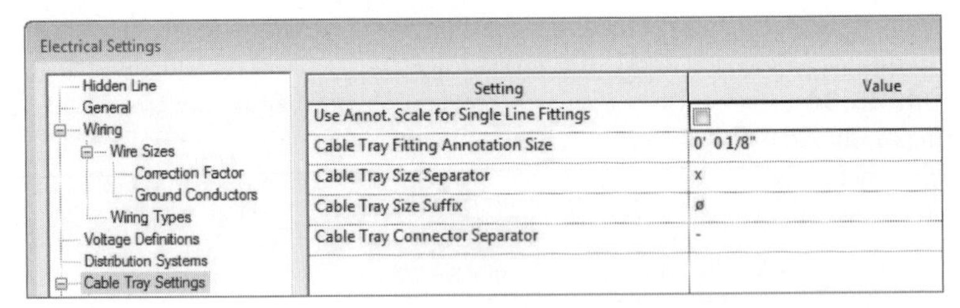

Rise/Drop graphics can be defined for both cable tray and conduit. Clicking the value cell activates the Select Symbol dialog box, which allows you to choose a graphic representation for a conduit or cable tray rise or drop, as shown in Figure 13.36. Graphics can be defined for both single-line and two-line representation. You cannot create your own symbols to use for this representation.

The Size settings for cable tray are very simple. You can input sizes available for cable tray and also choose whether certain sizes are available for use in your project by selecting the box in the Used In Size Lists column, as shown in Figure 13.37.

The Size settings for conduit are more detailed because of the difference in dimensions for various conduit materials. Sizes and a minimum bend radius can be defined for each conduit material type.

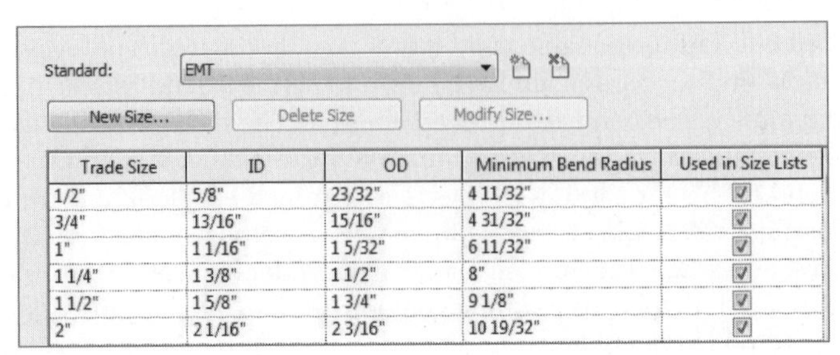

Trade Size	ID	OD	Minimum Bend Radius	Used in Size Lists
1/2"	5/8"	23/32"	4 11/32"	☑
3/4"	13/16"	15/16"	4 31/32"	☑
1"	1 1/16"	1 5/32"	6 11/32"	☑
1 1/4"	1 3/8"	1 1/2"	8"	☑
1 1/2"	1 5/8"	1 3/4"	9 1/8"	☑
2"	2 1/16"	2 3/16"	10 19/32"	☑

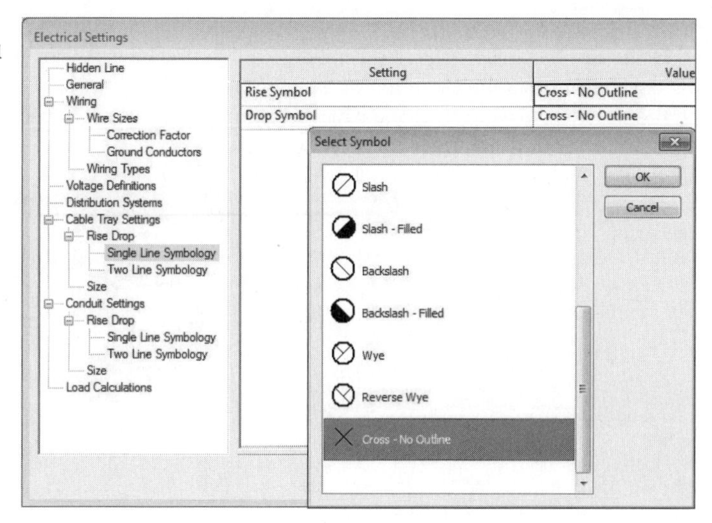

FIGURE 13.36 Rise/Drop symbol options

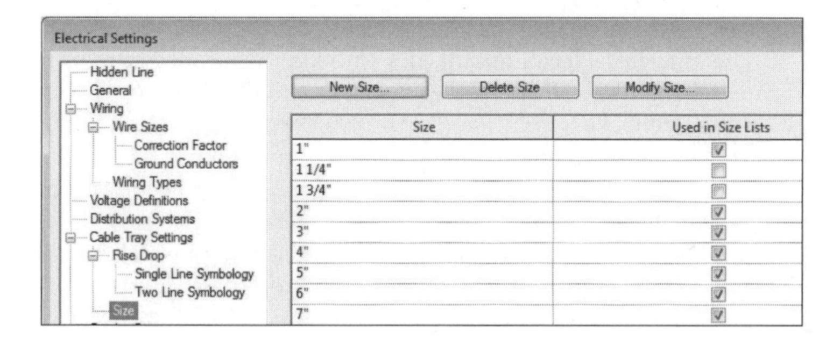

FIGURE 13.37 Cable tray size settings

You can create additional settings for materials by clicking the Add Standard button next to the drop-down at the top of the dialog box. A few standards are predefined, and you can use them as the base settings for creating custom standards. Figure 13.38 shows the predefined conduit standards available in Revit MEP 2011.

Whether a conduit size is available for use in your projects is determined by selecting or deselecting the check box in the Used In Size Lists column. Be sure to set your Project Units settings to coincide with the available sizes. For example, if you set the Rounding value of the Conduit Size units to the nearest 1/2″, then even if the 1 1/4″ conduit size in the electrical settings is set to be available in size lists, it will not display in the list. If you are using metric units, the Rounding setting must match the number of decimal places of the conduit or cable tray sizes in order for a conduit or cable tray size to display in the available sizes list.

The settings for conduit and cable tray that you use most often should be established in your project templates so you can begin modeling right away. The settings can be modified to meet any unique requirements of a project once the project file has been created.

Placing Conduit in a Model

Once you have defined the desired settings for conduit and cable tray, you can begin modeling in your projects. To model a run of conduit, click the Conduit button on the Home tab. Choose the desired style (with or without fittings) from the Type Selector in the Properties palette. The

Diameter drop-down on the Options Bar displays the list of available sizes based on the settings defined in the Electrical Settings and Project Units dialog boxes. The Offset drop-down on the Options Bar determines the elevation of the conduit above the level of the currently active view. Click in the drawing area to start the run of conduit. Move your mouse pointer in the desired direction, and click to finish the run or change direction.

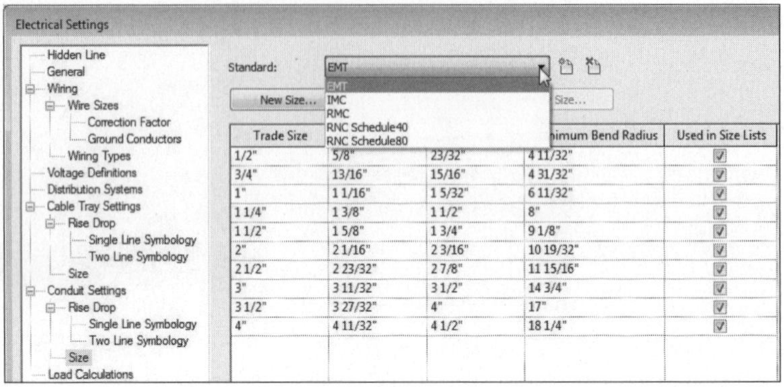

FIGURE 13.38
Conduit standards for size settings

The Modify | Place Conduit contextual tab also contains tools for placement. The Justification button allows you to define the point along the conduit that determines its elevation. If you choose to draw conduit with a vertical justification at the middle and the horizontal justification at the center, then the value given for the offset of the conduit will be directly in the center of the conduit. These settings allow you to draw conduit from the top, middle, or bottom to coordinate elevations with other model elements. The elevation of a conduit can be changed after it is drawn by selecting the conduit and editing the value in the Offset drop-down on the Options Bar or the Offset parameter in the Properties palette.

The Automatically Connect button on the contextual tab is set by default. This setting means that a fitting will be automatically placed when one conduit touches another in the model. The Tag On Placement button can be selected to automatically tag conduit as it is drawn. A conduit tag must be loaded into the project and settings for the tag placement are available on the Options Bar.

You can change the elevation in the midst of modeling a run by changing the Offset value in the drop-down and then continuing the run. Revit will automatically insert the proper fittings and vertical conduit to transition to the new elevation.

To connect conduit to equipment or a device, the object must have a conduit connector. When you select the object, you can right-click the conduit connector and choose the option to draw conduit from the connector, as shown in Figure 13.39. This connector is defined as a face connector, which means that conduit can be connected anywhere on the face of the equipment. It also means that multiple conduits can be connected to the face.

Selecting this type of connector allows you to choose the position of the connection prior to drawing the conduit. Figure 13.40 shows the conduit connection point on the face of the equipment. The connection point can be dragged to any location on the face, or you can edit the temporary dimensions to locate the connection point. Once you have determined the location, click the Finish Connection button on the Surface Connection contextual tab. You can now draw the conduit from the connector. Be sure to assign an appropriate Offset value prior to drawing. Once the conduit has been drawn, you must disconnect it from the connector point on the equipment if you want to move the location of the connection. Figure 13.40 shows plan and section views of conduit drawn from the top of an electrical equipment object.

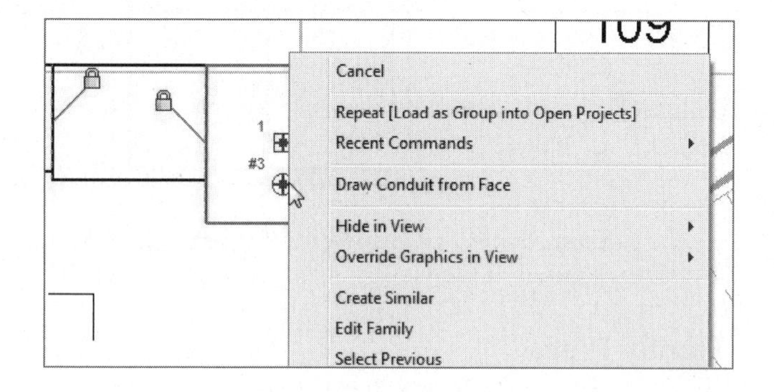

FIGURE 13.39
Drawing conduit from a connector

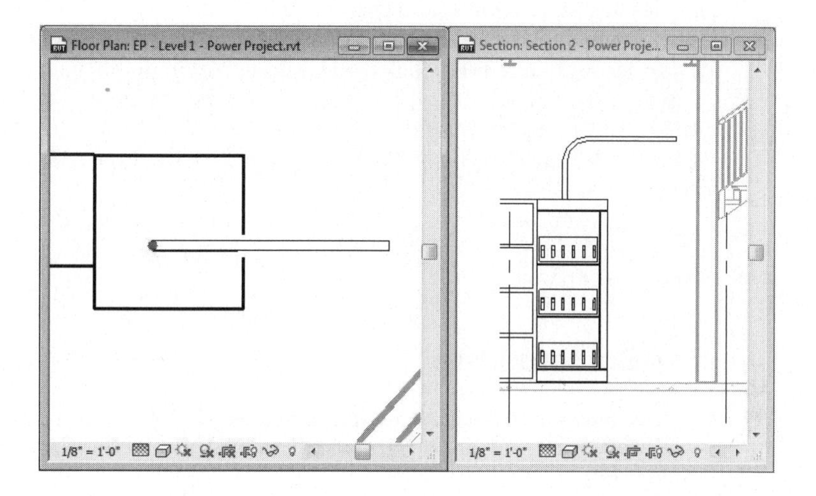

FIGURE 13.40
Conduit drawn in a model

Individual connectors have a static location that can be modified only by editing the family. Only one conduit can be drawn from this type of connector.

Placing Cable Tray in a Model

The process for modeling cable tray is the same as for conduit. There is a drop-down on the Options Bar for the Width and Height settings of the cable tray, as defined in the Electrical Settings dialog box. Cable tray can be connected to equipment or devices that have cable tray connectors. Cable tray connectors have a static location and cannot be moved along the face of the family without editing the family.

You can connect conduit to cable tray by snapping your mouse pointer from the cable tray edge when drawing the conduit. The conduit connects to the center of the cable tray, so if you are using the Fine detail level, setting the center line of the conduit will display beyond the connected edge of the cable tray.

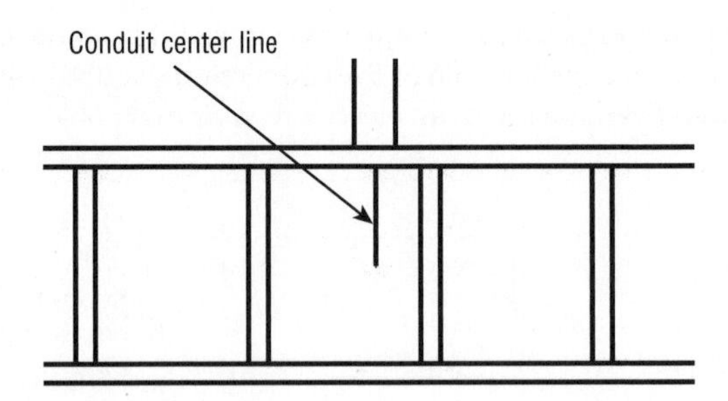

You can remedy this by turning off the center line of conduit in the Visibility/Graphics Overrides settings of the view.

Family Types

You can create unique family types for conduit and cable tray system families. Select a conduit or cable tray in the model, and click the Edit Type button in the Properties palette, or you can double-click a conduit or cable tray family in the Project Browser to access the Type Properties dialog box for that family. Click the Duplicate button in the dialog box to create a new family type. Give the new family type a descriptive name that defines the type. You can then assign fittings or parameter values that make the type unique from other types within the family. The newly created family type will now be available in the Type Selector when placing conduit or cable tray.

The types of conduit and cable tray that you most commonly used should be defined in your project templates for easy access. Additional fittings can be loaded to create unique family types if the project requires.

The Bottom Line

Place power and systems devices into your model Creating electrical plans that are not only correct in the model but also on construction documents can be achieved with Revit MEP 2011.

Master it Having flexibility in the relationship between model components and the symbols that represent them is important to create an accurate model and construction documents. Is it possible to show a receptacle and its associated symbol in slightly different locations to properly convey the design intent on construction documents? If so, how?

Place equipment and connections Electrical equipment often requires clearance space for access and maintenance. Modeling equipment in your Revit project allows you to coordinate clearance space requirements.

Master it Interference between model components can be detected by finding components that occupy the same space. Explain how you can determine whether an object interferes with the clearance space of an electrical equipment component.

Create distribution systems Proper setup of distribution systems is the backbone of the intelligence of your electrical design. It helps you to track the computable data within your project.

Master it Because your project may contain multiple distribution system types, explain the importance of assigning distribution systems to your electrical equipment and naming your equipment.

True or false: You cannot create a power riser diagram with Revit MEP 2011.

Model conduit and cable tray Large conduit and cable tray runs are a serious coordination issue in building designs. Revit MEP 2011 has tools that allow you to model conduit and cable tray in order to coordinate with other model components.

Master it Conduit and cable tray can be modeled with two different styles. One style uses fittings, and one does not. Does this mean that no fittings need to be assigned to the style that does not use fittings? Explain how this affects scheduling of the components.

Chapter 14

Circuiting and Panels

The purpose for putting information into the components that you use to build your Revit MEP models is so that you can use that data to confirm the integrity of your design and improve coordination. The computable data in a Revit MEP model is dependent on the systems that you establish. Revit MEP recognizes many types of systems for air flow and piping. For electrical, the systems are the types of circuits you create to establish the relationship between a fixture or device and its associated equipment. This relationship is typically conveyed on construction documents by schematic lines representing the wiring. Circuit types that can be created in Revit MEP 2011 include power, data, communications, security, and others.

With the parametric and connective nature of Revit components, you have the ability to maintain the relationship between your model elements and the schematic wiring associated with them. Wires can be used to extract information about the circuits by using tags or schedules. Even if you do not choose to show wiring on your construction documents, you still have the ability to create and manage electrical circuits.

In this chapter, you will learn to

- Establish settings for circuits and wiring
- Create circuits and wiring for devices and fixtures
- Manage circuits and panels
- Use schedules for sharing circuit information

Electrical Settings

The electrical settings for your project will determine your ability to connect devices and equipment and also define how wiring and electrical information is displayed. You can define the types of voltages available and also the distribution system characteristics. This allows you to properly connect devices and prevents you from accidentally wiring objects to the wrong panel. You can set the visibility behavior of tick marks to show wire counts and how wire tags will display the electrical information. All of these settings are project specific, so you can create a standard setup in your project template based on company or project type requirements.

You can access the electrical settings at any time by typing **ES** or by clicking the MEP Settings button on the Manage tab.

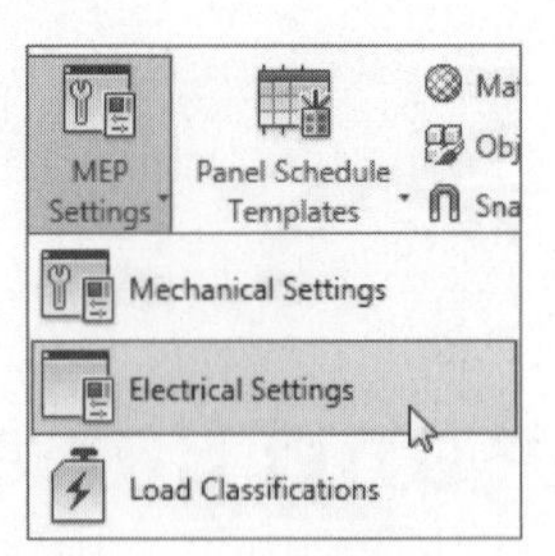

The General settings in the Electrical Settings dialog box are for how the electrical data is displayed for devices and the format for describing the electrical characteristics. These settings do not affect the behavior of electrical devices or circuits; they affect only how information is displayed in the Electrical Data parameter of devices. There are also settings for circuit naming if you name your circuits by phase.

Wiring Settings

In the Wiring settings section of the dialog box, you can define the ambient temperature to be used in order to apply a correction factor to the load of a circuit. The default setting is 86°F (30°C), which applies a correction factor of 1 for any of the three wire temperature ratings. In this section, you can also define the size of the gap that Revit MEP 2011 displays when wires cross. The value you input into this setting is for the print size of the gap on a full-size print.

Tick marks for wiring are also defined in the Wiring section of the Electrical Settings dialog box. You can create an annotation family to use as a tick mark if the defaults do not comply with your standards. The family must be loaded into your project in order for it to be assigned to represent hot, neutral, or ground wires. You can choose to show a slanted line across the tick marks to represent the ground conductor no matter what families you use. There are three options for when your tick marks will be displayed, as shown in Figure 14.1. Changing the display setting for tick marks will affect all wires in your project.

Always With this setting, tick marks will display on any wire when it is drawn.

Never With this setting, tick marks will not display on any wire when it is drawn.

Home Runs With this setting, tick marks will display only on wiring home runs and not on wires between devices on the same circuit.

New to Revit MEP 2011 are the settings for defining the maximum voltage drop for branch circuits and feeders.

Show Tick Marks	Always
Max Voltage Drop For Branch Circuit Wire Sizing	2.00%
Max Voltage Drop For Feeder Circuit Wire Sizing	3.00%

The Wire Sizes section of the Electrical Settings dialog box allows you to define the sizes of wires you want to use for different circuit ampacities. You can set the wire sizes for aluminum and copper wire. You can even create a different wire material if your project requires it. You can vary the wire sizes for the three standard wire temperature ratings and for different types of insulation. This gives you the ability to control what types of wires are used for different circuits because you can remove ampacity ratings for different wire temperature and insulation combinations.

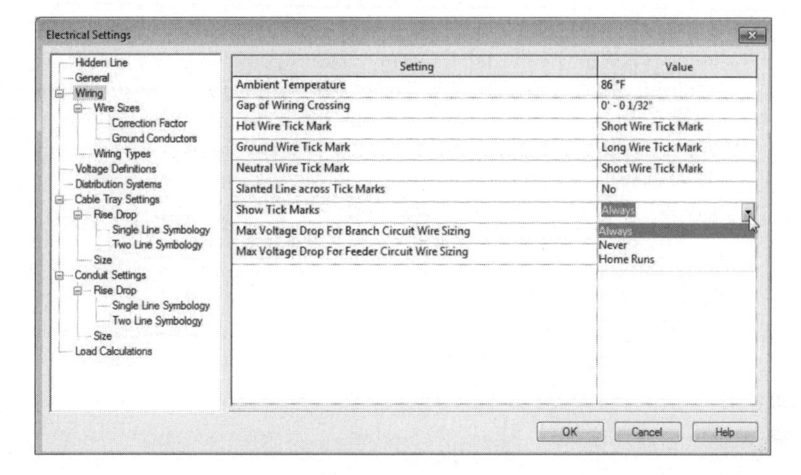

FIGURE 14.1 Wiring section of Electrical Settings dialog box

The Used By Sizing column allows you to establish which wire sizes are available for Revit MEP 2011 to use when wire sizes are calculated. You can assign only one wire size per ampacity, so if, for example, you want to use #12 wire for 15 amp and 20 amp circuits, you would have to delete the 15 A ampacity and set the size for the 20 A ampacity to #12. This does not mean you cannot use a 15 A circuit in your project. Revit MEP will use the smallest wire size available in your list for an ampacity smaller than the smallest in your list. Figure 14.2 shows an example of wire sizes for a project. Any 15 amp circuits in the project will have to use #12 wire.

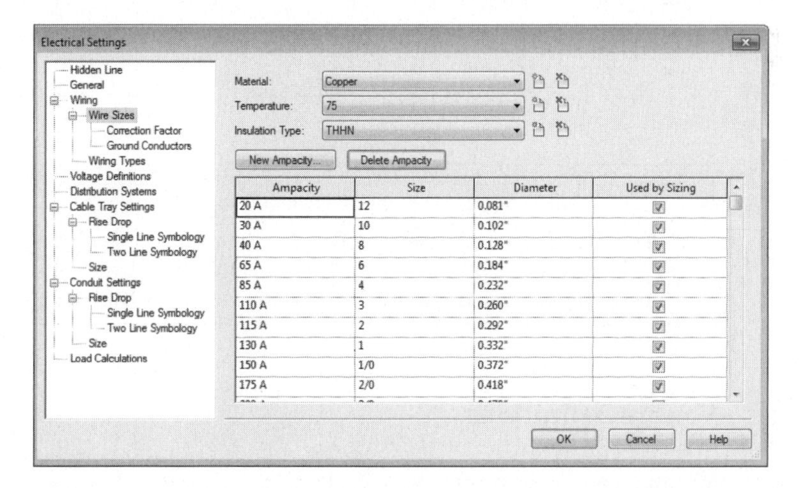

FIGURE 14.2 Wire size settings

The Correction Factor section of the Electrical Settings dialog box is for setting a correction factor based on temperature to be used in load calculations. You can set correction factors for the three wire temperature ratings for both aluminum and copper wire. The Ground Conductors section allows you to set the equipment grounding conductor size for different ampacities of different wire materials. So, for wire sizing, you have the ability to control several factors that determine the wire sizes that are calculated for your circuits.

The Wiring Types section of the Electrical Settings dialog box is where you can create different types of wires for different uses. By creating different wire types, you have the ability to assign all the desired settings to a type, and like component families, you can switch between wire types in the model. Creating wire types also provides you with an easy way to control the visibility of certain wires by using a view filter. You can give your wire types unique names that define their use or some characteristic about them. You can set the material, temperature rating, and insulation used by a wire type. You can also set the maximum wire size for a type. If a circuit requires a larger wire size than the maximum, Revit MEP 2011 will automatically create parallel sets of wires that do not exceed the maximum size to accommodate the load. You can establish whether your wire type requires a neutral conductor and, if so, set the neutral size to equal the hot conductor size or set it as an unbalanced circuit. After the neutral conductor is sized, the value you input for the neutral multiplier is assigned. Although Revit MEP 2011 has conduit tools for modeling, you can set the conduit material for your wire type, to be used for voltage drop calculations. Figure 14.3 shows some examples of wire types created for a project.

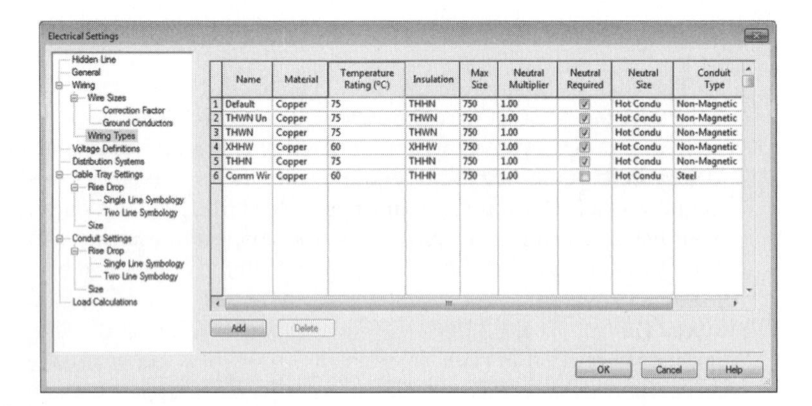

FIGURE 14.3
Wire types

BY DEFAULT

If you use the `default.rte` project template, there is a wire type named Default. If you do not create any wire types, the Default wire type will be used for all circuits. This wire type is aluminum, which may result in unexpected wire sizes.

Voltage Definitions

In the Electrical Settings dialog box, the Voltage Definitions section is for establishing the minimum and maximum values for the voltages used in your project. This allows for different ratings on devices or equipment. These voltages will be used to establish different distribution system definitions. You can add voltages to your project by clicking the Add button at the bottom of the dialog box and giving it a name and minimum and maximum values. To remove a voltage from your project, select the voltage, and click the Delete button at the bottom of the dialog box. You will not be able to delete a voltage if it is used by a distribution system definition in your project. You must delete the distribution system first, and then the voltage can be deleted. Setting up standard voltage definitions in your project template will save you time when setting up a project. Figure 14.4 shows some standard voltage definitions.

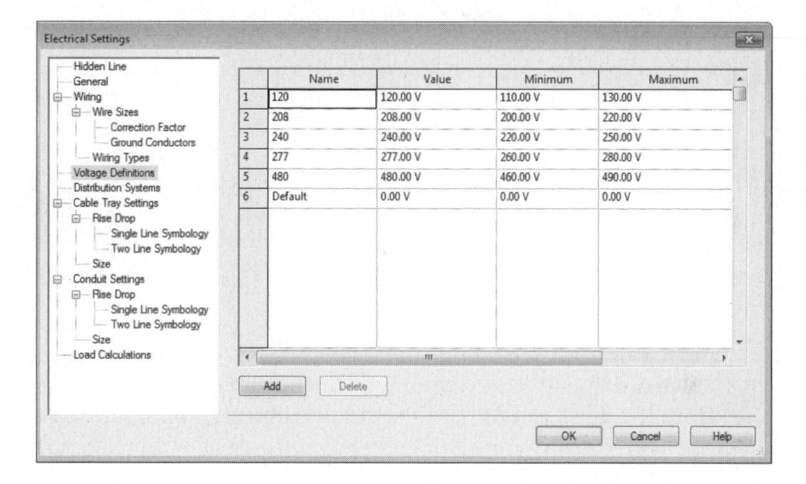

FIGURE 14.4
Voltage definitions for a project

Distribution Systems

You can define the distribution systems to be used in your project in the Distribution Systems section of the Electrical Settings dialog box. The connectors in your electrical families need to coincide with the systems you define so that you can assign devices and equipment objects to a system. You can create single or three-phase systems in delta or wye configurations and establish the number of wires for the system. The line-to-line and line-to-ground voltages for a system can also be defined. It is important to note that Revit will let you create systems that do not actually exist in the industry. For example, you could create a three-phase system and set the line-to-line voltage to 120 volts and the line-to-ground voltage to 277 volts. Use caution when creating distribution systems because they are very important for creating circuits for devices and equipment. Voltages can be assigned only to line-to-line or line-to-ground if they have been defined in the Voltage Definitions section. Figure 14.5 shows some examples of distribution systems created for a project.

Load Calculations

In the Load Calculations section, you can establish demand factors and load classifications. The load classifications of the connectors in your electrical families determine which demand factor they will take on in your project. In this section, you can deselect the check box so that calculations for loads in spaces will not be done, which can improve the performance of your project.

When you click the Load Classifications button in this section, the Load Classifications dialog box opens. In this dialog box, you can create new load classifications and define the demand factor used for them and a lighting or power load class for use with spaces. The available demand factors appear in the drop-down list, and the Demand Factors dialog box can be accessed by clicking the button next to the Demand Factor drop-down, as shown in Figure 14.6.

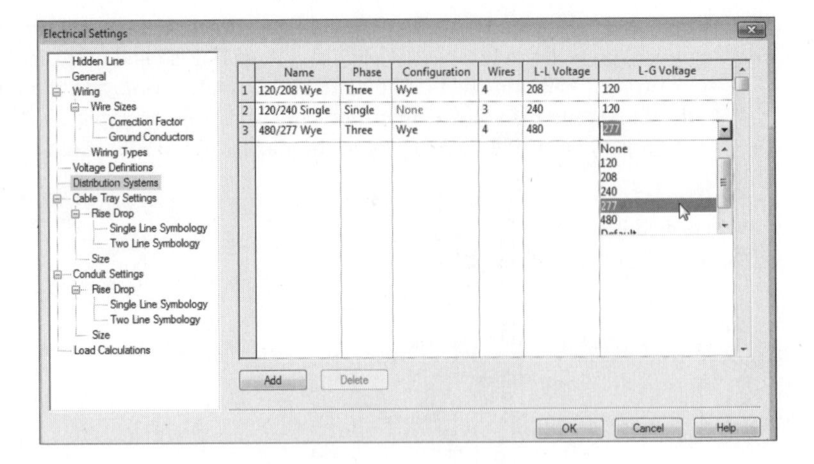

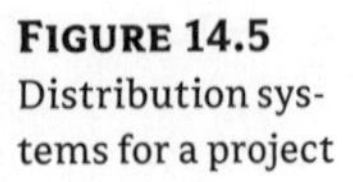
FIGURE 14.5
Distribution systems for a project

It is best to keep your load classification names simple and descriptive because you will have to assign them to the connectors in your electrical families. When working in the Family Editor, you can use the Transfer Project Standards tool on the Manage tab to transfer load classification settings from your project or template file into the family to ensure coordination.

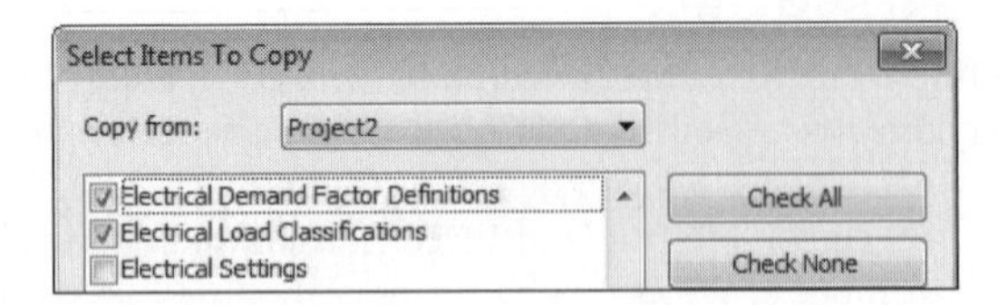

There are many load classifications that come with the default electrical templates. You can create your own using the buttons below the list, in the lower-left corner of the dialog box. The Spare classification cannot be assigned a demand factor and cannot be transferred to family files for use on connectors. It exists so that spares can be accounted for in your panel schedules.

In the Demand Factors dialog box, you can define the demand factors that are assigned to load classifications. The default electrical templates come with an extensive list of demand factors, and you can create your own using the buttons in the lower-left corner of the dialog box. Figure 14.7 shows the Demand Factors dialog box. The Calculation Method section is where you determine how the demand will be calculated. Additional load can be added to the calculated result using the check box at the bottom of the dialog box.

There are three methods that can be used for calculation:

Constant You can set a demand factor for all objects with the load classification that the demand factor is assigned to using the Constant calculation method.

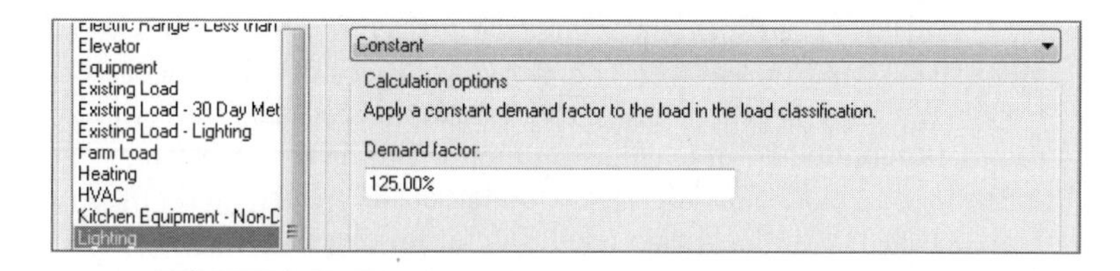

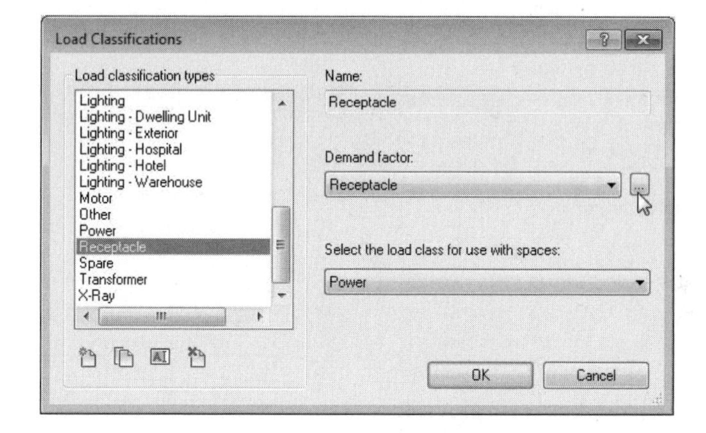

FIGURE 14.6
Load Classifications dialog box

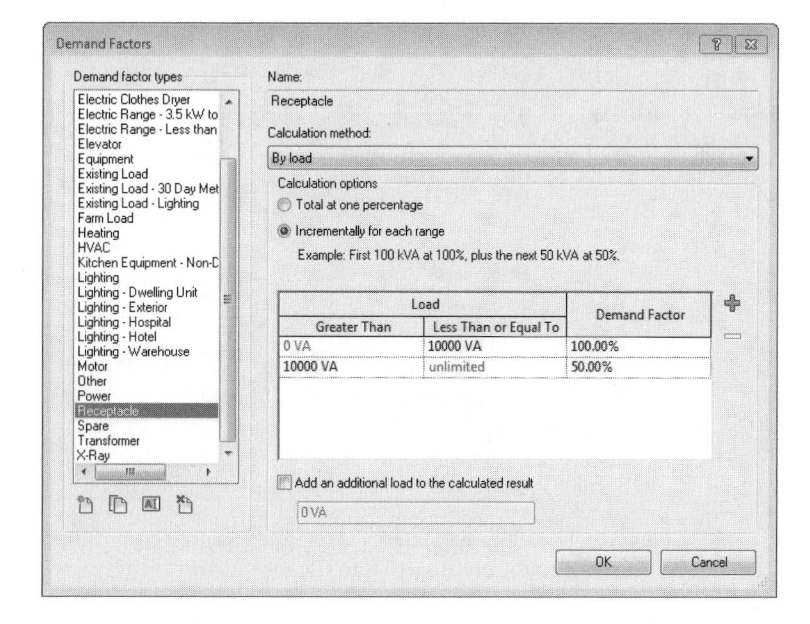

FIGURE 14.7
Demand Factors dialog box

By Quantity The By Quantity method is used to assign a demand factor to multiple items. You can assign different factors to ranges of items or the same factor to items depending on how many there are. Figure 14.8 shows this method used for a motor load classification. The calculation option is set to calculate incrementally for each range. The ranges defined result in the largest motor load being calculated at 125 percent and any other motors at 100 percent. Figure 14.9 shows this option used for clothes dryer load classification. The option is set to assign the demand factor to all objects within a range that defines the quantity of objects. These settings will calculate the first four dryers at 100 percent, the next dryer at 85 percent, the next at 75 percent, and so on.

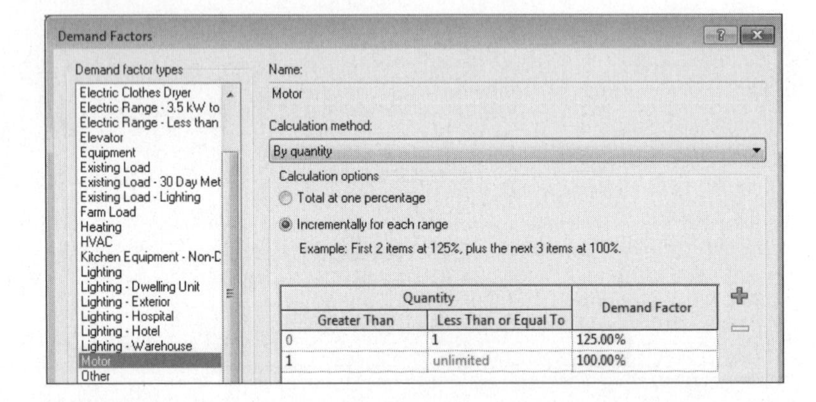

FIGURE 14.8 Incremental demand factor settings using the By Quantity method

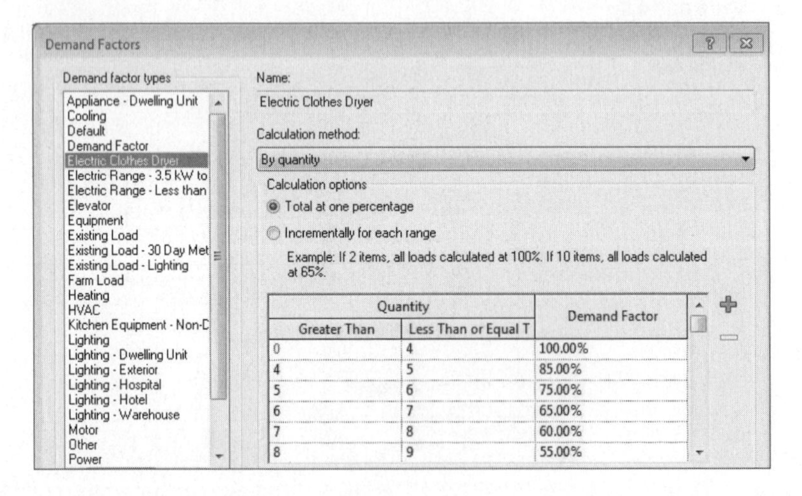

FIGURE 14.9 Total demand factor settings using the By Quantity method

By Load The By Load method is used to assign a demand factor to different ranges of loads. You can assign different factors to ranges of loads or the same factor to the total load connected to a panel. Figure 14.10 shows the settings using this method incrementally for each range defined. These settings will calculate the first 10,000 VA at 100 percent and anything greater at 50 percent. The option to calculate the total at one percentage would be used if you wanted to set the demand factor for a total load range. For example, you could set the factor at 100 percent for 10,000 VA, meaning that if the total load on the panel was 10,000 VA, then all loads would be calculated at 100 percent.

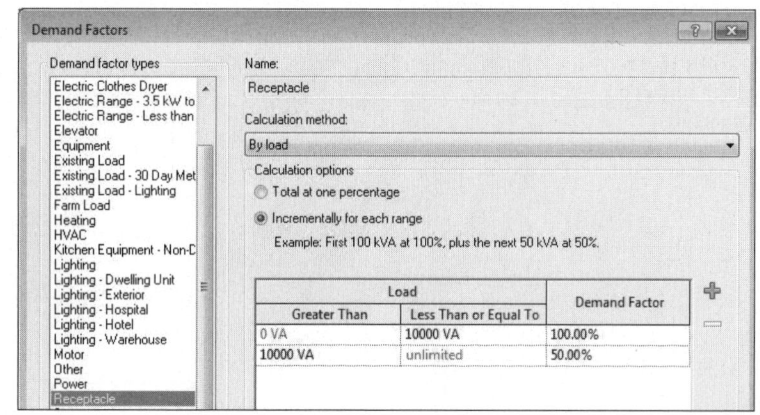

FIGURE 14.10
Incremental demand factor settings using the By Load method

The electrical settings established for your projects are the backbone of creating circuits and wiring. Properly set up in your project template, they can make creating and managing circuits easy and efficient.

SETTINGS FOR TEMPLATES

Duncan and Abbey Engineering has recently broadened its range of project types. Making the transition to Revit MEP 2011 has opened up opportunities to pursue projects that the company normally would not do. It recognized the power of having a Revit template for its projects and quickly set up templates with electrical settings for the different project types. Duncan and Abbey Engineering is now doing work in different countries that require unique settings for voltage and distribution systems. Having templates with preset electrical settings for each of its client's unique standards saves the company time and reduces the possibility for error.

Creating Circuits and Wiring for Devices and Fixtures

With Revit MEP 2011, you can create circuits for devices or equipment to keep track of the loads within your panels. Circuits are the "systems" that Revit MEP 2011 recognizes for electrical design. You can create a circuit for devices or fixtures without selecting a panel so that at a minimum you have removed them from the default system and therefore your project file performance will not suffer. It is important to realize that when you are working in Revit MEP 2011, circuits and wires are not the same thing. Circuits are the actual connection between elements, while wires are simply a symbolic representation of the connection.

The type of circuit that you can create for a device depends on the properties of the connector in the device family. The most important property is System Type, which defines what kind of circuit can be created for the device. Figure 14.11 shows the properties of a connector in a receptacle family and the various types of systems that can be assigned to the connector. Other properties of the connector also define the type of circuit that can be created for the family such as voltage and number of poles.

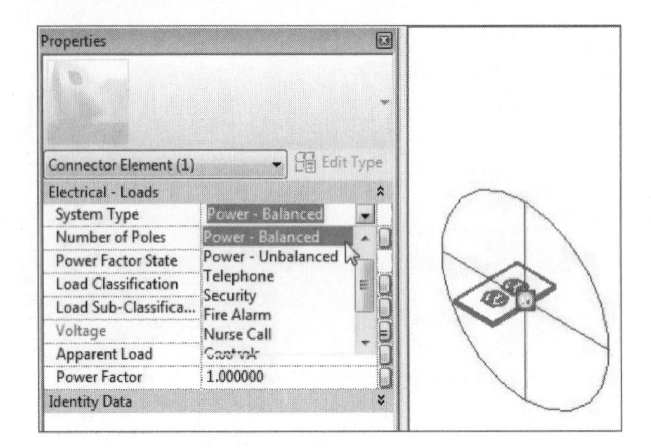

FIGURE 14.11 Device connector properties

When you select a device in your model, the Create Systems panel of the contextual tab displays a button to create the type of circuit that matches the properties of the connector in the selected family. If there are multiple connectors in the selected device, a button is available for each type of connector, as shown in Figure 14.12, where a floor box containing a power and data connector is selected.

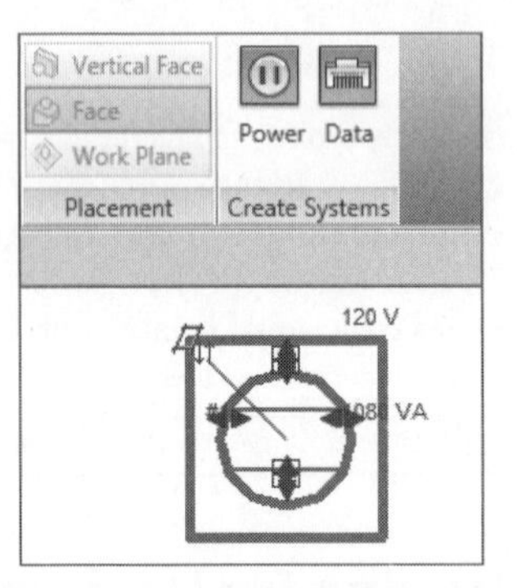

FIGURE 14.12 Create Systems panel of the contextual tab for a selected device

Clicking a button on the Create Systems tab creates a circuit for the selected device and activates the Electrical Circuits contextual tab. On this tab, there are tools to edit the circuit or select a panel for the circuit. The properties of the circuit can be seen in the Properties palette when this tab is active. The System Selector drop-down will show the circuit as <unnamed> until you select a panel for it. Click the Select Panel tool to choose a panel for the circuit. You can select the panel by clicking it in the drawing area, or you can use the drop-down on the Options Bar and select the panel by name. Only panels with a distribution system that matches the connector properties of your family will be available in the drop-down list. That includes the secondary side of transformers, so it is important to name your equipment and panels so they are easily identified, as shown in Figure 14.13.

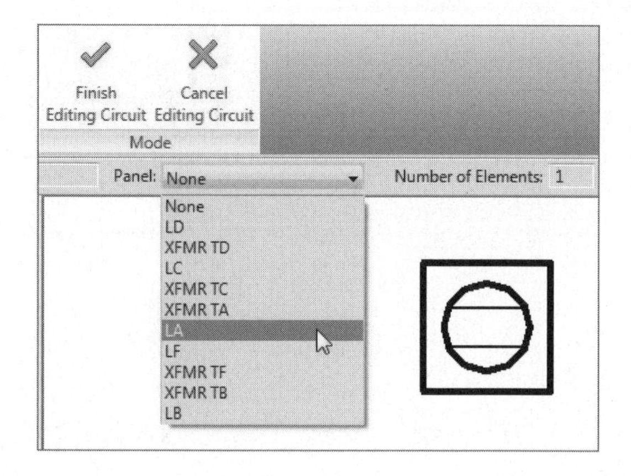

FIGURE 14.13
Panel drop-down on the Options Bar when selecting a panel for a circuit

When you select a panel for the circuit, a red, dashed box will appear in the drawing area that encompasses all components of the circuit, and a dashed red line with an arrow indicates the home run to the panel. Clicking one of the buttons on the Convert To Wire panel of the contextual tab will convert the indicated home run to wire. You have two choices for the graphical type of wire that you can convert to, arc or chamfered wiring. Arc wiring can be used for straight lines between components if that is your standard. Figure 14.14a shows the indicated home run, and Figure 14.14b shows the result of clicking the Arc button on the Convert To Wire panel.

When you select multiple devices or fixtures with the same connector type, the button to create a circuit is available on the contextual tab. This allows you to select all the elements for a circuit and create the circuit in one step. An icon will appear in the drawing area to convert the indicated wire to arc or chamfer type wiring, as shown in Figure 14.15. This keeps your cursor in the drawing area for improved efficiency.

Some of your elements may have more than one connector of the same type. When you select an element with multiple connectors and click the button to create a circuit, you will get a dialog box that allows you to select which connector you are creating the circuit for, as shown in Figure 14.16 for a receptacle with two power connectors. Once you have chosen a connector, you can complete the process of creating a circuit. When you select the device again, you will still have the option to create a circuit for the remaining connector.

When you are creating circuits for lighting fixtures, it is important that the properties of the connector in the switch family match those of the connector in the lighting fixture family, except for load. If you use a Controls system type for the connector in your switch families, you cannot add it to the circuit for the light fixtures, but you can still draw the wiring from the lights to the switch. If your switch does not have a connector, then you will have to use detail or model lines to represent the wiring from a fixture to the switch. If you use the Wire tool to draw from a fixture to the switch or from the switch to a fixture, a home run will be created when you connect to the fixture.

Editing Wiring

If the wiring that is automatically generated is not exactly how you want to show the connection of the devices, you can edit it after it is created. When you select a wire, there are grips that allow you to change the arc and location of the wire.

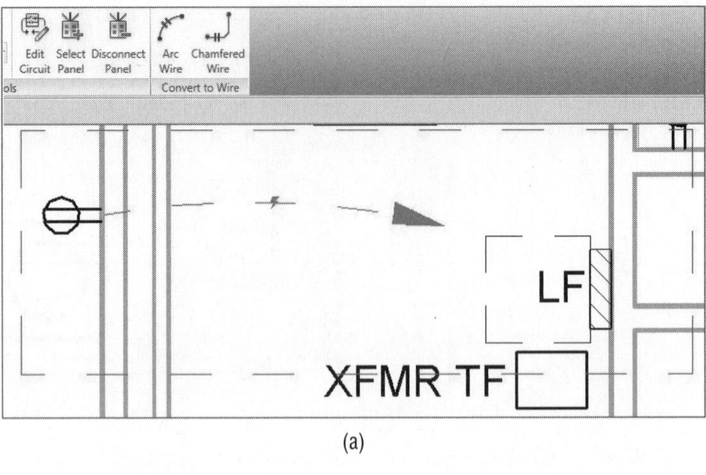

(a)

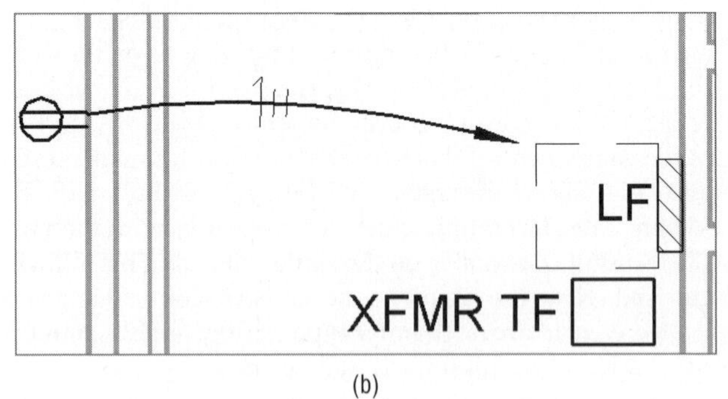

(b)

FIGURE 14.14
(a) Indicated home run for a circuit; (b) wire automatically generated by clicking Arc button on Convert To Wire panel

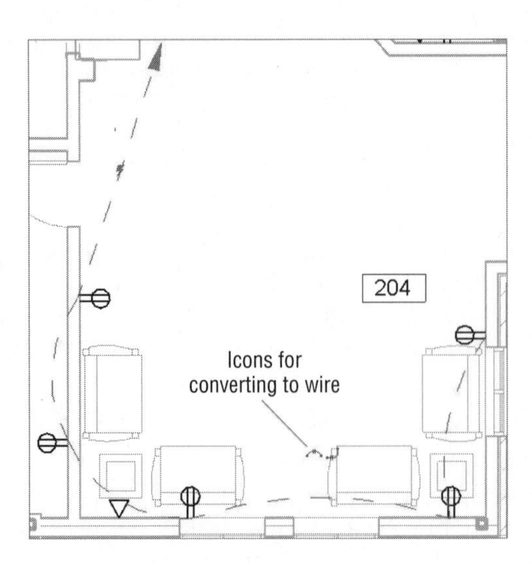

FIGURE 14.15
Multiple items selected for a circuit

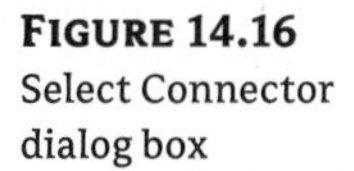

FIGURE 14.16
Select Connector dialog box

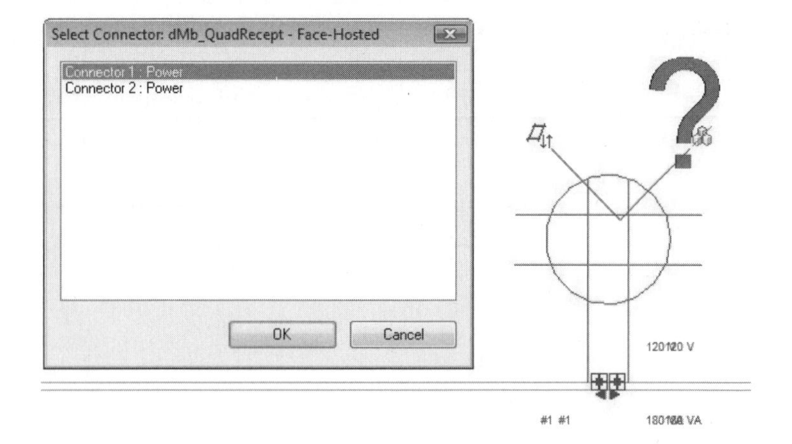

> **WHAT IS WIRE?**
>
> In Revit MEP 2011, wire is a special type of symbolic line. Wires are not model components, and they are visible only in the view in which they are drawn.

The small blue circles that appear near the ends of the wire are grips for changing the endpoints of the line that represents the wire. These Change End Offset grips do not represent the point where the wire is connected, only the graphical representation of the wire. The actual point of connection is indicated by the connector symbol. The Change End Offset grips allow you to show the wire from any point on the device or fixture regardless of the location of the connector. Occasionally the connection point grip and the Change End Offset grip are in the same place. You can use the Tab key to toggle between them for selection.

The larger blue circle with two lines tangent to the endpoints of the wire is for changing the arc of the wire. The + and – grips that appear near the wire are for adding or removing hot conductors. These are discussed later in this chapter. Figure 14.17 shows the editing grips for a wire.

FIGURE 14.17
Wire editing grips

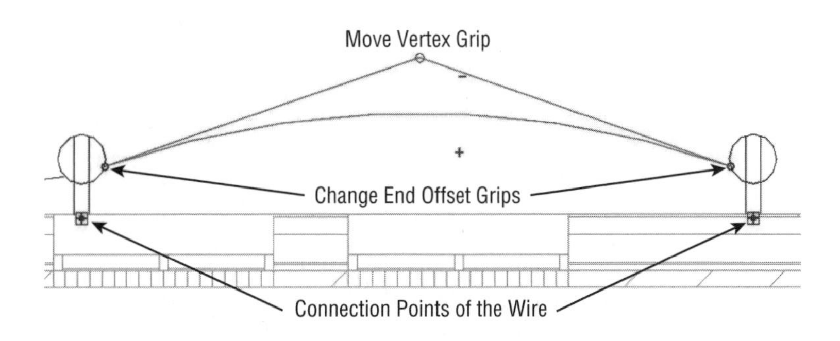

> **GET A GRIP**
>
> Thick symbolic lines can sometimes make it difficult to see the wire-editing grips. Setting your view to Thin Lines makes it easy to locate the grips.

The wire shown in Figure 14.18a can be edited with the grips so that it is drawn from the symbols correctly and does not interfere with other devices that are not part of the circuit, as shown in Figure 14.18b. The end offsets are changed, and the arc of the wire is reversed for a cleaner-looking drawing.

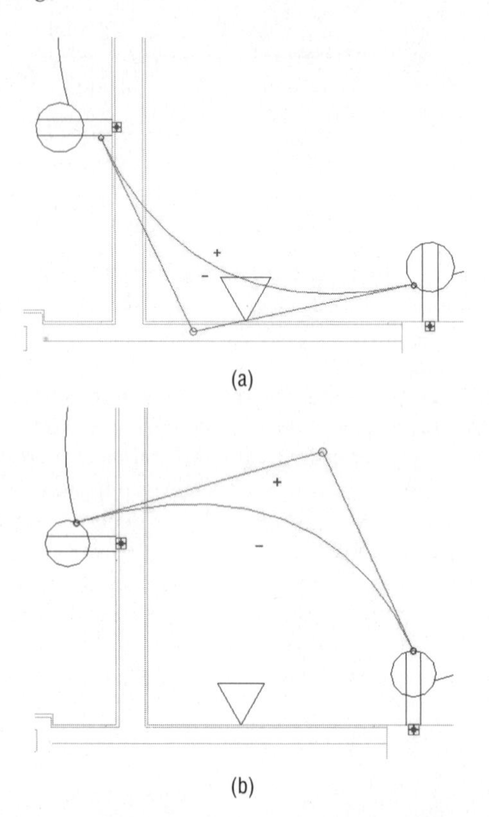

FIGURE 14.18
(a) Wire that needs to be edited; (b) the results of using the grips to change the appearance of the wire

You can use the connection point grip to change what object a wire is connected to. Click and drag the grip to a new object to connect the wire to it. If you click the connector grip and drag it off an object, a home run wire will be created. When changing the connection point of a wire, it is a good practice to make sure you snap to the connection point of the object you are connecting to. This will ensure that you do not have any erroneous home runs in your project. Figure 14.19a shows a wire moved from between two fixtures, and Figure 14.19b shows it connected to another fixture by dragging the connection point grip.

The type of wire that is created by using the Convert To Wire button is the last type of wire that was drawn in your project, or the default wire type is used if none has been manually drawn. You can change a wire type by selecting it and choosing a new type from the Type Selector in the Properties palette. You can change all the wires for a circuit by editing the properties of that circuit.

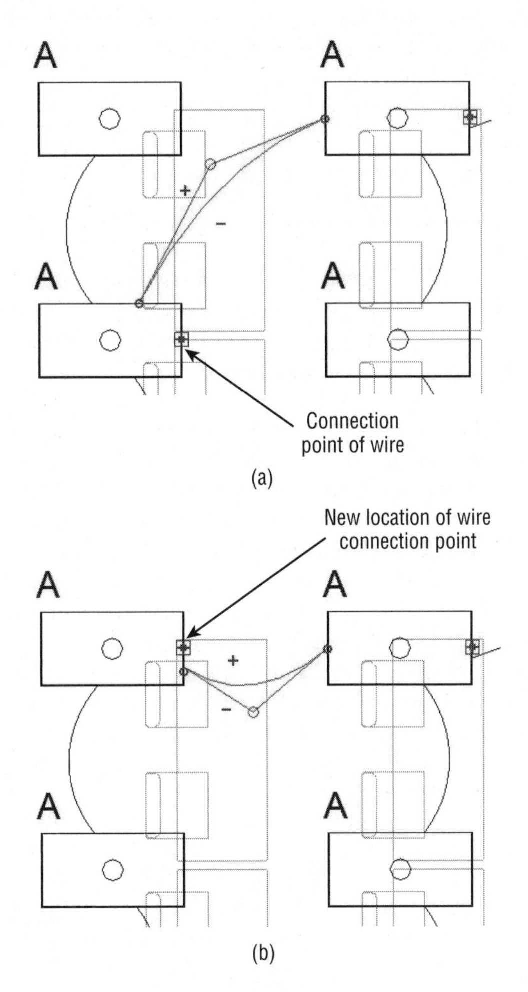

FIGURE 14.19
(a) Wire connected between two fixtures; (b) the result of dragging the connection point to another fixture

Editing Circuits

When you select a device or fixture that has a circuit, the Electrical Circuits contextual tab appears with tools for editing and managing the circuit. This tab is similar to the tab that appears when you create a circuit. If the device you have selected contains multiple connectors, the System Selector drop-down allows you to choose which circuit you would like to edit.

Click the Edit Circuit button to add or remove elements from the circuit. When you click the Edit Circuit button, the Edit Circuit contextual tab appears, and all items that are not part of the circuit turn to halftone in the drawing area, including annotation and wiring, while elements of the circuit remain as normally displayed. The Add To Circuit button is automatically selected when you click the Edit Circuit button. You can click any device or fixture in your view to add it to the circuit. You can also select devices or fixtures from other locations if you have multiple views of the model open. If you select an element that does not have a connector that matches the distribution system of your circuit, you will receive a warning dialog box, and the element will not be added to the circuit, as shown in Figure 14.20 where the receptacle was selected to be added to the lighting circuit.

FIGURE 14.20
Circuit warning

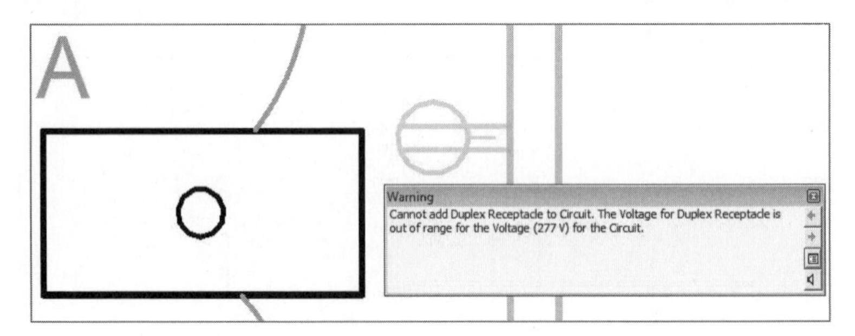

When you select a device or fixture with a connector that matches the distribution system of the circuit you are adding to, the display of the device or fixture changes from halftone to normal. You can click the Remove From Circuit button on the Edit Circuit contextual tab to remove items from the circuit. Once you have finished adding or removing items from a circuit, click the Finish Editing Circuit button to complete your changes. If you have wiring drawn to show items on a circuit and you remove some of the items, Revit MEP will not remove the wiring.

Another method for adding or removing objects is to select the object and right-click its connector grip. Select a device or fixture to add to a circuit, and right-click the connector grip. If you do not have your cursor over the connector grip when you right-click, you will get a menu with tools for visibility control. Right-clicking the connector will give you the same menu but with additional tools for circuiting. Select the Add To Circuit option from the menu, and click any device or fixture that is part of the circuit you want to connect to. You also have the option to create a new circuit from the right-click menu.

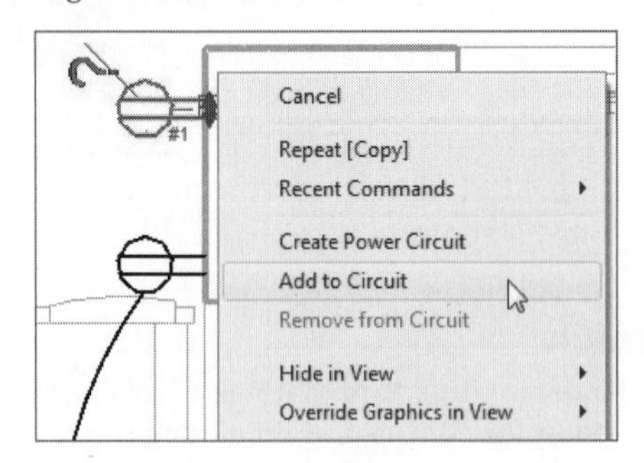

When you select a element on the circuit that you want to add your device or fixture to, you will be given the option to automatically generate wiring to show the added device as part of the circuit.

Selecting a device that is already part of a circuit and right-clicking its connector will give you a menu with the option to remove the element from the circuit. Removing a fixture or device from a circuit will not delete any wiring shown to it, but the wiring may change depending on the location of the removed element in the circuit. Figure 14.21 shows how wiring can be affected when a device is removed from a circuit. The receptacle in the lower-right corner was removed from the circuit. The receptacle in the upper-right corner is still a part of the circuit, so the wiring has changed to a home run. The wire from the unconnected receptacle to the one on the left has changed to a home run also, because it is not part of the circuit but has a device downstream that is part of the circuit. In any case, removing elements from a circuit will usually require some cleanup of the wiring.

FIGURE 14.21
Wire changed by removing a device from a circuit

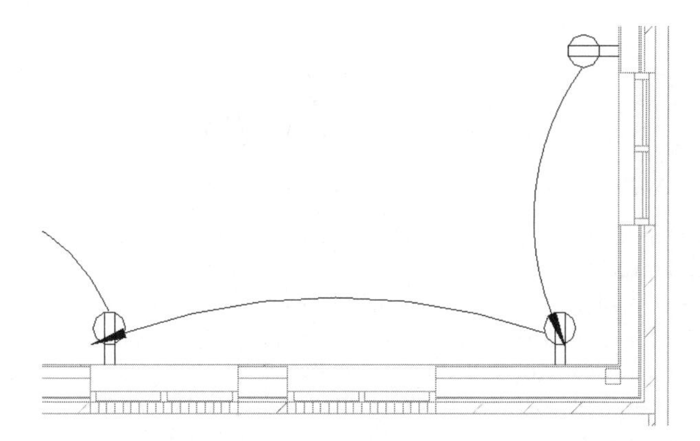

The workflow for creating circuits can be easily learned and can become second nature to your design and modeling processes. Practice creating circuits by doing the following:

1. Open the `Ch14Circuits.rvt` file from `www.wiley.com/go/masteringrevitmep2011`.
2. Select the four receptacles in Office 104. Click the Power button on the Create Systems panel of the Modify | Electrical Fixtures tab.
3. Click the Select Panel button on the System Tools panel of the Electrical Circuits tab. In the Panel drop-down on the Options Bar, select panel LA.
4. Click the Arc Wire button on the Convert To Wire panel of the Electrical Circuits tab.
5. Click the Tag by Category button located on the Tag panel of the Annotate tab. Deselect the Leader box on the Options Bar, and select the home run created for the circuit. Click the Modify button on the ribbon to exit the Tag command.
6. Repeat steps 1–5 for the three duplex receptacles in Office 103.
7. Select the quadruplex receptacle in Office 103. Right-click one of the connector grips. Select the Add To Circuit option from the menu. Click one of the duplex receptacles in Office 103. Click the Arc type wiring grip that appears. Delete one of the wires that is generated. Click the quadruplex receptacle again, and right-click the other connector. Select the Add To Circuit option from the menu. Click one of the duplex receptacles in Office 103. Hit the Esc key to finish.
8. Open the EL – Level 1 view.

9. Create a circuit for the lights in Office 104. Select panel HA from the Panel drop-down on the Options Bar. Tag the home run created.
10. Repeat step 9 for the lights in Office 103 and 102.
11. Select the home run for the Office 102 circuit, and using the vertex grip, drag the wire up until it connects with the left light fixture in Office 103.
12. Select the double home run from the light fixture in Office 103, and use the vertex grip to drag the wire up to the left light fixture in Office 104. Notice that a multicircuit run of wiring has been created.
13. Select the switch in Office 102, and right-click its connector. Select the Add To Circuit option from the menu, and select one of the light fixtures in the room. Click the arc-type wiring grip to create the wire. Use the wire-editing grips to adjust the display of the wire.
14. Repeat step 13 for Office 103 and 104.

Drawing Wires Manually

It is not necessary to only use the automatic wiring capability of Revit MEP 2011 to create the wiring for your circuits. You can quickly and easily draw wires manually. In some cases, it may be easier to draw the wire than to make adjustments to automatically generated wiring.

To draw wiring, click the Wire button located on the Home tab. The bottom half of the button is a drop-down with options for the style of wire you can draw. You can draw arc, chamfer, or spline wires. These are only options for how the wire is drawn and have nothing to do with wire types or the electrical properties of the circuit or devices. Once you have selected a wiring style, you can choose a wire type from the Type Selector in the Properties palette. You can use the Tag On Placement button on the contextual tab to place a tag on the wire as you draw it.

You can draw wire anywhere in your plan views, so it is important that you snap to the component geometry of the elements you are showing the wire for. If you draw a wire between two symbols, it may appear that they are connected by the wire, but if the wire is not "attached" to the device, then when the devices move, the wire will not move with them. To ensure that your wire is connected to an element, you need to snap to a point on the component geometry, not the symbol that represents the component. As long as your cursor is over some part of the component when you start your wire, the wire will connect to the component at the electrical connector. Figure 14.22a shows a wire being drawn from a mechanical unit by clicking the edge of the unit, and Figure 14.22b shows the wire attaching to the electrical connector of the unit.

The unit shown in Figure 14.22 has an electrical connector so the equipment connection symbol on the electrical drawing does not have a connector or any electrical properties. The wiring to this unit can be modified to be shown from the disconnect switch symbol while maintaining its connection to the unit, as shown in Figure 14.23.

Arc- and chamfer-style wires are drawn by clicking a start point for the wire, a middle point, and an end point. To draw wire between two components, start your wire at one component, click somewhere between the two items, and click the second component to end the wire. Spline style wires allow you to click several times between elements. To draw a wire home run, you simply start a wire at the connector of a device or fixture and leave the other end of the wire unconnected. If the component you draw from is assigned to a circuit, the wire will show an arrowhead, indicating a home run. If the component is not assigned to a circuit, the wire will display without an arrowhead until the element has been assigned to a circuit.

FIGURE 14.22
(a) Drawing a wire from a unit; (b) the wire connected to the unit at the electrical connector

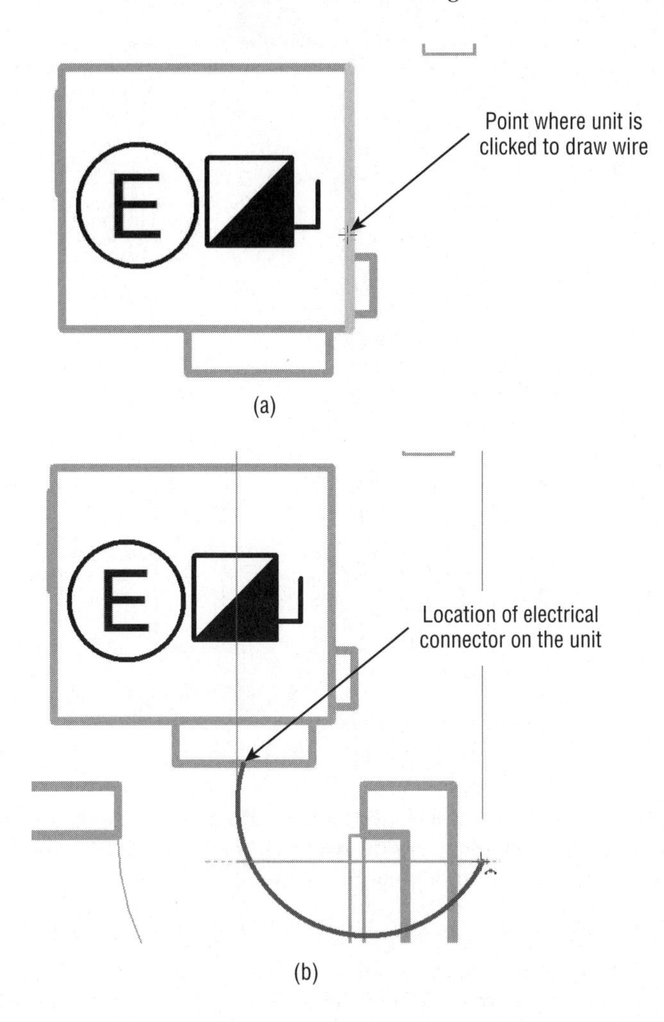

Drawing a wire between components does not mean that they are on the same circuit. If you draw a wire from a component that is on a circuit to another component that is not on a circuit, the second component will not be added to the circuit. You can draw wire between components that are not on the same circuit, creating a multicircuit run of wiring. A home run wire will be displayed between components that are on different circuits, as shown in Figure 14.24.

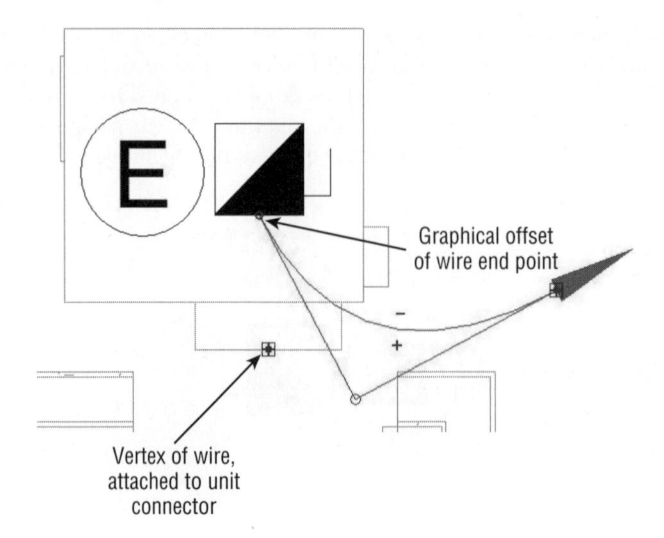

FIGURE 14.23
Modified wire graphics

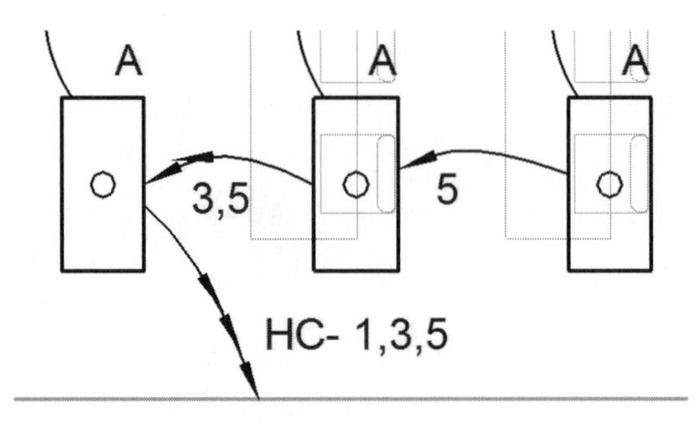

FIGURE 14.24
Multicircuit run of wiring

You can show multiple home runs from a single component that has multiple connectors. If you use the automatic wire option, it will work only for the first circuit that you create. You will have to draw the second home run manually.

1. Select the component, and click the button on the Create Systems panel to create a circuit.
2. Choose a connector from the dialog box that appears, select a panel for the circuit, and choose a wire style from the Convert To Wire panel on the contextual tab.
3. Select the component again to create a circuit for the next connector.
4. Select a panel for the circuit, but do not choose a wire style to be drawn. Instead, select the wire that was automatically created, and make note of the location of the connector.
5. Click the Wire button on the Home tab, and manually draw a wire from the component, starting the wire at the connector not occupied by the existing home run.

You can use a wire tag to confirm that the home runs are not connected to the same connector, as shown in Figure 14.25.

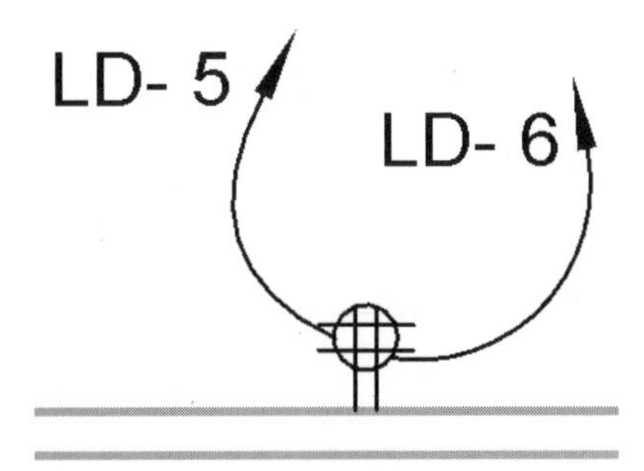

FIGURE 14.25
Multiple home runs from a single device

Wire Display

Using a filter to control the display of wires in your project is an effective method for distinguishing between different wire types. As mentioned earlier, creating wire types in the electrical settings of your project makes it easy to isolate specific wires for visibility control. This allows you to show wire such as underground, overhead, or low-voltage wire in the same view and be able to identify the different types.

To create a filter for wires, click the Filters button located on the Graphics panel of the View tab. In the Filters dialog box, select the New button in the upper-left corner. Give the filter a name that clearly identifies the purpose of the filter or what types of elements it applies to. In the Categories section of the dialog box, select the Wires box to apply the filter to wires. In the Filter Rules section of the dialog box, choose the Type Name parameter from the Filter By drop-down list. Set the condition to Equals in the second drop-down list, and choose the name of the wire type that you want to apply the filter to in the third drop-down list. This creates a filter that affects only wires with the type name chosen, as shown in Figure 14.26.

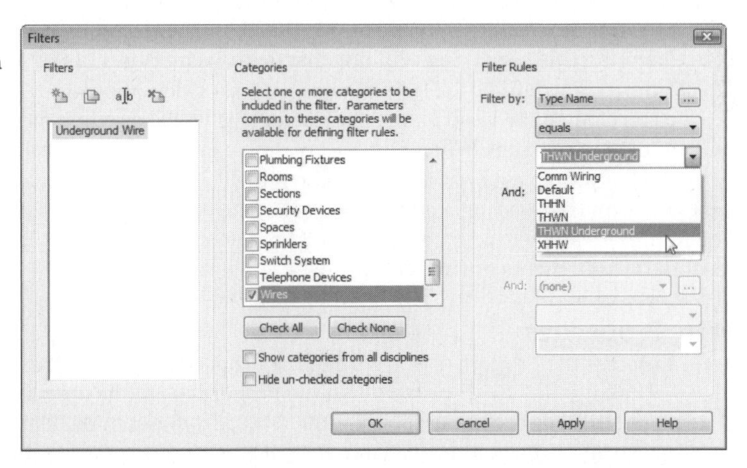

FIGURE 14.26
Filter settings for a wire-type filter

To apply the filter to your view, access the Visibility/Graphic Overrides settings for the view, and select the Filters tab. Click the Add button in the lower-left corner, and choose the filter you created for the wires from the Add Filters dialog box. Click the Override button in the Lines column for the filter. Set the Weight, Color, or Pattern overrides for the filter to display the wire as desired. Figure 14.27 shows floor boxes with underground wire in the same view as receptacles with wire as it is normally displayed.

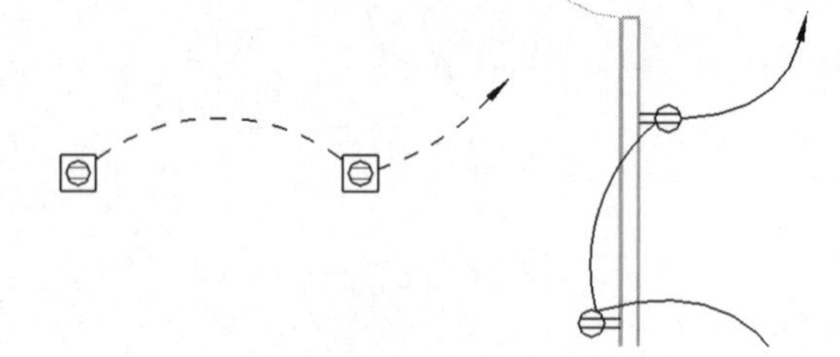

FIGURE 14.27 Different wire types displayed in the same view using a filter

Managing Circuits and Panels

By creating circuits for your devices, fixtures, and equipment, you have the ability to manage the properties of the circuits and the panels to which they are connected. You can also manage the location of circuits within your panels to balance loads and track the total electrical load for your project. The electrical connectors in your families determine some of the properties of the circuit to which they are connected.

Circuit Properties

To access the properties of a circuit, select an element connected to it, and select the Electrical Circuits contextual tab. If the selected component has multiple circuited connectors, you can choose which circuit properties to access by selecting the circuit from the System Selector drop-down. The properties of the circuit are displayed in the Properties palette. Click the Properties button on the ribbon to access the Properties palette. Because most of the data comes from the electrical characteristics of the connector and the distribution system, many of the parameters of the circuit are not editable and grayed out. Revit takes the load information from the components on the circuit and calculates the apparent load and current as well as the voltage drop and wire size. You can change the type of wire used for the circuit by editing the Wire Type parameter. Changing this parameter will not change the wire types of the wires you have drawn for the circuit; it only defines the type of wire used for calculations.

You can change the Load Name parameter to give the circuit a more descriptive name that will appear in the panel. When you are creating a circuit for components and the load exceeds 80 percent of the circuit breaker rating, Revit MEP 2011 will give you a warning. It does not prevent you from overloading a circuit, however. The default breaker size for circuits is 20 amps. The Rating parameter indicates the circuit breaker size for the circuit and can be changed to meet the requirements of your circuit.

Wire Properties

You can access the properties of a wire in the Properties palette when you select a wire in the model. Clicking the Edit Type button in the Properties palette opens the Type Properties dialog box, which displays the parameters for the wire as defined in the electrical settings for the project. You can add information such as a description or type mark to the wire type.

The instance properties of the wire control the display of tick marks and the number of conductors for the circuit. There are three options for the Tick Marks parameter:

Calculated This is the default value for wires and indicates that the number of conductors is determined by the circuit and distribution system properties regardless of whether the Tick marks are displayed.

On Selecting On displays tick marks on the selected wire.

Off Selecting Off turns off the display of tick marks on the selected wire.

The number of conductors can be increased or decreased by editing the values of the Hot, Neutral, and Ground Conductors parameters. You can change the number of hot conductors using the + and – editing grips that appear when you select a wire. The location of the tick marks on a wire can be changed by clicking the solid blue dot that appears at the center of the tick marks when a wire is selected.

Panel Properties

In the instance properties of a panel, you can provide information about it that will appear in the panel schedule. New to Revit MEP 2011 are the Schedule Header Notes and Schedule Footer Notes parameters. Click the Edit button in these parameters to add notes to the header or footer of a panel schedule. The Mains parameter is for indicating the size of the main circuit breaker and does not affect any load calculations. The Max #1 Pole Breakers parameter allows you to assign the number of poles in the panel. You can also indicate the short-circuit rating, enclosure type, and any modifications by editing the respective parameters.

You can define how the circuit tags will display in your drawings by editing the Circuit Naming, Circuit Prefix, and Circuit Prefix Separator parameters. The standard wire tag is a label that displays the circuit number. There are three options for circuit naming:

Panel Name This option places the name of the panel in front of the circuit number when a wire or device is tagged. You can use a separator such as a dash between the panel name and circuit number by using the Circuit Prefix Separator parameter.

Prefixed This option places the prefix defined in the Circuit Prefix parameter in front of the circuit number when a wire or device is tagged.

Standard This option does not add any additional information to tags that display the circuit number of a selected wire or device.

Figure 14.28 shows three circuits to panels that use different naming conventions. The panel for the circuit in the lower left uses the Panel Name option with a hyphen as a separator. The panel for the receptacle on circuit 2 uses the Standard circuit-naming option, and the panel for the circuit on the right uses a prefix "PWR" and a slash as a separator.

When you select a panel, the Modify | Electrical Equipment contextual tab appears. The Edit Panel Schedule button on this tab allows you to make changes to where the circuits are connected in the panel. In this view of the panel, you can change the fonts used in the cells to match your standards. Select a cell or multiple cells, and right-click to change the font. In the Edit Font dialog box, you can set the text size, font style, and color, as shown in Figure 14.29.

You can change the name of a circuit by simply editing the value in the cell. This will update the value of the Load Name parameter of the circuit. Once you have made changes to circuit names, the Update Names button becomes active when you select a cell that has been edited. Clicking this button will return the circuit name to the default, which includes the load classification name and the location of the object.

You can change the size of the circuit breaker for a circuit by editing the cell in the panel schedule. This will update the value of the Rating parameter of the circuit. If you edit the breaker to a size that causes the circuit to be overloaded, you will receive a warning dialog box.

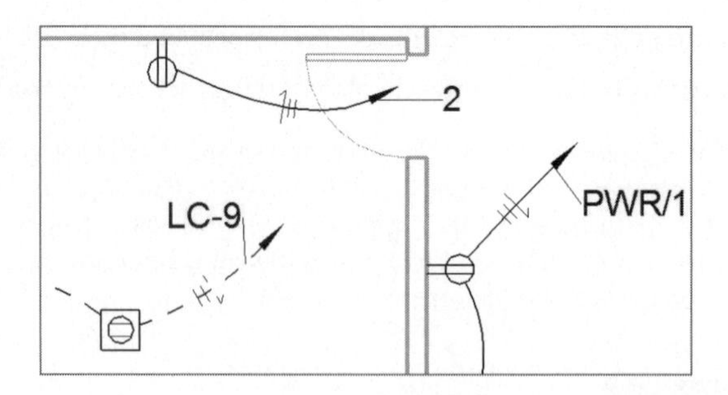

FIGURE 14.28
Wire tags using different circuit naming options

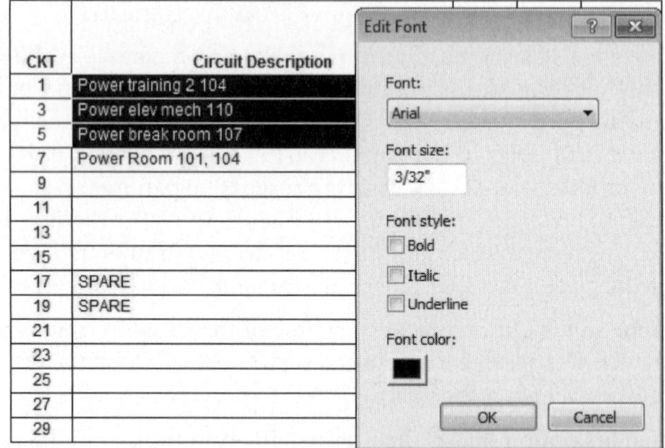

FIGURE 14.29
Edit Font dialog box for a panel schedule

To balance the loads in a panel, you can click the Rebalance Loads button on the ribbon. This will move circuits around in the panel to achieve the most balanced configuration possible across the phases. This may cause unwanted results, so use caution with this feature. For example, you may have a multicircuit run that occupies circuits 1, 3, and 5, but after rebalancing the loads, the circuits end up as 1, 4, and 18. It may also move multipole circuits to locations where they normally would not be placed. To avoid this scenario, you can use the Group/Ungroup button that becomes active when you select multiple circuits. This will maintain the order of the circuits and move them as a single unit. Grouped circuits are indicated with a blue dashed box around them. Selecting any cell within a group of circuits and clicking the Group/Ungroup button will remove all the circuits from the group, not just the selected cell or cells.

CKT	Circuit Description	Trip	Poles
1	Power training 2 104	20 A	1
3	Power elev mech 110	20 A	1
5	Receptacle-Rm 107	15 A	1
7	Power Room 101, 104	20 A	1

You do have the option to relocate circuits manually instead. To change the position of a circuit in a panel, click the circuit number. This will activate the move buttons on the Circuits panel of the tab. You can move the circuit up or down the side of the panel it is on. When you move a circuit up or down to a space that already contains a circuit, the two circuits will change places. Circuits can also be moved from one side of the panel to the other. As with moving up or down, moving a circuit to an occupied space on the other side of the panel will cause the two circuits to swap locations.

When you select a panel for a circuit, Revit MEP 2011 finds the first available circuit in the panel and places the circuit there. Spares and spaces can be added to occupy circuits in the panel. Because you cannot choose the circuit number for a circuit when it is created, when you add a circuit to the panel, the circuits occupied by spaces or spares will not be used.

To insert a spare or space, select an unoccupied circuit number, and click the Assign Spare or Assign Space button on the Circuits panel of the tab. A light brown background is applied to the cells for the circuit number. This background is a visual indication that the circuit is locked. A locked circuit cannot be moved in the panel. You can edit the name of a locked circuit and also change its breaker size. Any occupied circuit in the panel can be locked using the Lock/Unlock button on the Circuits panel of the tab. The only way to remove a spare or space is to select it and click the Remove Spare/Space button on the Circuits panel of the tab.

PARTIALLY FILLED PANELS

Your panel schedule templates define how many slots are shown for the panel. If you assign a number of one-pole breakers that is greater than the amount shown for the panel, you will receive a warning. Consider creating panel schedule templates for various sizes of panels based on the number of slots available.

Other Panels and Circuits

Circuits can be created for any of the categories of electrical systems. Wire types can be created to distinguish these systems visually if wiring is shown. The panels for these systems can be used to manage the circuit locations. You can create panel schedule templates for the various types of panel schedules that you use in your projects. Figure 14.30 shows the Edit Circuits dialog box for a telephone terminal block used as the panel for the telephone circuits in a project.

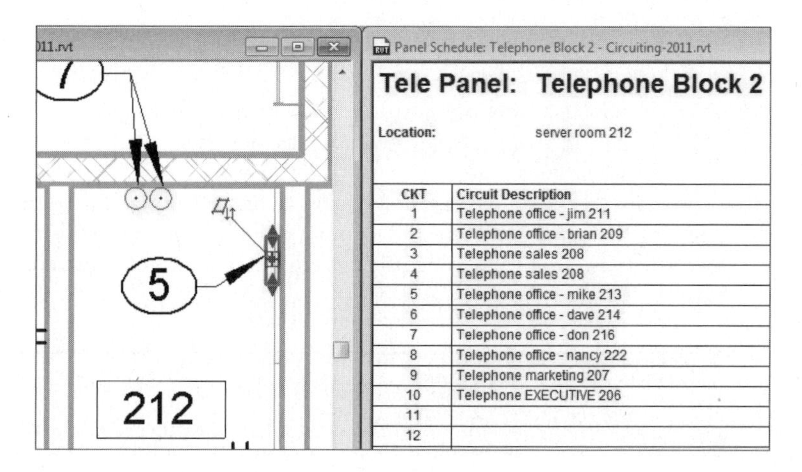

011.rvt
Panel Schedule: Telephone Block 2 - Circuiting-2011.rvt
Tele Panel: Telephone Block 2
Location:
server room 212
CKT
Circuit Description
1
Telephone office - jim 211
2
Telephone office - brian 209
3
Telephone sales 208
4
Telephone sales 208
5
Telephone office - mike 213
6
Telephone office - dave 214
7
Telephone office - don 216
8
Telephone office - nancy 222
9
Telephone marketing 207
10
Telephone EXECUTIVE 206
11
12
5
212

FIGURE 14.30
Telephone circuits

Although telephone wiring is not normally shown on a typical building project, creating these systems provides you with information that can help in making design decisions. This type of information could later be modified to represent the as-built conditions of your project, which could then be useful for facilities maintenance. It is about putting the *I* in your BIM project.

Using Schedules for Sharing Circuit Information

Panel schedules are a special type of schedule that is automatically generated by Revit MEP from the data within the panel objects. When you select a panel, the Create Panel Schedule button appears on the Modify | Electrical Equipment contextual tab. Clicking this button allows you to select a template to be used for the panel schedule.

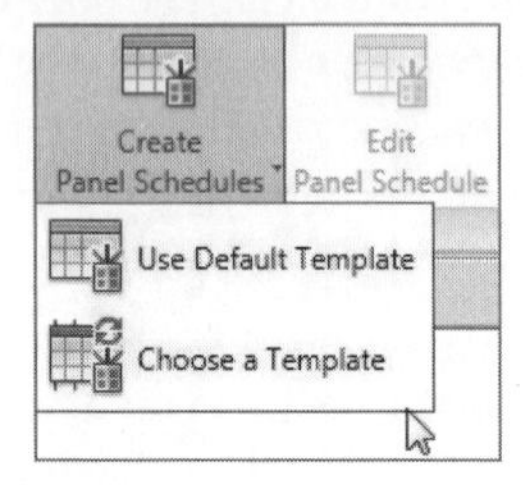

Once you have chosen a panel schedule template for the panel, the panel schedule appears in the Project Browser, as shown in Figure 14.31.

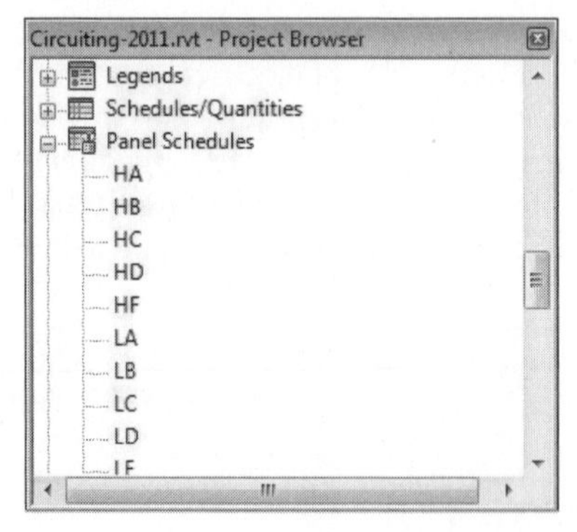

FIGURE 14.31 Panel schedules shown in the Project Browser

You can access the panel schedules by double-clicking them in the Project Browser or by selecting the panel in a view and clicking the Edit Panel Schedule button. You can change the template used for a panel schedule by right-clicking it in the Project Browser and choosing the Change Template option.

PANEL SCHEDULE TEMPLATES

See Chapter 5 for information on creating and managing panel schedule templates.

You can place panel schedules on your sheets by dragging and dropping them from the Project Browser. When you place a panel schedule onto a sheet, it will be formatted according to the settings defined in the panel template. Changes made to the circuits in your model or in the Edit Circuits dialog box of the panel will automatically show up in the panel schedule. Changes made to the panel schedule template that define the size and appearance of the schedule will be automatically applied to the schedule on a sheet. You can snap to the border of a panel schedule in order to align your schedules on a sheet.

Scheduling Panels and Circuits

You can create schedules for panels and circuits that can be used to extract information about your project. Organizing the data contained in your project regarding electrical loads can help you make design decisions and keep your design coordinated with other disciplines. Schedules can also be created to supplement the panel schedules generated by Revit MEP.

Circuits hold a host of useful data that can be scheduled for project coordination and design. Figure 14.32 shows a schedule of electrical circuits that displays information that can be used to verify circuit locations and length, breaker size, and load, as well as wire size. This type of schedule may be useful because you can display and manage information about the circuits that may not appear in your panel schedules. Having this information in a schedule makes it easy to check your design and make any necessary changes.

FIGURE 14.32
Electrical circuits schedule

Electrical Circuit Schedule

Circuit Number	Load Name	Voltage	Length	Voltage Drop	Breaker Size	Apparent Current	Wire Size
HA							
1	Lighting Room 103, 104, 105	277 V	288' - 5 3/4"	5 V	20 A	10 A	1-#8, 1-#8, 1-#8
2	Exterior Lighting - Entrance	277 V	311' - 3 5/32"	2 V	20 A	2 A	1-#12, 1-#12, 1-#12
3	Site Lighting	277 V	78' - 3 1/2"	0 V	20 A	1 A	1-#12, 1-#12, 1-#12
4	Site Lighting	277 V	57' - 8 5/8"	0 V	20 A	0 A	1-#12, 1-#12, 1-#12
5	SPARE	277 V	0' - 0"	0 V	20 A	0 A	1-#12, 1-#12, 1-#12
6	SPARE	277 V	0' - 0"	0 V	20 A	0 A	1-#12, 1-#12, 1-#12
7,9,11	ELEVATOR	480 V	28' - 11 1/16"	1 V	50 A	27 A	3-#6, 1-#6, 1-#10
8,10,12	TA	480 V	7' - 0 17/32"	0 V	50 A	14 A	3-#6, 1-#6, 1-#10
HB							
1	Lighting STAIR 3 S3	277 V	82' - 2 23/32"	0 V	20 A	1 A	1-#12, 1-#12, 1-#12
2	Lighting STAGE 115	277 V	103' - 1 7/32"	1 V	20 A	1 A	1-#12, 1-#12, 1-#12
3	Lighting AUDITORIUM/THEAT	277 V	232' - 9 3/16"	4 V	20 A	5 A	1-#12, 1-#12, 1-#12
4	Exterior Lighting	277 V	76' - 6 25/32"	0 V	20 A	0 A	1-#12, 1-#12, 1-#12
5	Lighting Room 102, 100, S3, 1	277 V	279' - 3 3/16"	5 V	20 A	5 A	1-#12, 1-#12, 1-#12
6	Site Lighting	277 V	101' - 3 27/32"	0 V	20 A	0 A	1-#12, 1-#12, 1-#12
7	Lighting Room 112, 113, 102,	277 V	131' - 6 7/16"	1 V	20 A	2 A	1-#12, 1-#12, 1-#12
8	Lighting Room 118, 117, 116	277 V	57' - 7 17/32"	0 V	20 A	2 A	1-#12, 1-#12, 1-#12
15	SPARE	277 V	0' - 0"	0 V	20 A	0 A	
17	SPARE	277 V	0' - 0"	0 V	20 A	0 A	
19,21,23	TB	480 V	6' - 6 1/8"	0 V	50 A	11 A	3-#6, 1-#6, 1-#10
20	SPARE	277 V	0' - 0"	0 V	20 A	0 A	

The Bottom Line

Establish settings for circuits and wiring Proper setup of the electrical characteristics of a project is important to the workflow for creating circuits and wiring. Settings can be stored in your project template and modified on an as-needed basis.

Master It The distribution systems defined in a project make it possible to connect devices and equipment of like voltages. Do you need to have voltage definitions in order to create distribution systems? If so, why?

Create circuits and wiring for devices and fixtures Circuits are the systems for electrical design. Wiring can be used to show the connection of devices and fixtures in a schematic fashion.

Master It Circuits can be created for devices or equipment even if they are not assigned to a panel. Circuits can then be represented by wiring shown on construction documents. Give two examples of how you can add a device to a circuit that has already been created.

Manage circuits and panels With the relationship between components and panels established, you can manage the properties of circuits and panels to improve your design performance and efficiency.

Master It While checking the circuits on a panel, you notice that there are only 14 circuits connected, but the panel has 42 poles. How can you reduce the amount of unused space in the panel?

Use schedules for sharing circuit information Panel schedules can be used on construction documents to convey the load information. Schedules can also be created for use as design tools to help track electrical data.

Master It The information in Revit panel schedules may not meet the requirements of your document or design standards. Describe how you can use the data within your Revit model to provide the required information.

Part 4

Revit MEP for Plumbing

Chapter 15

Plumbing (Domestic, Sanitary, and Other Piping)

Routing plumbing piping has come a long way from drawing circles and lines on paper. Over the past 20 years, tools such as the straight edge, 30/60 triangle, and Timely template have been replaced by CAD systems. With more owners requiring BIM, a plumbing designer has to become a virtual pipe installer. Instead of just drawing circles and lines, you have to understand more about how fittings go together to construct your piping design. This is where Revit MEP 2011 excels; it can help you create your designs more accurately and efficiently.

In this chapter, you will learn to

- Customize out-of-the-box Revit plumbing fixtures for scheduling purposes
- Use custom plumbing pipe assemblies to increase speed and efficiency in plumbing layouts
- Adjust and use the plumbing pipe settings
- Select and use the best pipe routing options for your project
- Adjust pipe fittings
- Adjust the visibility of pipes

Plumbing Fixtures

Plumbing fixtures are as important to the look of an architectural design as granite countertops or marble tile. Plumbing fixtures, when properly selected, will not only enhance the visual design but will also promote cleanliness and hygiene. Plumbing fixtures normally are placed by the architect during schematic design to coordinate usability and meet the requirements of governing codes. From a plumbing design point of view, there are some different criteria that must be examined. What are the water conservation guidelines? Are the plumbing fixtures required to meet LEED standards? Other questions should be asked during design, such as do you as the designer want to use the plumbing fixtures that the architect has used in their model to connect your piping to, or will you substitute them with your company standard? If so, do you need to apply shared parameters that reflect the design standards required? For example, in the United States, there are two major plumbing codes: Uniform Plumbing Codes and International

Plumbing Codes. To further complicate matters, some states will adopt one of these two codes and then add their own amendments, creating their own state code.

With Revit MEP 2011, you can apply this information through the use of parameters. This can be done in a couple different ways. First you can edit the information in the family itself by selecting a plumbing fixture family and editing the information through type properties (refer to Figure 15.1).

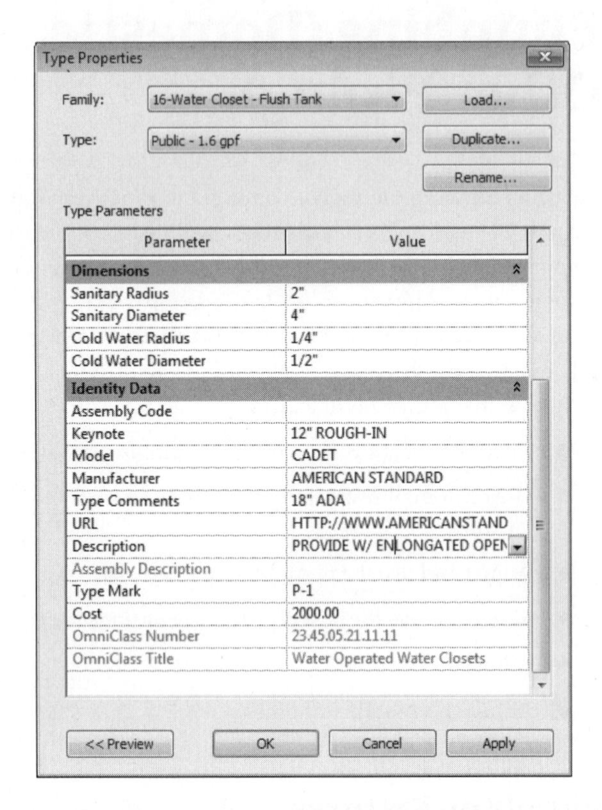

FIGURE 15.1
Editing the type properties

The second way to edit the information is to create a type catalog. Using type catalogs allows the user to easily produce more information about different types or models of the same family. For example, most manufacturers of bathrooms use model numbers to show the differences in finishes, rough-in locations, handle locations, and handicap accessibly information. Also, this will allow for the plumbing design to be easily changed from one manufacturer to another.

The easiest way to create a type catalog is to first open a plumbing fixture family and then review and make note of the type properties you want to be able to modify. Then create a `.txt` file that will populate the information. To create this `.txt` file, do the following:

1. Open Microsoft Excel. (If you do not have Excel, then you can use Notepad or OpenOffice.org to achieve to same goal. OpenOffice can be downloaded from `www.openOffice.org`.) Save the spreadsheet as a comma-separated values (CSV) file, making sure to name it the same as the plumbing fixture's family name.

2. Now leave cell A1 blank; this is necessary for the type name of the family, such as Kohler or American Standard, to be listed in column A properly.

3. Next you will add the parameter name followed by the parameter unit in row 1. For this example, you will be using the following system parameters from the identity data located in the plumbing fixture family:

Keynote	Keynote##other##
Model	Model##other##
Manufacturer	Manufacturer##other##
Type Comments	Type Comments##other##
URL	URL##other##
Description	Description##other##

4. Add the information for each row that you want to be able to schedule (refer to Figure 15.2).

FIGURE 15.2
Creating a `.csv` file for type catalogs

	A	B	C	D
1		Keynote##other##	Model##other##	Manufacturer##other##
2	American Standard	12" ROUGH-IN	3421.012	AMERICAN STANDARD
3	Kohler	12" ROUGH-IN	9334434.1	KOHLER
4	Zurn	12" ROUGH-IN	201100.11	ZURN
5	Crane	10" ROUGH-IN	2200.19	CRANE
6				

5. Once everything has been input, save the file. Make sure when the file is created, it is located in the same directory as the family it references.

6. Next go to the directory where you saved your file, and rename the extension from `.csv` to `.txt`. To review what the `.txt` file looks like or to make quick edits with converting back to a `.csv` file, you can open this file with Notepad (refer to Figure 15.3).

If you created your file properly, you should see a catalog of information when inserting the family (refer to Figure 15.4).

TYPE CATALOG REFERENCES

Type catalogs can reference by using system parameters and by using shared parameters. The following list is some of the types of parameters and parameter unit that can be used in creating a type catalog `.txt` file:

Text	parameter name##OTHER##
Integer	parameter name##OTHER##
Number	parameter name##OTHER##
Length	parameter name##LENGTH##FEET
Area	parameter name##AREA##SQUARE_FEET
Volume	parameter name##VOLUME##CUBIC_FEET
Angle	parameter name##ANGLE##DEGREES
Slope	parameter name##SLOPE##SLOPE_DEGREES
Currency	parameter name##CURRENCY##
URL	parameter name##OTHER##
Material	parameter name##OTHER##
Yes/No	parameter name##OTHER##
<Family Type>	parameter name##OTHER##

One note, the parameter in the family must be filled out for the type to load the value from the catalog properly; if not, the type catalog will not work.

Now that you have information added to your plumbing fixture family and you have placed the plumbing fixtures into the plan, you will want to schedule that information.

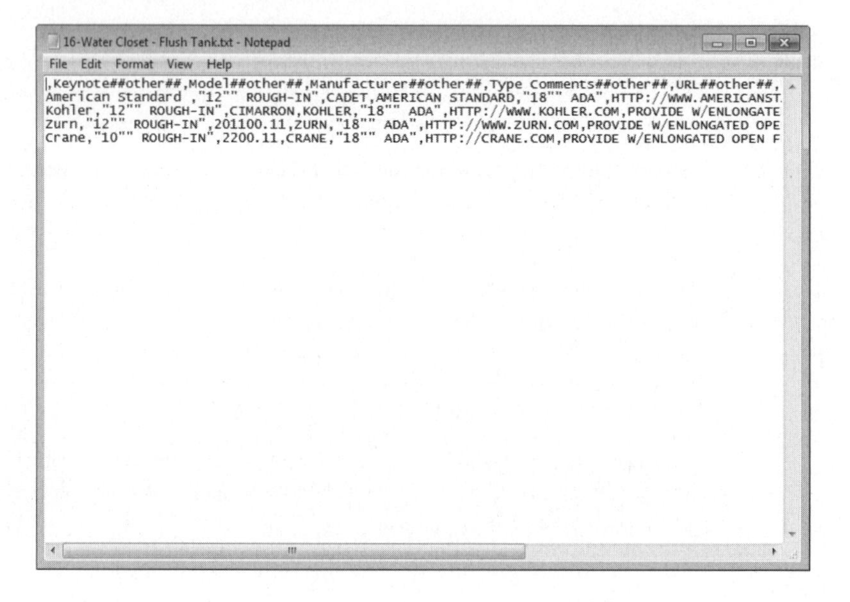

FIGURE 15.3
Opening the `.txt` file in Notepad

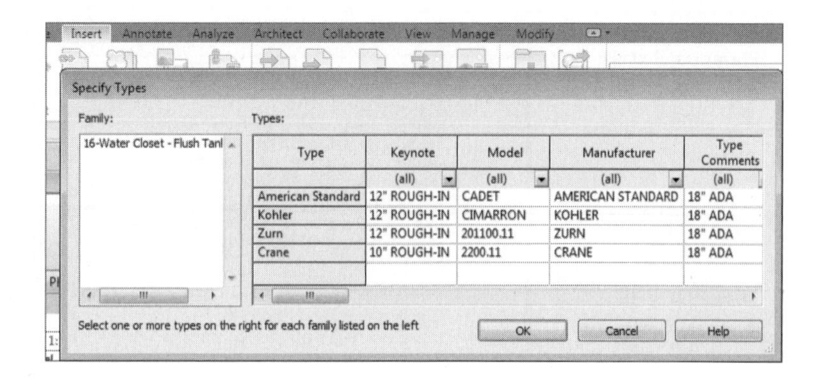

FIGURE 15.4
Type catalog

To accomplish this, go to the Analyze tab on the ribbon, and then select Schedule/Quantities. This will open the New Schedule dialog box. Select Plumbing Fixtures from the Category group, and then click OK (refer to Figure 15.5).

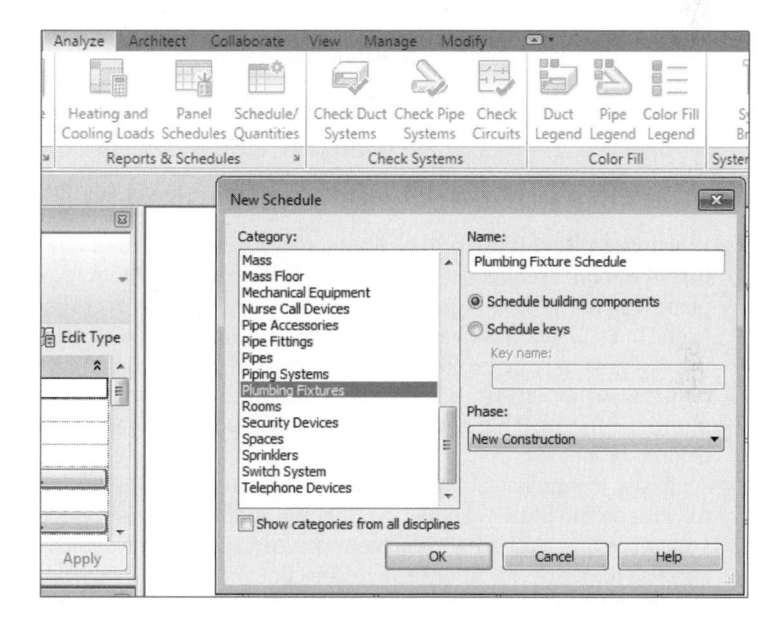

FIGURE 15.5
Select Plumbing Fixtures from the Category group

Next select the information from the Available Fields dialog box and add it to the Schedule fields (in order) dialog box. Then click OK, which will create your schedule (refer to Figure 15.6).

Now that you have the Plumbing Fixture schedule created, you may not want to see duplicate information or blank information, so you will need to sort the information. To do this, go to the Properties palette, select Sorting/Grouping, change the Sort By pull-down menu to Type Mark, and then deselect the Itemize Every Instance check box. Now your schedule will show only the items that have information (refer to Figure 15.7).

FIGURE 15.6
Plumbing Fixture schedule

Plumbing Fixture Schedule					
Type Mark	Description	Type Comme	Manufacture	Model	Keynote
P-1	PROVIDE W/ENLONGATED OPEN FRONT SEAT	18" ADA	AMERICA	CADET	12" ROUGH-IN
P-2	PROVIDE W/ENLONGATED OPEN FRONT SEA	18" ADA	KOHLER	CIMARRO	12" ROUGH-IN
P-3	PROVIDE W/ENLONGATED OPEN FRONT SEA	18" ADA	CRANE	2200.11	10" ROUGH-IN
P-4	PROVIDE W/ENLONGATED OPEN FRONT SEA	18" ADA	ZURN	201100.1	12" ROUGH-IN
P-4	PROVIDE W/ENLONGATED OPEN FRONT SEA	18" ADA	ZURN	201100.1	12" ROUGH-IN
P-4	PROVIDE W/ENLONGATED OPEN FRONT SEA	18" ADA	ZURN	201100.1	12" ROUGH-IN
P-4	PROVIDE W/ENLONGATED OPEN FRONT SEA	18" ADA	ZURN	201100.1	12" ROUGH-IN

FIGURE 15.7
Sorted schedule

Plumbing Fixture Schedule					
Type Mark	Description	Type Comme	Manufacture	Model	Keynote
P-1	PROVIDE W/ENLONGATED OPEN FRONT SEA	18" ADA	AMERICA	CADET	12" ROUGH-IN
P-2	PROVIDE W/ENLONGATED OPEN FRONT SEA	18" ADA	KOHLER	CIMARRO	12" ROUGH-IN
P-3	PROVIDE W/ENLONGATED OPEN FRONT SEA	18" ADA	CRANE	2200.11	10" ROUGH-IN
P-4	PROVIDE W/ENLONGATED OPEN FRONT SEA	18" ADA	ZURN	201100.1	12" ROUGH-IN

Working with Architectural Linked in Plumbing Models

When using Revit MEP for plumbing, there is a gray area of how to coordinate plumbing fixtures between architectural linked in models and plumbing models. The main thought is to place the plumbing designer's edited plumbing fixtures over the architectural plumbing fixtures. In fact, Autodesk has added a new feature to Revit MEP 2011 that will allow the plumbing designer to copy/monitor the architectural plumbing fixtures. The designer has the choice of using the architectural-supplied plumbing fixtures or replacing them with the designer's plumbing fixtures that have already been edited with proper connectors and scheduling information.

To use this method of copy/monitor, the plumbing designer must turn off the plumbing fixtures on the linked architectural model so that the plan does not show double fixtures. Following this method without constant coordination review can lead to costly mistakes because the Copy/Monitor feature does not automatically update to show the new fixtures that the architect may have added. We'll talk more about this later in the chapter.

The other option is to create custom pipe assemblies. Custom pipe assemblies are fittings preassembled or modeled to line up in the locations of the linked architectural plumbing fixtures. These are created using the plumbing fixture families.

Creating Custom Pipe Assemblies

Custom pipe assemblies can be represented one of two ways in the Family Editor. First they can be represented by using sweeps to represent the p-trap, wye, and associated piping, which creates a smaller file size and will reduce the size of the overall plumbing model but is not as accurate for quantity takeoffs. Refer to Figure 15.8.

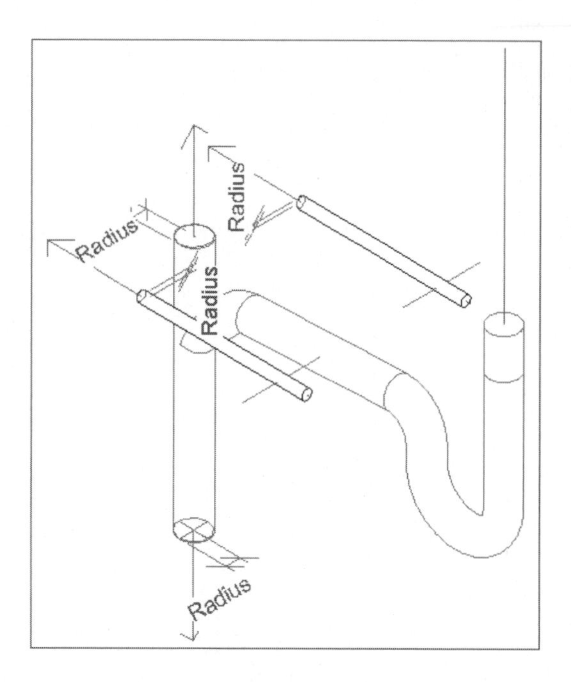

FIGURE 15.8
Representation of a pipe assembly created with sweeps

The second is assembling nested families, which can allow for better quantity takeoffs for all the fittings, can create more accurate dimensional information when supplied by manufacturers, and can be easier for the plumbing designer to create. The downside is that it will produce a larger family file. The second option will help you achieve more of the building information modeling status while helping to increase productivity. Refer to Figure 15.9.

Now let's examine how nested piping assemblies are put together and some key areas to be mindful of:

1. Open the `PR-Sinks and Lavs.rfa` file found on `www.wiley.com/go/masteringrevitmep2011`.
2. Several different modified pipe fitting families are nested into this family that make up the assembly. They are `Trap P - PVC-DWV.rfa`, `Tee Sanitary-PVC-DWV.rfa`, `PVC-DWV Pipe Section.rfa`, `Plug-PVC-DWV.rfa`, `Elbow -Copper Type L.rfa`, and `Copper Type L Pipe Section.rfa`. These can all be found on `www.wiley.com/go/masteringrevitmep2011`, or you can get the originals from the Pipe Fitting directory located in the Imperial Library.
3. When placing the fittings together, make sure to align, lock, and dimension each fitting together. If not, the fittings will pull apart.Refer to Figure 15.10.
4. When creating your pipe assembly, make sure to set a measurement from your Sanitary piping to the reference plane Front/Center. Next click the Reference plane, go to Properties, and select the Wall Closure box located under Construction. This will allow the assembly to act as if it is a wall- or face-hosted family without all the issues that come along with those family types. The wall closure will also come in handy later in this chapter for another purpose. Refer to Figure 15.11.

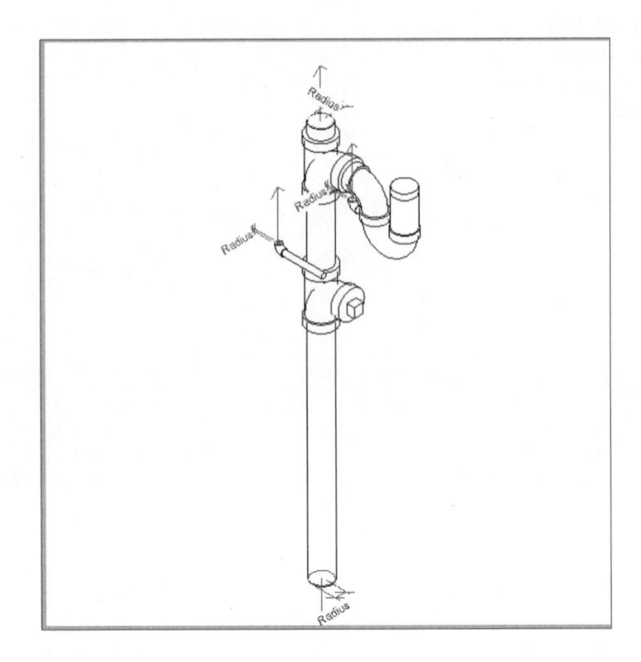

FIGURE 15.9
Nested pipe assembly

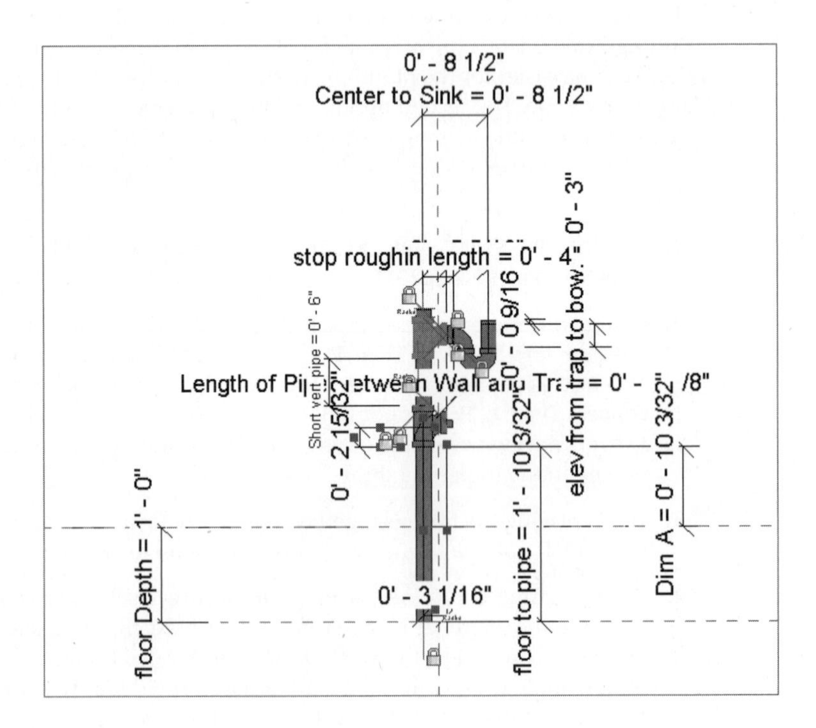

FIGURE 15.10
Aligning, locking, and dimensioning to lock nested families down

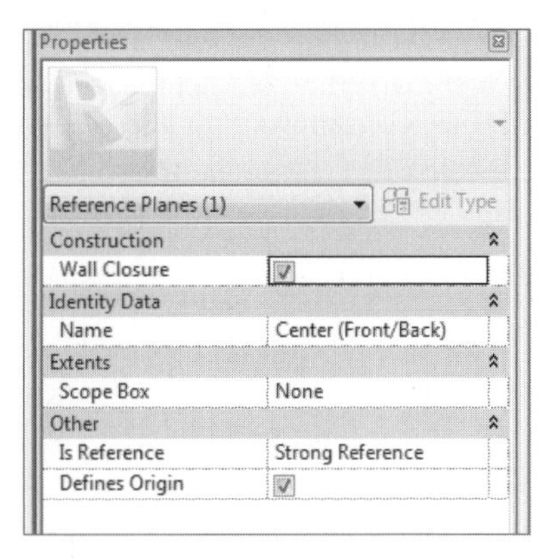

FIGURE 15.11
Wall closure

PLUMBING FIXTURE FAMILIES ARE NOT CREATED EQUAL

Plumbing fixture families are not always created equal. Rather than starting the family modeling front facing the front, some modelers will just model in whatever view is available. This has always been a problem in Revit, and as more families become available, the issue will increase. This can cause a few issues when aligning or when you replace one fixture with another. As your skills grow and you start creating your own content, please remember simple modeling etiquette. The model should be modeled in the proper orientation. The Front, Back, Top, Bottom, Left, and Right views should match the orientation of the product being modeled. This will increase productivity when aligning other items such as pipe assemblies under plumbing fixtures.

5. You can add parameters to flex your piping to align it under a sink or lavatory or to give a certain depth to account for floor slab. You can make these as complex or as simple as you want. Refer to Figure 15.12.

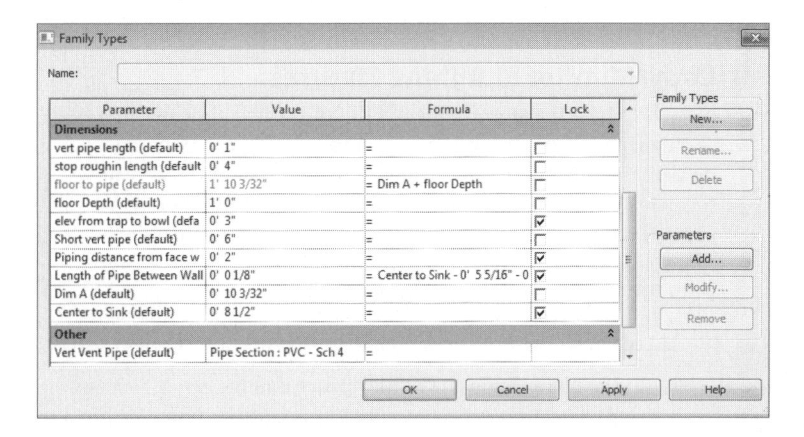

FIGURE 15.12
Using parameters to flex the piping assembly

6. Make sure when adding a water piping connection to a rough-in pipe to add an elbow pointing in the direction of the pipe routing such as from the floor or from the ceiling. This will help with autorouting and manual routing. Refer to Figure 15.13.

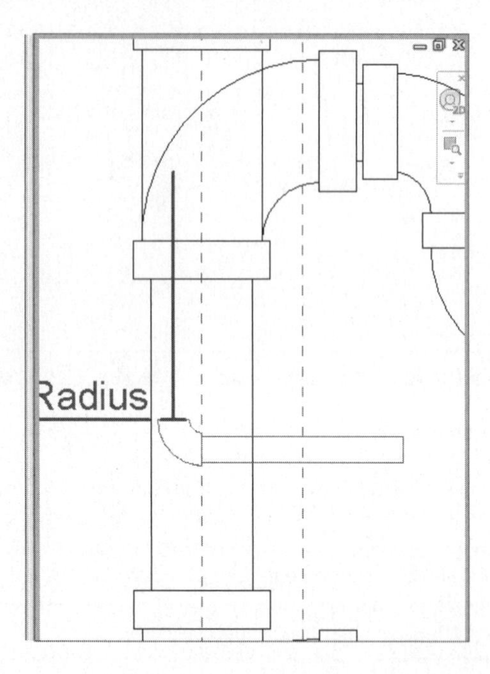

FIGURE 15.13 Adding a simple elbow pointed toward direction of flow

Now that you have reviewed some of the items that make up a pipe assembly, you can place the pipe assembly family in your plumbing model and align it with the architectural plumbing fixtures without having to turn them off. If the architect adds a plumbing fixture, you now have the capability to visibly see the added fixture. By following this workflow, you can have the intelligent plumbing fixture on the architectural model and still tag and schedule through the architectural link.

Copy/Monitor Plumbing Fixtures

Now let's take it a step further. If the plumbing fixtures that the architects are using match the same orientation of your pipe assemblies, you can copy/monitor the pipe assemblies directly behind the plumbing fixtures. Now when your architect moves a plumbing fixture, you will receive a warning that you need to coordinate your view. You will also have the capability to copy and change all of the plumbing fixtures on multiple levels all at one time. This can be a huge time-saver.

To use copy/monitoring, do the following:

1. Select Collaborate ➢ Copy/Monitor, and then select Current Link.
2. In the Copy/Monitor panel, select Batch Copy. Click Specify Type Mapping Behavior & Copy Fixtures. Once you click this button, you will be taken to a Coordination panel that has Category and Behavior selection tables.

3. Select Type Mapping, and replace the plumbing fixtures from the architects with the ones you created.

If you created your piping assemblies with Wall Closure selected and proper pipe locations, you will be able to host these with nonhosted, face-hosted, and wall-hosted fixtures. Refer to Figure 15.14.

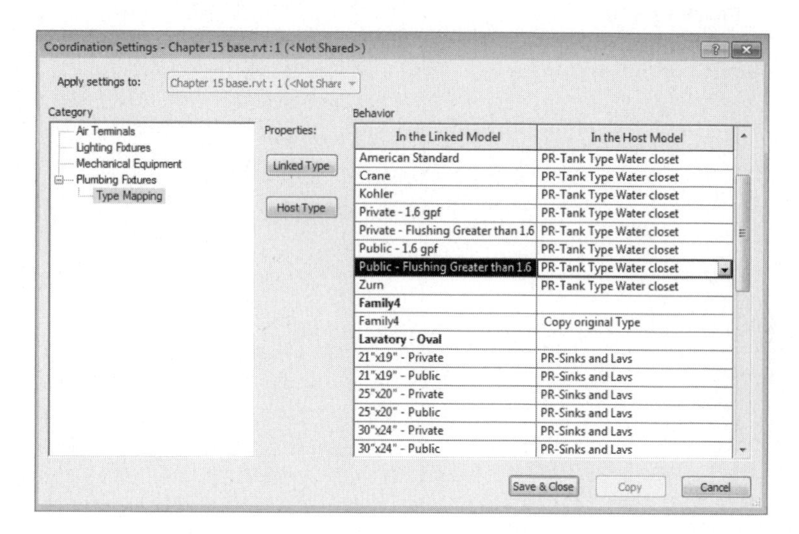

FIGURE 15.14
Type mapping

4. Next select Copy, and select the plumbing fixtures you want to copy. Refer to Figure 15.15.

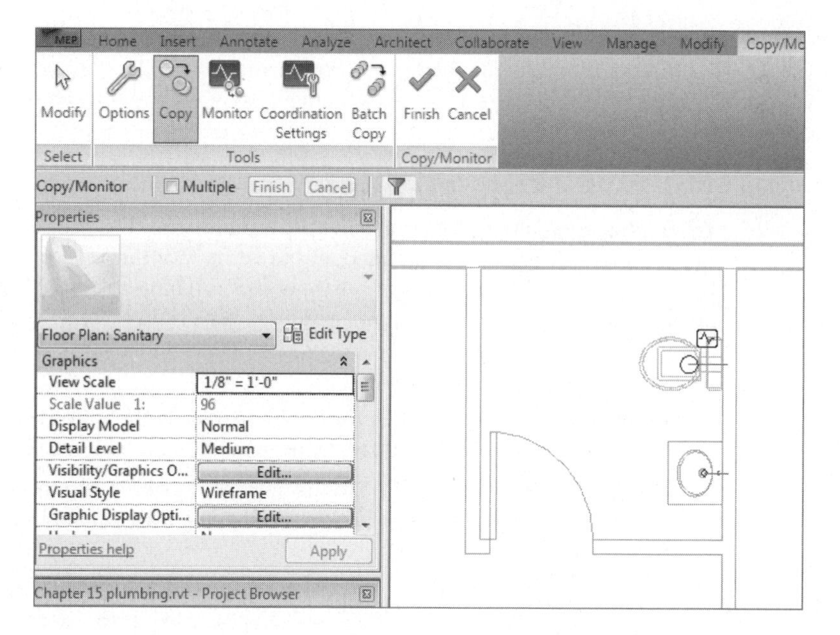

FIGURE 15.15
Copying and replacing fixture types

5. If you are doing multiple floors, you can select all levels from the elevations view and then filter and select only the plumbing fixtures. Then select Finish.

If you highlight the assembly, you should see the monitoring symbol on the assembly. Refer to Figure 15.16.

FIGURE 15.16
Monitoring symbol shown on the assembly

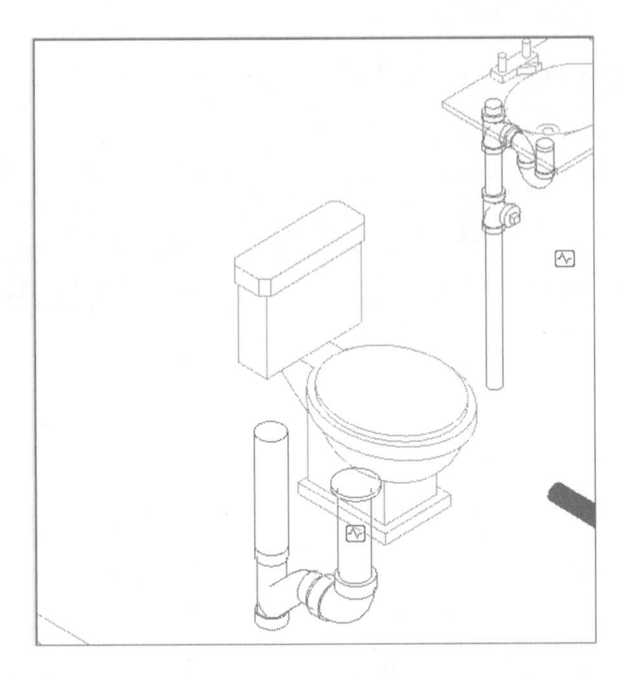

To fully take advantage of copy/monitoring, you will need to coordinate all of your plumbing fixtures with your pipe assemblies, but you will find it is time well spent.

Plumbing Pipe Settings

When setting up piping to route, you will want to apply the proper pipe material so that quantity takeoffs can be easily scheduled and to verify that the proper material is being used for the proper system. There are several areas you will have to adjust to set this up properly. These areas are system pipes, fittings, pipe material, pipe sizing tables, and fluids table, each of which is described here in more detail. Once you set these areas, you can then concentrate on autorouting and manual routing of pipe.

System pipes These are the pipes that are hard-coded into Revit. You have a limited amount of freedom to adjust parameters for these system families.

Fittings These can be applied to the parameters of the system pipes, which will allow for fittings to populate the model automatically. The fitting must be loaded into the model for them to work.

Pipe material This is set by selecting Mechanical Settings ➢ Sizes. This allows you to duplicate a pipe material, rename it, and apply the piping specification of material and roughness of pipe wall.

Pipe sizing table This is set by selecting Mechanical Settings ➤ Sizes. This allows you to duplicate and adjust the inside pipe diameter, outside pipe diameter, and pipe sizes to manufacturers' specifications.

Fluids table This is set by selecting Mechanical Settings ➤ Sizes. This allows you to duplicate, rename, and adjust the fluid type and also the viscosity, temperature, and density.

Creating System Pipes

To modify and create new system pipes, go to the Project Browser, and select Families ➤ Pipes ➤ Pipe Types. Then right-click the Standard pipe, and select Duplicate. This will create the additional pipe types you will need for the project. Next right-click the duplicated pipe types, and rename it to the pipe type you require. We prefer to rename them to the pipe material type for takeoff purposes (refer to Figure 15.17).

FIGURE 15.17
Renaming pipe types

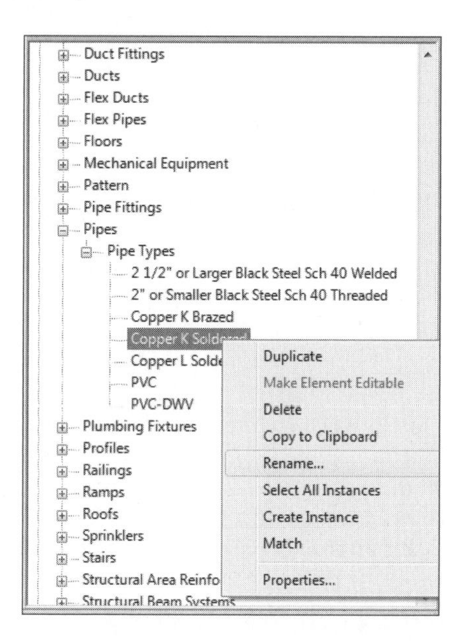

> **NAMING PIPE TYPES**
>
> Try to avoid naming your pipe types as system type names (such as *cold water pipe*). This makes it harder if you use your model for an integrated project delivery where you are partnered with a general contractor and subcontractors. The contractors will want to take the information and use as much of it for shop drawings as possible. Being able to quickly filter and schedule piping by material types is essential in the plumbing BIM world.

Now that you have your pipe types created, you will want to change some of the parameter options. First right-click the pipe you want to edit, and select Properties. This will open the Pipe Type parameters. Under Mechanical, you will find Material, Connect Type, and Class parameters. Change these settings to the appropriate types (refer to Figure 15.18).

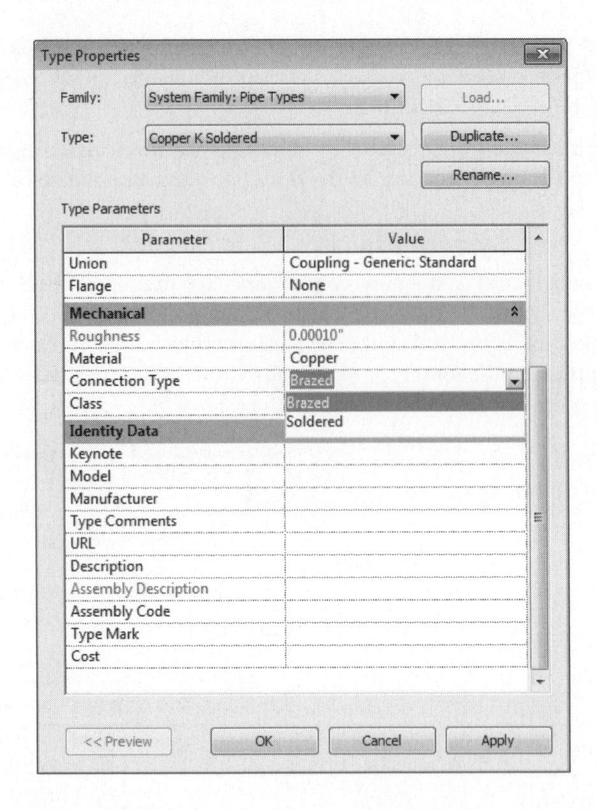

FIGURE 15.18 Mechanical parameters

Under the Identity Data parameter group, the following parameters are available: Keynote, Model, Manufacturer, Type Comments, URL, Description, Assembly Description, Assembly Code, Type Mark, and Cost.

If you have a certain manufacturer, model, or other special note that you may want to denote on the plans, you can use these settings to further describe your pipe type (refer Figure 15.19).

Creating Fittings for System Pipes

Under the Fitting parameters, you'll find Elbow, Preferred Junction Type, Tee, Tap, Cross, Transition, Union, and Flange. Before you can adjust the rest of your system pipe type parameters, you need to create the fittings that go with your system pipe types. To accomplish this, you must go to the Project Browser and select Families ➤ Pipe Fittings. Once you are under Pipe Fittings, you'll want to select the fittings that will be required for your pipe type parameters. You will right-click each of the family fittings and duplicate standard generic Elbow, Cross, Coupling, Tee, and Transition. After duplicating each one of these, rename them to the associated material of each pipe that you previously duplicated to create your pipe types (refer Figure 15.20).

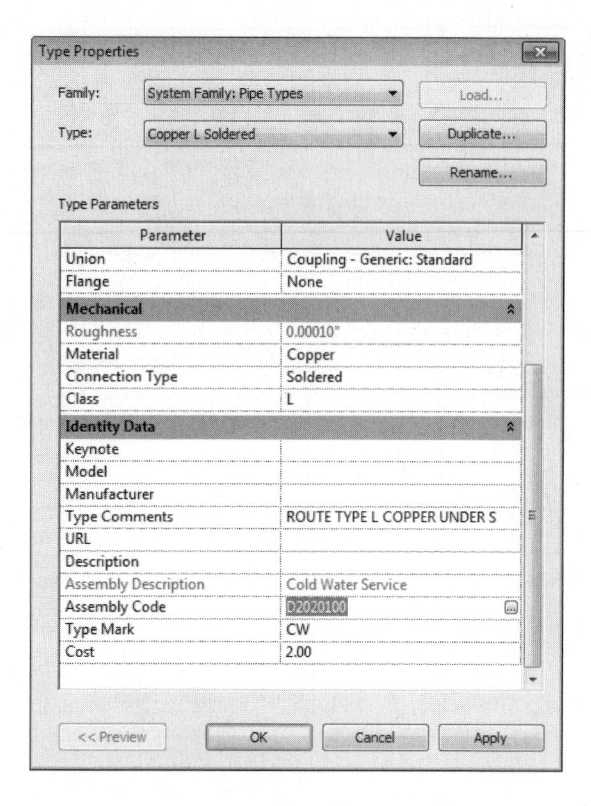

FIGURE 15.19 Identity Data parameters

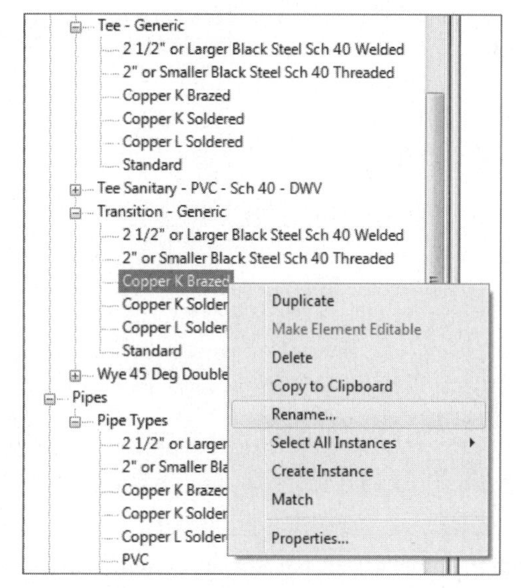

FIGURE 15.20 Duplicating and renaming fittings to match pipe types

Size Matters

When possible, use fitting families that are true dimensions or that closely resemble the dimensions from the pipe fittings you have in your specifications. For example, you would want to use the PVC-DWV pipe fittings that come with Revit rather than duplicating a generic fitting for sanitary piping system because the physical dimensions are so different because of the flow characteristics of the pipe itself.

Once you have all your pipe fittings, go to the Mechanical parameters to find the Material, Connect Type, and Class parameters. Change these settings to the appropriate types. Also, you may need to modify the Loss Method parameter to K Coefficient From Table. Also verify that the table is set to the appropriate setting (refer to Figure 15.21).

Figure 15.21
K Coefficient From Table setting

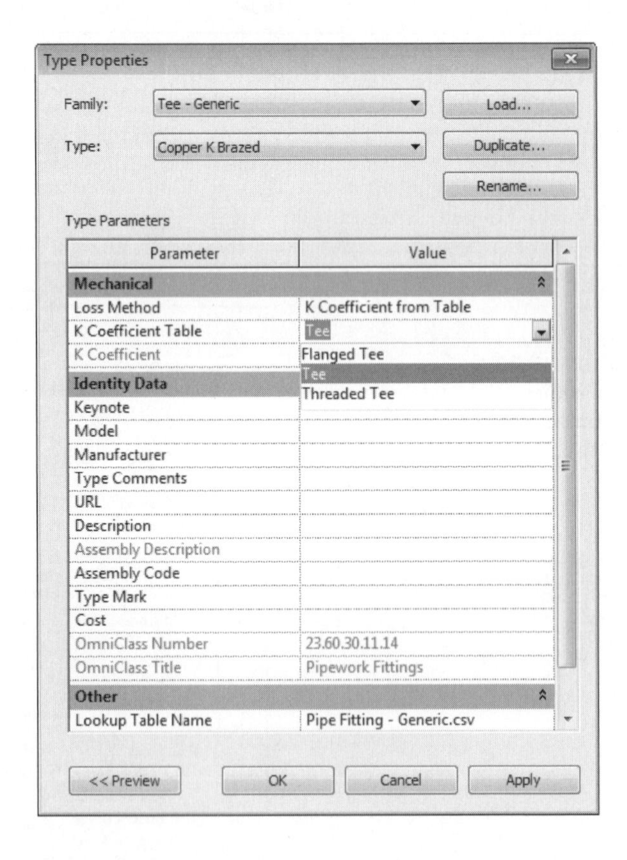

Once the pipe fittings have been created, go back to the Pipe type parameters, and under Pipe Fitting, select the proper pipe fitting for your pipe type. Doing this will allow you to be able to filter and schedule piping for quantity takeoff purposes to achieve more accurate pricing (refer to Figure 15.22).

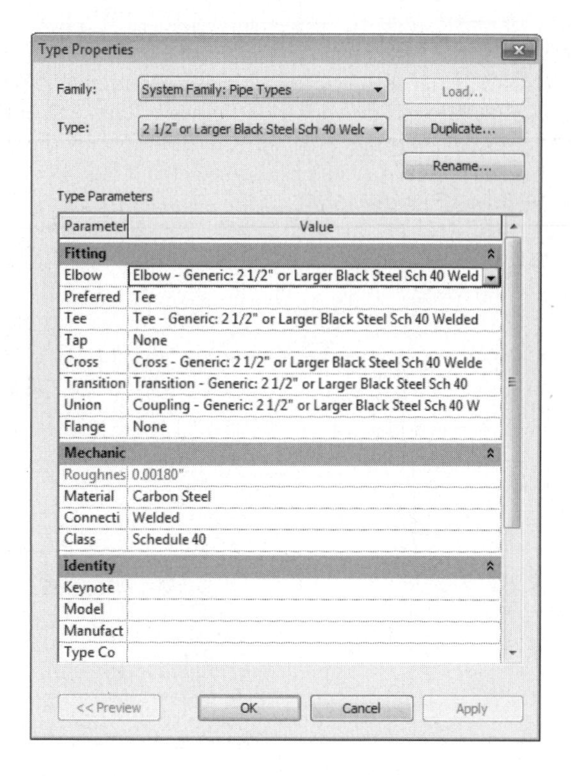

FIGURE 15.22 Selecting the proper pipe fittings for pipe types

Pipe Material

To get to the Pipe material settings, select Home ➤ Plumbing And Piping, and then select the small arrow in the lower-right corner of the panel. This will open the Mechanical Settings dialog box. Next select Pipe Settings ➤ Sizes. If you want to create a new pipe material, you can duplicate an existing pipe material and rename it to the new pipe material as required (refer to Figure 15.23). You can change the connection type and schedule of pipe.

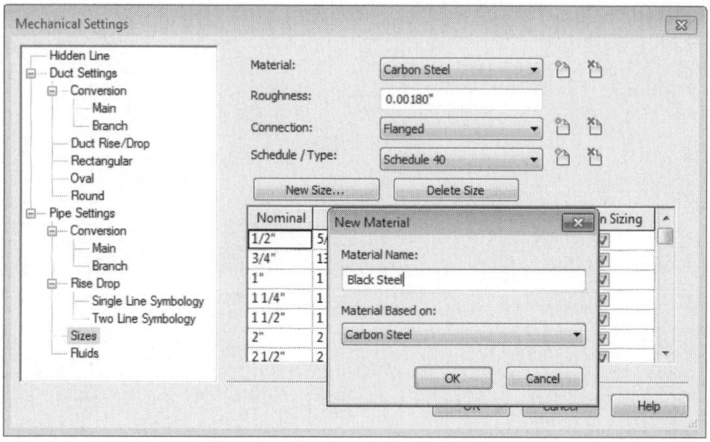

FIGURE 15.23 Duplicating and renaming pipe material

Pipe Sizing Table

If you want to adjust the sizing table, select Home ➤ Plumbing And Piping, and select the small arrow in the lower-right corner of the panel. This will open the Mechanical Settings dialog box. Next select Pipe Settings ➤ Sizes. You can duplicate the schedule of pipe and apply the pipe wall thickness as required. You can also select and deselect the piping sizes to match your design standards (refer Figure 15.24).

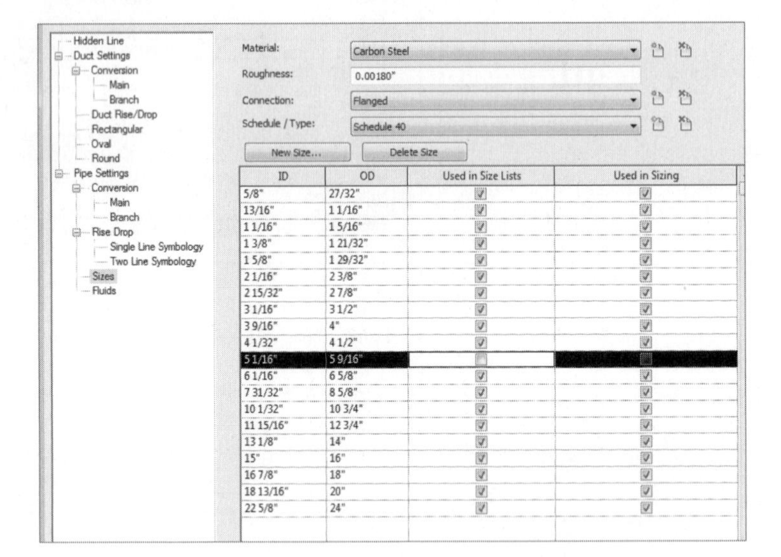

FIGURE 15.24
Pipe sizing

Fluids Table

You can add to the fluids table by selecting Home ➤ Plumbing And Piping and selecting the small arrow in the lower-right corner of the panel. This will open the Mechanical Settings dialog box. Next select Pipe Settings ➤ Sizes. Then you can duplicate one of the fluid categories that is closest to your needs and modify it as required (refer to Figure 15.25).

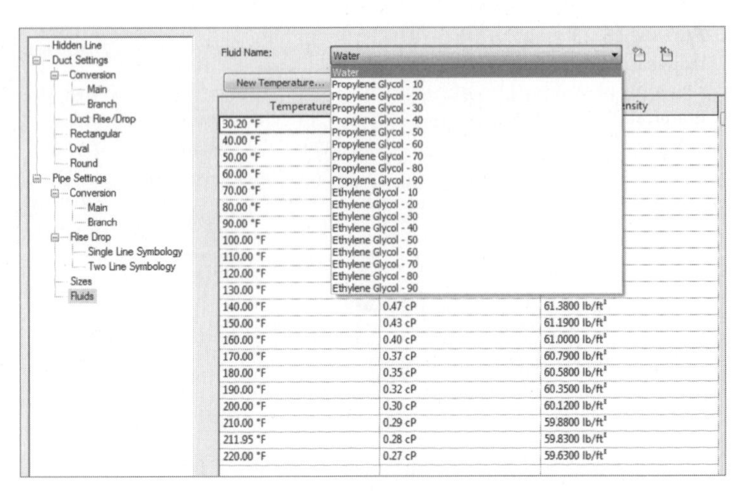

FIGURE 15.25
Fluids table settings

TYPE CATALOG REFERENCES

Although each of these options can be modified, use caution when changing these variables. If you are going to create new variables in these tables, always duplicate and do not edit the original tables. Most of the sizing variables are not 100% accurate because certain plumbing design principles have not been applied to the program.

Pipe Routing Options

Routing piping above ceilings, in walls, in chases, and under slab is always a concern for a plumbing designer. In fact, we used to focus so much on routing piping that we would feel like the whole project would fail if one section of piping was not routed just right. We were complaining that the structural engineer's concrete beams were too big in the first-floor ceiling space and how that would affect the gravity flow of the sanitary piping. He retorted that gravity affects buildings and people can die if the building falls. But we sleeved his beam to route our piping anyway. If the building falls, no one will call about your piping design, but if the toilet won't flush, then everyone knows your name. Now using Revit MEP 2011, you can show your concerns with color-coded visual coordination and interference checking — tools that can help you avoid putting a sleeve through a structural beam. There are a couple of routing options when you set out to design your piping model: the autoroute option, the manual routing option, and the sloping pipe option. Each of these is described in the following sections.

Auto Pipe Routing

We hope you have everything set up in your pipe types to begin routing piping. The smaller a system, the more beneficial the autoroute feature. If you have a large system you are designing, then manual routing will benefit you more because of the nature of designs changing more often. To start, open `Chapter 15 plumbing.rvt` and `Chapter 15 base.rvt` found on `www.wiley.com/go/masteringrevitmep2011`. Next do the following:

1. Select one of plumbing fixtures that you have added to the plumbing system, and press the Tab key until the autorouting features highlight its suggested path.
2. Select Generate Layout ➤ Modify Mechanical Equipment (refer to Figure 15.26).

You have four options to select for generating the layout: Network, Perimeter, Intersections, and Custom, and each one has several routing solutions to choose from that consists of a main (blue) and branches (green):

Network This solution creates a bounding box around the components selected for the piping system and then bases several solutions on a main segment along the center line for the bounding box, with branches coming from the main segment.

Perimeter This solution creates a bounding box around the components selected for the system and proposes several potential routing solutions. You can specify the Inset value that determines the offset between the bounding box and the components. Inset is available only when the Perimeter option is selected.

Intersections This solution bases the potential routing on a pair of imaginary lines extending from each connector for the components in the system. Perpendicular lines extend from the connectors. Where the lines from the components intersect are potential junctions in the proposed solutions along the shortest paths.

Custom The solution becomes available once you begin to modify any of the other solutions.

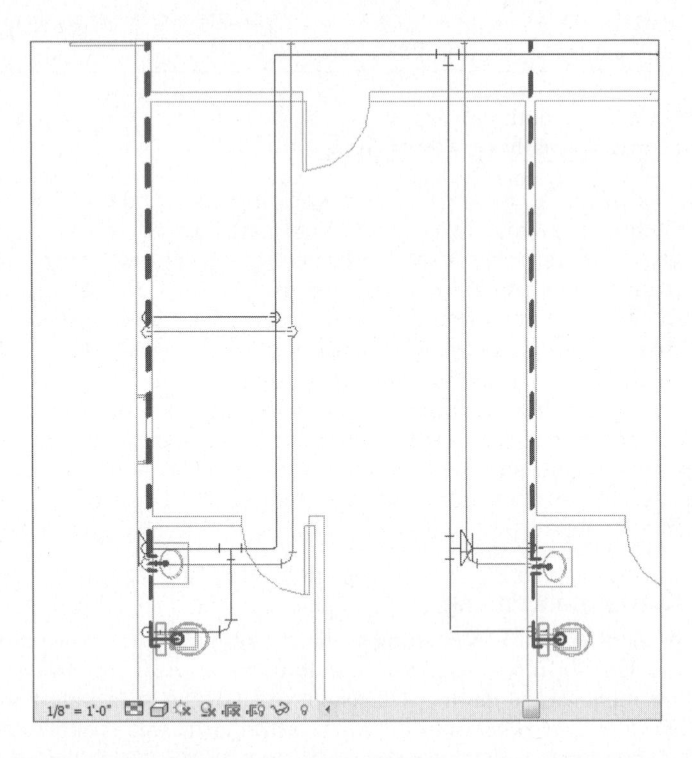

FIGURE 15.26
Generating layouts from systems

SELECTING FIXTURES WHEN AUTOROUTING

When using the autorouting feature, the number-one mistake that everyone makes is to try to select every fixture at one time, this will always lead to failure. The process is to use the Add And Remove From System ribbon of the General Layout ribbons.

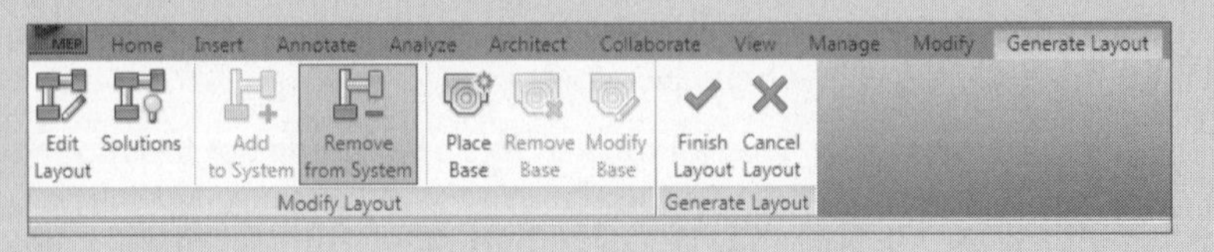

3. By making the autorouting smaller, you can actually modify the runs better. Do this by removing all the fixtures from the Path Layout tool except for the few you want to connect.
4. Place a base to show the end of the run. Make sure this offset is at the same height as your piping, which helps ensures that you are routing your piping at the correct offset.

5. Once you have everything routed, connect the smaller piping systems together to make the overall system.

Manual Pipe Routing

Manual routing of piping is the next method. The advantage of using this option is that the designer has total control over workflow. When routing manually, start your piping out at the elevation that you know will be most likely out of the way of other disciplines. Cutting sections when routing piping can really improve the success of your coordination of your Revit model. Normally we will route our mains first so we can make sure that most of the piping will fit before connecting all the branches to the mains. To route piping, select Home ➤ Pipe ➤ Placement Tools, and be sure to select Automatically Connect.

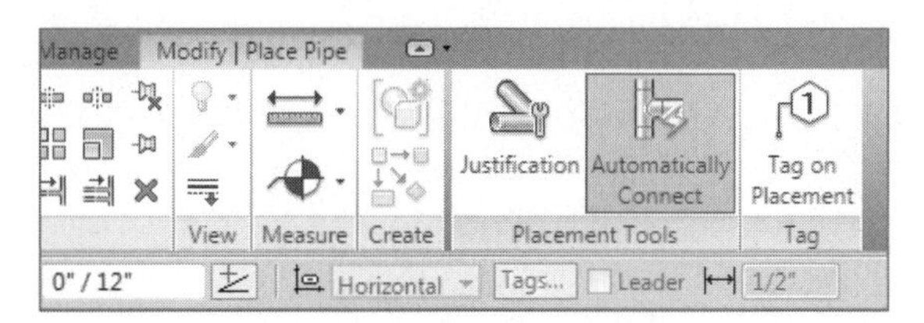

Before repeating this process for all of your plumbing systems, take some time to review the structural, mechanical, and architectural models. This can really cut down on modeling time. After coordinating the pipework and verifying the design flow rates, the layout is complete.

Whether you created your plumbing piping layout with autorouting or manual routing, you need to see whether your piping is reading flow. By using the Pipe Inspector, you can verify if your piping is well connected. To use this tool, click a piece of pipe that has been routed in a complete system. Once the pipe is selected, you will see the Modify Pipes ribbon. From this, select the System Inspector. With the System Inspector selected, pick any section of the pipe, and you will notice the GPM of the system and a direction of flow (refer to Figure 15.27).

Now you are assured your piping is well connected.

SYSTEM INSPECTOR

If your system is not well connected, you will not be able to use the System Inspector. To help troubleshoot the problem, use the Check Pipe Systems command located on the Analyze tab of the ribbon. This will flag possible problems. If any system components are flagged, disconnect your system components, and use the Connect Into tab on the ribbon located under Modify Mechanical Equipment. If that does not work, review your family to make sure it is created properly.

Sloping Pipe

A lot of people think sloped pipe applies only to sanitary sewer. But in most specifications, almost all piping calls for a slight slope to be in the pipe so if systems are drained down, they are less likely to trap water. Vent piping has a slope because it is open to the atmosphere because of the vent having to penetrate the roof. Storm sewer is one of the most critical systems to make sure you have the proper slope because large rain water loads have been known to make roofs collapse. Now that we have you thinking about checking your specifications to see whether you need to slope more piping than just your drainage, let's look at how you will slope your pipe.

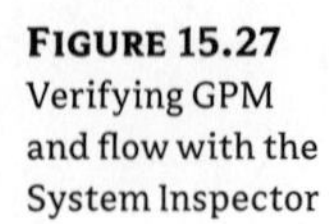

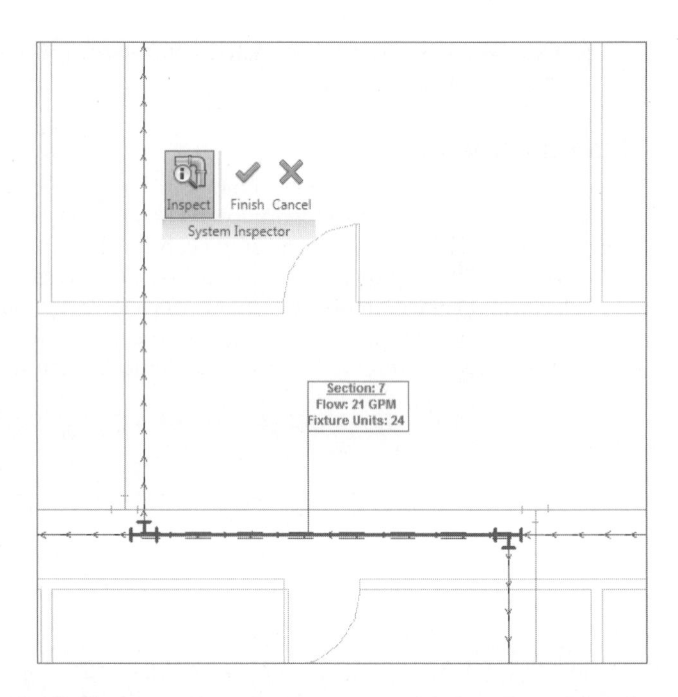

FIGURE 15.27
Verifying GPM and flow with the System Inspector

First you select the method of how you are going to run your piping using either the auto-routing or the manual routing feature. When we lay out sanitary sewer, we always like to manually route our main first to establish where the sanitary is going to be leaving the building and make sure it has been coordinated with the civil engineer.

For sloping pipe to work properly, it needs to have a starting point and an ending point. To set up sloping pipe, open `Chapter 15 plumbing.rvt` and `Chapter 15 base.rvt`. Also download `sanitary point of connection.rfa` and `wall cleanout.rfa`. All of these files are found on `www.wiley.com/go/masteringrevitmep2011`. Next do the following:

1. Make sure you have your view range set properly because you are going to be routing below 0′-0″. To change the view range, type **VP**, which will open the View Properties dialog box. Select View Range, and change Bottom and View Depth to Unlimited (refer to Figure 15.28).

FIGURE 15.28
View Range settings

View Range
Primary Range
Top: Associated Level (Level 1) Offset: 12' 0"
Cut plane: Associated Level (Level 1) Offset: 4' 0"
Bottom: Unlimited Offset: -6' 0"
View Depth
Level: Unlimited Offset: -6' 0"
OK Cancel Apply Help

2. Next look at your routing choices for your main. Then locate your end run fixture, floor cleanout, or wall cleanout.
3. Next locate your sanitary point of connection outside the building locating the sanitary piping at -4′0″ below finished floor.
4. Next select the Pipe toolbar located on the Home tab of the ribbon. Before you start routing your pipe, be sure to set the 1/8″ /12″ slope setting, and make sure you have the justification set to your preference.

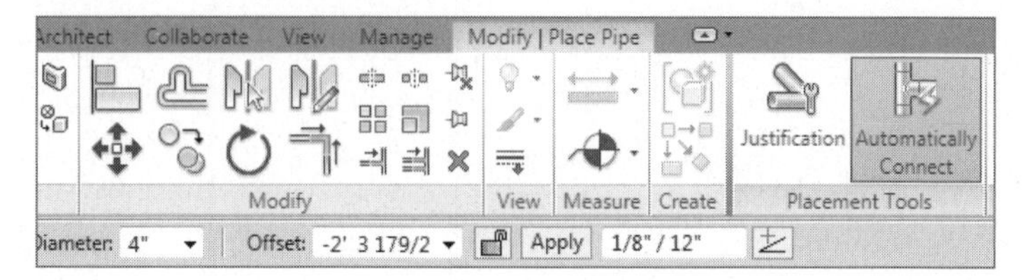

5. Now start your run from the sanitary point of connect, and route into your last fixture or cleanout. You now have a main trunk to build your system off.

> **SLOPED PIPE SYSTEMS ARE LIKE TREES**
>
> When you are routing sloped pipe systems, think about sketching a tree starting from the base of the tree and then from the base to the branches and then from the branches to the leaves, or in this case a water closet. The secret to successfully modeling sloped pipe is to draw your piping from the main first.

6. Select the Pipe toolbar, and then select the point on the main you want to route from. Once you do that hit the spacebar one time; this will make the piping connect at the same elevation as your pipe. Then route out to the water closets or any other components that take 1/8″ slope.
7. Route all of your sloped pipe that contains the same degree of slope first. Then route all of the next matching slope. Having a redundant pattern will add to your efficiency and productivity. Now you should have a sloped system (refer to Figure 15.29).

If you need to track the slope, use an elevation tag. This will allow you to find the invert of the pipe along any point of the pipe. It can be used only in elevations, sections, and 3D views. It does not account for insulation of pipe.

Fittings

Without fittings, piping would not be worth a whole lot. Fittings help shut off, help regulate, help open up, and help save lives. In Revit, most fitting families have the following functions:

End cap These can be placed only at the end of pipe.

Tee, tap, wye, or cross These can be placed anywhere along pipe runs.

Transitions, couplings, or unions These can be placed only at the end of pipe. They are used to join a smaller, larger, or same-size pipe.

Flange These can be placed at the end of pipe or face to face with an another flange.

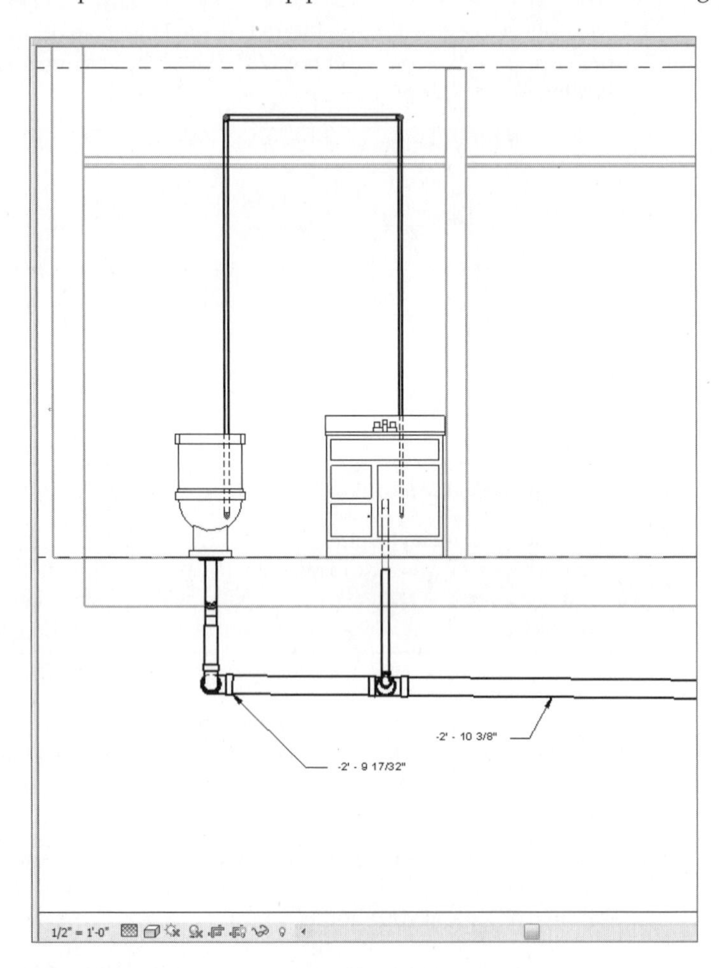

FIGURE 15.29 Sanitary layout with sloped piping

Using Pipe Fitting Controls

Understanding pipe fitting controls can really make life easier if you are routing a lot of piping. There are several fittings that have this ability. When you are laying out your piping, turn 90 degrees to create a elbow. If you click the elbow, you will notice a plus (+) sign. If you click that sign, it will change from an elbow to a tee, allowing you to add more piping and continue your pipe routing. If you select the minus (–) sign, it will downgrade the fitting. When you see the symbol on a fitting, it allows you to rotate the fitting, and the symbol allows you to flip the fitting.

Placing Fittings

When you need to add valves to your piping, select the Home tab, and click Pipe Accessories. Use the Type Selector to select the type of valve you want to use. Most valves are based on "break into" type parameter, so you can place them into the pipe type, and they will break into the piping and connect (refer to Figure 15.30).

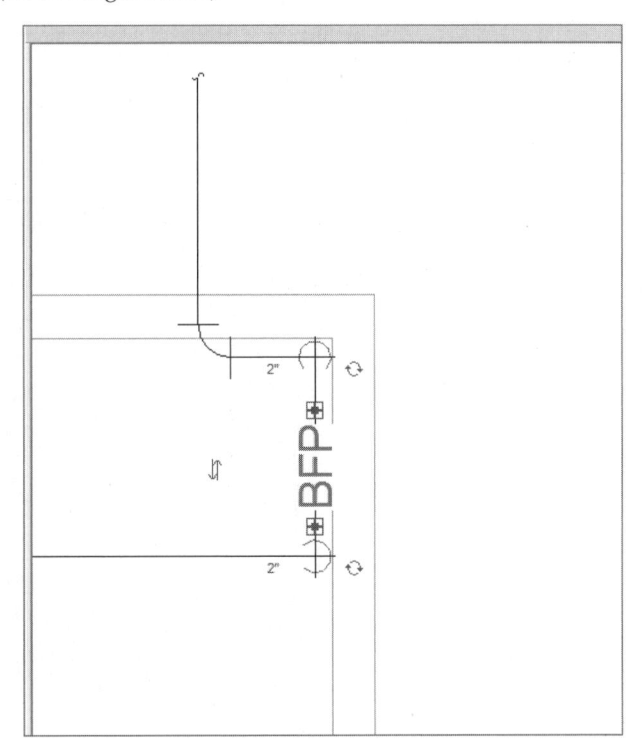

FIGURE 15.30
Fitting breaks into piping system

Real World Scenario

A FITTING END

In early 2008, an AEC firm in Knoxville, Tennessee, was working on a hotel project in Pigeon Forge, Tennessee. This 90-room hotel had already gone over the estimated projected design hours because of constant changes of field conditions, and it was the first hotel produced in Revit for this firm.

The building was a conventional plumbing sanitary with traditional multiple vent stacks. The project went out for bid, and when the bids came back, the building was slightly over the estimated budget, and the design team was asked to figure a way to get the cost back into budget. After carefully reviewing the project, the plumbing designer and engineer had decided to use a PVC version of a So-Vent plumbing system; only two families would have to be created for Revit: an Aerator and a De-aerator fitting. The cast-iron version of the system has been used in northern parts of the United States in high-rise design for years successfully.

The problem faced with using a system like this was that no one had installed such a system in East Tennessee. Plumbing code officials had not seen the system installed, and they stated it did not meet traditional methods of vent design. By using a section under codes that allowed the use of engineered plumbing systems, the design team proceeded to design the system. Because the existing layout was already in Revit, it was the logical choice to redesign the system in it.

Within two working days, the model and contract documents were revised with the new layout and then rebid. One-third of the vent system was reduced, which led to fewer holes being cut, fewer man-hours, and fewer fire stopping costs. Since the plumbing contractor had never installed a So-Vent system before, being able to use printouts of the 3D model helped speed up the installation of the system and assure it was installed properly.

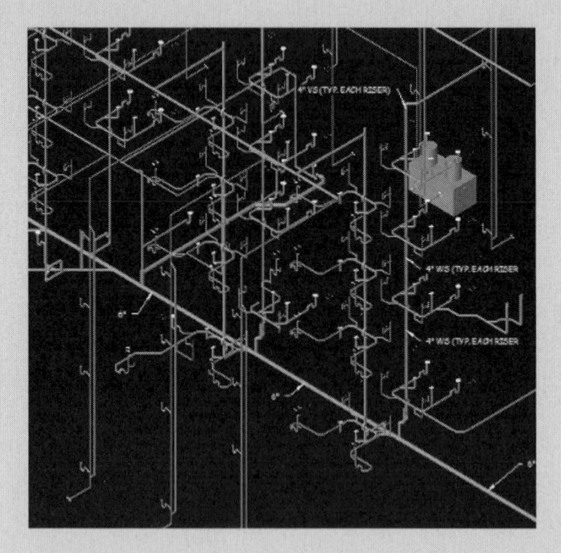

Although the project may have lost money for the firm, the hotel owner was happy with knowing that his construction costs were within budget and that his hotel could open on time.

Visibility of Piping

Now that you have created systems and added the piping, you need to be able to display them in different views, and you may also want them to display with different colors and line weights. If you created your systems correctly, you can go to Filters located on the View tab. Here you are able to create new filters by duplicating and modifying any of the existing ones (refer to Figure 15.31).

Once the filters have been created, go into each view where you want the different disciplines to show up, and type **VG** to bring up the Visibility/Graphics. Click the Filters tab, and add the newly created filters. At this point, you can control the line type, colors, and patterns of the filtered objects to match your company standards. Refer to Figure 15.32.

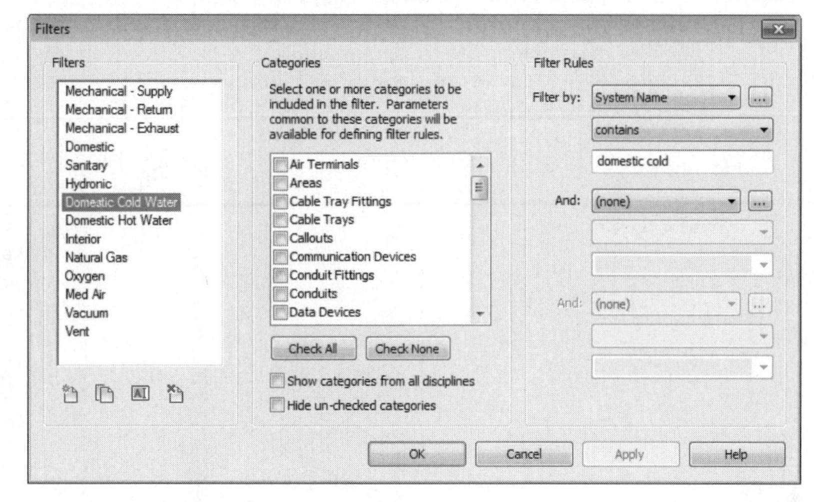

FIGURE 15.31 Modifying the filter settings

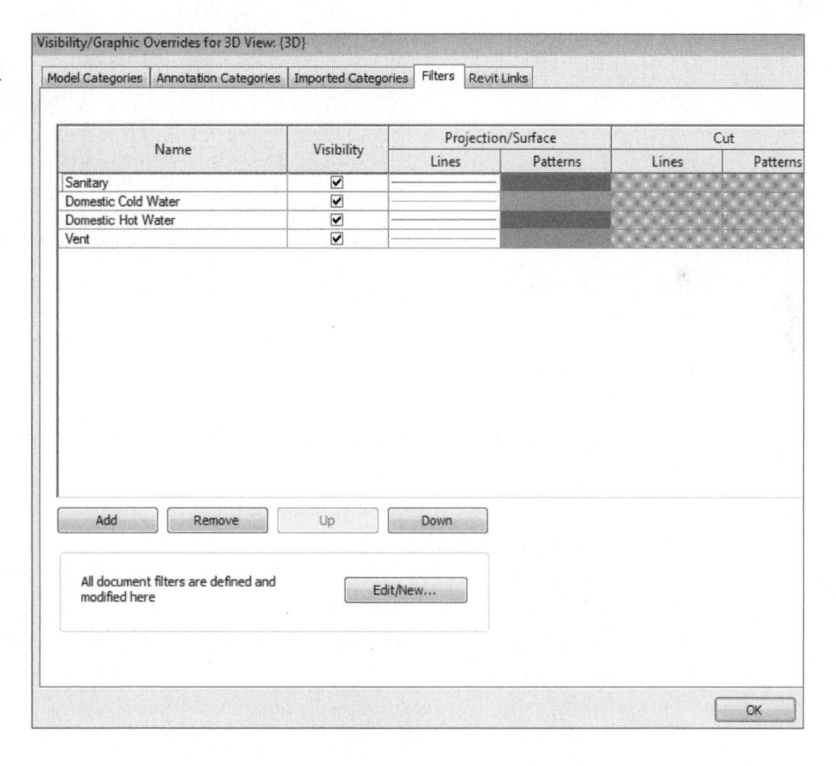

FIGURE 15.32 Adjusting color and line types through Visibility/Graphics

Now that your plans are coordinated and displaying the way you require, you are well on your way to completing your documentation. Also consider using the same filters to display 3D views, which can enhance the design, making it easier to understand and further reducing errors. Refer to Figure 15.33.

FIGURE 15.33
Color-coded 3D views for review

The Bottom Line

Customize out-of-the-box Revit plumbing fixtures for scheduling purposes Learning how to customize existing plumbing fixtures can help with productivity and provide more robust building information.

Master It What are the two types of parameter information that can be scheduled and used in type catalogs?

Use custom plumbing pipe assemblies to increase speed and efficiency in plumbing layouts Sometimes you are required to think outside the box and learn how to use new tools to get the most benefit out of them. By utilizing the new Copy/Monitor feature when creating custom pipe assemblies, you can take production to the next level.

Master It Why are nested pipe assembly families better to use than the modeled pipe assembly for a BIM project ?

Adjust and use the plumbing pipe settings Piping settings are crucial to the ability to have Revit MEP model your plumbing layout, the way it will look, and the way it will perform.

Master It Do fitting parameters have to be set up in the system pipe types?

Select and use the best pipe routing options for your project When using Revit MEP 2011 for your plumbing layouts, you must understand the functions of auto pipe routing, manual pipe routing, and sloping pipe. Once these functions are mastered, then the user can lay out any type of piping system.

Master It A plumbing designer has just been asked to lay out a sloped plumbing system and has only a day to pipe up a clubhouse. Where should he start his pipe route first?

Adjust pipe fittings Pipe fittings are needed in systems to make the systems function properly and to produce documentation for construction. Being able to add or modify fittings can increase productivity.

Master It You have just finished your modeled layout and given it to your employer for review. He's just came back and now has asked you to remove a couple of elbows and replace them with tees for future expansion. What would be your method to accomplish it quickly?

Adjust the visibility of pipes Being able to adjust the visibility gives the plumbing designer or user the ability to set up multiple views and control the graphics for documentation.

Master It There are too many systems showing up in your views. What would you do to show only one piping system in that view?

Chapter 16

Fire Protection

Fire protection is probably one of the least mentioned features in Revit. Fire protection designers use a variety of methods and software to lay out fire protection systems. There are considerable benefits to doing this process in Revit MEP 2011, including the coordination and clash detection with other services and building elements.

In this chapter, you will learn how to:

- Place fire protection equipment
- Create a wet fire protection system
- Route fire protection piping

The Essentials of Placing Fire Protection Equipment

Proper planning of placing fire protection equipment is essential when trying to create a productive layout with Revit MEP 2011. You should plan to have most of your equipment spotted during the schematic design phase of the project, which helps with productivity and coordination with other disciplines. You will need to use proper design methods to verify whether a fire pump is required on a project.

Although pump manufacturers are starting to provide Revit content, they are still few and far between. If you look under Fire Protection ➪ Imperial Library, you will find several components that can be used out of the box for fire protection; others can be found as Mechanical components or under Piping. For example, the backflow preventer is located under Imperial Library ➪ Pipe ➪ Valves ➪ Backflow Preventers.

Point of Connection

You will want to start your model by understanding where your water supply starts. Normally the architects will provide location details, and you can then display this information in your design model by either creating a water meter family or modifying an end cap family. To modify an existing family to indicate the water inlet point, do the following:

1. Open the `Chapter16_Dataset.rvt` file found on `www.wiley.com/go/masteringrevitmep2011`.
2. In the Project Browser, scroll down to the Families section, expand the Pipe Fittings category, and select Cap – Generic.

3. Right-click the family, and select Edit. This opens the Family Editor.
4. Click the Revit Home button, and select Save As ➪ Family. Save this family as **`Fire Protection Point of Connection.rfa`** in your My Documents folder.
5. Edit the newly created family by changing the family type from Pipe Fitting to Mechanical Equipment, located under Modify ➪ Properties ➪ Family Categories And Parameters.
6. Once the family is complete, select Family Types, and add three new parameters: Static Pressure, Residual Pressure, and Gallons Per Minute.

When creating these new parameters, be sure to use the piping discipline and appropriate units. Also, determine whether shared parameters would be applicable in this situation (see Chapter 19 for more information). You can leave the end cap the way it is modeled, or you can use model lines with an ellipse to create a break line to show up in single line piping, as in Figure 16.1.

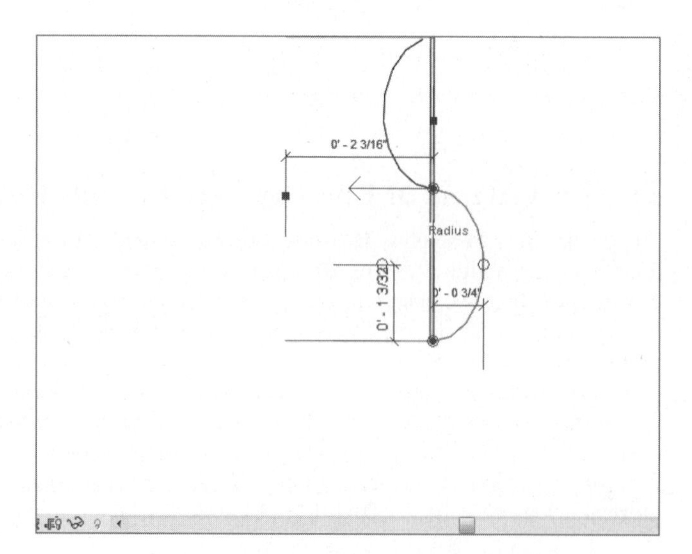

FIGURE 16.1
Fire protection point of connection

7. Select the connector, and change its type from Fitting to Fire Protection Wet.
8. Any existing line work can either be deleted or be changed to Invisible.

Fire Pump Assembly

You should try to preassemble as much of the fire protection components as possible to help with production time. Figure 16.2 shows a fire pump preassembled so one would only have to change out certain components, such as changing the pump for a smaller or larger pump depending on what the calculated fire flow demand calls for.

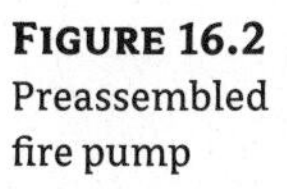

FIGURE 16.2
Preassembled fire pump

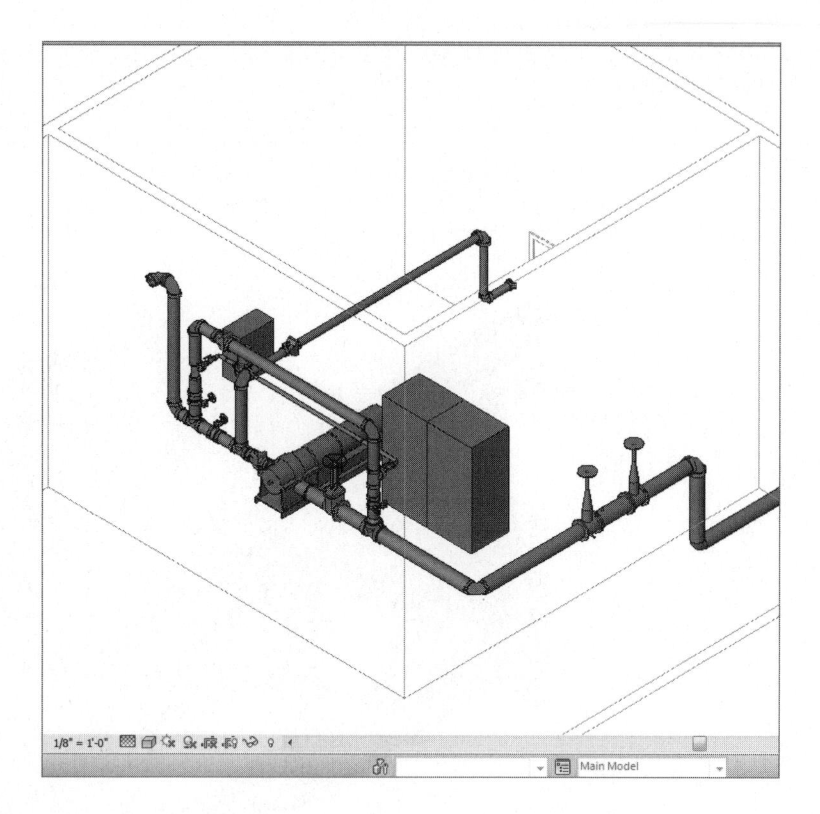

To create a preassembled fire pump, do the following:

1. Save a Revit file named `pump assembly model`.
2. Connect as much of the pump assembly as possible, taking into account where most of the components will likely be placed. You can use a split case pump that comes with Revit MEP as your base fire pump. This will normally give a large enough footprint once every piece has been assembled, but always verify the size of the equipment with the manufacturer's cut sheets to keep from making a costly mistake.

You can use an inline pump to represent a jockey pump, because you will find it matches closely in size. The inline pump is located under Imperial Library ➪ Mechanical Components ➪ Water Side Components ➪ Pumps.

Some of the harder items to find information about are the control panels because of proprietary components. For these, you can create a family using an Electrical Equipment family type. Another option for showing the control panel is to show it as an in-place component (see Figure 16.3).

This would be used as a placeholder to assure that the control panel is accounted for and so that electrical information may be added later through a electrical connector located by the electrical engineer.

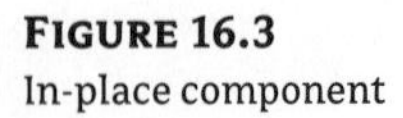

FIGURE 16.3
In-place component

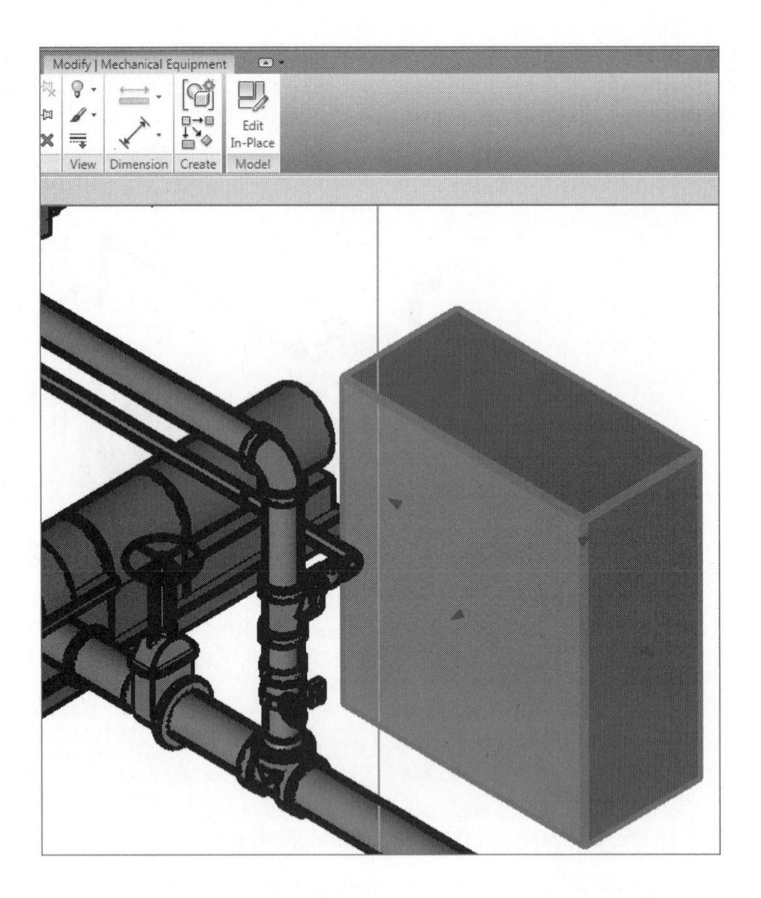

OPTIONS FOR ADDING CONNECTORS

When using in-place components, there are options for adding different types of connectors such as electrical connectors. Be aware these options may not work properly because of a known bug in the software. Only mechanical equipment and plumbing fixture families work properly. Because of this, it is recommended to use actual mechanical equipment families instead of in-place components if at all possible.

3. Once you have your layout the way you want it, you can save the model for future use. Select the elements required, and click Create Group. When prompted for a name, be sure to give it a suitable one — for example, `Fire Alarm Pump Set`, *not* `Group 130`.
4. Here the group insertion point can be modified and then saved as a library group. This can then either be loaded as a model group or be linked and bound.

Using the link method will give far more flexibility when positioning this object — as long as it is subsequently bound to allow access to the connections. Be aware that any hosted families used in this process will lose their associated host and will require rehosting.

The copy-and-paste method will give you a warning (see Figure 16.4), while the link-and-bind method will not. It is left to the user to check elements for hosting, although they retain their location properties correctly.

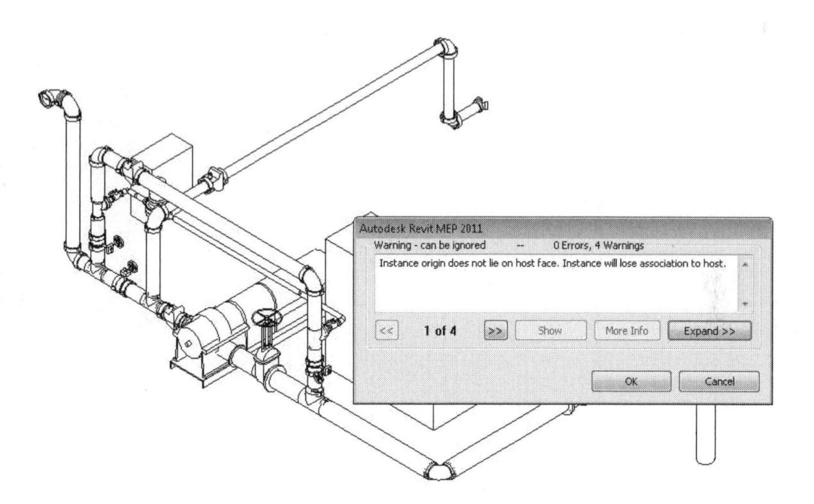

FIGURE 16.4 Copying and pasting fire pump into a new project

Creating a Fire Riser Assembly

Fire risers for most small projects are typically assembled in the same manner most of the time. The ideal way to handle assemblies like this would be to create the family as Mechanical Equipment. The reason for this is that during schematic design, placing the fire riser is crucial for understanding where the fire line will need to be routed (see Figure 16.5).

This family is created by nesting pipe fitting and pipe accessories families. To review what components make up this family, do the following:

1. Open `6 Inch Fire Riser.rfa` located at `www.wiley.com/go/masteringrevitmep2011`.
2. There are several different pipe fittings and pipe accessories families that are nested into this family and that make up the riser. They are as follows: `Pipe Elbow.rfa`, `Pipe Tee.rfa`, `Alarm Pressure Switch.rfa`, `Ball Valve - 2.5-6 Inch.rfa`, `Check Valve - 2-12 Inch - Flanged.rfa`, `Double Check Valve - 2.5-10 Inch.rfa`, `Multi-Purpose Valve - Angle - 1.5-2.5 Inch - Threaded.rfa`, and `Plug Valve - 0.5-2 Inch.rfa`. You can insert these from the Pipe Fitting and Pipe Accessories directory located in the Imperial Library.
3. When placing the fittings together, make sure to align, lock, and dimension each fitting together. If you leave out this step, the fittings will pull apart if you add parameters to make them flex. Refer to Figure 16.6.

FIGURE 16.5
Fire riser assembly

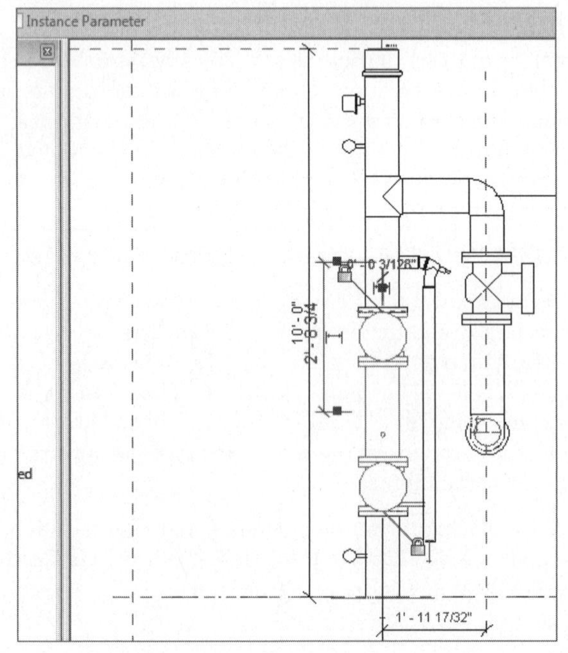

FIGURE 16.6
Fittings aligned, locked, and dimensioned to keep nested families in their place

4. Since system piping cannot be routed in the Family Editor, you will need to create extrusions for the piping sections and then just add the fire protection connectors. When adding the connectors, make sure the arrows are pointed in the direction of connection.

By assembling the riser together, you can coordinate the location it was be installed in and then start planning how to route your piping.

Sprinkler Heads

Since you now know about fire pump assemblies and how to create a standard fire riser, you will start planning for the type fire protection sprinkler heads you will need to use for your model. Within Revit MEP 2011, there are several different types of sprinkler heads to choose from. The different family types of sprinkler heads are hosted and nonhosted.

Hosted sprinkler heads are normally face-based families. When using these types of families, you will need to locate them on a surface. These locations depend upon the installation and the type of sprinklers, which could be Wall, Ceiling, Slab, or Soffit mounted. These surfaces can be part of the linked architectural model or walls/ceilings that have been created in your file using the tools located on the Architects tab.

A Strategy for Sprinkler Heads

You can use the reflected ceiling of a linked architectural model to coordinate the location of your sprinkler heads and to provide a surface for them to mount to by creating a fire protection ceiling plan. Type **VG** while in the model view or select View ⇨ Graphics ⇨ Visibility/Graphics, and then select the Revit links and change By Host View to Linked View. Now your view reflects what the architect has in their reflected ceiling plan.

Nonhosted sprinkler families must have the offset height parameter set to locate the heads at the proper elevation (see Figure 16.7).

Figure 16.7
Nonhosted sprinkler heads

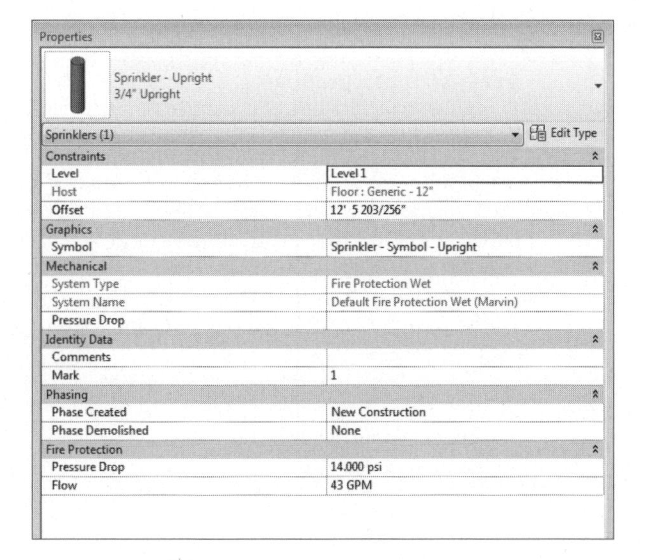

Upright sprinkler heads normally are nonhosted because they are located in spaces that do not have ceilings such as storage rooms or mechanical closets. If you do not set the offset height, then the heads will come in at a default of 0′, which could locate the heads on or below the floor level.

Creating Fire Protection Systems

There are several options for the type fire protection systems that can be created. They are as follows:

Fire protection This can be used for the building sprinkler piping, or it can be used for the utility fire protection coming into your building to connect the base of your fire protection riser.

Wet fire protection This pipe system type normally is used for the layout of the piping from the riser to the sprinkler head layout.

Dry fire protection This pipe system is used for layout from the fire riser to the sprinkler head or standpipe to keep the system from freezing.

Preaction This pipe system can also be used for a deluge system.

Fire protection other This pipe system can be used for a glycol antifreeze system and also can be used for a chemical suppression system.

You can also refer to piping systems in Chapter 8 for more information. When creating a fire protection system, one thing to remember is that the system will not calculate and autosize like domestic water systems will. The main reason is that fire protection systems have no true way of selecting and calculating which heads are in the highest demand.

Strength in Numbers

John's employer has just came back from a meeting with a high-profile client and has sold their ability to produce a fire protection model in BIM. His employer has stated that the client wants to see the total of flow for the highest demand on the system so their building can pass the fire inspection requirements as mandated by the local fire marshal. John has already calculated the system load so he knows where the highest demand shall be located. John decides that the easy way to accomplish this is to add `calc-gpm` in the `comments` parameters of the sprinkler heads and then filter these heads through a schedule with a grand total of GPM. The inspector reviewed the schedule and approved the client's building, and a satisfied client paid the bill.

To replicate what John did, you can do the following:

1. Select the sprinkler heads that need to be modified, and add `calc-gpm` into the `comments` parameters.

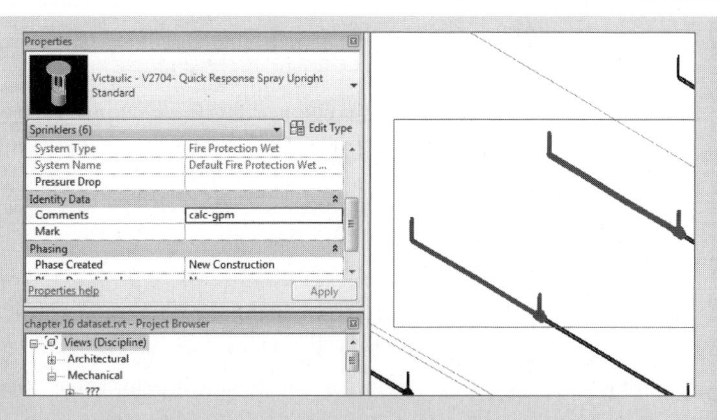

2. Go to the Analyze tab on the ribbon, and then select Schedule/Quantities, which will open the Schedule dialog box. Select Sprinklers from the Category group, and then select OK.

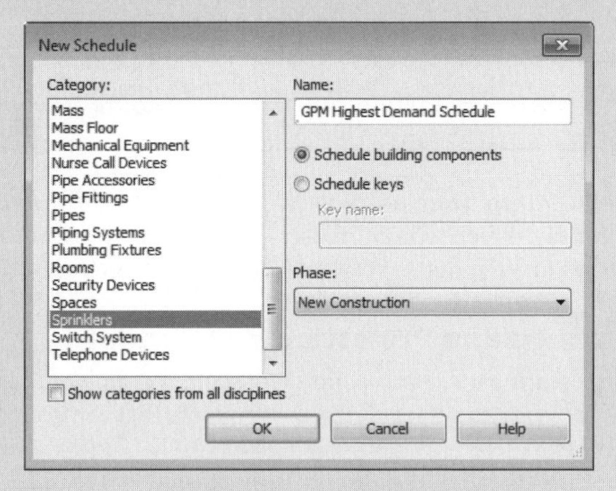

3. Next select Flow And Comments from the Available Fields dialog box, and add it to the Schedule Fields (In Order) dialog box. Then click OK, which will create your schedule.

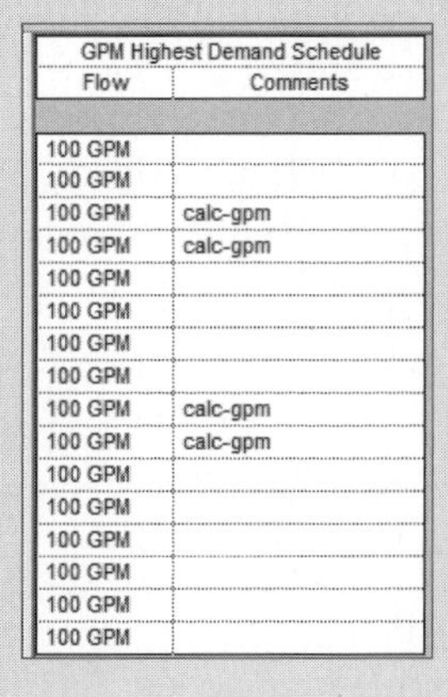

GPM Highest Demand Schedule

Flow	Comments
100 GPM	
100 GPM	
100 GPM	calc-gpm
100 GPM	calc-gpm
100 GPM	
100 GPM	
100 GPM	
100 GPM	
100 GPM	calc-gpm
100 GPM	calc-gpm
100 GPM	
100 GPM	
100 GPM	
100 GPM	
100 GPM	
100 GPM	

4. Now that you have the sprinkler schedule, you want to total only the sprinklers with comments. To do this, there are several parameters you have to set.
   a. First go to Properties browser, select Filter tab, and then change the filter to Comments.
   b. Set the parameter equals to `calc-gpm`.
   c. Select Sorting/Grouping, and change the Sort By pull-down menu to Comments.
   d. Select Itemize Every Instance.
   e. Select Grand Totals while selecting Totals Only.
   f. Select the Formatting tab, highlight Flow, and check Calculate Totals. Next highlight Comments, and select Hidden Fields.

Now your schedule will show only the items you want to show up, including the total GPM flow.

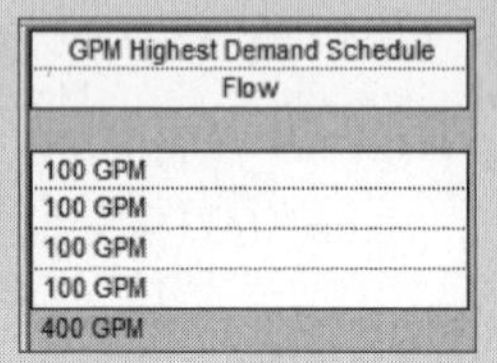

GPM Highest Demand Schedule
Flow
100 GPM
100 GPM
100 GPM
100 GPM
400 GPM

If the system were to try to calculate by GPM, it would account for every sprinkler head on the system, which would grossly oversize the system. Also, the fire protection system at this time has no effective way of calculating the water pressure as it goes higher in elevation.

Creating a Fire Protection Wet System

When creating a fire protection wet system or one of the systems previously mentioned, you will first want to select all the components that are going to be associated with that system (see Figure 16.8).

In case a system has already been started, you can add to the system by selecting a component on the system. Click Piping Systems ➪ Edit System ➪ Add To System, and then window all the items you want to add to the system. If this is done correctly, you should see all the items in the System Browser under the system you created (see Figure 16.9).

Filtering Fire Protection Systems

Once your fire protection system is created, you will need to make sure the piping will filter correctly. Using filters correctly (that is, by assigning colors and line types to piping) will help keep your systems organized. To create a fire protection filter, do the following:

1. On the View tab, select Visibility/Graphics ➪ Filters. The default keyboard shortcut VG will take you to the same location.
2. Click Edit/New, which will open the Filter Settings dialog box. You should notice that there are a number of filters already created. Select the Domestic Hot Water filter, right-click, and select Duplicate. Right-click again, and rename the filter to the type piping system you are filtering. In the right corner, you will see a dialog box named Filtering Rules. Change the rule from System Type to System Name. Refer to Figure 16.10.

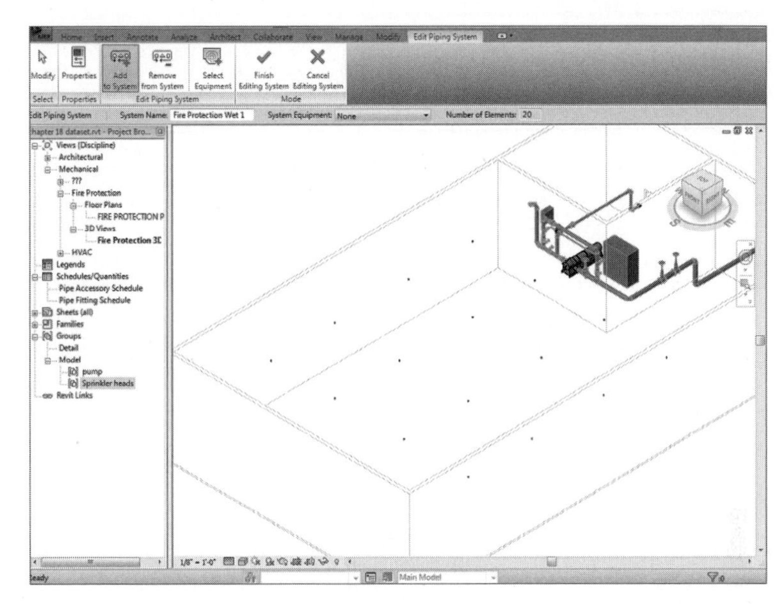

FIGURE 16.8
Adding sprinkler heads to a system

FIGURE 16.9
Fire protection system in System Browser

chapter 18 dataset.rvt - System Browser

Systems	Flow	Size	Space ...	Space ...
Mechanical (0 systems)				
Piping (1 systems)				
Fire Protection Wet				
Fire Protection Wet 1	N/A			
Fire Department Inlet Con...	N/A	4"		
Fire protection point of c...	1500 GPM	8"		
Inline Pump: 1.5"	N/A	1 1/2"		
Split Case Pump - Horizo...	N/A	8"		
Victaulic - V2704- Quick ...	100 GPM	1/2"		
Victaulic - V2704- Quick ...	100 GPM	1/2"		
Victaulic - V2704- Quick ...	100 GPM	1/2"		
Victaulic - V2704- Quick ...	100 GPM	1/2"		
Victaulic - V2704- Quick ...	100 GPM	1/2"		
Victaulic - V2704- Quick ...	100 GPM	1/2"		
Victaulic - V2704- Quick ...	100 GPM	1/2"		
Victaulic - V2704- Quick ...	100 GPM	1/2"		
Victaulic - V2704- Quick ...	100 GPM	1/2"		
Victaulic - V2704- Quick ...	100 GPM	1/2"		
Victaulic - V2704- Quick ...	100 GPM	1/2"		
Victaulic - V2704- Quick ...	100 GPM	1/2"		
Victaulic - V2704- Quick ...	100 GPM	1/2"		
Victaulic - V2704- Quick ...	100 GPM	1/2"		
Victaulic - V2704- Quick ...	100 GPM	1/2"		
Victaulic - V2704- Quick ...	100 GPM	1/2"		
Electrical (0 systems)				

3. Rename Domestic Hot Water 1 to Fire Protection (or whatever you named your system), select Apply, and close. Now you will see the Visibility Graphics dialog box with the Filters tab. Select Add, and you should see the newly created filter. Refer to Figure 16.11.

FIGURE 16.10
Verifying the system name

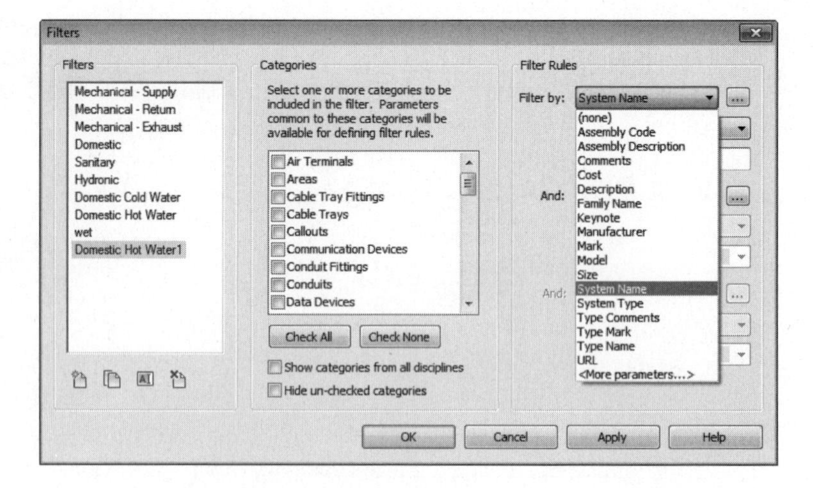

FIGURE 16.11
Matching the system name in the filter to the created system name in the model

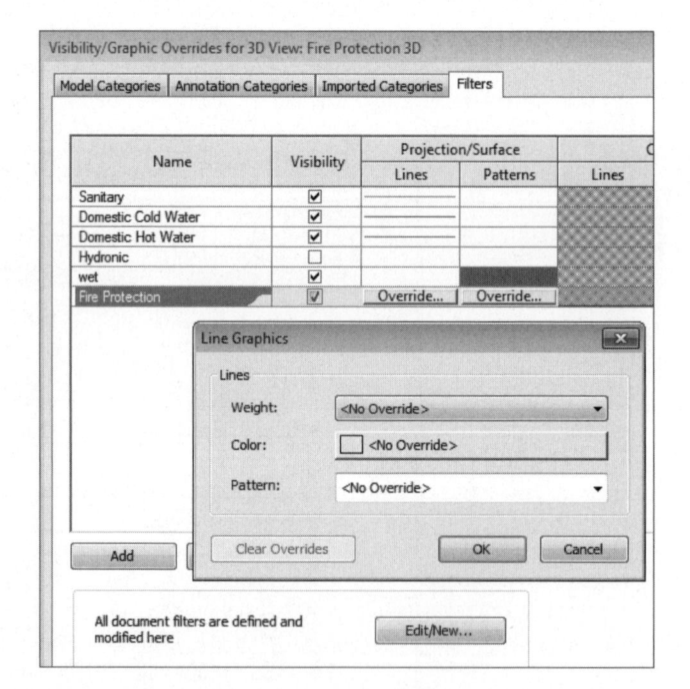

NAME FILTERS AND SYSTEMS THE SAME

Always remember to name your filter exactly the way you named your system, because filters are case sensitive. Incorrect naming because of misspelling or incorrect case are two of the main reasons for filters not working.

Deciding what colors and line types you want for the fire protection will most likely come from existing CAD standards. Once the filters are created and applied to the views, save them as view templates, and load them into the main templates. This will add to the ease of completing a design in an efficient manner.

Fire Protection Piping

When setting up piping to route, you will want to apply the proper pipe material. There are several areas you will have to adjust to set this up properly. These areas are system pipes, fittings, pipe material, pipe sizing tables, and fluids table. Once you set these areas, you can then concentrate on the autorouting and manual routing of pipe.

System pipes These are the pipes that are hard-coded into Revit. You have a limited amount of freedom to adjust parameters for these system families.

Fittings These can be applied to the parameters of the system pipes, which will allow for fittings to populate the model automatically. The fitting must be loaded into the model for them to work.

Pipe material This is set by selecting Mechanical Settings ➪ Sizes. This allows you to duplicate a pipe material, rename it, and apply the piping specification of material and roughness of pipe wall.

Pipe sizing table This is set by selecting Mechanical Settings ➪ Sizes. This allows you to duplicate and adjust the inside pipe diameter, outside pipe diameter, and pipe sizes to the manufacturer's specifications.

Fluids table This is set by selecting Mechanical Settings ➪ Sizes. This allows you to duplicate, rename, and adjust the fluid type, as well as the viscosity, temperature, and density.

So, what do you do if you require special fittings? It's quite common to see mechanical joints required on fire protection systems. Since they do not exist in the out-of-box content of Revit MEP 2011, you are stuck with three choices.

First, you can use regular fittings and then copy and rename them to the type fittings you need, as was demonstrated the fire pump. Then you can just use schedules to count the number and make of fittings.

Second, you could create your own, that is, if you have the budget to create every fitting you need (even though it is worth it in the long haul).

Third is to find a manufacturer that has already developed their content. As luck would have it, Victaulic has most if not all of their products in Revit on its website at `www.victaulic.com`. You can download them and load the fittings you need for your layout. Once they are loaded into your model, duplicate your system piping material, and apply the fittings as needed (refer to Figure 16.12).

Copying the Lookup Tables

When downloading or developing your own fittings, be sure to copy the lookup tables into the lookup table folder that the `revit.ini` file calls for; if not, the auto adjustment of fittings may not work properly.

FIGURE 16.12
Applying fitting parameters

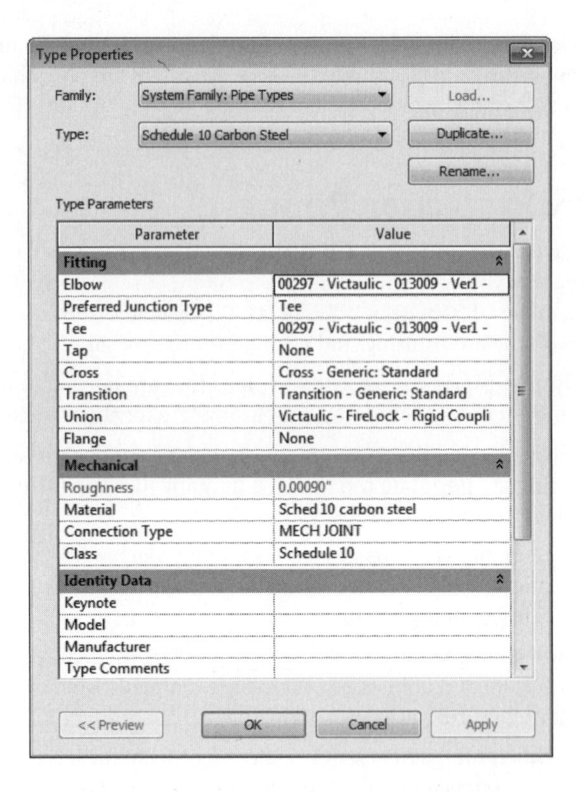

Auto Pipe Routing

Now you should be ready to route piping. Select one of the sprinkler heads that you have added to the fire protection system, and then press the Tab key a couple of times until the autorouting features highlight its suggested path. Or you can select Generate Layout under Modify Sprinklers. (Refer to Figure 16.13.)

You have four options to select for generating the layout: Network, Perimeter, Intersections, and Custom. Each one has several routing solutions to choose from.

Network This solution creates a bounding box around the components selected for the piping system and then bases several solutions on a main segment along the center line for the bounding box, with branches coming from the main segment.

Perimeter This solution creates a bounding box around the components selected for the system and proposes several potential routing solutions. You can specify the Inset value that determines the offset between the bounding box and the components. Inset is available only when the Perimeter option is selected.

Intersections This solution bases the potential routing on a pair of imaginary lines extending from each connector for the components in the system. Perpendicular lines extend from the connectors. Where the lines from the components intersect are potential junctions in the proposed solutions along the shortest paths.

Custom The solution becomes available once you begin to modify any of the other solutions.

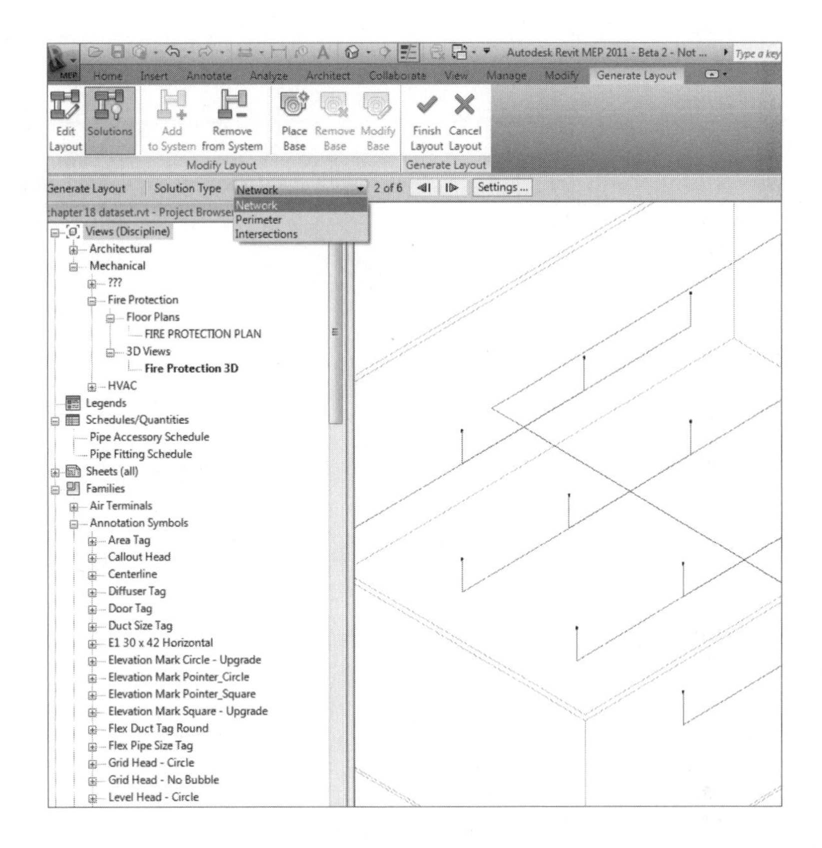

FIGURE 16.13
Autorouting path

Autorouting is great for generating layouts very quickly, but it may not give you ideal results and could lead to substantial rework. There are a variety of options for splitting and redirecting main runs, but these tools do require some practice.

Manual Pipe Routing

The manual routing of piping will give you the missing control. When routing, manually start your piping at the elevation that you know will be most likely out of the way of other disciplines. Cutting sections when routing piping can really improve the success of your coordination of your Revit model. Lay out the mains first, ensuring that most of the piping fits. Then work on connecting the heads to the mains. The adage of "measure twice, cut once" still applies. Taking a few minutes to review structural, mechanical and architectural models can really cut down on modeling time. After you coordinate and verify your routing, finish your layout. (Refer to Figure 16.14.)

If your systems were configured accurately, your piping will be filtered to the colors and line types selected.

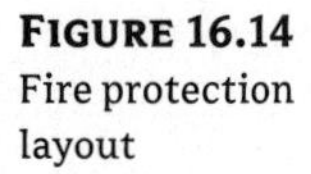

FIGURE 16.14
Fire protection layout

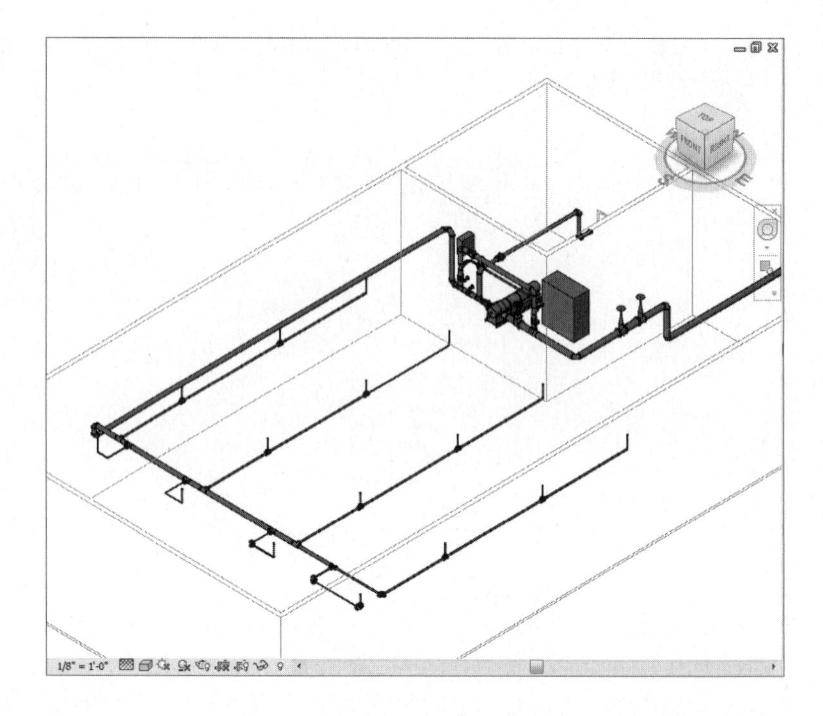

The Bottom Line

Placing fire protection equipment When starting a fire protection model, placing the equipment can make or break your design. The ability of Revit to verify your layouts early, through the coordination of this equipment with other disciplines, can set the pace for a successful project.

Master It What is a method to help speed up production when using a standard fire riser on multiple buildings?

Create fire protection systems Creating proper fire protection systems is essential to the performance and behavior of the fire protection model. Properly created fire protection systems also help with being able to coordinate with other disciplines during design.

Master It Marty has just created a fire protection system name called Wet1, and he has created a filter system type named wet1. Now Marty is in a presentation, and his system is not filtering properly. What should he look at first? What should he do if there is a second problem?

Route fire protection piping Fire protection piping can be routed by a couple different methods. It can be set up with different materials to help with takeoffs and specifications. Once piping has been routed, it can be coordinated with other disciplines to reduce errors and omissions.

Master It What are some of the methods to deal with fittings that may not be supplied with Revit MEP 2011?

Part 5

Managing Content in Revit MEP

Chapter 17

Solid Modeling

It may seem like being able to efficiently and effectively create solid models would be more appropriate for disciplines such as architecture and structural engineering; however, with these skills, you can create the types of components needed to accurately convey your design. MEP families such as pumps, condensers, valves, and even lighting fixtures can have very complex shapes or structure, and knowing how to create these objects with a minimum amount of effort is key to being productive while remaining accurate with your design.

Although MEP components are often very complex, it is important to keep your families as simple as possible while still making them recognizable and useful. This does not mean that all of your objects should be boxes and cylinders, but it also does not mean that you should be modeling the rivets, screws, handles, and hinges either. As your modeling skills develop, you will find the balance that works best for your workflow.

Knowing how to create solid model objects that are accurate to the specified components is equally important as making them recognizable. Allowing for parametric changes will make your families universally applicable to your projects and design standards.

In this chapter, you will learn to

- Model solids and voids
- Orient geometry using reference planes and lines
- Ensure the parametric change capability of models
- Determine optimal visibility settings for solids

Solids and Voids

The first step to success in creating solid model objects is to be able to think in 3D and to visualize the object you are attempting to create. It may be helpful to first sketch out the basic shapes needed to build the solids and voids that will make up a family. This will help you determine where to start and what the relationships between multiple solids will be. It can also help you understand how the object needs to be modeled in order to be used properly in your projects.

How you approach building a solid model may depend on the type of family you are creating. Face-hosted families are commonly used for MEP components, and when creating solids or voids in a face-hosted family, you need to consider their relationship to the host extrusion. The same is true for element-hosted families such as wall- or ceiling-mounted equipment. Building solid geometry in nonhosted families is similar when considering how the family will be placed into a project. Because the tools are the same in any type of family, a nonhosted environment will be used for most of the examples in this chapter except when discussing relationships to hosts. Much of the work described in this chapter is done in the Family Editor environment.

IN-PLACE FAMILIES

Although Revit MEP 2011 gives you the ability to create a family directly in your projects, this feature tends to be more useful for the architectural and structural disciplines.

The availability of this feature can lead to display and feature anomalies if you happen to be collaborating with an architect/structural engineer. A good example of this is if the architect models a wall in-place. This cannot be copy/monitored into the MEP file and can have detrimental effects on the performance of all models. If possible, talk to the project architect and try to figure out a better solution for the project.

MEP families are usually distinct in their form and function and component families can easily be found or created to represent equipment or fixtures, so the need to model in place is reduced for MEP disciplines.

Extrusions

While working in the Family Editor, the Home tab of the ribbon contains the tools necessary for creating component families. The Forms panel of the Home tab holds the tools for building solid model geometry. In essence, all the tools are for creating extrusions; they just vary in their functionality. The Extrusion tool is the most commonly used because it is the most basic method for generating a solid. The idea behind creating a solid with the Extrusion tool is that you are going to sketch a shape and then extrude the shape to a defined depth.

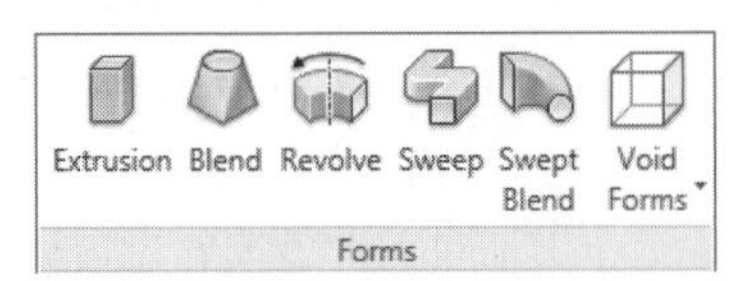

When you click the Extrusion button on the Forms panel, the Modify | Create Extrusion contextual tab appears on the ribbon. This tab contains the tools needed to create the shape of the extrusion. The same drafting tools that are available for any common drafting task are available when sketching the shape for solid geometry. When working in sketch mode, the reference planes and any other graphics in the view will be dimmed, and line work will be magenta as a visual indicator that you are working in sketch mode.

The sketch you draw for the shape of an extrusion can be as simple as a circle or as complex as you can imagine. The most important thing is to create a closed loop with your line work where the lines do not intersect. You will receive an error message when you attempt to finish a sketch that does not form a closed loop. The error dialog box allows you to continue sketching so that you do not have to start over, or you can quit the sketch, which will discard the work done while in sketch mode and return you to the Family Editor.

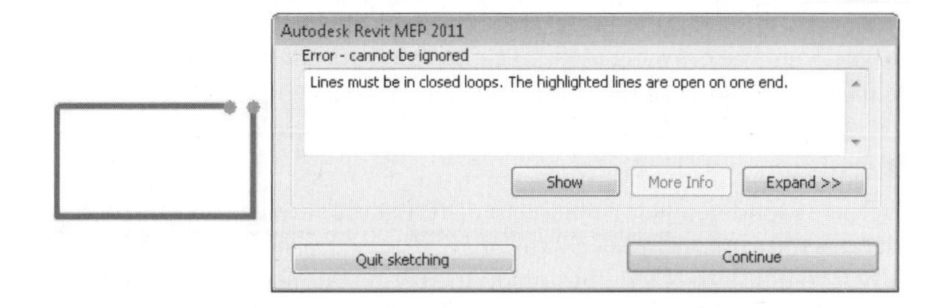

The sketch for an extrusion does not have to be a single continuous set of lines. You can draw several shapes for a single extrusion as long as each shape is drawn in a closed loop. You cannot draw shapes that intersect each other. When you draw one shape inside another, Revit will extrude the area between the two shapes, as shown in Figure 17.1 where one rectangle is drawn inside another. This is a very useful method for reducing the need for void geometry in a family, because having a lot of voids in a family can adversely affect file performance.

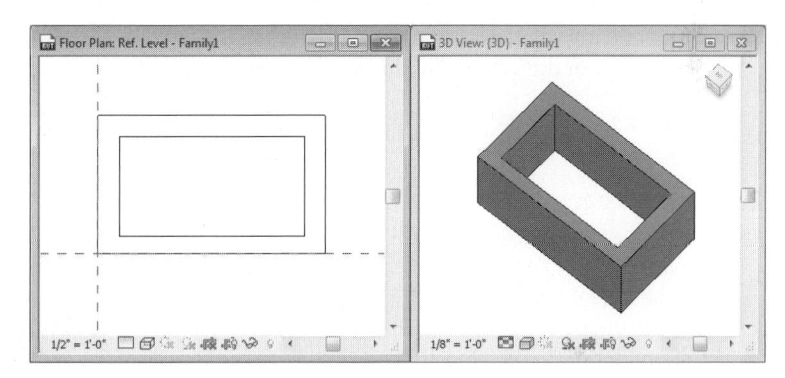

FIGURE 17.1
Multishape extrusion

As you are working in sketch mode, you can define the properties of the extrusion prior to completing the sketch. Your sketch will define the shape of the extrusion, but you also need to determine the distance of the third dimension for your solid geometry. The Depth setting on the Options Bar allows you to set the distance for the extrusion, or you can use the Extrusion End and Extrusion Start parameters visible in the Properties palette while sketching. These values can be changed after the sketch is completed and the extrusion is made, but it is helpful to know the value prior to completing the sketch. The default value is 1′-0″ (300 mm), but once you change it, the value you input will be used for the next extrusion until you exit the Family Editor.

When you have completed the sketch for an extrusion, you must click the green check mark button on the Mode panel of the contextual tab in order to exit sketch mode and return to the Family Editor. If you want to cancel the creation of the extrusion, you can click the red X button on the Mode tab.

You can edit the shape of an extrusion by clicking it and dragging the grips that appear at each shape handle of the solid geometry. This includes changing the depth of the extrusion, which can be done by editing it in a view perpendicular to the view in which the extrusion was sketched. Using grips for editing is an easy but also inaccurate method for changing the shape of an extrusion. To apply specific dimensions, you can edit the sketch by clicking the Edit

Extrusion button on the contextual Modify | Extrusion tab that appears when you select the extrusion in the drawing area. Clicking this button activates sketch mode and allows you to make changes to the shape handles of the geometry using dimensional input.

Blends

A *blend* is solid geometry that has a different shape at each end. The approach to creating a blend is the same as for an extrusion, with an extra step for creating the shape for each end. When you click the Blend button on the Forms tab, sketch mode will be activated, and you can begin creating the shape for the base of the blend. The base of the blend is the face of the solid geometry that is drawn on the reference plane that the blend is associated with. This is also referred to as First End in the properties of the blend. The sketch for the base geometry of a blend can contain only one closed loop.

Once you have completed the sketch for the base, you must click the Edit Top button on the Mode panel of the contextual tab. This will keep you in sketch mode, and the base geometry will be grayed out. You can sketch the geometry for the top of the blend anywhere in the drawing area. The solid geometry will be extruded to connect the base to the top along the depth distance and any distances in the X or Y direction. Figure 17.2 shows the result of a top sketch that is offset from the location of the base geometry.

FIGURE 17.2
Blend with offset top geometry

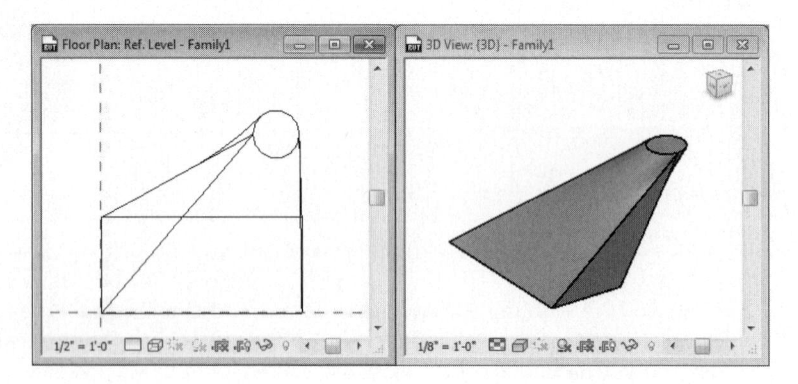

When you select a blend, the Modify | Blend contextual tab contains buttons that allow you to edit either the top or the base sketch by clicking the applicable button on the Mode panel. When you enter into sketch mode, you have the option to edit the other sketch if necessary by clicking the button on the Mode panel. You can also edit the vertices that are formed by the transition from the base shape to the top by clicking the Edit Vertices button.

The tools on the Edit Vertices tab of the ribbon can be used to change how the transition occurs from the base shape to the top shape.

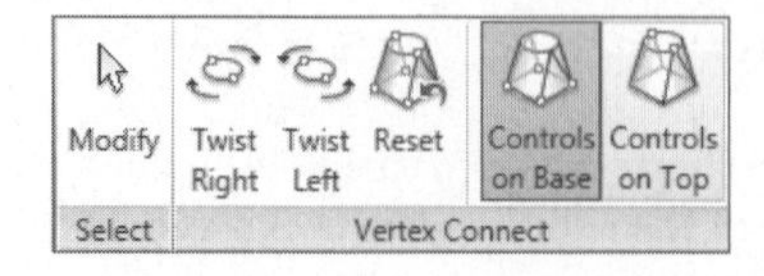

Depending on the shapes you have drawn, there will be different options for the vertices that define the transition. When editing the vertices of a blend, you can twist the vertices left or right. It is possible to twist too far, but you will not be warned until you exit sketch mode. Grips appear on the blend that allow you to manually edit vertices. The open circle grip is for adding a vertex, while the solid grip exists on a vertex that can be removed by clicking the grip, as shown in Figure 17.3. It is helpful to work in a 3D view when editing vertices because you can see more clearly the effect on the solid geometry. The Reset button on the Edit Vertices tab will return the orientation of the vertices to their original format.

FIGURE 17.3
Vertex editing grips

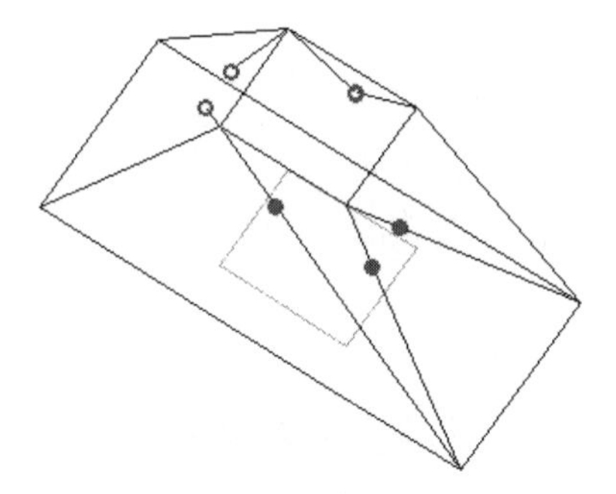

Once you are finished editing the vertices of a blend, you must click the Modify button on the Edit Vertices tab to return to the sketch mode tab. As always, you must then click the green check mark button to exit sketch mode.

Revolves

A *revolve* is an extrusion that follows a circular path around a specified axis. Using the Revolve tool allows you to create spherical solid geometry. You can start creating a revolve by clicking the Revolve button on the Home tab in the Family Editor. You can begin either by sketching the shape of the revolve or by selecting or creating the axis of rotation that the shape will revolve around.

The axis of rotation will always be perpendicular to the shape that you create, so it helps to first determine the orientation within the family of the shape you are creating. Once that has been determined, you will know where the axis needs to be and can switch to the appropriate view. For example, if you wanted to create a hemispherical solid that would lie flat when placed into a plan view, you would draw the axis perpendicular to the plan view. Switching to an elevation view in the Family Editor would allow you to draw the axis perpendicular to plan.

Clicking the Axis Line button on the Modify | Create Revolve contextual tab activates sketch mode with two drawing tools available in the Draw panel of the tab. You can either draw the axis or pick an existing line. The reference planes that determine the insertion point of a family are good choices for an axis line. Once you have drawn or selected an axis, you will remain in sketch mode, but the Draw panel will populate with the usual drawing tools for creating the shape of the revolve.

The shape of the revolve can be drawn anywhere in the view that the axis is drawn in. When you are sketching the shape of the revolve, you must create a closed loop or multiple closed loops. You can sketch shapes away from the axis to create an interior circular space within the solid geometry, as shown in Figure 17.4.

FIGURE 17.4
Revolve shape drawn away from axis

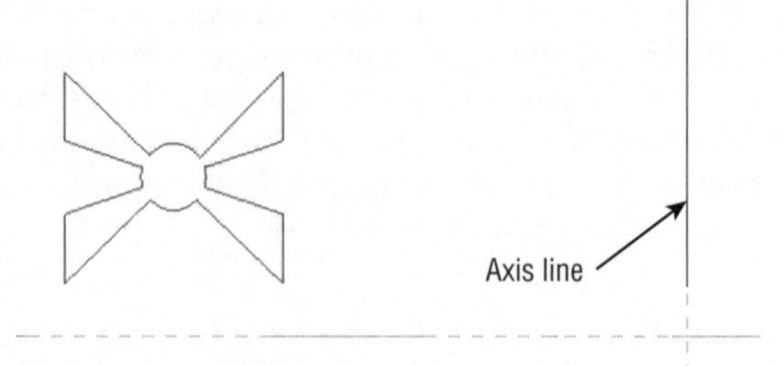

When you have completed the shape, you can click the green check mark button to exit sketch mode. The shape will be extruded in a circular path around the axis, forming the solid geometry. Figure 17.5 shows the resulting solid geometry formed by the shape and axis shown in the previous figure.

FIGURE 17.5
Revolve solid geometry

To create a spherical shape with a revolve, draw an arc shape adjacent to the axis line. Because the shape must be a closed loop, you will have a shape line at the axis, but this line will not be seen because it will be in the interior of the solid geometry. Figure 17.6 shows the sketch of a simple hemisphere solid.

FIGURE 17.6
Hemisphere revolve sketch

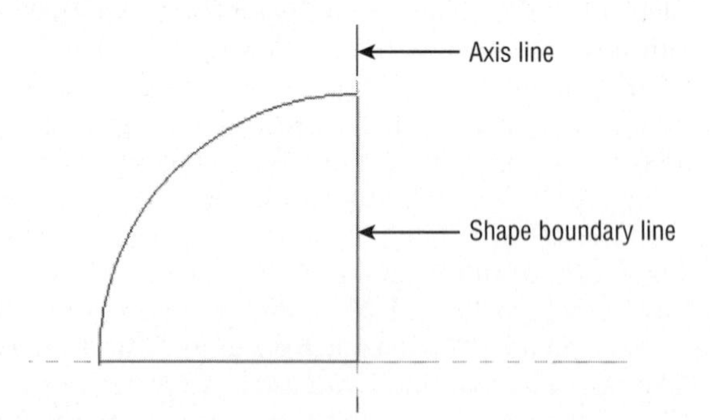

Unlike extrusions and blends, a revolve solid does not have a depth parameter. The depth or height of a revolve would be determined by the sketch dimensions. A solid revolve has parameters that allow you to control how far around the axis the solid geometry is extruded.

Adjusting the Start Angle and End Angle parameters adds further complexity to a solid revolve shape. Figure 17.7 shows the effects of values used in the parameters on a solid hemispherical revolve. Once you edit the parameters in the Properties palette, you can use the grips to pull the faces of the geometry around the axis. You can also select the temporary dimension that appears and input a value.

FIGURE 17.7
Revolve parameters

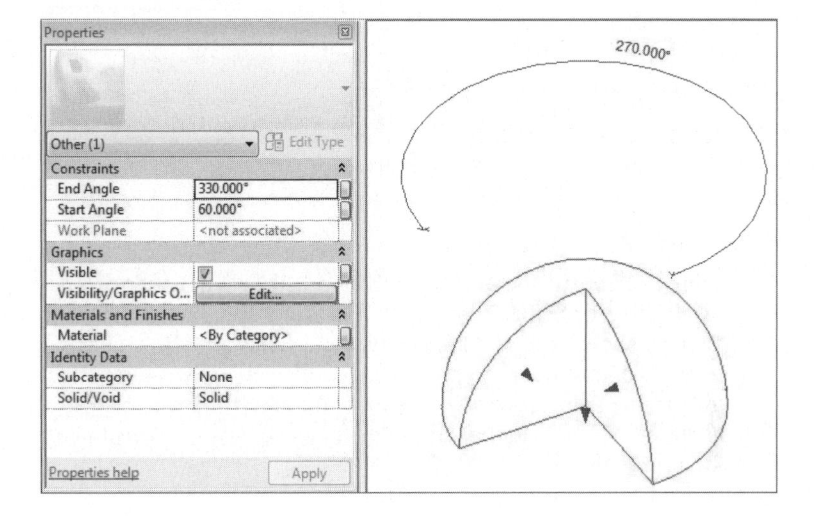

Sweeps

A *sweep* is an extrusion that follows a specified path. As you have just seen, a revolve is a special kind of sweep that follows a circular path around an axis. Using the Sweep tool allows you to specify the path for a shape to follow.

As with any solid geometry, it is important to first decide the orientation of the solid geometry within a family so that you can determine the location of the path for a sweep. When you click the Sweep button on the Home tab in the Family Editor, sketch mode will be activated, and the contextual Modify | Sweep tab appears on the ribbon.

The first step in creating a solid sweep is to define the path of the extrusion. You can either sketch the path using the standard drawing tools, or you can pick existing lines in the family. You cannot pick reference planes as a path for a sweep; however, you can pick reference lines because they have a set length.

After creating the path for the sweep, you must click the green check mark button to exit sketch mode for the path. This will return you to sketch mode for the sweep geometry. With the path drawn, a profile location indicator is shown on the path, as shown in Figure 17.8. The profile location indicator can be moved anywhere along the path while sketching the path. This allows you to locate the profile parallel to a standard view plane, which will make sketching the profile easier. This is especially useful when creating paths that contain arcs because the profile may end up in the middle of an arc at an angle that would be difficult to draw at. The location of the profile plane will not have any effect on predefined profile families.

FIGURE 17.8
Sweep path and profile location

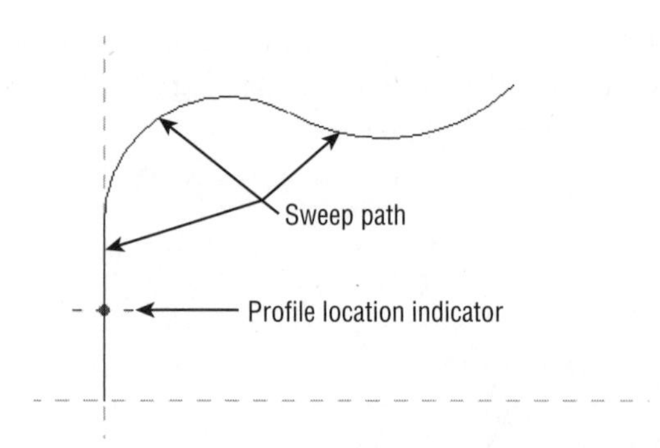

You can create the profile for the sweep by sketching, or you can select a predefined profile by clicking the Select Profile button on the Sweep panel of the Modify | Sweep contextual tab. When you click this button, any loaded profiles will be available in the drop-down on the Sweep panel. If no profiles are available in the file you are working in, you can click the Load Profile button on the Sweep panel. The content library that is loaded when you install Revit MEP 2011 contains a Profiles folder. From this folder, you can choose a predefined shape for use in your sweep geometry.

PROFILE FAMILIES

You can build a library of profiles that can be used when generating family geometry. This will allow you to maintain consistency in the way sweep geometry is generated. You can use the `Profile.rft` family template to create a profile family.

The profile for a sweep should be drawn in a view that is parallel to the profile location plane or perpendicular to the plane of the path. You can draw the profile in a plane that is perpendicular to the path but is not parallel to the profile plane. This will cause an error when you attempt to complete the sweep by exiting sketch mode. The best option is to draw the profile in a plane that is both perpendicular to the path plane and parallel to the profile plane. When you select a predefined profile, it will automatically be placed in the profile plane.

Once you have chosen or created a profile for the sweep, you can click the green check mark button to exit sketch mode and complete the solid geometry. To make changes to a sweep, select the solid geometry, and click the Edit Sweep button. You can modify the path by clicking the Sketch Path button, or you can modify the profile by clicking the Select Profile button and then the Edit Profile button.

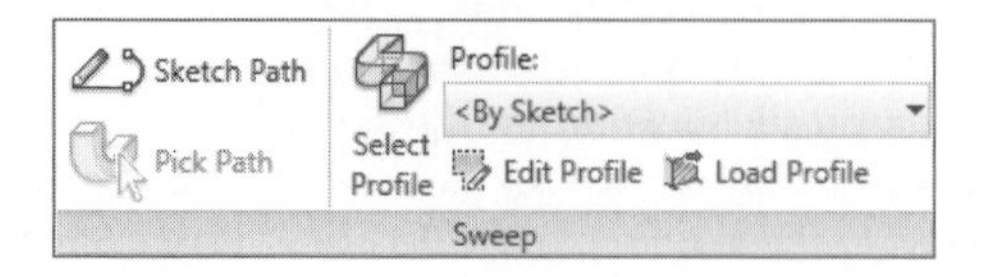

You cannot change the location of the profile plane once the profile has been created, so it is best to determine its location when creating the path for the sweep.

Swept Blends

A *swept blend* is a combination of a sweep solid and a blend solid. This tool allows you to create a sweep that has two profiles, one at each end of the path. The solid geometry will transition from the shape of the first profile to the shape of the second profile along the path.

The process for creating a swept blend is similar to that of creating a sweep, with an extra step for defining the shape of the second profile. When you click the Swept Blend button on the Home tab in the Family Editor, sketch mode is activated, and you can draw or pick a path for the extrusion to follow. When you draw a path line, the first profile location will be at the start of the line, and a second profile location will be placed at the end of the line.

You cannot use multiple lines to create a path for a swept blend, as shown in Figure 17.9. Since the profile plane for the second profile occurs at the end of the first line drawn, any additional lines drawn afterward cannot be included in the path. This is true if you use the Pick Path option also. You can select only one line for the path.

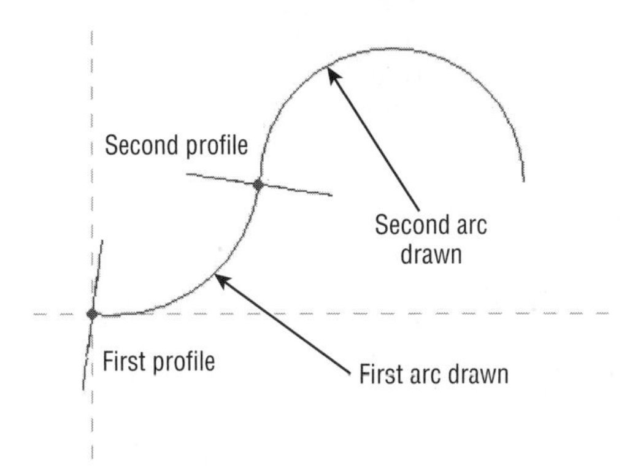

FIGURE 17.9
Multiple lines for a swept blend path

Instead of using multiple lines, you can use the Spline drawing tool for creating a complex path with a single entity. Figure 17.10 shows how a spline can be drawn to represent the same shape as in the previous figure.

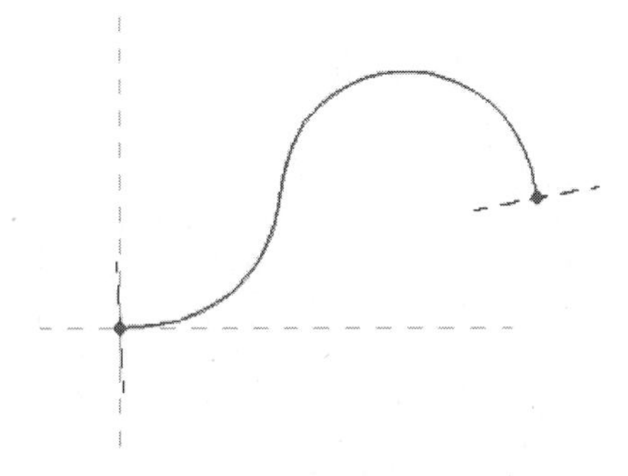

FIGURE 17.10
Spline path for a swept blend

When you click the green check mark button to exit sketch mode for the path, the Swept Blend panel provides tools for defining each of the profiles. Each profile can be drawn, or a predefined profile family can be used. When you click one of the profile buttons, you must click the Edit Profile button to draw the shape. The Go To View dialog box will appear so that you can choose a view where the profile sketch can be drawn. Once you have finished the profile shape, you can exit sketch mode by clicking the green check mark button. You can then create the second profile by clicking the Select Profile 2 button and then the Edit Profile button.

Because you are defining each end of a swept blend solid with a profile shape, you can create a twist in the solid geometry by locating the profile sketches at different elevations from the plane of the path. Figure 17.11 shows an example of a swept blend where the profiles are drawn above and below the plane of the path, creating a solid that not only changes shape from one end to the other but also changes elevation.

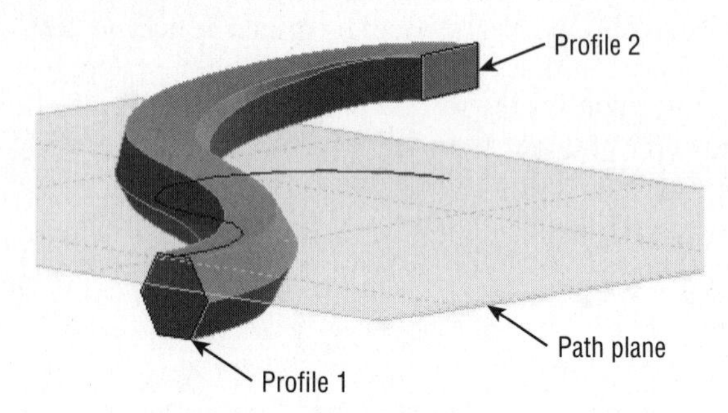

FIGURE 17.11 Swept blend with profiles offset from path plane

Joining Geometry

In some cases, it may be easiest to create a solid form by creating multiple individual solids and then combining them. When you select an extrusion in the Family Editor, the Geometry panel on the contextual tab contains tools for joining geometry.

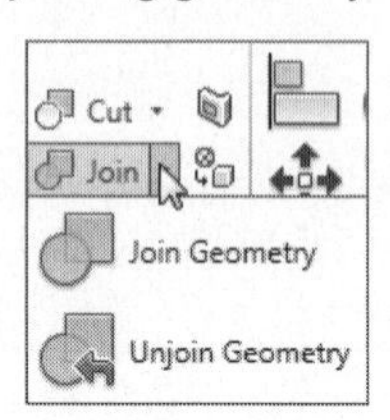

When you join geometry in the Family Editor, it creates a union between the selected solids. When geometry is joined, you can select all the forms by placing your mouse pointer over one of the forms and using the Tab key to select any forms joined to it. This makes for easier editing in the Family Editor, and the joined geometry can be assigned a material property, or the visibility of the joined multiple objects can be controlled with one setting.

Each individual form within a set of joined geometry can be edited normally. Solid forms do not even need to be touching each other to be joined. If you want to break the relationship of forms that are joined, you can select one of the solid forms and click the Unjoin Geometry option on the Join button. You will then be directed to select the solid geometry that you want to be unjoined from other geometry.

Voids

To this point, all the discussion on modeling tools has been for creating solid geometry. Sometimes it is necessary to create a void form. Voids can be used to cut shapes out of solid geometry, and in the case of hosted families, they can be used to cut the host. Some solids are easier to create by modeling a form and then using a void form to remove a portion of the solid geometry.

The same tools for creating solid geometry are used for creating void geometry. The Void Forms button on the Home panel of the Family Editor is a drop-down list of the tools for creating a void.

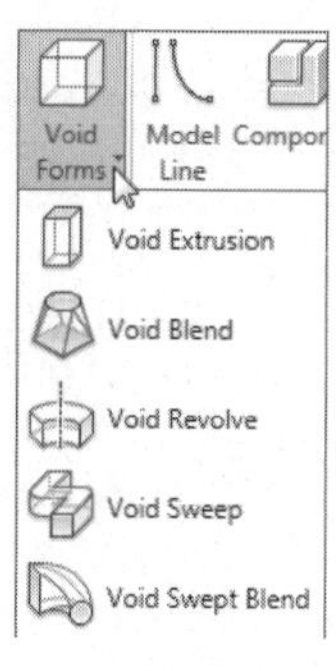

Because the form tools are similar, you can even create solid geometry and then change it to a void by editing the Solid/Void parameter in the properties of a solid form. Voids can also be changed to solid geometry by editing the same parameter.

In an earlier example, you saw a multishape extrusion where one rectangle was drawn inside another to create a solid with its center hollowed out all the way through (Figure 17.1). If you did not want the open space in the center to pass all the way through the solid, you could create multiple solids that result in the desired form, or you could use a void form to cut out the desired space in the solid, as shown in Figure 17.12.

FIGURE 17.12
Void geometry in a solid form

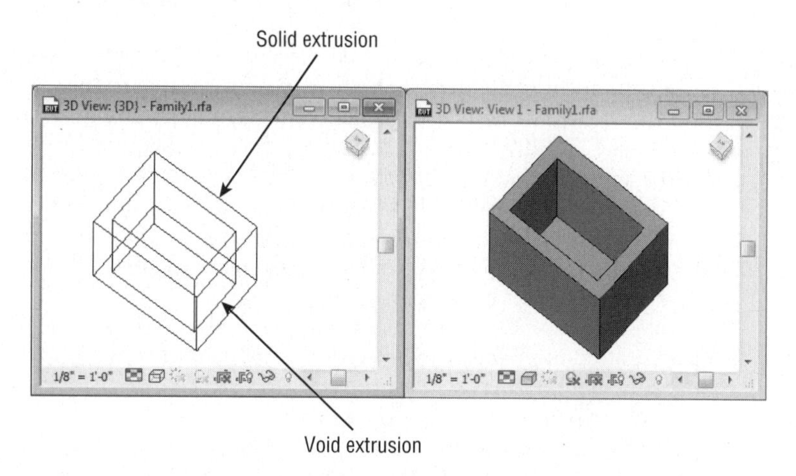

When you create a void form while in the Family Editor, it will appear in the 3D views as a transparent form and as orange lines in plan and elevation views as long as it is not cutting any solid geometry. The void will not automatically cut any solid geometry unless it is drawn overlapping the solid geometry. So if you were to draw a void form independent from any other geometry in the view and then move the void so that it overlaps solid geometry, the solid would not be cut.

You can tell the void form which solid geometry to cut by selecting it and then clicking the Cut button on the Geometry panel of the contextual tab. You can select the void first or the solid geometry and then select the other to establish the cut relationship.

STATUS BAR

The status bar at the bottom of the user interface will guide you through the steps for cutting or joining geometry when you are using the tools on the Geometry panel.

You can establish the cut relationship with a void and a solid before they are overlapping. Once you use the Cut tool and create the relationship, the void will cut the solid when it is moved to overlap the solid. You can create a cut relationship between one void and multiple solid objects, as shown in Figure 17.13. The void form has been highlighted in this image to illustrate that it is a single void cutting multiple solid objects.

FIGURE 17.13
Void cutting multiple solids

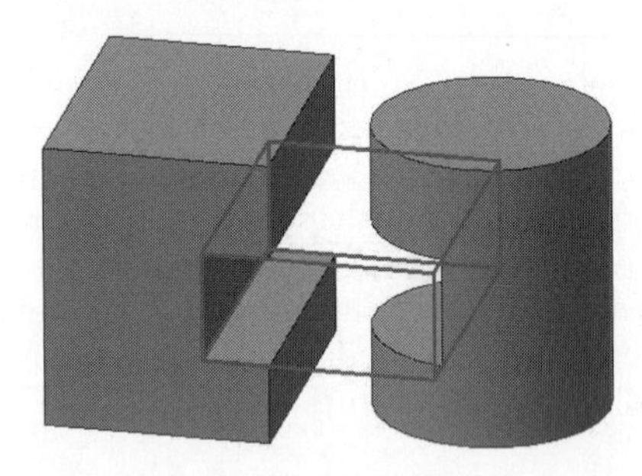

In the case of a hosted family, you can use the Cut tool to create a cut relationship between a void and the host geometry. This is useful for lighting fixture and air terminal families when you want to show that the component requires an opening in its host. This type of relationship can be established within face-hosted or with model-hosted families; however, when using face-hosted families in a project, the void will not cut a linked host face.

Voids should be used only when necessary to define the shape of a solid. Consider all the other solid modeling tools to create forms before resorting to using voids. It has been documented that having many voids in families, or having many families with voids in a project, can negatively affect file performance.

For some families, it may be tempting to use void forms to define the required clearance space around mechanical and electrical equipment. There are other methods to define these spaces without burdening your projects with void geometry. For more information on creating clearance spaces within families, see Chapter 20.

Mastering the tools for creating solid geometry will make content creation an efficient part of your Revit MEP 2011 implementation. Practice creating different types of solids by completing the following exercise:

1. Open the `Ch17_Solid Geometry.rfa` file found on `www.wiley.com/go/masteringrevitmep2011`.
2. Click the Extrusion button on the Forms panel of the Home tab. Select the Circumscribed Polygon drawing tool from the Draw panel.

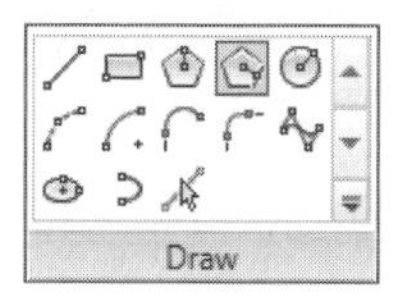

3. On the Options Bar, set the depth of the extrusion to 2′-0″ and the number of sides to 4.
4. Click at the intersection of the reference planes to start the extrusion. Drag your cursor to the right along the horizontal plane until the radius is 1′-0″, and then click to complete the sketch.

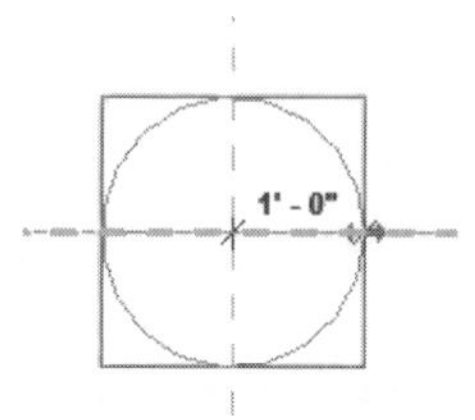

5. Select the Circle tool from the Draw panel. Click the intersection of the reference planes, and drag your cursor to the right along the horizontal reference plane until the radius is 0′-7″. Then click to finish the sketch.
6. Click the green check mark button on the Mode panel of the Modify | Create Extrusion contextual tab to finish creating the extrusion. Open the default 3D view to verify that the extrusion is a solid rectangle with a cylindrical opening in the center.

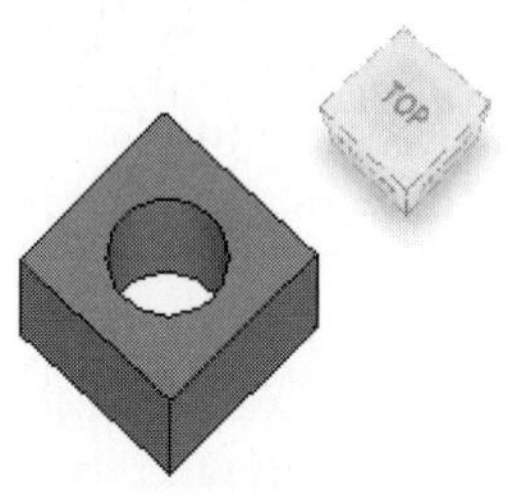

7. Open the Front elevation view. Click the Revolve tool on the Forms panel of the Home tab.
8. Click the Axis Line button on the Draw panel, and choose the Pick Lines tool from the Draw panel. Click the vertical reference plane in the view to set the axis of the revolve.
9. Click the Boundary Line button on the Draw panel. Draw a vertical line 1′-0″ long from the horizontal reference plane and 1′-5″ to the left of the axis.

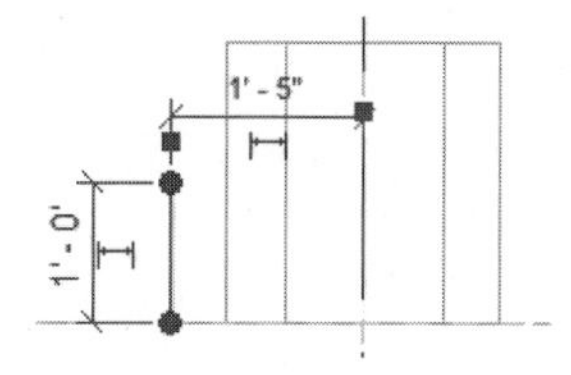

10. Draw a 1′-0″ line from the bottom end point of the vertical line to the left along the horizontal plane.
11. Select the Start-End-Radius Arc tool from the Draw panel. Choose the end point of the horizontal line drawn in step 10 as the start, and the end point of the vertical line drawn in step 9 as the end of the arc. Drag the radius until it snaps tangential to the lines.

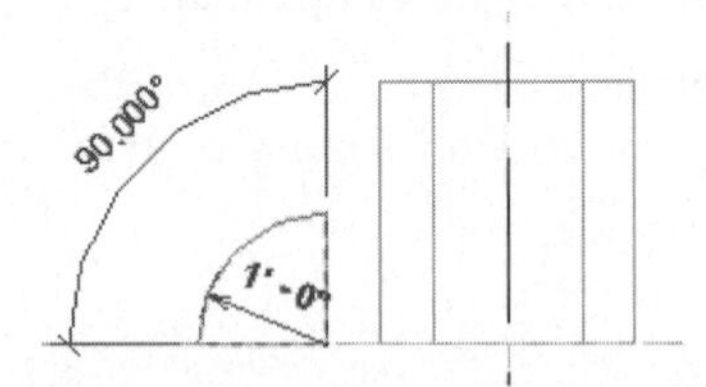

12. Click the green check mark button on the Mode panel of the Modify | Create Revolve contextual tab to finish creating the revolve. Open the default 3D view to verify that the solid geometry revolves completely around the axis.

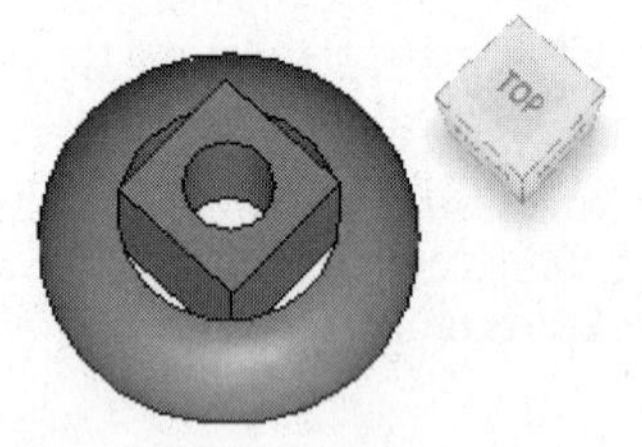

13. Open the Ref. Level floor plan view. Click the Sweep tool on the Forms panel of the Home tab. Click the Sketch Path button on the Sweep panel of the Modify | Sweep contextual tab.
14. Select the Start-End-Radius Arc tool from the Draw panel. Start the arc 2′-0″ from the top edge of the rectangular extrusion at the vertical reference plane. Drag your cursor down along the vertical axis a distance of 6′-0″, and click to establish the arc end point. Drag your cursor to the right until the radius dimension is 3′-0″.

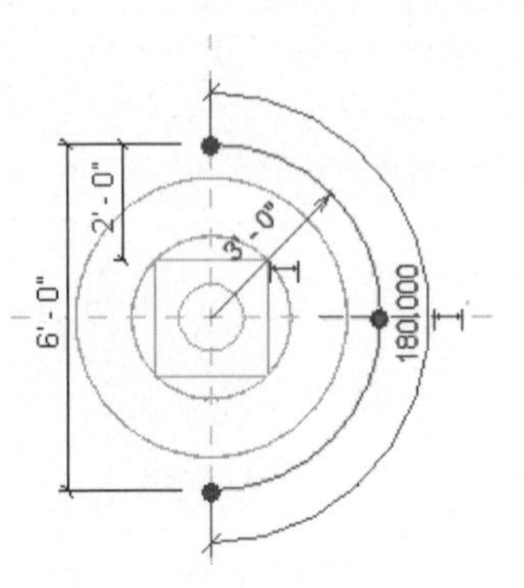

15. Click the Modify button on the Select panel of the contextual tab to close the arc tool. Click and drag the profile location point along the arc toward the top until the location point snaps to the end of the arc at the vertical plane.

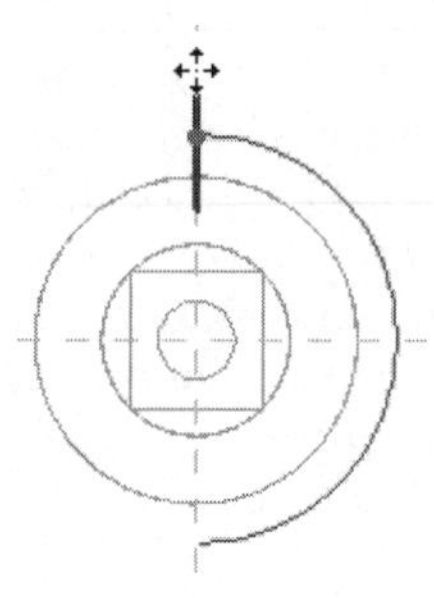

16. Click the Modify button on the Select panel of the Reference Planes contextual tab to finish editing the location of the profile plane. Click the green check mark button on the Mode panel of the Modify | Sweep Sketch Path contextual tab to complete the path for the sweep.
17. Open the Left elevation view. Click the Edit Profile button on the Sweep panel. (If the Edit Profile button is not active, click the Select Profile button to activate it.)
18. Select the Inscribed Polygon tool from the Draw panel. On the Options Bar, set the number of sides to 3. Click the profile location point to start the profile sketch. Move your cursor up until the radius is 0′-3″, and then click to create the profile sketch.

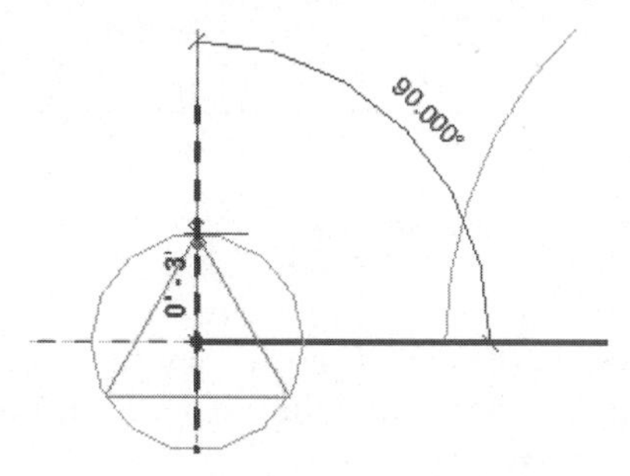

19. Click the green check mark button on the Mode panel to complete the profile. Click the green check mark button on the Modify | Sweep contextual tab to complete the sweep. Open the default 3D view to confirm that the triangular profile is extruded along the arc around the other solid geometry.

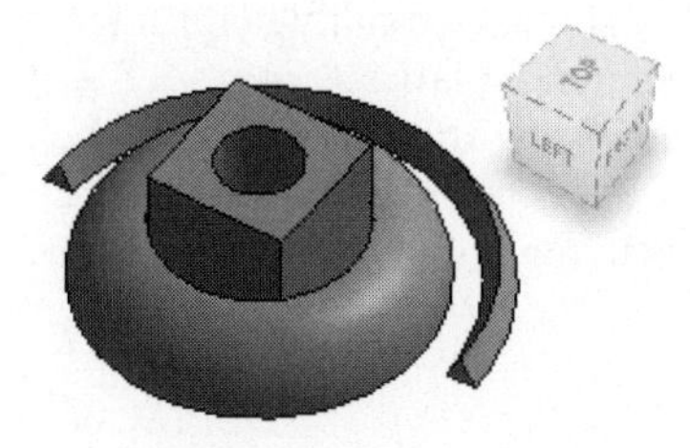

Reference Planes and Lines

When you begin to create a solid, it is important to understand how the current view will affect the orientation of the solid you create. When you click one of the buttons on the Form panel of the Home tab, you will be taken into sketch mode for the solid. A contextual tab appears on the ribbon with tools for generating the sketch, or shape, of the solid. The view that you are in determines the plane for the sketch, and the depth of the solid will be perpendicular to the sketch plane. If you are working in a file that contains multiple planes that are parallel to the current view, you can select a plane to associate the extrusion to by clicking the Set button on the Work Plane panel of the contextual tab. When you click the Set button, the Work Plane dialog box will appear. You can choose the desired plane from the drop-down list, as shown in Figure 17.14.

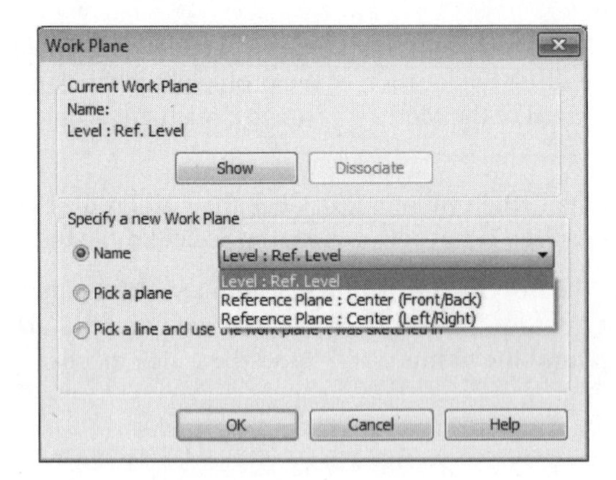

FIGURE 17.14 Work Plane dialog box

If you choose a reference plane that is not parallel to the current view, the Go To View dialog box will appear when you click OK in the Work Plane dialog box. The Go To View dialog box offers views that exist in the family file that are parallel to the selected reference plane. You can also choose a 3D view to work in if you are more comfortable working in 3D to generate solid geometry. You can click the Show button on the contextual tab for a visible reference of the plane chosen for the sketch. This is especially helpful when creating a sketch in a 3D view.

You do not have to set the reference plane to begin sketching the shape of the solid. Sketching it directly in the view will associate the solid geometry with the plane of that view, such as the reference level of a family.

When you are using reference planes to build solid geometry, it is a good idea to give any custom planes a name so that they can be easily identified and selected from the list in the Work Plane dialog box. This can be especially helpful when creating families with multiple solids where you can give reference planes names such as "Top of Unit" or "Face of Device" that aid in associating other geometry within the family to the plane.

You can control how a reference plane in a family is accessed when the family is loaded into a project by editing the Is Reference parameter. You can set the parameter to associate the plane with the orientation of the family by choosing any of the directional choices such as Front, Back, Bottom, Center (Left/Right), and others. You can also choose whether the reference plane is used for dimensioning when the family is loaded into a project. These options are as follows:

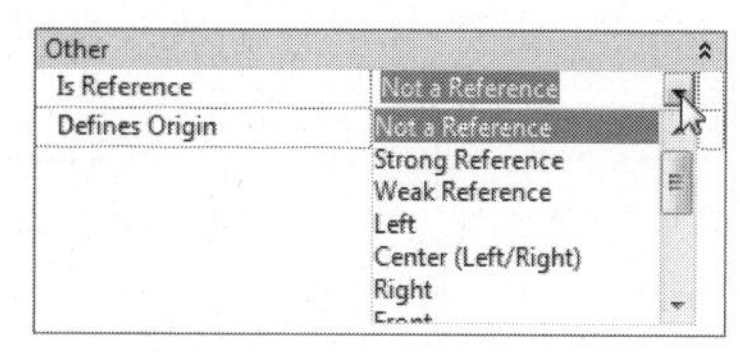

Not A Reference When placed into a project, the plane will not be available for dimensioning or alignment. This is a good setting for planes created in a family that are used only for association within the family. It will prevent unwanted selection or highlighting of a family when your mouse pointer passes over the plane.

Strong Reference When the family is placed into a project, the plane will be the first choice for temporary dimensions when placing the family into a project. This setting is best used for planes in a family that define the portions of a family that you would dimension or align to when using the family in a project.

Weak Reference This setting for the parameter is for when you want to be able to dimension or align to the plane but do not want temporary dimensions applied to the plane when placing the family in a project.

For example, if you have a family that you want to place at its edges but also want to dimension or align to its center, you would define reference planes at its edges as strong references and a reference plane at the center as a weak reference. This is typical for families such as lighting fixtures or air terminals.

The Defines Origin parameter allows you to set a reference plane as one of the planes of the origin of a family. Two planes must be set as defining the origin, and the planes must intersect. The intersection point of the planes in plan view will determine the insertion point of the family.

Reference lines are useful in creating solid geometry when you do not want to create an infinite plane of reference. Unlike reference planes, they have a start point and an end point. When you draw a straight reference line, two planes are formed at the line. One plane is parallel to the plane that the line is drawn in, and the other is perpendicular to that plane. This allows you to use the reference line in views parallel and perpendicular to where it is drawn. Figure 17.15 show a reference line that has been selected to show the planes associated with the line. Arced reference lines can be used for reference but will not create planes.

Reference lines can be dimensioned to when a family is placed into a project and they also have a parameter that allows you to set them as a strong or weak reference or as not a reference. You cannot use reference lines to define the origin of a family.

Solid geometry can be aligned to reference lines in the same way as to reference planes. Using reference lines and planes is the most effective way to create parametric behavior of solid geometry within a family.

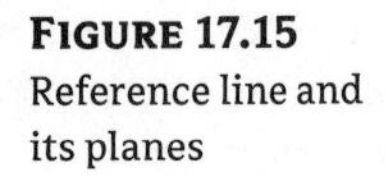

FIGURE 17.15
Reference line and its planes

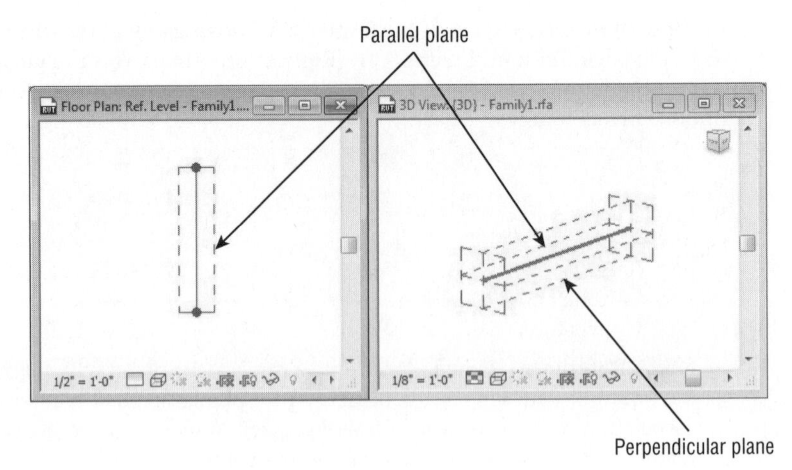

Constraints and Dimensions

Making the solid geometry in your families parametric will give you the ability to create multiple types within a single family and offers a higher level of management and control of the properties of components.

The key to making your solid geometry parametric is to constrain the geometry to reference planes and lines. This enables you to apply the parametric behavior to the planes and lines, which allows for multiple solid forms to react to changes to the parameters. Although you do have the ability to assign parametric constraints directly to the solid geometry, it is recommended that you assign it to reference planes or lines so that changes to a solid that affect other solids within the family are more easily achieved and managed.

Geometry can easily be constrained to a reference plane by using the Align tool on the Modify tab, or you can simply drag the edge of a solid to a reference plane or line, and it will snap into alignment. Once aligned, the padlock grip appears, allowing you to lock the alignment.

Some solid forms do not need a reference plane or line to be parametrically managed. Whenever you are sketching a circle and want to control the radius with a parameter, you can apply the parameter directly to the sketch. This is done by activating the temporary dimension that indicates the radius when sketching the circle. Clicking the dimension grip will change the temporary dimension to a permanent one, which can then be assigned to a parameter.

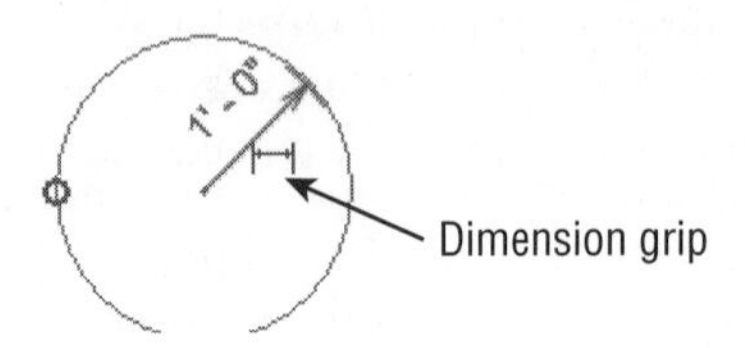

When you complete a sketch that contains dimensions within the sketch, the dimensions will not show unless you are in sketch mode. You can constrain sketch lines to reference planes while working in sketch mode, but if you are using dimensional constraints, it is best to put the dimensions directly in the family so that they will be visible while you are working on the family. It can be very frustrating to place a dimension only to find that one already exists in the sketch of a solid.

Creating angular constraints is often necessary for solid geometry. When you need to create angular parametric behavior for a family, it is best to use reference lines instead of reference planes. The location of the end point of a reference line can be constrained so that the line can be rotated with the end point serving as the axis of rotation. An angular dimension can be used to create the parametric behavior of the reference line, as shown in Figure 17.16. The padlock grip indicates that the reference line has been locked to the horizontal reference plane, although it is not necessary to lock the end point to the reference plane if the reference line is drawn connected to the plane.

FIGURE 17.16
Reference line with angular parametric behavior

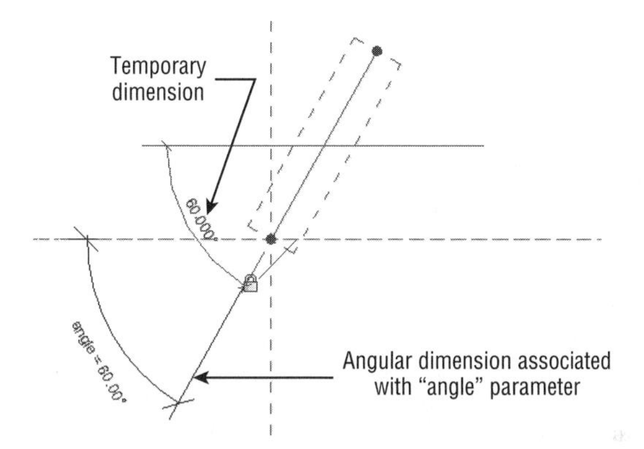

This type of angular constraint is useful for rotating solid geometry within a family. Figure 17.17 shows an extrusion that was modeled in the vertical plane of the reference line. Because the plane of the line was used, the extrusion is associated with the line so when the angle of the line changes, the solid will stay with it. The dashed lines indicate the original location of the line prior to changing the angle parameter to 120°.

Though reference lines work well for this type of constraint and parametric behavior, they can cause undesired results when the angle of solid geometry is supposed to change while the geometry remains in a fixed location. The solid geometry shown in Figure 17.18 is a sweep with a rectangular profile. The path of the sweep is an arc that is locked to the reference line and to the reference plane. As the angle parameter is modified, the length of the sweep increases. A radius parameter has also been applied for the sweep path.

This type of parametric relationship will behave as expected up to a certain point. Flexing the family reveals that at larger angles the geometry does not stay associated to the reference line, causing an incorrect representation of the solid, as shown in Figure 17.19. Notice that the solid geometry has gone past the reference line and appears to be aligned with the witness line of the angular dimension.

Different angles input into the parameter result in different undesired behavior of the solid geometry. The purpose of this example is not to point out shortcomings of the software, but to demonstrate that one choice for creating an angular constraint may not work for all situations.

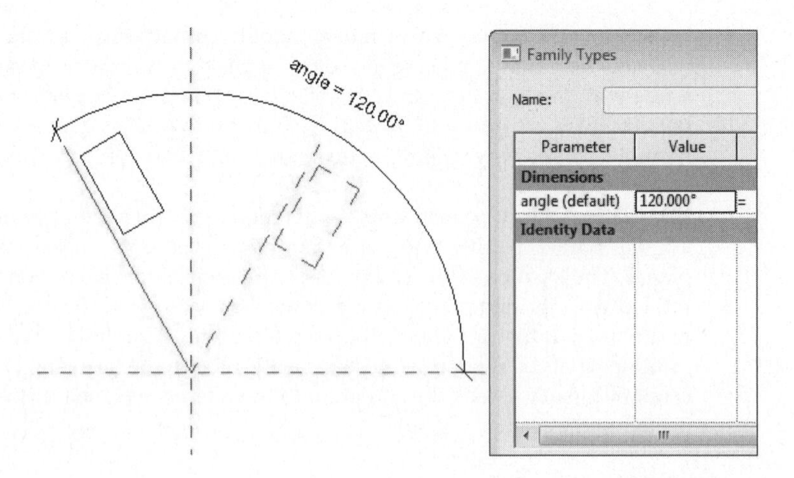

FIGURE 17.17
Extrusion constrained to reference line

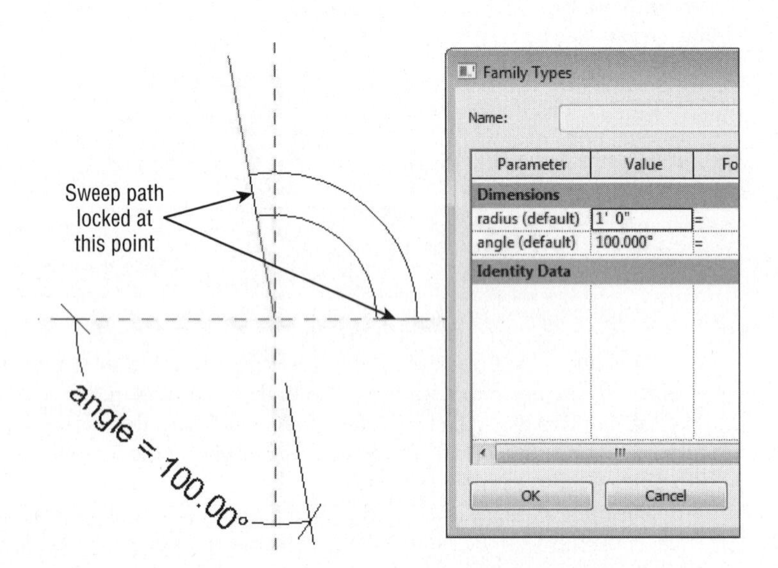

FIGURE 17.18
Sweep constrained to reference line

An alternative solution for this scenario would be to use an angular dimension of the sweep path itself instead of constraining to a reference line. Figure 17.20 shows the dimension parameters applied to the path of a sweep. The angular dimension was created by selecting the path and then activating the temporary dimension, which was then associated to the angle parameter.

With the path of the sweep being defined by the angle instead of its association to a reference line, the desired results for the solid geometry can be achieved, as shown in Figure 17.21. A dimension has been added to verify the angle of the solid geometry for display purposes only.

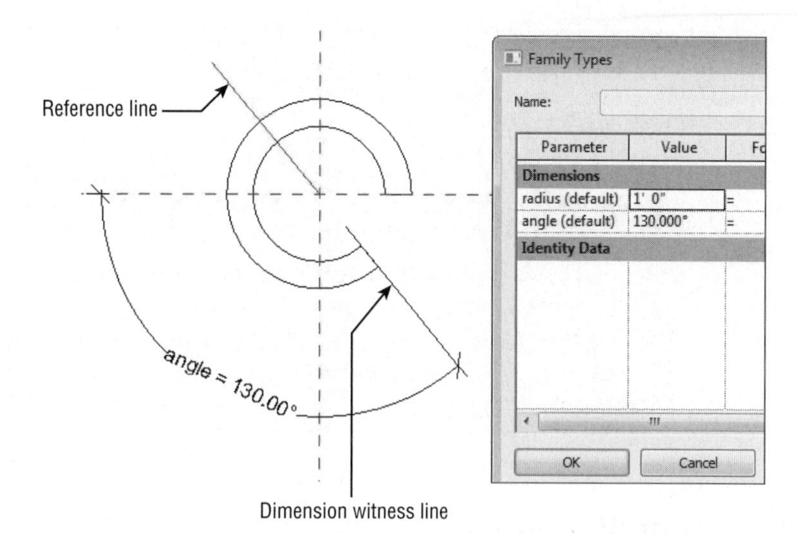

FIGURE 17.19 Undesired sweep geometry from angular constraint

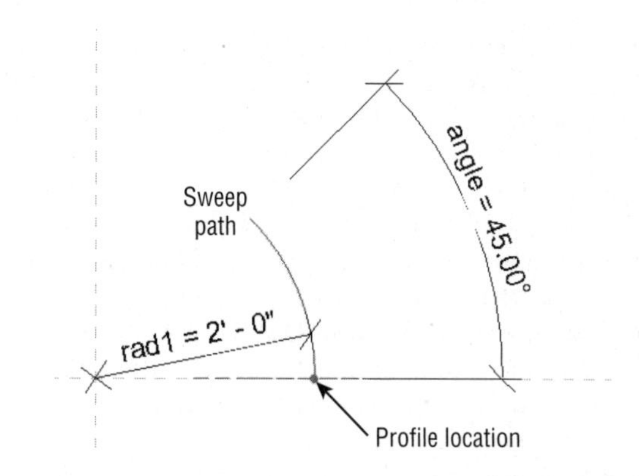

FIGURE 17.20 Sweep path with parametric constraints

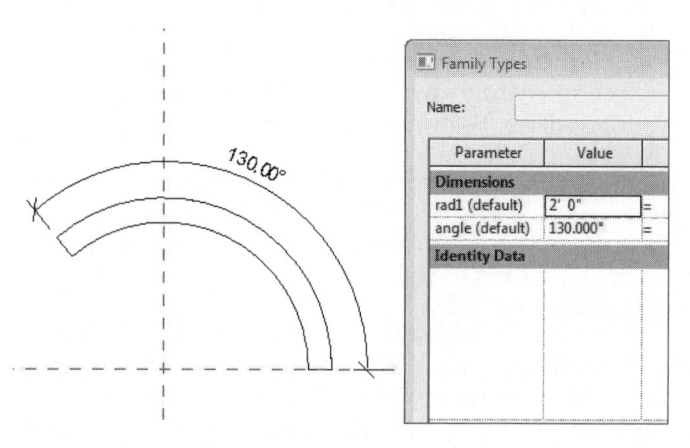

FIGURE 17.21 Sweep with an angular constraint for parametric behavior

CIRCLE OF LIGHT

Catherine is working on a project that has a rotunda and requires lighting that follows the curve of the walls. Knowing that there will be many changes before the final decision is made on this high-profile area, she creates a lighting fixture that represents a custom linear LED strip.

She creates a sweep with an arced path, provides a parameter to control the length of the arc by adjusting its angle, and provides a parameter for the length of its radius.

As her architect works with the client to finalize the design, Catherine can easily keep up with the changes by editing parameters instead of having to re-create the fixture each time the model changes.

Visibility Control

The visibility parameters of solid geometry provide useful means for controlling the behavior of solids in a family that is used in a project. For many MEP discipline families, the actual solid geometry is not shown in plan views, but instead a symbol is used. This is especially true for electrical devices and even some light fixtures.

If you are concerned about file performance, it is a good practice to limit the visibility of solid geometry to only the types of views where it is necessary to be seen. Symbolic or model lines can be used to represent the geometry in views where the solid does not have to be shown, such as plan views or reflected ceiling plans. Setting the visibility of solid geometry to display in 3D views, sections, and elevations is important because it helps keep your project coordinated by allowing you to locate items when a section or elevation view is created, without having to draw additional line work in the section or elevation view.

In some cases, you may also want to control the visibility of solid geometry based on the detail level setting of a view. For example, you may set the visibility of the solid geometry in a power receptacle family so that the geometry shows only in views set to Fine detail. This enables you to quickly see the actual location of the receptacles by switching the view detail level for instances where you are trying to coordinate exact locations. It also benefits other disciplines, such as architectural, which may want to see the receptacle geometry in a section or elevation.

To set the visibility of solid geometry in the Family Editor, select the geometry, and click the Visibility Settings button located on the Mode panel of the contextual tab that appears on the ribbon.

In the Family Element Visibility Settings dialog box, you have options for where the solid geometry will be visible when the family is used in a project. Figure 17.22 shows the dialog box and the available options.

All solid geometry is visible in 3D views when the category of the family is visible. The View Specific Display options allow you to select other types of views where the solid geometry will be visible. Keep in mind that for face-hosted families, the Front/Back setting will display the geometry in plan view when the geometry is hosted by a vertical surface. The Left/Right setting is for section and elevation views taken from the side of the geometry.

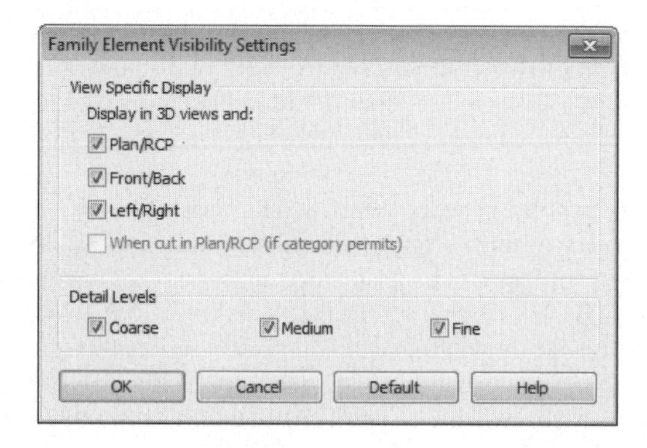

FIGURE 17.22 Family Element Visibility Settings dialog box

One of the options is to display the geometry when the cut plane of a view intersects the geometry. This option is not available for all categories of families and is typically used for architectural and structural types of families such as doors or windows.

In the Detail Levels section of the dialog box, you can select the detail levels in which the geometry will be visible. When using symbolic or model lines to represent the geometry in plan and RCP views, consider what detail level will be used for section, elevation, and 3D views so that the geometry will display when desired. When creating duct and pipe fitting and accessory families, be sure to set the Detail Levels settings within the Family Element Visibility Settings dialog box to match the behavior of duct and pipe system families so that the solid geometry will display in Medium detail for duct and Fine detail for pipe-related families. Any model lines used to represent the objects should only display in Coarse detail for duct and Coarse and Medium for pipe.

In some cases, you may want to represent an object with a very simple shape for Coarse detail–level plans and a more realistic shape in smaller scale views using Fine detail. The box shown in Figure 17.23 is set to display at Coarse detail, while the more detailed solid geometry displays at Medium and Fine detail levels.

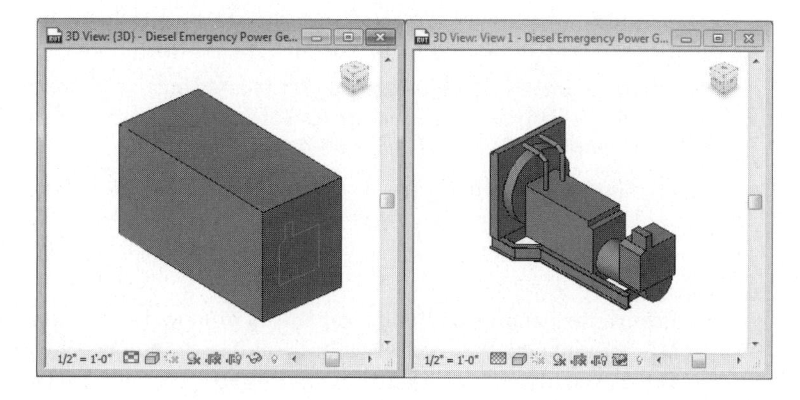

FIGURE 17.23 Geometry display at different detail levels

The visibility settings you apply will not affect the geometry while you are working in the Family Editor. Solid geometry will remain visible in the Family Editor so that you can work on the family. In 3D views within the Family Editor, solid geometry will appear halftone if the visibility settings of the geometry would cause the object to not be visible in the same type of project view. In other words, if you set an object to display only in Fine detail and your 3D view in the Family Editor is set to Medium detail, the solid will be halftone in the Family Editor 3D view. This is a helpful visual aid when working in a family to understand the visibility behavior of the geometry.

The visibility of solid geometry can also be controlled by the Visible parameter in the properties of the geometry. This is a Yes/No parameter that simply determines whether the geometry is visible. This parameter can be associated with another parameter defined in the family for control of the solid visibility when the family is used in a project.

Another option for the visibility of solid geometry is to assign a material to it using the Material parameter. You can create a custom material type with settings that are desired for use in your projects. One example is to create geometry in a family that represents the required clearance space for equipment. Along with visibility control settings, a material can be applied to the solid geometry to make it semitransparent when viewed in a 3D project view. See Chapter 20 for more details.

When you are considering the visibility of solid geometry in a family, keep in mind that the visibility of the entire family is controlled by its category. In some cases, you may want control individual components independently from the family category. You can create subcategories within a family by clicking the Object Styles button on the Manage tab in the Family Editor. Once you have created a subcategory, you can assign solid geometry to the category using the Subcategory parameter in the properties of the geometry.

The Bottom Line

Model solids and voids Being able to efficiently model will decrease the time you spend creating content and give you more time to spend on design decisions. Solid geometry is crucial for the physical coordination of components to achieve a design that will result in fewer changes during construction, where changes are the most expensive.

Master It Several tools are available to create the shapes needed to represent MEP discipline components. Each tool generates an extrusion in a unique way. Describe the difference between a swept blend and a regular sweep.

Orient geometry using reference planes and lines Reference planes and lines are the most effective way to define the orientation of solid geometry within a family. Reference planes define how an object will be inserted into a project.

Master It Knowing the resulting orientation of an extrusion prior to creating it will save lots of time by not having to duplicate modeling efforts. Nothing is more frustrating than taking the time to create a solid only to find that it is in the wrong plane. Describe the process for creating an extrusion that is associated with a custom reference plane.

Ensure the parametric change capability of models Building solid geometry to represent MEP discipline components is good. Building the geometry with parametric change capabilities is even better.

Master It Solid geometry can be defined by parameters that can change the size or shape of the geometry. Reference planes and lines are an important part of creating parametric behavior. Why?

Determine optimal visibility settings for solids The visibility behavior of solid geometry plays an important part in the creation of consistent, coordinated construction documents.

Master It It is important to know how a family will be used in a project to determine the visibility settings required for the solid geometry in the family. Why is it important to set the Detail Level visibility settings for pipe- and duct-related families?

Chapter 18

Creating Symbols and Annotation

Many of the components of an MEP design are represented on drawings as symbols. Having a well-stocked library of symbols can improve your efficiency in creating device families and reduce the time spent on drafting tasks. Symbols can also be used to create project legends that can be utilized on different project types when required.

Companies have spent many hours and significant funds to develop and maintain their drafting standards. Many think it is important to create construction documents that are recognizable as their work. Annotation styles and symbols can be created in Revit format in order to make the transition to Revit MEP 2011 without losing the signature look of your company construction documents.

Having a good understanding of how annotative objects can be used in Revit MEP 2011 will allow you to easily optimize content and create documents that align with your company standards.

In this chapter, you will learn to

- Create symbolic lines and filled regions
- Use symbols within families for representation on drawings
- Work with constraints and parameters for visibility control
- Use labels to create tags

Drafting Tools in Revit

It is easy to assume that because Revit MEP 2011 is a BIM solution that it does not have very good tools for drafting tasks. The truth is the drafting tools available in Revit MEP 2011 are quite useful for creating symbolic line work, hatch patterns, and annotation objects. Becoming proficient with these tools will greatly reduce your need to rely on a CAD application for drafting tasks. The more work you do in Revit MEP 2011, the less work you'll have to do in multiple applications, which will increase efficiency and will make managing standards and content easier.

The drafting tools available for use when drawing in a project view are the same as those available in the Family Editor. These tools appear on the Draw panel of the contextual tab when you click the Detail Lines button, Line button, Symbolic Lines button, or Model Lines button depending on what type of file you are working in. The image shown here is of the Draw panel because it appears on the Modify | Place Detail Lines contextual tab when the Detail Lines button on the Annotate tab is clicked.

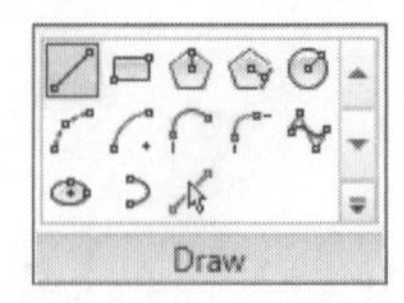

You can use the buttons in this panel to draw lines or shapes. Clicking a button for the type of line or shape to be drawn will activate different options for setting conditions for the line work on the Options Bar, depending on what is to be drawn. The Line Style panel contains a drop-down that is populated with line styles that are defined in the file.

When you click the Line button on the Draw panel, the option to draw a chain of lines appears on the Options Bar. This option will continue the Line command from the last point you click to end a line. This can save extra clicks because you do not need to click the start point for each new line you draw. If you draw a series of lines, they will be joined at their end points. If you deselect the Chain check box on the Options Bar, you will need to click in the drawing area to establish the start point of each new line you draw.

By holding down the Shift key while you are drawing a line, you will force it to snap to the orthogonal axis. The temporary dimensions allow you to see the length and angle of the line you are drawing. You can type in a length to automatically set the length of a line. You change the angle of a line by selecting it and editing the temporary angle dimension. You can also use the Drag Line End grips that appear when you click a line to change the length and/or angle.

When lines are drawn with Revit and they are connected at their end points, the end points remain connected when either of the lines is moved. There is no "stretch" command in Revit because lines maintain their connections at end points. You can select the Rectangle button on the panel to click a start point and opposite corner for an end point of a rectangle. Once drawn, the rectangle is treated as four separate lines, not a single entity. When you select any one of the lines and move it, the adjacent lines will adjust to maintain the shape of a rectangle.

There are two buttons on the panel for drawing polygons, one for a polygon inscribed in a circle and one for a polygon circumscribed around a circle. Either one gives you options on the Options Bar for the number of sides and to draw at an offset. You can select the Radius box on the Options Bar to set the radius of the imaginary circle that determines the size of the polygon.

To move or edit a shape as a whole, you must select all the lines that make up the shape. This is easily done by a selection window or by placing your mouse pointer over one of the lines and then pressing the Tab key until all the lines are highlighted for selection. Once you have selected a shape, the grips at the end points can be used to edit the shape without breaking the connection of the lines. Figure 18.1 shows two rectangles that have been edited. The rectangle on the left was edited by selecting the line on the right side and dragging the line's end point. The one on the right was edited by selecting the entire shape and dragging the end point in the upper-right corner.

The Circle button allows you to draw a circle by selecting the center point and then dragging your cursor to set the radius. You can type in a length for the radius after you click to set the center point, and the circle will be drawn automatically with a radius of your chosen length.

The temporary dimension for the radius remains on the screen after you click to set the radius. You can then click the temporary dimension to edit the radius of the circle. When you click a circle that has been drawn, the Drag Line End grip that appears can be dragged to set the size of the circle. The Options Bar also has a radius option to set the size of the circle prior to selecting a center point.

FIGURE 18.1
Editing options for line work

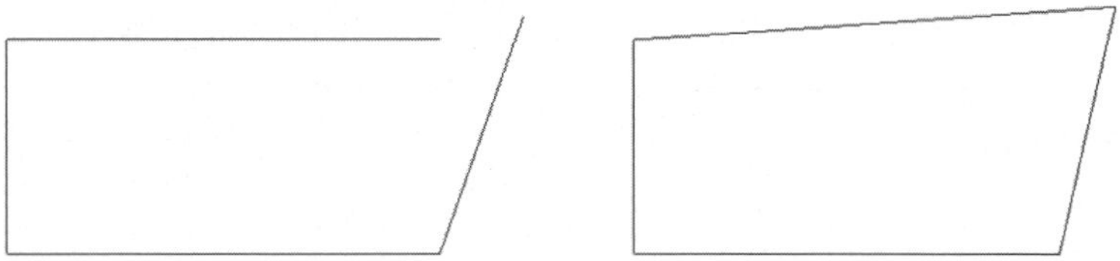

TEMPORARY DIMENSIONS

Temporary dimensions are useful for more than just modifying items as you draw. You can click a temporary dimension and click the icon that appears to change it from temporary to an actual dimension annotation in the view.

Drawing Arcs

There are four buttons on the Draw panel for drawing arcs. The first two are for drawing arcs whether or not they are connected to any other type of line work, while the other two are for drawing specific types of arcs that relate to lines.

The Start-End-Radius Arc button is used for drawing an arc by selecting a starting point and an end point and then dragging your mouse pointer to define the radius. You can preset the radius on the Options Bar. When you set the radius prior to drawing, you need to click to set the start and end points. The third click will determine the direction of the arc instead of the radius. You can input a length for the temporary dimension while drawing to establish the distance between the start and end points and then again to establish the radius. The grips that appear when you select an arc are for changing the radius or the arc length.

The Center-Ends Arc button is used for drawing an arc by first clicking to establish the center point of the arc and then clicking to set the first end point. From this end point, an arc will be drawn with a radius that is the distance from the center to the selected end point when you click to establish the second end point. You can draw an arc only at a 180-degree angle using this tool. When dragging your mouse pointer to establish the end point and moving it past 180 degrees from the start point of the arc, the direction of the arc will change.

The Tangent End Arc button can be used to draw an arc from the endpoint of a line. You must have an endpoint to start drawing this arc. Once you click the endpoint to begin drawing the arc, you can drag your mouse pointer, and an arc will appear so that the line remains tangential to it. This tool is useful for transitioning from a straight line to a curve without having to draw separate items and then adjust tangency. You can edit the radius of the arc after it is drawn by editing its temporary dimension. Because the line is tangent to the arc the larger you make the radius, the shorter the line becomes.

The Fillet Arc button is for drawing an arc that creates a filleted corner between two lines. Unlike most CAD applications that apply a fillet using a command, Revit requires that you draw the fillet arc. Clicking the button will prompt you to select the lines for the fillet. Dragging your mouse pointer will reveal the arc to be created between the two lines. You cannot type a radius length while dragging the arc, but you are able to edit the temporary dimension that appears after the fillet arc is drawn, or you can preset the radius on the Options Bar.

The Spline button is used for drawing free-form lines. Each click point creates a vertex that can be edited to change the curve of the line at that point. The Drag Line End grips that appear at the end points of a spline will lengthen or shorten the entire spline when they are dragged to new locations. You cannot type any lengths while drawing a spline, and there are no temporary dimensions to edit once the spline is complete.

The Ellipse button can be used to draw an ellipse by first selecting the center point and then defining the length of each axis. Once drawn, you can use the grips that appear at each quadrant to edit the size and shape of the ellipse. You can also edit the temporary dimensions that appear to adjust the distance from the center to the quadrant (half of the axis length).

The Partial Ellipse button is used for drawing half an ellipse. This is different from the arc tools because it creates a parabola instead of an arc with a uniform radius. When using this button, you click to establish the start and end points and then drag your mouse pointer to define the height of the parabola. You can enter dimensions while drawing, or grips and temporary dimensions are available for editing once the partial ellipse is drawn.

The Pick Lines button is used to create Revit line work by selecting lines that already exist in the view. You can select lines from a linked or imported CAD file, and a Revit line will be drawn in the same location. This tool is very useful for duplicating CAD details or symbols. You cannot use window selection to pick multiple lines, but you can use the Tab key to select multiple lines that are connected. The Offset option on the Options Bar allows you to create lines that are offset a specified distance from the lines you select.

CREATE SIMILAR

The Create Similar tool in Revit MEP 2011 is very useful for duplicating a model component, but when it is used on line work, it will activate the Line command only, not the specific button for the shape selected. For example, if you click a circle and use the Create Similar tool, you will still need to click the Circle button on the Draw panel.

Filled Regions

Filled regions are very useful in the symbols that represent model components. They can be used to represent different types of an object by controlling their visibility. There are two types of filled region patterns, one for use on model elements and one for use in details and drafting views. The Drafting pattern types should be used when creating filled regions for symbols. Drafting patterns maintain their density despite changes in the view scale.

To create a filled region within an annotation family, click the Filled Region button located on the Detail panel of the Home tab. Filled regions cannot be created in component families. When you click the Filled Region button, a contextual tab appears containing the Draw panel.

The same tools discussed earlier can be used to define the boundary of the region. The type of line used for the region boundary can be chosen from the drop-down on the Line Style panel. You can use invisible lines for the region boundary in order to avoid duplication of line work. Figure 18.2 shows two symbols with filled regions. The region on the right has invisible lines for its boundary, so only the pattern is displayed. The boundary lines of the symbol on the left are thicker than the line work of the symbol, causing the region to appear to be too large.

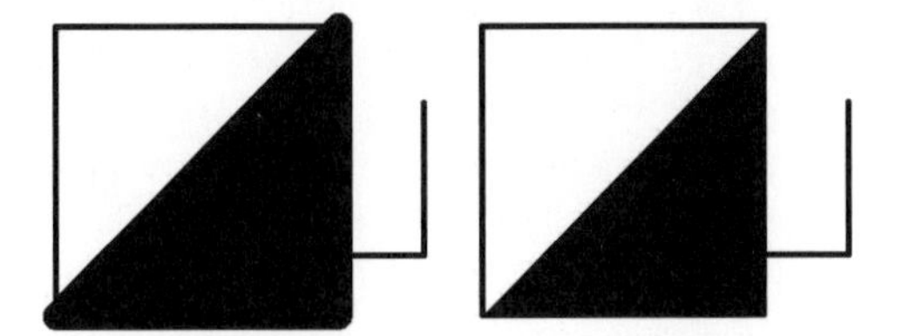

FIGURE 18.2 Filled region boundaries in a symbol

Building a Symbol Library

As with most CAD systems, it is helpful to have a library of symbols that are used repeatedly on projects and in details and diagrams. You can create a library of Revit symbols that matches your CAD library and reduce your dependence on importing CAD data into your projects. These symbols can be nested into component families in order to create construction documents that show your design as you normally would, but with the benefit of 3D model information also.

Generic Annotations

Annotation families are typically tags, but there is one type that can be used to create symbols that can be nested into component families or used directly in your project. The `Generic Annotation.rft` template can be used to create an annotation family that is not categorized as a tag. To create a new generic annotation, click the New link in the Families section of the Recent Files screen and browse to the Annotations folder. You can also click Annotation Symbol under New on the Application menu, as shown in Figure 18.3.

The generic annotation family contains a horizontal reference plane and a vertical reference plane that intersect to define the insertion point of the family. There is also a text object that serves as a reminder to assign the family to a category. When creating a symbol for use in a component family, there is no need to change the category of the annotation family. Delete the text object immediately so as not to forget to do so later. Annotation families are created at a scale of 1:1, so it is important to draw your symbol at its printed size. When used in a project or nested into a component family, the annotation will scale according to the view scale of the project or family.

If you already have a library of CAD symbols, you can use them to create your Revit symbols by importing them into the annotation family and then duplicating the line work with Revit lines. CAD files cannot be linked into families, so you must use the Import CAD button on the Insert tab when working in the Family Editor.

When importing a CAD file, the layers are imported as subcategories, so if you subsequently explode the imported CAD file, any line types defined will also be added with a prefix of IMPORT-. These subcategories are unnecessary and only add extra weight to your files. Both the line patterns and subcategories will remain in your file even if you delete the imported CAD file.

FIGURE 18.3 Creating a new annotation symbol from the Application menu

There is a workflow that will allow you to convert your existing CAD symbols to Revit without creating unwanted baggage within the Revit family. Line work that is created in an annotation family can be copied and pasted to another annotation family. You can start by creating a generic annotation family that you will import your CAD symbols into. Once you have created the line work in Revit, you can delete the imported CAD line work and copy the Revit line work to your clipboard. You can then create a new generic annotation family and paste the Revit line work into it. By doing so, you will have an annotation family that does not contain any of the imported subcategories or line patterns.

Subcategories

There is a single annotation category for all generic annotations in a project. To control the visibility and appearance of your symbol lines, you need to create subcategories for them. Otherwise, all symbols for all families would have the same appearance throughout your project. Make subcategories that are specific to the kind of symbol you are creating so you can control those specific symbols independently from other generic annotations. Be as specific with the names of subcategories as you like to avoid any confusion as to what the subcategory is.

Access the Object Styles settings in the family file from the Object Styles button on the Manage tab. Click the New button in the Object Styles dialog box, and give the subcategory a name. You can set the line weight, line color, and line pattern for the subcategory. These settings will carry through when the annotation family is loaded into a project or component family file. You can choose to use the default settings for the subcategory and override them in the project or family file that the annotation is loaded into. When you draw line work or a filled region in the annotation family, set the line style to the newly created subcategory in the drop-down on the Line Style panel. Figure 18.4a shows a subcategory created in an annotation family, and Figure 184.b shows how the subcategory appears when the annotation family is loaded into a project.

FIGURE 18.4
Subcategory settings displayed in (a) an annotation family and (b) a project

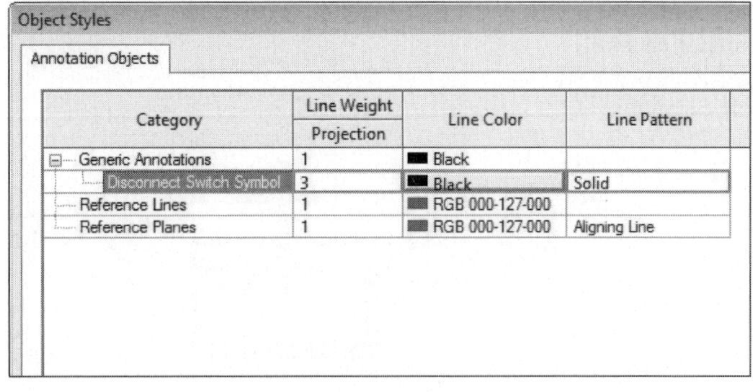

(a)

Object Styles

Model Objects | Annotation Objects | Imported Objects

Category	Line Weight Projection	Line Color	Line Pattern
Flex Duct Tags	1	Black	Solid
Flex Pipe Tags	1	Black	Solid
Generic Annotations	1	Black	Solid
Centerline	4	Black	Solid
Disconnect Switch Symbol	3	Black	Solid
Medium Lines	4	Black	Solid
North Line	8	Black	Solid
Thin Lines	2	Black	Solid
Wide Lines	6	Black	Solid
Generic Model Tags	1	Black	Solid
Grid Heads	1	Black	Solid
Keynote Tags	1	Black	Solid

(b)

With a CAD file from your symbol library imported into a new annotation family file, you can use the Pick Lines button on the Draw panel to duplicate the CAD line work. If you have defined line styles with colors, it may be helpful to invert the colors of the CAD file during import so that it is clear which lines are CAD and which are Revit. As stated earlier, the CAD file should be deleted when all the line work has been duplicated. Save the annotation to your own Revit MEP 2011 library in the appropriate discipline folder within the Annotations folder. You may find it best to separate your custom families from the Autodesk ones, making it easier to upgrade your families the next time you upgrade. The annotation family can now be used in a project file by loading it into your project and clicking the Symbol button on the Annotate tab.

Your annotation family can also be loaded into a component family for use as a schematic symbol representation of the component. Open the component family, and click the Load Family button on the Insert tab. Browse to the annotation family that you want, and click Open. On the Annotate tab, click the Symbol button to place the loaded annotation family into the view. You can only place an annotation symbol into a family in floor plan or ceiling plan views, not elevations or 3D views. Place the symbol so that its insertion point matches the insertion point of the component family. When one family is loaded into another, the loaded family is considered to be "nested." Figure 18.5 shows a thermostat annotation nested into a thermostat component family.

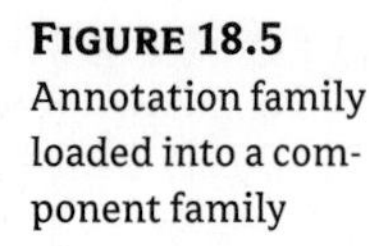
FIGURE 18.5
Annotation family loaded into a component family

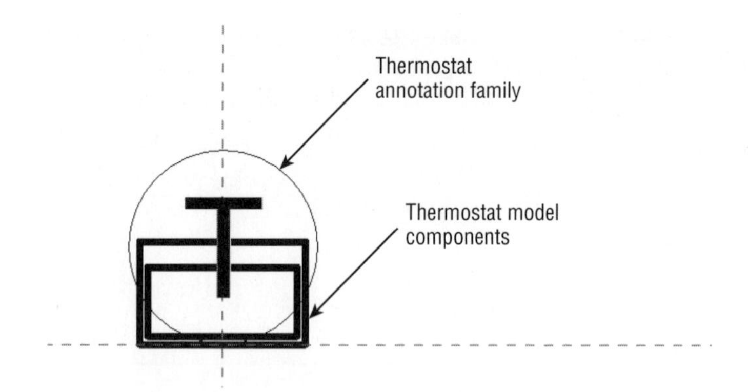

Text and Labels

You can use text or labels in your symbols along with line work and filled regions. Text is simply text within your symbol that does not change. For example, if your thermostat symbol always only contains the letter *T*, then you can use text in the annotation family. The only way to change the text is to modify it directly in the family by clicking it to change its value. If you want to be able to change the value of a text object in the family through the family's properties, you need to use a label. Labels act as family attributes and display the value of a parameter to which they are assigned.

To place text into your annotation family, click the Text button on the Home tab within the Family Editor. Select the desired text style from the Type Selector in the Properties palette. If the text style you want to use does not exist, you can create it. When you create a text style in a family and load the family into another family or a project file, the text style does not carry over into the file you are loading into. This means that if you want to use the text style in the project, you need to create it within the project also. It is possible to have a text style in a family with the same name as a text style in your project while both have different settings, such as font or text height. This can be particularly confusing when it comes to fonts. Your default font and default text heights are important considerations for your implementation. If you want to use anything other than the default Arial, you will not only need to set up your project template with all the necessary styles but also create your own family templates.

This is not as daunting a task as it first seems. Here are the steps:

1. Make a copy of all the default family templates, and place them into a folder with a custom name, such as your company name.
2. Rename all the templates with a prefix that denotes that they are yours.
3. Rename them again to *all* have the file extension `.rfa`. The template files are now all Revit families.
4. You can now edit each one to make sure the text and label styles all match your company standards.
5. Rename them one final time back to `.rft`. You now have a set of bespoke family templates!

This method can also be used to add specific shared parameters directly to the family template files, meaning you will not have to add them each time you create a new family.

Although this task can take a while to do, it's well worth doing, and if you know how, you can do some of the repetitive part by using the family upgrade tools and editing them for your own purposes with extracts from journal files. This can be a bit of trial and error, so make sure anything you do is backed up and always work offline, but once you get it right, your whole library can become personalized.

When you place a text object into your family, it is important to consider the orientation of the text when the family is used in your project. Text objects have an instance parameter called Keep Text Readable that allows you to set the orientation of the text to be readable, which is defined as text that is read from left to right or from bottom to top. Annotation families also have a family parameter called Keep Text Readable that applies to any text within the family. You can find this parameter by clicking the Category And Parameters button on the Properties panel of the ribbon, as shown in Figure 18.6. It is not necessary to set both parameters to Readable in order for the text to remain readable when the family is inserted into a project.

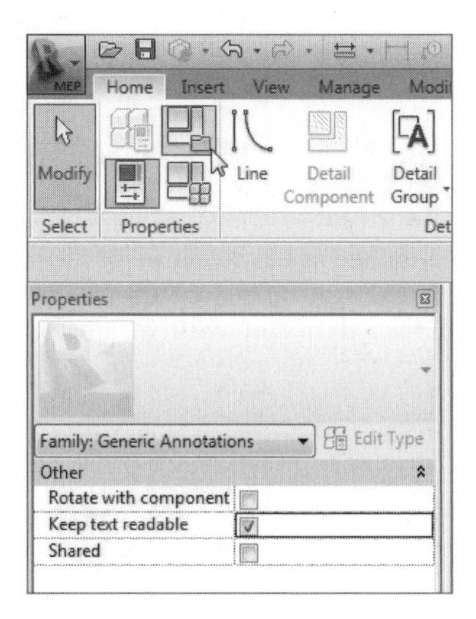

FIGURE 18.6
Keep Text Readable parameter of an annotation family

Detail Components

Although you cannot place an annotation family in any view other than a plan view, you can use a detail component family in lieu of an annotation family. This is useful for adding symbol lines or regions to a face-hosted family that does not have the Maintain Annotation Orientation parameter, such as a light fixture or electrical equipment.

A filled region is often used for electrical panels to represent panels with different voltages. You cannot create a filled region directly in the component family, and creating one in an annotation family will not work because it needs to display in the panel's Front or Back elevation view. The Back elevation view is what is displayed when a face-hosted family is hosted to a vertical surface. To show a filled region in the front or back elevation of a face-hosted family, you must create a filled region within a detail component and nest the detail component into the face-hosted family. The nested detail component can be placed in either the Front or Back elevation view to display in a project plan view.

To create a detail component, click the New link in the Families section of the Recent Files screen, or choose New ➤ Family from the Application menu. Select the `Detail Component.rft` template, and click Open.

You may want your detail component to change size with the component family that it represents, so you need to create parameters for its size. In the example of a filled region for an electrical panel, you need to make the length and width of the region parametric so that it can change size with the panel model component. Figure 18.7 shows a filled region created in a detail component that has instance parameters that define its length and width. The boundary lines for the filled region are invisible lines.

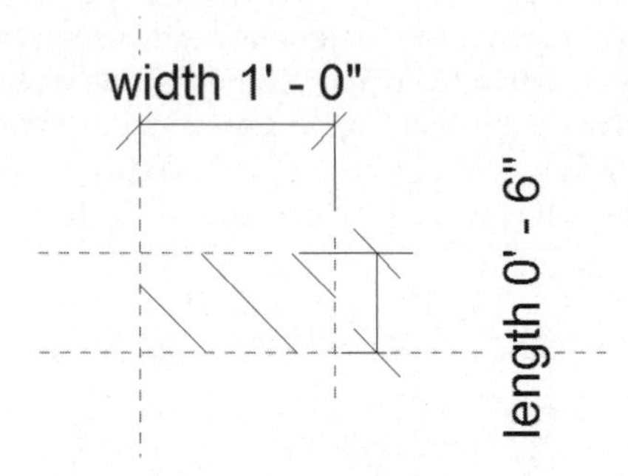

FIGURE 18.7
Filled region in a detail component family

The detail component family can then be loaded into the panel family, and the boundaries of the filled region can be constrained to the geometry of the panel object, as shown in Figure 18.8. The boundaries of the filled region are able to be stretched and locked to the reference planes in the panel family because of the length and width instance parameters in the detail component family.

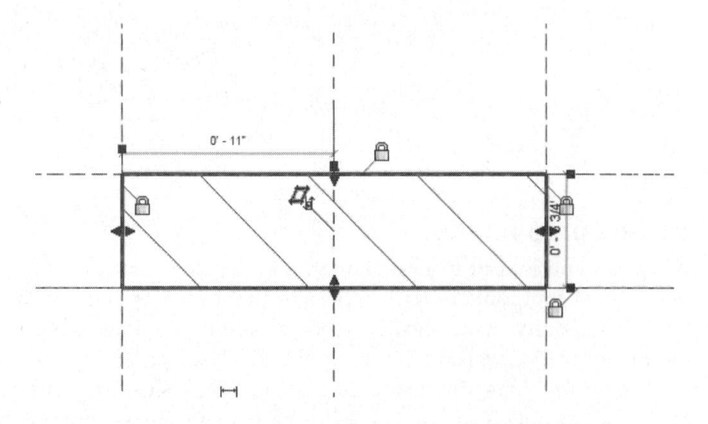

FIGURE 18.8
Detail component filled region in a panel family

When the panel family is used in a project, the filled region will adjust to the different sizes of the panel, creating the desired representation for the panel in a plan view, as shown in Figure 18.9.

Unlike annotation families, detail components are not drawn at their print size. When you create a detail component family for use as a symbol such as for a wall-mounted light fixture, you must draw the symbol to scale. The detail component will not adjust to changes to the view scale.

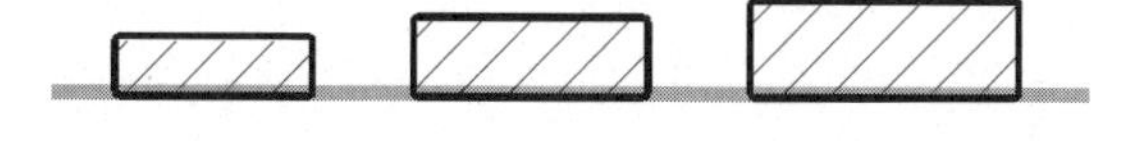

FIGURE 18.9
Various sizes of the same panel family

Real World Scenario

MAKING THE TRANSITION

It can be difficult to adopt an entirely new software solution when you have so much reusable content for your current drafting application. Using a new application does not mean you cannot create your construction documents to look the same as they do with what you currently use.

The ability to create the symbols that you use for documentation and to place them in the model components can improve the quality of your projects. Consider taking the time to create a library of annotation families that matches what you currently have. You will be surprised at how quickly your library will grow!

Visibility Control of Lines, Regions, and Annotation

Elements in a family have unique visibility control options, and parameters can also be used so that you can control what lines, regions, or text is visible for the various types within your family. This functionality allows you to have multiple symbols drawn within the same annotation family that can be nested into a component family containing multiple types. By doing so, you are able to avoid having numerous separate families for the same kind of component.

Visibility Parameters

Lines, regions, and text have a parameter called Visible that allows you to designate whether the item can be seen. It is not likely that you will create line work in your family only to turn off its visibility, but you can turn it on or off by associating its Visible parameter with a Yes/No type parameter in the family. This allows you to show certain lines, regions, or text for one family type and then turn them off to show others for another type. The receptacle annotation in Figure 18.10 has two filled regions. The vertical region is displayed to represent a GFI receptacle, while the half-circle horizontal region is for a countertop receptacle.

The visibility of each region can be set to the yes or no (selected or deselected) value of a parameter defined in the family. This is done by associating the value of the parameter with another by clicking the small button at the far right of the parameter value field. When the annotation family is loaded into a component family, its parameters can be associated with parameters in the component family that define the type of component and therefore which region to display. Figure 18.11 shows that the visibility of the vertical region is associated with a parameter in the annotation family called GFI.

Figure 18.12 shows that when the annotation is used in a component family, the visibility of the regions depends on the family type used. The process of nesting annotation families into component families is described in more detail in Chapter 22.

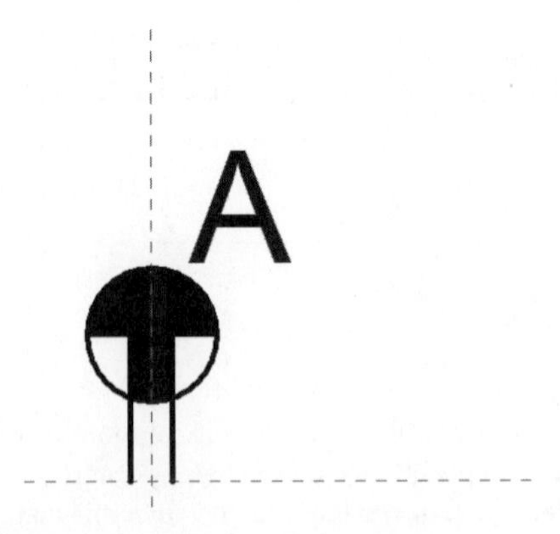

FIGURE 18.10
Annotation family with multiple filled regions

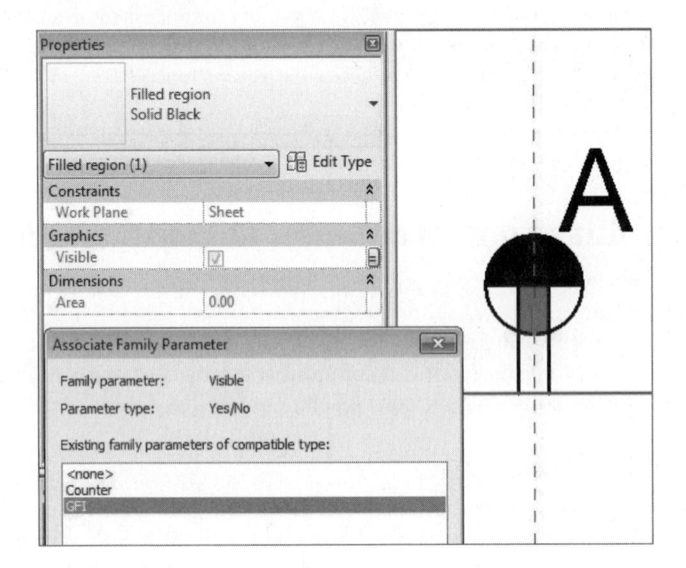

FIGURE 18.11
Visibility of a region associated to a parameter

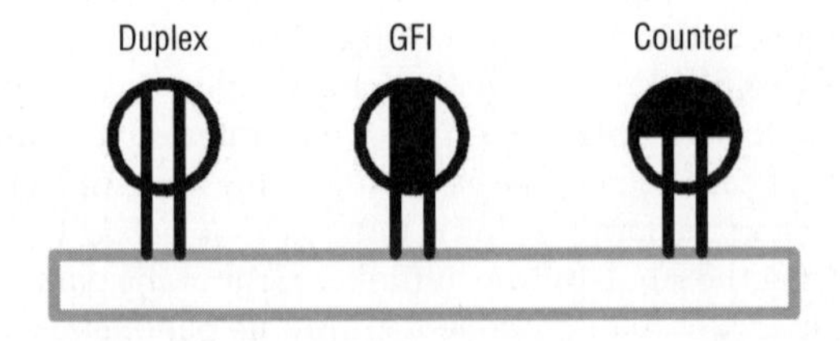

FIGURE 18.12
Display of an annotation family based on component type

Another visibility control option for annotation families is the orientation of the symbol. As stated earlier, the orientation of text within an annotation family can be controlled with the Keep Text Readable parameter. However, it may be your standard or preference to keep the text reading from left to right regardless of the symbol position. This can be achieved with a combination of text objects and visibility parameters.

You can copy a label or text in your annotation family and rotate it so that when the family is rotated, the label or text reads from left to right. The visibility of each label or text object can be set to an instance parameter allowing you to control which label or text to display at each instance of the component. Figure 18.13 shows a thermostat annotation family with two instances of a label. One has been rotated, and its visibility has been associated to a parameter in the family.

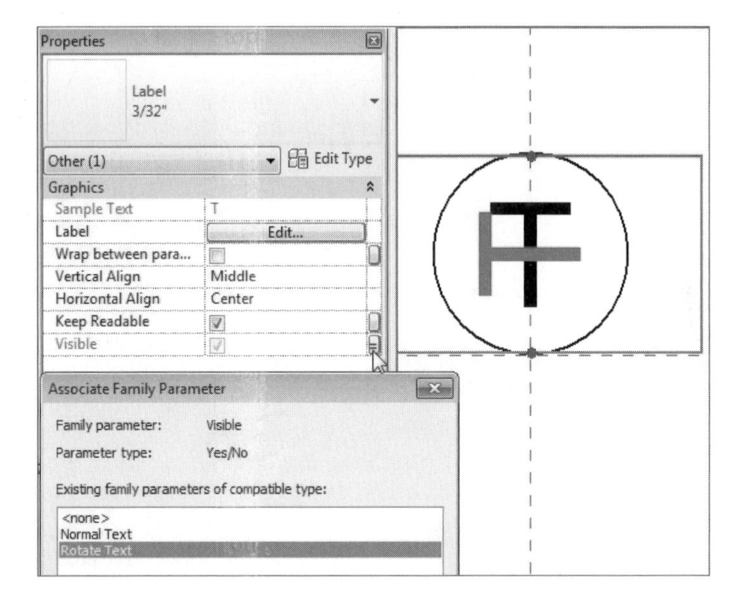

FIGURE 18.13 Annotation family with multiple instances of the same label

When used in a component family, the parameter controlling the visibility of each label can be associated with a Yes/No type parameter of the component. The result is that when the component is used in your model, you can control the visibility of the label to display the annotation symbol in the desired orientation, as shown in Figure 18.14 where the top thermostat has been modified to display the rotated label in the annotation family while the thermostat on the bottom is displayed as it normally would be.

This functionality can also be applied to the line work in your annotation family. It may be necessary to create a separate annotation family for the symbol that is to be rotated so that when you insert the families into a component family, the insertion point will be the same. Figure 18.15a shows an annotation family for a light switch, and Figure 18.15b shows an annotation family for a light switch symbol that is to be rotated when inserted into a component family.

The two annotation families can be placed into a light switch component family with the normal switch annotation placed into the family at the appropriate insertion point and the switch annotation for the rotated symbol placed into the family and rotated 180 degrees. Yes/No parameters are used to control the visibility of each nested annotation. The result is a switch family that can be displayed as desired depending on its orientation in your model. Figure 18.16 shows the switch family mounted to a wall. The switch on the right has been modified to show the rotated (or "flipped") symbol.

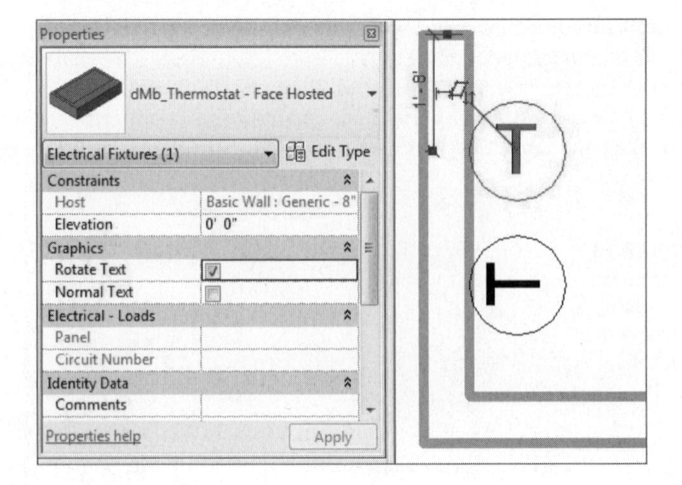

FIGURE 18.14
Multiple display options for a label within a family

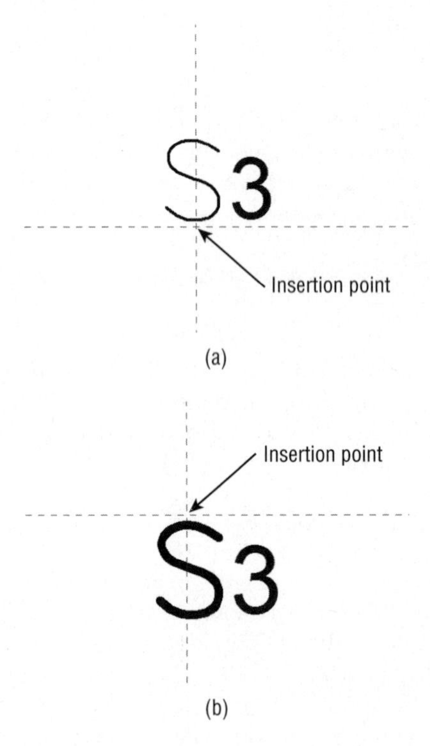

FIGURE 18.15
(a) Light switch annotation;
(b) light switch annotation to be rotated

Using Constraints

You can create reference lines in your annotation families to constrain the line work to. This allows you to make the annotation family parametric, giving you the freedom to move the annotation symbol independently from the component when using nested annotations in component families. This is most useful when working with face-hosted families.

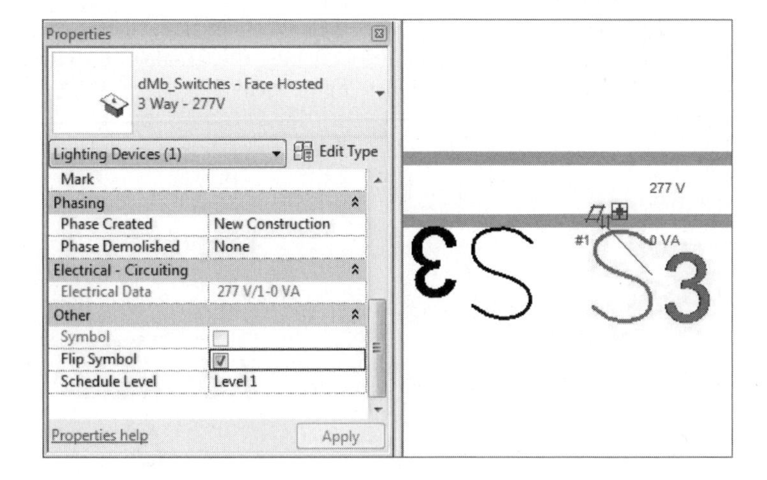

FIGURE 18.16
Display options for light switch symbol

As stated in Chapter 13, it is sometimes necessary to offset the symbol for a component so that it does not interfere with other symbols. For face-hosted items, an offset to the left or right can be created in the component family once the annotation is nested. The offset to show the symbol away from its host needs to be created directly in the annotation family.

To create an offset for a symbol that will show the symbol offset from its host, do the following:

1. Open the `Ch18_Duplex Annotation.rfa` family found at `www.wiley.com/go/masteringrevitmep2011`.
2. Click the Reference Line button on the Datum panel of the Home tab in the Family Editor.
3. Draw a horizontal reference line above the horizontal reference plane that defines the insertion point. The reference line only needs to be long enough to dimension to.
4. Dimension from the horizontal reference plane to the reference line by clicking the Aligned button on the Dimension panel of the Home tab.
5. Click the dimension, and select the Add Parameter option from the Label drop-down on the Options Bar.

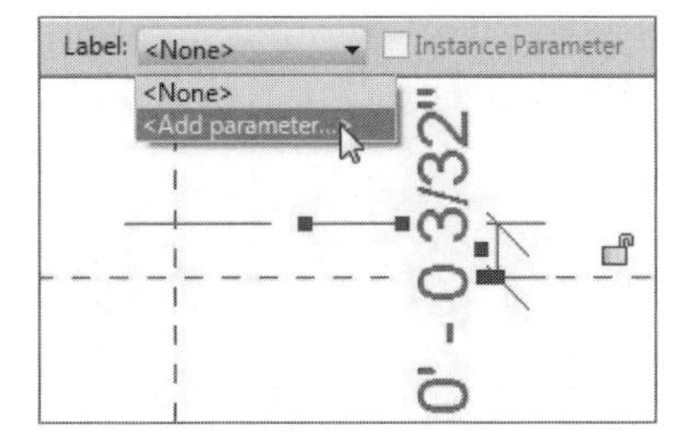

6. In the Parameter Properties dialog box, name the parameter **Symbol Offset**, choose Graphics for the group, and make it an instance parameter. Click OK.

7. Select the symbol line work and label. Do not select the reference planes or reference line. Group the selected items by clicking the Create Group button on the Create panel of the contextual tab. Click OK to name it Group 1.

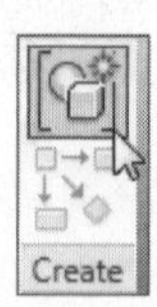

8. Click the Align button on the Modify tab, and align the group to the reference line by selecting an endpoint of one of the vertical lines in the receptacle symbol. Lock the alignment by clicking the padlock icon.

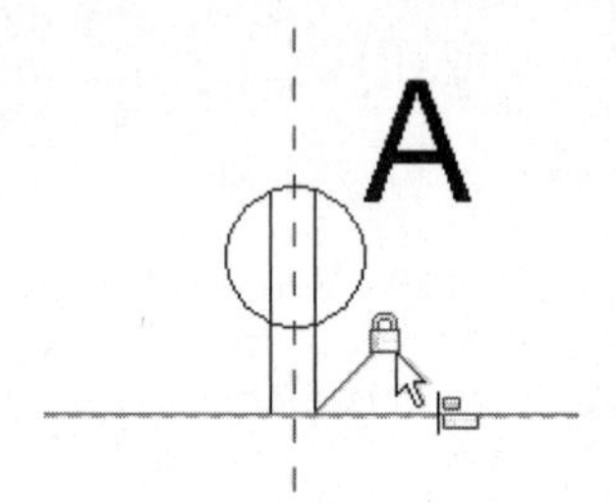

9. Click the Family Types button on the Properties panel of the ribbon. Input 0 for the value of the Symbol Offset parameter. Click OK.
10. Open the `Ch18_Duplex Receptacle.rfa` file found at `www.wiley.com/go/masteringrevitmep2011`.
11. With the duplex annotation family current in your drawing area, click the Load Into Project button on the Family Editor panel of the ribbon.
12. Place the annotation symbol at the intersection of the reference planes in the receptacle family.

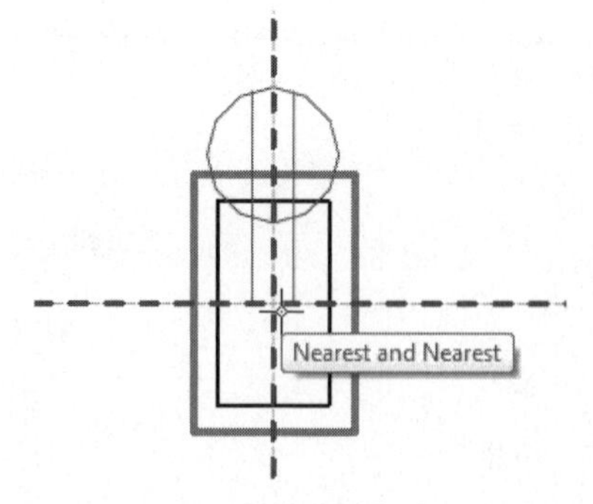

13. Select the annotation symbol, and click the small box to the far right of the Symbol Offset parameter in the Properties palette (click the Properties button to turn on the palette if

needed). In the Associate Family Parameter dialog box, select the Symbol Offset From Wall parameter. Click OK in all open dialog boxes. Save the family.

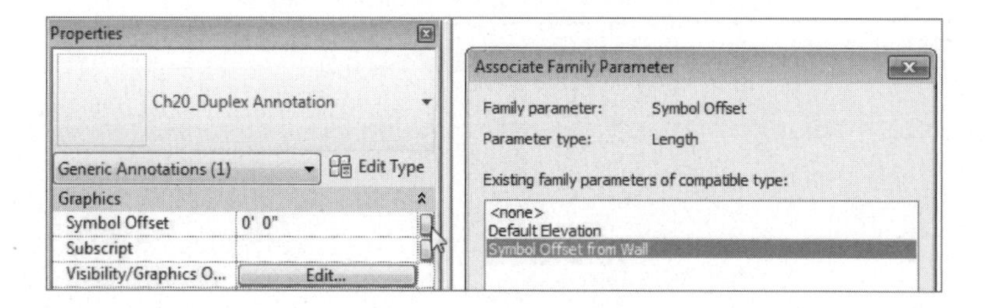

14. Load the receptacle family into a project file. Place a receptacle on a vertical host. Select the receptacle, and input a value of 1/4" (0.6 cm) for the Symbol Offset From Wall parameter. Set the Detail Level setting of your view to Fine. Notice that the symbol is offset from the wall while the component remains hosted inside the wall.

Note that you must use a value for the offset that relates to the actual size of the symbol because the symbol size in the view is determined by the view scale.

Labels and Tags

The most common type of annotation family is a tag. Tags can be created for any category of Revit model element to report information about the element on your sheet views. A tag family can contain a combination of labels and line work. The `Generic Tag.rft` family template can be used to create a tag, which can then be categorized for a specific element category. There are also category-specific tag templates that can be used, and you can even use the `Generic Annotation.rft` template and change its category to a tag. Tags are much more useful than plain text because they will update automatically when the parameters of an element are changed. This is done by using labels in your tag families. Using tags in lieu of text objects will save you editing time and improve coordination.

You can place a label into your annotation family by clicking the Label button on the Text panel of the Home tab. When you click in the drawing area to place the label, the Edit Label dialog box will appear. When using the generic annotation template, any parameters that exist in the family will be listed in the box on the left side. If none exists, you can create one by clicking the Add Parameter button in the lower-left corner of the dialog box. After giving the parameter a name and defining its type and whether it is an instance or type parameter, it will appear in the list.

When you use the generic tag family template, the parameters that are available in the Edit Label dialog box (see Figure 18.17) are parameters common to all Revit model components.

If you categorize your annotation family as a tag for a specific element category, the parameters for that element type will appear in the Edit Label dialog box. Figure 18.18 shows a list of parameters for an electrical equipment tag, including the common parameters. Any shared parameters that you have created for the element category can be added also. See Chapter 19 for more information about shared parameters.

FIGURE 18.17
Parameters available in a generic tag

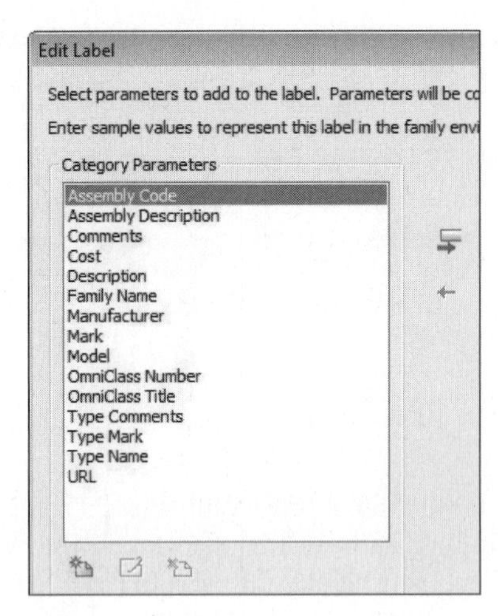

FIGURE 18.18
Parameters for an electrical equipment tag

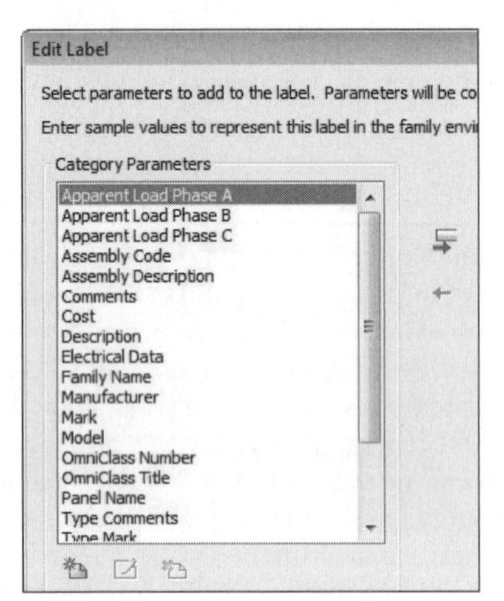

Label Format Options

You can assign a parameter to your label by selecting it from the list and clicking the Add Parameter(s) To Label button (with the green arrow).

You can apply a prefix or suffix to the label. These will appear at every location of the tag in your project. The Sample Value column allows you to input a value that will be seen when you are editing the tag family. If a value is given for the parameter while you are editing the tag family, the sample value will be overwritten.

It is possible to add multiple parameters to a single label. Doing so will list the parameter values in the same label object. Selecting the box in the Break column of the Edit Label dialog box will create a hard return after the value of the parameter when it is displayed in the label. This enables you to have a label with multiple lines of text, as shown in Figure 18.19.

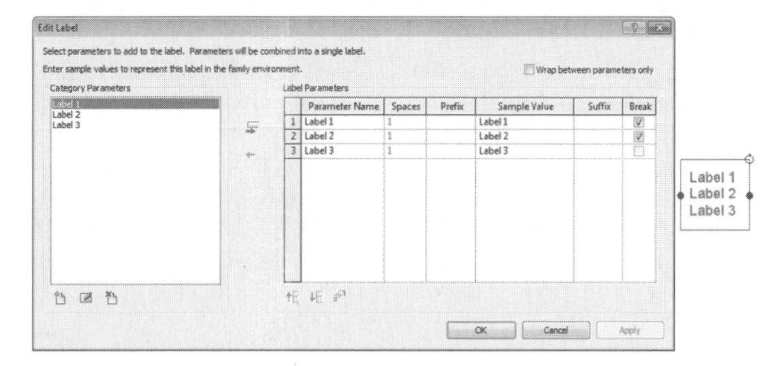

FIGURE 18.19 Single label with multiple parameters

If you do not use the Break option, the parameter values will display on a single line. If the single line of parameter values exceeds the bounding box of the label, the values will wrap in the same manner as a text object. You can select the box in the upper-right corner of the Edit Label dialog box to wrap the values between parameters only. This prevents values of a single parameter from ending up on multiple lines. Figure 18.20 shows a multiparameter label on a single line and the same label with its bounding box shortened and the Wrap Between Parameters Only option selected. Notice that the value for the Description parameter did not wrap.

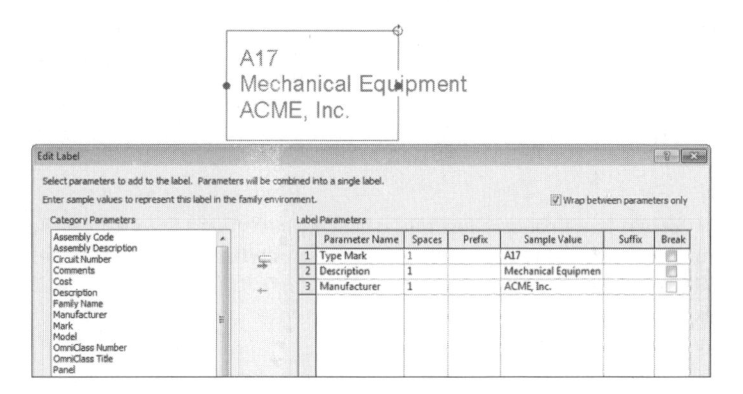

FIGURE 18.20 Multiparameter label

When you click OK to close the Edit Label dialog box, the label will appear with the sample values. The label properties can be edited to display the parameter values as desired. Select the label, and access the Properties palette to edit the alignment of the text and its readability instance properties. The type properties of the label define the text used to display the parameter values. You can assign a color and line weight to the text and define whether the background is transparent or opaque. A tag with a label that has an opaque background will mask out any elements that it is placed over in your project views.

In the Text group of a label's type properties, you can assign a font and text size. You can also make the text bold, italicized, or underlined. You can apply a width factor to the text as well.

FOLLOW THE LEADER

You cannot define the type of arrowhead for a tag family leader when you are editing the family. The leader arrowhead definition is a type property of a tag family after it is placed into a project.

Labels and Line Work

Your tag families can be a combination of labels, text, and line work. It is important to position these items around the family's insertion point. Improper placement of items in a tag family will cause inaccuracies in the location of leader lines when the tag is placed into a project. The same drafting tools used to create an annotation family can be used in tag families.

Figure 18.21 shows a duct tag that contains labels and line work. There are two labels so they can be positioned independently. A suffix was added to each label. The line work is positioned at the intersection of the reference planes for proper leader orientation when the tag is used in a project.

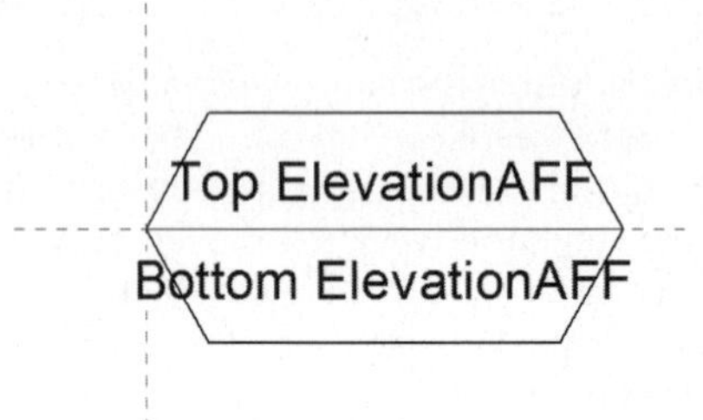

FIGURE 18.21
Sample duct tag family with line work and labels

> **OH, SNAP!**
>
> You cannot snap to text objects or labels, so you have to position them manually.

When you properly locate the graphics in your tag families, their leaders will maintain connection with the tag graphics regardless of which direction the leader is pointing, as shown in Figure 18.22.

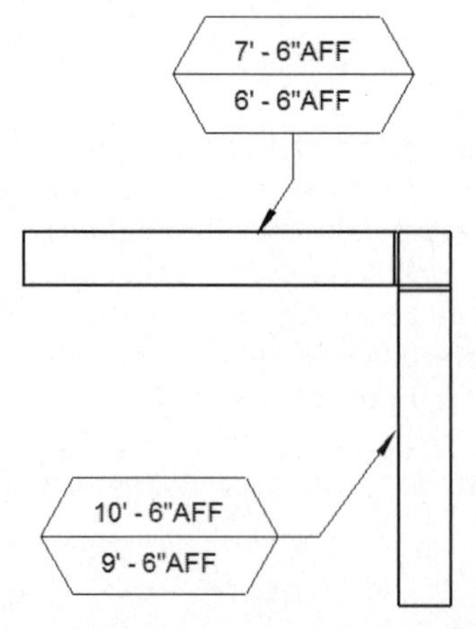

FIGURE 18.22
Tags with leaders

When you place a tag into your project view, there is an option on the Options Bar to orient the tag vertically or horizontally. If you want your tag to align with its associated item, you can select the box in the Rotate With Component parameter. You can find this by clicking the Family Category And Parameters button on the Properties panel of the ribbon when editing the family. Setting your tag family to rotate with its associated component disables the option to set the tag to horizontal or vertical because the tag will align with the component, as shown in Figure 18.23.

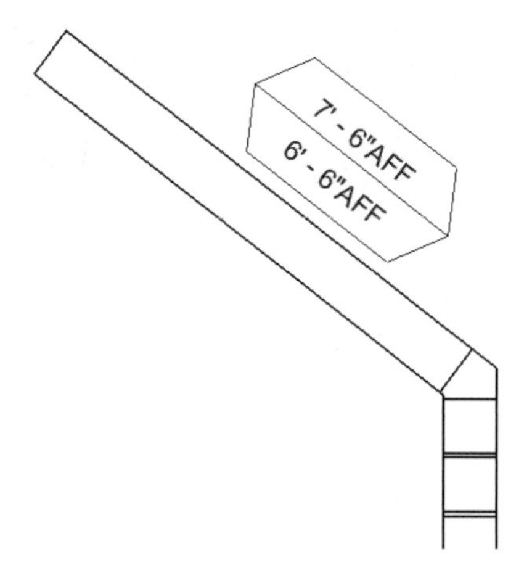

FIGURE 18.23
Tag aligned with angled duct

The Bottom Line

Create symbolic lines and filled regions Not only is Revit MEP 2011 a modeling application, but it contains the tools necessary to accomplish drafting tasks.

Master It Having a good command of the tools available for creating symbols will help you create families that represent your design elements exactly the way you want to see them. What line tool is best suited for duplicating the line work of an imported CAD symbol?

Use symbols within families for representation on drawings Given the schematic nature of MEP plans, symbols and annotation objects are important parts of your Revit MEP 2011 workflow, allowing you to represent your model components per your company standards.

Master It By having annotation symbols nested into your component families, you can create an accurate 3D model that is displayed schematically on your construction documents. Explain the importance of creating subcategories for the graphics in your annotation families.

Work with constraints and parameters for visibility control The parametric modeling capabilities of Revit MEP 2011 make it a powerful BIM solution. These capabilities can be used in annotation families as well.

Master It A common scenario for a Revit project is to link consultant files into your project file. Because of this, face-hosted families are often used. Face-hosted components can be attached to either a vertical or horizontal host, so the ability to separate the annotation symbol from the host needs to be created in the annotation family.

When using a length parameter to define the offset of a symbol from its host, what value should be input for the parameter when the component family is in a project?

Use labels to create tags Tags are the most commonly used annotation families in a project. They are used to report information about objects in a Revit model.

Master It The use of labels is a much more effective method than using text objects for keeping documents coordinated.

If your project requires that you show DIA (diameter) after each pipe size tag on your construction documents, how can you accomplish this globally throughout your project?

Chapter 19

Parameters

Revit is sometimes referred to as a parametric modeling application. Parameters are the very core of what makes Revit MEP 2011 such a powerful design and modeling tool. Parameters hold the computable data that defines the properties of not only model components but everything that makes up a Revit project. They are the characteristics of all elements of a Revit project that ultimately determine behavior, appearance, performance, and information.

Parameters and properties are often considered synonymous, but it is *parameters* that determine the properties of a component. Properties may be static, yet parameters allow for change to be propagated throughout the project. For example, you can have an object with a length property, but it is a length parameter that allows you to change the length as needed.

There are three basic kinds of parameters in Revit MEP 2011. Some parameters are hard-coded into the software. The values of these parameters can be edited as needed, but the parameters themselves cannot be removed or modified. In this chapter, these will be referred to as *coded parameters*.

Family parameters are used to build and define graphical structure and engineering data within component families. These parameters can be customized as needed to enhance the capabilities of component objects and to extract and analyze data.

Shared parameters are useful to help maintain consistency within families and to coordinate information within a project. They are the most useful kind of parameter because they can be used in component families, schedules, and tags and annotations.

When you realize the power of parameters in Revit and understand the types of things you can achieve with them, you will have a better understanding of why Revit can improve your work-flow processes and the efficiency of your design projects.

In this chapter, you will learn to

- Manipulate the properties of parameters
- Work with parameters in families
- Use shared parameters
- Use parameters in project files

Parameter Properties

Before you can understand how to use parameters to drive the properties of your objects, you need to understand the properties of parameters. When you create a parameter to hold some form of computable data, you want to define the way in which it will do so. Figure 19.1 shows the Parameter Properties dialog box accessed from within the Family Editor. Other versions of this dialog box that contain additional settings are discussed later in the chapter. This dialog box is the first step when adding a parameter to either a family, a project, or a schedule.

FIGURE 19.1
Parameter Properties dialog box

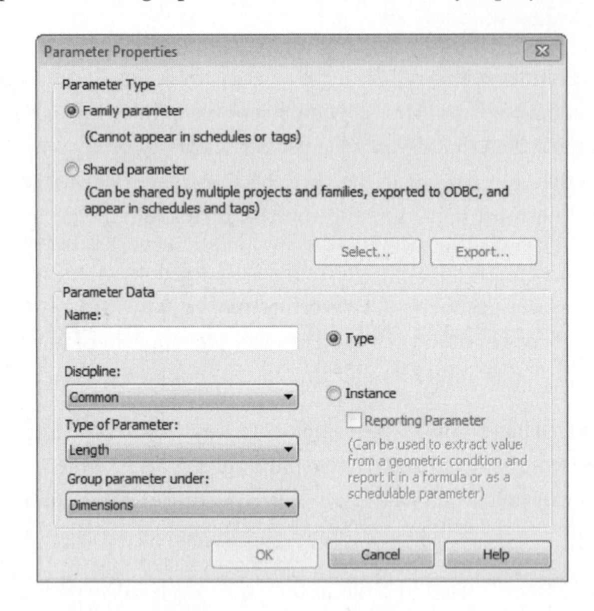

The first decision to make is what type of parameter to create. Family parameters can be created when working in the Family Editor, and they are limited to being used only within the family. The information that they hold cannot be used in schedules or reported by a tag or annotation. Shared parameters can be used the same way that family parameters are used, but they are unique in that they can also be used in schedules or tags. Parameter types are discussed in further detail later in this chapter.

Parameter Naming

When you choose to add a parameter, it is likely that you have a specific purpose for it. That may sound like an obvious statement, but it is important to consider when you decide what to name the parameter. There is no harm in naming your parameters in a descriptive manner, especially when you are working with others who need to understand the purpose of a parameter. However, it is possible to be too verbose. Long parameter names can cause the annoyance of having to resize columns within dialog boxes in order to read the name.

Consistency is the key to good parameter naming. It can be frustrating to go from one family to another and see different names for parameters of the same information. Using descriptive words is also helpful, especially when you have similar parameters within the same object, such as a component that is made up of multiple shapes with each requiring a width parameter. Having W1,

W2, W3, and so on, as parameter names can be difficult to work with, whereas using names like Housing Width, Lens Width, and Bracket Width make it easier to make adjustments or changes when working in the Properties palette or Type Properties dialog box of the object.

If you intend to abbreviate measurements such as length, height, or radius, be sure to use a consistent format. Will you use the abbreviation as a prefix or suffix to the descriptive portion of the name? Will punctuation such as dashes or parentheses be used? Decide up front whether you will use *height* or *depth* to describe the third dimension of an object.

Parameter naming is important because when you refer to a parameter in a formula, calculated value, or filter, Revit is case sensitive and content sensitive. Spelling and capitalization accuracy are critical, so develop a naming convention that is as simple as possible while still being easily understood.

Type Parameters

Type parameters are the reason that you can have multiple variations of a family within one file. Family types are driven by one or more type parameters. When you are creating a parameter, it is important to decide whether the parameter will be used to define a type within the family.

Type parameters can cause the most damage when misused because they enact changes to every instance of the family type to which they belong. For this reason, you will receive a warning when editing a type parameter in a schedule view, and accessing a type parameter in a model view requires an extra mouse click.

ALWAYS DOUBLE-CHECK

When editing a type parameter of a family in a project, it is a good idea to double-check that you do not need to create a new family type. It seems there is always that one instance of the family somewhere else in the model that gets changed when it should not have.

Type parameters do not always have to define a family type. In some cases, you may want to define a parameter as a type parameter so that changes can be made to a family everywhere the family exists in a project. For example, if you are creating a parameter for the finish color of an object, you could make it a type parameter so that when you change the value of the parameter, it changes all instances of the object.

Type parameters do require that if you need to change just one or a few instances of an object, you will have to create a new family type. This can lead to having several types within a family, causing your Type Selector to be cluttered and confusing. If you are creating a type parameter that will define a family type, it is best to name the family type as it relates to the value of the type parameter(s). A light fixture family defined by its width and length parameters would likely have family types with names such as 2 4 (600 1200) and 1 4 (300 1200), for example.

Instance Parameters

Instance parameters provide the most flexibility for editing an object. They are easily accessed via the Properties palette when an object is selected. Create instance parameters for values that you want to be able to change for just the selected object. Using the finish color example again, if the color is an instance parameter, then you could have one family type that could vary in color, without having to create a separate family type for each color option.

The drawback to instance parameters is that they apply only to the selected objects, so if you want to change an instance parameter value for all instances of an object, you will have to select each object and change it individually or use the tools that allow you to select all instances of a family.

An instance parameter can be set to be a reporting parameter. This is a new feature for Revit MEP 2011. A reporting parameter will hold a value that can be used in formulas for other parameters or to drive the behavior of another parameter. These are most useful in families that are wall or ceiling hosted because you can use a reporting parameter to recognize the thickness of the host wall, for example. Some portion of a dimensional reporting parameter must be associated to the host in order to be used in a formula.

Once you have defined a parameter as either instance or type, you can change it if required for the desired behavior. However, switching a parameter from type to instance, or vice versa, can cause problems if the parameters are used in formulas, so it is best to know up front what kind of parameter to create.

Parameter Discipline, Type, and Grouping

The Discipline drop-down list in the Parameter Properties dialog box contains the different disciplines that can be assigned to a parameter. Parameter discipline is important for defining the measurement units that the parameter value will have. Figure 19.2 shows the drop-down list of available disciplines.

FIGURE 19.2
Parameter discipline options

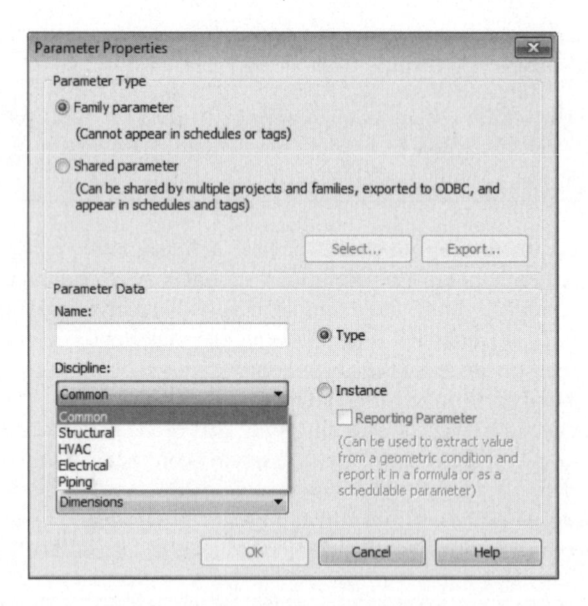

The Type Of Parameter option in the Parameter Properties dialog box is directly related to the chosen discipline. Each discipline has a unique set of parameter types that relate to the various units of measurement for that discipline.

Figure 19.3 shows the Type Of Parameter options for the Common discipline. Notice that many of the types are the same as in the Project Units settings for the Common discipline of a project such as Length, Area, and Volume.

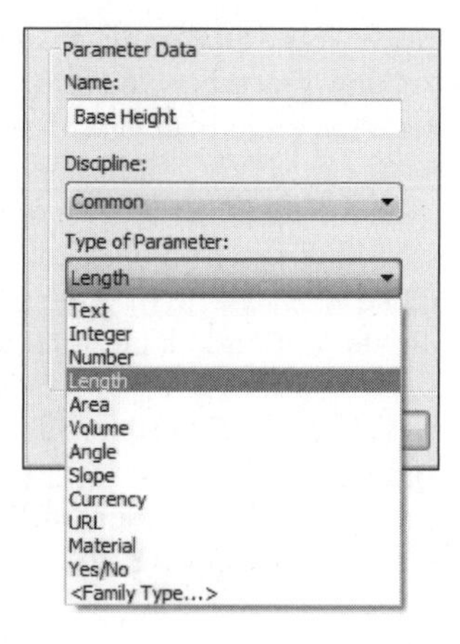

FIGURE 19.3
Parameter type options for the Common discipline

There are additional options for parameter values that are not a unit of measurement. The Text option allows you to input anything for the value of the parameter. This is the most versatile option, but from an engineering standpoint, it offers the least amount of intelligence because a text string provides only information, not computable data. If you are creating a parameter that is scheduled and want the ability to input either numbers or text or a combination of characters, then the Text option is best.

The Family Type option for a parameter is a useful tool when using multiple nested families within a family. You can create a Family Type parameter to toggle between all the nested families. When the host family is loaded into a project, the parameter can be modified to display any of the nested families by selecting from the list in the Family Type parameter. Figure 19.4 shows an annotation family for graphic scales with several nested annotations loaded. A Family Type instance parameter has been created to allow for use of any of the nested families.

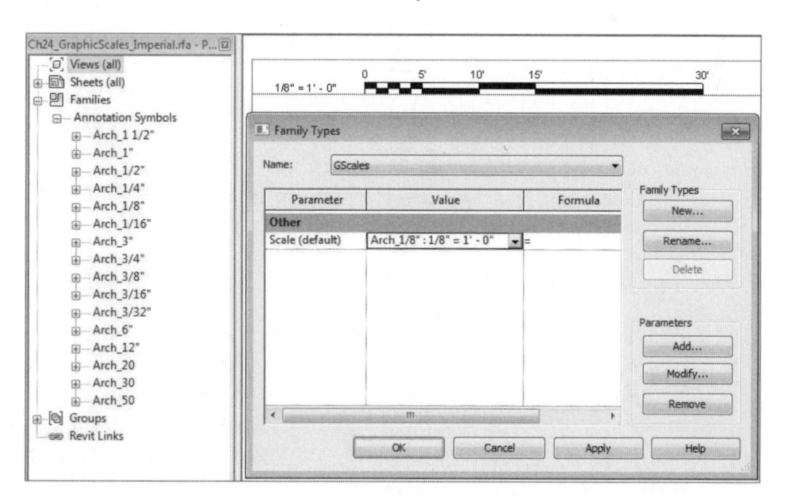

FIGURE 19.4
Family using Family Type parameter

You can use the Yes/No option if your parameter requires a simple yes or no value. The value for this type of parameter appears as a check box, which can be used to control the visibility of objects or to verify that a condition exists.

Other disciplines have options for Type Of Parameter that relate to units of measurement for that discipline. It is important to know that when you select a specific Type Of Parameter setting the value used for that parameter must be consistent with the unit of measurement. For example, if you choose the Air Flow option for the HVAC discipline, you could not input a value of anything other than a number consistent with the unit of measurement you are using for air flow. This can cause problems with schedules when an object does not have a value for this parameter and you want to use something like N/A or — to indicate that the value is not actually 0.

By default, the Type Of Parameter option will be set to Length. This is new to Revit MEP 2011, whereas in older versions the default was Text. In older versions, it was easy to overlook this setting because of the versatility of the Text type, but with Length being the default now, it is important to set the proper type.

You can determine where the parameter will show up in the Properties palette or Type Properties dialog box of an object, as shown in Figure 19.5.

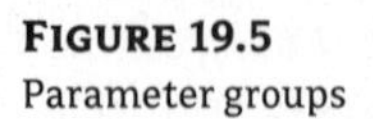

FIGURE 19.5
Parameter groups

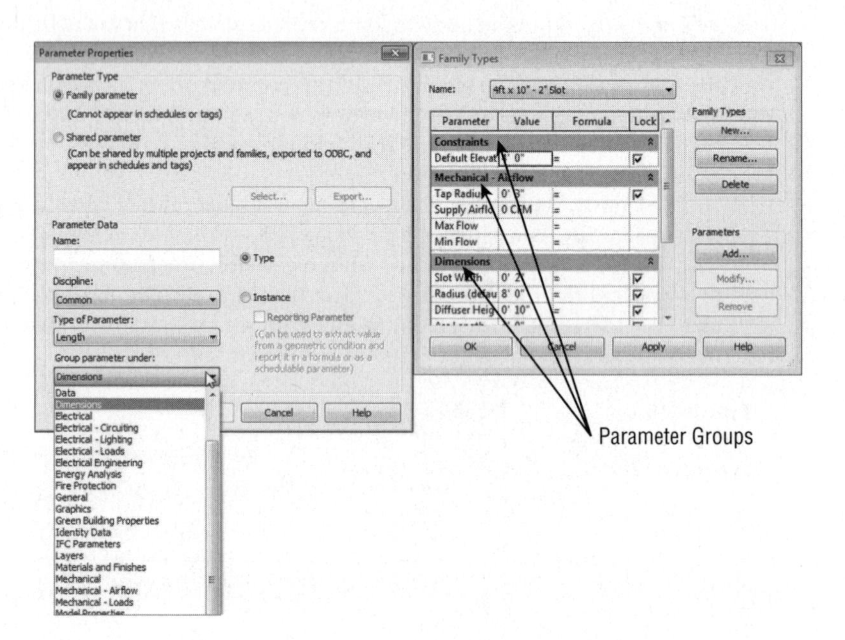

The option for grouping parameters is sometimes confusing to people because they think that it is related to the Type Of Parameter setting. This setting does not have any bearing on the Type Of Parameter or Discipline settings, so you could have a Duct Size parameter that is placed in the Identity Data group. To make it easier to keep parameters organized, new functionality has been added that automatically associates a parameter type with an appropriate group. So if you create a Length parameter, it will be placed in the Dimensions group, Air Flow parameters are placed in the Mechanical-Airflow group, and so on. You still have the option to place a parameter in any group you choose, but now the defaults are more intuitive.

Parameter grouping is another area where being consistent is important to improved workflow and efficiency. You want to be able to find your parameters in the same location for each family while editing in your model.

Parameters in Families

Parameters are primarily created when working in the Family Editor. As content is created or edited, it becomes clear what type of data is needed for either analysis or reporting or to drive the geometry. The new functionality in Revit MEP 2011 for editing parameters in the Family Editor makes it easier to work with dimensional parameters in the drawing area without having to access the Family Types dialog box.

Dimensional Parameters

You now have the ability to lock dimensional parameters in the Family Editor so they cannot be changed while working on the geometry of a family. There is now a Lock column in the Family Types dialog box with a check box for each dimensional parameter. What is nice about this feature is that if you do not lock a parameter, you can change its value while working on the geometry, eliminating the need to stop and access the parameter to change its value manually. An object that is constrained by a dimension can be moved, and the parameter's dimension value will adjust. This eliminates the pesky Constraints Are Not Satisfied warning dialog box that appeared in previous versions. However, this warning will appear if a dimensional parameter is locked and an object is moved.

You also have the ability to edit the value of a dimensional parameter in the drawing area by clicking the text, just as you would edit a dimension object. This can be done whether the parameter is locked or not. Locking only prevents the accidental dragging of an object while working in the drawing area.

Parameter Types

When creating type parameters in a family, there will be a set value for each family type. When the family is inserted into a project, the values established in the family will remain until the family type is edited. Instance parameters can be given a default value when created in a family. These parameters are easily identified in the Family Types dialog box via a suffix of (default). This is the value that the parameter will initially have when placed into a project. If you edit a family that exists in your project and then load it back in, the instance parameter values will not change from their existing states in the project. Fortunately, the dialog box that warns you that the object already exists in the project gives you the option to overwrite the family and its parameter values, which will change any instance parameters to the value as it exists in the family.

As mentioned earlier, Yes/No parameters are great for controlling the visibility of objects. After creating a Yes/No parameter, you can select the desired object and set its Visibility parameter to the value of the Yes/No parameter using the small box at the far right of the Visibility parameter value. Figure 19.6 shows the settings used to associate the visibility of line work with a Yes/No parameter.

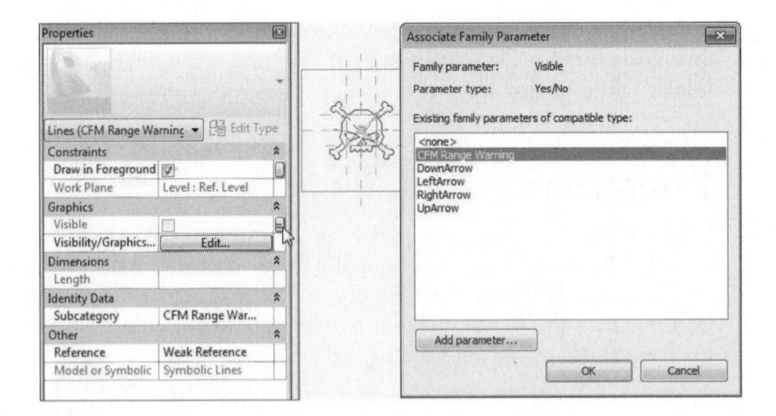

FIGURE 19.6
Yes/No parameter used in a family for visibility

Real World Scenario

YES/NO PARAMETERS IN NESTED ANNOTATION FAMILIES

To control the number of annotation families that are nested into a family, Yes/No parameters within an annotation are very useful. One example is for electrical receptacles. Many receptacles look essentially the same in the model, but the symbol used for each type may vary. Instead of creating a separate annotation symbol for each type of receptacle, consider using Yes/No parameters within one annotation to account for each symbol. The image shown here is an annotation family with all of the line work and regions required to represent each type of symbol used.

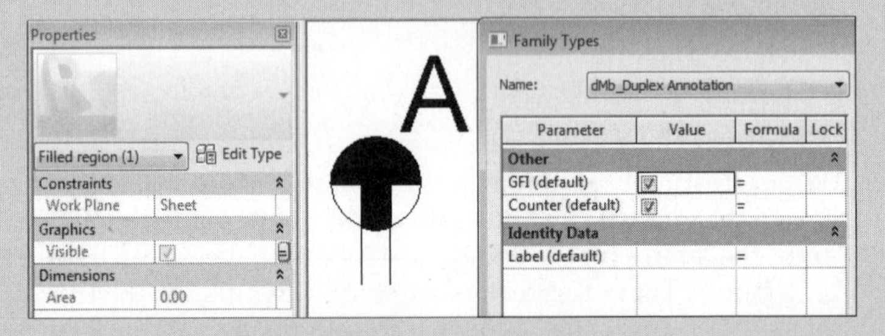

The visibility of the lines and regions are associated with Yes/No parameters so that when this annotation is nested into a family, the Yes/No parameters can be associated with parameters in the family. The result is a single receptacle family that can represent multiple types of receptacles, as shown here.

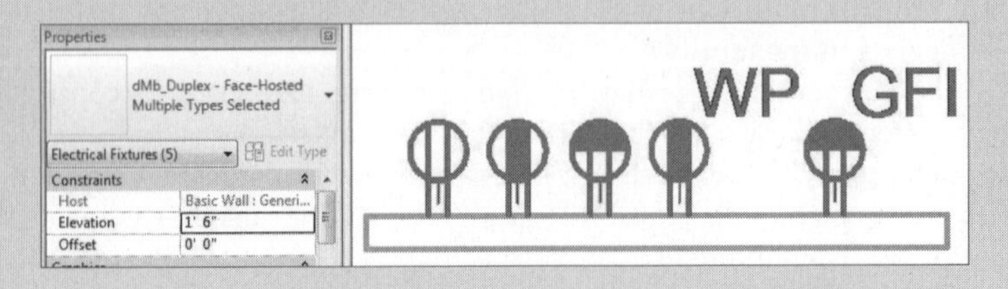

Type Catalogs

Families can sometimes become very crowded with many types. The number of type parameters used to define a family type will increase the number of family types. When these families are loaded into a project, all of the family types are loaded. This can quickly cause your project to be overloaded with unused family types. One way to remedy this scenario is to create type catalogs for families that contain many types.

A type catalog is a `.txt` file that contains values for the type parameters of a family. Having a type catalog associated with a family allows you to select only the family types you want to load when you insert the family into a project. You can create a type catalog for a family by creating a `.txt` file with the same name as the family in the same folder as the family file. Because type catalogs are `.txt` files in comma-delimited format, it is easier to edit them using a spreadsheet program such as Microsoft Excel.

To create a type catalog, do the following:

1. Start a new spreadsheet file. The first column of the spreadsheet will be a list of all the family types. Each column after that will be a type parameter within the family. The type parameters must exist in the family file and have some value in order to be used in the type catalog.

You cannot use instance parameters in a type catalog. If you leave a parameter value blank in the family, the type catalog will ignore the parameter when the family is loaded. So even though the parameters in the family are essentially placeholders for the type catalog, they need to have something input for their value. Figure 19.7 shows an example of a type catalog for a motor connection family.

FIGURE 19.7 Sample type catalog

	A	B	C	D	E
1		Horsepower##Other##	Device Voltage##Other##	Device Load##Other##	No. of Poles##Other##
2	120V - 1/6HP	0.17	120	506	1
3	120V - 1/4HP	0.25	120	667	1
4	120V - 1/3HP	0.33	120	828	1
5	120V - 1/2HP	0.5	120	1127	1
6	120V - 3/4HP	0.75	120	1587	1
7	120V - 1HP	1	120	1840	1
8	120V - 1 1/2HP	1.5	120	2300	1
9	120V - 2HP	2	120	2760	1
10	120V - 3HP	3	120	3910	1
11	120V - 5HP	5	120	6440	1
12	120V - 7 1/2HP	7.5	120	9200	1
13	120V - 10HP	10	120	11500	1
14	208V - 1/6HP	0.17	208	520	2
15	208V - 1/4HP	0.25	208	686	2
16	208V - 1/3HP	0.33	208	853	2
17	208V - 1/2HP	0.5	208	1165	2
18	208V - 3/4HP	0.75	208	1643	2
19	208V - 1HP	1	208	1914	2
20	208V - 1 1/2HP	1.5	208	2392	2
21	208V - 2HP	2	208	2870	2
22	208V - 3HP	3	208	4077	2
23	208V - 5HP	5	208	6698	2
24	208V - 7 1/2HP	7.5	208	9568	2
25	208V - 10HP	10	208	11960	2

It is easy to see why a type catalog for this family is used; otherwise, all 23 family types would be loaded when this family is inserted into a project. Notice that the format for the type parameters is ##parameter name##group name##. The group name is relatively unimportant because the parameter is already grouped in the family file, so Other can be used. For consistency, however, it is best practice to use the actual name that the parameter is grouped under. It is not a necessity, but this format is important because otherwise the columns cannot be parsed when the family is inserted into a project. Use decimals for measurement values. The units of a dimensional parameter should be defined in the column heading, such as ##DuctSize##Other##Inches.

2. Once you have finished creating the spreadsheet, you can save it as a `.csv` (comma-delimited) file. If you receive a warning that some of the features of the file may not work when saved to `.csv` format, you can click Yes to save the file.

Once the file is saved, you can browse to it and rename the file with a `.txt` extension to convert it. If you receive a warning that the file might become unusable when the extension is changed, click Yes.

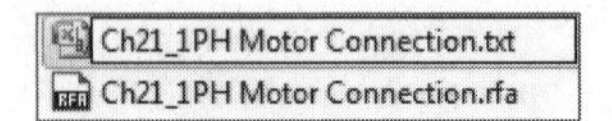

When a family with an associated type catalog is inserted into a project, a dialog box will appear that allows you to see the parameter values for each type and choose which family types you would like to load.

3. Choose family types by selecting them from the list. You can select multiple types, as shown in Figure 19.8. You can filter the dialog box by specific parameter values using the drop-down list under the heading of each column.

FIGURE 19.8
Specifying types from a type catalog

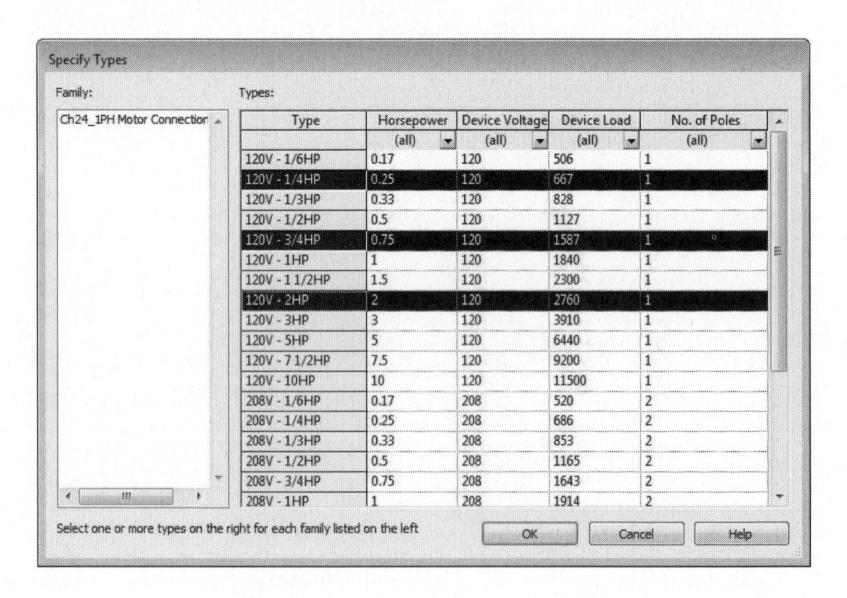

Type	Horsepower	Device Voltage	Device Load	No. of Poles
	(all)	(all)	(all)	(all)
120V - 1/6HP	0.17	120	506	1
120V - 1/4HP	0.25	120	667	1
120V - 1/3HP	0.33	120	828	1
120V - 1/2HP	0.5	120	1127	1
120V - 3/4HP	0.75	120	1587	1
120V - 1HP	1	120	1840	1
120V - 1 1/2HP	1.5	120	2300	1
120V - 2HP	2	120	2760	1
120V - 3HP	3	120	3910	1
120V - 5HP	5	120	6440	1
120V - 7 1/2HP	7.5	120	9200	1
120V - 10HP	10	120	11500	1
208V - 1/6HP	0.17	208	520	2
208V - 1/4HP	0.25	208	686	2
208V - 1/3HP	0.33	208	853	2
208V - 1/2HP	0.5	208	1165	2
208V - 3/4HP	0.75	208	1643	2
208V - 1HP	1	208	1914	2

DRAG AND DROP FAMILIES WITH A TYPE CATALOG

The type catalog functionality of a family will work only if you use the Load Family option from the Insert tab of the ribbon. Dragging and dropping a family file that has a type catalog into your project will load only the default type. To access the type catalog after a family has been loaded into your project, you can locate the family in the Project Browser, right-click, and select Reload.

Formulas

Not all parameters in a family are used to directly drive the geometry. Many are used to hold data that results from the creation of the geometry, which will be used for driving additional geometry or spatial relationships. This is done by creating a formula for a parameter. One of the most common occurrences of this is when a diameter measurement is required. Revit will only dimension the radius of a circle, so a diameter parameter must be created by using a formula. The radius of the geometry is dimensioned and applied to a parameter, which is then used in a formula to generate the diameter measurement.

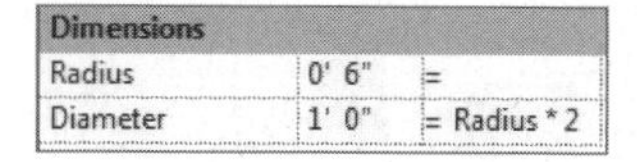

Dimensions		
Radius	0' 6"	=
Diameter	1' 0"	= Radius * 2

One nice feature of formulas is that you can change the value of the parameter that is created by a mathematical formula, and the value of the parameter used in the formula will update accordingly. Mathematical operators and Boolean functions can all be used. Placement of parentheses, proper units, case, and context sensitivity are all very important for your formulas to work properly. A warning will appear if the result of a formula does not match the units for a parameter as well as if a parameter name is misspelled.

Formulas can even be used for parameter types such as a Yes/No parameter. Figure 19.9 shows a Boolean formula for a Yes/No parameter to determine when the box should be selected. The formula indicates that the box is either deselected or selected when the conditions of the formula are true.

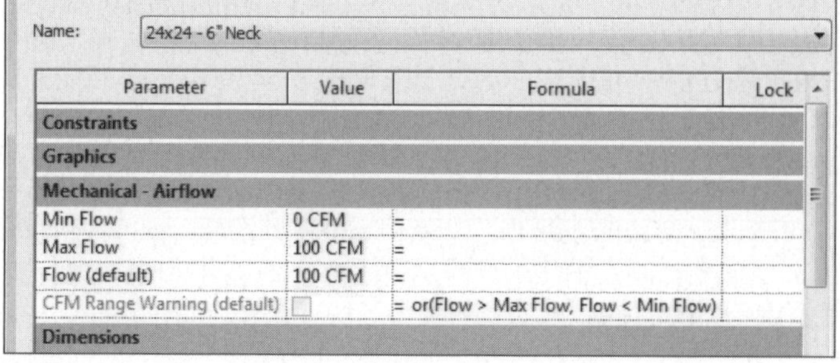

FIGURE 19.9
Boolean formula for a Yes/No parameter

Family Types

Name: 24x24 - 6" Neck

Parameter	Value	Formula	Lock
Constraints			
Graphics			
Mechanical - Airflow			
Min Flow	0 CFM	=	
Max Flow	100 CFM	=	
Flow (default)	100 CFM	=	
CFM Range Warning (default)	☐	= or(Flow > Max Flow, Flow < Min Flow)	
Dimensions			

Formulas using `if` statements are very powerful for providing exact conditions and variations in parameter values based on other parameter values. The format for an `if` statement is as follows:

```
if(condition, result, default result)
```

The default result is the value given to the parameter when the condition is not met. You can use other parameters to define the condition. For example, if you want a Width parameter to equal the Length parameter under certain conditions, you could write this formula:

```
if(Length>2' 0"(600mm), Length, 1' 0"(300mm))
```

This formula would cause the Width value to equal the Length value when the Length is greater than 2′-0″ (600mm); otherwise, the Width would be 1′-0″ (300mm).

CROSSING PARAMETER TYPES IN FORMULAS

When referencing a parameter in a formula, it is important to know that you cannot use an instance parameter in a formula for a type parameter.

Coded Parameters

When you create a family, certain parameters exist by default. These parameters are hard-coded into the software and cannot be removed. You can use these parameters where they apply to avoid having to create custom parameters.

These parameters vary depending on the category you choose for the family. The coded parameters under the Identity Data group are common to component families and are also included in system families within a project.

Identity Data
Keynote
Model
Manufacturer
Type Comments
URL
Description
Assembly Code
Cost

There are, however, a few that do not appear in the Family Types dialog box when working in the Family Editor but do appear in the Type Properties dialog box once the family has been loaded into a project. The most notable of these is the Type Mark parameter. This parameter is often used to identify a component with a tag or in a schedule. Because this parameter does not exist in the family file, you cannot use it in a type catalog. One way to avoid creating a custom parameter that does essentially the same job as the Type Mark parameter is to have your type catalog create family types named with the same value you would use for the Type Mark. Then you can tag or schedule the Type parameter instead of the Type Mark parameter.

Lookup Tables

One coded parameter that is unique is the Lookup Table Name parameter that exists in families that can utilize a lookup table for dimensions. These are primarily duct, conduit, and pipe fitting families. When you install Revit MEP 2011, an extensive library of lookup tables is installed that can be referenced by families.

Lookup tables are `.csv` files that work like a type catalog. They provide values for dimensions based on other dimensions within the family. The data in lookup tables can be driven by design codes or manufacturing standards to ensure the graphical accuracy of your components. Pipe fittings, for example, have a nominal diameter that is used to identify the size, but the actual outside diameter is slightly different especially for different pipe materials. A lookup table can provide the outside diameter dimension for each nominal diameter that exists in the table.

The Lookup Table Name parameter is used to identify which `.csv` file the family is referencing. The location of your lookup tables is defined in your `Revit.ini` file. When you type in the name of a lookup table, you do not need to include the full path to the file, only the name and file extension. As with parameter names, referencing a lookup table name is case and context sensitive.

Once you have referenced the lookup table with the Lookup Table Name parameter, you can access the data in the table by using a formula for the value of a parameter. The formula using lookup table data is as follows:

```
text_file_lookup(Lookup Table Name, "Column Name", Value if not found in table,
Value found in table)
```

The result of this formula will apply the value found in the table to the parameter or the defined value given in the formula if none is found in the table. The image shown here is the formula used to determine the value of the Fitting Outside Diameter parameter of a pipe fitting family. The FOD column is searched for a value that coincides with the value given for the Nominal Diameter and applies it to the Fitting Outside Diameter parameter. If the Nominal Diameter value given in the family does not match one in the table, then the Nominal Diameter + 1/8″ is used for the Fitting Outside Diameter.

Dimensions		
Tick Size (default)	115/256"	= Fitting Outside Diameter * 0.4 * tan(Angle / 2)
Nominal Radius (default)	1/2"	=
Nominal Diameter (defa	1"	= Nominal Radius * 2
Insulation Radius (defaul	9/16"	= Fitting Outside Radius + Insulation Thickness
Fitting Outside Diameter	1 1/8"	= text_file_lookup(Lookup Table Name, "FOD", Nominal Diameter + 0' 0 1/8", Nominal Diameter)
Center to End (default)	13/16"	= Center Radius * tan(Angle / 2)
Angle (default)	90.000°	=
Other		
Lookup Table Name	Pipe Fitting - Generic.csv	=

For more information on working with parameters in family files, see the Families Guide, which can be accessed from the Documents On The Web section of the Revit MEP 2011 Help menu.

Shared Parameters

Shared parameters are the most versatile parameters you can use, but they also require the most management. Used properly, shared parameters can help ensure that your schedules are coordinated and that your construction documents are reporting the correct information. Shared parameters can be type or instance parameters that are used in families or as project parameters. The main advantage to using shared parameters is that the data they hold can be exported or reported in tags and schedules.

Shared parameters are parameters that are created with their settings stored in a .txt file. It may help to think of this file as a library of parameters, similar to a library of model components.

Shared Parameters File

Do not attempt to edit your shared parameters' .txt files with a text editor. They should be edited through the Revit interface only.

When you need to add a parameter to a family or project, you can use one from your shared parameters file. This helps with the management of your content and project standards because you can be consistent in your use of parameters. It also helps with maintenance by allowing you to avoid duplication of parameters, which can cause coordination issues. Multiple parameters with the same name can show up as available for use in a schedule, and you will not be able to tell which one is the correct one to use.

You can create a shared parameter by doing the following:

1. Click the Shared Parameters button on the Manage tab. When you create a parameter in the Family Editor or add a project parameter, you have the option for it to be a shared parameter. Selecting this option will activate the Select button in the Parameter Properties dialog box.
2. In the Edit Shared Parameters dialog box, you must first create a shared parameters file. Click the Create button to select a location for the file.

You can have multiple shared parameters files, so it is a good idea to create a folder in a common location that you and others can access. You can access these files by clicking the Browse button in the Edit Shared Parameters dialog box, which is shown in Figure 19.10.

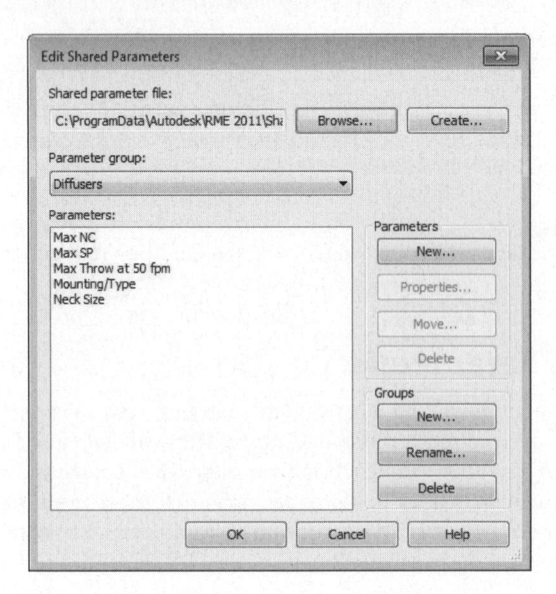

FIGURE 19.10 Edit Shared Parameters dialog box

There are two components to a shared parameters file. The parameter group is a level of organization that you can establish to group parameters together. This is not the same group as shown in the Parameter Properties dialog box for defining where parameters will be listed. These groups are so that you can keep your shared parameters organized. A typical method of organization is to create groups based on family categories. Figure 19.11 shows an example of parameter groups created for parameters that apply to specific family categories.

FIGURE 19.11
Sample parameter groups

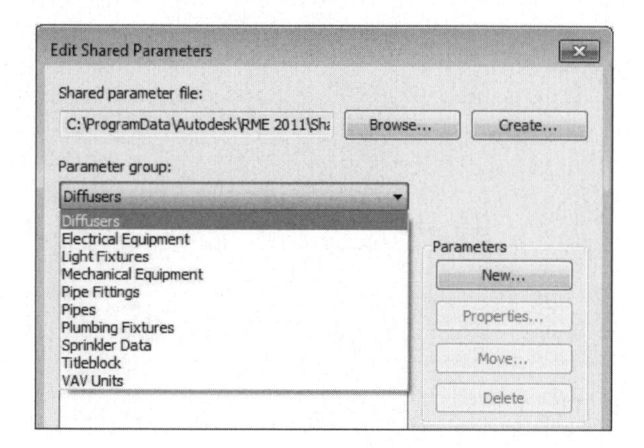

Notice that a group has been created for VAV units. Although it is possible to create a group with any name, keep in mind that the parameters must be assigned to a category. In the case of these parameters, they will be applied to all mechanical equipment if used as a project parameter even though they are specifically designed for VAV units. The parameters in this group could be added to a lighting fixture family if chosen. Parameter groups can be renamed or deleted using the buttons at the lower right of the dialog box. You cannot delete a group until all parameters have been removed from it.

3. After you have established a group in your shared parameters file, you can begin to create parameters. Click the New button in the Parameters section of the Edit Shared Parameters dialog box to open a Parameter Properties dialog box. This is not the same dialog box that you get when you create a parameter in the Family Editor or in a project. This is a very simple dialog box because all you need to define for a shared parameter are the Name, Discipline, and Type Of Parameter settings, as shown in Figure 19.12. The parameter that you create will be added to whatever group you have active when you click the New button.

The Name, Discipline, and Type Of Parameter settings are the same options as when creating a family parameter.

4. Once a parameter has been created, you can select it and click the Properties button to view its settings. The settings cannot be changed once a parameter has been created. If you want to change a shared parameter, you have to delete it and re-create it with the new settings. If you do so, you will have to add the parameter back to any object that had the parameter being replaced.

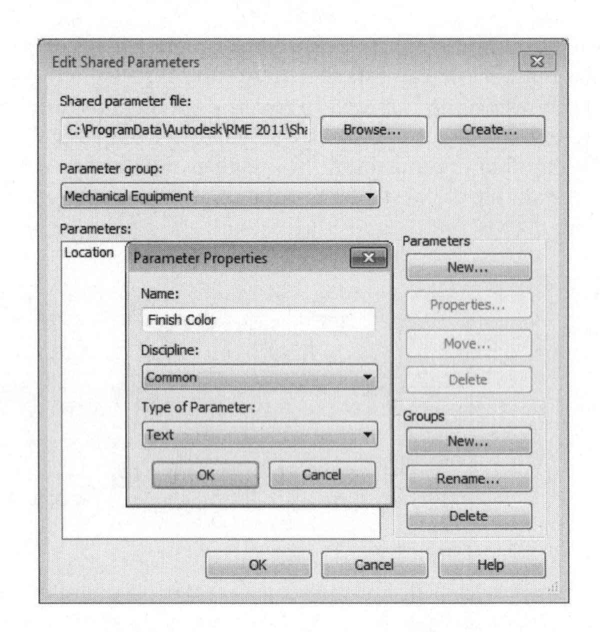

FIGURE 19.12 Parameter Properties dialog box for a Shared Parameter

Choose and establish shared parameter settings very carefully. A parameter that is deleted from your shared parameters file will remain in any families or projects, but you will not be able to add it to anything new.

You can add a shared parameter to a family by doing the following:

1. Select the Shared Parameter option in the Parameter Properties dialog box. This will activate the Select button, which opens the Shared Parameters dialog box.
2. Choose the group that contains the desired parameter, and then select the parameter from the list.
3. Once you click OK, you still need to define whether it will be a type or instance parameter and where it will be listed. These are the only two settings that can be modified once the parameter is added to the family.
4. Click the Edit button in the Shared Parameters dialog box to open the Edit Shared Parameters dialog box, where you can browse to another shared parameters file or make changes to the active file.
5. Once you exit this dialog box, you still need to select the parameter from the Shared Parameters dialog box to add it to the family. Once a shared parameter is added to a family, it can be used as a constraint or in formulas like any other parameter.

Managing shared parameters should be treated with the same importance as managing your content library. Because these parameters provide intelligence that carries through from a family all the way to your construction documents, it is important that they are maintained and used correctly.

One category where shared parameters can become cumbersome is the Mechanical Equipment category. Typically, many of the characteristics of a mechanical unit are required to be scheduled, so shared parameters are necessary. Some of these characteristics are the same unit of measurement but for a different component of the unit. Since you cannot add the same shared parameter to a family more than once, you may need to make multiple parameters of the same type. It is best to try and keep your shared parameters as simple as possible for this category. Naming parameters specifically for their use is helpful in keeping track of them, and developing a standard for where these parameters are grouped in the families will help you avoid confusion when editing the properties of a family.

Even though it may require a bit of work initially, adding these parameters directly to your Mechanical Equipment families rather than as project parameters will go a long way in keeping your families from becoming overcrowded with unused parameters. Consider keeping a document such as a spreadsheet that lists all of your custom parameters, whether they are family or project parameters, what parameter group they exist in if they are a shared parameter, where they are grouped in the properties of a family, and whether they are used as a type or instance parameter. Having this document open will be very helpful when creating new content because you will know what parameters already exist and how to use them. As new parameters are created, the document can be updated. If you work in an environment with multiple users, it is best to keep only one copy of this document in a common location.

Using Parameters in Projects

Parameters are typically handled at the component level for building objects, but there are also parameters for noncomponent objects such as views, sheets, and annotations. There may be a need for you to create custom parameters for system families that cannot be edited in the Family Editor. These parameters can be added to designated categories within your project so as to assign them to system families. Your projects themselves can have parameters that convey project-specific information. Understanding how to use parameters in a project is the key to getting the most benefit from constructing an intelligent model with computable data.

Project Parameters

The only way to add a parameter to a system family is to add it by creating what is known as a *project parameter*. This allows you to customize the information you want from elements within the model. Space objects can be given a lot of useful data to help make design decisions and to analyze the model performance. Project parameters make it possible to add this data to spaces and other elements that cannot be physically edited.

You can add a project parameter by clicking the Project Parameters button on the Manage tab. In the Project Parameters dialog box, you can see a list of any parameters that have been added to the project. Clicking the Add button opens the Parameter Properties dialog box, as shown in Figure 19.13. This is the same dialog box as in the Family Editor but with additional settings to assign the parameter to a category.

The settings for creating a parameter or adding a shared parameter are the same for project parameters. The only difference is the Categories section of the dialog box. This is where you can select the Revit category that the project parameter you are creating or the shared parameter you are loading is applied to in the project.

Project parameters are a great way to use shared parameters in families without having to edit each family individually. The check box in the lower-left corner indicates that the parameter will be added to all elements in the project that belong to the selected category. So if there are parameters that you want to use on a particular category of elements, such as light fixtures, you can create the parameters as shared parameters and then load them as project parameters into your project template file. Then whenever you load a family belonging to that category into your project, it will have the desired parameters, which can be used for scheduling or tagging.

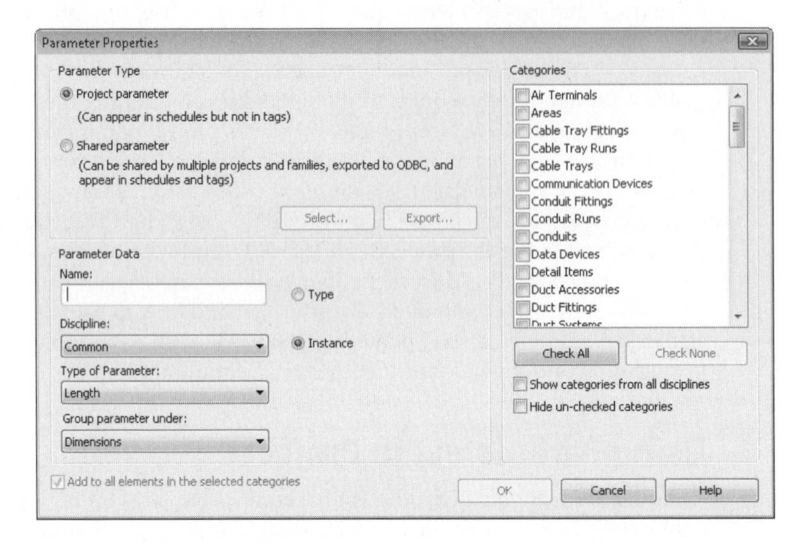

FIGURE 19.13 Parameter Properties dialog box for a project parameter

The Categories section lists all MEP-related categories that the parameter can be applied to. Like the Visibility/Graphics Overrides dialog box, you can select the box below the list to show all Revit categories. Project parameters can be added to multiple categories, allowing you to quickly add information to multiple objects.

DUPLICATE PARAMETERS

It is possible to create a project parameter with the same name and settings as a parameter that already exists in the elements to which it is being applied. This may not pose a problem for objects already loaded into your project, but keep in mind that a component may be loaded that already has a parameter equal to one of your project parameters. Managing project parameters is as essential as managing shared parameters and content.

When you add a parameter to your project and it is applied to elements in the chosen category, the parameter will not have a value. The parameters can be easily given values by creating a schedule of the elements and including the project parameter(s) in the schedule.

Yes/No type parameters will be set to Yes (checked) by default and appear to be uneditable when they are viewed in a schedule or the properties of an element. You can click the grayed-out check box once to make it editable.

Parameters in Schedules

Family parameters and some coded parameters cannot be used in schedules. If you do not want to use shared parameters in your families, you can create project parameters for scheduling information about components. In the Schedule Properties dialog box at the center of the Fields tab is an Add Parameter button, as shown in Figure 19.14. Clicking this button opens the Parameter Properties dialog box just like when working in a family, but instead of the Family Parameter option, you have the Project Parameter or Shared Parameter option (see Figure 19.15).

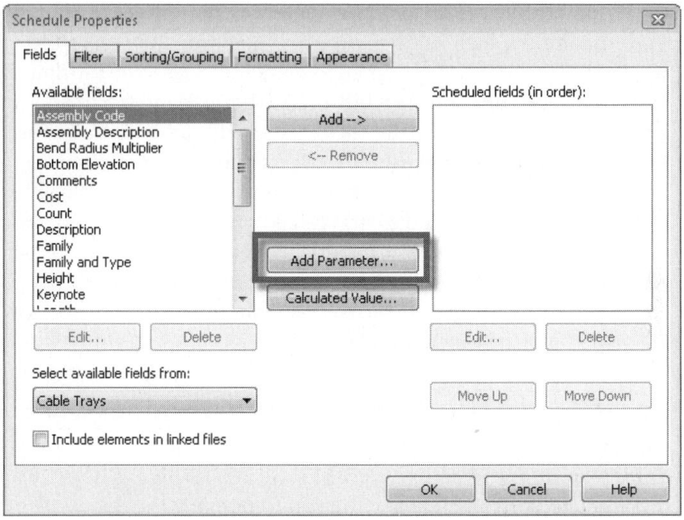

FIGURE 19.14
Schedule Properties dialog box for adding parameters

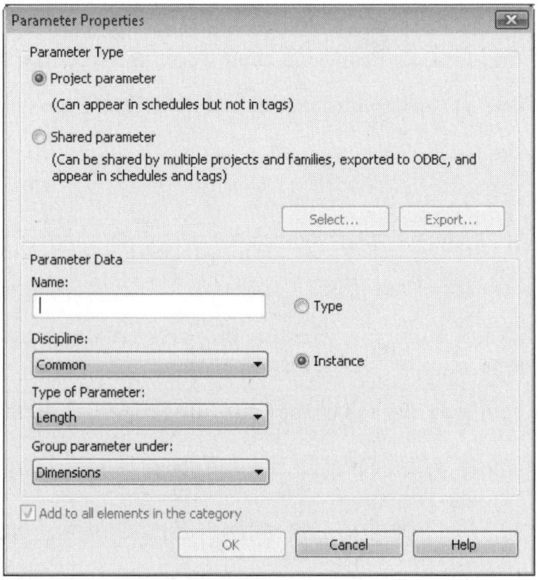

FIGURE 19.15
Parameter Properties dialog box

As you can see in Figure 19.15, there is no choice for a category in this dialog box because the parameter will be applied to the category of the schedule. All elements within the category will get the parameter even if they are not in the schedule. For example, if you are creating a Mechanical Equipment schedule that will only include VAV units and you create a parameter within the schedule, it will be added to all mechanical equipment.

Use of the Calculated Value feature of a Revit schedule will create a parameter to hold the value, yet this parameter is not added to the elements in the schedule.

Whether they are shared parameters or not, project parameters are required for scheduling system families such as duct, pipe, or cable tray. It is easiest to create these parameters when you are building the schedule for such elements. Once the parameter is created, you can access it from the Project Parameters button on the Manage tab to add it to other categories or make any necessary changes. Creating these parameters in schedules within your project template will ensure that they are consistently used from project to project.

One useful type of project parameter to create is for the schedule type of an element. This parameter can be applied to any category that is scheduled and is very useful for filtering your schedules. There is no need for this to be a shared parameter since it is not information that you will be tagging in your construction documents. This kind of parameter should be included in your project template to ensure consistent use and coordination with any preestablished schedules you have in your template(s).

Creating and Using Parameters in Families and Schedules

Understanding how to use parameters to get the information you require is the key to successfully reaping the benefits of a BIM solution. Knowing where and when to use certain types of parameters will make it easy for you to manage the data within your Revit projects.

1. Download the `Ch19_Project.rvt` and `Ch19 Shared Parameters.txt` files found at `www.wiley.com/go/masteringrevitmep2011`.
2. Open the `Ch19_Project.rvt` file. Open the VAV SCHEDULE view. Access the Fields tab of the Schedule Properties dialog box, and click the Add Parameter button.
3. Create a Project parameter called **Schedule Type** with the following settings:

   Instance

   Discipline: Common

   Type Of Parameter: Text

   Group: Identity Data
4. Click OK to exit the Schedule Properties dialog box. For each VAV listed in the schedule, enter a value of **VAV** for the schedule type.
5. Access the Filter tab of the Schedule Properties dialog box. Apply the following settings:

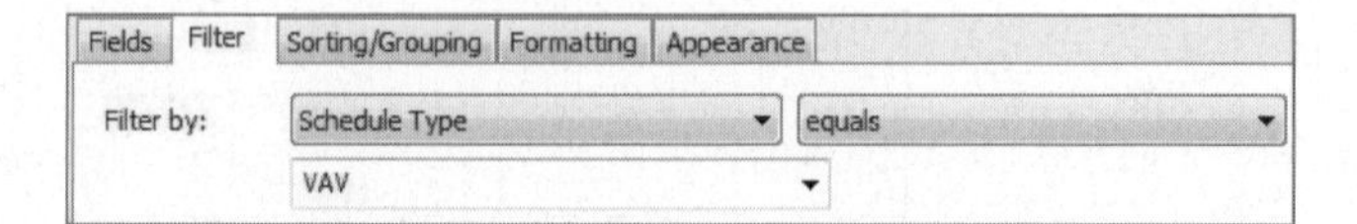

6. Click OK to exit the Schedule Properties dialog box. Notice that now only the VAVs are listed in the schedule. Click the Project Parameters button on the Manage tab of the ribbon. Notice that the Schedule Type parameter is now a project parameter that will be added to any mechanical equipment loaded into the project. Exit the Project Parameters dialog box.
7. Open the 1 – Lighting ceiling plan view. Click one of the light fixtures in the view, and click the Edit Type button in the Properties palette. Notice the Ballast Voltage parameter in the Electrical group. Click OK to exit the Type Properties dialog box.
8. Open the LIGHTING FIXTURE SCHEDULE view. Access the Fields tab of the Schedule Properties dialog box. Notice that the Ballast Voltage parameter that exists in the lighting fixture family is not listed in the Available fields list, because it is a family parameter.
9. Click the Add Parameter button on the Fields tab of the Schedule Properties dialog box.
10. In the Parameter Properties dialog box, select the Shared Parameter option, and click the Select button.
11. Click the Edit button in the Shared Parameters dialog box. Click the Browse button in the Edit Shared Parameters dialog box, and browse to the downloaded `Ch19 Shared Parameters.txt` file location. (Note: If you have never selected a shared parameters file, you will be asked to select one. Browse to the location of the downloaded file.)
12. Set Parameter Group to Lighting Fixtures. Click the New button under Parameters on the right side of the dialog box.
13. Create a new parameter named Fixture Voltage. Set Discipline to Electrical. Set Type Of Parameter to Electrical Potential. Click OK to exit the Parameter Properties dialog box. Click OK to exit the Edit Shared Parameters dialog box.
14. In the Shared Parameters dialog box, set the parameter group to Lighting Fixtures. Select the Fixture Voltage parameter from the list, and click OK to exit the Shared Parameters dialog box.
15. In the Parameter Properties dialog box, select the check box in the lower-left corner to add the parameter to all elements in the category. Set the parameter as a type parameter, and click OK to exit the dialog box.
16. Notice that the Fixture Voltage parameter is now in the Lighting Fixture schedule. You can now input values for voltage into the parameter.

View and Sheet Parameters

Views and sheets are system families, so you need to use project parameters to include additional information or functionality in them. This type of information may be necessary for construction documentation or simply for organizing your project for more efficient workflow. Depending on their use, these parameters may need to be shared parameters.

One of the more common parameters for views is the Sub-Discipline parameter. This parameter allows you to assign a sub-discipline value to the properties of any view to establish a secondary level of organization within the Project Browser. This parameter is already established in the default template files provided with Revit MEP 2011.

Much of the information that appears on a sheet border is common to every sheet. If the parameters that hold this information were unique to each sheet, it would be very time-consuming to make certain changes. Project parameters that are applied to the Project Information category can be included in your titleblock so that the value can be changed in one location and globally updated to all sheets.

One example is for total sheet count. Including this information on a sheet as an editable parameter would require creating a shared parameter. A label of this parameter could then be placed within the titleblock family. The shared parameter could then be added as a project parameter applied to the Project Information category, as shown in Figure 19.16.

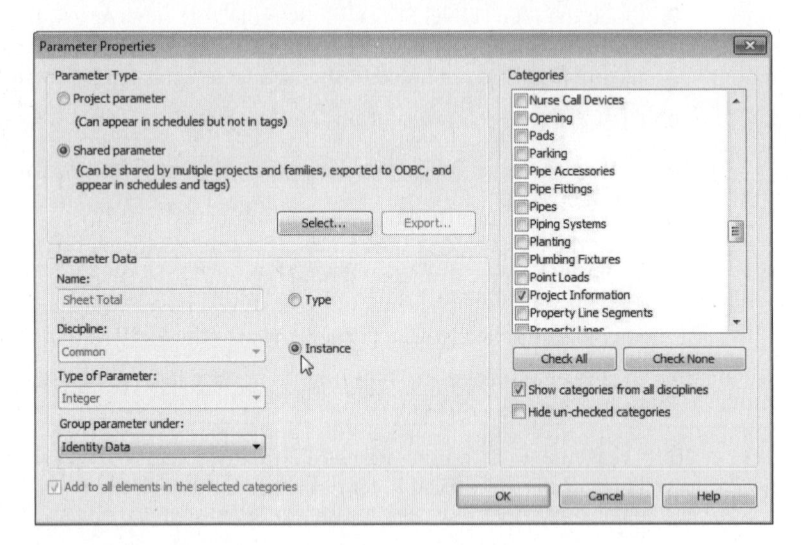

FIGURE 19.16
Project parameter for sheet total

Notice that the parameter is an instance parameter. You cannot create type parameters for the Project Information, Sheet, or View categories.

When the project information is edited, the value will be updated on every sheet in the project. Figure 19.17 shows this example on a titleblock in a sample project.

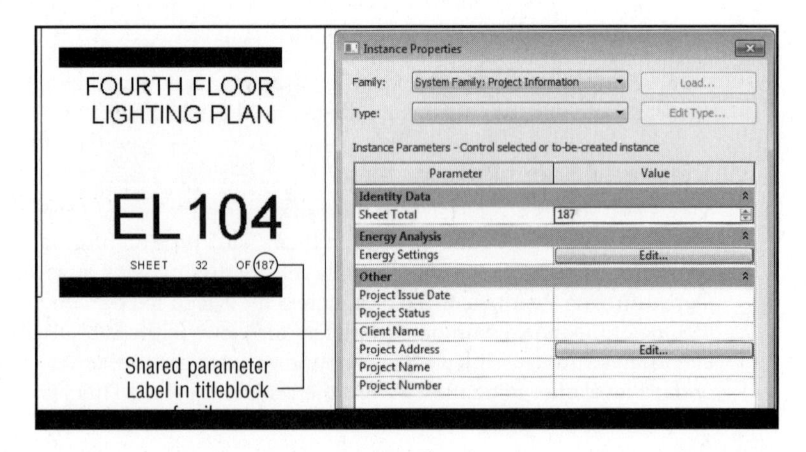

FIGURE 19.17
Shared parameter used on a sheet

Combining the power of shared and project parameters can give you the ability to report, tag, or schedule any data within your model or about your project in general. Once you understand how to use parameters effectively, the real work becomes managing them for consistency and accuracy within your projects.

The Bottom Line

Manipulate the properties of parameters The parameters used to define the properties of elements have properties of their own that define the behavior of the parameters and how they can be used.

Master It It is important to know when and where parameters can be used for extracting data from the model or project. It is also important to understand how instance and type parameters are used. Describe how the use of instance and type parameters affects the way data is changed in a family.

Work with parameters in families Parameters created in family files are useful for defining the geometry of a family and also for assigning engineering data to the family.

Master It Certain families can have multiple family types. If a family has many types, all of them will be loaded into a project when the family is loaded. What can be done to limit the number of family types that are loaded when a family is inserted into a project?

Work with shared parameters Shared parameters are very useful because they can be used in schedules and in annotation tags. Shared parameters can be applied directly to families or added as project parameters.

Master It Managing shared parameters is as important as managing your component libraries. Explain the importance of keeping a common shared parameters file for multiple-user environments.

Use parameters in project files The use of parameters is not limited to component families. Parameters can be added to any element that makes up a Revit project.

Master It Parameters can be added to system families only by creating project parameters. When you create a project parameter, it will be added to all the elements in the chosen category. Explain why managing project parameters is important to using them in schedules within a project.

Chapter 20

Creating Equipment

The equipment families used for the design of your MEP engineering systems are important not only for coordinating the physical model but also for establishing the components that serve as the equipment for your engineering systems. The properties you assign to your equipment families will define how they can be used in relationship to other components within a system.

Many mechanical and electrical equipment items require clearance space for maintenance and installation. With Revit MEP 2011, you can define the required clearances for equipment families either directly in the family files or as a separate component. This gives you the ability to fully coordinate your model to avoid costly conflicts during construction.

Creating equipment families that are generic enough for use early on in the design process, yet parametrically changeable, will help you transition through design phases and changes to systems smoothly and efficiently. The physical properties of MEP equipment are similar enough in most cases to use simple geometry that can be sized according to specified equipment.

Whether you are creating equipment for mechanical, plumbing, or electrical systems, knowing how the equipment will be used from an engineering standpoint as well as a modeling standpoint will help you create the types of families that fit your workflow and processes best.

In this chapter, you will learn to

- Create MEP equipment families
- Add connectors to equipment for systems
- Create clearance spaces for equipment
- Add parameters to equipment

Modeling MEP Equipment

With the rising popularity of building information modeling (BIM), more and more manufacturers are providing their products for use in a virtual model environment. This can be very useful when you get to the stage in your design where you can specify the exact equipment to be used. Early in the design, however, you may not yet know what equipment will be used. Furthermore, you may not yet have done calculations that would determine the size of the equipment needed.

All of this boils down to a need for MEP equipment families that are realistic in size and function but also parametrically editable to compensate for changes in the design of both the project and the systems used. These families need to be flexible enough to handle the seemingly constant change that occurs early in the design process but also need to have the functionality to accurately represent their intended purpose.

Some of the resistance to adopting a BIM approach to designing a project is that too much information is required early on that is not known in a typical project environment. Having usable equipment families can alleviate some of that concern and allow you to move forward with your design processes.

Hosting Options

The first thing to consider when creating an equipment family is how the family will be placed into a project model. Is the equipment in-line with duct or pipe? Does the equipment require a building element to be attached to? Knowing the answers to these types of questions will help you start with the right family type so that the equipment you create can be used properly in your projects.

Equipment that sits on the floor of a room or at the exterior of a building does not need to be created as a hosted family. When you place a nonhosted family into your project, it will be associated with the level of the view in which it is placed. However, if the equipment requires a pad for housekeeping or structural support, you may want to make the family face hosted so that it can be attached to the face of the pad. This will allow for coordination with changes to the pad, assuming that the pad is provided by the structural engineer or architect. If the equipment you are creating always requires a pad, you can build it directly in the equipment family.

Face hosted families should be used for equipment such as control equipment or panels that are typically mounted to walls or on the side of large equipment. This will save you time in making changes when walls are moved and allow you to see when changes have a negative effect on equipment location.

Once you have determined the hosting behavior of an equipment family, you can build the solid geometry of the equipment in relation to the host object or reference plane that represents how the family will be placed into a model.

Family Categories

There are essentially two choices for categorizing your equipment families. You can categorize an equipment family as either Mechanical or Electrical equipment. This may seem very limiting, but keep in mind that with Revit you make any family whatever category you choose. It may be necessary to categorize an equipment family as some other model category, depending on how it is used in your projects. This is typically done for visibility control or scheduling purposes.

Systems Equipment

If you were to build an equipment family and categorize it as something other than Mechanical or Electrical equipment, you would not be able to assign it as the equipment for a system when you create MEP systems in your project.

You can set the category of a family by clicking the Family Category And Parameters button on the Home tab in the Family Editor.

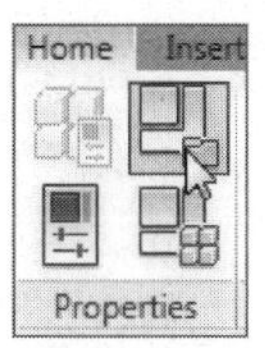

When you categorize a family as Mechanical equipment, you can set its behavior with the Part Type parameter, as shown in Figure 20.1. The Normal option is for equipment that is simply placed into the model and stands on its own, while the Breaks Into option is for equipment that is in-line with ductwork or pipe. The Breaks Into option allows you to insert the family directly into a run of duct or pipe without having to first create an opening for the equipment.

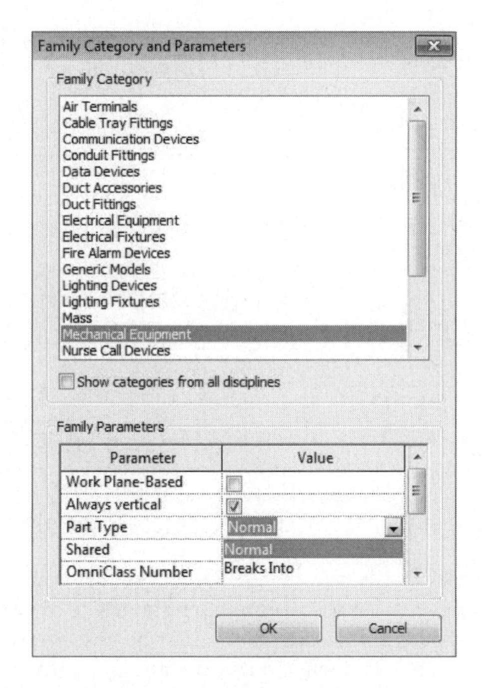

FIGURE 20.1
Part Type options for Mechanical equipment

The Part Type parameter for Electrical Equipment families also has options for defining the equipment behavior, as shown in Figure 20.2. These options let you define how the equipment is used in an electrical system.

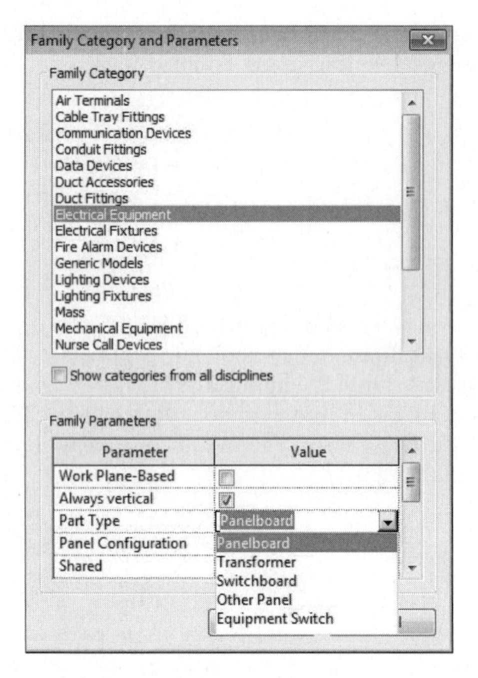

FIGURE 20.2
Part Type options for Electrical equipment

If you choose the Panelboard option, you can set the configuration of the breakers and circuit numbering in the panel. This corresponds to the template used for panel schedules.

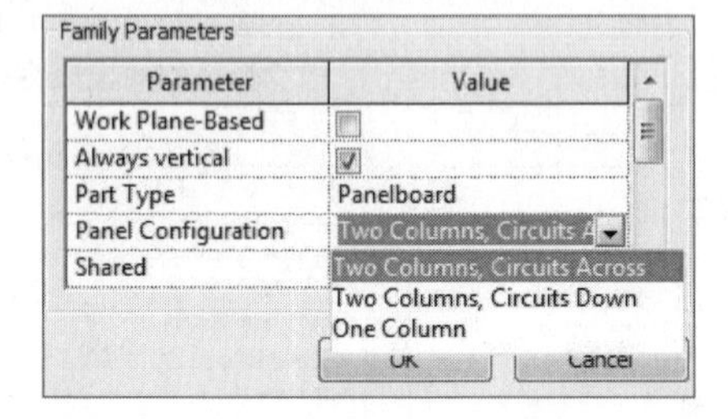

Another family parameter that you should consider is the Always Vertical option. Selecting this option will cause the equipment family to always appear vertical at 90° no matter how the host is sloped. This setting is used for hosted families. If you are creating a nonhosted family and you want to be able to place the family in any orientation, you can set the Work Plane-Based option. This allows you to place a nonhosted family onto any work plane defined in your project. Otherwise, if you place a nonhosted family into a project and it is associated with the level of the view placed, you will not be able to rotate the family in section or elevation views.

Setting the family to Work Plane-Based does not mean you can rotate the family in section or elevation, but you can create a reference plane that is at the desired rotation in the project and associate the family with that plane, as shown in the section view in Figure 20.3. When placing a Work Plane-Based family into a project, the option to place the family onto a face is available on the Placement panel of the contextual tab.

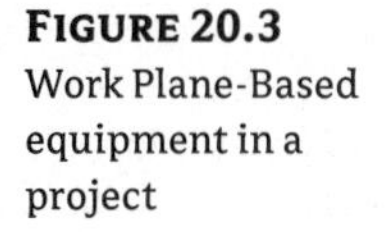

FIGURE 20.3 Work Plane-Based equipment in a project

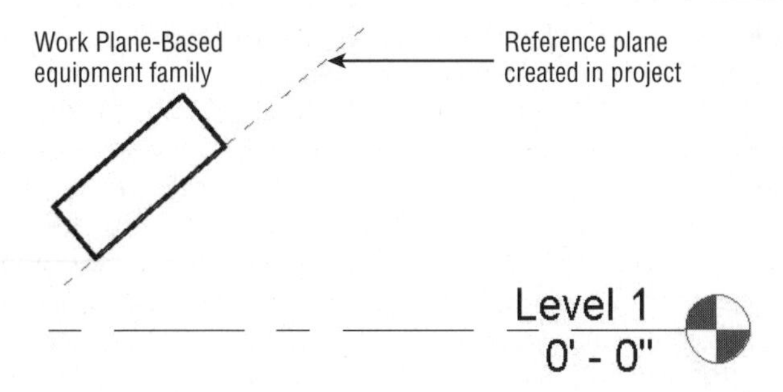

This functionality is very useful because it allows you to use nonhosted families in a face-hosted fashion.

FACE-HOSTED AND NONHOSTED CHOICES

Some companies create a nonhosted and face-hosted version of the same families for use in different project scenarios. Consider using the Work Plane-Based option wherever possible to eliminate the need for duplicate family versions.

Detail Level

When you are modeling your equipment families, consider the amount of detail required to represent the equipment. Mechanical components can be very complicated structures that include many intricate parts. It is not necessary to model your equipment families to a high level of detail for them to function properly in your projects. If you need to show equipment with a lot of detail, you can use model lines or symbolic lines to represent the equipment in views where the high level of detail is needed, while keeping the family geometry simple.

Most equipment families can be modeled using very simple geometric forms such as cubes and cylinders. If the equipment family you are creating requires more complex geometry, you will need to use the other modeling tools available in the Family Editor, such as the Sweep, Blend, and Revolve tools.

For instances where you only need to show the amount of space required for equipment, you can model it as a box that is visible in Coarse detail views. More detailed geometry can be modeled to show in views with a finer level of detail. The important thing with using this technique is that the connection points for ductwork or pipe should be located on the outer edges of the box geometry.

Geometry for Connection Points

MEP discipline equipment families can include connectors for ducts, pipes, conduit, and even cable tray if necessary. These connectors are added to define the function of the equipment from an engineering standpoint. Connectors are added by one of two methods, either by placing them on a face of the solid geometry or by associating them with a work plane within the family. When you associate a connector with a work plane, it can be located anywhere in that plane.

Connectors placed with the Face option on the Placement panel of the contextual tab will automatically attach to the center of the face.

Using the Face option is useful when you know that the connection point is always at the center of the geometry because if the geometry changes size, the location of the connector will adjust, or if the geometry changes location within the family, the connector will move with it. This method requires less constraint within the family for the location of the connector, but it may be necessary to model extra geometry to provide a properly located face for a connector.

Creating geometry for the location of connectors is an easy way to manage not only connector locations but also connector sizes. The size of a connector can be associated with the size of its geometry host face. This allows for accurate modeling of connection points of different family types within a family.

Geometry used for connection points may be more easily dimensioned than connectors themselves because you can dimension to the edges of geometry, where connectors can be dimensioned only to their center points. For example, if you have a rectangular duct connection that varies in size yet needs to maintain a certain distance from the edge of the unit, you can model geometry that is constrained to the dimensional requirements and host the connector to that face. Figure 20.4 shows an example of an extrusion used in this manner. The 0′-6″ and 0′-4″ dimensions are locked to hold the extrusion in place when the UnitHeight and UnitWidth parameters are changed and when the SupplyHeight and SupplyWidth parameters change.

FIGURE 20.4
Extrusion used for connector host

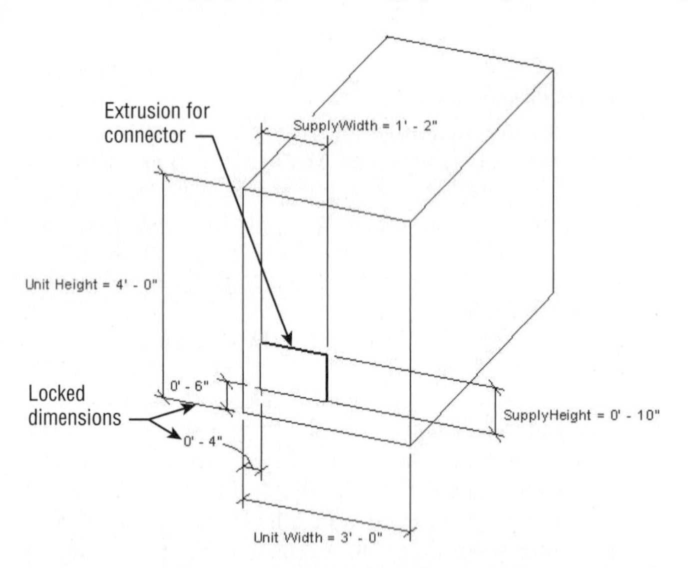

Using extrusions for connection points can make it easier to identify where to connect pipes, duct, or conduit to your equipment when working with the family in a project. More information on connector geometry is provided in the next section of this chapter.

Equipment Pads

When you are creating an equipment family that requires a pad to be mounted on, it may be best to include the pad geometry directly in the equipment family. This will enable you to ensure that the proper dimensions are used for the pad without having to make any changes in another file such as a structural or architectural link. If the equipment does not always require a pad, you can control whether the pad is used with a parameter or with family types.

The key to creating pad geometry in an equipment family is to provide a reference for the top of the pad so that the equipment geometry can be modeled in the correct location and maintain a proper relationship with the pad. If you are creating a face hosted family, the pad should be modeled so that it is associated with the host extrusion, as shown in Figure 20.5.

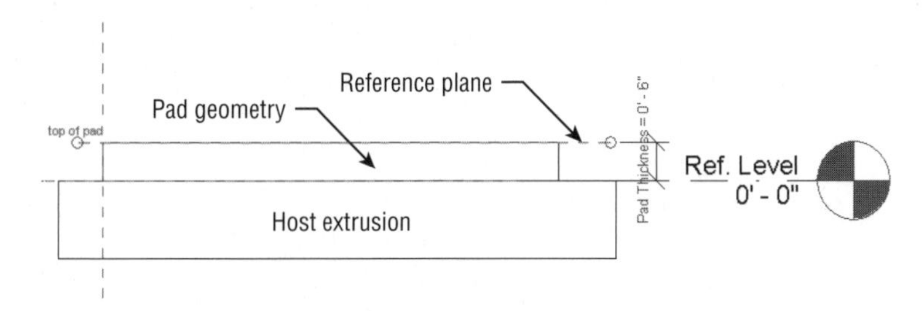

FIGURE 20.5
Pad modeled in equipment family

Real World Scenario

PAD COORDINATION

If you work in the same company with a structural engineer or perhaps have a close working relationship with a structural consultant, you have an opportunity to utilize the power of BIM by collaborating on the design and use of equipment pads.

Most MEP engineers know the basic size and shape of the pads that their equipment requires, but in cases of heavy equipment and unique design requirements, a structural engineer is often used to design an equipment pad that meets the applicable codes for construction.

A foundation family can be created that can be nested into an equipment family. This family can contain all of the parameters and data that a structural engineer might use to represent the pad in their design model, and the size parameters can be associated to parameters in the equipment family. Formulas can even be applied to adjust the properties of the nested pad when changes in size are made to the equipment.

Of course, not everyone has the luxury of having other engineers design portions of their equipment families, but this example is just another one of the many ways that Revit and BIM are changing how things can be done in the design industry to improve project quality.

Adding Connectors to Equipment Families

The connectors that you add to an equipment family will ultimately determine its use in your Revit projects. Connectors can be added for the completion of air, fluid, and electrical systems when the family is selected as the system's equipment in your project.

In most cases, the location of connectors on equipment families is easily handled by geometry modeled specifically for the connector. However, it is not always necessary to have connector geometry. Some equipment families are simply used for equipment location and do not need to be modeled to any level of detail. Early on in the design of a project you may want to use "placeholder" equipment families until specified equipment can be chosen based on calculations. It is still helpful for these types of families to have connectors so that duct, pipe, and conduit can be connected and system analysis can be performed.

If you do not need extra geometry to define the location of a connector, you can dimension directly to the connector by selecting it and then activating the temporary dimension to the center of the connector. Once activated, you can associate the dimension to a parameter. This can be done only if you use the Work Plane option for placing the connector. Using the Face option constrains the connector to the center of the face so its dimensions cannot be manually changed. Figure 20.6 shows how a Work Plane–placed connector location can be dimensioned and parametrically managed.

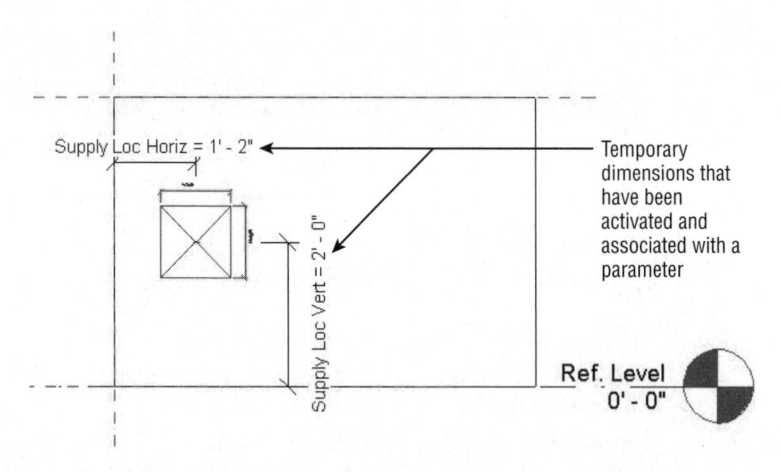

FIGURE 20.6
Connector location dimensions

In this example, if you want to keep the right and bottom edges of the connector static when the connector changes size, you would need to write a formula for the connector dimension parameter. For example, with a parameter that defines the connector width, such as SupplyWidth, the left edge of the connector can be made to stay at 0′-8″ (200mm) from the left edge of the equipment by using a formula for the Supply Loc Horiz parameter, as shown in Figure 20.7.

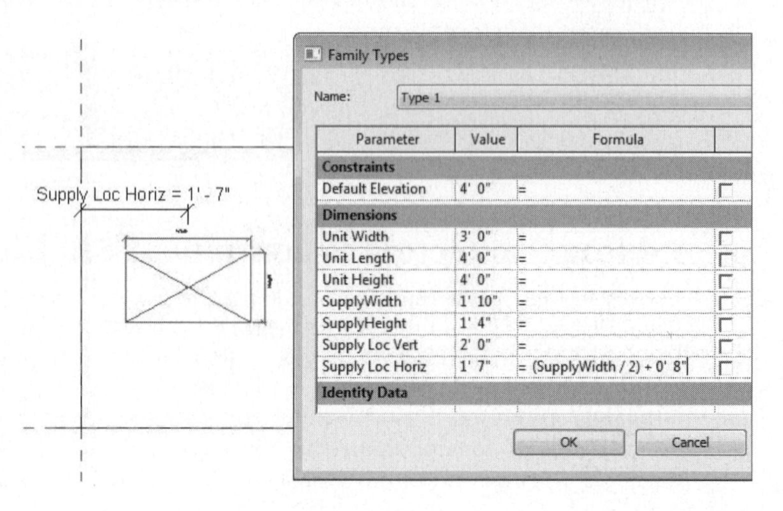

FIGURE 20.7
Connector location using formula

When you have equipment that has several pipe connectors, it is helpful to create geometry for the connectors. Otherwise, you would have to define a reference plane for each connector unless they all connected at the same plane. With geometry used for a connector, you can dimension and constrain the geometry and then use the Face option for connector placement. This does not eliminate the need for dimensions or constraints; it only takes them off the connector object itself.

When you sketch a circle to create a cylinder, the point at which you draw the center can be dimensioned while you are in sketch mode. This dimension can be used for the location of the connector, as shown in Figure 20.8. The radius of the sketch can also be dimensioned and associated with a parameter for easy size adjustment.

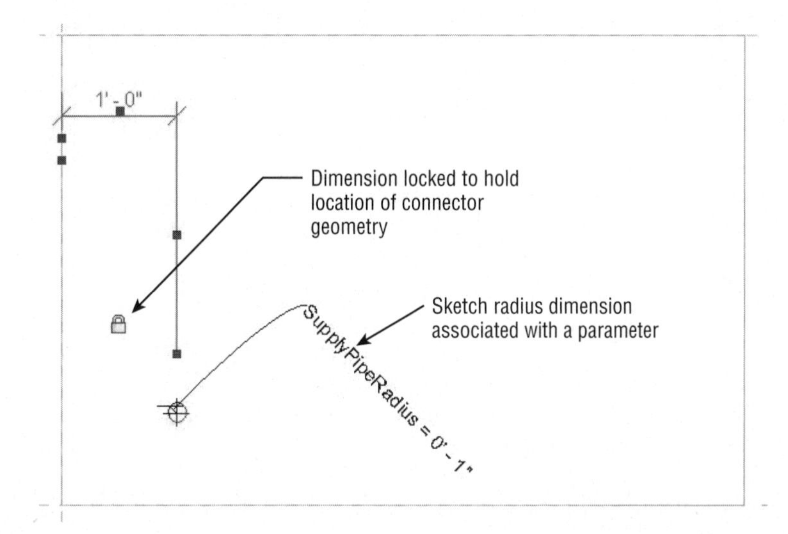

FIGURE 20.8
Geometry for cylindrical connector

> **PIN IT**
>
> To hold the location of connector geometry, you can pin it in place instead of using dimensional constraints. Connectors themselves can be pinned if no geometry is used for their location. Pinning a connector will have no effect on its ability to change size parametrically.

With the geometry in place, a connector can be added using the Face option for placement, as shown in Figure 20.9. You can adjust the connector properties to associate the size of the connector with the parameter created for the geometry radius.

Using the Face option is the easiest way to place connectors into a family. The direction of the connector can be easily flipped using the blue arrows grip that appears when you select the connector. The direction arrow of a connector indicates the direction that pipe, duct, conduit, or cable tray will be drawn from the connector. The drawback to using the Face option is having to create additional geometry if the location of a connector needs to be parametrically managed.

FIGURE 20.9
Connector added to face of geometry

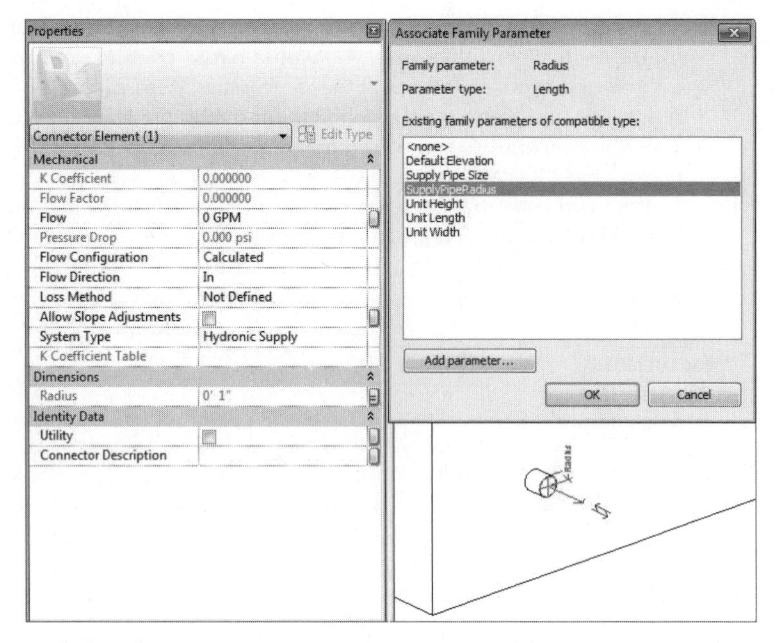

Connectors that are placed using the Work Plane option may not be facing the right direction when you place them. The blue arrows grip will not work to flip the direction, so you will have to rotate the plane that the connector is associated with or redraw it in the opposite direction. The benefit of using the Work Plane option is that you can locate a connector anywhere on the plane.

> **PLANE DIRECTION**
>
> **If you draw a reference plane from left to right in a plan view, connectors attached to it will point up, while a reference plane drawn from right to left will cause the connectors to point down.**

The decision to use the Face or Work Plane option for connectors depends on how you want to control the location of the connector. Whatever method you choose, once a connector is placed, you can adjust its properties so that the equipment family will behave as desired in your projects.

Duct Connectors

You can select the system type of a duct connector from the Options Bar when you click the Duct Connector button on the Home tab in the Family Editor.

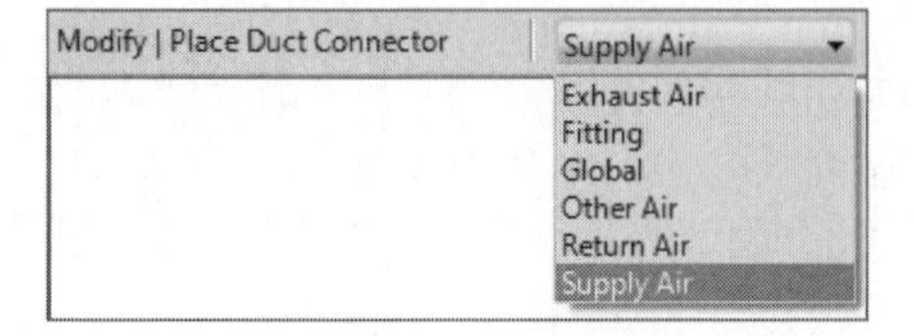

The Global option will enable creating air systems depending on flow direction when the family is used in a project. A Global connector with a flow direction of In will enable a Supply system to be created, while Out can be used for Return or Exhaust. Bidirectional flow gives you options for Supply, Return, and Exhaust. Fitting connectors do not have parameters to define air system behavior and are typically used on duct fitting families to establish connectivity.

For air systems equipment, your duct connectors should be set with the properties that coincide with the behavior of the equipment that the family represents. Figure 20.10 shows the properties of a duct connector selected that has just been added to a family.

FIGURE 20.10
Duct connector properties

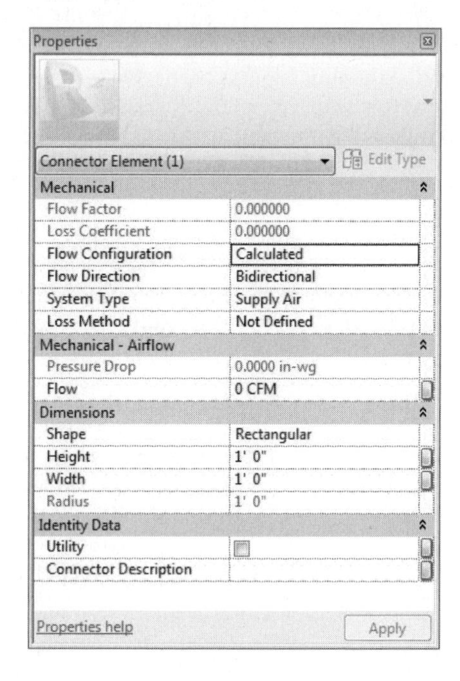

Flow Configuration The Flow Configuration parameter is used for setting how the flow is determined at the connector point.

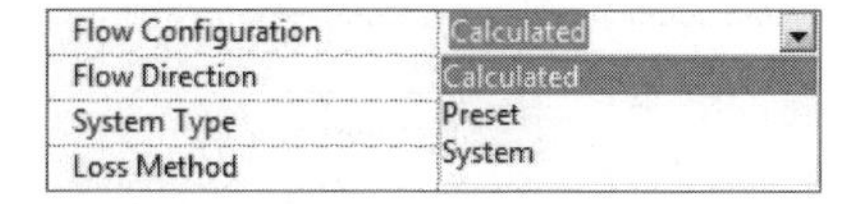

Calculated This option will add up all the flow values from objects downstream in the system.

Preset This option is for direct input of the flow value at the connector point.

System This option can be used if the flow at the connector point is a percentage of the entire flow for a system. With the System option, you will have to provide a value for the Flow Factor parameter. For example, if the equipment you are creating is used in a system with other equipment in your project and it provides 30 percent of the flow, the Flow Configuration would be set to System, and the Flow Factor would be set to 0.33.

Flow Direction The Flow Direction parameter is used to define the direction of the air at the connector point.

IN OR OUT?

Do not confuse Flow Direction with the direction of the arrow shown on the connector itself. You may have a connector whose arrow points away from the equipment, indicating the direction of duct from that point, while the direction of the air is into the equipment. Flow direction is important because when you use the equipment in a project, the flow direction must coincide with the objects that the equipment is connected to. If your connector flow direction is IN, the object it is connected to should have a flow direction of OUT. Making a connection between connectors with the same flow direction will produce analysis errors in your projects.

Loss Method The Loss Method parameter allows you to assign either a Coefficient, by placing a value in the Loss Coefficient parameter, or a Specific Loss by placing a value in the Pressure Drop parameter. Keep in mind that these values are for the selected connector. If you have additional connectors in the family, you can assign unique values to them also.

Shape Use the Shape parameter to define the connector as round or rectangular. The dimensions of a connector can then be assigned in the connector properties. Placing values for the Height, Width, or Radius of a connector will set the size of duct drawn from that connector. If you want to change the size of the connector in the family properties, you can associate its dimensions with parameters created in the family, as shown in Figure 20.11.

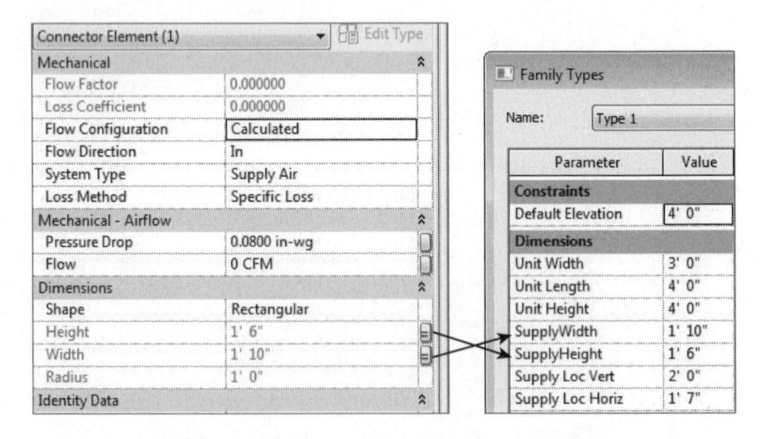

FIGURE 20.11
Connector dimensions associated to family parameters

Utility The Utility parameter for a connector is used to define whether the connector indicates a utility connection. This allows for coordination when your project is exported for use in Autodesk Civil 3D. The connection points defined as utilities will be available in the exported file for coordination of things such as service entrance locations and invert elevations.

The dimensions of a connector can be associated to a parameter by clicking the + sign grip that appears near the connector label when the connector is selected in a view.

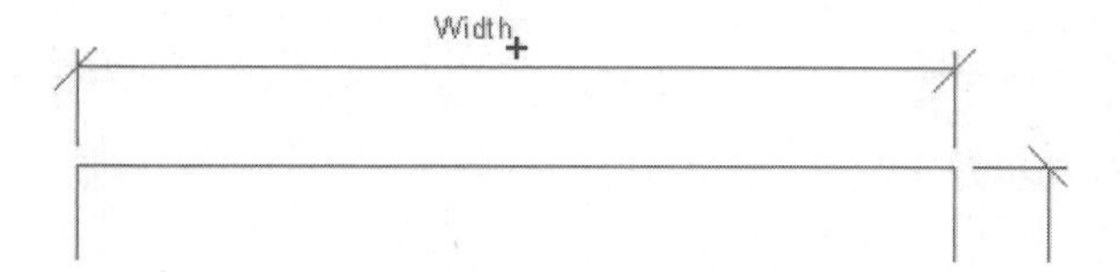

It is important to establish the orientation of the Width and Height dimensions of a connector on your equipment. Otherwise, duct or duct fittings that are connected to the connection point may not be oriented properly. You can rotate a connector placed with the Face option by clicking and dragging it; however, this is an inaccurate method of rotation. Using the Rotate tool allows you to set a specific angle of rotation. If you use the Pin tool to lock the location of a connector, you will also not be able to rotate it. This can be useful if you find that you are accidentally clicking and dragging the rotation of your connectors when working in the Family Editor.

Pipe Connectors

Connectors for pipes can be placed on your equipment families in the same manner as duct connectors. You can find options for setting the system type on the Options Bar when you click the Pipe Connector button on the Home tab in the Family Editor.

Modify | Place Pipe Connector
Hydronic Supply
Domestic Hot Water
Domestic Cold Water
Fire Protection Wet
Fire Protection Dry
Fire Protection Pre-Action
Fire Protection Other
Fitting
Global
Hydronic Return
Hydronic Supply
Other
Sanitary

The properties for a pipe connector are similar to that of a duct connector except that they deal with the flow of liquid instead of air. Because liquid flows differently, there are some different choices for the same parameters you would find in a duct connector.

Flow Configuration The Flow Configuration for a pipe connector has options depending on the system type. Systems related to plumbing pipes have an additional Flow Configuration option called Fixture Units. This establishes that the flow at the connector is based on the value given to the Fixture Units parameter, as shown in Figure 20.12. Notice that the Flow parameter is not active when using the Fixture Units option for Flow Configuration.

Radius The size of the connector is determined by the Radius parameter. Unless the connector does not need to change size with different family types, you should associate the Radius parameter with a family parameter for easy editing. Since pipe sizes are usually given as the nominal diameter of the pipe, you can also create a parameter for pipe size and use a formula to derive the radius from the size given. That way, your input is consistent with industry standards.

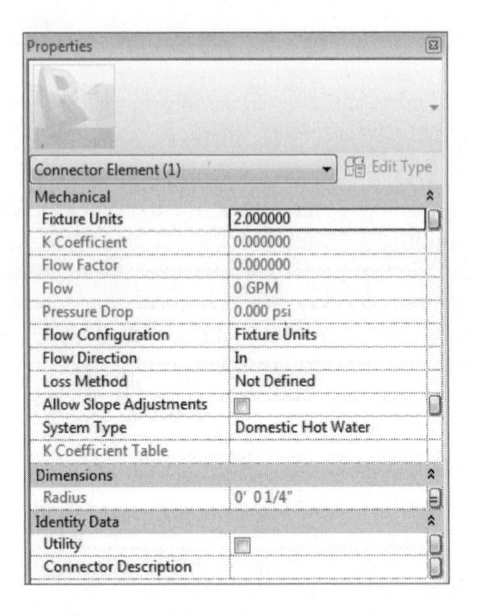

FIGURE 20.12
Properties of a pipe connector for plumbing pipe

PipeRadius	0' 0 1/4"	= Pipe Size / 2
Pipe Size	0' 0 1/2"	=

Allow Slope Adjustments The Allow Slope Adjustments parameter lets you establish whether pipe attached to the connector can be sloped. If you do not select the box, then when you apply a slope to pipe that is connected to the connector, you will receive an error message that the angle between the elements is too great, and the pipe will become disconnected from the equipment.

If you draw sloped pipe from a connector that allows for slope adjustment, the Pipe Connector Tolerance angle defined in the MEP settings of your project will determine the maximum angle that a sloped pipe can enter a connector. If the angle is exceeded, a straight run of pipe and an elbow fitting will be drawn from the connector prior to the sloped pipe.

Electrical Connectors

There are three kinds of connectors for electrical systems: conduit connectors, cable tray connectors, and electrical connectors.

Conduit connectors These can be added by clicking the Conduit Connector button on the Home tab in the Family Editor. The Options Bar provides the two choices available for conduit connectors.

Individual connector This option is for placing a single connection point in a specific location on the equipment. These connectors have properties for adjusting the angle and setting the radius of the connector.

Surface connector The surface connector is a unique and powerful option because it allows you to establish an entire surface of the equipment geometry as a connection point for conduit. Figure 20.13 shows a conduit surface connector that has been added to the top of an electrical panel family. This allows for the connection of multiple conduits at any location on the top of the panel. The Angle parameter can be set to allow for adjustments in the angle of the surface that the connector is attached to. This parameter is typically used in fitting families but may have application in an equipment family.

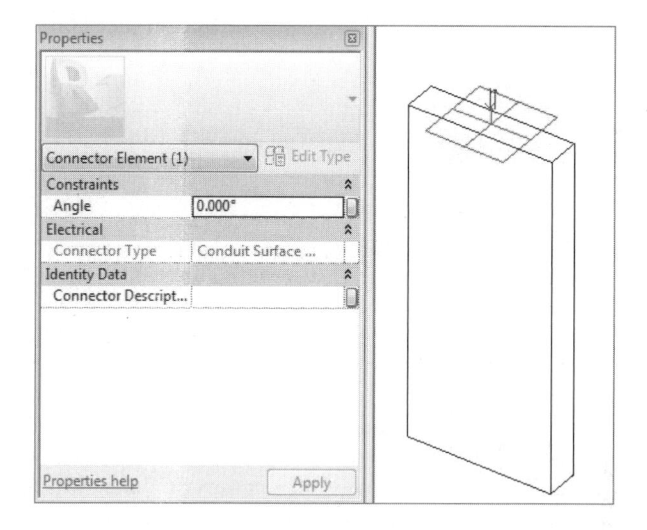

FIGURE 20.13 Conduit surface connector on the top surface of a panel family

When an equipment family with a surface connector is used in a project, conduit can be drawn to or from the surface using the Conduit tool on the Home tab. The surface containing the connector will highlight when you place your cursor over it. Selecting the surface will activate the Surface Connection mode, which allows you to drag the connector location to any point on the surface. This process can be repeated several times for the same surface connector, allowing you to model multiple conduits from your equipment, as shown in Figure 20.14. See Chapter 13 for more information on modeling conduit.

Cable tray connector The Cable Tray Connector button on the Home tab in the Family Editor allows you to place a point of connection for cable tray. These connectors are similar to the individual conduit connector, having properties for the angle and for setting the height and width of the connector.

Electrical connector Clicking the Electrical Connector button on the Home tab in the Family Editor allows you to place a point of connection for wiring, which will also define what type of system the equipment can be used in. The Options Bar allows you to define the system of the connector prior to placement.

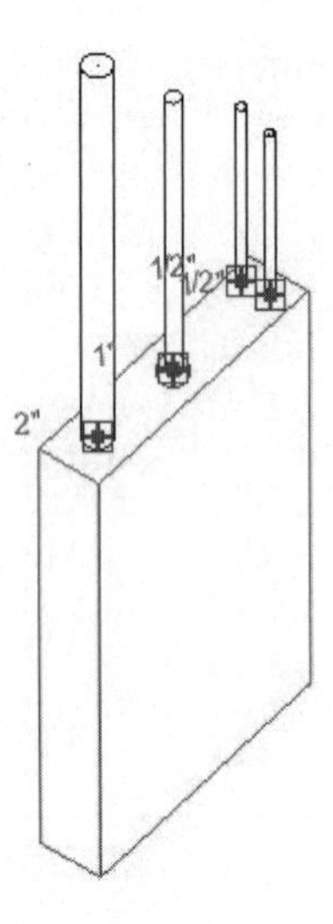

FIGURE 20.14
Multiple conduits from a surface connector

> **FACE OPTION**
>
> When placing electrical connectors on equipment families, the Face option for placement is usually best. Electrical connectors do not have a direction property, and the wiring can be drawn to any point on the equipment. The need for using the Work Plane placement option would be if you had to move the connector for easier access or because it interferes with another connector.

Many of the properties for an electrical connector can be associated with family parameters for easy management and creation of family types. The Voltage parameter will determine what distribution systems in your project the equipment can be used in, which must be coordinated with the value of the Number Of Poles parameter. The Load Classification parameter enables you to assign a load classification to the connector that will determine the demand factor when used in a project. If you create a new load classification in the family, it will be available in the project when the family is loaded.

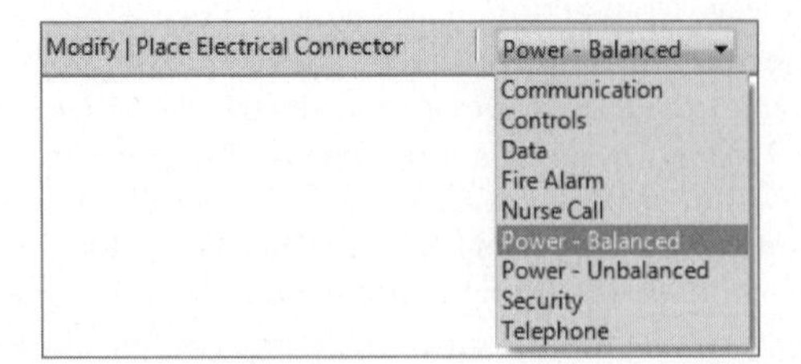

Multiple Connectors in Families

Many MEP equipment families have multiple types of connectors. Your work environment may determine the connectors that you add to equipment families. For example, if your projects contain all MEP systems in one model, you may want to have electrical connectors in your HVAC equipment families that require electricity. If you have a separate model for each discipline, there is no need to have an electrical connector on any mechanical equipment families because the connection cannot be made through a linked file.

It is possible to link connectors in a family. This is useful for when you want the system behavior to pass through from one connector to the other, such as when you have an air systems equipment family that has a connector with supply air flowing into the equipment and a supply air connector with air flowing out. Linking the connectors propagates system information through the equipment, so air flow in is related to air flow out of the equipment. You can link connectors by selecting one of the connectors and clicking the Link Connectors button on the Connector Links panel.

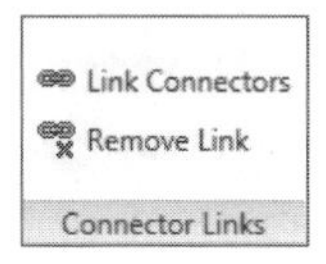

Linked connectors will take on the Flow Configuration value of the primary connector. So even if you have a connector in the family that has a Flow Configuration set to Preset, if it is linked to a primary connector set to Calculated, the behavior of the connector when in a project will be as if the connector is set to Calculated. You can tell whether connectors are linked in a family by selecting one of the connectors in the Family Editor. Red arrows will appear between the connector and the one it is linked to, as shown in Figure 20.15.

FIGURE 20.15
Linked connectors in an equipment family

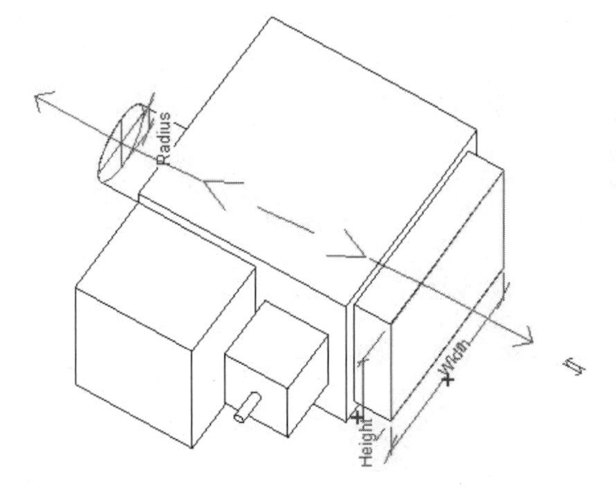

The primary connector in a family is the first connector of each type added to the family and is indicated by crosshairs in the connector graphics. Figure 20.16 shows an equipment family with multiple types of connectors. You can change which connectors are the primary ones by clicking the Re-assign Primary button that appears on the Primary Connector panel of the contextual tab when you select a connector. Clicking the button allows you to select another connector, which will become the primary for that type.

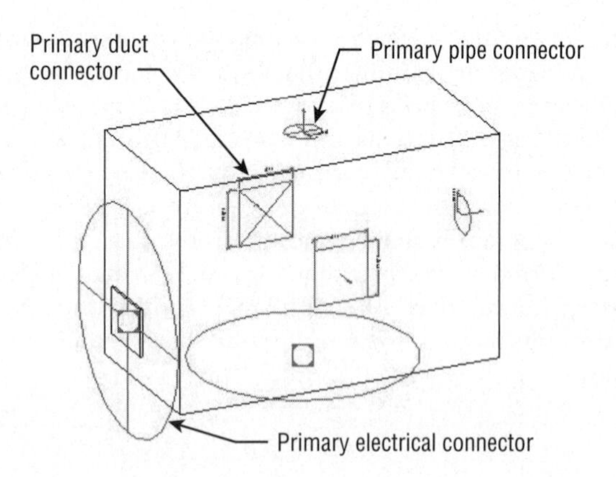

FIGURE 20.16
Equipment family with primary and secondary connectors

Naming connectors in a family makes it easier to understand which one you are connecting to when working in a project. The Connector Description parameter in a connector's properties is where you can name the connector. When using the Connect Into tool in a project on a family with multiple connectors, the connectors will appear in a list with their descriptions so you can choose the proper one.

Now that you have learned some ideas about creating an equipment family, practice some of the basic skills by completing the following exercise. No matter what your discipline of expertise, the steps in this exercise will help you become more familiar with the process of creating equipment families.

1. Open the `Ch20_equipment.rfa` file found at `www.wiley.com/go/masteringrevitmep2011`.
2. In the Ref. Level view, the intersection of the reference planes will be the corner of the equipment pad. Create a vertical reference line to the right of the vertical plane and a horizontal reference line above the horizontal plane. These lines will be used to define the length and width of the pad.

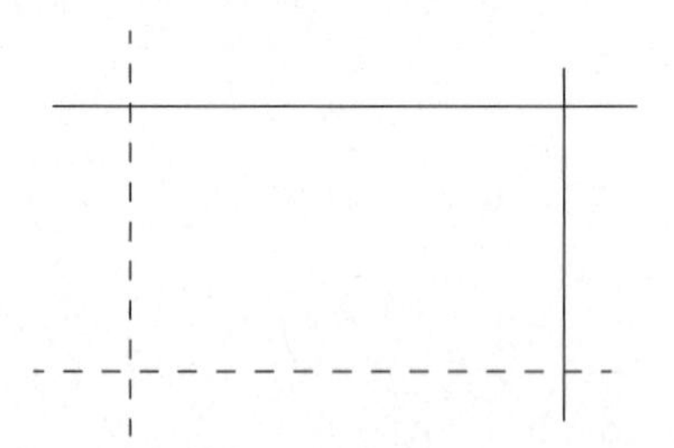

3. Create a dimension from the vertical reference plane to the vertical reference line. Select the dimension, and click the Label drop-down on the Options Bar. Select the <Add parameter...> option. In the Parameter Properties dialog box, name the parameter **PadLength**. Set the parameter to an instance parameter. Click OK. Repeat the process for the horizontal reference line, using **PadWidth** for the parameter name.

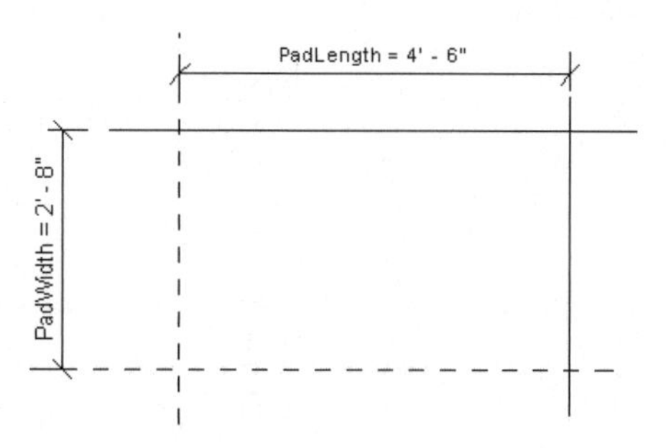

4. Open the Front elevation view. Draw a horizontal reference plane above the Ref. Level. Select the plane, and name it **TopOfPad** in its element properties. Create a dimension from the Ref. Level to the horizontal plane. Create an instance parameter for the dimension called **PadDepth**.

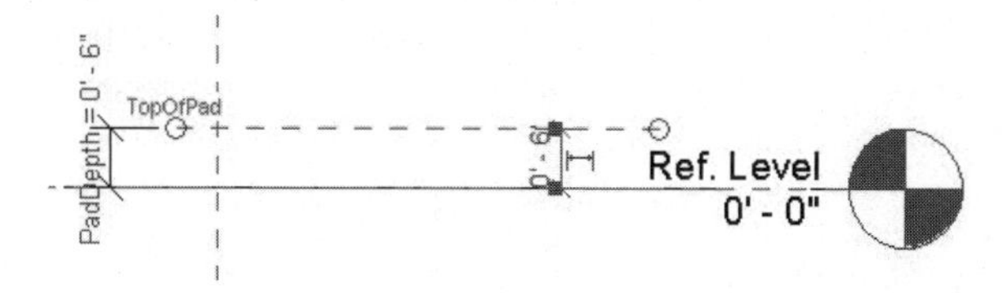

5. Open the Ref. Level view. Click the Extrusion button, and sketch a rectangle. Align and lock the sketch lines to the reference lines and planes. Click the green check button to complete the sketch. Open the Front elevation view. Align and lock the top of the extrusion to the TopOfPad reference line.

6. Click the Family Types button on the Properties panel of the Home tab. Create a Yes/No Instance parameter called **Pad**, grouped under Graphics. Create another parameter with the same settings called **No Pad**. Click OK to exit the Family Types dialog box.

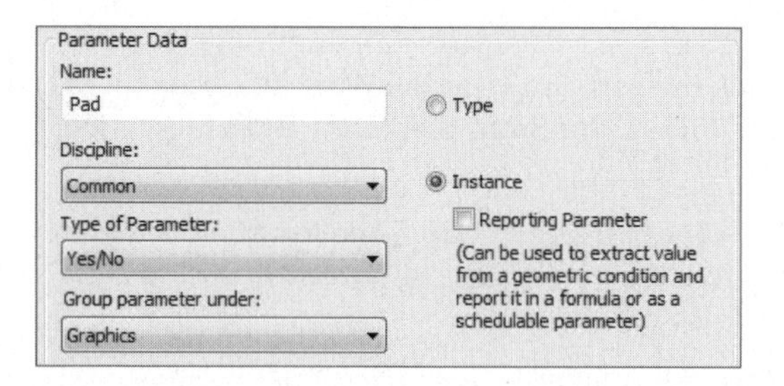

7. Open the Ref. Level view. Create a vertical reference plane 4″ to the right of the vertical plane. Create another vertical plane 4″ to the left of the vertical reference line drawn in step 2. Create a horizontal reference plane 4″ above the horizontal plane and another 4″ below the horizontal reference line drawn in step 2.

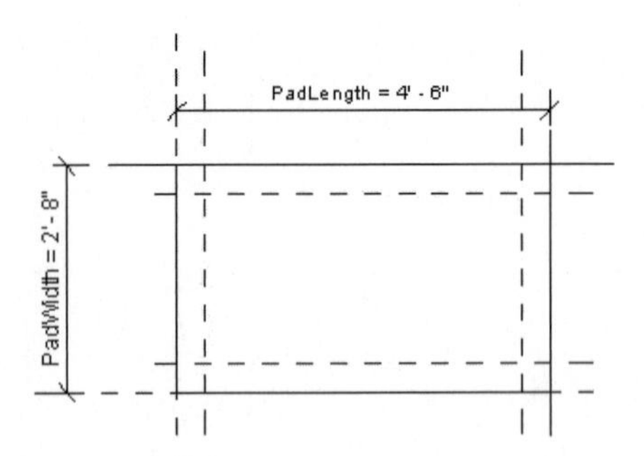

8. Dimension between the two vertical planes drawn in step 7, and create an instance parameter for the dimension called **UnitLength**. Repeat the process for the two horizontal planes drawn in step 7, with **UnitWidth** as the parameter name. Add a dimension for the 4″ distance from the initial vertical plane and the vertical plane drawn in step 7. Repeat the process for the 4″ distance between the initial horizontal plane and the one drawn in step 7. Be sure to lock the dimensions using the padlock grip that appears when the dimension is placed.

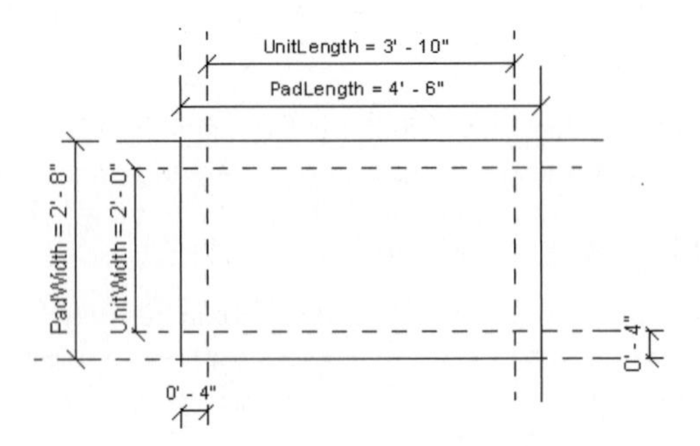

9. Click the Set button on the Work Plane panel of the Home tab. Select the TopOfPad reference plane from the Name drop-down list in the Work Plane dialog box. Click OK.
10. Click the Extrusion button, and sketch a rectangle. Align and lock the sketch lines to the reference planes drawn in step 7. Click the green check button on the Mode panel to finish the sketch.
11. Open the Front elevation view. Draw a reference plane from left to right above the extrusion created in step 10. Name the reference plane **TopOfUnit**. Add a dimension between the TopOfPad reference plane and the TopOfUnit plane. Create an instance parameter for the dimension called **UnitHeight**. Align and lock the top of the extrusion created in step 10 to the TopOfUnit plane.

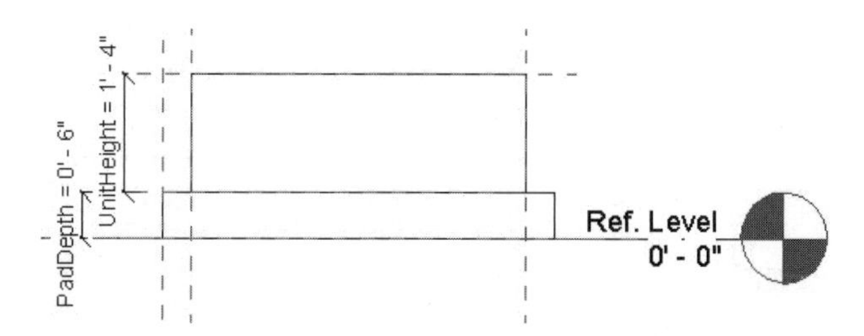

12. Click the Family Types button on the Properties panel on the Home tab. Create an instance parameter called **SupplyHeight** and one called **SupplyWidth**, using the settings shown here. Give each of the parameters a value of 10″.

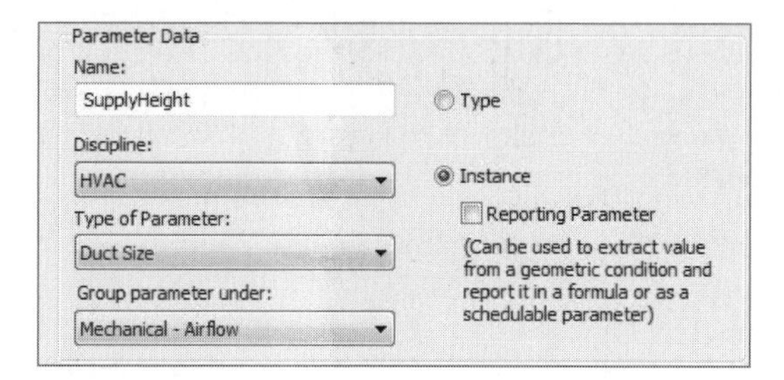

13. Repeat step 12, creating parameters called **ReturnHeight** and **ReturnWidth**. Click OK to exit the Family Types dialog box.
14. Open the default 3D view. Click the Duct Connector button on the Home tab. From the drop-down on the Options Bar, select Supply Air for the system. Place your cursor at an edge of the top extrusion to highlight the end face. You can use the Tab key if the desired face will not highlight when your cursor is on one of its edges. Click to select the face, and place the connector.

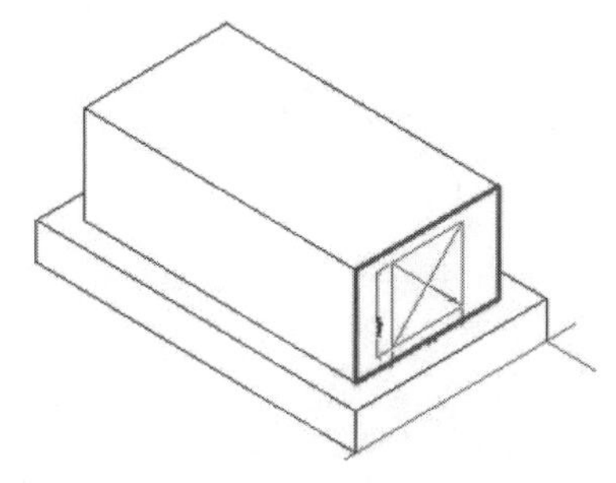

15. Rotate the view to show the opposite end of the extrusion. From the drop-down on the Options Bar, set the system type to Return Air. Select the face opposite from the Supply Air connector, and click to place the connector.

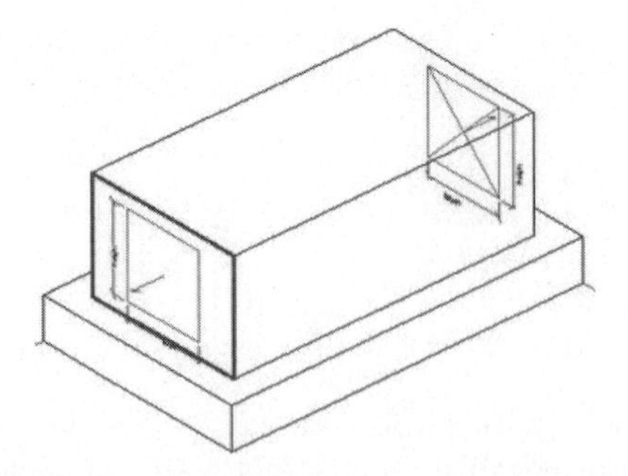

16. Click the Modify button or use the Esc key to finish placing connectors. Select the Supply Air connector, and access its element properties. Confirm that the Flow Configuration parameter is set to Calculated. Set the Flow Direction parameter to Out. Click the small rectangle to the far right of the Height parameter to associate it with the SupplyHeight family parameter, as shown in Figure 20.17. Click OK to close the Associate Family Parameter dialog box.

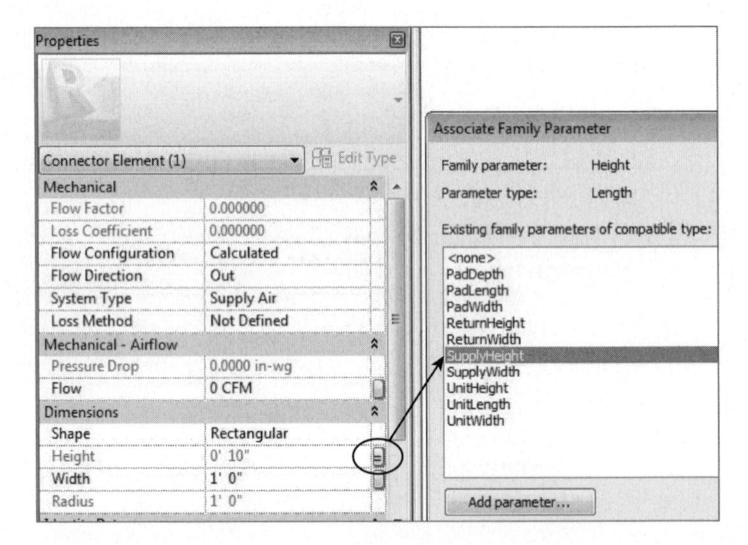

FIGURE 20.17
Associating a connector parameter to a family parameter

17. Associate the Width parameter of the Supply Air connector to the SupplyWidth family parameter. Click OK.

18. Select the Return Air connector in the view, and access its element properties. Set the Flow Configuration parameter to Preset. Set the Flow Direction parameter to In. Associate the Height and Width parameters to the ReturnHeight and ReturnWidth family parameters.

19. Select the Supply Air connector. Click the Link Connectors button on the Connector Links panel of the contextual tab. Click the Return Air connector. Click either of the connectors to confirm they are linked by the display of arrows between them. With a connector selected,

click the Remove Link button on the Connector Links panel of the contextual tab. The arrows will disappear when the link is removed.

20. Click the Electrical Connector button on the Home tab. From the drop-down list on the Options Bar, set the system type to Power – Balanced. Place the connector on one of the side faces of the equipment extrusion. Click the Modify button or use the Esc key to finish placing connectors.

21. Select the electrical connector, and access its element properties. Set the Number Of Poles parameter to 3. Set the Load Classification parameter to HVAC by clicking the small box that appears in the value cell and choosing HVAC from the Load Classifications dialog box.

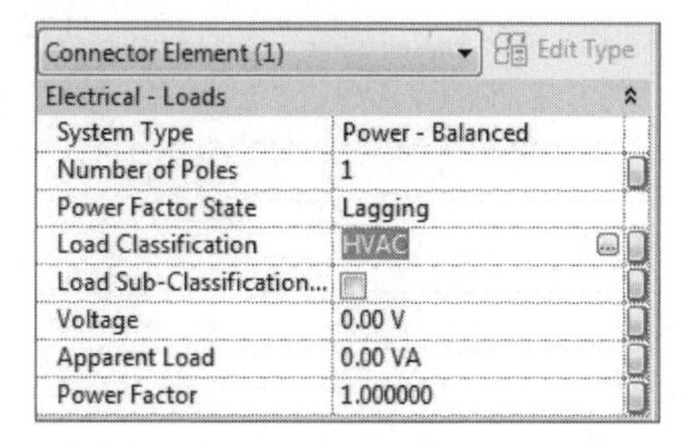

22. Click the small rectangle to the far right of the electrical connector's Voltage parameter to associate it with a family parameter. Since no family parameter is available, click the Add Parameter button at the bottom of the Associate Family Parameter dialog box. Create an instance parameter named **UnitVoltage**. Because the Voltage parameter was selected, the other properties of the parameter are already selected. Click OK to close the Parameter Properties dialog box. With the UnitVoltage parameter highlighted in the list, click OK to close the Associate Family Parameter dialog box.

23. Repeat step 22 for the Apparent Load parameter of the connector. Name the family parameter **UnitLoad**.

24. Click the Family Types button on the Properties panel of the Home tab. Set the value of the UnitVoltage parameter to **480**. Set the value of the UnitLoad parameter to **5000**. Click OK to close the dialog box. Select the electrical connector in the view, and access its element properties. Confirm that the values have been associated to the connector parameters.

You can save this family and use it to practice techniques for creating equipment. Load the family into a test project to test its behavior and usability.

Creating Clearance Spaces

Most MEP equipment requires some sort of clearance space for safety or maintenance reasons. Modeling the clearance spaces directly in your equipment families is useful for model coordination. Because the space is part of the family, any objects encroaching on the space will show up when you run an interference check of your model. You can choose to keep the spaces visible during modeling for visual reference to avoid such interferences. Even if you choose to turn the visibility of the spaces off, any interferences will still be detected.

Figure 20.18 shows an electrical panel family with clearance spaces modeled for the front of the panel and the space above the panel.

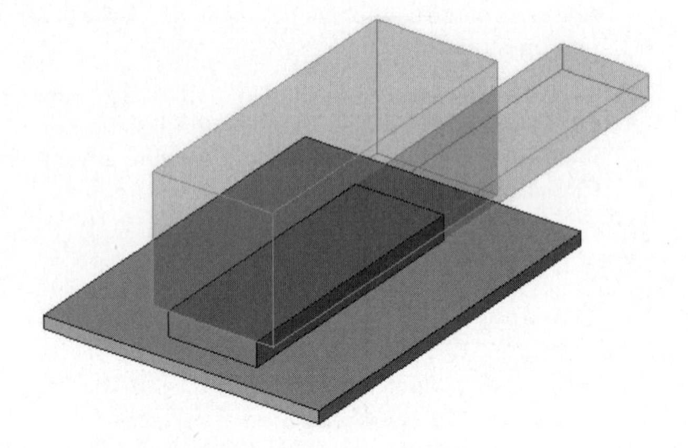

FIGURE 20.18
Clearance spaces in an equipment family

You can create clearance spaces by modeling an extrusion in the family that is associated with a reference plane at the appropriate face of the equipment. You can use other reference planes to control the size of the clearance space and constrain it to the equipment. In the example of the electrical panel, reference planes were created to constrain the top of the panel so that the front clearance space will always be 6″ above the top of the panel. The dimension for the space height was done in the extrusion sketch, as shown in Figure 20.19, which shows the plan view of the face-hosted family.

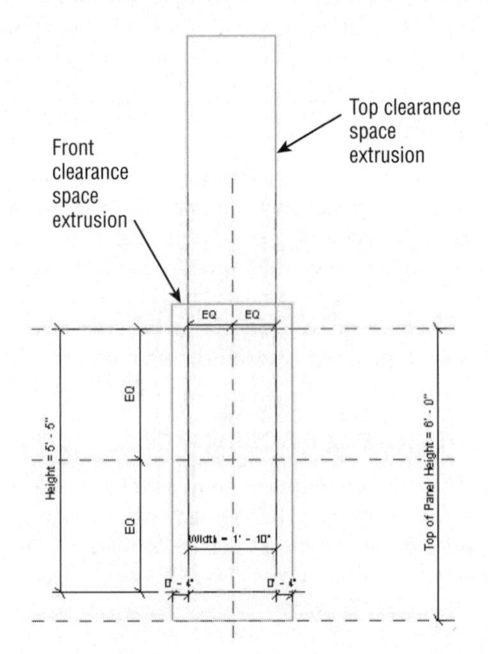

FIGURE 20.19
Clearance space reference planes in an equipment family

Once you have modeled a clearance space extrusion, there are properties you can apply to it for visibility. In the panel example, a unique material type was created and applied to the extrusions for their semitransparent display. This keeps them from blocking out the actual equipment when shown in a model view.

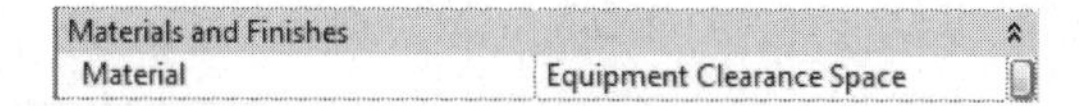

A subcategory was created for the clearance space extrusions so that they can be turned off in model views without turning off the equipment category. The subcategory is given a color for the edge lines of the clearance space extrusions to stand out from other solid geometry.

Object Styles

Model Objects | Annotation Objects | Imported Objects

Category	Line Weight		Line Color	Line Pattern
	Projection	Cut		
Electrical Equipment	1		Black	
Clearance Area Lines	1		Black	Hidden
Clearance Space	1		RGB 255-128-064	Solid
Hidden Lines	1		Black	Dash

Although having solid geometry in an equipment family to represent clearance spaces is useful for interference coordination, you may not want to have the solid geometry displayed in plan views. Model or symbolic lines can be used to represent the clearance space for plan views. The visibility of these lines can be controlled by a Yes/No type parameter or by a subcategory. Figure 20.20 shows clearance space lines in the Front elevation view of the panel family. The lines exist in the Front elevation view so that they will be visible in plan views when the panel is attached to a vertical host.

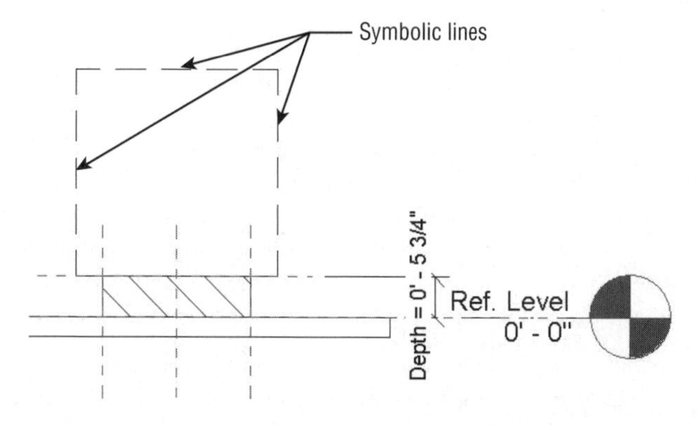

FIGURE 20.20
Lines representing clearance space in a family

Another option for adding clearance spaces to your equipment is to create unique clearance space families. These families should be categorized as Mechanical or Electrical equipment families, but you could categorize them as any of the available subcategories depending on how you want to control their visibility. In a project file with worksharing enabled, a workset could be created for equipment clearance spaces, allowing for easy visibility control. There is a downside to using worksets for visibility control, however. Worksets cannot appear in a view template.

Clearance space families can be face hosted families for easy attachment to the solid geometry of your equipment families. Consider making individual families with the dimensions for each unique type of clearance or for specific types of equipment.

Adding Parameters and Constraints

The key to having equipment families that are usable in the early stages of your designs is to have families that are easily modified to meet the demands of frequent changes to your project model. Using parameters gives you the flexibility to change not only the size of equipment but also the values for connectors.

Most equipment can be defined by three simple dimensions: length, width, and height. By using these parametrically, you can be confident that your equipment families occupy the space required for them in the model even if they do not look exactly like the object they represent. Be consistent with your use of and naming of these parameters. They do not need to be shared parameters unless the information is to be included in schedules, but you could create them as shared parameters for consistent naming. Because there may be other dimensions in your families, it is best to give these parameters a descriptive name, such as Unit Length or Equipment Width, to distinguish them as the overall size. Of course, having these parameters in your families is not enough; you need to have the family geometry associated to them.

MEP equipment schedules vary in size and information not only from one company to another but even in the same company, depending on the project standards or client demands. Mechanical equipment schedules are particularly difficult to manage without the proper use of parameters in your equipment families. The use of shared parameters is necessary for scheduling the systems information required from equipment schedules. The problem is that all mechanical equipment families are in one category.

To avoid having families with unnecessary parameters, it may be best to apply your shared parameters directly in the equipment families instead of using project parameters. That way, you can be sure that you have only the parameters required for scheduling of the specific families in your mechanical equipment schedules.

Connectors have parameters that you may need to change based on the performance of your engineering systems. Creating parameters in your families that the connector parameters can be associated to allows you to change the connector values without having to open, edit, and reload your families. The main thing to remember when creating these parameters is that they must match the units of the connector parameter in order for them to be associated. For example, you cannot associate a connector's Voltage parameter with a number or integer parameter.

For more information on parameters and schedules, see Chapter 19 and Chapter 5.

The Bottom Line

Create MEP equipment families The ability to create the types of equipment families needed for accurate modeling of components and systems is a major factor in the success of your Revit projects.

Master It MEP equipment can be quite complex in its structure. Complex geometry can have an adverse effect on model performance. What are some ways to model equipment in the most simple form yet still convey the proper information on construction documents?

Add connectors to equipment for systems Adding connectors to equipment families will make them functional for use in the design of engineering systems.

Master It It is important to know how your equipment families will be used in your projects from an engineering standpoint as well as for model coordination. Explain how connectors determine the behavior of an equipment family.

Create clearance spaces for equipment Space for safety and service of equipment is crucial to building design. The ability to coordinate clearances around equipment improves project quality and can reduce construction and design cost.

Master It Equipment families with built-in clearance spaces allow you to quickly and easily determine whether the equipment will fit into your project model. Describe some options for controlling the visibility of clearance spaces so that they are not shown when not needed.

Add parameters to equipment Parameters in your equipment families can be useful for creating schedules in your Revit projects that report data directly from the equipment used in the design. Family parameters can enable you to make equipment families that are changeable without having to create new families.

Master It Shared parameters must be used in your equipment families if you want to schedule the data they provide. If you are creating parameters for parametric behavior of the solid geometry, do they also need to be shared parameters?

Chapter 21

Creating Lighting Fixtures

Lighting fixtures can be anywhere in a building. They hang from ceilings and structure; are mounted in floors, walls, and stairs; and even stand on their own. With Revit MEP 2011, you can make any shape or form into a lighting fixture family so that if your project requires it, you can create a unique fixture. However, the majority of lighting fixtures used in building design are of a few basic shapes. Unless your project requires it for visualization, there is no need to go into great detail when creating the geometry of your lighting fixture families. Keeping their design simple will enable you to focus more on the computable data within your fixture families.

Lighting fixtures are unique Revit MEP families because not only are they used to create a coordinated 3D model, but they also can be used as a design tool. They have the ability to display the output of light for a specific fixture type by using `.ies` files. The photometric data for a specific lighting fixture can be applied to a lighting fixture family regardless of the geometry of the family. This gives you the freedom to build families for the basic size, shape, and mounting options of lighting fixtures and then apply manufacturer-specific data to them for an accurate lighting design. Lighting fixture families can also contain electrical data, which allows for connection to distribution systems.

There are many options for the level of detail and data you can put into your lighting fixture families.

In this chapter, you will learn to

- Create different types of lighting fixture families
- Use a light source in your lighting fixture families
- Create and manage fixture types and parameters
- Use lines and symbols to represent lighting fixtures

Types of Lighting Fixture Families

Prior to creating a lighting fixture family, it is important to know how it will be used in your projects. The first thing to consider is how the fixture will be hosted in your model. Lighting fixture families can be made to be hosted by specific building elements such as walls or ceilings or by any three-dimensional face within your model. You can also create lighting fixture families that do not require a host. Each hosting option determines how the geometry of the fixture is to be oriented in the family file; it is also important to realize that a family created with one type of hosting template cannot be exchanged for another, even though they may be of the same category. For this reason, planning, naming conventions, and training for staff are all very important.

There are family template files (`.rft`) for different hosted-type lights. You can also use one of the generic family templates and later categorize the family as a lighting fixture. When you use a lighting fixture template, a reference plane for the elevation of the light source is already included. Using a generic template will require you to add a reference plane for the light source if needed.

Nonhosted Lighting Fixtures

Nonhosted lighting fixture families are useful in areas that do not have a ceiling or for free-standing lights. They can be given an offset to the level at which they are inserted to show them above the floor. They are also useful for wall-mounted lights when you need to show a symbol on your drawings instead of the actual fixture. The symbol can be nested into your fixture family without concern for the annotation orientation because the fixture will always be vertical. One example of this use is a wall-mounted exit light. Figure 21.1 shows an exit light family that has a nested annotation. The geometry of the fixture is modeled to appear correctly when placed adjacent to a wall, while the annotation is nested to display in plan view.

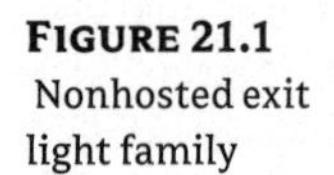

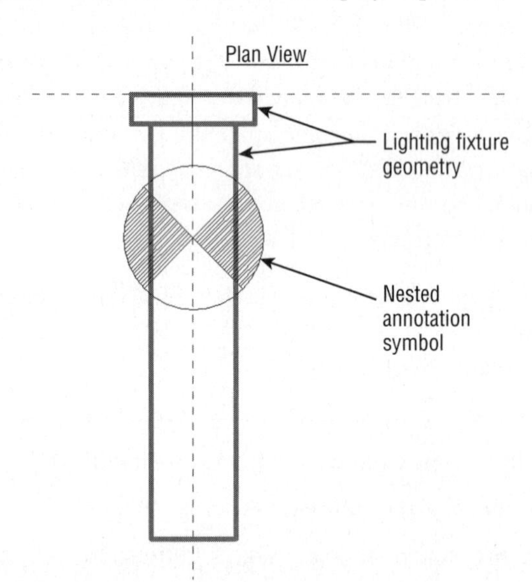

FIGURE 21.1 Nonhosted exit light family

The drawback to this scenario is that although you will show the fixture adjacent to a wall, if the wall moves, the fixture will not move with it. This can cause inaccuracies on your drawings and therefore requires manual editing of the fixture locations, so carefully consider what types of lights to use this method for.

As with any family you create, the insertion point is an important consideration when building the geometry of a lighting fixture. When you create a lighting fixture with the `Lighting Fixture.rft` or `Generic Model.rft` family template, Revit will open four views automatically, with the floor plan view active.

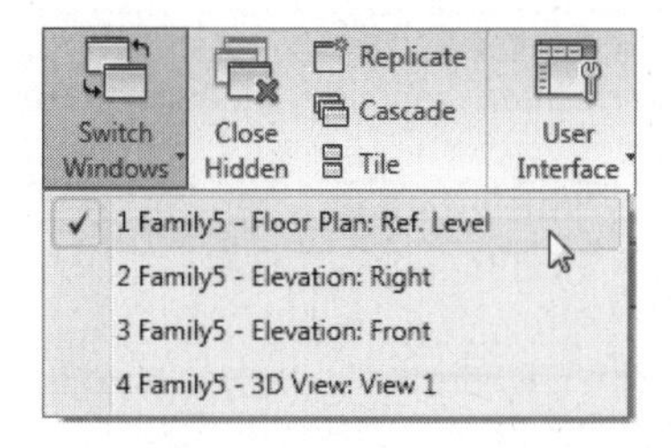

The intersection of the reference planes in the view defines the insertion point of the family. Be sure to build the fixture geometry around the insertion point in a manner that makes it easy to place the fixture into your models. Fixtures that need to be adjacent to elements are easier to insert when their geometry is properly oriented with the insertion point. Also take into consideration how the fixture may be rotated in your model.

Because of the potential complexities of creating a new family and maintaining company standards and consistency, it can be useful to have a standard "family planning" sketchpad, where design, parameters, and reference planes can all be drawn out, before committing yourself to an actual family. Figure 21.2 shows an example of a form used for planning the construction of a Revit family.

Creating geometry in the plan view will place it at the reference level. The reference level in a family is a placeholder for the project level when you insert the family into a project. That is how a light fixture inserted on the second level of your project is associated with the second level. Therefore, the elevation or offset that you apply to the fixture is relative to the level at which it is placed. If you are creating a nonhosted lighting fixture that you want at a certain elevation when you insert the family, you can build the geometry at the desired elevation. Doing so can cause some confusion when using the Offset parameter.

Real World Scenario

OFFSET VS. ELEVATION

Grant has created a lighting fixture modeled at 8′-0″ above the reference level. When he inserts the family into his project, it has an Offset value of 0′-0″. In the breakroom area of the project, the lights need to be at an elevation of 10′-0″, so he changes the Offset parameter value to 2′-0″ for each of the fixtures in the lounge.

Later that day his project manager Lucy is reviewing the Revit file and selects one of the light fixtures to view its properties. She notices the offset of 2′-0″, and assuming that is the elevation of the lighting fixture, she changes the value to 10′-0″. Concerned with quality control, she checks and changes all of the fixtures in the breakroom area. It is not long before she receives an email from Grant wondering why the lighting fixtures in the lounge are 18′-0″ above the ground.

You can control the elevation of a lighting fixture by creating a reference plane that will define the fixture elevation. The plane can be dimensioned and associated with an instance parameter called Fixture Elevation to easily edit and manage the elevation of the fixture. Draw a reference plane in an elevation view when working in the Family Editor. Be sure to give the plane a name so that you can select it later as the host for your fixture graphics.

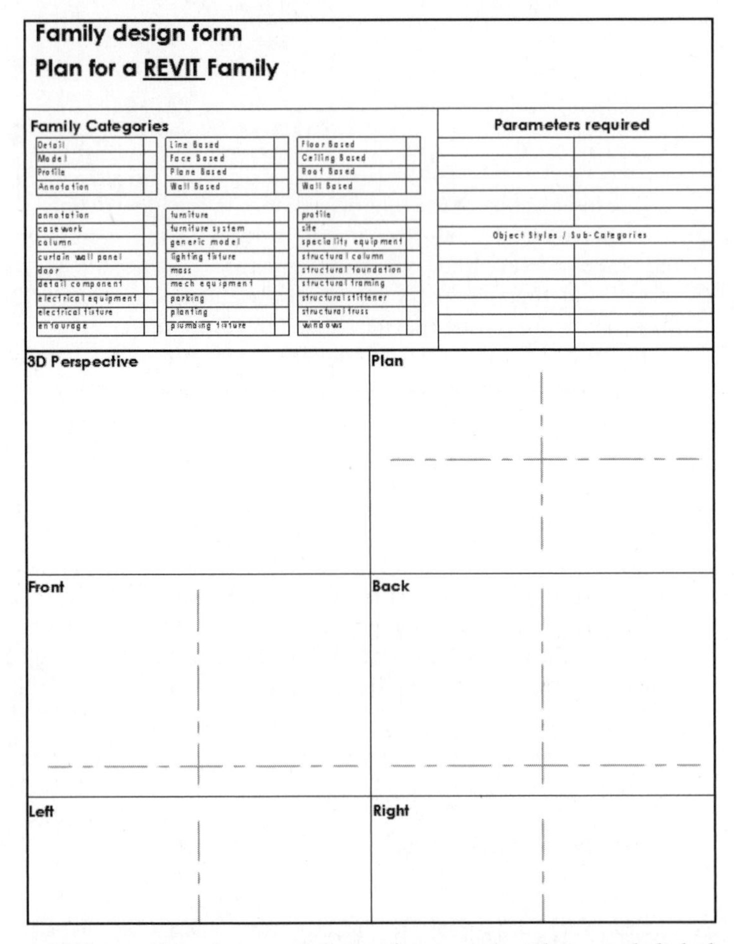

Family design form

Plan for a REVIT Family

Family Categories

Detail	Line Based	Floor Based
Model	Face Based	Ceiling Based
Profile	Plane Based	Roof Based
Annotation	Wall Based	Wall Based

annotation	furniture	profile
casework	furniture system	site
column	generic model	speciality equipment
curtain wall panel	lighting fixture	structural column
door	mass	structural foundation
detail component	mech equipment	structural framing
electrical equipment	parking	structural stiffener
electrical fixture	planting	structural truss
entourage	plumbing fixture	windows

Parameters required

Object Styles / Sub-Categories

3D Perspective

Plan

Front

Back

Left

Right

FIGURE 21.2 Family planning design form

Once you have created the graphics for your lighting fixture, select them, and click the Edit Work Plane button on the Work Plane panel of the contextual tab. In the Work Plane dialog box, click the Name radio button, and choose the plane you created for the fixture elevation from the drop-down list (see Figure 21.3).

You can change the work plane of an item only to a parallel plane, so select those items that have been created on the same plane or a parallel plane. If an extrusion is created while in an elevation view, it is associated with a vertical plane and cannot be moved to a horizontal plane. However, you will be able to align and lock it to a horizontal plane.

Figure 21.4 shows an elevation view of a nonhosted fixture family. A reference plane has been created to define the fixture elevation. The rectangular solid extrusion was created in plan view, and its work plane was changed to the elevation plane. The arced void extrusion was created in the left elevation view and aligned and locked to the fixture elevation plane. The dimension has been associated to an instance parameter for fixture elevation.

FIGURE 21.3
Choosing a new work plane for family geometry

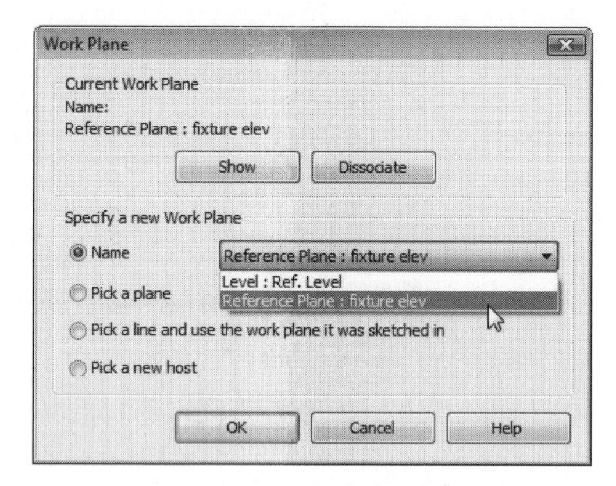

FIGURE 21.4
Nonhosted family with an elevation plane

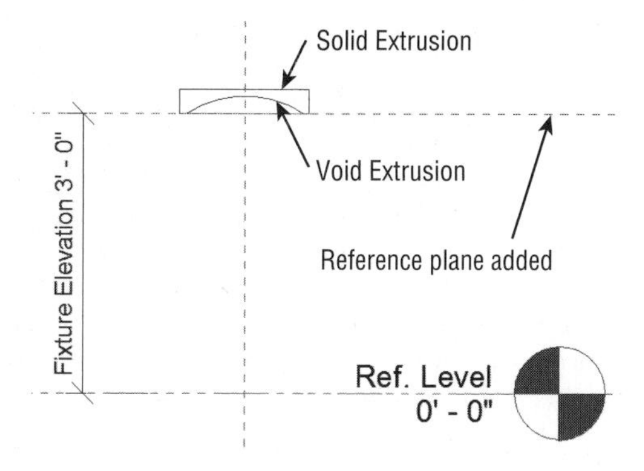

Creating a fixture with an elevation parameter does not disable the functionality of the Offset parameter, so be sure that you and your users understand the difference. When your family is placed into a project, the offset is applied to the entire family, while the elevation is applied to the geometry within the family.

Face-Hosted Lighting Fixtures

Creating face-hosted fixture families adds another level of coordination to your projects because the families will move with their associated hosts. This keeps your fixtures at the proper elevations when ceiling heights change. Wall-mounted fixtures will move with changes to wall locations also. Face-hosted families can also be hosted by reference planes within the project. In an area with no ceiling, a reference plane or level can be created to host lighting fixtures at a specific elevation.

To create a lighting fixture family that is face-hosted, you can use the `Generic Model face based.rft` family template. Once you've opened it, you can change the category of the family by clicking the Category And Parameters button on the Family Properties panel of the Create tab. When you change the category to Lighting Fixtures, the Light Source parameter becomes available. Select the box if your fixture will have a light source to be used for rendering or lighting calculations (see Figure 21.5).

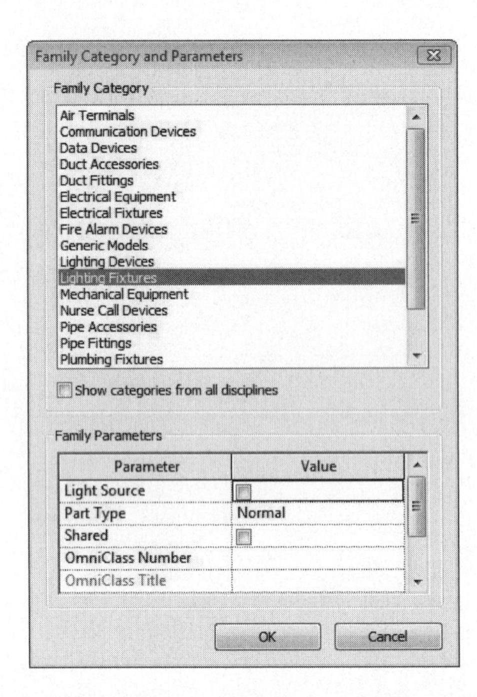

FIGURE 21.5
Family Category And Parameters dialog box

The face-based template contains an extrusion that is used as a reference for the host when the family is placed into a model. You can change the size of the extrusion, but you cannot delete it. The extrusion is necessary in order for your family graphics to know how to attach to their model host. It also allows you to indicate how void extrusions will cut their host.

> **CUT THROUGH A LINK?**
>
> Families with a void designed to cut the host will not cut the face of a linked file. This does not affect light output when using families with a light source.

Face-hosted families also contain a reference-level plane at the face of the placeholder extrusion. This face represents the face of a ceiling or wall host when the family is placed into a model. The placeholder extrusion face is locked to this reference plane and cannot be unlocked, so although you *can* associate the extrusion to another plane, it is not a good idea. Doing so will change only the thickness of the extrusion, which is not important.

To build a recessed lighting fixture, you want your extrusion(s) to be inside the placeholder extrusion. Although it will appear that the fixture is upside down when working in the family file, it will have the correct orientation when hosted by a ceiling or reference plane in your project file. Surface- or pendant-mounted lights should be modeled adjacent to the face of the placeholder extrusion.

If you want your lighting fixture families to follow the slope of a surface, such as a sloped ceiling, you need to deselect the Always Vertical parameter. This setting is found in the Family Categories And Parameters dialog box, which you can access from the Home tab.

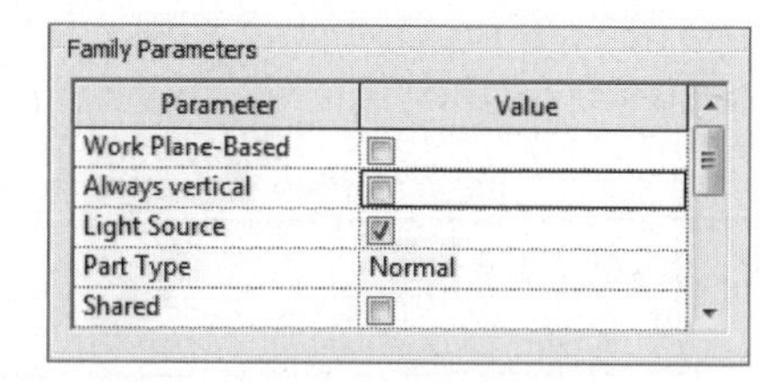

Setting the family to Always Vertical will cause the fixture to attach to the host in its normal orientation even if the host surface is sloped. Deselecting the parameter will cause your fixture family to follow the slope of its host.

Face-Hosted Families for Wall-Mounted Lights

When creating a face-hosted family for a wall-mounted light fixture, you need to treat the placeholder extrusion as though it were a wall. When working in the Ref. Level plan view of the family, it is as though the wall has been laid flat on the ground and you are looking down at it. It is easy to assume that the distance from the edge of the placeholder extrusion to your geometry will be the mounting height of the fixture. That is not the case (see Figure 21.6).

FIGURE 21.6
Face-hosted family for a wall-mounted light

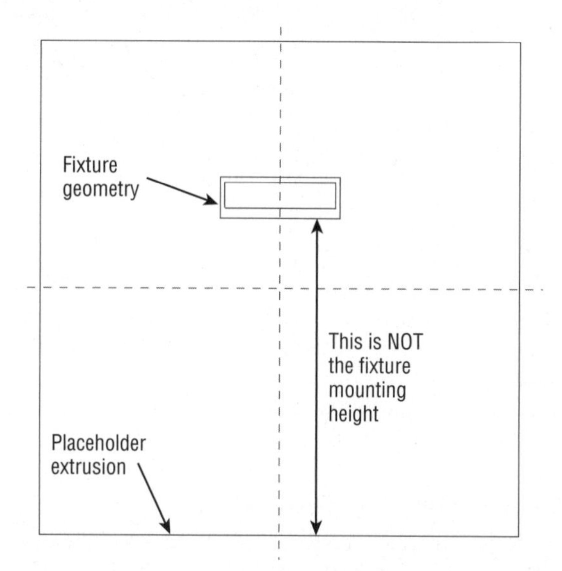

Face-hosted families have a Default Elevation parameter for when the family is placed on a vertical face. When they are placed into a model on a vertical face, the Elevation Instance Parameter becomes active and is used to set the mounting height. The Default Elevation parameter is simply a starting point until an elevation is assigned after the placement of the family. You can set the Default Elevation within the family file to any value that you want for the mounting height when the fixture is first placed into your project. The Elevation parameter is active only when the family is placed on a vertical host. When a face-hosted family is placed on a horizontal plane, the Elevation parameter is inactive and cannot be edited. Figure 21.7 shows a wall-mounted light that was placed into the model at an elevation of 4′-0″ (Default Elevation) and then given a mounting height of 7′-6″ (Elevation) after placement.

FIGURE 21.7
Wall-mounted fixture with an elevation

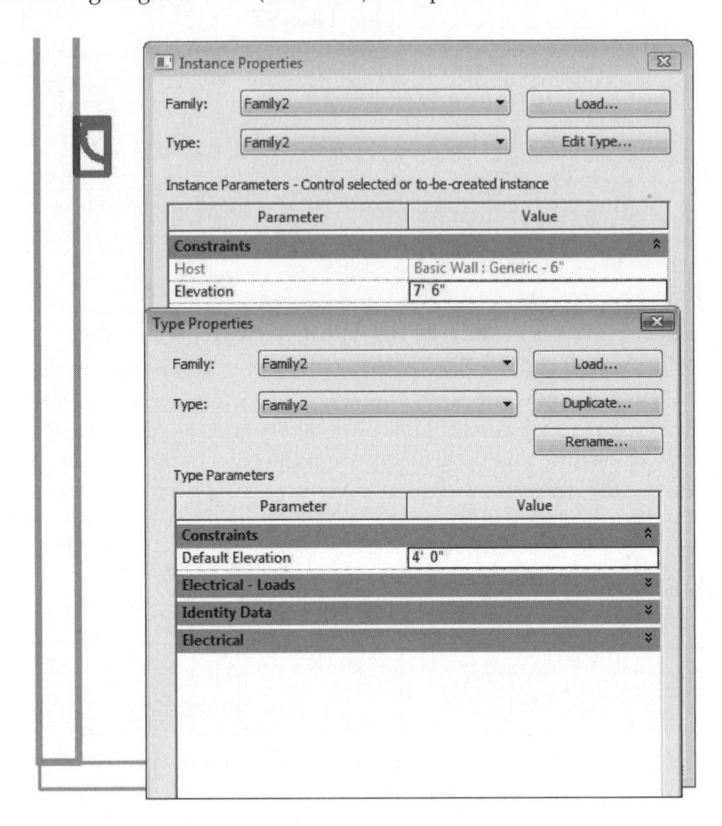

When creating an extrusion to represent the body of a light fixture that is to be hosted by a vertical face, it is often easier to create it in a left or right elevation view instead of the Ref. Level plan view. By doing so, the extrusion you create will be associated with the Center Left/Right reference plane. Additional reference planes can be added to create constraints for fixture length, depth, and width.

Fixture Types Based on Dimensions

When you are creating lighting fixtures, you first need to decide what type parameters are going to be used to determine the different types within that family. One common practice used is to create fixtures based on their dimensions. This is a good starting point for developing a fixture library, and additional parameters can be added later to create more types within the family for more specific purposes.

For example, you can start by creating a recessed fluorescent troffer; then, creating parameters for the length and width of the fixture allows you to make types in the family such as 2×4, 2×2, or 1×4. This is a very generic method for creating fixtures but is useful for general modeling where lighting analysis or circuiting is not required. The depth of the fixture should also be considered because it is a very important dimension for coordination within a model.

This can be done for any kind of fixture you may use in a project such as downlights, surface mounted, and pendant fixtures. A library of generic or "placeholder" fixtures is very useful in the preliminary design stages of a project. Early on you may not know exactly what fixtures will be used, but you can use these generic fixture families to represent the basic layout and design intent. Because they are categorized as lighting fixtures, you can easily replace them with the actual fixtures to be used when the decision is made.

Fixture Types Based on Fixture Performance and Lighting Characteristics

Having fixture families based on size and shape is a good start, but you will also need fixtures that are more defined to their specific uses and performance. This is especially true if you intend to use your Revit model for lighting analysis. It is a good practice to develop lighting fixture families that meet the design requirements of your Revit projects without having to create a separate fixture family for every light fixture you use.

You can create families based on their performance and how they look. Not only will a parabolic fixture perform differently than a lensed fixture, but it also looks different, so creating separate families for each may be necessary. Appearance is important only if you are concerned with modeling to a level of detail that requires a visible difference between the two fixtures, but having separate families may be easier to manage than having one family with several types.

With these fixtures, the dimensions are important for defining different types, but they do not need to define the family itself. You can create a parabolic troffer family with types such as 2×2 or 2×4, but you also want parameters to define what the voltage, load, or number of lamps is. Each type parameter that you add to a fixture family will determine a fixture type within that family. This allows you to use a unique `.ies` file for each type.

There are a large number of lighting manufacturers to choose from when specifying your fixtures in a project. It is not necessary to have a library that contains each and every option. You can make a set of lighting fixture families that cover the basic fixture types and then modify them to the manufacturer criteria specified as needed. Choose an `.ies` file that meets the basic requirements for the fixture type as the baseline for that type of fixture. When you use the fixture in a project, you can get an idea of its performance prior to making a decision on which specific manufacturer and model number to be used for your project. Some people have resisted using a BIM solution such as Revit MEP for their projects because they think they have to make too many decisions early on in the project that they are not able to make. Using a baseline fixture type will allow you to move forward with the project and make the more specific decisions when necessary.

Figure 21.8 shows the different types of a parabolic 2×4 light fixture family where the Number Of Lamps and Ballast Voltage parameters were used to determine the family types. A unique Photometric Web File is used for each variation in the number of lamps but is not required for each voltage.

There is no right or wrong answer to how many light fixture families you should have. How you use the software and your design standards will help determine what types of fixture families you require. A minimum number of families that can be easily adjusted to the specifics of your project will be easiest to manage and maintain.

However you decide to create a fixture, it is important to build your fixture family so that when it is placed in a model, it can associated with a Space object in order to provide the desired engineering data from calculations. A light fixture that is not inside or touching a Space object will not generate an average estimated illumination for that Space object. Figure 21.9 shows a group of fixtures placed higher than the upper limit of an engineering Space object; therefore, there is no illumination calculated for the Space object.

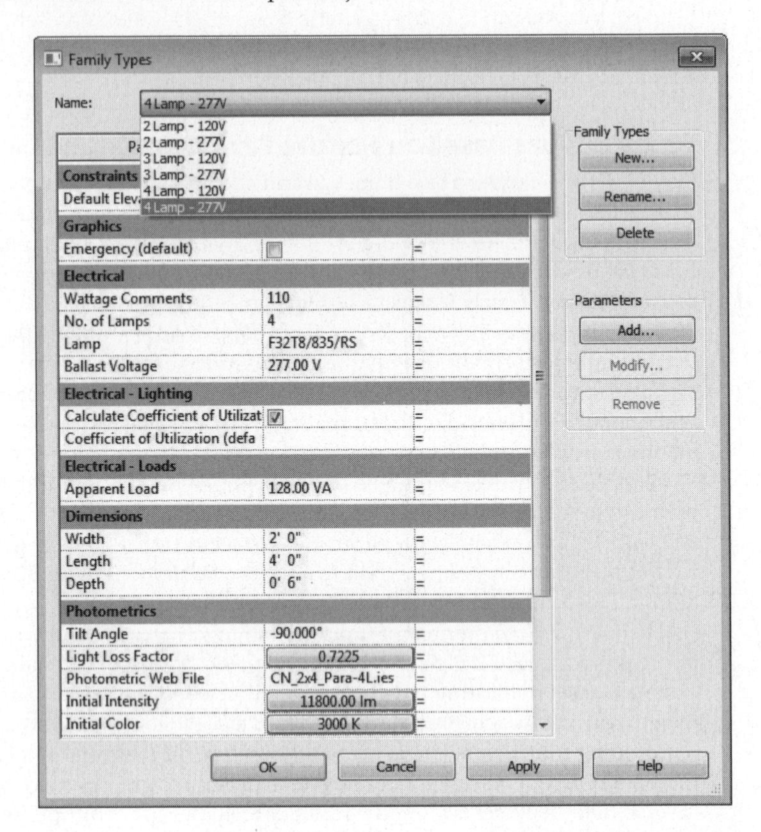

FIGURE 21.8
Lighting fixture family types

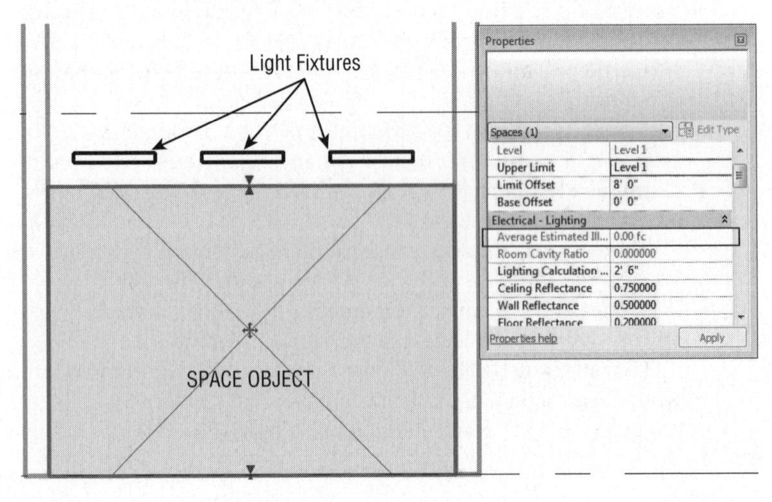

FIGURE 21.9
Fixtures outside the limits of a Space object

Naming Conventions

A standard naming convention will go a long way in helping you maintain and manage your lighting fixture families. When you create a lighting fixture or modify an existing family, it is best to distinguish it from families that come installed with Revit MEP. This will prevent the files from being overwritten when your library is updated at the installation of a new release of the software.

A common practice is to prefix the family name with your initials or your company's initials. This not only distinguishes them as unique to you or your company but also keeps them organized because they will all be listed together alphabetically.

The naming of `.ies` files is also important to management and organization. Most `.ies` files that are provided by manufacturers are named with a convention unique to whoever is providing them. It can be difficult to determine what type of fixture an `.ies` file is associated with if it is not named in a manner that indicates the fixture type or performance. When you acquire an `.ies` file, consider renaming it to indicate its characteristics so that those using the file will be able to easily see what type of fixture it represents.

Figure 21.10 shows that the Photometric Web File parameter for a parabolic 2×4 fixture has been named so that it clearly indicates what type of fixture it should be used for.

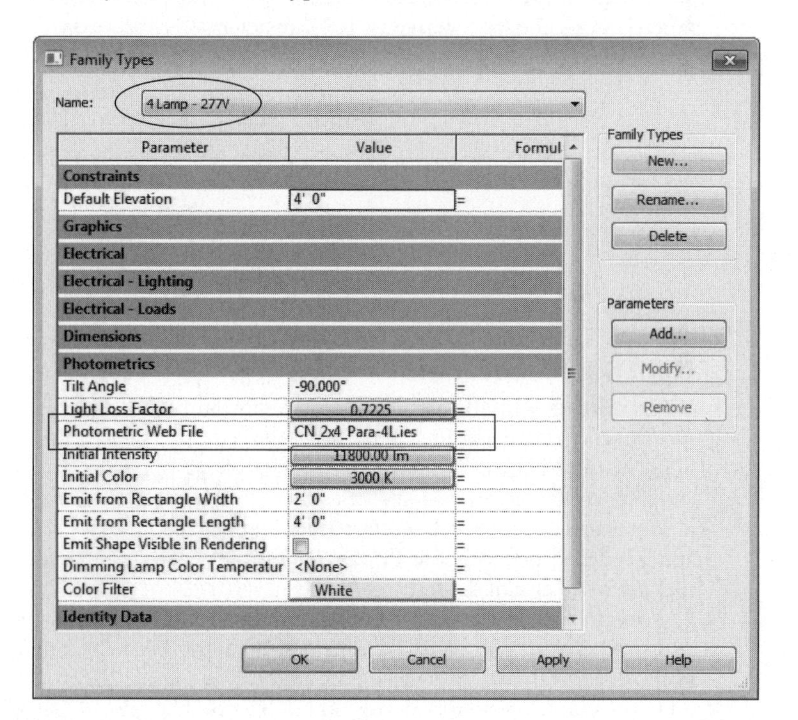

FIGURE 21.10 Photometric Web File parameter named to match fixture type

Lighting Analysis

Your lighting fixture families require a light source if they are to be used for rendering or lighting calculations. A light source is a unique feature of a lighting fixture family that acts as the part that emits light. This does not necessarily mean a lightbulb because you can define the shape of the light source to how light is seen coming from the fixture. When you look at a light fixture that has a lens, the light appears to come from the shape of the lens.

GARDEN OF LIGHT

In lighting design circles, it is often said that a *bulb* is something you plant in your garden; light fixtures have *lamps*.

You can also define the light distribution from the light source. Setting how the light is thrown from a fixture will give you a more accurate representation of the behavior of the fixture for lighting calculations and rendering. To give your family the exact photometric characteristics of a specified fixture, you can designate a photometric web file for the light source and also define settings such as color and intensity.

Lighting fixture families have one light source by default. It is possible to have a fixture family with multiple light sources, such as track lights or a chandelier. Doing this requires creating a separate family that defines the light source and then nesting it into your fixture family. The nested family must be shared in order to act as a light source. You can set a family to be shared by selecting the Shared box in the Family Category And Parameters dialog box.

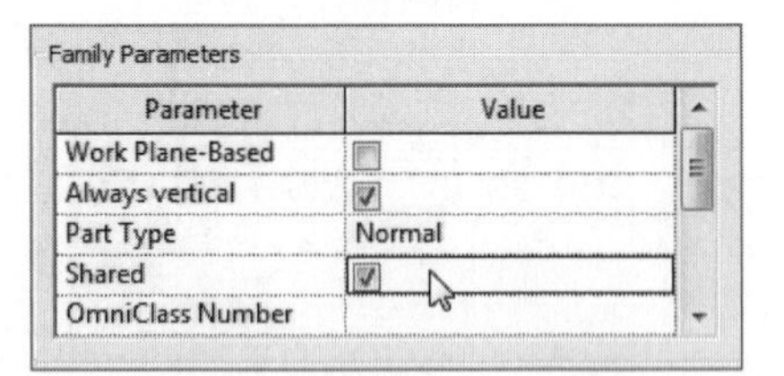

Light Source Location

When starting with the `Generic Model.rft` template and categorizing the family as a light fixture with a light source, you will see the light source as a yellow object in the drawing area. Yellow can be a difficult color to work with on a white background, so you may want to change the color by accessing the Object Styles settings within the family file. The light source appears at the insertion point on the reference plane. You can move the light source to anywhere in 3D space within the family file. Figure 21.11 shows a light source for a site light pole that has been moved to the face of the fixture.

You can change the size of the light source symbol by editing the Light Source Symbol Size and Emit From Line Length parameters. Click the Types button on the Family Properties panel of the Create tab to access these parameters (see Figure 21.12).

Light source objects have axes that allow you to align and lock them to the fixture geometry or to a reference plane. If you place the light source inside solid geometry, you will receive a warning that no light will be emitted when the family is in a rendered view. The light source can be located on the face of your fixture geometry or within a void inside the geometry for a more realistic representation of the fixture. Figure 21.13 shows a downlight family with a light source located inside the geometry of the fixture. The light distribution that appears to pass through the sides of the fixture will be blocked by the solid geometry when shown in a rendering. A material can be applied to the geometry to affect how the light will be reflected inside the fixture.

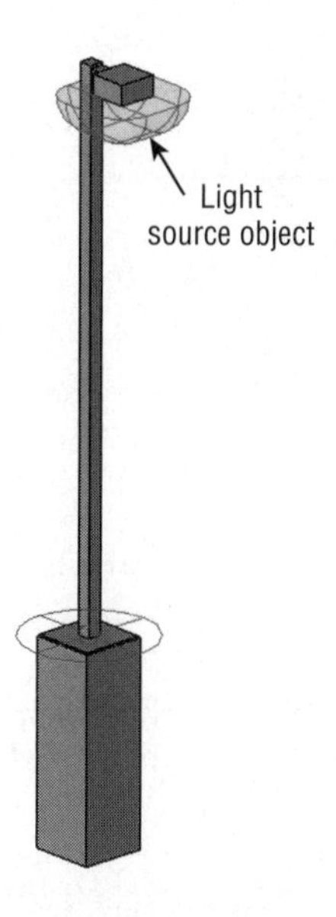

FIGURE 21.11
Light source of a pole light family

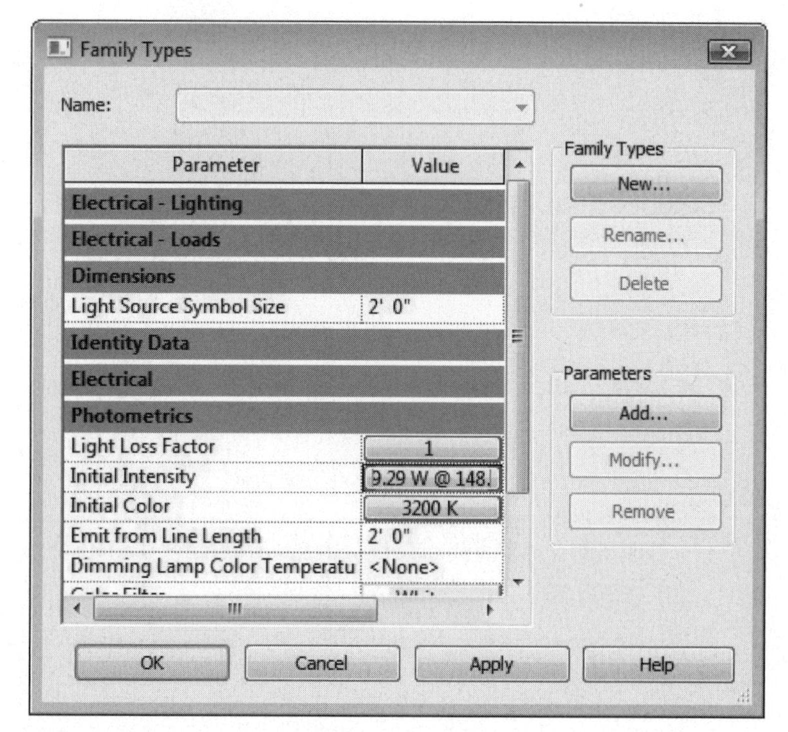

FIGURE 21.12
Light source size parameters

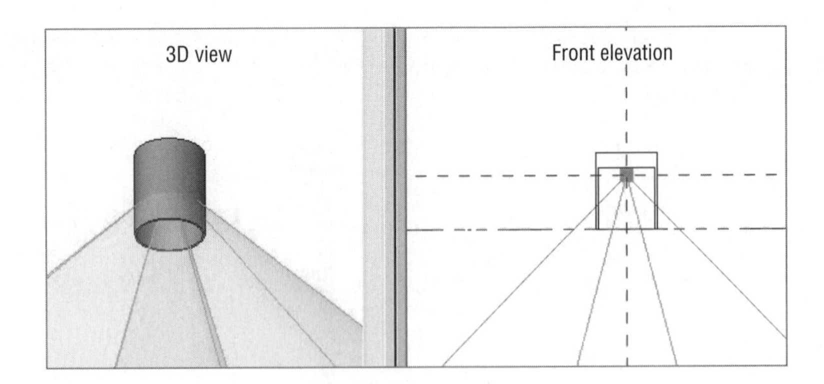

FIGURE 21.13
Light source inside fixture geometry

Light Source Definitions

You can define the shape of the light source and its distribution pattern by clicking the Light Source Definition button on the Lighting Panel of the contextual tab that appears when you select the light source. You can choose from four shapes: point, line, rectangle, and circle. You can combine any of these shapes with one of the four light distribution patterns located at the bottom of the Light Source Definition dialog box. Pattern options are spherical, hemispherical, spot, and photometric web (see Figure 21.14).

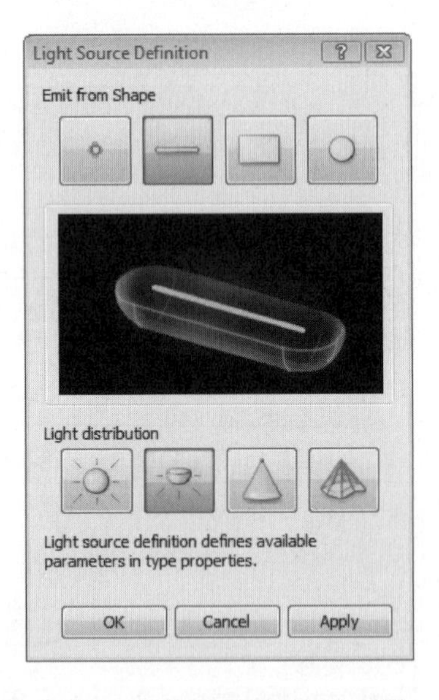

FIGURE 21.14
Light Source Definition dialog box

The light distribution patterns have different properties. The spherical and hemispherical distributions have the basic settings for size and photometric data, while the spot distribution has additional settings for the tilt angle and beam spread. The photometric web distribution has a parameter for tilt angle and one to define the light source by a photometric (`.ies`) file. The properties of the light source distribution are all associated with type parameters of the fixture family. Figure 21.15 shows the properties of a light source with a spot distribution. Notice that the parameter values are mapped to family parameters (indicated by the = at the right side of the parameter value).

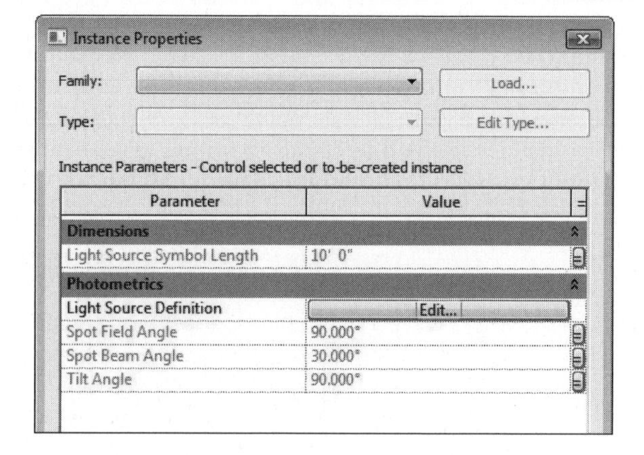

FIGURE 21.15
Light source properties

Light Source Parameters

The family parameters that control the properties of the light source are type parameters in the Photometrics parameter group. The parameters vary depending on the light distribution style chosen. These parameters have more to do with rendering appearance than with lighting calculations since the `.ies` file associated with a fixture will drive the calculation values.

You can apply a light loss factor to the fixture family by clicking the button in the Light Loss Factor parameter cell. This opens a dialog box with two methods for applying a light loss factor. The Simple method allows you to use the slider in the dialog box to assign a total light loss factor to the family, while the Advanced method provides more specific options. You can input values for losses and depreciation manually or use the sliders. The combination of settings will generate a total, as shown in Figure 21.16.

FIGURE 21.16
Advanced light loss factor settings

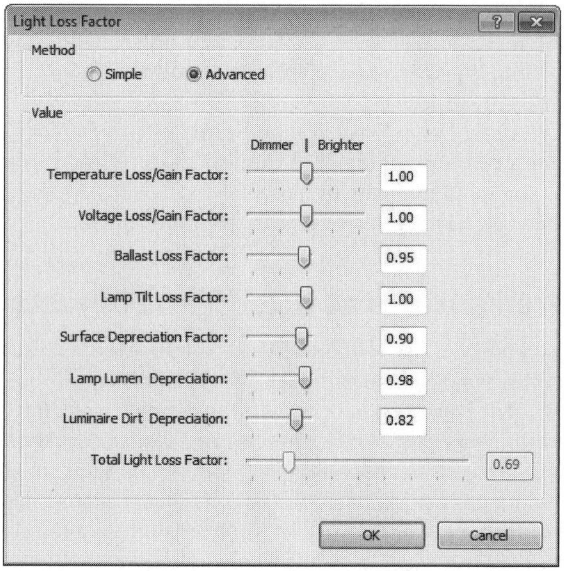

You can set the Initial Intensity parameter by any of the four values shown in the dialog box that appears when you click the button in the parameter cell. You can choose any of the value options to input. This information is often found in a lamp catalog or may be stored in the `.ies` file associated with a fixture. You do not need to know the value for each option. When you input a value, the other options will populate according to your input. Figure 21.17 shows an example where an input was given for Luminous Flux and the other values populated accordingly.

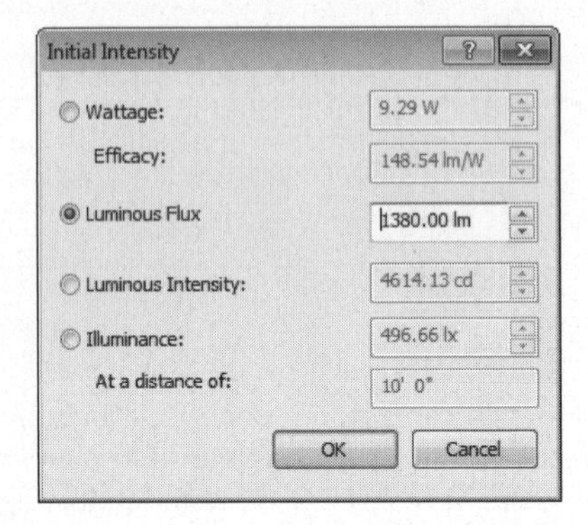

FIGURE 21.17
Initial Intensity dialog box values

VIEWING THE INFORMATION IN A PHOTOMETRIC WEB FILE

Many lighting manufacturers that provide photometric files of their products also offer viewers that allow you to see the information contained within the file. These are very useful tools for inputting engineering data into your Revit lighting fixture families.

Clicking the button in the Initial Color parameter cell opens a dialog box that lets you choose the color temperature of the light. The Color Preset drop-down offers settings based on various common lamp types and also offers the option to input a custom color temperature value, as shown in Figure 21.18.

The Dimming Lamp Color Temperature Shift and Color Filter parameters are also for rendering the appearance of the light from the fixture. You can apply a predefined lamp curve to affect the color and intensity of lights when they are dimmed. You can also apply a color filter to change how the light appears coming from the fixture.

Fixture Families as Intelligent Objects

The parameters you use will provide the intelligence within your lighting fixture families, making them easily modifiable and useful for calculations.

Use family parameters to constrain the geometry of a fixture or to add data to the fixture. Family parameters cannot be included in a schedule or tagged in views, so be sure to only use them for data that is not necessary to be shown on your construction documents.

As mentioned previously in this chapter, dimensions are good parameters to have in your fixture families because they allow you to customize a family to the exact dimensions of fixtures specified for your project without having to create an entirely new family. These dimension constraints are also useful for controlling the symbolic representation of fixtures, which is discussed later in this chapter.

Another useful parameter is one for electrical load. This gives you a parameter to associate the connector parameter with so that when fixtures from different manufacturers are chosen, you can edit the family parameter and update the load of the connector.

Some types of lighting fixtures such as site lighting are available in different voltages that require a different number of poles for connection to a circuit. Having a parameter that defines the number of poles allows for the easy modification of the connector through the use of parameter association. A fixture family with this parameter can have a type for 208V single phase (2 poles), 208V three phase (3 poles), or 277V single phase, for example.

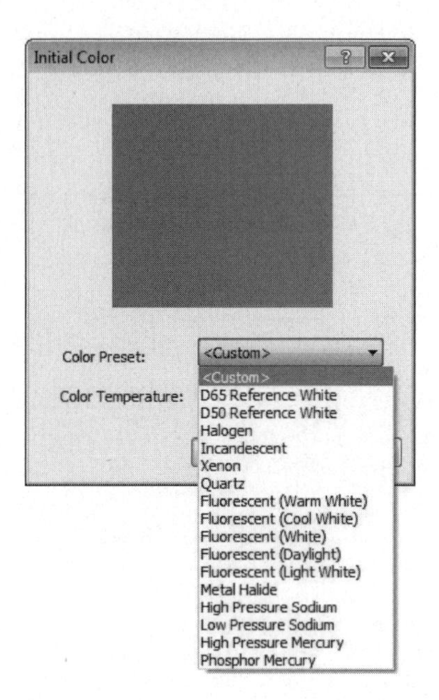

FIGURE 21.18 Initial Color dialog box

Adding Connectors

One feature of your lighting fixture families that will make them useful for the design of your projects is an electrical connector. Adding connectors to your families will allow you to connect them to electrical circuits in order to manage your panel loads and also will allow you to use wiring objects that maintain a connection to the fixtures when they are moved in the model.

The location of a connector in a lighting fixture family is not as important as on other types of electrical objects because the graphical representation of the wiring will stop at the edge of the fixture. The easiest placement method for adding a connector is to choose the Face option. This places the connector in the center of whatever 3D face you select. Click the Electrical Connector button on the Connectors panel of the Home tab, and then choose the Face option on the Placement panel of the contextual tab that appears.

Next, choose the type of connector from the drop-down list on the Options Bar. You can change the type later by accessing the properties of the connector if needed. If you are using the connector to circuit your lights to a power panel, choose one of the power type connectors.

As you place your cursor over the fixture geometry, the 3D faces will highlight indicating where the fixture can be placed. You can use the Tab key to cycle through available faces that have a common edge. Once you have highlighted the desired face, click to place the connector. There is no need to dimension or constrain the connector because it will always be in the center of the selected 3D face.

If you want to control the location of a connector, you can use the Work Plane method of placement. This method requires that you select a plane on which to place the connector. You can use any reference plane within the fixture family. Choosing the Work Plane option will open a dialog box with options for specifying the desired plane. You can pick the plane from a list or choose to manually select the plane. Figure 21.19 shows the dialog box and the various planes available in a particular lighting fixture family.

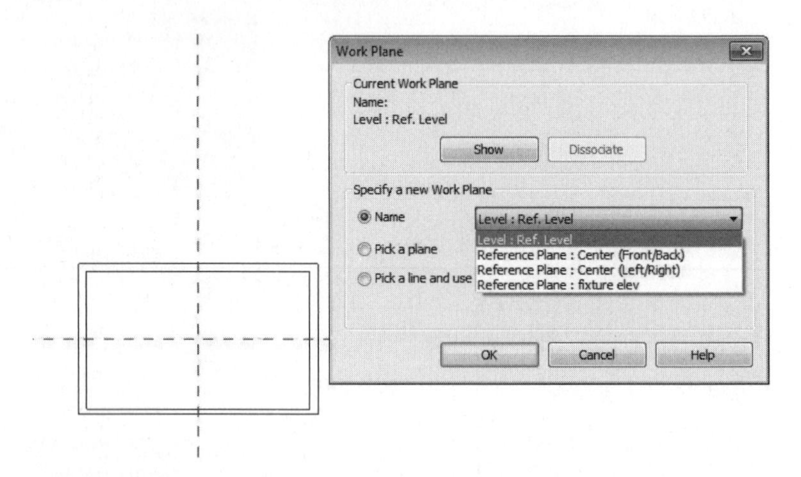

FIGURE 21.19
Work Plane connector placement dialog box

Once you have selected a plane for placement, the connector will appear at the insertion point of the family. You can then move the fixture by selecting it and using the Move command on the Modify Connector Element contextual tab. It is important to use the Move command because attempting to click and drag the connector will only rotate the orientation of the connector object.

After you have moved the connector to the desired location, you can convert the temporary dimensions to actual dimensions in order to constrain the location of the connector if necessary.

The properties of a connector determine how the fixture family behaves electrically. The parameters of a connector can be associated with parameters of the fixture family so that they can be changed parametrically in a project or be given different values for different types within the family. For example, you can have a fixture family with multiple voltage options. The Voltage parameter of the connector can be associated with the Voltage parameter of the family by clicking the small square to the far right of the connector Voltage parameter. This opens a dialog box that allows you to choose the family parameter to associate the connector parameter to. Only family parameters that match the discipline and type will be available, so you cannot associate a Voltage parameter with something that is not a voltage.

Representing Light Fixtures on Construction Documents

As with most of an electrical design, a lighting plan can be represented symbolically. There is no real need to show the actual fixtures in the model on construction documents. Using symbolic lines and symbols to show lighting fixture locations enables you to represent fixtures that are alike with a common symbol. This reduces the amount of symbols required, simplifying your construction documents. For example, though your project may have several types of 2×4 recessed fixtures, they can all be represented with the same symbol in your plan views.

Using symbolic lines and symbols will also help improve the performance of your Revit model. 2D graphics can be more easily processed and regenerated than 3D graphics. Although it is recommended that you model your light fixtures in the simplest form possible, you may receive fixture families from manufacturers or other sources that are modeled to a more complex level. Having many light fixtures in a project and showing their 3D graphics can significantly affect how well your model performs.

Another reason for using symbolic lines or nested annotation symbols in your lighting fixture families is that it gives you another level of visibility control for your fixtures. Through the use of parameters, a light fixture can display different configurations of symbolic lines based on variations within the fixture, without having to create another family or family type.

A good example of a light fixture using a nested annotation symbol is an exit light. These fixtures are commonly shown on construction documents as a symbol, instead of showing the actual light fixture; however, the symbol may be shown in different configurations depending on how many faces or direction arrows are on the fixture. Figure 21.20 shows an exit light family in both plan and 3D view. Notice that a symbol is used to represent the fixture in ceiling plan view and that the symbol indicates the location of the face of the fixture with a filled region.

Creating this type of functionality is easily achieved by using a nested annotation in the lighting fixture family. The annotation family contains lines and filled regions whose visibility is controlled by Yes/No parameters.

To create an exit light family with multiple display options, do the following:

1. Open the `Ch21_Exit Light Annotation.rfa` family found at `www.wiley.com/go/masteringrevitmep2011`.
2. Draw a circle with the center point at the intersection of the reference planes. The radius of the circle should be 3/32″ so that the annotation will be the correct size at different view scales.
3. Draw a line from one quadrant of the circle to the opposite quadrant. Rotate the line 45 degrees using the center of the circle as the axis, and mirror it using one of the reference planes as the axis, creating four equal quadrants within the circle.
4. Create a filled region by tracing over one of the quadrants. Use invisible lines for the border of the filled region and Solid Fill for the pattern. Click the green check box button on the contextual tab to finish creating the filled region.
5. Mirror the filled region using the reference plane perpendicular to the region to create another filled region directly opposite.
6. Click the Family Types button on the Properties panel of the Home tab. Create a new parameter by clicking the Add button at the right of the dialog box. Name the parameter FACE1. Set the parameter as an instance parameter, and set Type Of Parameter to Yes/No. Click OK.

7. Repeat step 6 to create another parameter named FACE2. Click OK to exit the Family Types dialog box.
8. Select one of the filled regions. Click the small box to the far right of the Visible parameter in the Properties palette. Select FACE1 from the list, and click OK. Select the other filled region and repeat, choosing FACE2 from the list. Save and exit the family.
9. Open the `Ch21_Ceiling Exit Light.rfa` family found at `www.wiley.com/go/masteringrevitmep2011`.
10. Create a 4″ square extrusion centered at the intersection of the reference planes. Set the depth of the extrusion to 1″. Create another extrusion that is 2-1/2″ wide and 12″ long. In the Properties palette, set Extrusion Start to 1″ and Extrusion End to 11″. Center the extrusion at the intersection of the reference planes.
11. Select both extrusions, and click the Visibility Settings button on the Mode panel of the contextual tab. Remove the check mark from the box next to Plan/RCP, and click OK. This will keep the extrusions from being visible in plan or reflected ceiling plans.
12. Click the Family Types button on the Properties panel of the Home tab. Create a new parameter by clicking the Add button at the right of the dialog box. Name the parameter Show Face 1. Set the parameter as an instance parameter, and set Type Of Parameter to Yes/No. Group the parameter under Graphics. Click OK.
13. Repeat step 6 to create another parameter named Show Face 2. Click OK to exit the Family Types dialog box.
14. Click the Load Family button on the Insert tab. Browse to the annotation family you created, and click Open. Click the Symbol button on the Annotate tab, and place the annotation symbol at the intersection of the reference planes.
15. Select the annotation symbol, and click the small box to the far right of the FACE1 parameter in the Properties palette. Choose Show Face 1 from the list, and click OK. Repeat for the FACE2 parameter, choosing Show Face 2 from the list. Save and exit the family.
16. Open the `Ch21_SampleProject.rvt` file found at `www.wiley.com/go/masteringrevitmep2011`.
17. Click the Load Family button on the Insert tab. Browse to the newly created exit light fixture family, and click Open. Click the Lighting Fixture button on the Home tab. Select the Place On Face option from the contextual tab, and place the fixture on the ceiling. Notice that only the annotation symbol is displayed and not the extrusion graphics.
18. Click the light fixture and deselect the box for the Show Face 2 Parameter in the Properties palette. Notice that the filled region for FACE2 is no longer displayed.

Not all light fixtures are represented by an annotation symbol. Some fixtures need to be shown as their actual size for coordination purposes. If a nested annotation is used for these types of fixtures, the annotation will change size with the scale of the view, which would be an inaccurate representation. Symbolic lines can be used in a fixture family to represent the outline of a fixture without having to show the model graphics within the family.

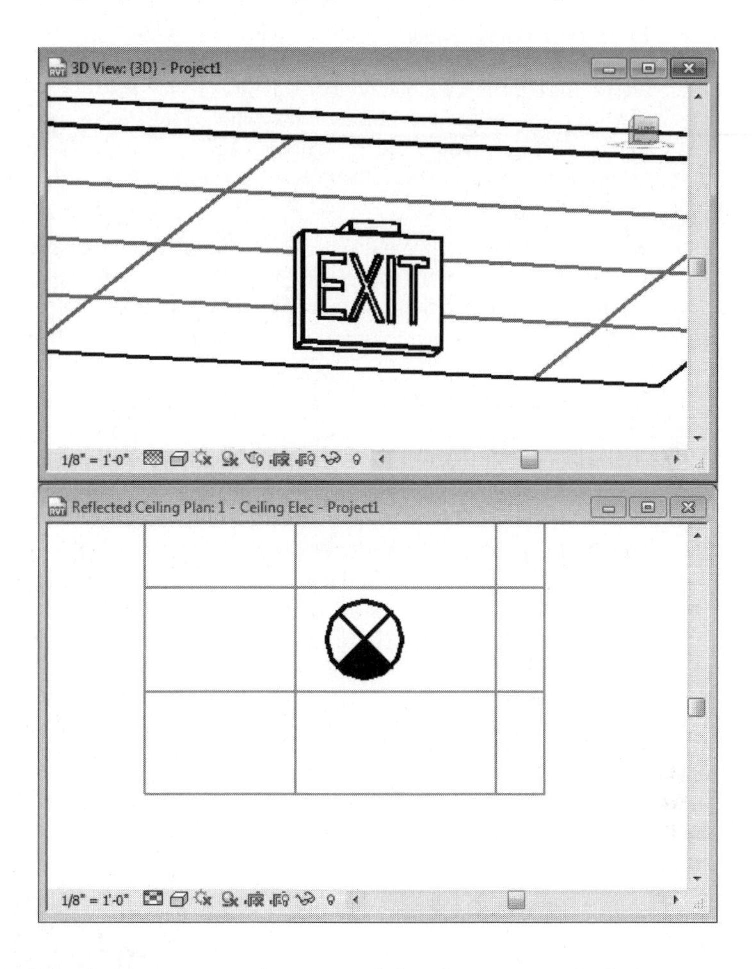

FIGURE 21.20
Exit lighting fixture in plan and 3D view

Symbolic lines will display in any view that is parallel to the view in which they are created. When adding symbolic lines to a lighting fixture family, they need to be drawn in the Ref. Level Floor Plan view or Ref. Level Ceiling Plan view for ceiling-mounted fixtures. If the fixture you are creating is wall mounted, draw the symbolic lines in either the Front or Back elevation view.

Symbolic lines will not change size in different scale views so they can be drawn to match the size of the fixture. You can constrain symbolic lines to reference planes or model graphics so that they will move when the fixture is changed.

When working in a lighting fixture family, any symbolic lines will belong to the Lighting Fixtures subcategory by default. You can create a unique subcategory for the lines to allow for greater visibility control within the model.

TURN OUT THE LIGHTS

Wesley is a lighting designer who uses symbolic lines in his lighting fixture families. He shares his lighting layout with his architect Buck, who is required to submit a reflected ceiling plan in his construction documents. During a coordination meeting, Buck points out that all the wall-mounted lights are showing up in the reflected ceiling plan. He obviously cannot turn off the Lighting Fixture category in the view.

Wesley has drawn the symbolic lines in his wall-mounted lights on a subcategory called Wall Lights – Linework. This allows Buck to turn off the subcategory while leaving on the Lighting Fixtures category. With the subcategory turned off, the reflected ceiling plan now shows only ceiling lights.

Although you can draw symbolic lines in a lighting fixture family, you cannot create a filled region. Filled regions are useful for showing a portion of the fixture filled in to denote an emergency lighting fixture. If you create a filled region in an annotation family and then nest that annotation into a fixture family, the annotation cannot be resized to match the fixture if its dimensions change. However, you can use a detail component family instead.

You can create a detail component family with parameters for length and width. A filled region can be drawn in the detail component family that is constrained to the parameters. When nested into a lighting fixture family, the parameters of the detail component can be associated to the parameters within the fixture family so that the detail component will match the size of the fixture. Figure 21.21 shows a detail component family designed to indicate when a lighting fixture is an emergency type.

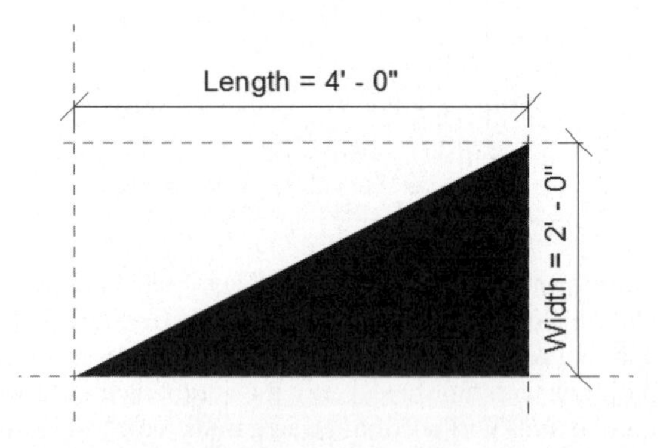

FIGURE 21.21
Detail component family for an emergency lighting fixture

With the detail component nested into the fixture family, its visibility can be associated with a YES/NO parameter so that it can be turned on or off as desired. This should be an instance parameter so that the same fixture type can be shown as normal or as emergency. Figure 21.22 shows two instances of a light fixture, one of which has been set to an emergency light.

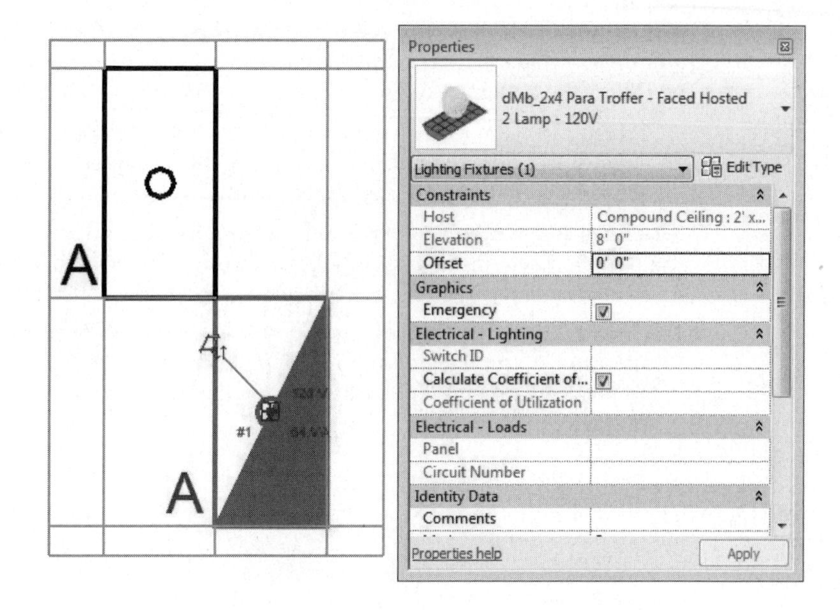

FIGURE 21.22 Lighting fixture family with a detail component for emergency lighting display

Detail components can be more useful than annotation symbols because they can placed into an elevation view within a fixture family where annotation symbols cannot. When you create a wall-mounted fixture that is face hosted, the Front or Back elevation view is what you will see when the fixture is hosted by a vertical face, so any symbolic representation must be placed in the Front or Back elevation view of the family.

The Bottom Line

Create different types of lighting fixture families Many different types of lighting fixtures are required for various applications within a building. With Revit MEP, you can create any type of lighting fixture and include any data associated with the fixture.

Master It Knowing how a lighting fixture will be used in a Revit model is important to determining the kind of family to create. True or false: A face-hosted lighting fixture can be replaced by a nonhosted fixture using the Type Selector.

Use a light source in your lighting fixture families Lighting fixtures can be used in making design decisions because they not only represent the fixture as a model but also contain photometric data from real-world lighting fixtures for use in lighting analysis.

Master It Photometric web files can be obtained from lighting fixture manufacturers. These files provide the lighting distribution characteristics of a fixture when added to a family. How can you be sure that the `.ies` file you are using is appropriate for the type of fixture it is being used in?

Create and manage fixture types and parameters The parameters of a lighting fixture family are what makes it an intelligent object. They can be used to create multiple types within the same family or for managing the electrical characteristics of a fixture.

Master It Connectors are what determine the electrical properties of a lighting fixture family. Describe the process of ensuring that a connector has the same load and voltage values that have been assigned to the fixture.

Use lines and symbols to represent lighting fixtures Some lighting fixtures are shown on construction documents as symbols, while others are shown as their actual size. Symbolic lines or annotation symbols can be used to eliminate the need to display model graphics.

Master It Annotation symbols nested into lighting fixture families can represent the fixture without having to show the model graphics. Is it possible to use a nested annotation family to represent a wall-mounted fixture in a face-hosted lighting fixture family? Explain.

Chapter 22

Creating Devices

Device families can be used by any MEP discipline to represent the types of components that are crucial to engineering systems but do not necessarily play a major part in the physical model of the systems. Components such as thermostats, switches, and receptacles are all important to how engineering systems are used, yet their size is generally not an issue when it comes to interferences with other model components.

Although devices do not require very detailed modeling, it is useful to have solid geometry that represents the devices in your projects. This will allow you to coordinate their locations in the situations where it is important for collaboration with other disciplines. It may seem unnecessary to model receptacles or switches, but since these types of items are shown in a set of construction documents anyway, you might as well show them correctly in the model for further coordination. With a good library of device families, there is no additional effort to adding model devices as opposed to adding symbols that hold no system information or model intelligence.

Device families can be given connectors that enable them to be included in the engineering systems that you create in your projects. This adds another level of intelligence to the systems and allows you to keep track of things such as circuits and device-component relationships.

Because there is no single standard for devices that is used by all engineers and designers, it is important that you are able to develop the device families that work in the way you design. Of course, it is also important that these components look the way you want them to on your construction documents.

In this chapter, you will learn to

- Model device geometry
- Use annotations to represent devices
- Add parameters and connectors

Modeling Device Geometry

Creating solid geometry for device families is similar to modeling equipment families only on a smaller scale. The level of detail for devices can be kept very simple to represent the components. In most cases, the geometry does not need to be parametrically controlled, but there are some families where parametric geometry is very useful. Junction boxes are a good example of this. You can create one junction box family with parametric geometry so that multiple types can be created by editing the dimensions of the geometry. This eliminates the need for separate families for each size of junction box.

Category and Parameters

As with any family you are about to create, you should first decide how the family will be used in your projects. For the most part, device families should be face-hosted families, because it is likely they will be placed in walls, floors, and ceilings. Of course, there are some exceptions such as junction boxes, which you may want to put into the model without a host. The hosting option you choose will determine the family template that will be used. The generic family templates can be used since you can categorize the family once you have started.

After you have chosen a template to create the family from, click the Family Category And Parameters button to assign the family to a category.

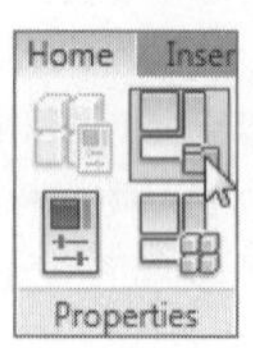

All of the device categories have the same parameters, but the options for the Part Type parameter vary depending on the category. Figure 22.1 shows the available part types for the Lighting Devices category. These are the same options for the Nurse Call and Security Devices categories. The other device categories do not have the Switch option. Using the Switch option allows you to specify the device as the switch for a system in your model.

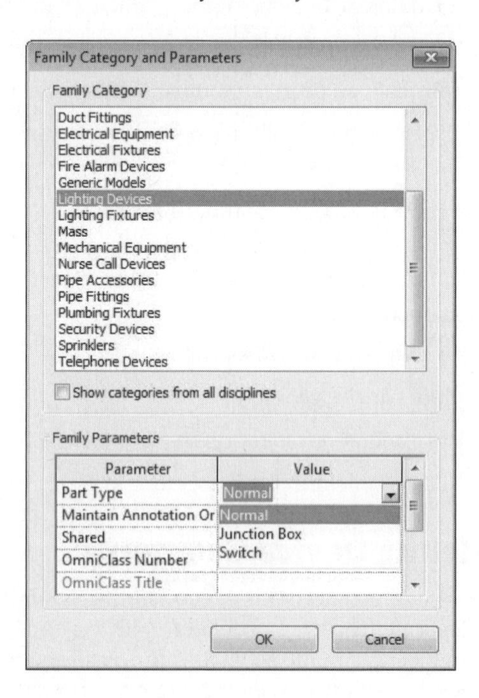

FIGURE 22.1
Part types available for a Lighting Devices family

The Maintain Annotation Orientation parameter is very important for device families that use a nested annotation symbol to represent the object in your model views. This parameter allows the annotation to be displayed regardless of the hosting of the device. This means that the annotation symbol will display whether the device is mounted to a vertical or horizontal surface. If the device is placed on a sloped surface, the annotation will not display.

Select the box for the shared parameter if you want to load your device family into another family and be able to account for the device individually when the host family is used in a project. With this functionality, you could create a family such as a typical systems furniture layout and have the outlets included by nesting them into the furniture family, for example.

The OmniClass parameters are for classifying a family for sharing on websites like Autodesk Seek. If you share a family, it can be identified or filtered based on its classification. You do not need to assign an OmniClass number to your families in order to share them.

Geometry and Reference Planes

With the family category and parameters established, you can begin modeling the geometry of your device family. It is a good idea to establish the category right away, because this will activate certain device-specific parameters, if applicable, and also because once you start working, it is easy to forget until you attempt to use the family in a project. It can be very frustrating to work on a family and load it into a project only to find that it is not available when you click the button to place it in your model.

The amount of solid geometry that you include in a device family depends not only on the object you are creating but also on how it will be used in your projects. If your intent is to only use devices so that their symbols are shown on your plans and you are not concerned with 3D visualization, you can use a simple box. This will at least give you the ability to see the location of the device in section and elevation views for coordination with things such as casework and openings.

Solid geometry is not required in order to place connectors into the family, but it gives you a good reference point for the connectors. The reference planes in the generic family templates define the insertion point of the family. The vertical plane is the center point between the left and right side, while the horizontal plane defines the front and back of the family. The insertion point will be at the intersection of these two planes. You can change the properties of a reference plane so that it will establish the insertion point by editing the Is Reference and Defines Origin parameters.

For face-hosted families, these planes only define the insertion point as it relates to the face of the host extrusion. If you place a device family into a model and attach it to a vertical face, you will need to give it an elevation. A default elevation is assigned in the type properties of the family.

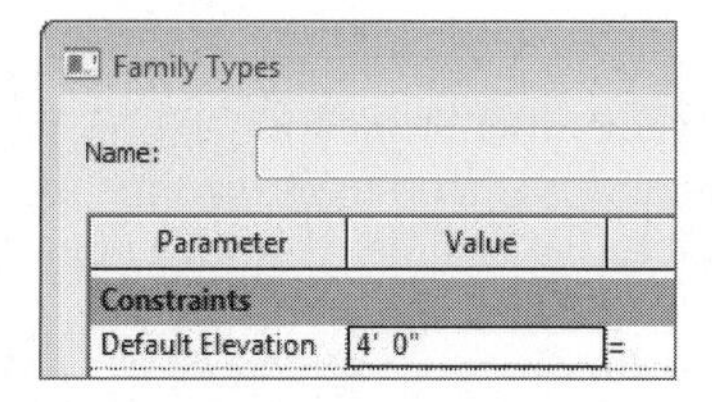

If you have determined that your device family should be parametric, it is best to use reference lines or planes to establish the dimensions that are to be editable. For the most part, the length, width, and depth dimensions should give you enough flexibility to create the various family types within your device family.

When using a reference line or plane to create a dimension, consider how you want the geometry to change upon input of data. Should the geometry grow or shrink in one direction? Does it need to grow equal distances from the center? These types of questions will help you determine the amount and placement of references. The order in which you draw reference planes will determine the direction that a parametric dimension between them will move. This behavior applies only to reference planes. In Figure 22.2, the vertical reference plane on the left was drawn first. When the Length parameter is updated, the reference plane on the right will be the one that moves.

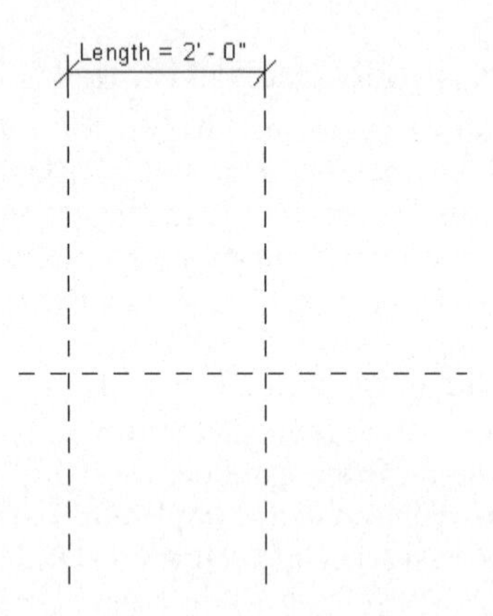

FIGURE 22.2
Reference planes with parametric dimension

Because the two default reference planes are already in the family when you begin, any reference planes dimensioned to them will move when parameters are updated. You can use the Pin button on the Modify | Reference Planes tab to lock the position of a reference plane. This will keep it from moving when a parametric dimension is updated. Obviously, you cannot pin both reference planes that have a parametric dimension between them; otherwise, you will get an error when the parameter value is changed.

PINNING REFERENCE LINES

Reference lines will always move when parametrically dimensioned to a reference plane. Even if you pin a reference line in place, it will move when pushed by a parametric dimension.

The majority of your device families can be modeled as simple boxes. It is important to create the solid geometry in the proper orientation to how it will be used in your projects. Face-hosted devices should be modeled so that the geometry is inside the host or lying on the surface of the host where applicable. For families that are not parametric, you can simply create the solid geometry as the appropriate size and position it relative to where you want the insertion point.

Set the visibility of your device geometry as desired by clicking the Visibility Settings button on the Mode panel when you select a solid form. Even though it is not likely that you will show this geometry in your plan views, it can be helpful to allow the geometry to be visible at some level of detail so that you can determine the exact location of the device when needed. The geometry should be kept visible in Front/Back and Left/Right views so that your devices can be seen in section and elevation. This will allow for coordination with architectural and structural features as well as the ability to edit the device location when working in section or elevation views. Figure 22.3 shows the visibility settings for device geometry that work well for standard model views while allowing for a view of the geometry in Fine detail views. Your results may vary depending on the level of detail used for your plan views at certain scales.

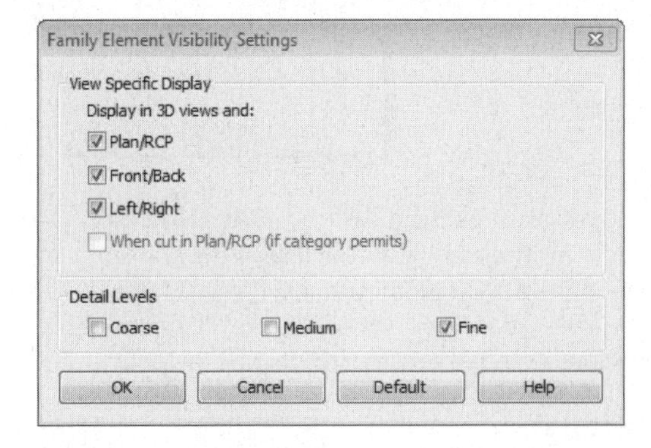

FIGURE 22.3
Device geometry visibility settings

FACE-HOSTED VISIBILITY

When you are creating a face-hosted device family, remember that the Ref. Level view is parallel to the front face of the host extrusion. If you want to set the visibility of device geometry so that it does not display in model plan views when hosted by a vertical face, you must deselect the Front/Back visibility of the geometry.

Another option for viewing device family geometry is to create a Yes/No parameter that controls the visibility of the solid forms in the family. The Visible parameter of a solid form can be associated with the Yes/No parameter. This enables you to turn on the geometry when needed for coordination in plan views. However, using this type of visibility control will turn on or turn off the geometry everywhere that it can be seen, so section and elevation views would be affected.

When creating geometry in a face-hosted family, be aware that if you model a solid form so that it is inside the host, the device will not be visible when the family is hosted by a floor unless the view range of the view includes the floor. Figure 22.4 shows a device family hosted by a floor in a project. The device is not visible in the plan view because the bottom of the view range is at the face of the floor. So technically, the device is not within the view range and is therefore not visible.

One way to avoid this scenario is to adjust the view range of your plan views so that the bottom of the view is slightly below the top face of the floor geometry. This could have an adverse effect on the overall look of your floor plan, depending on how the building is designed. Another option is to create the solid geometry in the family so that it protrudes from the host. This may not be a completely accurate representation of the device, but the protrusion can be minimal. Extending past the face of the host just 1/256″ (0.1mm) is enough to cause the device to display.

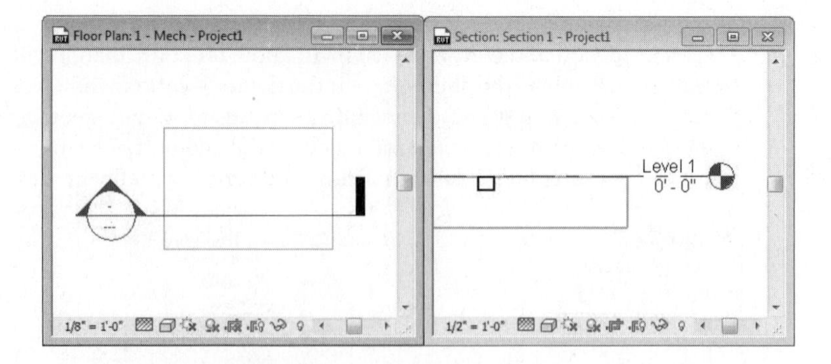

FIGURE 22.4
Device family hosted by a floor

If you are creating a device family for an object that requires a face plate, the plate geometry should be modeled to the surface of the host for a face-hosted family. This eliminates the need to extend the device geometry past the face of the host because the plate geometry will make the family visible when floor hosted. Figure 22.5 shows how adding face plate geometry to the family in the previous example causes it to display properly in the project plan view when hosted by a floor.

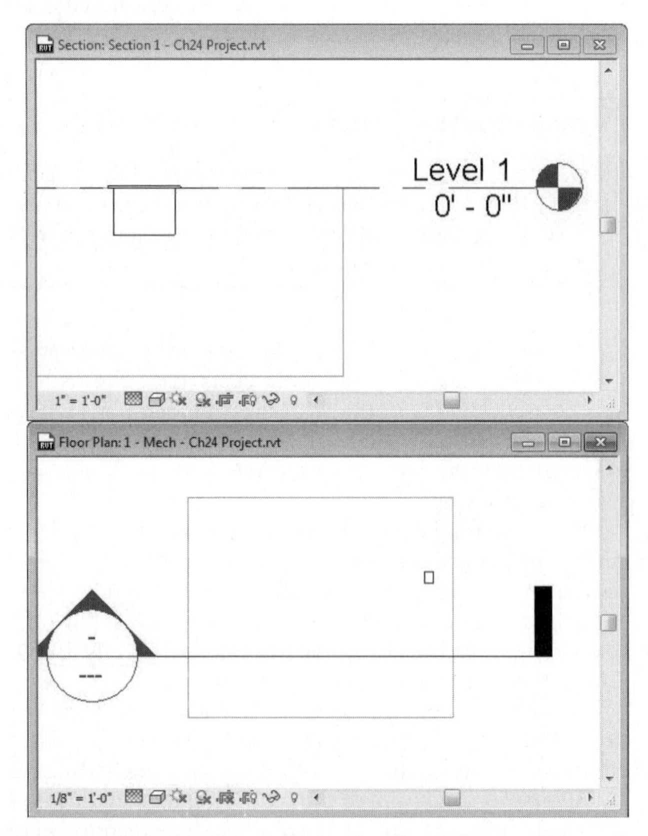

FIGURE 22.5
Device family with face plate geometry

Symbolic lines are useful for adding detail to a device family without adding a lot of complexity to the family. Symbolic lines drawn on the face plate of a device are useful for identifying the device when viewed in section or elevation views. These lines will be visible only in views parallel to the plane in which they are drawn, so they will display in plan views when the device is mounted horizontally only. Vertically hosted devices shown in plan views will not display the symbolic lines; however, the lines will show in section and elevation views. This is useful for interior elevations where devices need to be coordinated with casework or furniture. Can you tell which device is a power receptacle, which is a switch, and which is a telephone outlet in Figure 22.6?

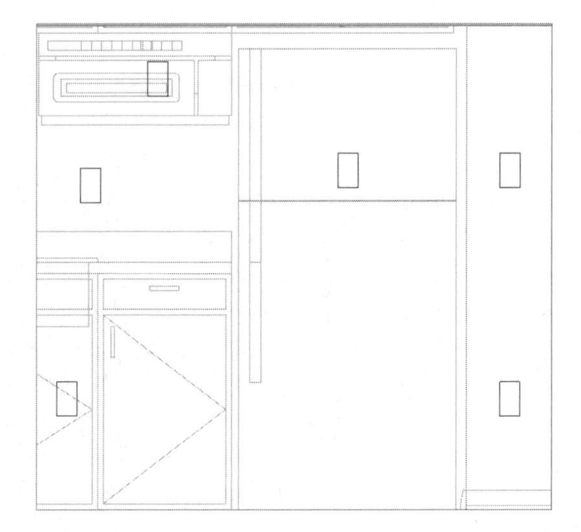

FIGURE 22.6
Devices shown in a section view

Without symbolic lines to identify the devices, additional information would have to be added to the view for the construction documents. Also, while working in the project, you would have to select the device to see what it is. With symbolic lines added to the families, it is much easier to determine what the devices are in section or elevation views. This can increase coordination of your construction documents and improve productivity during design. Figure 22.7 shows the same devices with symbolic lines added to identify them.

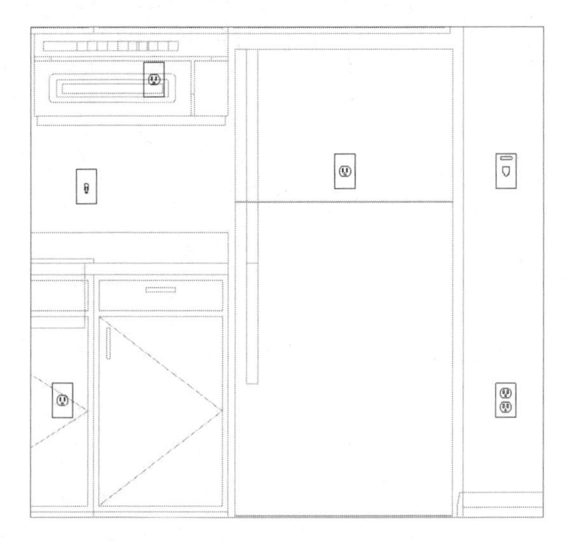

FIGURE 22.7
Devices with symbolic lines for identification

Annotations for Devices

Although the use of symbolic lines is helpful for identifying devices in section and elevation views, these objects are typically represented in plan views by symbols. Annotation families can be added to your device families to represent them in the same manner as you would for a traditional 2D CAD project. Nested annotations can be parametric as well, allowing you to control their orientation and visibility so they can be shown on your construction documents in an appropriate manner.

An annotation family is loaded into another family in the same way as loading a family into a project. Click the Load Family button on the Insert tab of the Family Editor to load an annotation family.

> **ANNOTATION LIBRARY**
>
> Having a well-stocked library of annotation families is very useful for creating devices. You can build a library of symbols that matches your library of CAD blocks. You can even use your CAD blocks to develop your Revit symbols by importing them and duplicating their line work, but be sure to remove the imported CAD data once you have created a Revit version of the symbol. See Chapter 18 for more information on creating annotation families.

For face-hosted families, this is where the Maintain Annotation Orientation parameter is important. When you place an annotation family into your device family, it will only appear in model views that are parallel to the view in which it is placed in the family. With the parameter selected, you can place the annotation in the Ref. Level view of the family, and it will display when the device is hosted vertically. Annotation families cannot be placed into an elevation view when working in the Family Editor.

Once you have loaded an annotation family into your device family, you can place it in the Ref. Level view by dragging it from the Project Browser or by using the Symbol button on the Annotate tab within the Family Editor. While placing the annotation, you can snap to reference planes or reference lines within the device family. The snap point within the annotation is its insertion point defined by the reference planes in the annotation family. You cannot snap line work or geometry to the line work within a nested annotation. Once the annotation is placed, you can use the Move tool to locate the annotation, or you can click and drag it to the desired location.

The annotation family that you use in a device family should be located in proper relationship to the geometry of the family so that when the device is placed into a project on a vertical face, the annotation symbol will be aligned with the device host. The insertion point of the annotation family is important to establish a proper relationship to the device geometry as well as the host of the device family. Figure 22.8 shows the location of a junction box annotation family nested into a device family. The insertion point of the annotation is the center of the circle. When the family is used in a project, the annotation does not properly align to the device family host.

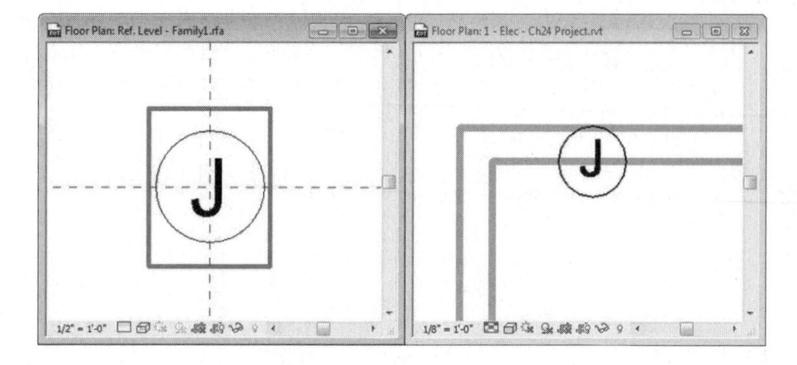

FIGURE 22.8
Annotation family location in a device family and behavior when used in a project

The alignment of the annotation with the vertical reference plane is correct, but regardless of where the annotation is located along that vertical plane, it will display in the project incorrectly. The solution to this problem is to have the insertion point of the annotation at the bottom quadrant of the circle. That way, when the annotation orientation is "maintained," it will display properly, as shown in Figure 22.9.

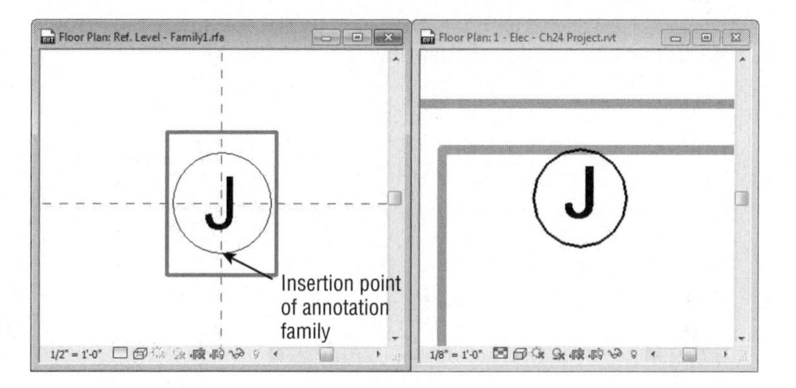

FIGURE 22.9
Annotation family with proper insertion point for use in a vertical host

The orientation of an annotation symbol may not be correct for every instance in your projects. In some cases, the annotation may need to be rotated to display as desired.

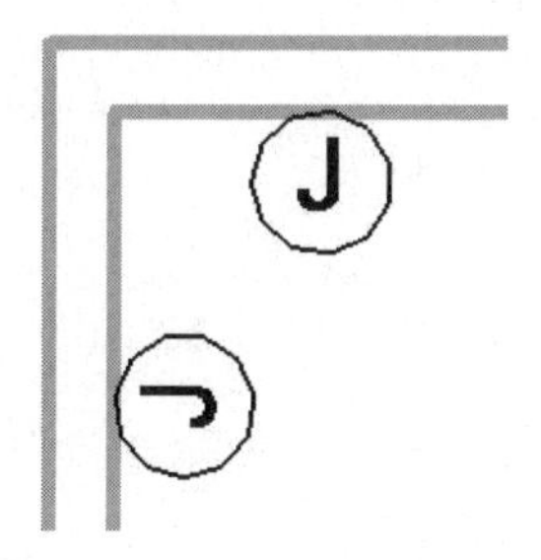

You cannot rotate the annotation separately from the device family once the device is placed into a project. If you edit the device family, rotate the annotation, and then load it back into the project, all instances of the device will have a rotated annotation.

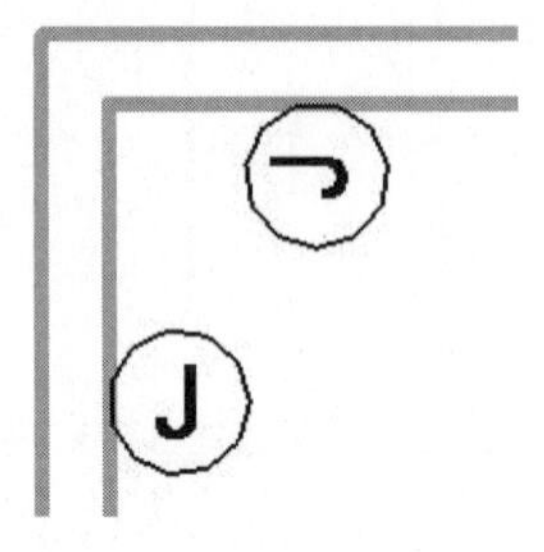

Consider using multiple annotations within a device family to allow for displaying a rotated annotation symbol when required. Yes/No parameters can be used to control the visibility of the individual nested annotations, giving you the ability to toggle between annotations when using the device family in a project. Figure 22.10 shows a device family with multiple annotations nested into it.

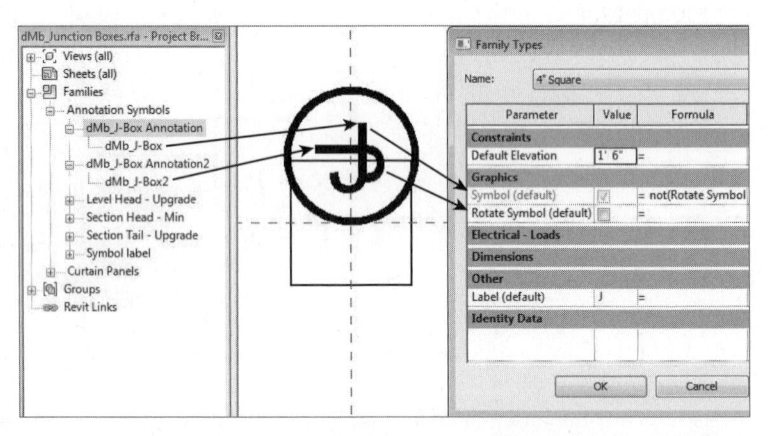

FIGURE 22.10
Device family with multiple nested annotations

The Symbol parameter controls the visibility of the annotation for normal orientation, and the Rotate Symbol parameter controls the visibility of the rotated annotation. A formula was used so that when working in a project, only one of the boxes can be selected for each instance, preventing accidental display of both annotations. With this type of behavior built into your device families, you can display devices in your projects as desired.

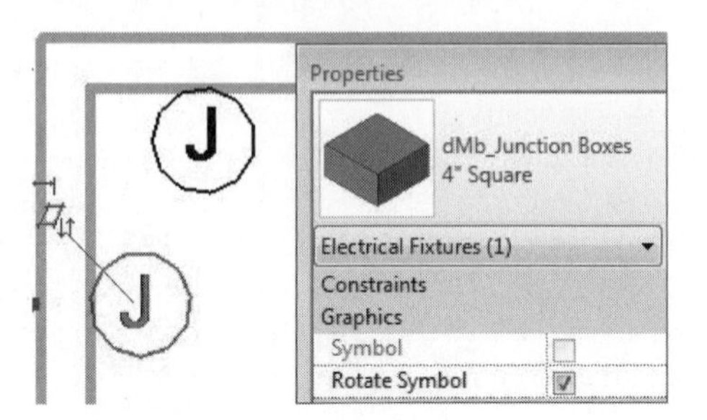

Another method for controlling annotation behavior is to apply an offset parameter so that the annotation symbol can be offset from the actual location of the device. This is useful for situations where two devices are next to each other and their annotation symbols interfere. This can be done directly in the device family by creating a reference plane or line parallel to the plane that defines the left and right sides of the device. A dimension between the plane and line or plane can be associated with an instance parameter that defines the offset distance, as shown in Figure 22.11. The annotation family is aligned and locked to the reference line so that it moves when the dimension is edited.

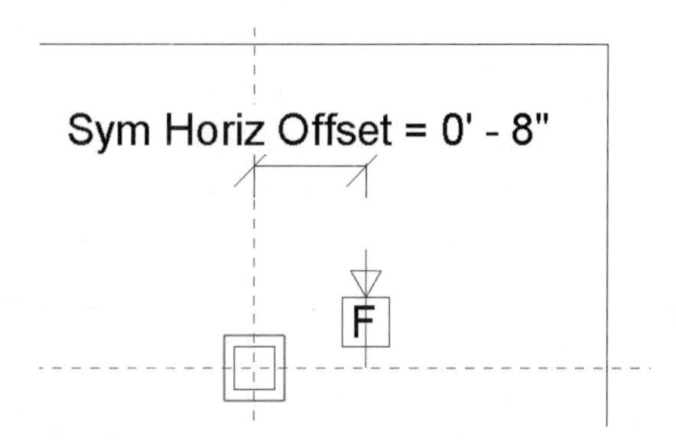

FIGURE 22.11
Reference plane used for annotation symbol offset

A default of 0′-0″ (0mm) can be used for the offset distance to allow for the normal display of the annotation. Using this type of parametric behavior gives you the freedom to locate the annotation symbol in an appropriate location without affecting the device geometry location. You will be able to move the symbol in only one direction, so it may be necessary to rotate the device if the symbol needs to be offset to the other side. Figure 22.12 shows how an offset parameter is used to achieve the desired display of the annotations. The items on the left have interfering symbols because of the locations of the device geometry. The items on the right are displayed properly without affecting the device locations. The device geometry is visible only to show the offset.

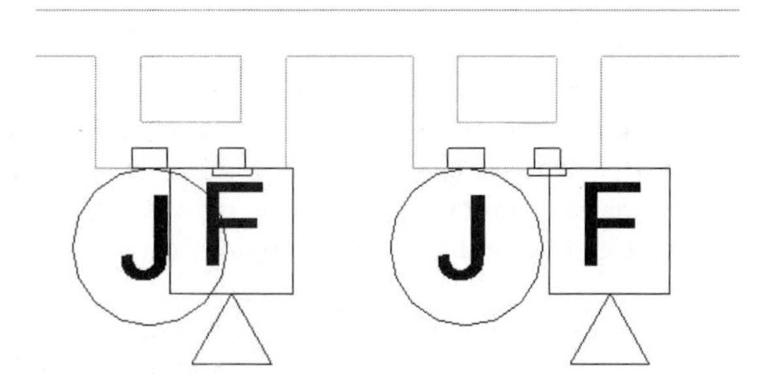

FIGURE 22.12
Symbol offset parameter used for proper device display

You can use this same functionality to create an offset that will pull the annotation away from the device geometry. The offset plane or line can be created directly in the device family only if the family is a nonhosted type. For a face-hosted family, you need to have an offset defined in the annotation family. The parameter that defines the offset can be associated to a parameter in the device family for controlling the offset. See Chapter 18 for how to create an offset within an annotation family.

Having device families that are functional for both your 3D model and your construction documents enables you to be more efficient with your design and project coordination. Now that you have learned about creating a device family with a nested annotation, practice the techniques to make a device family usable in a project by completing the following exercise:

1. Open the `Ch22_Device Family.rfa` file found on `www.wiley.com/go/masteringrevitmep2011`.
2. Click the Family Category And Parameters button located on the Properties panel of the Home tab.
3. In the Family Category And Parameters dialog box, set the category to Electrical Fixtures. Set the Part Type parameter to **Junction Box**, and select the box for the Maintain Annotation Orientation parameter. Click OK.
4. Select the extrusion located at the intersection of the reference planes. Click the Visibility Settings button located on the Mode panel of the Modify | Extrusion contextual tab. Remove the check mark from the Front/Back box in the Family Element Visibility Settings dialog box. Remove the check mark from the Coarse and Medium boxes also. Click OK.
5. Click the Symbol button located on the Detail panel of the Annotate tab. Confirm that the Ch22_J-Box Annotation is shown in the Type Selector, and click to place the annotation symbol in the drawing area, near the extrusion.
6. Click the Reference Line button located on the Datum panel of the Home tab, and draw a vertical reference line to the right of the vertical reference plane.

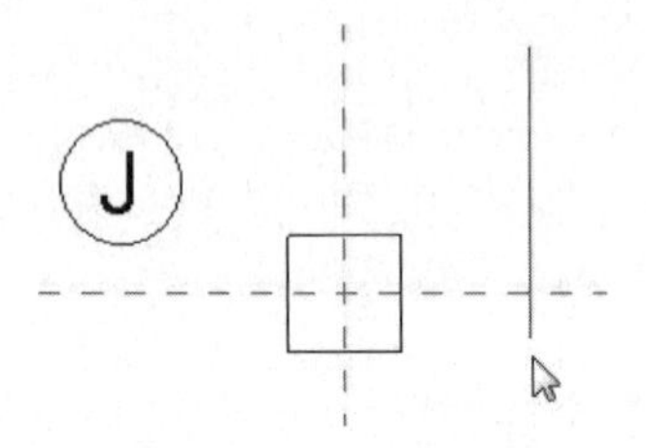

7. Click the Align button located on the Modify panel of the Modify tab. Select the vertical reference line drawn in step 6, and then move your mouse pointer over the center of the annotation symbol, highlighting the vertical reference within the annotation. Then click to align the symbol to the reference line. Click the padlock grip to lock the alignment.

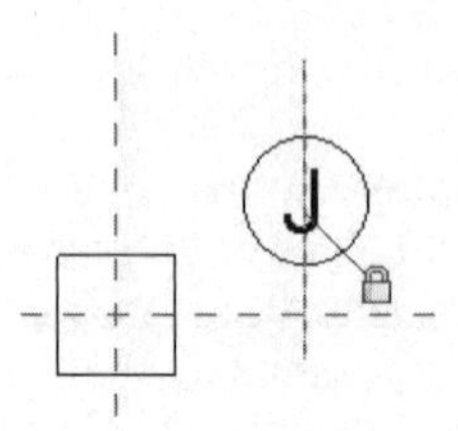

8. With the Align tool still active, click the horizontal reference plane, and then move your mouse pointer near the bottom quadrant of the annotation symbol, highlighting the horizontal reference within the annotation, and click to align the symbol to the reference plane. It is not necessary to lock this alignment.

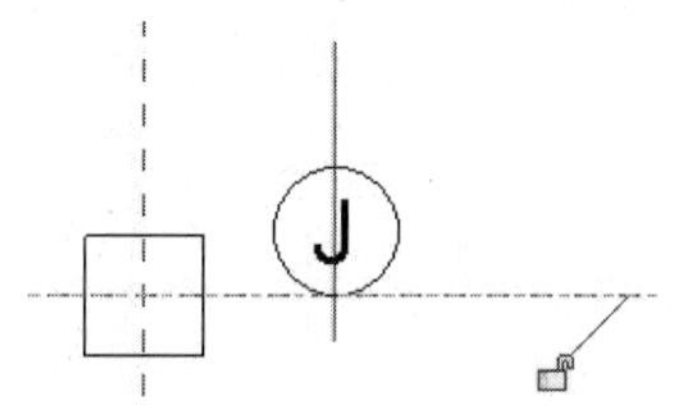

9. Click the Aligned button on the Dimension panel of the Annotate tab. Place a dimension between the vertical reference plane and the reference line drawn in step 6. Click the Modify button on the Select panel of the contextual tab.

10. Select the dimension created in step 9. Choose the <Add Parameter … > option from the Label drop-down on the Options Bar.

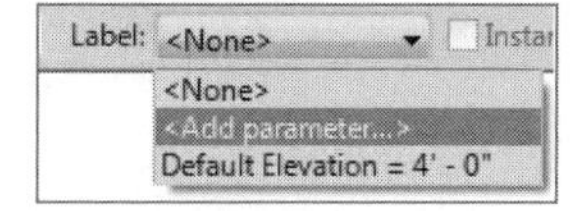

11. Create an instance parameter with the following settings:

Parameter Type: Family

Name: Symbol Offset

Group Parameter Under: Graphics

Click OK to exit the Parameter Properties dialog box.

12. Click the Family Types button on the Properties panel of the Home tab.

13. In the Family Types dialog box, set the Symbol Offset parameter value to 0′-0″, and click OK.

14. Save the family to a location that you can access to test its usability in a project.

Parameters and Connectors

The device families you create can do more than just indicate device locations on your drawings. Connectors can be added to device families for use with systems. The parameters you add to a device family can hold the engineering data required to connect devices to the appropriate system and account for their effect on the entire system. Adding connectors also allows you to use the wiring tools available in Revit MEP 2011. Revit wire objects that are attached to devices will move and stretch with edits to the device location, reducing the amount of work required to make a simple change.

Parameters for Labels

Using labels in device families seems like the logical solution to provide the information sometimes shown with devices. Although adding text labels to the annotation families that are used in your device families works well, there are some instances where the text needs to be moved for an instance of the device within a project. The best way to have the freedom to move text associated with a device is to create a parameter that holds the value and use a tag in your project that reads that parameter. This places the information directly in your project, not embedded in the device family, allowing you to place the tag wherever you want.

To use a project tag for device labels, you can use one of the coded parameters that exists for the family type. The Description parameter is a good candidate if you are not using it for other information. Keep in mind that this is a type parameter, so whatever value you use will apply to all instances of that type. This can result in having to create unique family types for each unique label. Using a shared parameter allows you to make it an instance parameter that can be a unique value for each instance of the device in your project. If you want to use a custom parameter for the tag, it must be a shared parameter. Figure 22.13 shows that a shared parameter has been added to a device family for its candela rating. There is no label in the device family for this parameter; it is simply there to give a value when the device is tagged in a project.

FIGURE 22.13 Shared parameter added to a device family for tagging

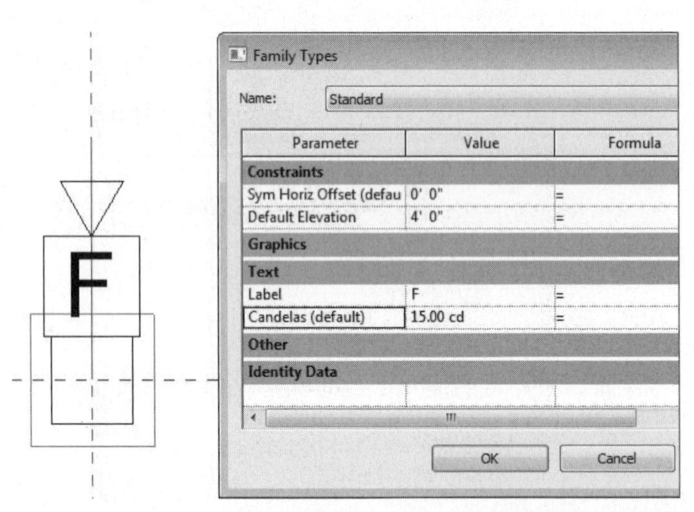

This type of functionality in your device families allows you to place the text associated with the device anywhere in the project view to avoid conflict with other graphics. A tag must be created that reads the parameter used. When creating the tag, be sure to categorize it as a tag for the appropriate type of device. Another benefit of using a tag for a device label is that you can use a leader on the tag when necessary. Figure 22.14 shows how the device with a custom parameter and a tag can be used in a project. The text shown is a Fire Alarm Device tag that was created to read the custom instance parameter called Candelas.

Adding Connectors

The connectors that you add to device families will most likely be electrical connectors, since these families are the types of building components for controlling systems or power distribution. There are several types of electrical connectors to choose from. The type you add should coincide with the function of the device in your projects.

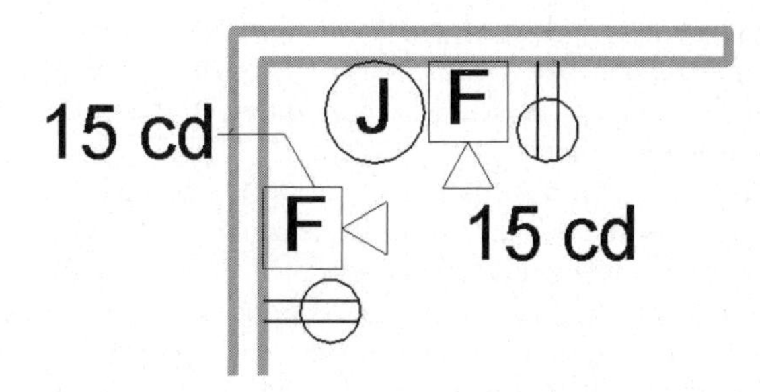

FIGURE 22.14 Sample device family with a tag for a label in a project

When you click the Electrical Connector button on the Home tab in the Family Editor, you can choose the connector type from the drop-down on the Options Bar.

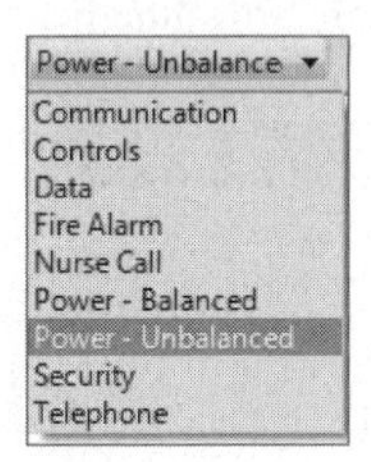

The connector can be placed on the face of any 3D geometry that exists in the device family. If you are creating a family without any 3D geometry, you can use the Work Plane placement option for the connector. You can find this option on the Placement panel of the Modify | Place Electrical Connector contextual tab.

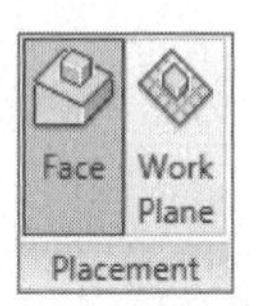

An equipment connection and a motor connection are examples of device families that do not have any 3D geometry yet still require a connector for use in the electrical systems defined in your projects. The properties of the connector are what define the system behavior of the device family. Most of the electrical connector system types do not have any parameters for defining electrical behavior. These connectors are simply used for keeping track of connectivity between the devices and their equipment, so their properties are very simple.

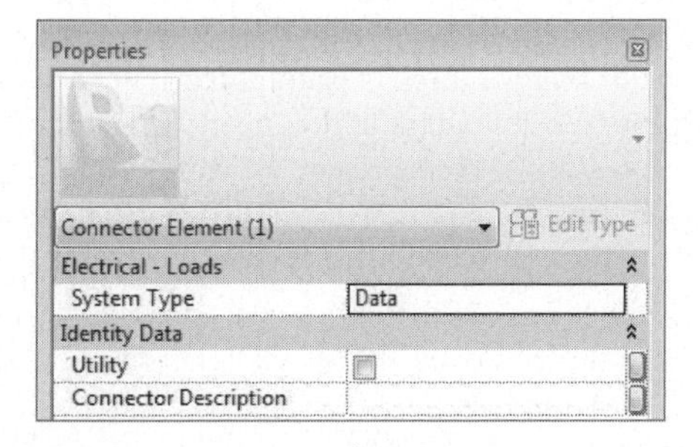

The Power-Balanced and Power-Unbalanced system options for a connector contain more properties that are specific to the system behavior. These properties define the electrical characteristics of the device, which then determine the distribution system it can be used with in your projects.

Since you cannot change the properties of a connector without editing the family it resides in, it is best to associate the connector parameters to parameters defined in the device family. This allows you to control the electrical characteristics of the device by editing its properties while working in your projects. A connector parameter can be associated to a family parameter by clicking the small box at the far right of the parameter. This opens a dialog box that lets you select the family parameter to associate the connector parameter to. Only family parameters of the same type will appear in the list. Figure 22.15 shows how the Apparent Load parameter of a connector is associated with a family parameter. Notice that only one parameter is available in the list because it is the only family parameter that is an Apparent Power type.

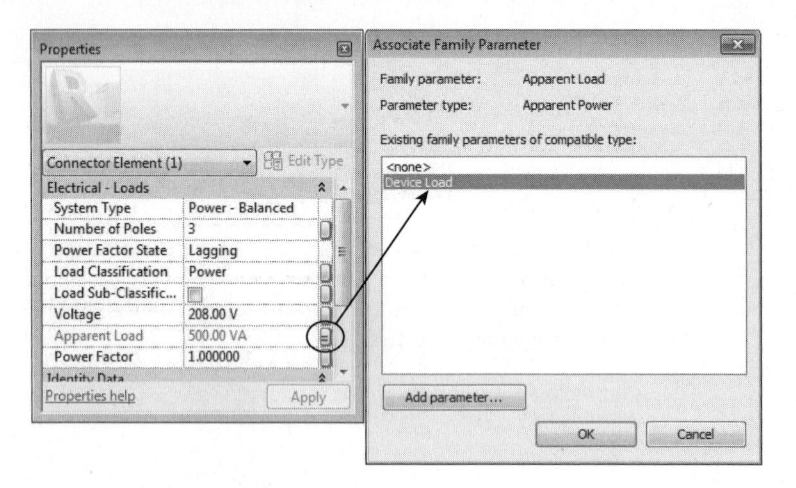

FIGURE 22.15
Associating a connector parameter to a family parameter

Although the Apparent Load parameter can be an instance parameter, the parameters that you use to define the device's voltage or number of poles should be type parameters. In the case of a motor connection device family where specific loads are used for each voltage, you could use a type parameter for the apparent load. With these settings in place, you can use a type catalog to create the various types of motor connections. Figure 22.16 shows a sample type catalog for a three-phase motor connection device family. The values for the parameters are taken from a code table. This ensures that the proper load values will be used when the devices are connected to systems in a project.

Even though the motor connection family has no 3D geometry, the connector is available when the device is used in a project. Although the location of an electrical connector is not that important in a family that contains 3D geometry, it is very important for a family without geometry. Be sure to locate the connector at the insertion point of the family; otherwise, you can end up with undesired results in your project, as shown in Figure 22.17.

For some device families, you may have the need for multiple connectors. This may be because of different systems within the device or because the device has multiple connection points.

FIGURE 22.16
Sample type catalog for a device family

Specify Types

Family: dMb_3 PH Motor Connection

Types:

Type	Horsepower	Device Voltage	Device Load
	(all)	(all)	(all)
208V - 50HP	50	208	51962
208V - 60HP	60	208	61315
208V - 75HP	75	208	76557
208V - 100HP	100	208	98727
208V - 125HP	125	208	124361
208V - 150HP	150	208	143414
208V - 200HP	200	208	191218
480V - 1/2HP	0.5	480	876
480V - 3/4HP	0.75	480	1275
480V - 1HP	1	480	1673
480V - 1 1/2HP	1.5	480	2390

FIGURE 22.17
Device family with improper connector location

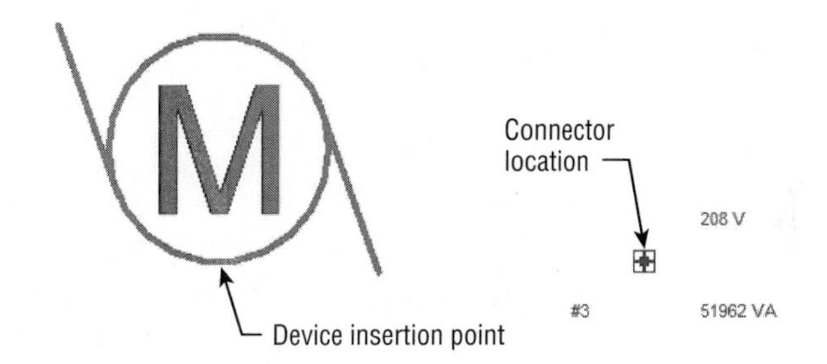

A floor box device family can be created that allows for the connection of all the types of systems within the box. With all the connectors added to the device family, the device can be connected to the appropriate systems when used in a project. Figure 22.18 shows a floor box family with multiple connectors in a project. The systems associated with the connectors can be created for the device in the project.

FIGURE 22.18
Device with multiple connectors in a project

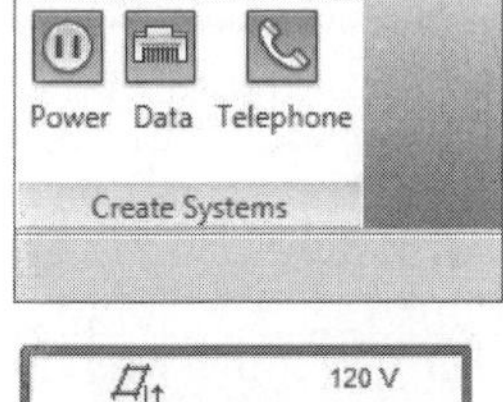

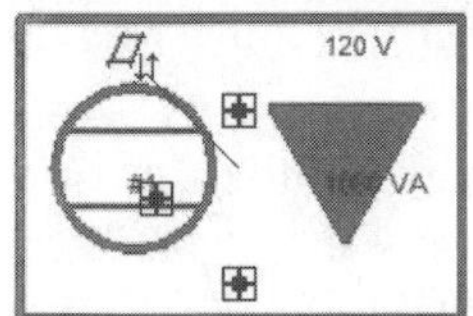

A junction box is another example of a device family that may require multiple connectors. In this case, the connectors are the same type. Their properties can be associated to the same family parameters or unique ones. Having multiple connectors of the same type within a device family allows you to connect the device to multiple circuits. Figure 22.19 shows how having two connectors in a junction box device family allows the device to be connected to multiple circuits in a project.

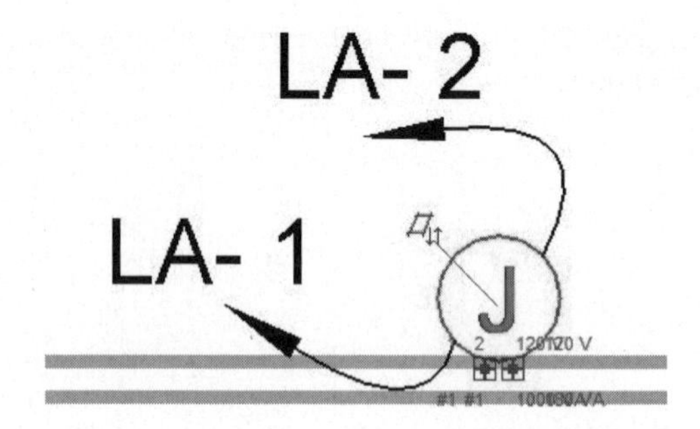

FIGURE 22.19
Junction box with multiple connectors

Real World Scenario

Wendy is working on a large office building project. The open floor plan requires systems furniture to be used. She is required to provide a junction box for each group of furniture, but each cubicle within the group needs its own power source. By adding connectors to her junction box family, she is able to show that multiple circuits are provided to each junction box location.

In another area of the building she is using quad receptacles. To limit the load on each circuit, she needs a unique circuit for each pair of outlets within a quad receptacle. She creates a quad receptacle device family that has two connectors. The connector voltages are associated to the device family voltage, and she creates two load parameters, one for each connector.

When using multiple connectors in a device family, it is a good idea to use the Connector Description parameter to identify the connectors. This will help with creating systems when working in a project. When a device has multiple connectors, a dialog box will appear with a list of available connectors when a system is created for the device.

The Bottom Line

Model device geometry The 3D geometry of a device family may not be its most important component, but there are things you can do to make the geometry useful for project coordination and accuracy.

Master It Device families are commonly face-hosted because of the way they are used in a project model. Describe why a device family that is hosted by a floor in a project may not display correctly in plan view.

Use annotations to represent devices Because of their size and simple shapes, device families are not typically shown on construction documents, but rather a symbol is used to represent the device. Annotation families can be nested into device families for display in project views.

Master It The orientation of an annotation family nested into a device family is important for proper display in project views. What is the device family parameter that allows for the display of the annotation in plan and section views? What hosting options in a project allow for this parameter to work?

Add parameters and connectors The connectors used in device families provide the engineering characteristics of the device so that it can be incorporated into systems within a project.

Master It The parameters of a connector cannot be edited while working in a project file. Describe how you can change the properties of a connector without having to edit the device family.

Appendix

The Bottom Line

Each of The Bottom Line sections in the chapters suggest exercises to deepen skills and understanding. Sometimes there is only one possible solution, but often you are encouraged to use your skills and creativity to create something that builds on what you know and lets you explore one of many possible solutions.

Chapter 1: Exploring the User Interface

Navigate the ribbon interface The ribbon is an efficient user interface style that works well in Revit. The ability to house numerous tools in a single area of the interface allows for maximum screen real estate for the drawing area.

Master It Along with the standard tabs available on the ribbon interface, contextual tabs also are available while you're working on a project. Explain what a contextual tab is and how it may differ throughout your workflow.

Solution A contextual tab is an additional tab that appears when an object is selected. It is an extension of the Modify tab that contains tools specific to the editing of the selected object.

Utilize user interface features Many features are available in the Revit MEP 2011 user interface that allow for quick access to tools and settings. The use of keyboard shortcuts can also improve workflow efficiency.

Master It It is important to workflow efficiency to know how to access features of the user interface. What tool can be used to activate or remove user interface features?

Solution The User Interface tool on the View tab can be used to turn on or turn off user interface features.

Use settings and menus Establishing settings for your user interface is another way to create a working environment that is the most efficient and effective for your use of Revit MEP 2011.

Master It The use of keyboard shortcuts has been part of design and drafting software for a long time. The ability to customize the shortcuts to best suit your workflow is key to improved efficiency. How can the settings for keyboard shortcuts be accessed? How can the settings be shared among users?

Solution The Keyboard Shortcuts dialog box can be accessed from the User Interface tab of the Options dialog box, from the User Interface button on the View tab, or by using the KS keyboard shortcut.

Chapter 2: Creating an Effective Project Template

Set up views and visibility The settings for views are crucial to being able to visualize the design and model being created and edited in a project. Establishing the default behavior for views and visibility of objects can increase not only the efficiency of working on a project but also the accuracy of design.

Master It The properties of a view determine how objects and the model will appear in the view. Along with Visibility/Graphics Overrides, what other view property determines whether items are visible in that view? For a floor plan view, describe the three major components of this property.

Solution View Range determines what elements can be seen in a view based on their location. For a plan view, the Top setting defines the elevation at which the model is being viewed from. The Bottom setting is the extent to which the model is being viewed from the Top. The Cut Plane setting is an imaginary plane that cuts through the architectural and structural elements.

Establish project settings Many project settings can be established in a Revit template to determine the display of objects in views and on construction documents. There are also settings that define the makeup of the project itself.

Master It Phase settings for a project are very important to defining what portions of a building design occur in certain phases. Explain why having phases established in a template might cause a need for a separate template file for phased projects.

Solution When phases are established in a project, it is important to assign a phase to each view so that objects modeled in a view will belong to the phase of the view. If you are creating a template with phases and preset views, you will want to have a view type for each phase. If you use this template for a project that does not require phasing, you will have many unnecessary views already established in your project.

Define preloaded content and its behavior The more items you have available for immediate use when a project begins, the more your focus can be on design decisions and less on loading required items. In a multiuser environment, preloaded content ensures that improper variations do not occur, causing inconsistencies in the project documentation.

Master It Having system family types defined in your template is just as important as having the appropriate components loaded. Explain why certain component families are required in order to create and define MEP system family types.

Solution To create and define duct, pipe, cable tray, and conduit system family types, you need fitting component families loaded. The system families cannot be modeled without their fittings defined.

Create sheet standards As with other template elements, standards for sheets are a useful component to have established.

Master It Having a predefined organization for drawing sheets in your template will ensure consistency from one project to the next. True or false: You must have all the required sheets for any project built into your template in order for them to be organized properly? Explain.

Solution False. You only need to establish the organization settings. As new sheets are created in a project, they will be organized according to the settings.

Chapter 3: Worksets and Worksharing

Create a central file by dividing the model into worksets Setting up your Revit project file correctly will help users visualize and coordinate their systems easily.

Master It You are working on a project with a mechanical engineer, a plumbing designer, and an electrical engineer. Describe the types of worksets that the model can be divided into to accommodate the different systems for each discipline.

Solution The model can be divided into HVAC and mechanical piping worksets for mechanical, domestic and sanitary water piping worksets for plumbing, and lighting and power worksets for electrical.

Allow multiple users to work in the same file Revit MEP provides functionality to set up your project in a manner that allows users to edit and manage their systems without conflicting with other systems in the model.

Master It Describe how to create a local file copy of a central file and how to coordinate changes in the local file with other users who are accessing the central file.

Solution Browse to the central file in the Open dialog box, and select the Create New Local box; alternatively, copy/paste the central file to a new folder, open the pasted file, and click Save.

Use the Synchronize With Central command to coordinate changes made in the local file with the central file.

Work with and manage worksets Working in a project with multiple users means it is likely that you will need to coordinate the availability of worksets.

Master It Describe how you would isolate a system in the model so that no other user could make changes to that system. What is the best way to release a system so that others can work on it?

Solution Create a workset for that system, and place any model elements on that workset. Take ownership of the workset to prevent changes from other users. Use the Relinquish All Mine command to release ownership of the workset, allowing others to have access to the components on that workset.

Control the visibility of worksets Visualization is one of the most powerful features of a BIM project. Worksets give you the power to control the visibility of entire systems or groups of model components.

Master It You are facing a deadline and need to add some general notes to one of your plumbing sheets. Because of the intricate design of the HVAC system, your project file is very large and takes a long time to open. What can you do to open the file quickly to make your changes?

Solution From the Open dialog box, select the Options drop-down, and choose the Specify option to open only the worksets required to complete the task.

Chapter 4: Best Practices for Sharing Projects with Consultants

Prepare your project file for sharing with consultants Taking care to provide a clean, accurate model will aid in achieving an integrated project delivery.

Master it Describe the importance of making worksets visible in all views when your file will be shared with consultants.

Solution If you do not make a workset visible in all views, the model elements on that workset will not be visible to consultants when they link in your file.

Work with linked Revit files There are many advantages to using linked Revit files in your project. Revit provides many options for the visibility of consultants' files, allowing you to easily coordinate your design.

Master it How would you turn off a model category within a Revit link while allowing that category to remain visible in your model?

Solution Select the Revit Links tab in the Visibility/Graphics Overrides dialog box. Click the By Host View button in the Display Settings column next to the appropriate linked file. Change the display to Custom. Select the appropriate tab within the RVT Link Display Settings dialog box, and turn off the desired category.

Coordinate elements within shared models Revit can alert you to changes to certain model elements within linked files. Managing these changes when they occur can reduce errors and omissions later in the project and help keep the design team coordinated.

Master it List the types of elements in a linked file that can be monitored for changes.

Solution For Revit Architecture: Levels, Grids, Columns, Walls and Floors.

For Revit MEP components: Air Terminals, Lighting Fixtures, Mechanical Equipment, and Plumbing Fixtures.

Working with non-Revit files Not all your consultants may be using Revit. This does not mean that you cannot use their files to develop and coordinate your design. You can also share your design by exporting your file to a format they can use.

Master it Describe the difference between linking and importing a CAD file and why the linking option is preferred.

Solution When a file is linked, it can be updated by reloading the link. An imported file will not update when the original CAD file is updated. Importing causes unnecessary information from the CAD file to be brought into Revit, which can increase file size and impede file performance greatly.

Set up a system for quality control As a BIM solution, Revit provides functionality to keep your design coordinated with your consultants.

Master it What functionality exists in Revit that could allow a design reviewer to comment on coordination issues within a project?

Solution Annotation symbols can be created and used throughout the model to comment on design issues. The annotations can be scheduled to organize and manage issues that need to be addressed.

Chapter 5: Schedules

Use the tools in Revit MEP 2011 for defining schedules and their behavior The capabilities of schedules in Revit MEP 2011 can increase your project coordination and the efficiency of your workflow. The ability to track items within a model can help you to better understand the makeup of your design.

Master It The information in schedules comes from information stored within the objects of a Revit model. Explain why editing the data of an object in a schedule will change the properties of the object.

Solution Schedule views are just another view of the model objects. So when you edit an object in a schedule, you are actually editing the properties of that object.

Schedule building components Scheduling building components is the primary use of the scheduling tools in Revit. Schedules are used on construction documents to provide additional information about components so that drawings do not become too cluttered.

Master It Understanding what information can be used in a schedule is important to setting up a specific component schedule within your Revit project. What types of parameters can be included in a schedule? What type cannot?

Solution Project parameters and shared parameters can be included in a schedule as well as some of those that are inherent in objects. Family parameters cannot be included in schedules.

Create schedules for design and analysis Scheduling can go beyond counting objects and tracking their information. You can also create schedules that assist in making design decisions by providing organized analytical information.

Master It The information stored in Space objects often comes from their relationship with other objects. Some of the data for analysis needs to be manually input. Explain how using a schedule key can assist in adding data to a Space object.

Solution A Schedule key allows you to assign a value to a parameter based on the key value, which ensures consistency throughout the project. Key values can be added to objects via a schedule without having to access the properties of each individual Space object.

Schedule views and sheets for project management Not only the components that make up a model, but also the views and sheets within your project can be scheduled. Specialized schedules for views and sheets are useful for project management.

Master It A Note Block schedule is a schedule of annotation family so a list of the information within the annotation can be generated. What are some of the benefits of using a Note Block instead of static text for plan notes?

Solution When a note is removed from the list in a Note Block, all the associated annotation instances are removed from the plan. Notes can be easily renumbered in the Note Block schedule.

Chapter 6: Details

Use Revit drafting and detailing tools efficiently Revit MEP 2011 has many tools for creating the details and diagrams needed to enhance your model and provide the necessary level of information for your projects.

Master It Although the drafting tools in Revit MEP 2011 may be unfamiliar at first, learning to use them efficiently and effectively will help you spend more time focusing on design decisions instead of drafting efforts. Describe how filled regions can be used not only to display a pattern but also to provide line information in a detail.

Solution The line styles used for the borders of a filled region can be used to convey information about the detail without having to overlap lines and regions.

Use CAD details in Revit projects Much of the detail information used in projects has already been drawn. When you transition to Revit, you can still use your CAD details.

Master It Using CAD details in a Revit project can be a quick way to complete your construction documents in a timely manner. However, using many CAD files for details can have a negative effect on file performance, so it is important to link CAD files whenever they are used. Explain why importing and exploding CAD files can adversely affect your project.

Solution Exploded CAD files cause many text and line styles to be created. These styles are unnecessary and could be used by mistake instead of the standard styles developed specifically for Revit projects.

Build a Revit detail library Having a library of details makes it easy to save time on projects by not having to spend time drawing details that have already been created.

Master It Revit drafting views can be saved as individual files for use on projects as needed. True or false? A drafting view will be added to your project when you use the Insert 2D Graphics From View option of the Insert From File tool. Explain.

Solution False. When you use this option, only the graphics drawn in the selected view will be placed into your project.

Create detail views of your Revit model Some details require the model to be shown in order to show installation or placement of objects.

Master It Callout views can be created from plan, section, and elevation views. Explain how detail views are different from drafting views.

Solution Detail views show the portion of the model from the callout tool. You cannot show model graphics in a drafting view.

Chapter 7: Sheets

Create a titleblock A titleblock can be the signature for your projects. Its design and layout can be an immediate indicator as to who has created the construction documents. A titleblock is also important for conveying general project and specific sheet information.

Master It To ensure that your Revit projects look the same as, or similar to, your other projects, it is necessary to have a titleblock family that looks the same as other file format titleblocks you use. Describe the process for creating a Revit titleblock family from an existing CAD format titleblock.

Solution The CAD titleblock can be imported into a new Revit titleblock family. The drafting tools can be used to create Revit lines that match the CAD lines. Labels and text can be placed in the same locations as CAD text and attributes.

Establish sheets in your project The sheets that make up your construction documents can be organized in your Revit projects for easy access and for management of project information.

Master It A Sheet List schedule is a very useful tool for managing the information shown on your construction documents as well as for organizing the order of sheets for your project. Is it possible to create parameters for sheets that can be used in the sheet list? Explain.

Solution Yes, custom parameters can be created for use in a sheet list. These parameters can be shared or project parameters.

Place views on sheets For a Revit project, the construction documents are created as a result of the model, whereas in traditional CAD environments the sheets are the main focus. You can put your construction documents together by placing the views you have created onto your sheets.

Master it Uniformity among sheets in a document set is important not only to the look of a project but also for ease in document navigation. Explain how guide grids can be used to place model views in the same location on individual sheets.

Solution Guide grids can be established so that column grids in the model views can be snapped to the same location on each sheet.

Print and export sheets Although we live in a digital age, the need to print sheets and views is still part of our daily workflow. With the ability to work with consultants from all over the world, the need to share digital information is crucial. Exporting sheets and views to a file format that is easily shareable increases our ability to collaborate with consultants.

Master It Printing sheets is often necessary for quality control checking of a project. How can you keep section and elevation marks of views that are not used on your sheets from printing?

Solution The Options area of the Print dialog box has check boxes for hiding unwanted graphics and view tags.

Chapter 8: Creating Logical Systems

Create and manage air systems Knowing how to effectively manage air systems can help productivity through organizing systems so that items can be easily interrogated to verify that the systems are properly connected.

Master It True or False: Outside air cannot be modeled because there is no default system type to select from.

Solution False. Although there is no default Outside Air system, the system can be created by using Supply Air and renaming the system to Outside Air.

Create and manage piping systems Being able to understand how to change and manage piping systems, the user can create and maintain different systems effectively.

Master It A designer has been asked by the engineer to create a grease waste system to accommodate a new kitchen that has been added to the project. What would be the quickest way to accomplish this feat?

Solution The designer can easily create a new system called Grease Waste and add all the plumbing components that are required to be on the system.

Manage display properties of systems Filters and Visibility settings can help the user show the intent of the layout.

Master It A plumbing designer has just created a Grease Waste system, and now the engineer has decided that the Grease Waste line should appear as a dashed line. How would the designer accomplish this?

Solution The designer would create a filter that referenced the name of the system, which in this case would be Grease Waste, and then add this to the Filters section of Visibility Graphics. Once this is added, then the designer can add any linetype, pattern, color, or line weight they desire.

Chapter 9: HVAC Cooling and Heating Load Analysis

Prepare your Revit MEP Model for analysis The key element to a successful building performance analysis is the proper accounting of all variables that will influence the results.

Master it Describe the relationship between rooms and spaces — are they the same element? Describe an essential tool that can be created to maintain and track space input data and building construction for a heating and cooling load analysis.

Solution A room is a region that can be occupied within the architectural building model. A space is the design region, based on an architectural room, within the MEP model. Be sure to check the Room Bounding option on your link to ensure that the same elements that define the room boundaries also define the respective space boundaries.

Developing a working Space Properties schedule, either by creating a custom schedule or by modifying an existing one, will enable you to track and modify space properties as needed to properly account for critical properties that affect the space loads.

Perform heating and cooling analysis with Revit MEP 2011 Before a piece of equipment can be sized or duct systems designed, the building heating and cooling performance must be known in order to accurately condition your spaces.

Master it How does project location affect building heating and cooling loads? Describe methods to determine project location in Revit MEP 2011.

What is a sliver space, and how does it affect the building performance?

Solution The location of the project building determines the environmental conditions surrounding the building. Conditions, such as ASHRAE climate zone data, determine minimum design conditions that the building construction must perform to. Two methods exist in Revit MEP 2011 to determine project location: selecting the design city from a list of default cities within Revit MEP or using an Internet mapping service and locating the exact coordinates of your design building.

A sliver space is a narrow space in the building bounded on two sides by occupiable interior spaces — pipe chases, HVAC shafts, wall cavities, and so on. The volume of the sliver space is added to the larger of the two adjacent interior spaces when the building performance is analyzed.

Export gbXML data to load-simulating software Often, to complete the building analysis, the Revit MEP model has to be analyzed in greater detail by a third-party simulation program.

Master it What is gbXML? Why is it necessary to export your Revit MEP project?

Solution gbXML is a Extensible Markup Language file that is able to be written and read by multiple programs. It allows interoperability between building modeling programs and energy-simulating programs.

It is necessary to export the Revit MEP model, via gbXML data format, to a third-party simulation program because Revit does not offer the user extensive customization of the spaces, zones, building constructions, or even the heating and cooling systems requirements. It also does not have the capabilities for energy analysis of the building, as required by LEED and other energy conscious organizations.

Chapter 10: Mechanical Systems and Ductwork

Distinguish between hosted and nonhosted components Deciding whether hosted or nonhosted components are used is crucial for the success of your project. It will play a large factor in performance and coordination with other companies.

Master It Should you choose hosted or nonhosted components for your project?

Solution A mixture may well be the best solution, but whichever you choose, consistency is the key. If you start with nonhosted, stick with that type of component during the life of the project because you will find it difficult to change between hosted and nonhosted.

Use the different duct routing options When using Revit MEP 2011 for your duct layouts, the user must understand the functions of automatic duct routing and manual duct routing. Once these functions are mastered, then the user can lay out any type of ductwork system.

Master It When asked to submit a design proposal for a multifloor office building, the HVAC designer needs to show a typical open plan office and the supply and extract ductwork. How should the designer start this process?

Solution The designer should start by laying out air terminals and, if required, VAV units. Then they should use the autorouting tools to generate an initial duct design.

Adjust duct fittings Duct fittings are needed in systems to make the systems function properly and to produce documentation for construction. Being able to add or modify fittings can increase productivity.

Master It You have just finished your modeled layout and given it to your employer for review. Your boss has just came back and has asked you to remove a couple of elbows and replace them with tees for future expansion. What would be your method to accomplish this quickly?

Solution Select the elbow, and click the + sign; this will change the elbow to a tee.

Chapter 11: Mechanical Piping

Adjust and use the mechanical pipe settings Making sure that the mechanical piping settings are properly set up is crucial to the beginning of any project.

Master It A designer has just been asked to model a mechanical piping layout, and the engineer wants to make sure the designer will be able to account for the piping material used in the layout. What steps must the designer take to the complete this request?

Solution First the designer must figure what piping materials are needed. Next the designer will duplicate the piping and associated fittings. Once the piping and fittings have been duplicated and renamed to the piping material as needed, the designer will assign the fittings to the proper pipe types. This will ensure that when the piping is routed, all of the fittings and piping can be accounted for.

Select and use the best pipe routing options for your project When using Revit MEP 2011 for your mechanical layouts, one must understand the functions of automatic pipe routing and manual pipe routing. Once these functions are mastered, then the user can lay out any type of piping system.

Master It The engineer has just come back from a meeting with the owner and architect, and it has been decided that the owner wants to have a hot water system and a chilled water system rather than a two-pipe hydronic system. How would you modify your hydronic layout to accommodate the change?

Solution First duplicate your mains. Rather than a two-pipe system (one supply, one return), you will have a four-pipe system (one hot water supply, one hot water return, one chilled water supply, one chilled water return). The designer will also have to change out the mechanical equipment to add chilled water coils. Additional chiller and pumps may also be required. Once you have all your equipment added and modified, add the mechanical equipment to the proper systems by changing the hydronic supply systems to hot water supply, hot water return, chilled water supply, and chilled water return. This will allow the pipes to be filtered properly.

Adjust pipe fittings Pipe fittings are needed in systems to make the systems function properly and to produce documentation for construction. Being able to add or modify fittings can increase productivity.

Master It You have printed off a check set for review and have noticed that there are no shutoff valves. Now you need to load the shutoff family. What directory should you look in for pipe fittings?

Solution You should first look in the imperial library for Pipe and then Valves. If you have a manufacturer that is creating proper Revit content, its fittings would be the preferred choice. Also, you can refer to Autodesk Seek, `Arcat.com`, and other sites for additional content.

Adjust the visibility of pipes Being able to adjust the visibility gives the mechanical designer or user the ability to set up multiple views and control the graphics for documentation.

Master It The engineer has just come back from a meeting with the owner and architect, and it has been decided that the owner wants to have a hot water system and a chilled water system. You have just modified your hydronic layout to accommodate the change. Now the owner wants the pipes color coded, so it's easier to visualize the changes. Describe how this would be done.

Solution First create filters that are named after the systems you created (Hot Water Supply, Hot Water Return, Chilled Water Supply, and Chilled Water Return) by duplicating hydronic piping. Change the system type to the system name, and then add the name of the systems, remembering that the text is case sensitive. Finally, add these to your filters, and apply the colors you want to use.

Chapter 12: Lighting

Prepare your project for lighting design The greatest benefit you can receive from a lighting model is coordination with other systems. Properly setting up the project file is key to achieving this coordination.

Master it Describe the relationship between ceilings and engineering spaces. How can you be sure that your engineering spaces are reporting the correct geometry?

Solution Ceilings can be set to define the upper boundaries of Room and Space objects.

The upper limit of a space should be set to the level above to ensure that room-bounding ceilings will define the actual height of the space.

Use Revit MEP for lighting analysis Though the design of electrical systems is usually represented schematically on construction documents, you can use the intelligence within the model to create a design tool that analyzes lighting levels.

Master it What model elements contain the data required to determine proper lighting layout?

Solution Spaces contain data that determines the average estimated illumination within them. Lighting fixtures contain photometric web file information to provide accurate light output from the fixture family.

Compare and evaluate hosting options for lighting fixtures and devices As a BIM solution, Revit MEP offers multiple options for placing your lighting model elements into your project. These options are in place to accommodate several workflow scenarios.

Master it What is the default hosting option for face-hosted families? Describe the limitations of representing wall-mounted lights with symbols and how they can be shown in a plan view.

Solution The default hosting option for face-hosted families is to place them on a vertical face.

Lighting fixtures cannot maintain the orientation of annotations within the family. In a plan view, wall-mounted fixtures must be represented by the actual graphics of the fixture or by model lines drawn in the fixture family.

Develop site lighting plans Creating a site lighting plan will allow you to coordinate with civil engineering consultants as well as with your architect. These plans are also useful for presentation documents and visual inspection of lighting coverage on the site.

Master it What is the benefit of using nonhosted lighting fixture families for site lighting?

Solution Nonhosted fixture families can be adjusted to match the elevation of the topographic surface if necessary. Hosted elements do not offer as much control when trying to coordinate with topography.

Chapter 13: Power and Communications

Place power and systems devices into your model Creating electrical plans that are not only correct in the model but also on construction documents can be achieved with Revit MEP 2011.

Master it Having flexibility in the relationship between model components and the symbols that represent them is important to create an accurate model and construction documents. Is it possible to show a receptacle and its associated symbol in slightly different locations to properly convey the design intent on construction documents? If so, how?

Solution Yes. You can add parameters that allow for an offset of the symbols that represent the model components in a plan view.

Place equipment and connections Electrical equipment often requires clearance space for access and maintenance. Modeling equipment in your Revit project allows you to coordinate clearance space requirements.

Master it Interference between model components can be detected by finding components that occupy the same space. Explain how you can determine whether an object interferes with the clearance space of an electrical equipment component.

Solution You can build the required clearance spaces into your equipment families or build a separate family to be used for clearance space. You can use Revit's interference-checking capabilities to find interferences with the clearance spaces.

Create distribution systems Proper setup of distribution systems is the backbone of the intelligence of your electrical design. It helps you to track the computable data within your project.

Master it Because your project may contain multiple distribution system types, explain the importance of assigning distribution systems to your electrical equipment and naming your equipment.

True or false: You cannot create a power riser diagram with Revit MEP 2011.

Solution Assigning a distribution system to electrical equipment allows for other items of the same distribution type to be connected to the equipment. Naming equipment makes it easy to locate when connecting elements.

False.

Model conduit and cable tray Large conduit and cable tray runs are a serious coordination issue in building designs. Revit MEP 2011 has tools that allow you to model conduit and cable tray in order to coordinate with other model components.

Master it Conduit and cable tray can be modeled with two different styles. One style uses fittings, and one does not. Does this mean that no fittings need to be assigned to the style that does not use fittings? Explain how this affects scheduling of the components.

Solution You must assign fitting families to be used by the style without fittings. The fittings must exist in order for the conduit or cable tray to be modeled. Scheduling styles without fittings enables you to report the total length of a run. Scheduling styles with fittings is done to report data about the components that make up a run.

Chapter 14: Circuiting and Panels

Establish settings for circuits and wiring Proper setup of the electrical characteristics of a project is important to the workflow for creating circuits and wiring. Settings can be stored in your project template and modified on an as-needed basis.

Master It The distribution systems defined in a project make it possible to connect devices and equipment of like voltages. Do you need to have voltage definitions in order to create distribution systems? If so, why?

Solution Yes, voltages must be defined to create distribution systems because they are used to establish the line-to-line and line-to-ground values for the system.

Create circuits and wiring for devices and fixtures Circuits are the systems for electrical design. Wiring can be used to show the connection of devices and fixtures in a schematic fashion.

Master It Circuits can be created for devices or equipment even if they are not assigned to a panel. Circuits can then be represented by wiring shown on construction documents. Give two examples of how you can add a device to a circuit that has already been created.

Solution You can right-click the device connector, select the Add To Circuit option, and then click a device that is on the circuit you want to connect to.

You can click a device that is part of the circuit and then click the Edit Circuit button on the Electrical Circuits contextual tab. You then click the Add To Circuit button and select the device to be added.

Manage circuits and panels With the relationship between components and panels established, you can manage the properties of circuits and panels to improve your design performance and efficiency.

Master It While checking the circuits on a panel, you notice that there are only 14 circuits connected, but the panel has 42 poles. How can you reduce the amount of unused space in the panel?

Solution Click the panel, and access its instance properties. Change the value of the Max #1 Pole Breakers parameter to a smaller value. Right-click the panel in the Project Browser, and select a template that utilizes fewer circuits.

Use schedules for sharing circuit information Panel schedules can be used on construction documents to convey the load information. Schedules can also be created for use as design tools to help track electrical data.

Master It The information in Revit panel schedules may not meet the requirements of your document or design standards. Describe how you can use the data within your Revit model to provide the required information.

Solution There is much more data contained in electrical circuits and panels than what may be shown in the Revit panel schedules. You can create a schedule of electrical circuits that shows the information you need.

Chapter 15: Plumbing (Domestic, Sanitary, and Other Piping)

Customize out-of-the-box Revit plumbing fixtures for scheduling purposes Learning how to customize existing plumbing fixtures can help with productivity and provide more robust building information.

Master It What are the two types of parameter information that can be scheduled and used in type catalogs?

Solution System parameters and shared parameters. Simple parameters that have been added to the family cannot be scheduled.

Use custom plumbing pipe assemblies to increase speed and efficiency in plumbing layouts Sometimes you are required to think outside the box and learn how to use new tools to get the most benefit out of them. By utilizing the new Copy/Monitor feature when creating custom pipe assemblies, you can take production to the next level.

Master It Why are nested pipe assembly families better to use than the modeled pipe assembly for a BIM project ?

Solution BIM stands for "building information modeling." With nested families, you can create more accurate information. If you are using Revit strictly as a modeling tool for drafting, then the modeled pipe assembly will help keep the file smaller, but you lose some of the *I* in BIM.

Adjust and use the plumbing pipe settings Piping settings are crucial to the ability to have Revit MEP model your plumbing layout, the way it will look, and the way it will perform.

Master It Do fitting parameters have to be set up in the system pipe types?

Solution Yes. If you do not set up the Fittings parameter properly, you will have mixed materials or fittings that may not be placed automatically when routing piping in your model.

Select and use the best pipe routing options for your project When using Revit MEP 2011 for your plumbing layouts, you must understand the functions of auto pipe routing, manual pipe routing, and sloping pipe. Once these functions are mastered, then the user can lay out any type of piping system.

Master It A plumbing designer has just been asked to lay out a sloped plumbing system and has only a day to pipe up a clubhouse. Where should he start his pipe route first?

Solution He should start from a point of connection outside the building and work his way in, from main to branch to fixture.

Adjust pipe fittings Pipe fittings are needed in systems to make the systems function properly and to produce documentation for construction. Being able to add or modify fittings can increase productivity.

Master It You have just finished your modeled layout and given it to your employer for review. He's just came back and now has asked you to remove a couple of elbows and replace them with tees for future expansion. What would be your method to accomplish it quickly?

Solution Select the elbow, and click the + sign; this will change the elbow to a tee.

Adjust the visibility of pipes Being able to adjust the visibility gives the plumbing designer or user the ability to set up multiple views and control the graphics for documentation.

Master It There are too many systems showing up in your views. What would you do to show only one piping system in that view?

Solution Type **VG**, go the Filters tab, and deselect all the systems you do not want to show up.

Chapter 16: Fire Protection

Placing fire protection equipment When starting a fire protection model, placing the equipment can make or break your design. The ability of Revit to verify your layouts early, through the coordination of this equipment with other disciplines, can set the pace for a successful project.

Master It What is a method to help speed up production when using a standard fire riser on multiple buildings?

Solution Create a nested family with all the components required on the fire riser.

Create fire protection systems Creating proper fire protection systems is essential to the performance and behavior of the fire protection model. Properly created fire protection systems also help with being able to coordinate with other disciplines during design.

Master It Marty has just created a fire protection system name called Wet1, and he has created a filter system type named wet1. Now Marty is in a presentation, and his system is not filtering properly. What should he look at first? What should he do if there is a second problem?

Solution Marty should review his naming first. His system starts with a capital *W*, and his filter has a lowercase *w*. Next he should review whether he selected System Name instead of System Type in the filter.

Route fire protection piping Fire protection piping can be routed by a couple different methods. It can be set up with different materials to help with takeoffs and specifications. Once piping has been routed, it can be coordinated with other disciplines to reduce errors and omissions.

Master It What are some of the methods to deal with fittings that may not be supplied with Revit MEP 2011?

Solution Use existing fittings for spatial restraints and visual coordination only, create your own fitting families, or use manufacturer-supplied content.

Chapter 17: Solid Modeling

Model solids and voids Being able to efficiently model will decrease the time you spend creating content and give you more time to spend on design decisions. Solid geometry is crucial for the physical coordination of components to achieve a design that will result in fewer changes during construction, where changes are the most expensive.

Master It Several tools are available to create the shapes needed to represent MEP discipline components. Each tool generates an extrusion in a unique way. Describe the difference between a swept blend and a regular sweep.

Solution A swept blend is used when the shape of the extrusion changes from one end to the other. A sweep is an extrusion of a consistent shape along a path.

Orient geometry using reference planes and lines Reference planes and lines are the most effective way to define the orientation of solid geometry within a family. Reference planes define how an object will be inserted into a project.

Master It Knowing the resulting orientation of an extrusion prior to creating it will save lots of time by not having to duplicate modeling efforts. Nothing is more frustrating than taking the time to create a solid only to find that it is in the wrong plane. Describe the process for creating an extrusion that is associated with a custom reference plane.

Solution You first must create a reference plane and give it a unique name. Then when you choose a form tool, you can set the reference plane that the form will be associated with by using the Set button on the Work Plane panel of the Home tab in the Family Editor.

Ensure the parametric change capability of models Building solid geometry to represent MEP discipline components is good. Building the geometry with parametric change capabilities is even better.

Master It Solid geometry can be defined by parameters that can change the size or shape of the geometry. Reference planes and lines are an important part of creating parametric behavior. Why?

Solution The dimensional parameters used to define geometry should be assigned to the dimensions associated with reference planes and lines that the geometry is locked to. This allows you to maintain the relationships of multiple solids when parameter values change.

Determine optimal visibility settings for solids The visibility behavior of solid geometry plays an important part in the creation of consistent, coordinated construction documents.

Master It It is important to know how a family will be used in a project to determine the visibility settings required for the solid geometry in the family. Why is it important to set the Detail Level visibility settings for pipe- and duct-related families?

Solution The pipe and duct system families have specific visibility behavior to show as a single or double line. Your pipe and duct component families should be set with the same visibility settings for consistency on construction documents.

Chapter 18: Creating Symbols and Annotation

Create symbolic lines and filled regions Not only is Revit MEP 2011 a modeling application, but it contains the tools necessary to accomplish drafting tasks.

Master It Having a good command of the tools available for creating symbols will help you create families that represent your design elements exactly the way you want to see them. What line tool is best suited for duplicating the line work of an imported CAD symbol?

Solution The Pick Lines tool allows you to select the line work of an imported file and creates a Revit line to match.

Use symbols within families for representation on drawings Given the schematic nature of MEP plans, symbols and annotation objects are important parts of your Revit MEP 2011 workflow, allowing you to represent your model components per your company standards.

Master It By having annotation symbols nested into your component families, you can create an accurate 3D model that is displayed schematically on your construction documents. Explain the importance of creating subcategories for the graphics in your annotation families.

Solution Symbols are created as generic annotations, which is a single category within Revit. A subcategory must be created to control the visibility of a specific symbol independently from the Generic Annotations category.

Work with constraints and parameters for visibility control The parametric modeling capabilities of Revit MEP 2011 make it a powerful BIM solution. These capabilities can be used in annotation families as well.

Master It A common scenario for a Revit project is to link consultant files into your project file. Because of this, face-hosted families are often used. Face-hosted components can be attached to either a vertical or horizontal host, so the ability to separate the annotation symbol from the host needs to be created in the annotation family.

When using a length parameter to define the offset of a symbol from its host, what value should be input for the parameter when the component family is in a project?

Solution You need to input a length that relates to the actual size of the symbol because the symbol size in a view is determined by the view scale.

Use labels to create tags Tags are the most commonly used annotation families in a project. They are used to report information about objects in a Revit model.

Master It The use of labels is a much more effective method than using text objects for keeping documents coordinated.

If your project requires that you show DIA (diameter) after each pipe size tag on your construction documents, how can you accomplish this globally throughout your project?

Solution When defining the label in your pipe size tag, you can add DIA as a suffix for the label.

Chapter 19: Parameters

Manipulate the properties of parameters The parameters used to define the properties of elements have properties of their own that define the behavior of the parameters and how they can be used.

Master It It is important to know when and where parameters can be used for extracting data from the model or project. It is also important to understand how instance and type parameters are used. Describe how the use of instance and type parameters affects the way data is changed in a family.

Solution Type parameters are used to define different family types within a family. Changes to a type parameter will affect every instance of that type. Instance parameters are unique to each instance of an element as it exists in the project. Changes to an instance parameter occur only at the selected instance of a family.

Work with parameters in families Parameters created in family files are useful for defining the geometry of a family and also for assigning engineering data to the family.

Master It Certain families can have multiple family types. If a family has many types, all of them will be loaded into a project when the family is loaded. What can be done to limit the number of family types that are loaded when a family is inserted into a project?

Solution A type catalog can be created to allow for the selection of specific family types when a family is loaded into a project.

Work with shared parameters Shared parameters are very useful because they can be used in schedules and in annotation tags. Shared parameters can be applied directly to families or added as project parameters.

Master It Managing shared parameters is as important as managing your component libraries. Explain the importance of keeping a common shared parameters file for multiple-user environments.

Solution Because a shared parameters file is a library of parameters, you want to keep just one location to be shared by multiple users. This will prevent the unnecessary duplication of parameters that contain the same type of data.

Use parameters in project files The use of parameters is not limited to component families. Parameters can be added to any element that makes up a Revit project.

Master It Parameters can be added to system families only by creating project parameters. When you create a project parameter, it will be added to all the elements in the chosen category. Explain why managing project parameters is important to using them in schedules within a project.

Solution It is possible to create a project parameter that has the same settings as a parameter that already exists in some of the families being scheduled. This would create confusion as to which parameter should hold the data for the schedule.

Chapter 20: Creating Equipment

Create MEP equipment families The ability to create the types of equipment families needed for accurate modeling of components and systems is a major factor in the success of your Revit projects.

Master It MEP equipment can be quite complex in its structure. Complex geometry can have an adverse effect on model performance. What are some ways to model equipment in the most simple form yet still convey the proper information on construction documents?

Solution The basic shapes of equipment can be modeled, and symbolic or model lines can be used to represent the actual geometry in plan views. The solid forms used to create the geometry can be set to show only in 3D views. Simple forms can be set to show at lower levels of detail, while more complex geometry can be reserved for finer detail.

Add connectors to equipment for systems Adding connectors to equipment families will make them functional for use in the design of engineering systems.

Master It It is important to know how your equipment families will be used in your projects from an engineering standpoint as well as for model coordination. Explain how connectors determine the behavior of an equipment family.

Solution Connectors have properties that define the behavior of the family, depending on the system type of the connector. Duct and pipe connectors determine the direction of flow that may affect how the equipment can be connected in project systems. Electrical connectors determine which distribution systems the equipment is used for.

Create clearance spaces for equipment Space for safety and service of equipment is crucial to building design. The ability to coordinate clearances around equipment improves project quality and can reduce construction and design cost.

Master It Equipment families with built-in clearance spaces allow you to quickly and easily determine whether the equipment will fit into your project model. Describe some options for controlling the visibility of clearance spaces so that they are not shown when not needed.

Solution The visibility of clearance space geometry can be controlled by creating a unique subcategory for the geometry or by using Yes/No type parameters to toggle the visibility.

Add parameters to equipment Parameters in your equipment families can be useful for creating schedules in your Revit projects that report data directly from the equipment used in the design. Family parameters can enable you to make equipment families that are changeable without having to create new families.

Master It Shared parameters must be used in your equipment families if you want to schedule the data they provide. If you are creating parameters for parametric behavior of the solid geometry, do they also need to be shared parameters?

Solution They need to be shared parameters only if they are to be used in a schedule. Family parameters cannot be scheduled.

Chapter 21: Creating Lighting Fixtures

Create different types of lighting fixture families Many different types of lighting fixtures are required for various applications within a building. With Revit MEP, you can create any type of lighting fixture and include any data associated with the fixture.

Master It Knowing how a lighting fixture will be used in a Revit model is important to determining the kind of family to create. True or false: A face-hosted lighting fixture can be replaced by a nonhosted fixture using the Type Selector.

Solution False. Face-hosted fixtures can be replaced only by other face-hosted families of the same category.

Use a light source in your lighting fixture families Lighting fixtures can be used in making design decisions because they not only represent the fixture as a model but also contain photometric data from real-world lighting fixtures for use in lighting analysis.

Master It Photometric web files can be obtained from lighting fixture manufacturers. These files provide the lighting distribution characteristics of a fixture when added to a family. How can you be sure that the `.ies` file you are using is appropriate for the type of fixture it is being used in?

Solution The information in an `.ies` file obtained from a manufacturer can be viewed using third-party software. Files should be renamed to indicate what type of fixture they are used for.

Create and manage fixture types and parameters The parameters of a lighting fixture family are what make it an intelligent object. They can be used to create multiple types within the same family or for managing the electrical characteristics of a fixture.

Master It Connectors are what determine the electrical properties of a lighting fixture family. Describe the process of ensuring that a connector has the same load and voltage values that have been assigned to the fixture.

Solution The load and voltage values of a connector can be associated to the load and voltage of a fixture by accessing the instance properties of the connector and using the box in each parameter value to associate it to the appropriate family parameter.

Use lines and symbols to represent lighting fixtures Some lighting fixtures are shown on construction documents as symbols, while others are shown as their actual size. Symbolic lines or annotation symbols can be used to eliminate the need to display model graphics.

Master It Annotation symbols nested into lighting fixture families can represent the fixture without having to show the model graphics. Is it possible to use a nested annotation family to represent a wall-mounted fixture in a face-hosted lighting fixture family? Explain.

Chapter 22: Creating Devices

Model device geometry The 3D geometry of a device family may not be its most important component, but there are things you can do to make the geometry useful for project coordination and accuracy.

Master It Device families are commonly face-hosted because of the way they are used in a project model. Describe why a device family that is hosted by a floor in a project may not display correctly in plan view.

Solution If the geometry of the device family is flush with the host extrusion, it will not appear in a plan view when hosted by a floor if the bottom of the view range is set to the face of the floor.

Use annotations to represent devices Because of their size and simple shapes, device families are not typically shown on construction documents, but rather a symbol is used to represent the device. Annotation families can be nested into device families for display in project views.

Master It The orientation of an annotation family nested into a device family is important for proper display in project views. What is the device family parameter that allows for the display of the annotation in plan and section views? What hosting options in a project allow for this parameter to work?

Solution The Maintain Annotation Orientation parameter allows for nested annotation families to be visible in plan and section views. The parameter displays the annotation when the device family is hosted to a vertical or horizontal surface.

Add parameters and connectors The connectors used in device families provide the engineering characteristics of the device so that it can be incorporated into systems within a project.

Master It The parameters of a connector cannot be edited while working in a project file. Describe how you can change the properties of a connector without having to edit the device family.

Solution The parameters of a connector in a device family can be associated to the device family parameters. The properties of a device can be modified directly in a project, therefore updating the connector properties.

Index

Note to the reader: Throughout this index **boldfaced** page numbers indicate primary discussions of a topic. *Italicized* page numbers indicate illustrations.

D

E

F

G

H

I

T

U

V

W

X

Y

Z

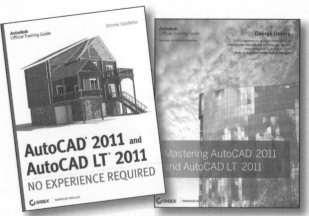

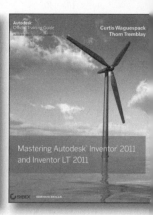